High Pressure Molecular Science

NATO Science Series

A Series presenting the results of activities sponsored by the NATO Science Committee. The Series is published by IOS Press and Kluwer Academic Publishers, in conjunction with the NATO Scientific Affairs Division.

A. **Life Sciences**	IOS Press
B. **Physics**	Kluwer Academic Publishers
C. **Mathematical and Physical Sciences**	Kluwer Academic Publishers
D. **Behavioural and Social Sciences**	Kluwer Academic Publishers
E. **Applied Sciences**	Kluwer Academic Publishers
F. **Computer and Systems Sciences**	IOS Press

1. **Disarmament Technologies**	Kluwer Academic Publishers
2. **Environmental Security**	Kluwer Academic Publishers
3. **High Technology**	Kluwer Academic Publishers
4. **Science and Technology Policy**	IOS Press
5. **Computer Networking**	IOS Press

NATO-PCO-DATA BASE

The NATO Science Series continues the series of books published formerly in the NATO ASI Series. An electronic index to the NATO ASI Series provides full bibliographical references (with keywords and/or abstracts) to more than 50000 contributions from internatonal scientists published in all sections of the NATO ASI Series.
Access to the NATO-PCO-DATA BASE is possible via CD-ROM "NATO-PCO-DATA BASE" with user-friendly retrieval software in English, French and German (WTV GmbH and DATAWARE Technologies Inc. 1989).

The CD-ROM of the NATO ASI Series can be ordered from: PCO, Overijse, Belgium

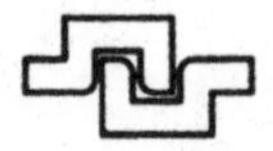

Series E: Mathematical and Physical Sciences – Vol. 358

High Pressure Molecular Science

edited by

Roland Winter

University of Dortmund,
Department of Chemistry,
Physical Chemistry,
Dortmund, Germany

and

Jiri Jonas

University of Illinois,
School of Chemical Sciences,
Urbana, Illinois, U.S.A.

Kluwer Academic Publishers

Dordrecht / Boston / London

Published in cooperation with NATO Scientific Affairs Division

Proceedings of the NATO Advanced Study Institute on
High Pressure Molecular Science
Il Ciocco, Italy
27 September – 11 October 1998

A C.I.P. Catalogue record for this book is available from the Library of Congress.

ISBN 0-7923-5806-6 (HB)
ISBN 0-7923-5807-4 (PB)

Published by Kluwer Academic Publishers,
P.O. Box 17, 3300 AA Dordrecht, The Netherlands.

Sold and distributed in North, Central and South America
by Kluwer Academic Publishers,
101 Philip Drive, Norwell, MA 02061, U.S.A.

In all other countries, sold and distributed
by Kluwer Academic Publishers,
P.O. Box 322, 3300 AH Dordrecht, The Netherlands.

Printed on acid-free paper

Printed in the Netherlands.

Table of Contents

Chemical, Biophysical, Biochemical, and Biomedical Aspects

PREFACE

This monograph contains the proceedings of the NATO-ASI on "High Pressure Molecular Science", held in Il Ciocco, Barga, from September 27 until October 11, 1998, and illustrates new developments in the field of high pressure science. In fact, for chemists, biochemists, physicists and materials scientists, pressure as an experimental variable represents a tool which provides unique information about the microscopic properties of the materials studied. In addition to its use as a research tool for investigating the energetics, structure, dynamics, and the kinetics of transformations of materials at a molecular level, the application of pressure is also being used to modify the properties of materials to preserve or improve their qualities. It is interesting to note how the growth of the high pressure field is reflected in the content of the ASI's dealing with this field. The ASI "High Pressure Chemistry" held in 1977 was followed by the ASI "High Pressure Chemistry and Biochemistry" held in 1986, by the ASI on "High Pressure Chemistry, Biochemistry, and Materials Science" held in 1992, and the coverage of the present ASI also includes new developments and applications to all high pressure fields, in particular to bioscience and biotechnology. In fact, the potential of high pressure techniques in biotechnology has only recently been fully recognized. On the experimental side we have seen advances in static and dynamic high pressure probes. Surely one of the most dramatic changes that has taken place in the last years has been, associated with the development of fast and capable computers, the immense rise of ab-initio molecular dynamics calculations, a means of peering into the ionic and electronic structure of condensed matter.
In view of the teaching character of the ASI, it is natural that the main contribution to this volume presents overviews of the different subfields of high pressure research. The lectures cover the main areas of high pressure applications to materials science, condensed matter physics, chemistry, and biochemistry. In addition, a few contributed papers offer more specialized aspects of various high pressure studies. The contributions to this volume make clear the impressive range of fundamental and applied problems that can be studied by high pressure techniques, and also point towards a major growth of high pressure science and technology in the near future. This ASI focused mainly on advances achieved in the six years since the previous ASI devoted to the high pressure field. We hope that these proceedings will help to establish promising directions for future research on effects of pressure on complex systems.
The editors gratefully acknowledge input of the organizing committee and express their thanks to all lecturers and contributors to this volume for their efforts in preparing their lectures and manuscripts, and to all the participants of the ASI whose enthusiasm convinced us of the importance and excitement of this interdisciplinary area. On behalf of all ASI participants, we express our gratitude for the generous financial support provided by the Scientific Affairs Division of the North Treaty Organization.

Dortmund, Germany
Urbana, Illinois, USA
March, 1999

Roland Winter
Jiri Jonas

HIGH-PRESSURE RAMAN SCATTERING STUDIES OF FLUIDS

JIRI JONAS
Beckman Institute for Advanced Science and Technology
University of Illinois at Urbana-Champaign
405 North Mathews Avenue
Urbana, IL 61801 USA

1. Abstract

The field of laser Raman spectroscopy of liquids and gases is reviewed. After introducing the importance of using pressure as an experimental variable in Raman studies of fluids, a brief overview of the experimental techniques is presented. Illustrative examples of specific high-pressure Raman studies deal with the following topics: reorientational motions in liquids, vibrational dephasing, collision-induced scattering, Fermi resonance and Raman frequency noncoincidence effect.

2. Introduction

Laser Raman and Rayleigh scattering experiments on liquids and gases at high pressure continue to provide important and unique information about dynamic processes and intermolecular interactions in liquids. In high-pressure Raman scattering studies of fluids carried out in our laboratory, the following phenomena were of main interest:

- Reorientational motions
- Vibrational dephasing
- Collision-induced scattering
- Fermi resonance
- Raman frequency noncoincidence effect

The great advantage of Raman experiments lies in the fact that the analysis of Raman lineshapes provides information about the detailed nature of the correlation function - one obtains the time dependence of the correlation function itself and not only an integral. It is only natural that the Raman experiment also has some limitations, namely, only Raman lineshapes of relatively simple molecules can be analyzed to yield unambiguous results. The general theory of light scattering [1] is

R. Winter and J. Jonas (eds.), High Pressure Molecular Science, 1–23.

well established and at this point a few pertinent comments about the Raman experiments are appropriate.

From the experimental polarized and depolarized Raman bandshapes one can obtain the isotropic scattering intensity $I_{iso}(\omega)$ and the anisotropic scattering intensity $I_{aniso}(\omega)$. Only vibrational (nonorientational) processes contribute to $I_{iso}(\omega)$, whereas both reorientational and vibrational processes contribute to $I_{aniso}(\omega)$. This provides the means of separating reorientational processes from vibrational processes and of calculating reorientational and vibrational correlation functions. Assuming vibrational relaxation to be the major nonreorientational broadening mechanism, one can show that

$$I_{iso}(\omega)=[I_{VV}(\omega)-4/3\ I_{VH}(\omega)]\ /\ \int[I_{VV}(\omega)-4/3\ I_{VH}(\omega)]\,d\omega \tag{1}$$

and

$$C_v(t)=\langle Q^{\upsilon}(0)Q^{\upsilon}(t)\rangle=\int I_{iso}(\omega)\ \exp(-i\omega t)\,d\omega \tag{2}$$

where $C_v(t)$ is the vibrational correlation function. Similarly, one may write

$$I_{aniso}(\omega)=I_{VH}(\omega)\ /\int I_{VH}(\omega)\,d\omega, \tag{3}$$

and

$$C_R(t)=\langle Tr\ \beta^{\upsilon}(0)\beta^{\upsilon}(t)\rangle=\frac{\int I_{aniso}(\omega)\ exp(-i\omega t)d\omega}{\int I_{iso}(\omega)\ exp(-i\omega t)d\omega} \tag{4}$$

where $C_R(t)$ is the reorientational correlation function.

During the past two decades, reorientational motions of molecular liquids and gases have been extensively studied by Raman lineshape analysis. High-pressure laser Raman scattering experiments provided conclusive evidence for the need to separate the effects of density and temperature on reorientational processes in liquids. In order to illustrate the large difference between isobaric and isochoric experiments, Figure 1 shows the rotational correlation functions for liquid methyl iodide (CH_3I), a symmetric top molecule. The correlation function describes the reorientation of the CH_3I about axes perpendicular to the main symmetry axis.

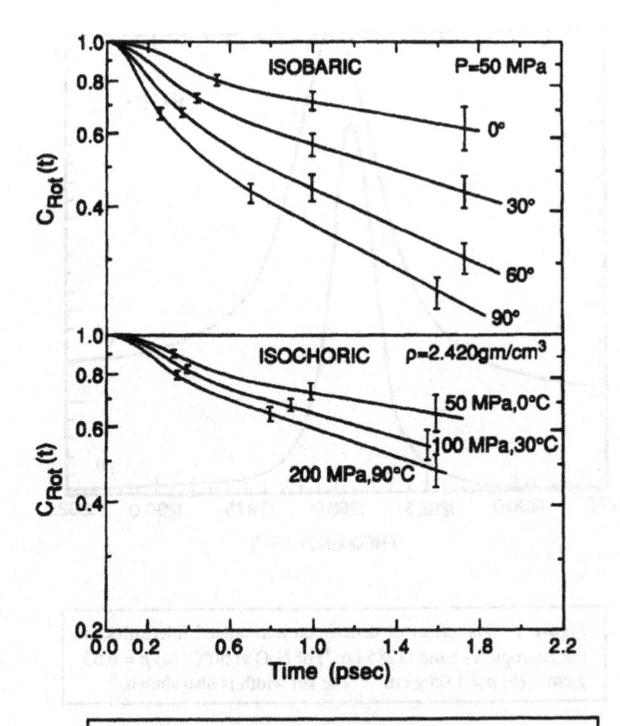

Figure 1. **Temperature dependence of the rotational correlation function for liquid methyl iodide (CH_3I) under isobaric and isochoric conditions. (Taken from [2])**

As an illustrative example of pronounced density effects on vibrational lineshapes, one can use Figure 2, taken from a study [3] of vibrational relaxation of N_2O which shows the density effects on the normalized intensities of isotropic linewidths of the ν_1 band of N_2O.

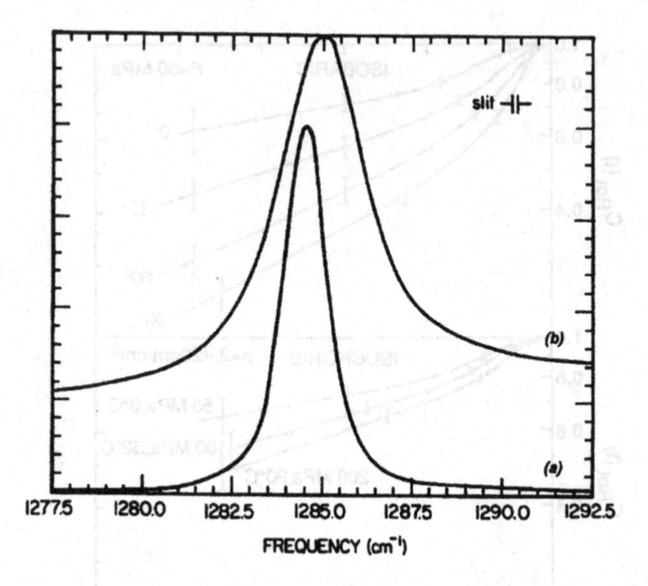

Figure 2. The effect of density on normalized intensities of the isotropic ν_1 band (1285 cm^{-1}) of N_2O at 50°C; (*a*) $\rho = 0.53$ g cm^{-3}, (*b*) $\rho = 1.05$ g cm^{-3}. The slit width is also shown.

The overwhelming majority of Raman studies of the dynamic processes in liquids dealt with the investigation of properties of individual molecules by studying the reorientational and vibrational relaxations which reflect only indirectly the influence of intermolecular interactions. Therefore, the problem of collision-induced scattering (CIS) has attracted both theoretical and experimental interest [4, 5]. It has been observed that collisions in dense liquids or gases produce depolarized Rayleigh spectra in fluids composed of atoms or molecules of spherical symmetry. Collision-induced Raman spectra, forbidden by selection rules (symmetry), have been investigated in polyatomic molecular liquids, as well as collision-induced contributions to the allowed Raman bands. The origin of these collision-induced spectra lies in the polarizability changes produced by intermolecular interactions. It is clear that studies of CIS can provide direct information about intermolecular interactions. However, the CIS represents a very difficult theoretical problem because the scattered intensity depends on the polarizability change in a cluster of interacting molecules, and the time dependence of this change is a function of the intermolecular potential.

Studies of intermolecular interactions and Fermi resonance [6] may serve as yet another example of the important role of high-pressure experiments. The Fermi resonance between the ν_1 and the first overtone of ν_4 have been studied in liquid ND_3 as a function of density and temperature [6]. Figure 3 shows the changes in relative intensities of these Fermi resonance-coupled bands for the extreme density range of our measurements. Since Fermi resonance parameters are sensitive to intermolecular potential, we can change them by varying temperature and pressure. The transition dipole moments of the $\nu_1 + 2\nu_4$ bands are found to vary and the Fermi resonance treatment enables us to estimate the changes in their relative magnitude. The high-pressure experiments provided the spectroscopic information needed for the theoretical analysis of intermolecular interactions in ND_3.

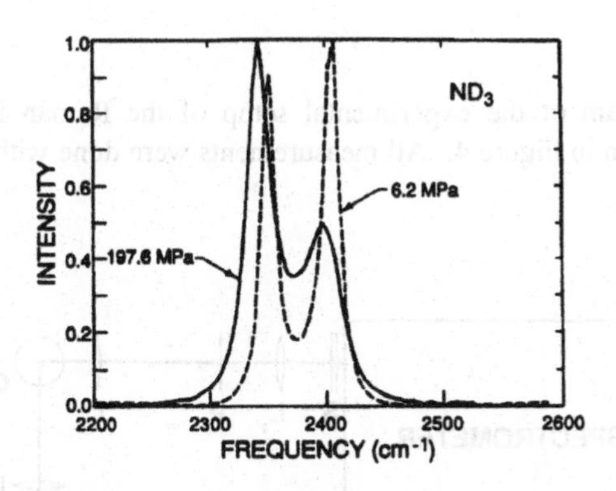

Figure 3. Density effects on the relative intensity of the isotropic $\nu_1 + 2\nu_4$ Fermi resonance lines in ND_3. The full line denotes density = 0.730 g cm^{-3} (T=0°C; P = 197 MPa), and the dashed line denotes density = 0.457 g cm^{-3} (T = 100°C; P = 6 MPa).

In the study of symmetric modes in polar liquids by Raman spectroscopy, considerable attention has recently been paid to the noncoincidence effect [5] because Raman measurements of the noncoincidence effect for polar liquid molecules can provide valuable information about the coupling mechanisms of short- and long-range orientational order which arise from the inter- and intramolecular interactions in dense fluid phases. The noncoincidence effect denotes the phenomenon that the peak wavenumbers of the isotropic and anisotropic components of a Raman band do not coincide. The noncoincidence value, $\delta\nu$, is defined as

$$\delta\nu = \nu_{aniso} - \nu_{iso} \quad (5)$$

where ν_{aniso} and ν_{iso} are the respective anisotropic and isotropic peak wavenumbers. Differences in peak wavenumber start from 0 cm^{-1} in the gaseous phase and increase gradually as the density increases. In the liquid state, normally the wavenumber difference is slight, but for some polar molecules the difference may be as large as 30 cm^{-1} [3]. The origin of the noncoincidence in liquid polar molecules is mainly due to the orientationally dependent intermolecular forces. The intermolecular forces modulate the vibration of a symmetrical mode, changing the oscillator force constant of this mode. The isotropic component is a measure of the spherically symmetric average of the potential while the anisotropic component reflects the angularly dependent portion of the potential. The noncoincidence effect is primarily associated with symmetric vibrational modes of polar molecules and is very pronounced for the modes which are both Raman active and strongly infrared active.

3. Experimental

The schematic diagram of the experimental setup of the Raman light scattering measurement is shown in Figure 4. All measurements were done with 90° scattering geometry.

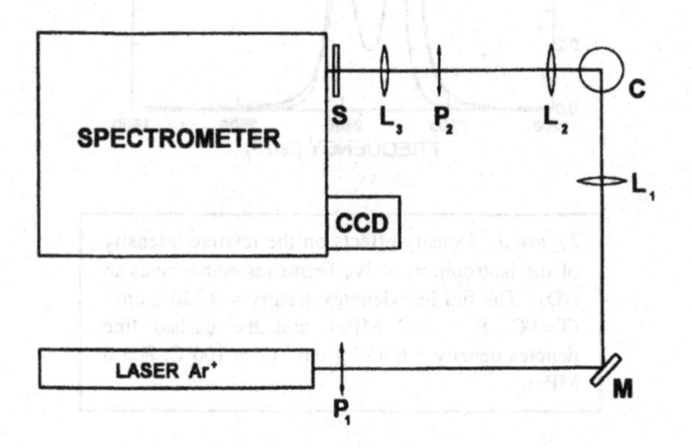

Figure 4. **Schematic diagram of the Raman light scattering experimental setup, where P1 is a polarizer, M is a reflection mirror, L1 is a focus lens, C is a Raman sample cell, L2 is a collection lens, P2 is a polarization analyzer, L3 is a focus lens, and S is a polarization scrambler.**

The incident radiation is the 4880 Å line from a Spectra-Physica Ar-ion laser. P1 is a polarizer; M is a mirror; L1 is a focus lens used to collect scattered light at 90 degrees. The collection cone angle was kept small (< 5°) in order to minimize the errors in the measurements of anisotropic spectra. P2 is a rotatable Glan-Thompson prism used as a polarization analyzer. The VV and VH spectra were obtained with this polarization analyzer set to parallel and perpendicular to the polarization of the incident beam, respectively. L3 (300mm f.1.) is a focus lens used to focus the collected light into the slit of the spectrometer. S is a crystalline-amorphous quartz compensated wedge scrambler which is placed immediately prior to the entrance of the spectrometer and is used to scramble the polarization of the light to ensure equivalent transmission of different polarized light through the spectrometer. (The need of this scrambler is due to the non-equivalent reflection efficiency of the grating mirrors for different polarized light.) The scattered light was analyzed by a spectrometer (a Spex1403 0.85m double monochromator) and then recorded by a liquid nitrogen-cooled CCD detector (the CCD chip contains 1024x1024 pixels with pixel size: 27μmx27μm) operated at −110°C (manufactured by Princeton Instrument, Inc.). The whole experimental setup was installed on a vibration isolated optic table (manufactured by Newport Corporation).

One important experimental aspect of high-pressure Raman experiments is related to possible problems with stress-induced birefringence by optical windows in the high-pressure Raman optical cell. To introduce the discussion of the stress-induced birefringence, Figure 5 gives a schematic drawing of a high-volume optical cell which can be used for laser scattering experiments on fluids at high pressure.

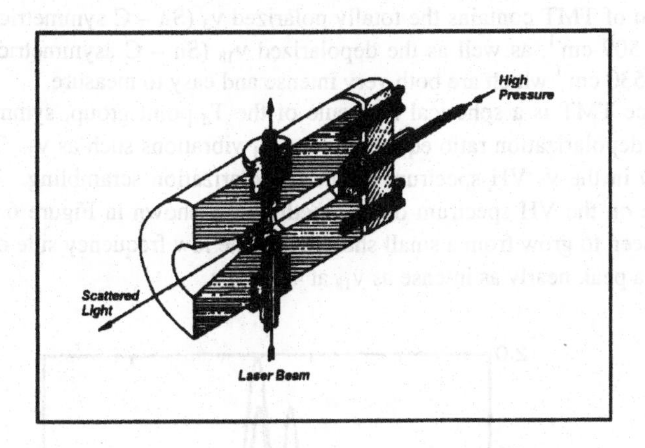

Figure 5. Schematic drawing of high volume optical cell for light scattering experiments at high pressure.

As indicated in the introduction, Raman bandshapes contain information on the orientational and vibrational dynamics of the molecular system. The theoretical basis by which this information may be extracted from the experimental spectra depends on the use of linearly polarized incident radiation and the ability to measure accurately the scattered light in two polarization directions. Lasers, of course, are an excellent source of intense, linearly polarized radiation. However, if the optical cell window material is birefringent, the linearly polarized light will become elliptically polarized, and the results of a bandshape analysis will become questionable. Fortunately, since the most frequently used cell materials such as fused silica and glass are not birefringent, this is usually not a problem for measurements at atmospheric pressure. However, the stress applied to the cell windows in a high-pressure experiment can lead to strain-induced anisotropies which result in a pressure-dependent scrambling of the polarization. Even though the scrambling is small, it can have a large effect on the I_{VH} band, especially for cases when the depolarization ratio is small. The effect has been known for several years and was studied earlier in our laboratory by Cantor, et al. [7].

Since neglect of the effects of polarization scrambling by optical windows during a high-pressure experiment can lead to erroneous results, the following example [8] stresses the importance of a careful analysis of each specific high-pressure light scattering experiment where polarization measurements are important. In principle, any compound exhibiting a totally polarized band with the depolarization ratio equal to zero can be used as a test liquid to measure the polarization scrambling of the windows. We found particularly useful spherical molecules of T_d symmetry for which the totally symmetrical modes have no anisotropic spectrum and therefore $\rho = 0$. There are many possible choices for the test liquid, but we used tetramethyltin (TMT) as an illustrative example. The spectrum of TMT contains the totally polarized ν_3 (Sn – C symmetrical stretch, A_1) band at 509 cm^{-1}, as well as the depolarized ν_{18} (Sn – C asymmetrical stretch, F_2) band at 530 cm^{-1} which are both very intense and easy to measure.

Since TMT is a spherical molecule of the T_d point group, symmetry demands that the depolarization ratio equal zero for A_1 vibrations such as ν_3. Therefore, any intensity in the ν_3 VH spectrum is due to polarization scrambling. The effects of pressure on the VH spectrum of tetramethyltin is shown in Figure 6. The ν_3 peak can be seen to grow from a small shoulder on the low frequency side of the ν_{18} at 50 MPa to a peak nearly as intense as ν_{18} at 400 MPa.

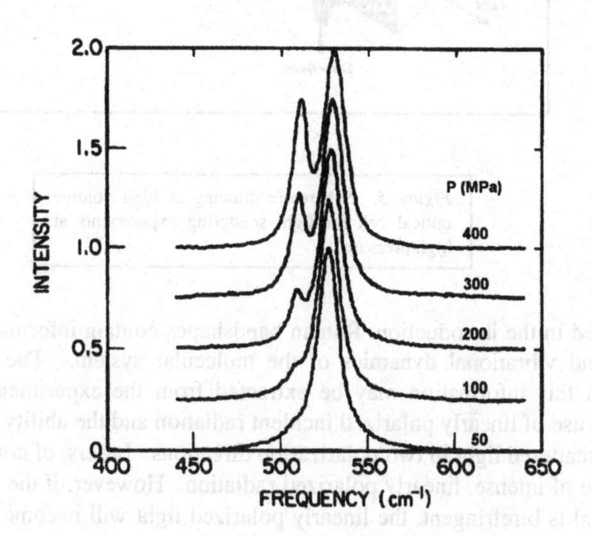

Figure 6. Depolarized spectra of tetramethyltin at different pressures showing the growth of the ν_3 intensity (509 cm^{-1}) produced by polarization scrambling of the float glass window in the high-pressure cell. The strong band at 530 cm^{-1} is the depolarized ν_{18} stretching mode. All measurements were carried out at 90°C.

4. Raman Studies of Reorientational Motions

As already show in Figure 1, the high-pressure Raman experiments provided conclusive evidence for the need to separate the effects of density and temperature on rotational motions in liquids.

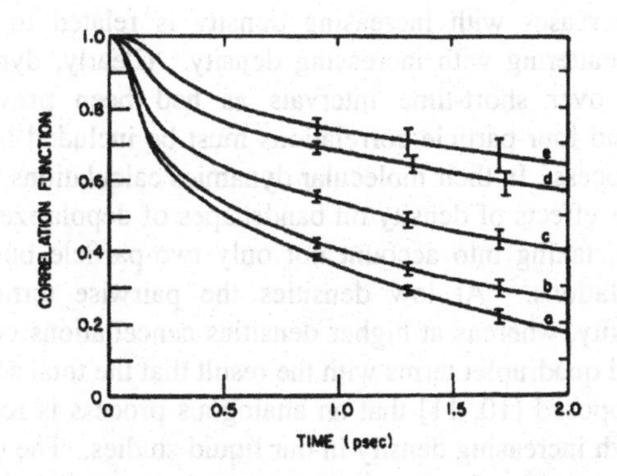

Figure 7. Reorientational correlation function for propyne in acetone solution at 50°C as a function of pressure: (a) 50; (b) 100; (c) 200; (d) 300; (e) 400 MPa.

Yet another example is shown in Figure 7 which illustrates the pressure effects on the reorientational correlation function of propyne in acetone solution [9, 10]. Propyne (methyl acetylene) is a symmetric top molecule and the lineshape of the ν_3 C≡C stretching mode at 2142 cm^{-1} was analyzed to obtain the correlation functions given in Figure 7. Here again, the changes with pressure are typical for liquids as with increasing pressure to 400 MPa the rotational process slows by a factor of about 5. Analogous behavior has been observed for other molecular systems.

At this point, it is worthwhile to point out the specific results of several such studies, which showed that the rotational second moment is strongly pressure/density-dependent. According to classical interpretation, the second moment $M_R(2)$ is related to the rotational kinetic energy of the molecule and should not change with density. Its value is constant at a given temperature, $M_R(2) = 6kT/I_\perp$. ($I_\perp$ is the molecular moment of inertia perpendicular to the symmetry axis.) In contrast to this classical prediction, we found that $M_R(2)$ is strongly density-dependent in liquid chloroform [11] and propyne [10]. In these cases we found that $M_R(2)$ decreases with increasing density.

For example, for the ν_1 (A_1) C-H stretching mode in liquid chloroform [11] at 303 K there is a decrease of $M_R(2)$ from 510 cm^{-2} with a density increase from 1.47 to 1.78 g cm^{-3}. The decrease of the second-moment with increasing density can be explained in terms of CIS which contributes to the wings of the Raman band. The fact that the CIS affects only the far wings of the band explains why this effect was very often neglected; the correlation functions and the correlation times are affected only slightly. However, second-moment analysis, which describes the short-time behavior of a correlation function and thus reflects the collision-induced high-frequency contributions to the spectral line rather strongly, is the most reliable method of investigating short-time intermolecular interactions. The reason why the second-moment decreases with increasing density is related to the decrease of collision-induced scattering with increasing density. Clearly, dynamic effects are not as inefficient over short-time intervals as had been previously assumed. Evidently, three- and four-particle correlations must be included in any description of the scattering process. In their molecular dynamics calculations, Alder, et al. [12, 13] investigated the effects of density on bandshapes of depolarized light scattering from atomic fluids, taking into account not only two-particle but also three- and four-particle correlations. At low densities the pairwise term dominates and increases with density, whereas at higher densities cancellations occur between the pairwise, triplet and quadruplet terms with the result that the total scattering intensity decreases. We proposed [10, 11] that an analogous process is responsible for the decrease of CIS with increasing density in our liquid studies. The observed density-dependence of the rotational second-moment makes questionable studies using Raman bandshapes to calculate torques in liquids.

5. Vibrational Dephasing in Liquids

Many different theories dealing with dephasing have been developed (for review, see [14]), examples of which are the isolated binary collision (IBC) model [15, 16], the hydrodynamic model [17], the cell model [18], and the model based on resonant energy transfer [19].

In the measurements of isotropic lineshapes of the C-H, C-D, and C-C stretching modes in a variety of simple molecular liquids, it was found that the IBC model [15, 16], which considers only the repulsive part of the intermolecular potential in calculating the dephasing rate, reproduces the general trends of the experimental data observed. It should be pointed out that this model and other models predict that the dephasing rate (line width) will increase with density at constant temperature; therefore, it was interesting to observe a very different behavior of the dephasing rate in liquid isobutylene ($CH_2 = C(CH_3)_2$) [20].

In order to understand this behavior, one has to realize the different character of these two vibrations of interest. First, the $C{=}CH_2$ vibration is strongly infrared active, while the $C\text{-}CH_3$ vibration is inactive. Secondly, the classical vibrational amplitudes show that the repulsive forces of the bath may affect the $C\text{-}CH_3$ vibration more.

The Raman lineshapes of the $\nu_4(A_1)$ symmetry $C{=}CH_2$ stretching mode of 1657 cm^{-1} and the ν_9 (A_1) symmetric $C\text{-}CH_3$ stretching mode of 807 cm^{-1} in isobutylene [20] were measured as a function of pressure from vapor pressure to 0.8 cm^{-3} over the temperature range from –25 to 75°C. Figure 8 shows the density-dependence of the experimental half-width $\Gamma = (2\pi c \tau_v)^{-1}$ where τ_v is the dephasing time and Γ is proportional to the dephasing rate. From Figure 8, we see that increasing density at constant temperature affects the bandwidths of the two vibrations in a qualitatively different way. The C-CH_3 stretching bandwidth increases with increasing density, a behavior found for many other models of liquids. This basic trend can be predicted in terms of an IBC model based on rapidly varying repulsive interactions. The most interesting result of this study was the observed decrease of the bandwidth of the C=CH_2 stretching mode (strongly infrared active band) with increasing density. To our best knowledge, this was the first experimental observation of a decrease in dephasing rate with increasing density in a liquid. It appears that this band is inhomogeneously broadened as it is affected by environmentally-induced frequency fluctuations. These fluctuations are due to dispersion, induction, and electrostatic forces that depend on the dipole and polarizability of the molecule. The decay of the inhomogeneous environment around a molecule results in motional narrowing. The correlation function modeling [21], which uses the Kubo stochastic lineshape theory [22], was in agreement with the experimental data. In further studies we found a similar decrease in bandwidth for the C=O stretching mode in liquid acetone [23, 24]. Again, the attractive interactions influence the dephasing process and are responsible for the density behavior of the bandwidth.

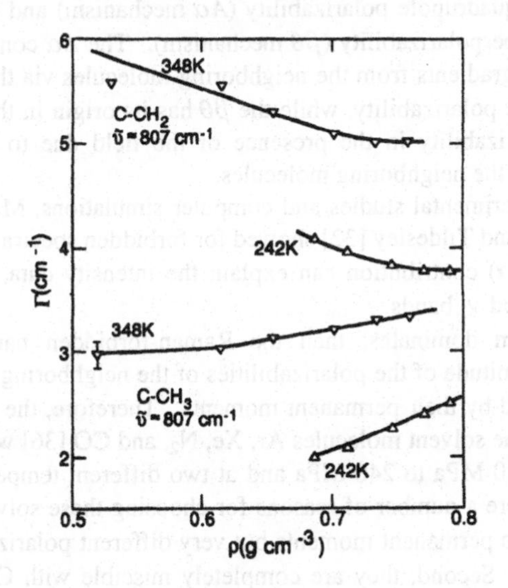

Figure 8. The density dependence of the experimental half-widths Γ for the C-CH_3 and C=CH_2 stretching modes in liquid isobutylene at 242 K and 348 K. (Taken from [20])

The Raman experiments on isobutylene and acetone have been discussed in a theoretical study of dephasing by Schweizer and Chandler [25], who analyzed in detail the relative of slowly varying attractive interactions and rapidly varying repulsive interactions on the frequency shifts and dephasing in liquids. Their theoretical model correctly predicted the isothermal density dependence of the $C=CH_2$ bandwidth in isobutylene and the C=O bandwidth in acetone.

6. Collision-Induced Scattering

The microscopic mechanism responsible for CIS has been the subject of many theoretical and experimental studies [5]. The depolarized Rayleigh wing of globular molecules [26-28], the depolarization ratio, as well as the appearance of the forbidden spectra of CS_2 [29-32] and CO_2, have been explained by the scattering mechanism involving multipolar polarizabilities [33-35].

The forbidden spectra of molecules are specially advantageous for the study of the CIS mechanism. The fact that forbidden spectra result completely from CIS and therefore no allowed portion of the spectra needs to be separated, outweighs the experimental difficulties related to their low intensity. Two mechanisms have been proposed to explain the forbidden spectra of linear molecules in the framework of multipolar and higher-order polarizability series expansion: the mechanism involving the dipole-quadrupole polarizability ($A\alpha$ mechanism) and the mechanism involving the first hyperpolarizability ($\beta\theta$ mechanism). The $A\alpha$ contribution arises from reradiated field gradients from the neighboring molecules via the vibrationally modulated quadrupole polarizability, while the $\beta\theta$ has its origin in the vibrationally modulated hyperpolarizability in the presence of the field due to the permanent multipole moments of the neighboring molecules.

By means of experimental studies and computer simulations, Madden and Cox [29-31] and Madden and Tildesley [32] showed for forbidden spectra of CS_2 that the dipole-quadrupole ($A\alpha$) contribution can explain the intensity data, as well as the lineshapes of the ν_2 and ν_3 bands.

If $A\alpha$ mechanism dominates, then the Raman-forbidden bands should be dependent on the magnitude of the polarizabilities of the neighboring molecules, but should be less affected by their permanent moments. Therefore, the ν_2 spectrum of CO_2 1:1 mixed with the solvent molecules Ar, Xe, N_2, and CO [36] was measured at pressure range from 30 MPa to 240 MPa and at two different temperatures, 313 K and 353 K. There were a number of reasons for choosing these solvents. First, Ar and Xe both have zero permanent moments but very different polarizabilities, α_{Xe} = 4.02 and α_{Ar} = 1.64. Second, they are completely miscible with CO_2. Third, no other Raman bands are found in the spectral region of interest (450-900 cm^{-1}). The total intensity of CO_2-Ar, I^{tot} (CO_2-Ar), and CO_2-Xe systems, I^{tot} (CO_2-Xe), were found increasing with the density as shown in Figure 9, indicating the domination of the two-body collision contributions. Furthermore, the intensity ratio I^{tot} (CO_2-Xe) / I^{tot} (CO_2-Ar) was constant over the densities considered at 313 K and its value was 2.2. Since the multipole moments of the two systems are the same, this density-independent intensity ratio implied that the dipole-quadrupole mechanism may be the dominant contribution to the ν_2 forbidden band of CO_2. Applying the method described by Cox and Madden [29-31], this intensity ratio due to the $A\alpha$ mechanism was calculated and its theoretical value of 2.0 was in surprisingly good agreement with the experimental value.

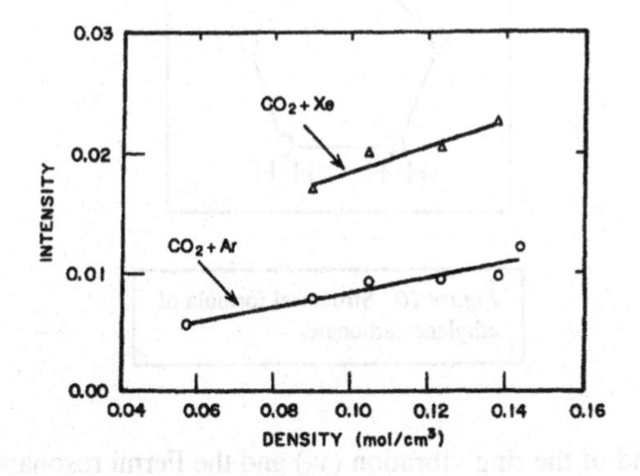

Figure 9. Relative intensities of the forbidden ν_2 band of CO_2 in AR (O) and Xe (Δ) mixtures plotted as a function of density at 313 K. (Taken from [36])

7. Fermi Resonance

The strong dependence of the Fermi resonance upon temperature and pressure is illustrated by the high-pressure Raman study of liquid ethylene carbonate [37]. Ethylene carbonate, 1,3-dioxolane-one (its structural formula is shown in Figure 10), was chosen because it is a molecule with large permanent dipole moment (4.9 *D*) of relatively high symmetry (C_{2v}) and its complete band assignment was available, showing two vibrations, ν_2 and ν_5, with large transition dipole moments. Moreover, there is a Fermi resonance between the ν_2 and the $2\nu_7$ vibration which is sensitive to changes of external parameters like pressure, temperature, and solvent. It will be shown how the analysis of the Fermi resonance enables estimation of the relative changes of the magnitude of the transition dipole moments for both coupled vibrations.

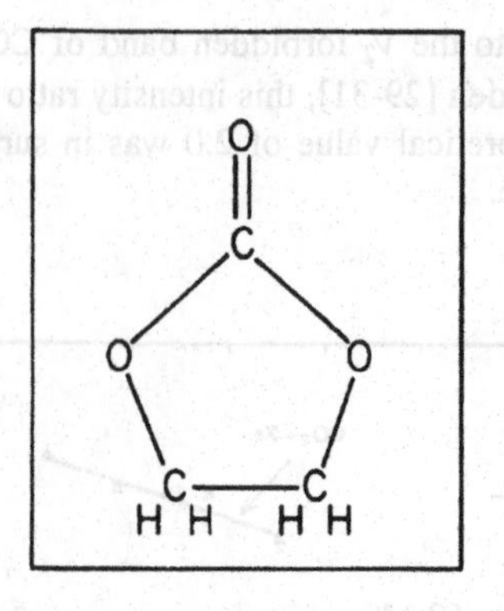

Figure 10. **Structural formula of ethylene carbonate.**

The Raman band of the ring vibration (ν_5) and the Fermi resonance between the carbonyl stretching (ν_2) and the first overtone of the ring breathing vibration ($2\nu_7$) were studied in liquid ethylene carbonate in the pressure range between 0.1 and 300 MPa and at temperatures from 40 to 160°C. The complex structure of Fermi resonance in the C=O stretching region of ethylene carbonate has been interpreted in Raman, as well as in infrared spectra. Below, we will give a short review of the general theory of Fermi resonance. It occurs when two molecular vibrational modes of the same symmetry characterized by wave function ψ_a^0 and ψ_b^0 and frequencies Ω_a and Ω_b are coupled by anharmonic term H^1 in the intramolecular potential function. In effect, the energy levels of both vibrations are changed and the Raman and/or infrared bands are observed at frequencies ν_a and ν_b. The separation between the observed bands $\Delta = \nu_a - \nu_b$ depends on the Fermi resonance parameter $W = \langle \psi_a^0 | H^1 | \psi_b^0 \rangle$ and on the separation between the unperturbed frequencies $\Delta^0 = \Omega_a - \Omega_b$,

$$\Delta^2 = (\Delta^0)^2 + 4W^2 \tag{6}$$

The perturbed wave functions can be expressed as a linear combination of harmonic wave functions

$$\begin{aligned} \psi_a &= s\psi_a^0 - t\psi_b^o \\ \psi_b &= t\psi_a^0 + s\psi_b^o \end{aligned} \tag{7}$$

("*t*" may be negative if $W > 0$). The relative contributions "*s*" and *t* can be expressed as

$$s^2 = \left(\Delta + \Delta^0\right) / 2\Delta$$
$$t^2 = \left(\Delta - \Delta^0\right) / 2\Delta \quad (8)$$

The mixing of the two states increases as the energy difference between the two states Δ^0 becomes smaller. Experimentally, the mixing leads to intensity transfer from one band to another. Therefore, the intensity ratio of the two bands in Fermi resonance R is a very important quantity and provides information on Δ, Δ^0, s, and t.

Following the arguments given in detail in the original study [37], the simplified expression for R is as follows:

$$R = \frac{\Delta - \Delta^0}{\Delta + \Delta^0} \quad (9)$$

The temperature- and pressure-dependence of the ratio R for the $\nu_2 + 2\nu_7$ isotropic bands are shown in Figure 11.

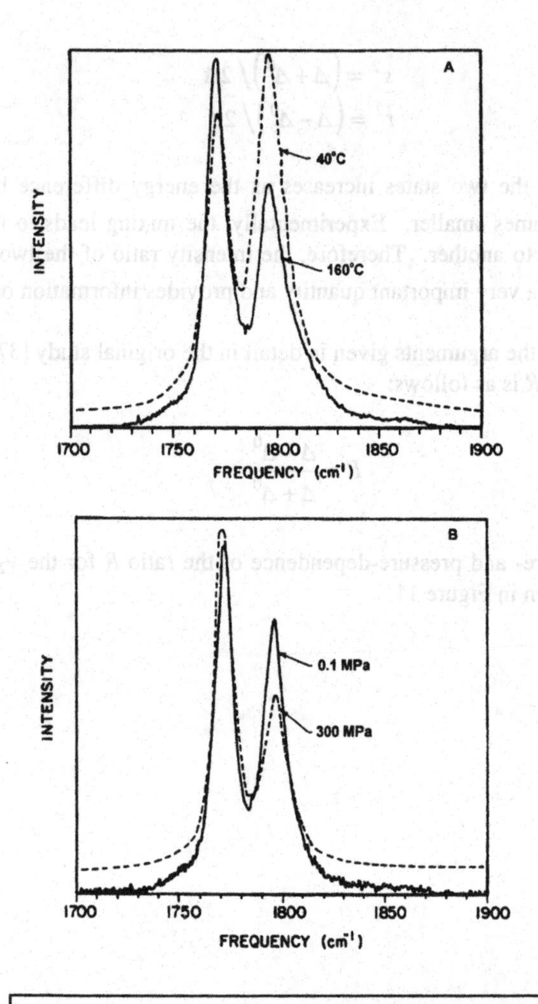

Figure 11. Change of the relative intensities of the two isotropic lines comprising the $\nu_2 + 2\nu_7$ bands, which are coupled by Fermi resonance. (A) Constant pressure (0.1 MPa) temperature is changed from 40 to 160°C; (B) at constant temperature (80°C) pressure is changed from 0.1 to 300 MPa.

The temperature- and pressure-dependence of Fermi resonance in the $\nu_2 + 2\nu_7$ region of ethylene carbonate was analyzed, and transition dipole moments for the coupled vibrations were obtained from the measurements of the intensity rations of the resonant bands. It appears that $\partial\mu/\partial q$ for the bands in Fermi resonance are density-dependent. The frequency splitting $\delta\nu$ between ANISO and ISO parts of $\nu_2 + 2\nu_7$ (which is also density-dependent) is shown to be proportional to $(\partial\mu/\partial q)^2$, indicating that transition dipole moment coupling is responsible for the observed effect. The frequency splitting, $\delta\nu$, for the ν_5 band is also attributed to the transition dipole moment coupling. It is shown that the frequency shifts $\Delta\nu$ of the ν_5 and the $\nu_2 + 2\nu_7$ bands can be explained qualitatively by assuming permanent dipole-dipole interactions.

8. Raman Frequency Noncoincidence Effect

Arising from the coupling of a vibrational mode with the same vibration in neighboring molecules, the Raman noncoincidence effect has been used as a measure of intermolecular interactions. Commonly associated with the Raman bands that are also strong IR active, this effect has been observed in various vibrational modes, most frequently in the C=O stretching modes of carbonyl compounds such as acetone [38], formamide [39], *N,N*-dimethylformamide [40, 41], *N,N*-dimethylacetamide [39], cyclic carbonates [37, 42], and methyl ethyl ketone [43]. The interaction between the vibrational modes of two different molecules leads to a frequency difference between the anisotropic and isotropic Raman bands.

Several theories have been developed to explain the noncoincidence effect, notably by Wang and McHale [44-46], Fini and Mirone [47, 48], and Logan [49-51]. Since the C=O stretching modes have large transition dipole moments, the noncoincidence effect in these bands is considered to arise from the resonant transfer of vibrational excitation via transition dipole-transition dipole coupling. The interactions between the transition dipoles of neighboring molecules have a different effect on the isotropic and anisotropic components of the Raman scattering tensor when there is local order present in the system. The magnitude of the noncoincidence splitting is generally considered to reflect the degree of interaction between two molecules or, more specifically, two vibrating dipoles.

Several studies have been performed employing temperature and/or pressure as experimental variables in order to investigate the sources contributing to the noncoincidence effect and their dependence on temperature and pressure. The temperature dependence of the noncoincidence effect is equivocal. For some systems, the $\Delta\nu$ values decrease with increasing temperature [49, 52, 53]. This has been interpreted to be due to the partial breakdown of regular short-range order of the liquid structure or faster reorientation of the molecules at higher temperature averaging out the interactions that are responsible for the noncoincidence effect. The opposite trend has been observed for the molecules where a resonance structure [39] or hydrogen-bonding structure [54] is present, which may be due to the effect of temperature on these structures. Regarding the effect of pressure, the noncoincidence generally increases with an increase in density as the molecular distances between neighboring molecules become smaller, increasing the intermolecular interaction which increases the coupling of the two transition dipoles.

The results of the recent high-pressure laser Raman scattering study [55] of the noncoincidence effect in a homologous series of alkyl benzoates represents a novel use of the Raman noncoincidence effect to investigate density effects on conformation of the molecules in the liquid state.

2-Ethylhexyl benzoate (EHB) has been studied extensively using NMR [56-60], infrared [61, 62], and Raman [63] spectroscopies as a model compound for elastohydrodynamic lubrication. A previous experiment in our laboratory [60] using

2D NOESY found cross-peaks appearing at the elevated pressure, indicating possible through-space dipolar coupling between the aromatic protons and the methyl protons on the ends of the alkyl chains in EHB. Furthermore, the intensity of the cross-peaks increased with pressure. These experimental results were explained by the approach of a methyl group to the ring at higher density by the bending of alkyl chain toward the aromatic ring. Simple force field calculations suggested that the energy penalty for bending the alkyl chain toward the ring is small [60].

When the alkyl chain of EHB is "folded" toward the benzene ring, it is expected to somewhat shield the carbonyl group of the benzoate from the carbonyl in neighboring molecules. Since the noncoincidence effect reflects the interaction between two vibrational modes, the noncoincidence measurements for the C=O bands allow one to estimate the degree of shielding of the carbonyl group by the folding of the alkyl side chain. We have measured the noncoincidence effect for a homologous series of alkyl benzoates: methyl benzoate (MB), ethyl benzoate (EB), propyl benzoate (PB), butyl benzoate (BB), and hexyl benzoate (HB), as well as EHB. By examining the noncoincidence effect of benzoates of various alkyl chains as a function of pressure, we hoped to obtain the information about the conformational preferences of the alkyl chain in these complex liquids. In addition, by varying the length of the alkyl chain of the benzoate, the effect of chain length on the conformational change can be investigated.

To compare directly the noncoincidence behavior for various alkyl benzoates, $\Delta\nu$ is plotted as a function of packing fraction (ρ^*) in Figure 12. The packing fraction is defined by

$$\rho^* = V_W/V_M \tag{10}$$

where V_W is the (molar) van der Waals volume or the net volume occupied by the molecules excluding free volume. V_W was calculated by Bondi's method [64] of additive molecular group volume increments. In Figure 12, the lines are drawn to show clearly the two regions of the effect of changes in packing fraction. Here the trends mentioned above are very obvious, and the transition points are clearly seen where two lines representing each regime cross. Moreover, the location of the transition point between the two regimes is found to be very similar for the straight chain alkyl benzoates. The packing fraction at which the transition occurs for each of the compounds is shown in Figure 13. The lines in the figure show the packing fraction for each compound as a function of density. The open symbols indicate the packing fraction at ambient temperature and pressure, and the solid symbols indicate the transition points. For EB, PB, BB, and HB, the transition between the two regions is observed at about $\rho^* = 0.64$. For EHB, the only branched chain benzoate in this experiment, the transition packing fraction is higher, $\rho^* = 0.67$.

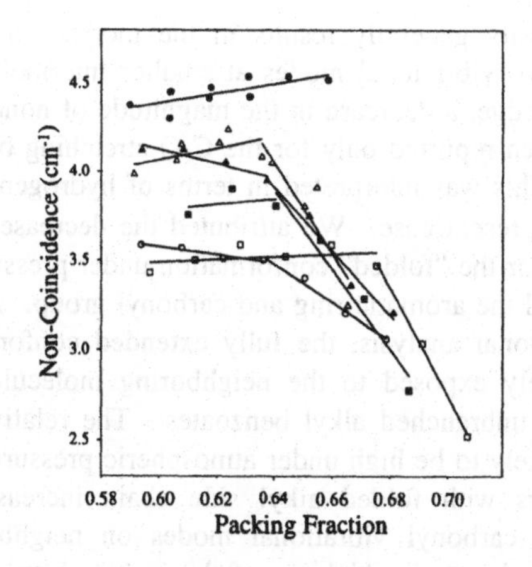

Figure 12. **Noncoincidence values of the alkyl benzoates as a function of packing fraction: MB (●), EB (O), PB (▲), BB (Δ), HB (■), and EHB (□).**

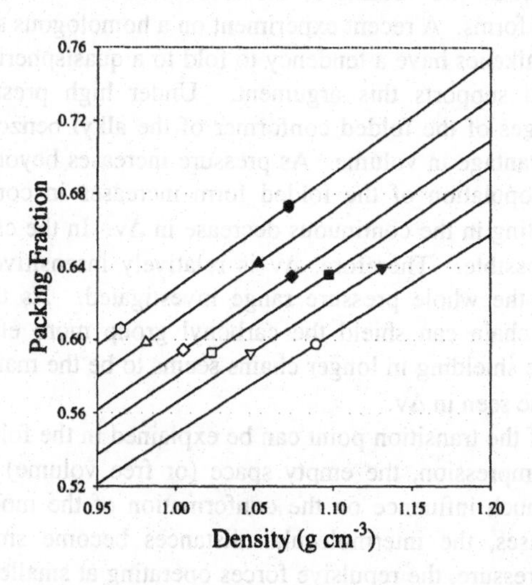

Figure 13. **Transition packing fraction of the alkyl benzoates. A detailed explanation is given in the text. MB (O), EB (∇), PB (□), BB (◇), HB (Δ) and EHB (O).**

The increase in density generally results in the increase in $\Delta\nu$ due to the stronger interaction of two vibrational modes at smaller intermolecular distances. To the best of our knowledge, a decrease in the magnitude of noncoincidence with increasing density has been reported only for the C-O stretching band of methanol [53] and ethanol [54]. This was interpreted in terms of hydrogen bonding, which cannot contribute to the present case. We attributed the decrease in $\Delta\nu$ at higher pressure to a preference for the "folded" conformation under pressure, in which the alkyl chain is bent toward the aromatic ring and carbonyl group. According to the results of the conformational analysis, the fully extended conformation with the carbonyl group completely exposed to the neighboring molecules is the lowest energy conformation for unbranched alkyl benzoates. The relative population of this conformer is most likely to be high under atmospheric pressure. If the relative population of conformers with folded alkyl side chain increases, the average interaction between two carbonyl vibrational modes on neighboring molecules would be reduced due to the steric shielding of the carbonyl group by the folded alkyl chain. This would result in a decrease of $\Delta\nu$.

Considering the fact that the relative molecular volume is the main factor affecting the relative population of the conformers, however, an increased population of the folded conformers at higher pressure should occur owing to the compact nature of the folded conformers. It is known that in hydrocarbon chains the volume of methylene unit is smaller in the gauche conformation than in the trans conformation. Since the folded side chains in alkyl benzoates are formed by a series of gauche conformations, the volume of the folded conformers would be smaller than that of extended forms. A recent experiment on a homologous series of alkanes [65] has shown that alkanes have a tendency to fold to a quasispherical shape under pressure, which also supports this argument. Under high pressure, the slight energetic disadvantages of the folded conformer of the alkyl benzoates are largely overcome by the advantage in volume. As pressure increases beyond the transition point, the relative population of the folded form increases in comparison to the extended form, resulting in the continuous decrease in $\Delta\nu$. In the case of MB, only one conformer is possible. Therefore, $\Delta\nu$ is relatively insensitive to changes in pressure throughout the whole pressure range investigated. As the chain length increases, the alkyl chain can shield the carbonyl group more effectively. This increasingly effective shielding in longer chains seems to be the main reason for the larger rate of decrease seen in $\Delta\nu$.

The existence of the transition point can be explained in the following way. At the first stage of compression, the empty space (or free volume) in the liquid is decreased without much influence on the conformation of the molecules. As the empty space decreases, the intermolecular distances become smaller, and with further increase of pressure, the repulsive forces operating at smaller intermolecular distances compel the molecules to change confirmation to the compact form. According to this interpretation, the point at which the noncoincidence starts to decrease corresponds to the density, or packing fraction, above which the intermolecular distances are so close that the interaction between molecules would induce the change of the conformation of the molecules. For all unbranched alkyl benzoates other than MB, the transition between the two regions was observed at about $\rho^* = 0.64$, as shown in Figure 13. It is interesting to note that 0.64 is the packing fraction of random close-packed spheres [66-68], which is the most compact packing possible for spheres without introducing crystalline order. For EHB, with a branched alkyl chain, the transition is at a higher packing fraction, which may be due to either a less spherical folded conformation or a smaller volume change upon folding, reducing the preference for the folded conformer.

9. Acknowledgments

This work was supported in part by the National Science Foundation under grant NSF CHE 95-26237.

10. References

1. Herzberg, G. (1945) *Molecular Spectra and Molecular Structure, II Infrared and Raman Spectra of Polyatomic Molecules*, D. Van Nostrand Co., New York.
2. Jonas, J. (1984) *Acc. Chem. Res.* **17,** 74.
3. Zerda, T.W., Song, X., and Jonas, J. (1985) *Chem. Phys.* **94,** 427.
4. Jonas, J. (1987) *J. Chem. Soc., Faraday Trans. 2* **83,** 1777.
5. Jonas, J. and Lee, Y.T. (1991) *J. of Phys.: Condens. Matter* **3,** 305.
6. Bradley, M., Zerda, T.W., and Jonas, J. (1984) *Spectrochim. Acta* **40A,** 1117.
7. Cantor, D.M., Schroeder, J., and Jonas, J. (1975) *Appl. Spectrosc.* **29,** 393.
8. Perry, S., Sharko, P.T., and Jonas, J. (1983) *Appl. Spectrosc.* **37,** 340.
9. Perry, S., Zerda, T.W., and Jonas, J. (1981) *J. Chem. Phys.* **75,** 4214.
10. Zerda, T.W., Perry, S., and Jonas, J. (1981) *Chem. Phys. Lett.* **83,** 600.
11. Schroeder, J., Schiemann, V.H., and Jonas, J. (1977) *J. Mol. Phys.* **34,** 1501.
12. Alder, B.J., Weis, J.J., and Strauss, H.L. (1973) *Phys. Rev. A* **7,** 281.
13. Alder, B.J., Strauss, H.L., and Weis, J.J. (1973) *J. Chem. Phys.* **59,** 1002.
14. Oxtoby, D.W. (1981) *Ann. Rev. Phys. Chem.* **32,** 77.
15. Fischer, S.F. and Labereau, A. (1975) *Chem. Phys. Lett.* **35,** 6.
16. Oxtoby, D.W. and Rice, S.A. (1979) *Chem. Phys. Lett.* **42,** 1.
17. Oxtoby, D.W. (1979) *J. Chem. Phys.* **70,** 2605.
18. Diestler, D.J. and Manz, J. (1977) *Mol. Phys.* **33,** 227.
19. Madden, P.A. (1982) *Raman Spectroscopy – Linear and Non-linear*, Wiley, New York.

20. Schindler, W. and Jonas, J. (1980) *J. Chem. Phys.* **72,** 5019; (1980) *J. Chem. Phys.* **73,** 3547.

21. Rothschild, W.G. (1976) *J. Chem. Phys.* **65,** 455.

22. Kubo, R. (1962) *Fluctuations, Relaxation, and Resonance in Magnetic Systems*, Plenum, New York.

23. Schindler, W. and Jonas, J. (1979) *Chem. Phys. Lett.* **67,** 428.

24. Schindler, W., Sharko, P.T., and Jonas, J. (1982) *Chem. Phys.* **76,** 3493.

25. Schweizer, K.S. and Chandler, D. (1982) *J. Chem. Phys.* **76,** 2296.

26. Buckingham, A.D. and Tabisz, J.C. (1977) *Opt. Lett.* **1,** 220; *Mol. Phys.* **36,** 583.

27. Penner, A.R., Meinander, N., and Tabisz, G.C. (1985) *Mol. Phys.* **54,** 479.

28. Posch, H.A. (1982) *Mol. Phys.* **37,** 1059; (1980) *Mol. Phys.* **40,** 1137; (1982) *Mol. Phys.* **46,** 1213.

29. Cox, T.I. and Madden, P.A. (1980) *Mol. Phys.* **39,** 1487.

30. Madden, P.A. and Cox, T.I. (1981) *Mol. Phys.* **43,** 287.

31. Cox, T.I. and Madden, P.A. (1981) *Mol. Phys.* **43,** 307.

32. Madden, P.A. and Tildesley, D.J. (1983) *Mol. Phys.*, **49,** 193; (1985) *Mol. Phys.* **55,** 969.

33. Holzer, W. and Ouillon, R. (1978) *Mol. Phys.* **36,** 817.

34. Amos, R.D., Buckingham, A.D., and Williams, J.H. (1980) *Mol. Phys.* **39,** 1519.

35. Madden, P.A. and Cox, T.I. (1985) *Mol. Phys.* **56,** 223.

36. Song, X., Frattini, R., and Jonas, J. (1988) *J. Chem. Phys.* **89,** 156.

37. Schindler, W., Zerda, T.W., and Jonas, J. (1984) *J. Chem. Phys.* **81,** 4306.

38. Bradley, M.S. and Krech, J.H. (1993) *J. Phys Chem.* **97,** 575.

39. Thomas, H.D. and Jonas, J. (1989) *J. Chem. Phys.* **90,** 4144.

40. Shelley, V.M. and Yarwood, J. (1991) *Mol. Phys.* **72,** 1407.

41. Shelley, V.M. and Yarwood, J. (1989) *Chem. Phys.* **137,** 277.

42. Sun, T.F., Chan, J.B., Wallen, S.L., and Jonas, J. (1991) *J. Chem. Phys.* **94,** 7486.

43. Scheibe, D. (1982) *J. Raman Spectrosc.* **13,** 103.

44. Wang, C.H. and McHale, J. (1980) *J. Chem. Phys.* **72,** 4039.

45. McHale, J. (1981) *J. Chem. Phys.* **75,** 30.

46. McHale, J. (1982) *J. Chem. Phys.* **77,** 2705.

47. Mirone, P. and Fini, G. (1979) *J. Chem. Phys.* **71,** 2241.

48. Giorgini, M.G., Fini, G., and Mirone, P. (1983) *J. Chem. Phys.* **79,** 639.

49. Logan, D.E. (1986) *Mol. Phys.* **58,** 97.

50. Logan, D.E. (1986) *Chem. Phys.* **103,** 215.

51. Logan, D.E. (1989) *Chem. Phys.* **131,** 199.

52. Mirone, P. and Fini, G. (1974) *J. Chem Soc., Faraday Trans. 2* **70,** 1776.

53. Thomas, H.D. and Jonas, J. (1989) *J. Chem. Phys.* **90,** 4632.

54. Zerda, T.W., Thomas, H.D., Bradley, M., and Jonas, J. (1987) *J. Chem. Phys.* **86,** 3219.

55. Slager, V.L., Chang, H.-C., Kim, Y.J., and Jonas, J. (1997) *J. Phys. Chem. B* **101,** 9774.

56. Adamy, S.T., Grandinetti, P.J., Masuda, Y., Campbell, D.M., and Jonas, J. (1991) *J. Chem. Phys.* **94,** 3568.

57. Jonas, J., Adamy, S.T., Grandinetti, P.J., Masuda, Y., Morris, S., Campbell, D.M., and Li, Y. (1990) *J. Phys. Chem.* **94,** 1157.

58. Walker, N.A., Lamb, D.M., Adamy, S.T., Jonas, J., and Dare-Edwards, M.P. (1988) *J. Phys. Chem.* **92,** 3675.

59. Zhang, J. and Jonas, J. (1994) *J. Phys. Chem.* **98,** 6835.

60. Adamy, S.T., Kerrick, S.T., and Jonas, J. (1994) *Z. Phys. Chem. (Munich)* **184,** 185.

61. Whitley, A., Yarwood, J., and Gardiner, D.J. (1993) *J. Chem. Soc., Faraday Trans.* **89,** 881.

62. Whitley, A., Yarwood, J., and Gardiner, D.J. (1991) *J. Mol. Struct. (THEOCHEM)* **247,** 187.

63. Whitley, A., Yarwood, J., and Gardiner, D.J. (1990) *Ber. Bunsen-Ges. Phys. Chem.* **94,** 404.

64. Bondi, A. (1968) *Physical Properties of Molecular Crystals, Liquids, and Glasses*, John Wiley & Sons, Inc., New York.

65. Brüsewitz, M. and Weiss, A. (1993) *Ber. Bunsen-Ges. Phys. Chem.* **97,** 1.

66. Bernal, J.D. and Mason, J. (1960) *Nature* **188,** 910.

67. Scott, G.D. (1960) *Nature* **188,** 908.

68. Berryman, J.G. (1983) *Phys. Rev. A* **27,** 1053.

PRESSURE EFFECTS ON NON-LINEAR OPTICAL PHENOMENA

H. G. DRICKAMER, Y. LI, G. LANG and Z. A. DREGER
School of Chemical Sciences, Dept. of Physics
and Frederick Seitz Materials Research Laboratory
University of Illinois
600 S. Mathews
Urbana, IL 61801-3792

1. Abstract

The effect of pressure has been measured for two non-linear optical phenomena: second harmonic generation (SHG) in organic crystals and one and two photon fluorescence excitations in a molecule dissolved in an organic polymer. Both sets of results are discussed in terms of theory.

2. Introduction

A major aspect of modern materials development concerns non-linear optical properties of crystals and powders. In this paper we present some exploratory studies of two aspects of this phenomenon: second harmonic generation (SHG) and one and two photon excitations involving organic crystals and powders for SHG and Rhodamine B in solution in polyacrylic acid (PAA) for the one and two photon investigation.

The basis for non-linear optical phenomena can be established in the following way.

The bulk polarization of a medium can be written:

$$P = P_O + \chi^{(1)} \cdot E + \chi^{(2)} \cdot E \cdot E + \chi^{(3)} \cdot E \cdot E \cdot E + \ldots.$$

where the $\chi^{(i)}$ susceptibility coefficients are tensors of order (i + 1) (e.g., $\chi^{(2)}$ has tensor elements $\chi^{(2)ijk}$). P_0 is the built-in static dipole of the sample. The electric field of a plane light wave is: $E = E_0 \cos(\omega t)$ so that for an arbitrary point in space the polarization can be written as:

R. Winter and J. Jonas (eds.), High Pressure Molecular Science, 25–46.

$$P = P_0 + \chi^{(1)} E_0 \cos(\omega t) + \chi^{(2)} E_0^2 \cos^2(\omega t) + \chi^{(3)} E_0^3 \cos^3(\omega t) + \ldots \quad (1)$$

Since $\cos^2(\omega t)$ equals 1/2 + 1/2 cos(2ωt), the first three terms can be written:

$$P = \left(P_0 + 1/2\chi^{(2)} E_0^2\right) + \chi^{(1)} E_0 \cos(\omega t) + 1/2\chi^{(2)} E_0^2 \cos(2\omega t) + \ldots \quad (2)$$

3. Second Harmonic Generation (SHG)

Certain classes of crystals (or powders), when exposed to a laser beam of, say 1000 nm (10,000 cm^{-1}) will transmit or reflect a certain fraction of the energy at 500 nm (20,000 cm^{-1}). (See, e.g. ref 1-6) For SHG to occur it is necessary that neither the crystal lattice nor the molecule or molecular ions which make up the crystal have a center of symmetry.

SHG is widely used for generating laser beams in the visible or near UV from IR lasers. In general, inorganic crystals are used commercially, but organic crystals can have very high efficiencies. They are, however fragile and subject to attack chemically. Nevertheless, properly contained, they may have practical applications, and an understanding of their behavior with pressure can contribute to our insight into the SHG process.

We present here studies on three organic crystals: 4-Br-4′ methoxychalcone (BMC), (7) 4-aminobenzophenone (ABP) (7) and 3-(4-chlorophenyl)-1-(3 thienyl) propen-1-one (CTPO). The structures are given in Fig. 1. BMC and ABP are monoclinic crystals with well established point groups. The refractive indices are given as a function of wave length along the x-y-z optical axes. (The optical axes do not necessarily correspond exactly to the crystallographic axes.) There is considerably less information about CTPO but the pressure results form an interesting counterpoint to those for BMC and ABP.

The details of the experimental apparatus and procedure appear in the original paper (7). For the crystals the laser beam is transmitted through the crystal, reflected off an aluminum foil and transmitted back out the front diamond. There may also be a modest amount of reflection off the surface. The data for the powder are entirely by reflection. Measurements were made as a function of the angle of polarization of the incident laser light, which was the 1053 nm line of a mode locked Nd:YLF laser. The measurements were made at 526.5 μm using appropriate filters to control the intensity and exclude light of the fundamental frequency.

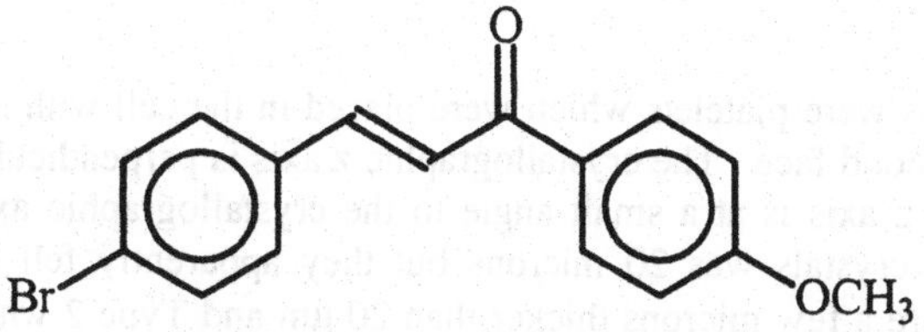

4-Br-4'methoxychalcone (BMC)

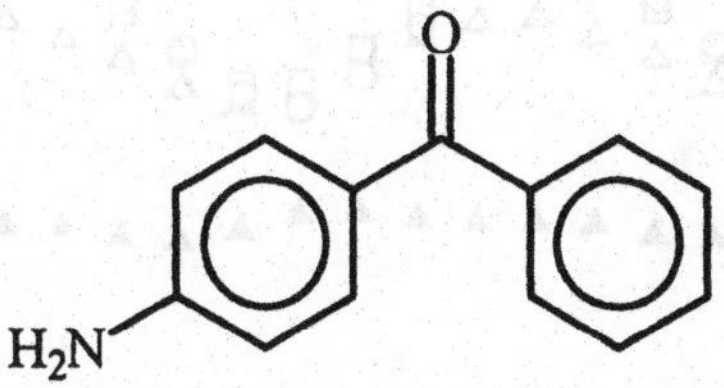

4-Aminobenzophenone (ABP)

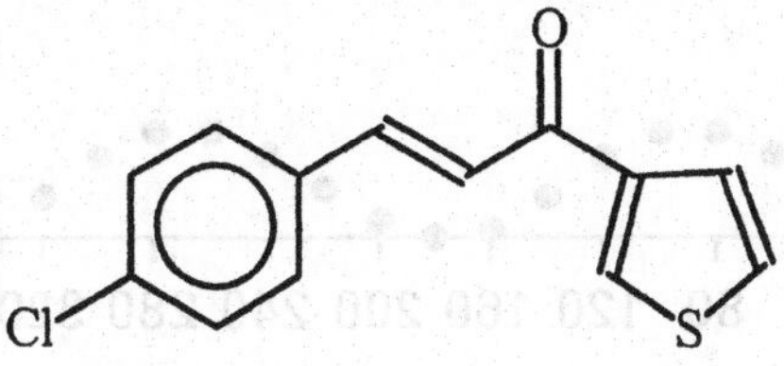

3-(4-Chlorophenyl)-1-(3-thienyl)propen-1-one

(CTPO)

Figure 1. Structure of molecules studied for SHG.

3.1 BMC

The BMC crystals were platelets which were placed in the cell with the plane surface parallel to the diamond face. The crystallographic z axis is perpendicular to the crystal face. The optical z axis is at a small angle to the crystallographic axis. The nominal thickness of the crystals was 20 microns but they apparently fell into two groups, Type 1, which were a few microns thicker than 20 μm and Type 2 which were thinner.

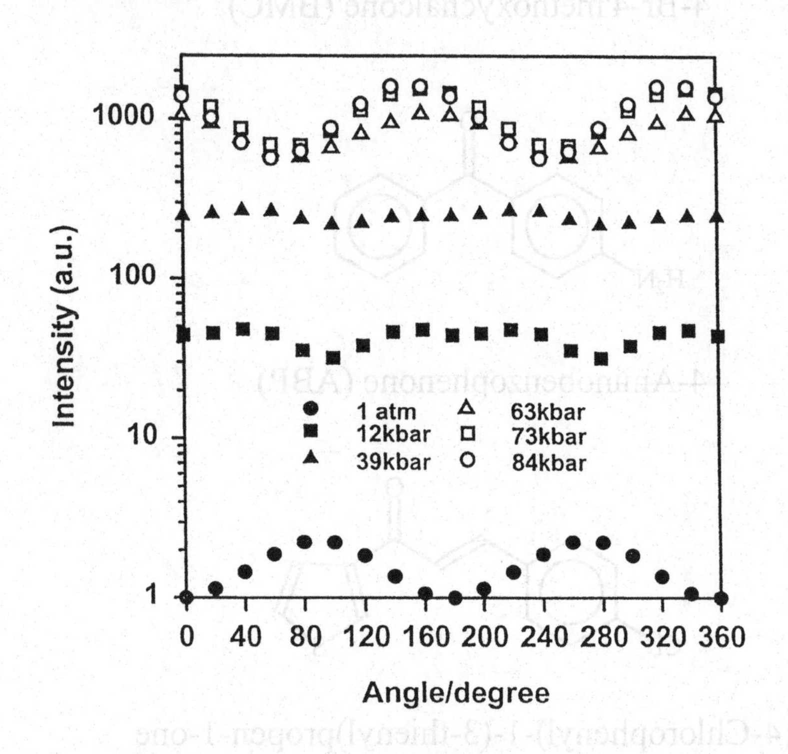

Figure 2. Typical SHG intensity as a function of polarization angle of the laser beam for BMC crystal for several pressures. (Reprinted with permission from L. Yi, G. Yang, Z. A. Dreger and H. G. Drickamer 1998 J. Physical Chemistry 102B, 5963-68. Copyright 1998 Amer. Chemical Society)

In Fig 2 we present the intensity as a function of laser polarization angle (here after referred to as "angle") for several pressures. It is evident that a phase transition initiates at ~ 35 kbar and above ~ 60 kbar there is clearly a different structure. It is our belief that it is a different point group of the monoclinic system. Fig 3 and 4 present the intensity as a function of pressure for Type 1 and Type 2 crystals measured at the angle for maximum change and for the angle at which the one atmosphere maximum occurs. The efficiency of SHG generation increases by several orders of magnitude. In Fig 5 we present the SHG efficiency for the powder (≤ 10 μm). The results are independent of angle and show a relatively small increase. The location of the phase transition is evident. The optical absorption data (not shown here) also give some evidence for a phase transition.

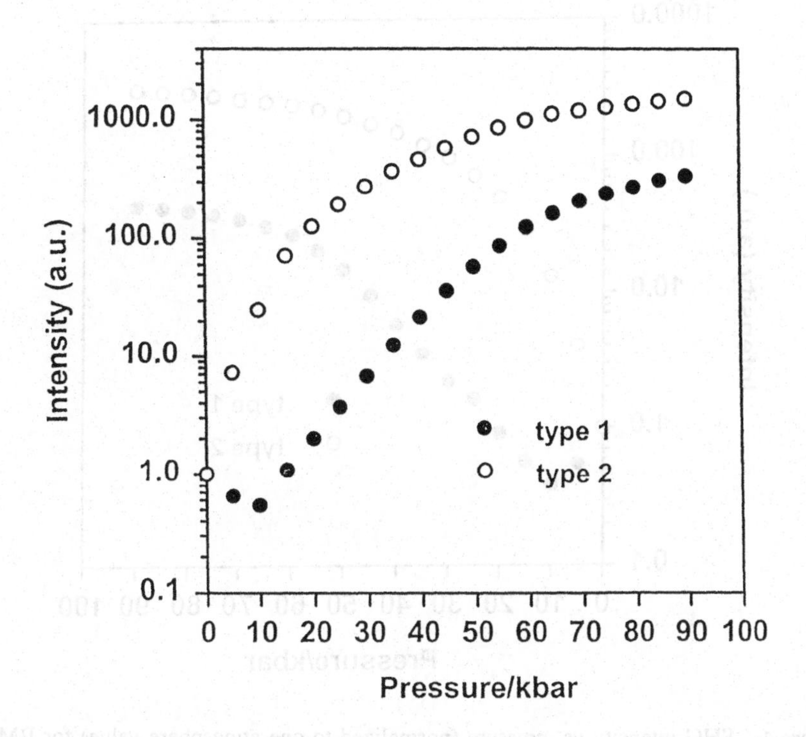

Figure 3. SHG intensity vs. pressure (normalized to one atmosphere value) for BMC at the angle giving maximum intensity increase – averaged over several loads for two crystal types. (Reprinted with permission from L. Yi, G. Yang, Z. A. Dreger and H. G. Drickamer 1998 J. Physical Chemistry 102B, 5963-68. Copyright 1998 Amer. Chemical Society)

Up to ~ 35kbar the intensity is completely reversible upon release of pressure both for crystal and powder. From 70-90 kbar the intensities relax to their value at ~ 40 kbar (just above the transition). There is some further relaxation over a period of 1-2 days, but never completely to the original value.

For BMC it is possible to give a fairly thorough, if somewhat qualitative analysis.

In general, one can write the non-linear polarizability in the form.

$$P_{ik}(t) = d_{ijk} E_i(t) E_k(t) \tag{3}$$

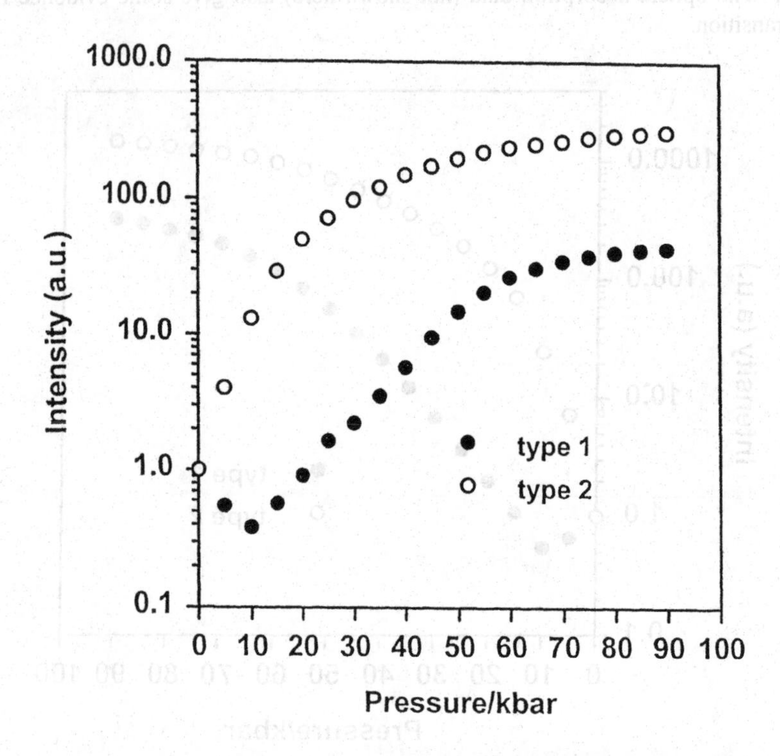

Figure 4. SHG intensity vs. pressure (normalized to one atmosphere value) for BMC at the angle for maximum intensity at one atmosphere – averaged over several loads for two crystal types. (Reprinted with permission from L. Yi, G. Yang, Z. A. Dreger and H. G. Drickamer 1998 J. Physical Chemistry 102B, 5963-68. Copyright 1998 Amer. Chemical Society)

Since the exchange of i and k has no physical significance, the coefficients (d) can be expressed as a 3 x 6 matrix. For all space and point groups which have the requirements to permit SHG the non zero components of (d) have been established (8).

SHG conversion efficiency in the plane-wave fixed-field approximation can be expressed as (9):

$$\frac{I^{2\omega}}{I^{\omega}} \approx \frac{d_{eff}^{2} L^{2} I^{\omega}}{(n^{\omega})^{2} n^{2\omega}} \left[\frac{\sin^{2}(\Delta kL/2)}{(\Delta kL/2)^{2}} \right] \tag{4}$$

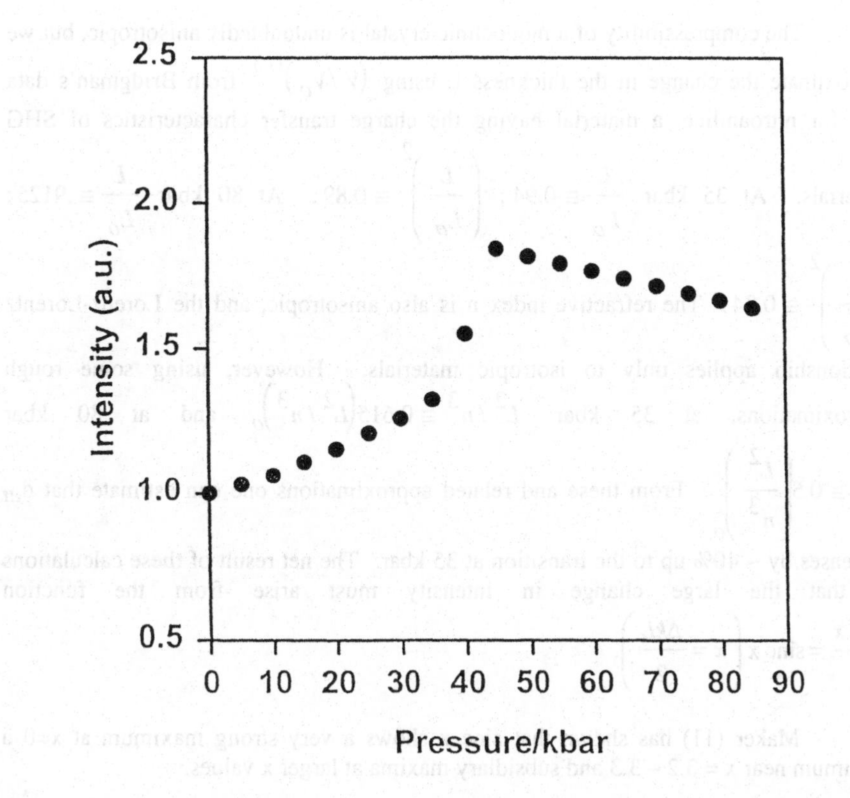

Figure 5. SHG intensity vs. pressure (normalized to one atmosphere value) for BMC powder averaged over several loads. (Reprinted with permission from L. Yi, G. Yang, Z. A. Dreger and H. G. Drickamer 1998 J. Physical Chemistry 102B, 5963-68. Copyright 1998 Amer. Chemical Society)

where $I^{2\omega}$, I^{ω} are the second harmonic and laser intensities, respectively. d_{eff} is the effective nonlinear coefficient.

L is the thickness of the nonlinear crystal in the direction of the incident beam.

$n^{2\omega}$, n^{ω} are the refractive indices, respectively for the second harmonic and fundamental waves.

Δk is the wave vector mismatch between the coupled fields:

$$\Delta k = \frac{4\pi}{\lambda}\left(n^{2\omega} - n^{\omega}\right) \tag{5}$$

The compressibility of a monoclinic crystal is undoubtedly anisotropic, but we approximate the change in the thickness L using $(V/V_o)^{1/3}$ from Bridgman's data (10) for nitroaniline, a material having the charge transfer characteristics of SHG materials. At 35 kbar $\frac{L}{L_o} \cong 0.94$; $\left(\frac{L}{L_o}\right)^2 \cong 0.89$. At 80 kbar $\frac{L}{L_o} \cong .9125$; $\left(\frac{L}{L_o}\right)^2 \cong 0.84$. The refractive index n is also anisotropic, and the Lorenz-Lorentz relationship applies only to isotropic materials. However, using some rough approximations, at 35 kbar $L^2/n^3 \cong 0.615\left(L^2/n^3\right)_o$ and at 80 kbar $\frac{L^2}{n^3} \cong 0.5\left(\frac{L^2}{n^3}\right)_o$. From these and related approximations one can estimate that d_{eff} increases by ~ 40% up to the transition at 35 kbar. The net result of these calculations is that the large change in intensity must arise from the function $\frac{Sin\,x}{x} = \text{sinc } x \left(x = \frac{\Delta kL}{2}\right)$.

Maker (11) has shown that sinc x shows a very strong maximum at x=0 a minimum near x = 3.2 – 3.3 and subsidiary maxima at larger x values.

If we assume as a first approximation that the optical and crystallographic axes coincide we can write the major contribution to the SHG intensity:

$$I^{2\omega} \approx d_{26}^2 Sinc^2\left[\frac{2\pi L}{\lambda}\left(n_y^{2\omega} - \frac{n_x^{\omega} + n_y^{\omega}}{2}\right)\right] \tag{6}$$

Using the one atmosphere refractive indices (9), depending on the initial thickness x = 3 -3.5. As can be seen from Fig 6, the thinner crystals would give the sharp continuous increase observed, while the slightly thicker crystals would give the initial small drop and then the increase. The decrease in L with pressure would combine with the "pressure tuning" of Δk to reduce the value of $x = \frac{\Delta kL}{2}$ by the appropriate amount.

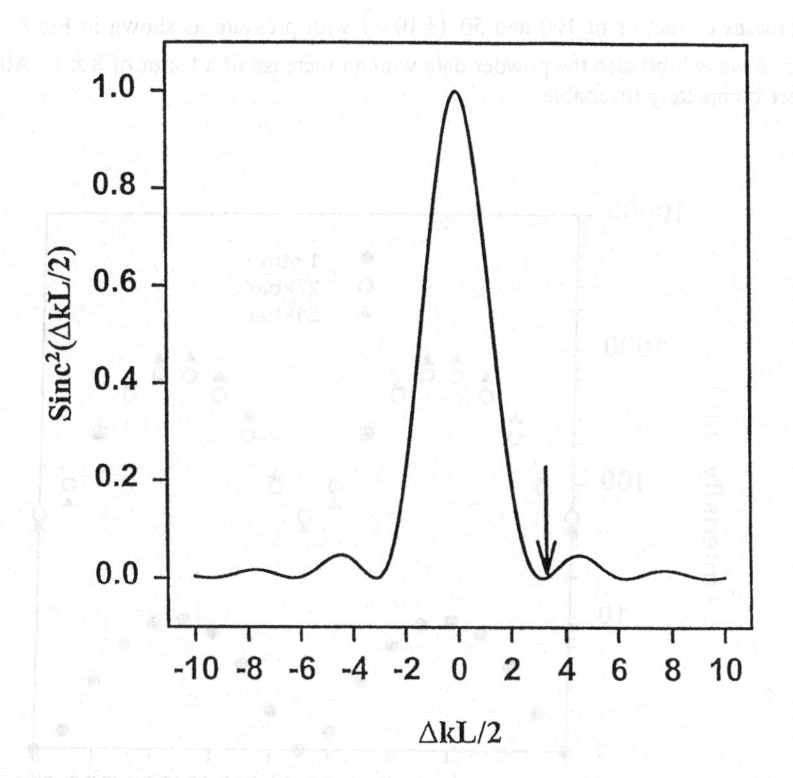

Figure 6. Sinc $\frac{\Delta kL}{2}$ vs. $\frac{\Delta kL}{2}$. The arrow indicates the approximate one atmosphere value for $\frac{\Delta kL}{2}$ for BMC. (Reprinted with permission from L. Yi, G. Yang, Z. A. Dreger and H. G. Drickamer 1998 J. Physical Chemistry 102B, 5963-68. Copyright 1998 Amer. Chemical Society)

3.2 ABP

ABP is also monoclinic, but it crystallizes as needles rather than as plates. The long axis is the y axis which was always oriented vertically, but there was no way to control the orientation in the x-z plane for such small needles. Fig 7 exhibits the SHG as a function of angle at several pressures, for a typical load. Although there was no control over the x-z orientation the data seemed to cluster around orientations giving increases of factors of 100 and 50 $(\pm 10\%)$ with pressure as shown in Fig 8. In this figure we exhibit also the powder data with an increase of a factor of 8 ± 1. All results were completely reversible.

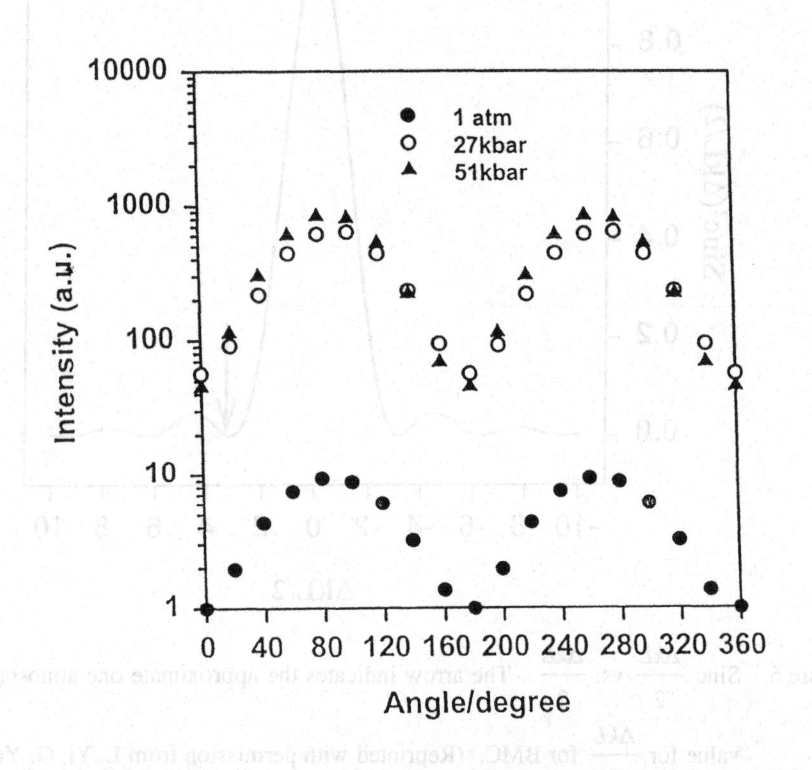

Figure 7. Typical SHG intensity as a function of polarization angle of the laser beam for ABP crystal at several pressures. (Reprinted with permission from L. Yi, G. Yang, Z. A. Dreger and H. G. Drickamer 1998 J. Physical Chemistry 102B, 5963-68. Copyright 1998 Amer. Chemical Society)

The crystal data give no indication of a phase transition, but it seems very probable that there is one near 20 kbar. The powder SHG which shows no angle dependence up to 20 kbar, at higher pressure shows a dependence 5-10% of that for the crystal which indicates some internal reorganization. The optical absorption date (not shown) also exhibits a modest discontinuity in the pressure region above 20 kbar. Any rearrangement must be minor.

Because of the uncertainty in the orientation it is not possible to make a calculation as clear as that for BMC. However, a rough calculation assuming various orientations in the x-z plane indicate values of $\frac{\Delta kL}{2}$ in the order of 2.5-3.5 which, with the estimated decrease in L and some tuning of Δk, would be very consistent with the observed results.

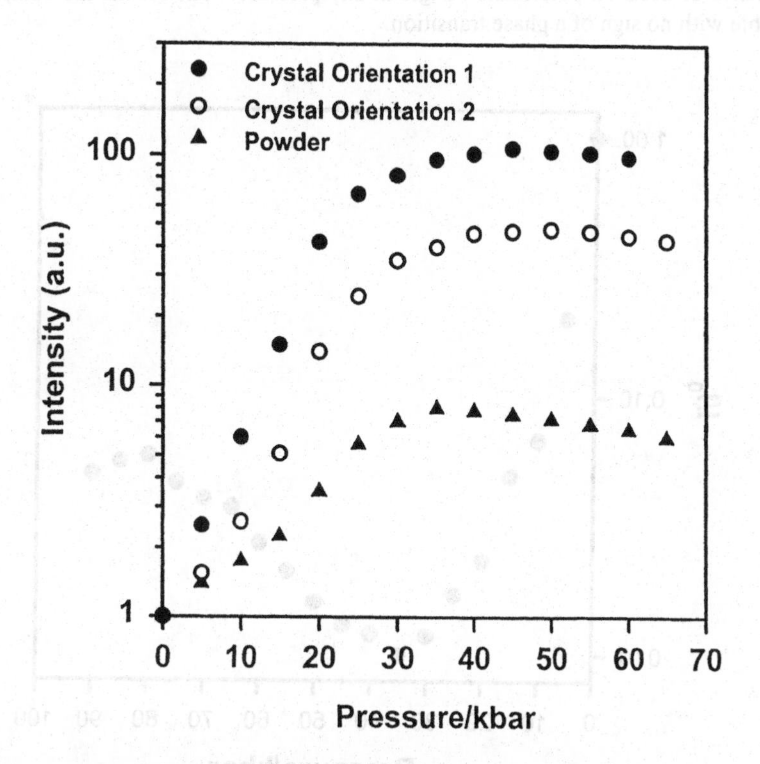

Figure 8. SHG intensity vs. pressure (normalized to one atmosphere value) for ABP (two crystal orientations) and the powder average value for several loads in each case. (Reprinted with permission from L. Yi, G. Yang, Z. A. Dreger and H. G. Drickamer 1998 J. Physical Chemistry 102B, 5963-68. Copyright 1998 Amer. Chemical Society)

3.3 CTPO

For CTPO the possible explanations of the results are considerably more tenuous because of lack of crystallographic and refractive index data, but the results are included here because they illustrate a significantly different behavior. The crystal data are presented in Fig 9 and the powder data in Fig 10. For the crystal the SHG efficiency drops by a factor of ~ 100 by 35 kbar and then increases by a factor of ~ 5 by 90 kbar. For the powder, the efficiency drops by a factor of ~ 2 by 55 kbar and then decreases more rapidly so that the 90 kbar the overall decrease is by a factor of 10 vis-à-vis one atmosphere. For the crystal the SHG dependence on the angle of polarization of the laser beam was essentially independent of pressure. For the powder there was no dependence of SHG on polarization angle at any pressure. The results are completely reversible with no sign of a phase transition.

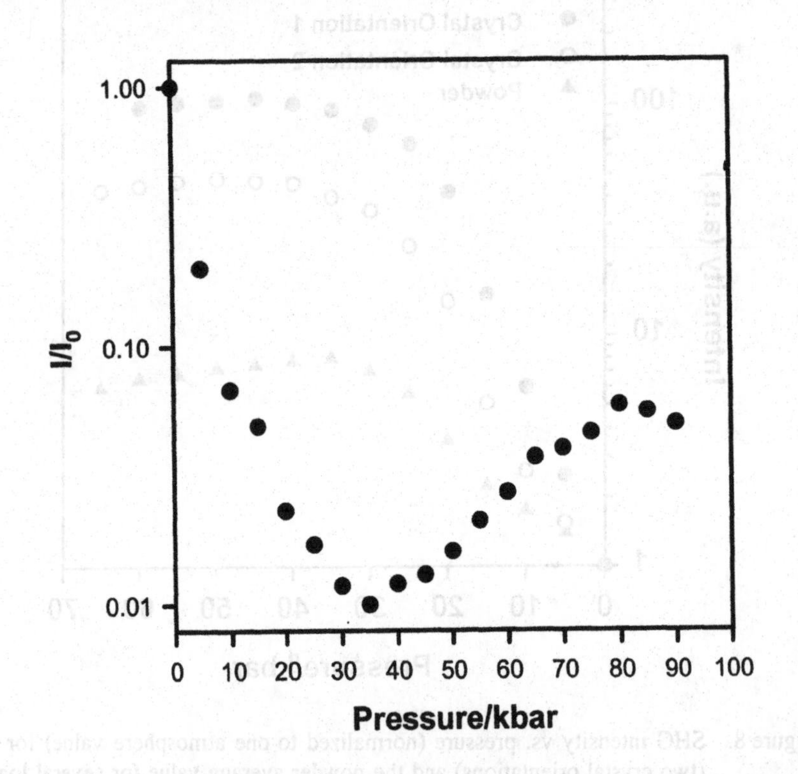

Figure 9. SHG intensity vs. pressure (normalized to one atmosphere value) for CTPO – averaged over several loads.

Little is known about the structure or refractive indices so the initial value of $\frac{kL}{2}$ is not known. If one uses the powder data as a rough correction for the coefficient of sinc $\frac{kL}{2}$ one can estimate values for sinc $\frac{\Delta kL}{2}$ vs pressure.

If the initial value of $\frac{kL}{2}$ is ~ at or near the first subsidiary maximum (See Fig 6) a decrease in $\frac{kL}{2}$ 15-20% in 90 kilobars would reproduce the value obtained closely.

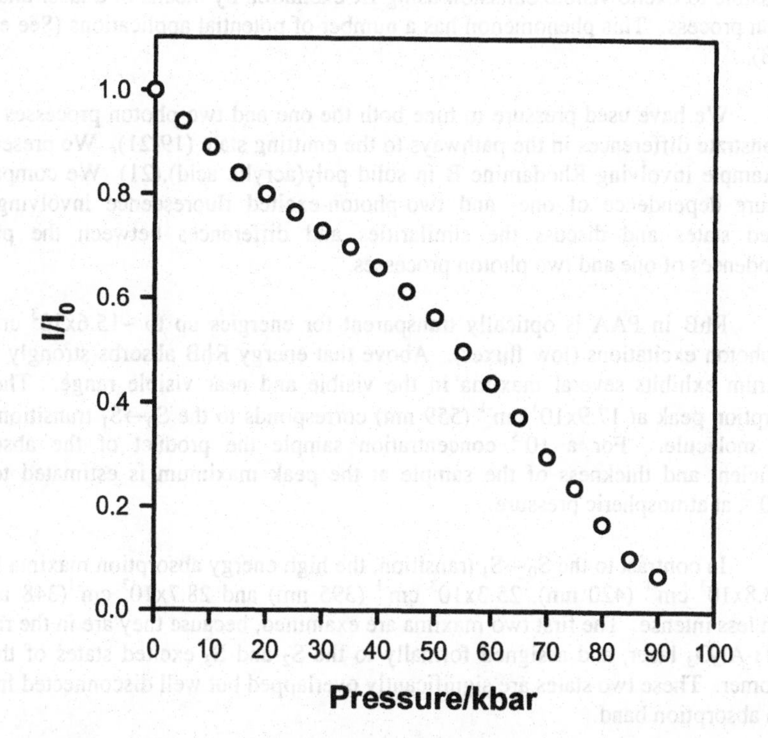

Figure 10. SHG intensity vs. pressure for CTPO powder normalized to one atmosphere value) – averaged over several loads.

These studies of SHG represent an exploratory investigation and can not be compared in a quantitative way with theory because it is not at present practical to measure the necessary changes in lattice parameter and refractive indices in the anisotropically compressible media. They illustrate the possibility of introducing rather spectacular changes in SHG with pressure which can be qualitatively explained and add to our understanding of the SHG process.

4. One and two photon processes

Normally fluorescence occurs by optical excitation via a Franck-Condon process to a vibrationally excited level of the S_1 electronic state, rapid relaxation to the lowest vibrational level of this state and then emission of the fluorescence at an energy lower than the absorption. The difference in energy between the absorption and emission maxima is known as the Stokes shift.

It has been shown that for some molecules with non-linear optical properties it is possible to excite visible emission using IR excitation by means of a laser and a two photon process. This phenomenon has a number of potential applications (See e.g. ref. 12-18).

We have used pressure to tune both the one and two photon processes and to demonstrate differences in the pathways to the emitting state (19-21). We present here an example involving Rhodamine B in solid poly(acrylic acid).(21) We compare the pressure dependence of one- and two-photon-excited fluorescence involving three excited states and discuss the similarities and differences between the pressure dependences of one and two photon processes.

RhB in PAA is optically transparent for energies up to $\sim 15.6 \times 10^3$ cm^{-1}, for one-photon excitations (low fluxes). Above that energy RhB absorbs strongly and its spectrum exhibits several maxima in the visible and near visible range. The main absorption peak at 17.9×10^3 cm^{-1} (559 nm) corresponds to the $S_0 \rightarrow S_1$ transition in the RhB molecule. For a 10^{-5} concentration sample the product of the absorption coefficient and thickness of the sample at the peak maximum is estimated to be ~ 6×10^{-2}, at atmospheric pressure.

In contrast to the $S_0 \rightarrow S_1$ transition, the high energy absorption maxima located at 23.8×10^3 cm^{-1} (420 nm), 25.3×10^3 cm^{-1} (395 nm) and 28.7×10^3 cm^{-1}(348 nm) are much less intense. The first two maxima are examined, because they are in the range of the Ti-Al_2O_3 laser, and assigned formally to the S_2 and S_3 excited states of the RhB monomer. These two states are significantly overlapped but well disconnected from the main absorption band.

The first excited state of RhB in PAA was reached either by a 1053 nm line (two-photon absorption) of the Nd-YLF laser (250 W peak power and 50 ps pulse duration, at 76 MHz) or its second harmonic at 526.5 nm (one-photon absorption). The excitation of higher energy states was by means of a titanium-sapphire (Ti-Al_2O_3) laser (13 kW peak power and 100 fs pulse duration, at 76 MHz) or its second harmonic generated with a BBO crystal. The beam of the laser was focused on a 7×10^{-2} mm^2 area of the sample. The fluorescence was corrected for the spectral sensitivity of the detection system. All measurements are made at room temperature. The details of the experimental setup have been described elsewhere (19-21).

In Fig 11 we present the peak location as a function of pressure for both the one and two photon processes at all excitation wave lengths. Similarly the lifetimes appear in Fig 12. Although the one photon processes involve three different excited states, and, as we shall discuss below, the two photon processes apparently involve still other pathways, the emission in all cases is $S_1 \rightarrow S_0$.

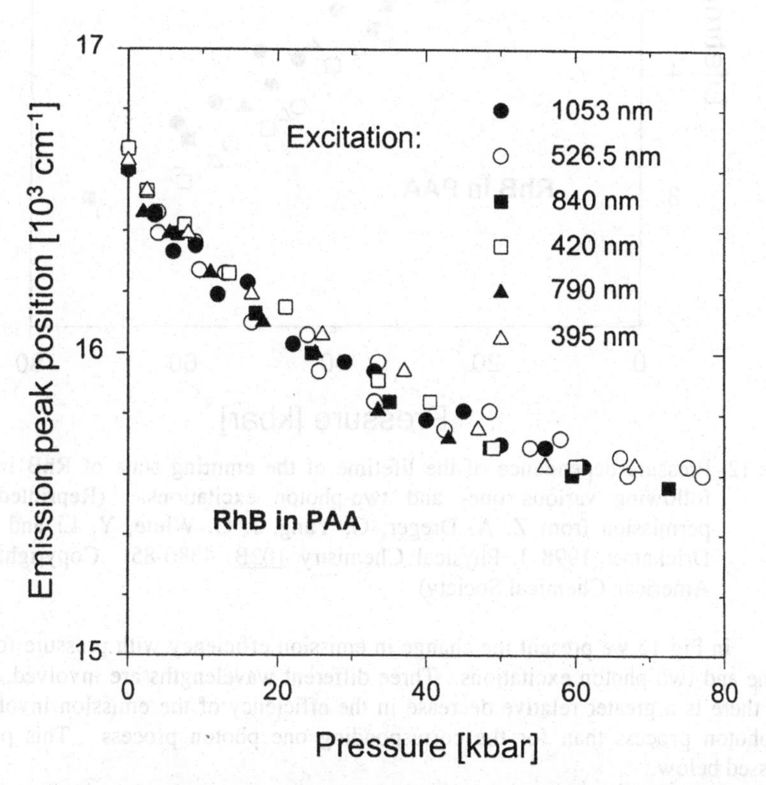

Figure 11. Pressure dependence of the energy of the emitting state of RhB in PAA following various one- and two-photon excitations. (Reprinted with permission from Z. A. Dreger, G. Yang, J. O. White, Y. Li and H. G. Drickamer 1998 J. Physical Chemistry 102B, 4380-85. Copyright 1998 American Chemical Society)

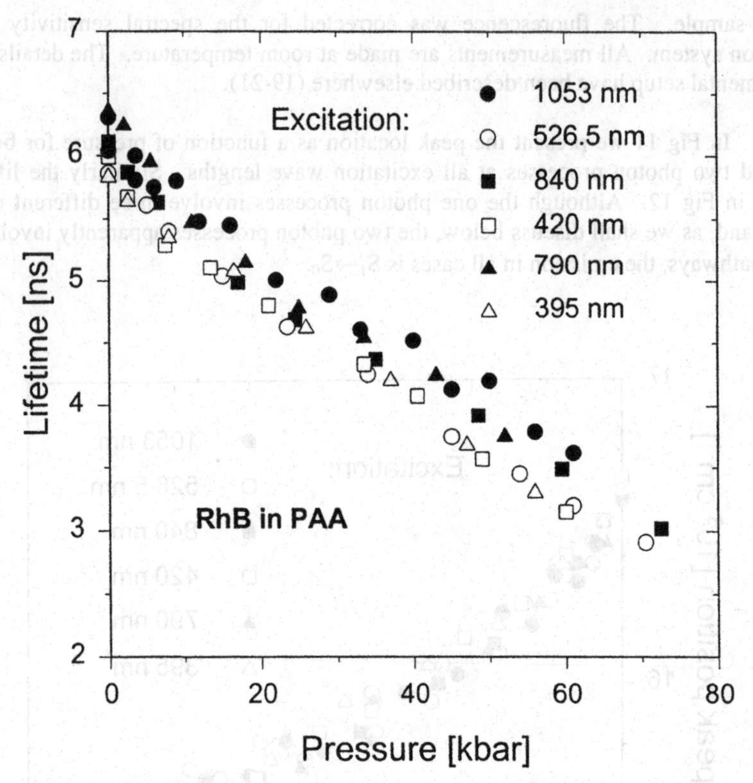

Figure 12. Pressure dependence of the lifetime of the emitting state of RhB in PAA following various one- and two-photon excitations. (Reprinted with permission from Z. A. Dreger, G. Yang, J. O. White, Y. Li and H. G. Drickamer 1998 J. Physical Chemistry 102B, 4380-85. Copyright 1998 American Chemical Society)

In Fig 13 we present the change in emission efficiency with pressure for both the one and two photon excitations. Three different wavelengths are involved. In all cases there is a greater relative decrease in the efficiency of the emission involving a two photon process than for the corresponding one photon process. This point is discussed below.

One can write the following equations for the emission intensities of the one photon process, one- and two-photon processes I_1' and I_2' relative to their one atmosphere values:

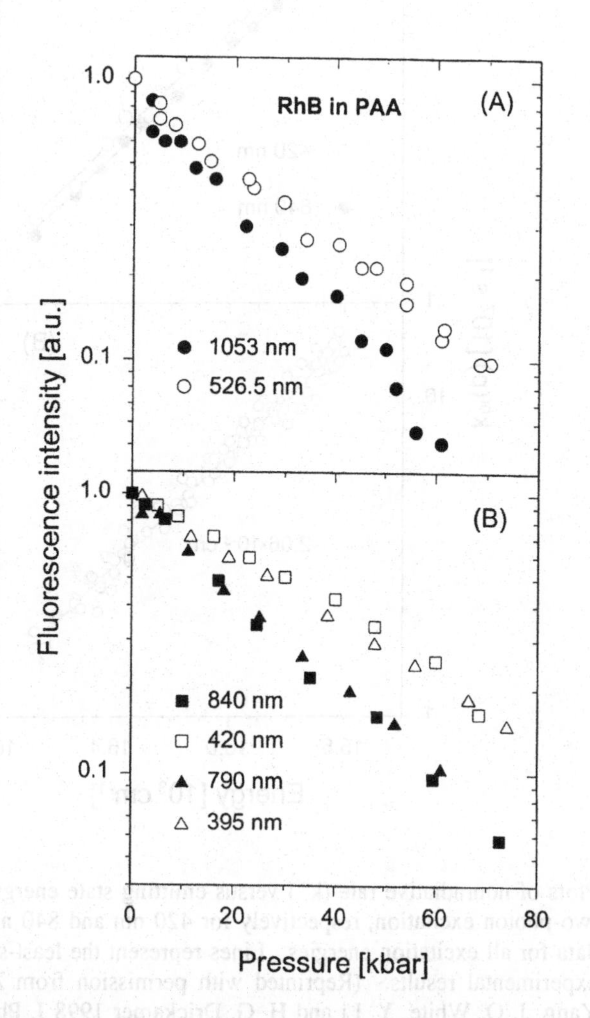

Figure 13. Pressure-induced fluorescence intensity change for RhB in PAA following one- and two-photon excitation. (A) Nd-YLF and (B) Ti-Al_2O_3 laser. (Reprinted with permission from Z. A. Dreger, G. Yang, J. O. White, Y. Li and H. G. Drickamer 1998 J. Physical Chemistry 102B, 4380-85. Copyright 1998 American Chemical Society)

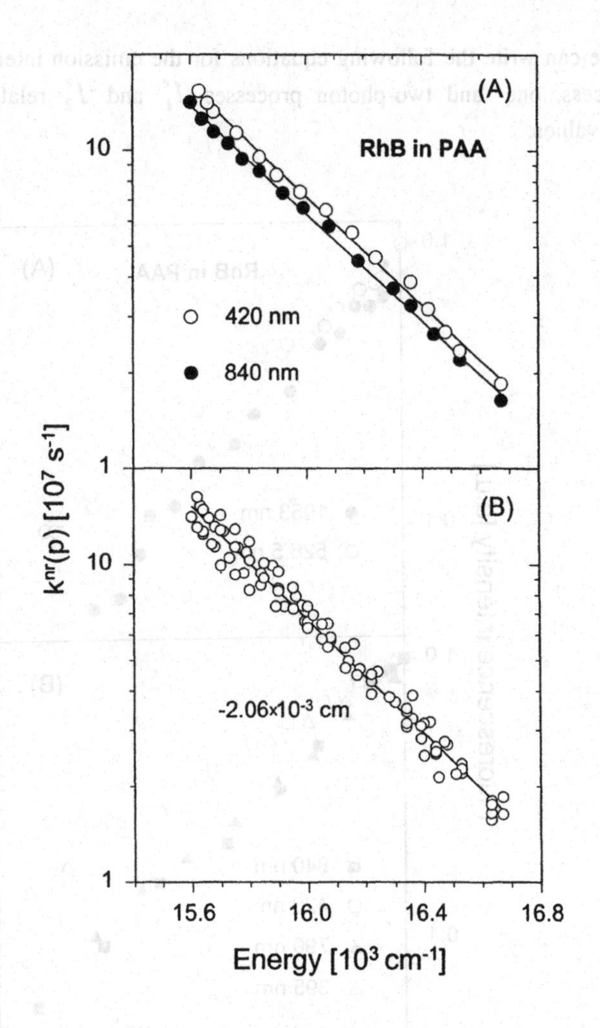

Figure 14. Plots of nonradiative rate (k^{nr}) versus emitting state energy. (A) one- and two-photon excitation, respectively for 420 nm and 840 nm, (B) collected data for all excitation energies. Lines represent the least-squares fits of the experimental results. (Reprinted with permission from Z. A. Dreger, G. Yang, J. O. White, Y. Li and H. G. Drickamer 1998 J. Physical Chemistry 102B, 4380-85. Copyright 1998 American Chemical Society)

$$I_1' = \frac{I_1(p)}{I_1(0)} = \frac{k_1^r(p)\tau_1(p)\sigma(p)}{k_1^r(0)\tau_1(0)\sigma(0)}, \tag{7a}$$

$$I_2' = \frac{I_2(p)}{I_2(0)} = \frac{k_2^r(p)\tau_2(p)\delta(p)}{k_2^r(0)\tau_2(0)\delta(0)} \tag{7b}$$

It has been established that in polymeric media k^r varies slowly compared with k^{nr} (22). We can then write for the k^{nr} (p) the rate of non-radiative dispersion of energy.

$$k^{nr}(p) = \frac{1}{\tau(p)} - \frac{\Phi(0)}{\tau(0)} \tag{8}$$

k^{nr} can involve either direct conversion to the ground state (internal conversion), indirect deactivation by intersystem crossing to a triplet state.

The nonradiative rate for internal conversion can be formulated within the theory of radiationless transitions in large organic molecules, given by Englman, Jortner, Freed. (23-24) Their treatment is based on the assumptions that molecular vibrations are harmonic and that normal modes and their frequencies are the same in the initial and final states. In the case when the electronic energy is dissipated by a single vibrational mode and the relative horizontal displacement of the potential energy surfaces in the initial and final electronic state is small (Stokes shift is comparable with the mean vibrational energy) they derive an analytical expression in the weak coupling limit.

This expression is of the form:

$$\ln k^{nr} \sim \Delta E \tag{9}$$

where ΔE is the energy of the thermally equilibrated excited state above the ground state. In Fig 14A we plot ln k^{nr} vs. emission peak location for the 420 nm and 840 nm excitation, and in 14B the results for all excitation wave lengths. The slopes are clearly linear and differ by less than 6%, certainly within experimental error. These results confirm that internal conversion controls the energy dissipation.

The RhB molecule in the ground state has relatively low symmetry, belonging essentially to the C_{2v} point group. Thus the S_0 ground state of the π-elections has the totally symmetric representation A_1 and the excited states are represented by either A_1 or B_2. The first excited singlet state is represented by B_2. The higher states, S_2 and S_3 are considered to be a A_1 symmetry (25). Both transitions to A_1 and B_2 are symmetry allowed for two-photon absorption. Since the $S_0 \rightarrow S_2$ and $S_0 \rightarrow S_3$ transitions take place between states of the same symmetry ($A_1 \rightarrow A_1$) it is not unexpected that absorption cross sections for these transitions, which take full advantage of the π-electron delocalization, are substantially stronger than for the $S_0 \rightarrow S_1$ transition ($A_1 \rightarrow B_2$). Moreover, two-photon transitions to higher excited states (S_2, S_3) can be enhanced by the contribution of S_1, participating as an additional intermediate state.

As pressure increases the fluorescence intensity decreases for both the one- and two-photon excitation. The effect is always stronger for two-photon excitation than for one-photon. Because for RhB in PAA the following relations are applicable: i) $\tau_1(p) = \tau_2(p)$ at all pressures and ii) $k_1^r = k_2^r$ with negligible pressure dependence, the ratio of the relative absorption cross sections for two- and one-photon excitation can be simply expressed from the intensity measurements:

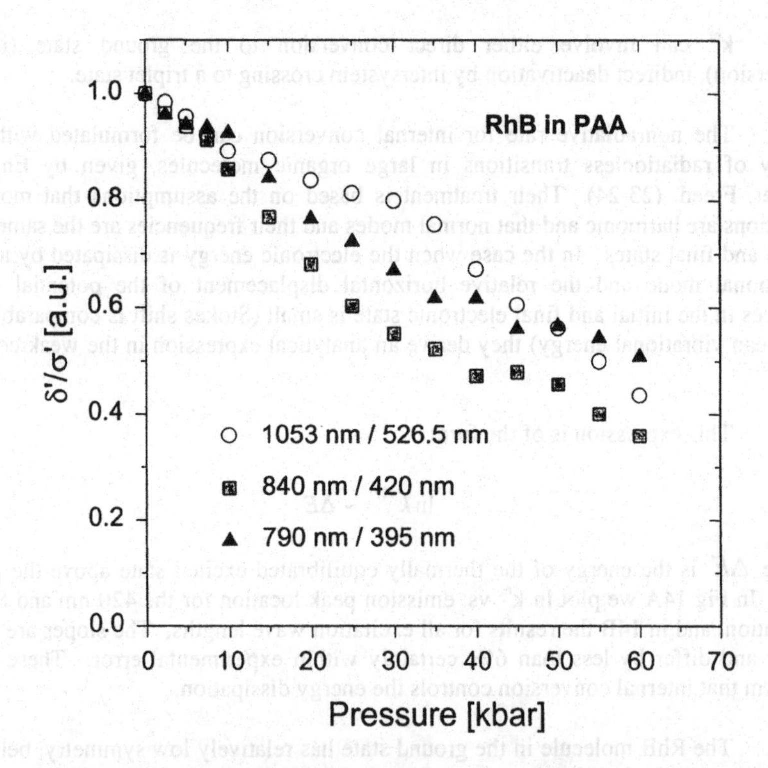

Figure 15. Pressure effect on ratio of the relative absorption cross section for two-photon excitation δ′ to relative absorption cross section for one-photon excitation σ′ for various excitations. (Reprinted with permission from Z. A. Dreger, G. Yang, J. O. White, Y. Li and H. G. Drickamer 1998 J. Physical Chemistry 102B, 4380-85. Copyright 1998 American Chemical Society)

$$\frac{\delta'}{\sigma'} \cong \frac{I'_2}{I'_1}, \tag{10}$$

As can be seen in Fig 15 the δ'/σ' ratio decreases gradually by a factor of two within 60 kbar. Moreover, the changes are approximately the same for three different excited states.

The most likely explanation is that the two photon processes involve intermediate states not relevant for the one photon processes and that the transfer of excitation from these states to the emitting state S_1 has an efficiency which is decreased with pressure more rapidly than the $S_2 \rightarrow S_1$ or $S_3 \rightarrow S_1$ process or than the internal conversion to S_0.

5. Conclusion

These high pressure studies of non-linear optical phenomena are exploratory in nature and require some approximations for their analysis. There are, however, very strong pressure effects and the results illustrate the power and versatility of pressure tuning spectroscopy.

6. Acknowledgement

The authors are pleased to acknowledge continuing support from the US Department of Energy, Division of Materials Science Grant DEF602-96ER45439 through the University of Illinois at Urbana-Champaign Frederick Seitz Materials Research Laboratory. Most of the research was performed in the MRL Laser Laboratory.

7. References

1. Marder, S. R., Sohn, J. E. and Stucky, G. D., eds., (1983) *Materials for Nonlinear Optics (Chemical Prospectives)*, ACS Symposium Series **455**, ACS Press: Washington DC.
2. Munn, R. W. and Ironside, C. N., eds., (1993) *Principles and Applications of Nonlinear Optical Materials*, Blackie Academic and Professional, London.
3. Kuhn, H. and Robillard, J. O., eds., (1992) *Non-Linear Optical Materials*, CRC Press: Boca Raton, FL.
4. Hahn, R. A. and Bloor, D., eds., (1991) *Organic Materials for Nonlinear Optics (II)*, The Royal Society of Chemistry: London.
5. Hahn, R. A. and Bloor, D., eds., (1993) *Organic Materials for Nonlinear Optics (III)*, The Royal Society of Chemistry: London.
6. Kurtz, S. K. and Perry, T. T., (1968) *J. App. Phys.*, **39**, 3798.

7. Li, Y., Yang, G., Dreger, Z. A., White, S. O. and Drickamer, H. G., *J. Phys. Chem.*
8. Yariv, A., (1975) *Quantum Electronics*, 2nd ed., J. Wiley & Sons: New York, **427**.
9. Dimitriev, V. G., Guradyan, G. G. and Nikogosyan, D. N., (1997) *Handbook of Nonlinear Optical Crystals*, ed., Springer-Verlog, Berlin.
10. Bridgman, P. W., (1948) *Proc. Amer. Acad. Arts Sci.*, **76**, 71.
11. Maker, P. D., Terhune, R. W., Nissenoff, M. and Savage C. M., (1962) *Phys. Rev. Lett.*, **8**, 21.
12. Bhawalkar, J. D., He, G. S. and Prasad, P. N., (1996) *Rep. Prog. Phys.*, **59**, 1041 and references therein.
13. Mukherjee, A., (1993) *Appl. Phys. Lett.*, **62**, 3423.
14. Qiu, P. and Penzkofer, A. (1989) *Appl. Phys. B* **48**, 115.
15. Tutt, L. W. and Boggess, T. F., (1993) *Prog. Quant. Electron*, **17**, 299 and references therein.
16. He, G. S., Bhawalkar, J. D., Zhao, C. F. and Prasad, P. N., (1995) *Appl. Phys. Lett.*, **67**, 2433.
17. Parthenopoulos, D. A. and Rentzepis, P. M., (1989) *Science*, **245**, 843.
18. Jutamulia, S. and Storti, G. M., (1995) *Optoelectronics*, **10**, 343 and references therein.
19. Dreger, Z. A., Yang, G., White, J. O. and Drickamer, H. G., (1997) *J. Phys. Chem. A*, **101**, 5753.
20. Dreger, Z. A., Yang, G., White, J. O., Li, Y. and Drickamer, H. G. (1997) *J. Phys. Chem. A*, **101**, 7948.
21. Dreger, Z. A., Yang, G., White, J. O., Li, Y. and Drickamer, H. G. (1998) *J. Phys. Chem.* **102**, 4380.
22. See for example, Offen, W. H. in *Organic Molecular Photophysics*, ed., J. B. Birks Wiley-Interscience, N. Y. ch. 3.
23. Englman, R. and Jortner, J., (1970) *Mol. Phys.* **18**, 145 and references therein.
24. Freed, K. F. and Jortner, J., (1970) *J. Chem. Phys.*, **52**, 6272.
25. Hermann, J. P. and Ducuing, J., (1972) *Opt. Commun.*, **6** 101.

PRESSURE EFFECTS ON THE INTRAMOLECULAR TWIST OF FLEXIBLE MOLECULES IN SOLID POLYMERS

H. G. DRICKAMER, Z. A. DREGER and J. O. WHITE
School of Chemical Sciences, Dept. of Physics and
Frederick Seitz Materials Research Laboratory
University of Illinois
600 S. Mathews
Urbana, IL 61801-3792

1. Abstract

There is a large class of "flexible" molecules which, when excited, can isomerize from a planar to a twisted form. In this paper we present both steady state and time resolved data for two such molecules in a solid polymer. We present a model consistent with these data and demonstrate that the steady state results can be predicted from the time resolved data.

2. Introduction

Disubstituted benzenes of the type D-Ph-A consisting of electron donor (D) and electron acceptor (A) group constitute a broad class of organic molecules with both technological and scientific interest. Such compounds, particularly in polymeric matrices, have potential applications in molecular electronics (e.g., charge carrier generation and transfer systems) and nonlinear optics (optical memories, frequency converters etc.).

It has been shown (e.g. 1-4) that these molecules, when excited, form an intramolecular charge transfer state. If the donor and acceptor moieties are connected by a flexible bond, in the excited state they may undergo a further change involving a twist or rotation around this bond. If, as in the present case, the twisted conformation is non-fluorescent, the probability of the twist occurring may limit the efficiency of emission.

The rate of these conformational changes is very sensitive to the properties of the medium, especially its rigidity, therefore the are a useful probe in studying internal processes in polymers (5-8).

R. Winter and J. Jonas (eds.), High Pressure Molecular Science, 47–57.

3. Results and Discussion

In this paper we present time-resolved measurements for two molecules, julolidinemalononitrile (JDMN) and p-N, N-dimethylaminobenzylidenemalononitrile (DMABMN), dissolved in polymethylmethacrylate (PMMA) at a concentration of 10^{-4} mol/mol of monomer (9). These molecules (Fig. 1) which belong to the same series of benzylidenemalononitriles, reveal fluorescence only from the untwisted conformation,

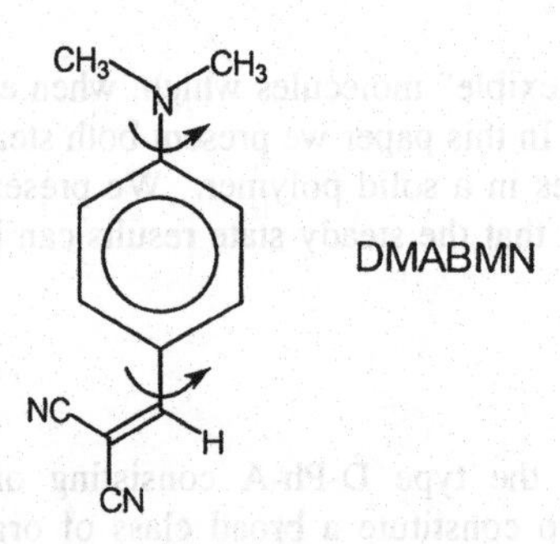

Figure 1. Molecular structure of JDMN and DMABMN. Arrows show the possible intramolecular-twists. (Reprinted with permission from Z. A. Dreger, J. O. White and H. G. Drickamer, 1998, *Chem. Phys Lett*, **290**, 399-404. Copyright 1998 Elsevier Pub. Co.)

and differ in the number of possible twists around single bonds. We show that these intramolecular reactions are governed by forward and backward transformation rates and are strongly limited by compression. The agreement between the pressure dependence of fluorescence intensities obtained in steady-state and transient conditions demonstrate the applicability of the intramolecular-twist concept for this type of flexible molecules. The samples are excited by 100 fs pulses from an 76-MHz mode-locked Ti:sapphire laser after frequency doubling to 420 nm with a BBO crystal. To assure the same excitation conditions at all pressures the intensity of the incident light is measured by a photodetector and adjusted by an attenuator. The front-side excited fluorescence is dispersed with a 0.25 m spectrometer and detected by a microchannel-plate photomultiplier-tube coupled to a single-photon time-resolved detection system.

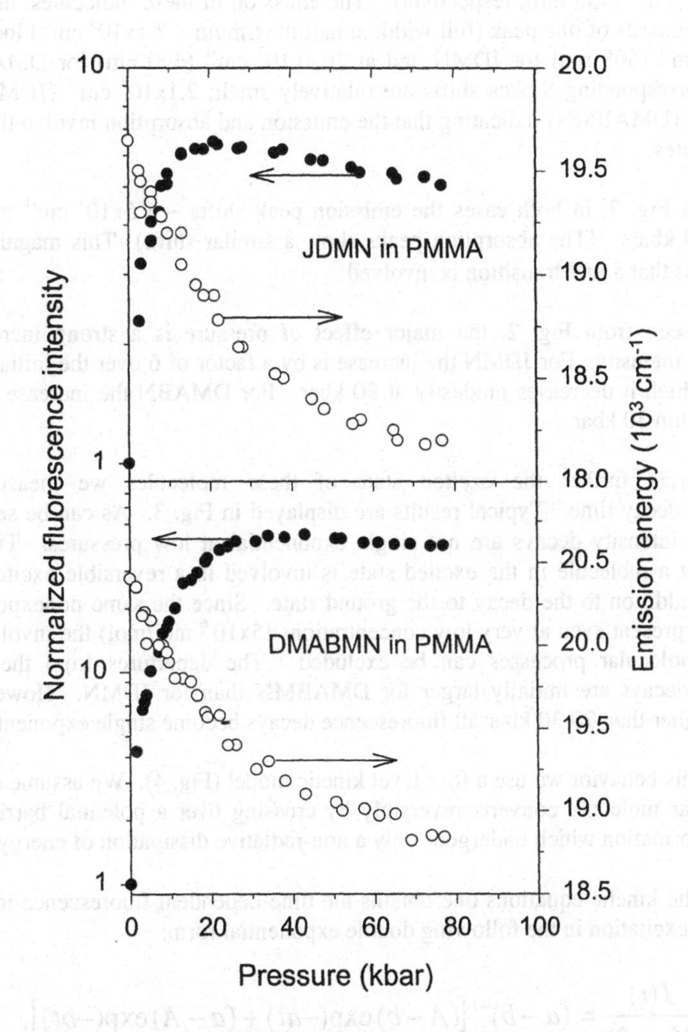

Figure 2. Pressure dependence of fluorescence intensity and energy of the emitting state for JDMN and DMABMN. (Reprinted with permission from Z. A. Dreger, J. O. White and H. G. Drickamer, 1998, *Chem. Phys Lett*, **290**, 399-404. Copyright 1998 Elsevier Pub. Co.)

A PC-based multichannel pulse-height analyzer is used to accumulate time-resolved data. The emission characteristics are corrected for the spectral sensitivity of the detection system. The details of the experimental setup can be found elsewhere (9-10).

At one atmosphere, both JDMN and DMABMN dissolved in PMMA exhibit an intense absorption band in the visible region with maxima located at 21.75x103 cm^{-1} (460 nm) and 23.0×10^3 cm^{-1} (435 nm), respectively. The emission of these molecules, like their absorption, consists of one peak (full width at half maximum ~ 2.1×10^3 cm^{-1}) located at 19.65×10^3 cm^{-1} (508 nm) for JDMN and at 20.4×10^3 cm^{-1} (490 nm) for DMABMN. Thus, the corresponding Stokes shifts are relatively small; 2.1×10^3 cm^{-1} (JDMN) and 2.6×10^3 cm^{-1} (DMABMN) indicating that the emission and absorption involve the same electronic states.

As shown in Fig. 2, in both cases the emission peak shifts ~ 1.5×10^3 cm^{-1} to lower energy in 80 kbars. (The absorption peaks show a similar shift.) This magnitude of shift indicates that a $\pi\pi^*$ transition is involved.

As can be seen from Fig. 2, the major effect of pressure is a strong increase in fluorescence intensity. For JDMN the increase is by a factor of 6 over the initial 10-20 kbar after which it decreases modestly at 80 kbar. For DMABN the increase is by a factor 50 within 30 kbar.

To characterize further the excited state of these molecules we measure the fluorescence decay time. Typical results are displayed in Fig. 3. As can be seen, the fluorescence intensity decays are not single exponential at low pressures. This fact indicates that a molecule in the excited state is involved in a reversible excited state reaction, in addition to the decay to the ground state. Since the same nonexponential decays were present even at very low concentrations (5×10^{-6} mol/mol) the involvement of any biomolecular processes can be excluded. The departures from the single exponential decays are initially larger for DMABMN than for JDMN. However, at pressures higher than 20-30 kbar all fluorescence decays become single exponential.

To explain this behavior we use a four level kinetic model (Fig. 4). We assume that the excited planar molecule converts reversibly by crossing over a potential barrier to a twisted conformation which undergoes only a non-radiative dissipation of energy.

By solving the kinetic equations one obtains the time-dependent fluorescence intensity upon pulsed excitation in the following double exponential form:

$$\frac{I(t)}{I(t=0)} = (a-b)^{-1}\left[(A-b)\exp(-at) + (a-A)\exp(-bt)\right], \qquad (1)$$

where,

$$a = \left\{A + B + \left[(A-B)^2 + 4k_{PT}k_{TP}\right]^{1/2}\right\}/2,$$
$$b = \left\{A + B - \left[(A-B)^2 + 4k_{PT}k_{TP}\right]^{1/2}\right\}/2,$$
$$A = k_P + k_{PT} \qquad B = k_T + k_{TP},$$

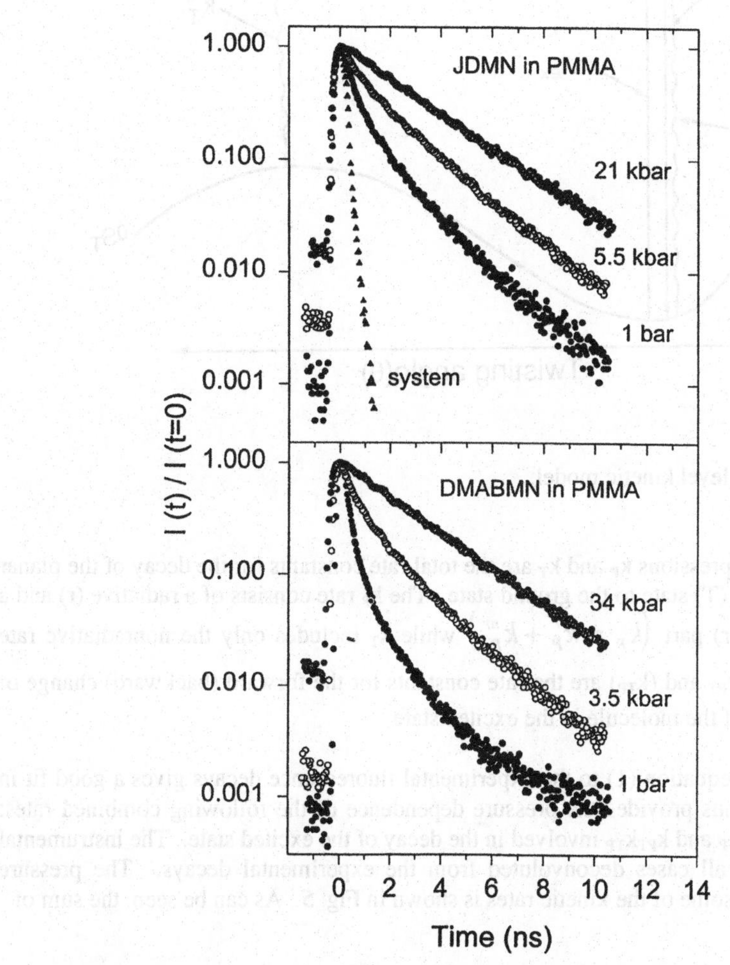

Figure 3. Fluorescence intensity decays for JDMN and DMABMN at several pressures. [System – denotes the instrumental response ~ 170 ps]. (Reprinted with permission from Z. A. Dreger, J. O. White and H. G. Drickamer, 1998, *Chem. Phys Lett*, 290, 399-404. Copyright 1998 Elsevier Pub. Co.)

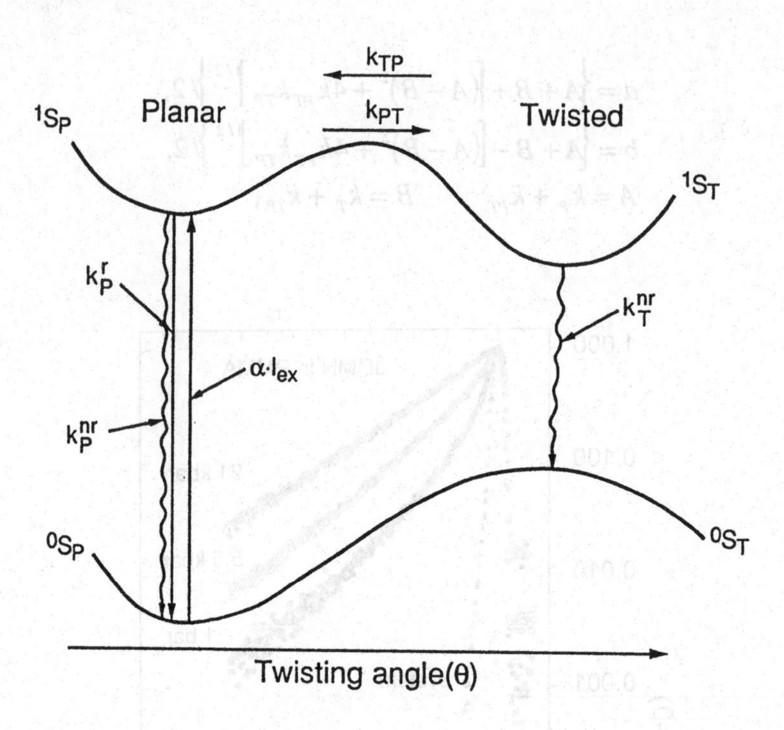

Figure 4. Four level kinetic model.

In the above expressions k_P and k_T are the total rate constants for the decay of the planar (P) and twisted (T) state to the ground state. The k_P rate consists of a radiative (r) and a nonradiative (nr) part $\left(k_P = k_P^r + k_P^{nr}\right)$ while k_T includes only the nonradiative rate $\left(k_T = k_T^{nr}\right)$. k_{PT} and (k_{TP}) are the rate constants for the forward (backward) change of conformation of the molecule in the excited state.

Application of equation (1) to the experimental fluorescence decays gives a good fit in all cases and thus provides the pressure dependence of the following combined rates: $k_P + k_{PT}$, $k_T + k_{TP}$ and $k_{PT}k_{TP}$ involved in the decay of the excited state. The instrumental response is in all cases deconvoluted from the experimental decays. The pressure dependence of some of the kinetic rates is shown in Fig. 5. As can be seen, the sum of

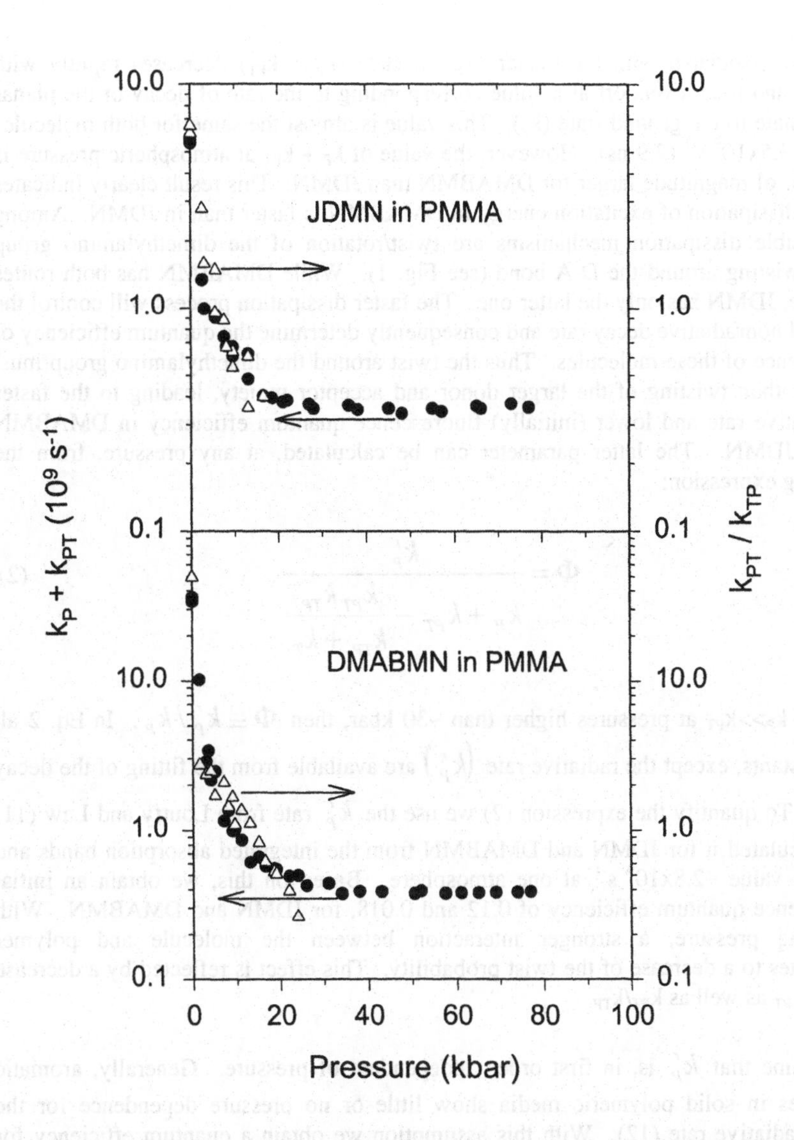

Figure 5. Pressure dependence of rate constants: $k_P + k_{PT}$ (•) the rates controlling the planar state, k_{PT} /k_{TP} (Δ) the ratio of the forward and backward rates for the intramolecular transformation in the excited state. (Reprinted with permission from Z. A. Dreger, J. O. White and H. G. Drickamer, 1998, *Chem. Phys Lett*, 290, 399-404. Copyright 1998 Elsevier Pub. Co.)

two rates associated with the planar excited state ($k_P + k_{PT}$) decreases rapidly with pressure and then levels off at a value corresponding to the rate of decay of the planar excited state to the ground state (k_P). This value is almost the same for both molecules and is ~ $3.5x10^9$ s^{-1} (2.9 ns). However, the value of $k_P + k_{PT}$ at atmospheric pressure is one order of magnitude larger for DMABMN than JDMN. This result clearly indicates that the dissipation of excitation energy in DMABMN is faster than in JDMN. Among the possible dissipation mechanisms are twist/rotation of the dimethylamino group and/or twisting around the D-A bond (see Fig. 1). While DMABMN has both routes available, JDMN has only the latter one. The faster dissipation process will control the observed nonradiative decay rate and consequently determine the quantum efficiency of fluorescence of these molecules. Thus the twist around the dimethylamino group must be faster than twisting of the larger donor and acceptor moiety, leading to the faster nonradiative rate and lower (initially) fluorescence quantum efficiency in DMABMN than in JDMN. The latter parameter can be calculated, at any pressure, from the following expression:

$$\Phi = \frac{k_P^r}{k_P + k_{PT} - \dfrac{k_{PT} k_{TP}}{k_{TP} + k_T}}. \qquad (2)$$

Because $k_P >> k_{PT}$ at pressures higher than ~30 kbar, then $\Phi \cong k_P^r / k_P$. In Eq. 2 all rate constants, except the radiative rate $\left(k_P^r\right)$ are available from the fitting of the decay curves. To quantify the expression (2) we use the k_P^r rate from Loutfy and Law (11) who calculated it for JDMN and DMABMN from the integrated absorption bands and obtained value ~$2.8x10^9$ s^{-1}, at one atmosphere. Based on this, we obtain an initial fluorescence quantum efficiency of 0.12 and 0.018, for JDMN and DMABMN. With increasing pressure, a stronger interaction between the molecule and polymer contributes to a decrease of the twist probability. This effect is reflected by a decrease of $k_P + k_{PT}$ as well as k_{PT}/k_{TP}.

We assume that k_P^r is, in first order, independent of pressure. Generally, aromatic molecules in solid polymeric media show little or no pressure dependence for the singlet radiative rate (12). With this assumption we obtain a quantum efficiency for fluorescence of ~0.8 for both molecules in PMMA at high pressure. This agrees exactly with the efficiency found at 77K and one atmosphere (13).

We can use the rate constants to calculate the changes in steady state emission efficiency as a function of pressure from the following expression.

$$\frac{I(p)}{I(p=0)}=\frac{\alpha(p)k_p^r(p)}{\alpha(0)k_p^r(0)}\frac{k_P(0)+k_{PT}(0)-\dfrac{k_{PT}(0)k_{TP}(0)}{k_{TP}(0)+k_T(0)}}{k_P(p)+k_{PT}(p)-\dfrac{k_{PT}(p)k_{TP}(p)}{k_{TP}(p)+k_T(p)}} \tag{3}$$

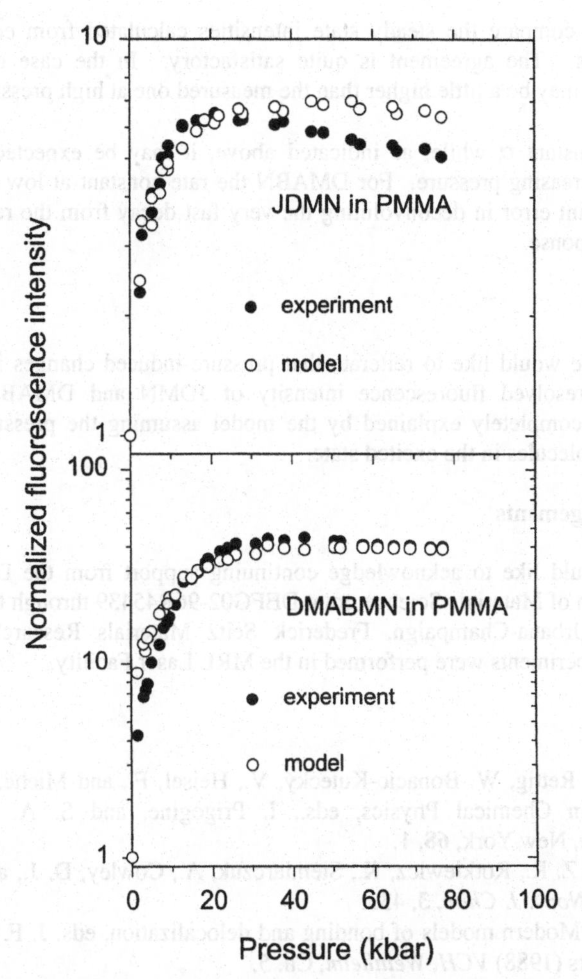

Figure 6. Comparison of the experimental and calculated fluorescence intensity change with pressure. (Reprinted with permission from Z. A. Dreger, J. O. White and H. G. Drickamer, 1998, *Chem. Phys Lett*, 290, 399-404. Copyright 1998 Elsevier Pub. Co.)

In the above expression α is the absorption coefficient at the excitation energy; the meaning of the remaining symbols is explained above.

In the first approximation we assume α is independent of pressure. However, it should be noted that the excitation (23.81×10^3 cm^1) is at the high energy side of the absorption bands (JDMN, maximum at 21.75×10^3 cm^{-1} and DMABMN at 23.0×10^3 cm^{-1}). Therefore, α may decrease slightly (more for JDMN than for DMABMN) with increasing pressure due to the shift of the absorption spectrum to lower energies.

In Figure 6 we compare the steady state intensities calculated from eq. 3 with the measured values. The agreement is quite satisfactory. In the case of JDMN the calculated value may be a little higher than the measured one at high pressure because

we assumed constant α while, as indicated above, it may be expected to decrease slightly with increasing pressure. For DMABN the rate constant at low pressure may involve significant error in deconvoluting the very fast decay from the relatively slow instrumental response.

4. Conclusion

In conclusion we would like to reiterate that pressure-induced changes in the steady-state and time-resolved fluorescence intensity of JDMN and DMABMN in solid PMMA can be completely explained by the model assuming the pressure-controlled twist of these molecules in the excited state.

5. Acknowledgements

The authors would like to acknowledge continuing support from the Department of Energy, Division of Materials Science grant DEFG02-96R45439 through the University of Illinois at Urbana-Champaign, Frederick Seitz Materials Research Laboratory (MRL). The experiments were performed in the MRL Laser Facility.

6. References

1. Libbert, E., Rettig, W. Bonacic-Kutecký, V., Heisel, F., and Miehé, J. A. (1987) Advances in Chemical Physics, eds., I. Prigogine, and S. A. Rice, *Wiley-Interscience*, New York, **68**, 1.
2. Grabowski, Z. R., Rotkiewicz, K., Siemiarczuk, A., Cowley, D. J., and Baumann, W., (1979) *Nouv. J. Chim.* **3**, 433.
3. Rettig, W. Modern models of bonding and delocalization, eds. J. F. Liebman and A. Greenbers (1988) *VCH*, *Weinheim*, Ch. 5.
4. Rettig, W. and Baumann, W. (1992) Photochemistry and Photophysics ed. J. F. Rabek, *C. R. Press*, Boca Raton, FL., **6**, 79.
5. Loutfy, R. O., (1981) *Macromolecules* **14**, 270.

6. Paczkowski, J. and Neckers, D. C., (1990) *Macromolecules* **24**, 3013.
7. Loutfy, R. O., Photophysical and photochemical tools in polymer science, ed. M. A. Winnek, *Reidel, Dordrecht* (1986) 429.
8. Itagaki, H., Horie, K., Mita, I. (1990) *Prog. Polym. Sci.* **15**, 361.
9. Dreger, Z. A., White, J. O., and Drickamer, H. G. (1998) *Chem. Phys. Lett.* 290, 399.
10. Lang, J. M., Dreger, Z. A. and Drickamer, H. G., (1994) *J. Phys. Chem.* **98**, 11308.
11. Loutfy, R. O. and Law, K. Y., (1980) *J. Phys. Chem.* **84**, 2808.
12. Offen, H., (1973) Organic Molecular Photophysics, ed. J. B. Birks, *Wiley-Interscience*, N.Y., 1, 3.
13. Loutfy, R. O., (1981), *Macromolecules* **14**, 270.

MOLECULAR DYNAMICS STUDIES OF HIGH PRESSURE TRANSFORMATIONS AND STRUCTURES

J.S. TSE and D.D. KLUG
Steacie Institute for Molecular Sciences
National Research Council of Canada
Ottawa, Ontario, Canada K1A 0R6

1. Introduction

The study of high-pressure transformations and the stability of dense materials has been an active area research for many decades. The change of the structure of a solid when pressurized has yielded many interesting new materials and many new phenomena such as insulator → metal transformations, important effects in high T_c superconductors, and pressure-induced amorphization that yields unique disordered materials. Ten years ago, the application of classical molecular dynamics techniques had already yielded results that provided important insight into materials under high pressure. This was made possible by the constant-pressure method developed by Anderson[1] and by Parrinello and Rahman[2,3] which allows the volume as well as the shape of the simulation cell to change. Another important aspect of these simulation methods was that systems could be studied at finite temperatures. These developments complemented the rapidly developing quantum mechanical methods being used in the solid state physics community to study the stability and structures of solids under pressure[4,5]. A major advance in the theoretical methods for the study of structural stability and dynamical properties of solids was made with the introduction of the Car-Parrinello method[6] for *ab initio* molecular dynamics (AIMD). A primary advantage of this method is the fact that a interatomic potential is not a part of the input. The Car-Parrinello method allows the calculation of all of the properties that can be obtained in the classical MD technique but with the important feature that interatomic forces are obtained from a quantum mechanical calculation. The

R. Winter and J. Jonas (eds.), High Pressure Molecular Science, 59–85.

AIMD method has more recently been extended to include Parrinello-Rahman variable cell dynamics[7,8] so this method has the capability to describe structural phase changes and stabilities from first-principles. Other variations of constant pressure MD have also been proposed and the basic ideas and capabilities are similar[9].

In this article, the basic ideas of classical and ab inbitio molecular dynamics as required for calculations of material properties under pressure are first summarized. There are currently a number of excellent reviews and articles that describe the details of each of the various methods. The application of these methods to phase transformations and the determination of structural stability of high pressure phases is illustrated by a variety of different materials from our research on molecular solids such a hydrogen, silicates, zeolites, and high pressure phases of ice.

2. Classical Molecular Dynamics

Classical molecular dynamics has been an important tool for the study of matter since its first application to a system of hard spheres more than 40 years ago[10]. The basic idea for this method is to solve the equations of motion for a collection of atoms or molecules in a simulation cell. An excellent recent review of the method is recommended[11]. The number of atoms is limited primarily by the size of the computer and in practice may vary from a few hundred to many thousands of atoms. The basic equations of motion are obtained from the Lagrangian for the system[12]

$$L = \frac{1}{2}\sum_{i=1}^{N} m_i \dot{\mathbf{r}}_i^2 - V(\mathbf{r}_1, \ldots, \mathbf{r}_N) \tag{1}$$

where m_i and $\mathbf{r}_i$ are the coordinate and mass of particle i, and $V(\mathbf{r}_1,\ldots,\mathbf{r}_N)$ is the interaction potential. In constant-pressure MD[2,3], the Lagrangian contains an extension for the kinetic energy of the variable volume cell and also a potential term for the externally applied pressure

$$L = \frac{1}{2}\sum_{i=1}^{N} m_i (\dot{\mathbf{s}}_i^T \mathbf{G} \dot{\mathbf{s}}_i) - V(\mathbf{r}_1, \ldots, \mathbf{r}_N) + \frac{1}{2} W \ \mathrm{Tr}\ \dot{\mathbf{h}}^T \dot{\mathbf{h}} - p\Omega \tag{2}$$

where W has dimensions of mass in the term which represents the kinetic energy of the MD cell. The matrix **h** defines the 3 × 3 matrix whose columns are the three vectors **a**, **b**, and **c** that define the edges of the MD cell with volume Ω. The real space coordinates of each particle is related to the scaled coordinates **s** by $\mathbf{r} = \mathbf{hs}$ where $0 < \mathbf{s}_i < 1$. The distance between particles is given in terms of the metric tensor $\mathbf{G} = \mathbf{h}^t\mathbf{h}$ with

$$(\mathbf{r}_i - \mathbf{r}_j)^2 = (\mathbf{s}_i - \mathbf{s}_j)^T \mathbf{h}^t \mathbf{h}(\mathbf{s}_i - \mathbf{s}_j) = (\mathbf{s}_i - \mathbf{s}_j)^T \mathbf{G}(\mathbf{s}_i - \mathbf{s}_j) \tag{3}$$

p is the hydrostatic pressure to be imposed on the system. The equations of motion that are obtained from Eqn. 2 and 3 are:

$$\ddot{\mathbf{s}}_i = -\frac{1}{m_i}\frac{\partial V}{\partial \mathbf{r}_i^\beta}(\mathbf{h}^T)^{-1}_{\beta\alpha} - \mathbf{G}^{-1}_{\beta\alpha}\dot{\mathbf{G}}_{\beta\gamma}\dot{\mathbf{s}}_i^\gamma \tag{4}$$

and

$$\ddot{\mathbf{h}}_{\alpha\beta} = \frac{1}{W}(\Pi_{\alpha\gamma} - p\delta_{\alpha\beta})\Omega(\mathbf{h}^T)^{-1}_{\alpha\beta}, \tag{5}$$

$$\mathrm{v}_i^\alpha = \mathbf{h}_{\alpha\gamma}\dot{\mathbf{s}}_i^\gamma \tag{6a}$$

$$\Pi_{\alpha\beta} = \frac{1}{\Omega}\left(\sum_i m_i \mathrm{v}_i^\alpha \mathrm{v}_i^\gamma - \frac{\partial \mathrm{V}}{\partial \mathbf{h}_{\alpha\beta}}\mathbf{h}^T_{\delta\gamma}\right) \tag{6b}$$

The Greek indices indicate components of vectors or matrices and implicit sums over repeated indices are implied. The molecular dynamics performed with the above Lagrangian has been shown to be equivalent to a constant enthalpy (N,P,H) ensemble[1].

The calculation of many useful time and spatial averages of various quantities of interest is relatively straightforward[13]. The dynamical properties of the system are calculated from the appropriate time correlation functions. For example the vibrational density of states is obtained from the Fourier transform of the single-particle velocity correlation function,

$$Z(\omega) = \int e^{-i\omega t} < \mathrm{v}(t) \cdot \mathrm{v}(0) > dt \tag{7}$$

In addition the infrared lineshape can be computed from the Fourier transform of the instantaneous total cell dipole moment, M(t) autocorrelation function,

$$I(\omega) = \int e^{i\omega t} < \mathbf{M(t)} \cdot \mathbf{M(0)} > dt \tag{8}$$

that is related to the infrared absorption cross-section. Another particularly useful quantity are the elastic constants which can be obtained from the strain fluctuations $\delta\varepsilon$ that are related to the adiabatic compliances Γ_{ijkl} as follows[14],

$$< \delta_{ij}\delta_{kl} >_{SHN} = (k_B T / V_0)_{ijkl} \tag{9}$$

where ε_{ij} is the strain tensor and V_0 is the equilibrium volume for the system. The

compliances can then be transformed to the conventional elastic constants using the Voight equations[15]. Another quantity that is particularly important for the analysis of the structures obtained from a molecular dynamics simulation at high pressures is the structure factor. This is almost essential for the precise determination of the crystal structures before and after a phase transformation. The contribution from Bragg scattering is the coherent structure factor that can be obtained from the density operator

n(q,t), [16,17]

$$S_c(\mathbf{q}) = \left|\langle \mathbf{n(q)} \rangle\right|^2 / N \tag{10}$$

and

$$\mathbf{n(q,t)} = \sum_i b_i e_i^{-i\mathbf{q}\cdot\mathbf{r}}(t) \tag{11}$$

where b_i is the scattering length of atom i and for x-ray diffraction, the atomic scattering form factor. The peaks in the coherent structure are at values of **q** that correspond to Bragg reflections from a periodic solid.

3. Ab Initio Molecular Dynamics

The molecular dynamics method was greatly advanced with the publication of the paper by R. Car and M. Parrinello[6] in 1985 that introduced *ab initio* molecular dynamics

(AIMD). This has become an essential tool in solid state physics and chemistry for the study of structural stability and phase transitions. The principle advantage over classical MD is the elimination of the requirement for an interaction potential as explicit input and usually obtained from the fitting of parameters to experimental properties. There are numerous excellent descriptions of the AIMD method[18-22] and we will therefore only outline the method in this section and discuss the aspects of the method that are particularly important for high pressure studies. In this method, the forces acting on the ions are obtained from a quantum mechanical method based on density functional theory. This is accomplished by considering a fictitious dynamical system associated with the ions and the electrons. The electronic wave-functions are introduced as classical fields. The dynamics of the ions follow that of the physical system with the electrons on the Born-Oppenheimer surface defined by

$$V(\{\mathbf{R}_{\mathrm{I}}\}) = \min_{\{\psi_i\}} E[\{\Psi_i\},\{\mathbf{R}_{\mathrm{I}}\}] \tag{12}$$

The generalized Lagrangian for the system is now defined as follows,

$$\mathrm{L} = \sum_i \mu\langle\dot{\psi}_i|\dot{\psi}_i\rangle - \mathrm{E}[\{\psi_i\},\{\mathbf{R}_I\}] + \sum_{ij}\Lambda_{ij}\left(\int d\mathbf{r}\psi_i^*(\mathbf{r})\psi_j(\mathbf{r}) - \delta_{ij}\right) + \frac{1}{2}\sum_I M_I\dot{\mathbf{R}}_{\mathrm{I}}^2 \tag{13}$$

where ψ_I are the electronic wave-functions, and $\mathbf{R}_I$ the ionic positions. The first term containing the parameter μ for the generalized mass of the electronic degree of freedom, represents the fictitious kinetic energy of the electronic wave-functions, the second term is the total electronic energy, the third term containing the Lagrangian multiplier Λ_{ij} contains the orthonormalization constraints. The equations of motion derived from this Lagrangian are:

$$\mu\ddot{\psi}_i(\mathbf{r}) = -\frac{\delta E}{\delta\psi_i^*(\mathbf{r})} + \sum_j \Lambda_{ij}\psi_j(\mathbf{r}) \tag{14}$$

$$\ddot{\mathbf{R}}_{\mathrm{I}} = -\frac{1}{M_I}\frac{\partial E}{\partial \mathbf{R}_{\mathrm{I}}}. \tag{15}$$

For Car-Parrinello MD at constant pressure, the Lagrangian contains the electronic wave-functions in addition to the terms for the variable cell. In this case, an additional term is included for the kinetic energy of the MD cell as well as a term for the external pressure with the same form as in Eqn. 2 for the classical limit. In addition the wave-functions need to be defined onto a scaled variable space as is done for the classical equations. There are recent successful examples of the application of this method to phase transformations[23,24] and it will undoubtedly continue to be an important technique in future high pressure materials science.

There are several methods of calculating the energy of solid phases within the context of the AIMD technique and the density functional formalism. In the calculation of the band-structure energy of a solid phase there is the requirement of adequate sampling of the k points over the Brillouin zone. This can be accomplished by either performing a calculation with a large enough simulation cell that there is adequate sampling do the folding of bands back onto the appropriate zone or by using a extended mesh of k points in the calculation[22,25]. The first example for hydrogen discussed below used the former method and the second example for a high-pressure form of ice we discuss employed a large k-point sampling method.

4. Examples of Classical Molecular Dynamics Studies Applied to High-Pressure Transformations and Structures

There are numerous systems where classical MD has been particularly successful in obtaining not only information on the structure and stability of high pressure phases but also has provided detailed pictures of the mechanisms of phase transformations. One system where constant pressure MD described in the first section has been particularly useful has been in the study of the mechanism of pressure-induced amorphization in SiO_2[26]. We will discuss this material in detail since it provides an example of the wide spectrum of properties and behavior that can be obtained from classical MD. We will focus on results from our laboratory and therefore the references to other work will not be complete.

The essential component of a classical MD calculation is the interatomic potential. Two excellent potentials[27,28] have been developed in the last decade which are able to reproduce the structural properties of the known phases of SiO_2. This was an important development since a potential for SiO_2 must be able to yield the correct structures and dynamical behavior for phases in which the Si atom is in either a 4- or 6-coordinated environment. Both potentials were tested[29] and in particular it was found that the potential suggested by van Beest et al.[28] could be used to reproduce the phase pressure-volume phase diagram for both 4- and 6-coordinated structures. The form of the potential is

$$\Phi_{ij} = q_i q_j / r_{ij} + A_{ij} \exp(-b_{ij} r_{ij}) - c_{ij} / r_{ij}^6 , \qquad (16)$$

where q_i and q_j are the charges on atoms i and j, r_{ij} the corresponding interaction distance, and A_{ij}, b_{ij}, and c_{ij} are the force field parameters. This potential has been used to calculate the pressure-volume equations of state for α-quartz, cristobalite, and 6-coordinated stishovite[29], the pressure-induced amorphization transformation in α-quartz[26], structural changes in densified amorphous SiO_2[30,31], as well as the as well as properties of related materials such as zeolites[32] and materials that are isostructural with α-quartz[33] such as $AlPO_4$. The experimental and calculated equations of state[26] using this potential for α-quartz, cristobalite, and stishovite at 300 K are shown in Figure 1. The agreement of the pressure-volume equation of state with the experimental data is very significant since the original potential was parameterized for the fourfold coordination that is found in the low pressure forms of SiO_2. It was believed that it would be necessary that a three-body potential would be required due to the strong covalent and directional bonding characteristics in crystalline forms of SiO_2. This two body potential has therefore overcome the difficulties that would have been present if a three-body potential were used.

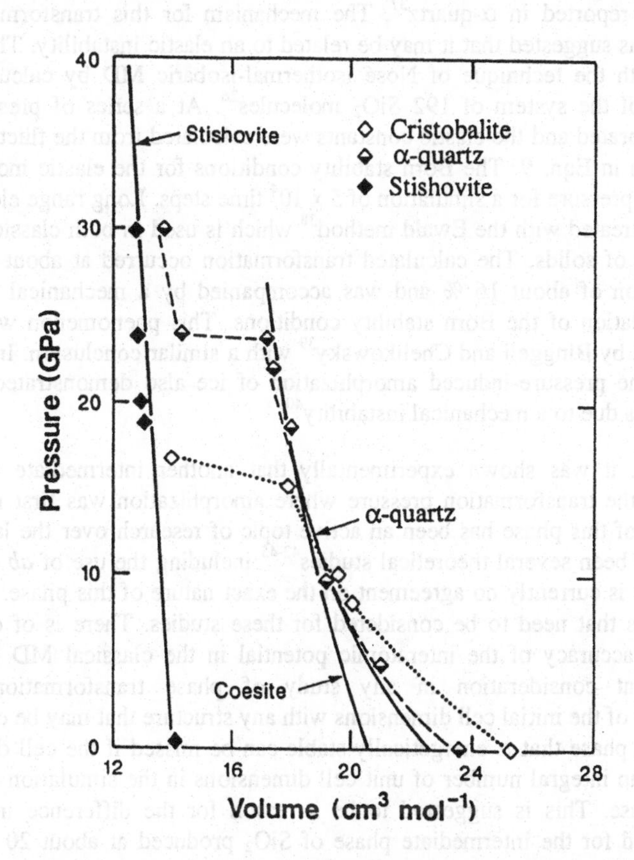

Figure 1. Comparison of the calculated and experimental pressure-volume behavior of 4- and 6-coordinated phases of SiO_2. The calculated points were obtained with the potential of van Beest et al.[28]

4.1 PRESSURE-INDUCED AMORPHIZATION

The phenomena of pressure-induced amorphization under hydrostatic conditions was first described in detail for ice Ih in 1984[34]. In this report, ice was shown to transform to a high-density amorphous solid that could be quench recovered at 77 K for further study by a variety of techniques. It was shown that the radial distribution function g(r) as determined by neutron diffraction[35] was that of an amorphous solid and it was suggested that the g(r) was similar to that of a quenched high-density liquid. The mechanism for the transformation was however unknown although a melting type of mechanism was suggested. There are however two different types of melting that could occur under such conditions. There is either true thermodynamic melting or mechanical melting[36] where there was a collapse of the crystalline structure due to a mechanical or elastic instability in the lattice. Soon after the report of pressure-induced amorphization in ice, a similar phenomenon was reported in α-quartz[37]. The mechanism for this transformation was unknown but it was suggested that it may be related to an elastic instability. This system was examined with the technique of Nosé isothermal-isobaric MD by calculating the elastic constants of the system of 192 SiO_2 molecules[26]. At a series of pressures, the system was equilibrated and the elastic constants were calculated from the fluctuations of the strain as given in Eqn. 9. The Born stability conditions for the elastic moduli were monitored at each pressure for a simulation of 5×10^4 time steps. Long range electrostatic interactions were treated with the Ewald method[38] which is used in both classical and ab initio MD studies of solids. The calculated transformation occurred at about 22.3 Gpa with a densification of about 16 % and was accompanied by a mechanical instability resulting in a violation of the Born stability conditions. This phenomenon was further examined in detail by Binggeli and Chelikowsky[39] with a similar conclusion. In addition, a MD study of the pressure-induced amorphization of ice also demonstrated that this transformation was due to a mechanical instability[40].

More recently, it was shown experimentally that another intermediate crystalline phase appears at the transformation pressure where amorphization was first reported[41]. The exact nature of this phase has been an active topic of research over the last several years. There have been several theoretical studies[42-45] including the use of *ab initio* MD methods but there is currently no agreement on the exact nature of this phase. There are several difficulties that need to be considered for these studies. There is of course the limitation of the accuracy of the interatomic potential in the classical MD technique. Another important consideration in any study of phase transformations is the commensurability of the initial cell dimensions with any structure that may be obtained at high pressures. A phase that is energetically stable can be missed if the cell dimensions do not allow for an integral number of unit cell dimensions in the simulation cell of the high-pressure phase. This is suggested to be a reason for the difference in different structures obtained for the intermediate phase of SiO_2 produced at about 20 GPa. In a recent report, a 4608 atom cell of α-quartz was examined using the Nosé isothermal-isobaric method[44]. Calculations were performed at a series of pressures in small pressure increments close to the expected phase transformation. The x-ray diffraction patterns were obtained as given in Eqn.10 from the density operator and the coherent part of the structure factor. The transformed structure was then analyzed using a method developed recently[46] to obtain the translationally equivalent unit cell and the correct space group of the new intermediate-pressure phase. An examination of the new phase showed that the structure has a lattice resembling a sheared α-quartz lattice. Analysis of the simulation cell yielded a triclinic cell with lattice parameters a = 4.07, b = 3.92 Å, c = 4.97 Å, α = 87°, β = 95.5°, and γ = 118.5°, and a calculated density of 4.32 g/cm^2. The optimized internal coordinates for the cell are listed in Ref. 45. A further test of the stability was done using *ab initio* MD as described above. This calculation showed that the structure was indeed stable and that the energy was comparable to that of α-quartz. The comparison of the theoretical and experimental diffraction patterns is shown in Figure. 2. There remain differences between these results and others that used a different (smaller) simulation cell. The use of a smaller simulation cell however can easily introduce an error if the new cell is not commensurate with the simulation cell or cannot form in the limited volume. The result that the intermediate pressure phase produced by pressurization of α-quartz is a sheared quartz structure is consistent with the earlier result[26] that there is a shear instability that is apparent before the first-order transition occurs.

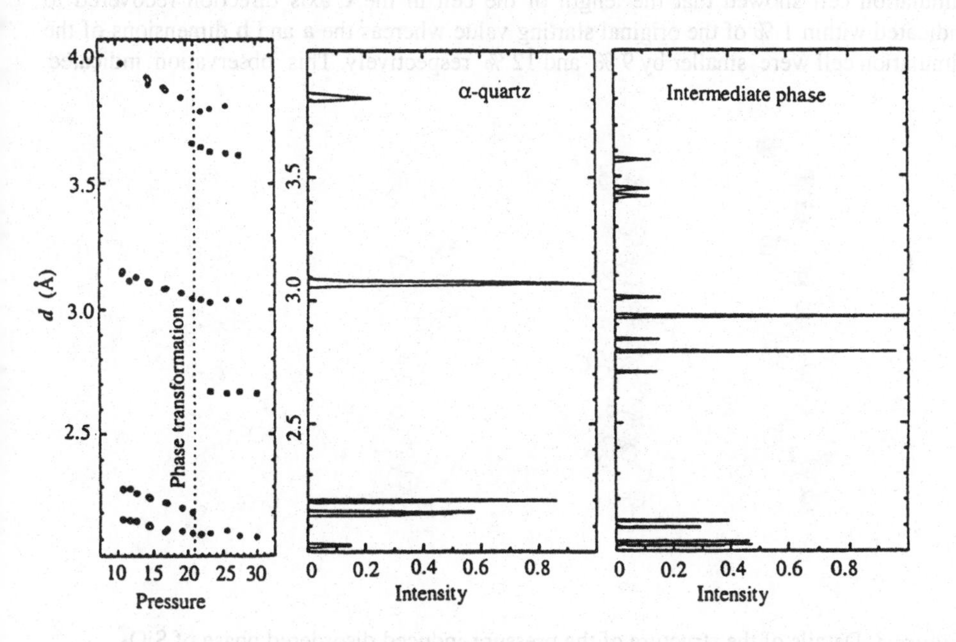

Figure 2. The experimental (a) and calculated diffraction patterns of α-quartz (b) and the phase of SiO_2 obtained at about 22 GPa (c).

4.2 ANISOTROPIC NATURE OF PRESSURE-INDUCED DISORDERED SYSTEMS

The characterization of an amorphous or disordered phase produced by pressure is important in order to provide an atomic level understanding of this phenomenon. In another study, the structure of the disordered phase of SiO_2 was examined and found to be anisotropic in a Brillouin scattering experiment[47]. It was observed that when α-quartz was increased to about 25 GPa and the densified solid was then recovered at low pressure, that the acoustic wave velocity parallel to the original **c** axis of the crystal was smaller than the acoustic velocity in the perpendicular direction. This was a reversal of the order of the sound velocities that are found in crystalline α-quartz. This was a significant observation since it demonstrated experimentally that the pressure amorphized solid had a structure that was quite different than the structure resembling a liquid at a similar density. Classical MD was able to provide a detailed understanding of the experimental results[48]. This was obtained by a series of calculations using the constant pressure method with the potential developed by van Beest et al. The elastic modulii were calculated and then converted to acoustic wave velocities and frequency shifts. The experimental pressure dependence of the sound wave velocities was reproduced and the material produced was disordered. At high pressures, the average coordination number of oxygen atoms about silicon atoms was more than 5. The increase in coordination number is in agreement with evidence provided by high-pressure x-ray diffraction results[49]. Upon release of pressure the coordination number decreased to a value of 4.3. Inspection of the simulation cell showed that the length of the cell in the **c** axis direction recovered to indicated within 1 % of the original starting value whereas the **a** and **b** dimensions of the simulation cell were smaller by 9 % and 12 % respectively. This observation indicated

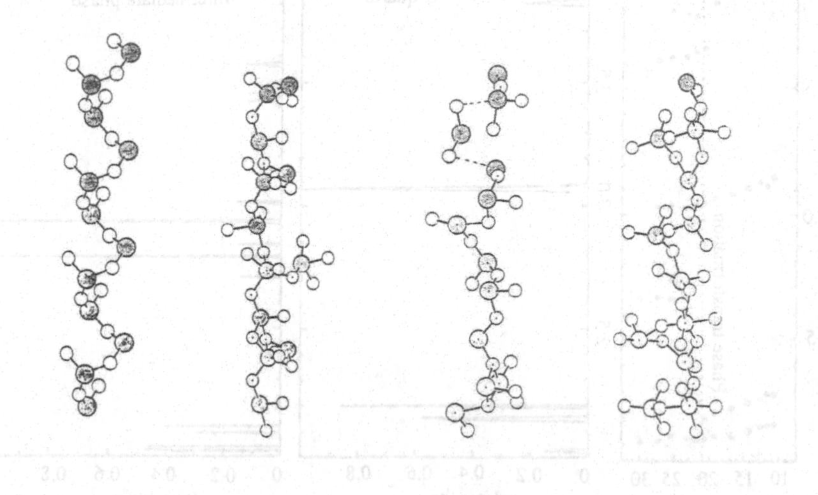

Figure 3. Details of the structure of the pressure-induced disordered phase of SiO_2 showing the anisotropic nature of this phase.

that the packing of the original helical chains had increased in the direction perpendicular to the chain axes. Apart from cross-linking between neighboring chains, the original helical network of a-quartz was largely preserved (Figure. 3). This observation provides an understanding of the reversal in the frequency shifts or acoustic velocities that occurs upon depressurization of the dense solid. The close agreement between experiment and theory showed in a very detailed manner that the pressure amorphized solid was significantly different that that of a quenched high-density liquid which of course would be isotropic.

4.3 HIGH PRESSURE CRYSTALLINE FORMS OF SiO_2

The success of this investigation implied that it may be possible to determine what structures of SiO_2 might be possible at pressures in the megabar range which is now obtainable experimentally in diamond anvil devices. Classical MD has also been applied to the effect of pressure on the amorphous phase with the goal of determining the possible structures of SiO_2 in the megabar pressure range[30,31]. These studies have been carried out with the van Beest et al. potential and it was found that a new phase related to a α-PbO_2 structure was obtained. This structure (Figure. 4) was shown to be energetically comparable to the to the 6-coordinated stishovite structure and more stable than the cubic

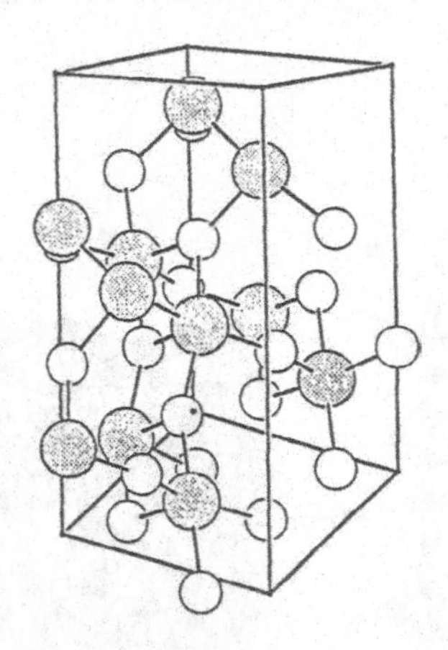

Figure 4. The structure of the monoclinic I2/a phase obtained at high pressures.

Pa3 structure which was considered as a candidate for high pressure SiO_2[50]. This was confirmed[51] by the application of first principles pseudopotential total energy calculations within the local-density approximation similar to those used in the ab initio studies described in the next section. More recently, it was shown that a large class of energetically similar high-pressure phases can be generated so it may be possible that a wide variety of metastable phases may eventually be found at high pressures[45].

4.4 "MEMORY" EFFECT IN HIGH-PRESSURE TRANSFORMATIONS

Another interesting phenomenon which has been examined using the potential of van Beest et al. is that of the memory effect or reversibility of order-disorder transformations that occurs in several materials. One early example of this effect was reported for $AlPO_4$, a material that is isostructural with α-quartz. It was reported that when a single crystal of this material was pressurized to about 20 GPa, it transformed to a disordered solid[52]. However, after release of the pressure the disordered sample recovered to the single crystal starting structure with the same orientation as before application of pressure. This effect has also been reproduced by classical molecular dynamics[33]. In Figure 5, the

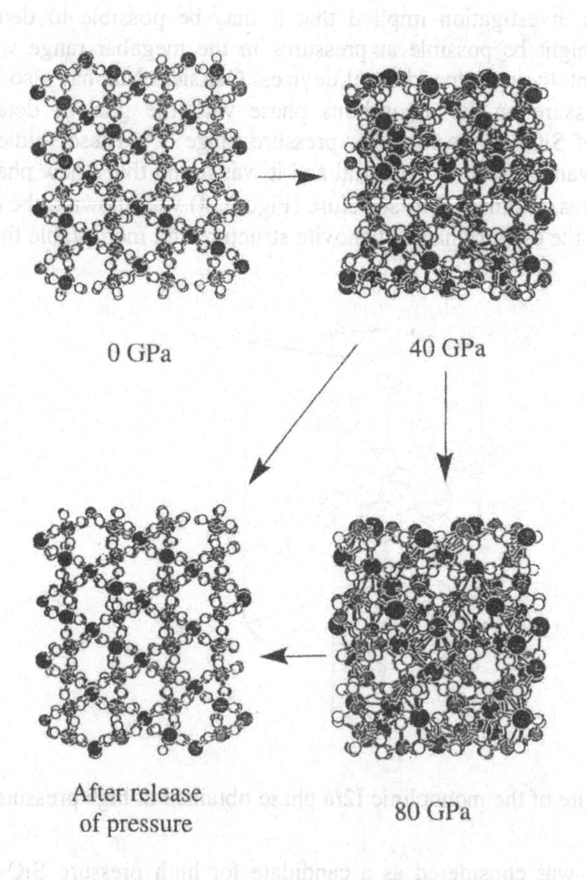

Figure 5. The calculated order → disorder → order transformation in $AlPO_4$ obtained with classical MD and the potential of Ref. 28.

calculated structure of $AlPO_4$ is shown as viewed down the crystal **c** axis at several pressures during the pressurization cycle. The important information that one obtains from this calculation provides the reason for the different behavior in this system than that observed in α-quartz. In α-quartz the densified solid is recovered at 1 bar and room temperature. In $AlPO_4$, the behavior of the AlO_4 and the PO_4 tetrahedra are quite different in that the Al-O coordination number increases and the O – Al – O bond angles distort more easily than the O – P – O angles with increase in pressure. The PO_4 tetrahedra are largely preserved at high pressure and therefore provide nucleation sites for the reconstruction of the original crystal.

An identical behavior has been observed and calculated in both clathrates hydrates[53] and in clathrasil dodecasil-3C[32], a new type of tectosilicate that has the same structure as that of a clathrate hydrate. These materials contain cage-like voids with guest molecules occupying the voids. The clathrasil can be prepared with empty cages whereas the clathrate hydrate is not stable without the guest molecules present. In both the clathrasil and in the clathrate, high-pressure experimental studies and MD simulations have been

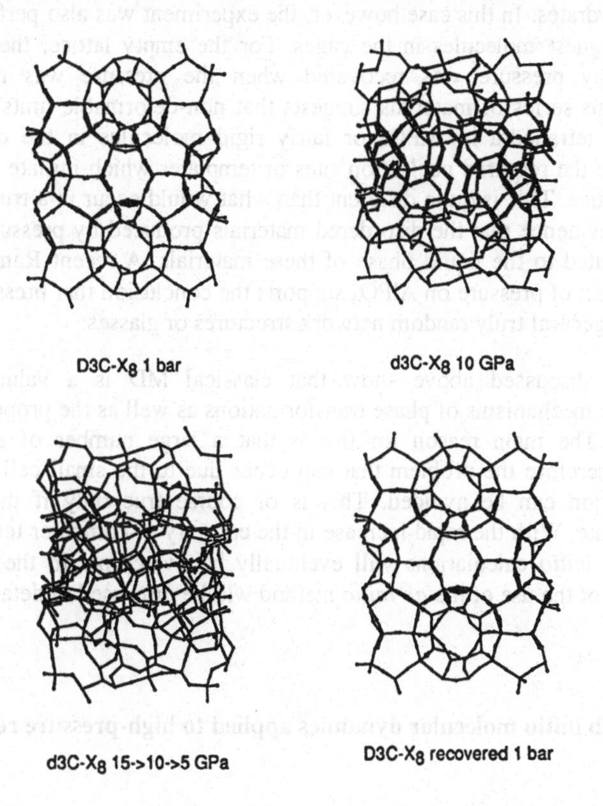

Figure 6. The order → disorder → order transformation in the clathrasil D3C-X_8 that contains Xe atoms in the cages.

performed. The transformation to a high-density disordered solid has been observed in both of these materials. In the case of the clathrate hydrates of tetrahydofuran and SF_6, sharp transformations with a decrease in volume were observed when the materials were compressed at 77 K to about 1.5 GPa. The materials recovered immediately to the starting crystalline structures when the pressure was released at 77 K. Molecular dynamics simulations of these transformations reproduced the experimental behavior and provided insight into the reasons for the recovery of the crystalline phase after release of pressure. The ice-like lattice collapses around the guest molecules and these guests provide centers of nucleation for the recovery of the starting structure. The behavior is illustrated in Figure 6 that shows the calculated radial distribution function for the oxygen atoms as a function of pressure. The fundamental change that occurs in the 25 - 40 kbar pressure range is evidence for the production of a disordered material since the second coordination shell broadens greatly. This change is similar to that for ice Ih and is a result of the displacement of water molecules into the open cavities upon pressurization. The radial distribution function of a structure almost identical to the starting material is obtained upon release of the pressure. The behavior of the clathrasils was similar to that of the clathrate hydrates. In this case however, the experiment was also performed on the material without guest molecules in the cages. For the empty lattice, the high-density solid produced by pressure was recovered when the pressure was released. The observations in this series of materials suggests that non-deformable units in the lattice such as the PO_4 tetrahedra in $AlPO_4$ or fairly rigid molecules in the clathrates and clathrasils provide the required nucleation sites or templates which initiate the formation of the original lattice. This is quite different than what would occur in a true glass and is therefore strong evidence that the disordered materials produced by pressure are unique structures not related to the liquid phase of these materials. A recent Raman scattering study[54] on the effect of pressure on $AlPO_4$ supports the conclusion that pressure-densified phases are not in general truly random network structures or glasses.

The examples discussed above show that classical MD is a valuable tool for understanding the mechanisms of phase transformations as well as the properties of high-pressure phases. The main reason for this is that a large number of atoms can be considered and therefore the problem that can occur due to the small cell size in a full quantum calculation can be avoided. This is of course true only if the interatomic potential is accurate. With the rapid increase in the capacity of computer technology, this limitation for *ab initio* calculations will eventually be overcome. In the next section, several examples of the use of the *ab initio* method will be discussed in detail.

5. Examples of ab initio molecular dynamics applied to high-pressure research

5.1 MOLECULAR HYDROGEN

The application of Car-Parrinello methods to solid hydrogen and Ice VIII provides examples of materials where classical methods may be deficient. The problem of the structure and electronic properties of hydrogen has been an active area of solid state physics for decades since hydrogen is a major component of the larger planets such as jupitor and also due to the apparent fundamental simplicity of this material. Since the early prediction of Wigner and Huntington[55], considerable effort has been made in attempting to produce metallic hydrogen and to predict the structure of both the metallic phase and the high pressure insulating phases. So far metallic hydrogen has not been made under hydrostatic conditions at pressures as high as 342 GPa at low temperatures[56]. A conducting phase has only recently been reported in shock wave experiments. Molecular hydrogen exists at 1 bar and low temperatures as a rotor phase solid (Phase I) where the H_2 molecules are rotating about their positions in a hexagonal close packed lattice. Above about 110 GPa the molecular rotations are frozen resulting in an ordered broken-symmetry phase (BSP or Phase II)[57]. At pressures above about 150 GPa another phase (Phase III) is observed but the structure of this phase has not been experimentally determined. Since this phase may remain stable up to the transformation, the determination of its structure is particularly important. There are several remarkable changes in the properties of molecular hydrogen when this phase appears and these properties can be used as a test of any model proposed for the structure. In particular, infrared absorption measurements[57] and Raman spectra[58] have revealed significant changes in the structure of solid hydrogen after the transformation to Phase III. In addition to discontinuous changes in the vibron frequencies there is a large increase in the absorptivity. There is an increase of about three orders of magnitude for vibrational excitations that are inactive for the isolated molecule.

Our first principles calculations[59] were based on local density functional theory and pseudopotential plane-wave approximations. The simulations were carried out on a simulation cell of 96 atoms with the H_2 molecules initially located at hexagonal close packed sites. The simulation temperature was maintained at 77 K. At pressures below 60 GPA, the hydrogen the hydrogen molecules rotated almost freely about the hexagonal lattice sites. As the pressure increased, a phase transformation occurred with the molecules aligned in layers parallel to the crystallographic ***a-b*** plane. This ordered structure remains stable up to about 150 GPa with the main effect being a slight shortening of the hydrogen bond length. At about 260 GPa, there is a transformation to another ordered structure consisting of hydrogen molecules with two distinct intramolecular hydrogen bond lengths in an exact ratio of 2:1. The appearance of this phase provides clear evidence for significant change in the electronic structure. The structures of the three phases are shown in Figure 7. Although the high pressure structure may not be precisely correct due to several limitations in these calculations such as the lack of a full Brillouin zone sampling or k-point convergence and the neglect of the effect of zero point motions on the stability of the various phases, a great deal of physical insight can be obtained from these results. It should be emphasized that the physical properties of the high-pressure phases have been reproduced and this provides an illustration of the ability of the ab initio MD method to uncover the origins of these properties. One such property is the vibrational density of states obtained from the Fourier transform of the velocity autocorrelation function. As the pressure is raised the calculated H-H stretching vibrations increases until the pressure reaches 200 GPa and then the frequencies drop as phase III is formed. The calculated trend in frequencies is in qualitative agreement with the experimental Raman and infrared results[57,58]. The lowering of frequencies is most likely due to the increasing intermolecular interactions at high pressure.

One experimental result that can possibly be understood from the calculated structure for phase III is the significant increase in infrared absorption intensity that is characteristic of this phase. As noted above, the high-pressure orthorhombic phase is characterized by two bondlengths for the hydrogen molecules and a unique grouping of the molecules with three molecules that were originally in the same plane to be alternately shifted along the c axis with an accompanied rotation of molecules out of the

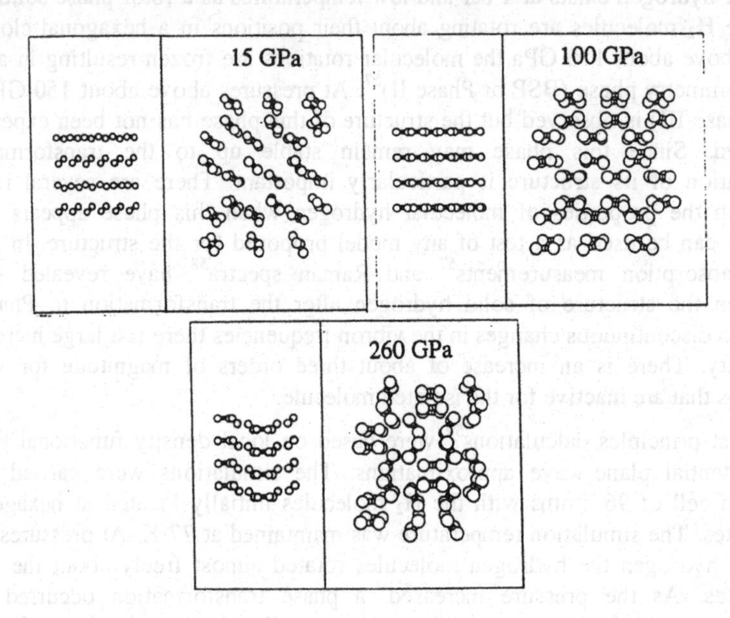

Figure 7. The three calculated crystal structures of molecular hydrogen at several pressures. The views on the left and right side of the figure are down the y axis and z axis, respectively, of the simulation cell.

a-b plane. The short bonded molecules are located at the ends of the third hydrogen molecule. Each component therefore consists of three molecules where two having a short bond are equivalent. This arrangement and grouping of the hydrogen molecules has an important effect on the electronic properties of this phase. Electronic band structure calculations indicate a small overlap between the hydrogen molecules. There is a small accumulation of electronic charge on the longer of the hydrogen molecules and a slight lowering of charge on the shorter bonded molecules. This is therefore a small charge-transfer interaction that is a result of this unique molecular arrangement in this phase. It has been shown that the electron structure can be understood using a molecular orbital (MO) diagram (Fig. 8) that consists of six sets of orbitals derived from the linear combination of three bonding and anti-bonding orbitals from each molecule. We review the arguments here since these types of arguments may be useful for the understanding of structural stability of other high-pressure phases of simple and complex materials. The ordering of the relative energies of these MOs can be given in accordance with the number of nodal planes at each level. At high pressures and high density the lowest energy symmetric (S) orbital in Figure 8 is stabilized and there will be a small effect on the asymmetric "nonbonding" orbital. The highest energy occupied symmetric orbital would be destabilized and its energy shifted upward as a result of out of phase interactions between orbitals at this level. The energy of the lowest energy empty orbital will decrease due to the in-phase interactions between molecules. The two highest unoccupied orbitals shown would shift to higher energies due to nonbonding interactions. The unoccupied asymmetric antibonding orbital that shifts to lower energies can interact

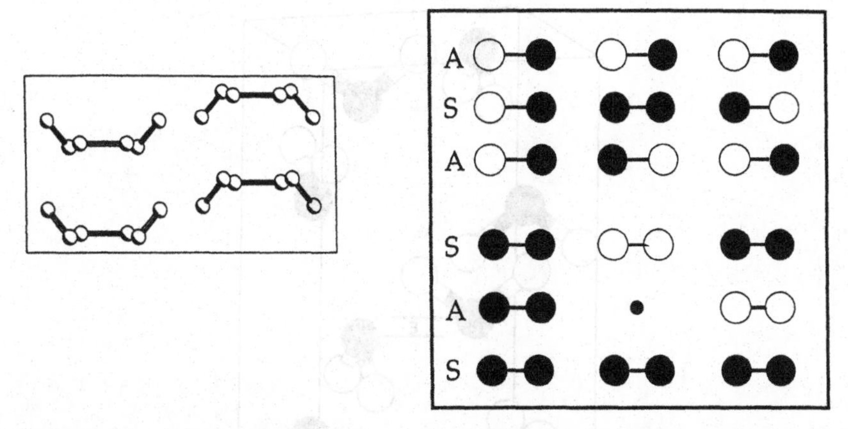

Figure 8. Structure of phase III of molecular hydrogen (left) and schematic MO diagram (right) for hydrogen in phase III. The open and shaded circles represent the positive and negative phases of the orbitals. The dot is a nodal plane.

with the occupied "nonbonding" orbital. This would result in a destabilization of the central hydrogen molecule caused by the antibonding character and would result in a donation of charge density from the two neighboring hydrogen molecules to the central one. The result is a lengthening of the bond of the central molecule and the formation of a weak charge-transfer complex. This charge transfer results in the enhancement of the infrared absorptivity. This simple picture resulting directly from the *ab initio* MD results provides a basic picture for the ideas that had been proposed earlier for the strong infrared enhancement in this system[58]. It should be noted that the band structure of this phase indicates that this phase of hydrogen is an insulator with an indirect band gap. A recent study based on density functional theory has also found that a movement of the hydrogen molecules away from the close-packed hexagonal sites at high pressures results in a polarized state that would give rise to high infrared absorptivity[61]. The calculation of the vibron frequency using the frozen phonon method fully supported the *ab initio* MD results. There is certainly a great deal of additional work required however in order to fully characterize the high-pressure phases of molecular hydrogen.

5.2 AB INITIO STUDY OF A STRUCTURAL TRANSFORMATION IN ICE VIII

A recent example which illustrates the current ability of first-principles quantum methods to characterize of subtle aspects of structural transformations is in the study of ice VIII[62]. Ice VIII is a high-pressure phase made from H_2O or D_2O that has a particularly simple structure[63] as shown in Figure 9. The structure is anti-ferroelectric consisting of two

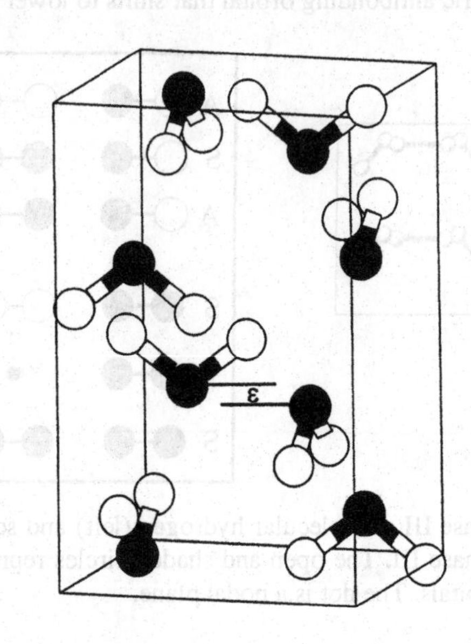

Figure 9. The crystal structure of ice VIII and the tetragonal distortion parameter ε.

interpenetrating cubic lattices with diamond structures for the hydrogen bonded oxygen atoms. This phase is stable at pressures above 2 GPa at temperatures below about 263 K but it can be recovered at ambient pressure if the sample is maintained at low temperatures. A recent accurate neutron diffraction study[64] of this material was carried out using a spallation source and a unique high-pressure apparatus that permitted measurements to be made on decompression to ambient pressure. The experimental measurements revealed an isostructural phase transformation in the pressure range below 2 GPa. This transformation was characterized by a rapid drop in the fractional coordinate $z(\mathrm{O})$ shown in Fig. 10 which is related to the distortion parameter ε by

$$\varepsilon = 2c[1/8 - z(\mathrm{O})] \tag{17}$$

where c is the lattice constant along the z axis of the tetragonal lattice. The parameter ε is therefore a measure of the tetragonal distortion of the lattice. There are also Raman and infrared measurements in the literature that indicate unusual frequency shifts at low pressures[64-66]. It was reported that there are somewhat large changes in the frequencies of several inter and intra-molecular vibrational modes which may also indicate a structural change. It was suggested that there was a new phase denoted ice VIII′ that occurs possibly as a result of a second-order isostructural phase transformation. There are only a few examples of these types of transformations in both metallic and molecular solids and it was apparent that ice VIII would be an excellent example for a theoretical study due to the large amount of structural information in the literature. The goal of our study was therefore to determine if an atomic level understanding of the transformation mechanism could be obtained and in particular, to characterize the thermodynamic order of the transformation.

Total energy calculations were carried out using the pseudopotential planewave method in density functional theory[25]. A gradient corrected functional was used for the exchange correlation[67]. The pseudopotential used was that of Vanderbilt[68]. Due to the apparent subtle nature of the transformation, special care was taken to have convergence in the calculated total energies and structural parameters. This was ensured by the use of a 48 k-point mesh obtained by the Monkhorst-Pack scheme[69] and the use of a 600 eV energy cutoff for the planewave basis set. The calculated O–H bondlength and H-O-H angle for an isolated water molecule were 0.971 Å and 104.0° respectively. The experimental values are 0.957 Å and 104.5°. The slight difference in the bond length can be attributed to minor features of the pseudopotential. A series of calculations were performed where both the internal coordinates and the c/a ratio were optimized. Frozen phonon calculations[70] were performed on vibrational modes to examine the pressure dependence of modes that could be considered as possible soft modes that may accompany the structural transformation.

The first result (Figure 10) reproduced completely the observed drop in z(O) with the calculated volume for the onset corresponding to a pressure of about 10 GPa which is slightly higher than in the experiment. This is most likely due mainly to the kinetic factors present in the experiment. The calculated drop is also more abrupt than seen in the experiment and this possibly a result of kinetic factors in the experiments. The characterization of the thermodynamic order of the transformation was first investigated by examination of the volume dependence of the total energy and the lattice parameters $\boldsymbol{a}$ and $\boldsymbol{c}$. In order for the transformation to be classified as a first order transformation, there would have to be a discontinuity in the volume dependence of these properties. As seen in Figure 11, there is no detectable discontinuity in either lattice parameter and this is in agreement with the experimental results. The transformation therefore does not appear to be first order. The calculated volume dependence of the A_{1g} lattice mode which can be described as the displacement of each interpenetrating lattice in opposite directions parallel to the c axis does not show a discontinuity within the accuracy of the frozen phonon calculation. This mode is a good candidate for a possible soft mode and therefore an indicator of a second-order thermodynamic transformation from Landau theory[71]. The A_{1g} intramolecular vibration which consists of symmetric stretching vibrations of the water molecules also does not show a discontinuity or a soft mode behavior.

Although the transformation in ice VIII does not appear to be a classical first or second order thermodynamic transformation based on the criteria discussed above, there are several details of the structural changes that provide insight for the mechanism of the transformation. The calculated volume dependence of the hydrogen bonded O-H - - O distances is distinctly different than that of the nearest neighbor nonbonded O - - O distances (Figure 12). The calculated nonbonded distance has a definite change in behavior

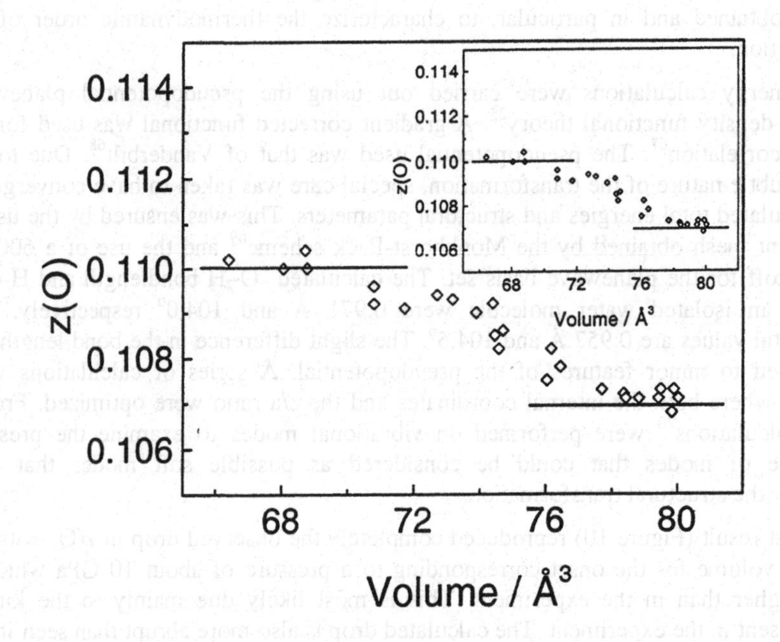

Figure 10. The calculated pressure dependence of the fractional coordinate z(o) for ice VIII and the experimental data of Ref. 4 (Inset). The solid lines are guides for the eye.

over the pressure range where z(O) changes rapidly. There is a smaller volume dependence over the transformation range. The hydrogen-bonded lattices have a more uniform pressure dependence. The uniform volume dependence as reflected in the c and a lattice parameters is consistent with the behavior of the O - - O distances in that it means that the compressibility is determined by the hydrogen bonds and that the relative motion of the nonbonded lattices can occur without affecting this property. An examination of the change in the structural properties of individual water molecules (Figure 13) shows that the H-O-H angle also changes less rapidly with pressure over the transformation range whereas the O-H intramolecular bond lengths change uniformly and at a rate that is in excellent agreement with experiment.

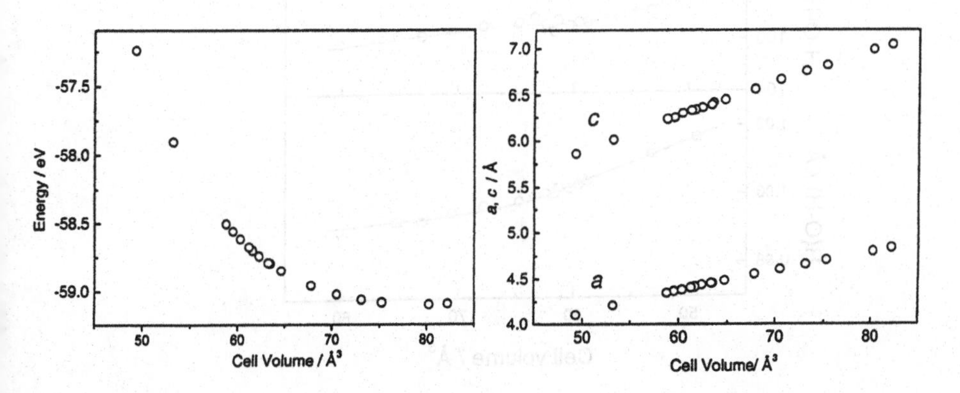

*Figure 11.*The calculated volume dependence of the (a) total energy and (b) lattice constants **c** and **a** in ice VIII.

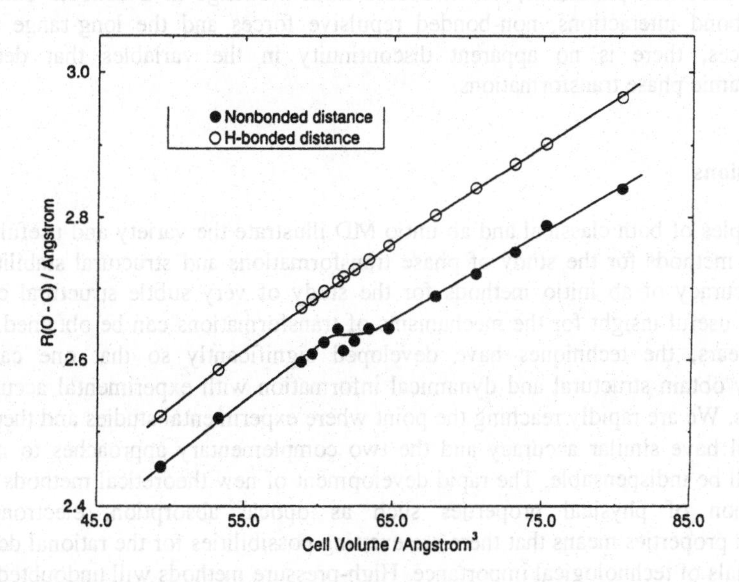

Figure 12. The calculated volume dependence of the internal structural parameters of the hydrogen-bonded and non-bonded nearest neighbor distances in ice VIII.

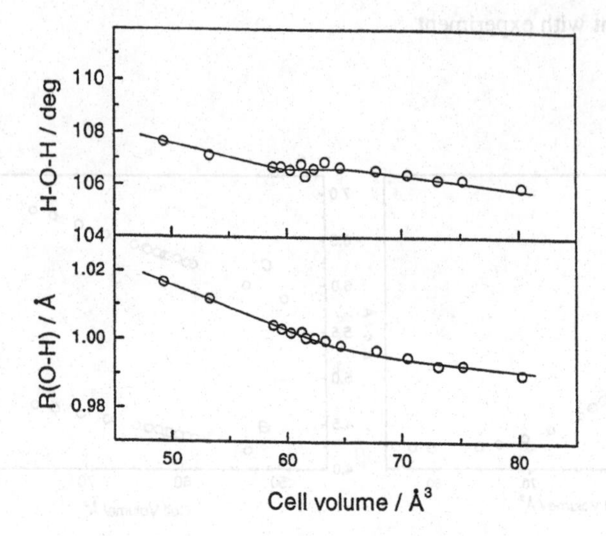

Figure 13. The calculated volume dependence of the O-H bond length (lower frame) and the H-O-H angles (upper frame) in ice VIII.

The results of the calculations show that although there is a rapid increase in the tetragonal distortion parameter, which occurs from a change in a delicate balance of hydrogen bond interactions, non-bonded repulsive forces and the long-range dipole-dipole forces, there is no apparent discontinuity in the variables that determine thermodynamic phase transformations.

6. Conclusions

This examples of both classical and ab initio MD illustrate the variety and usefulness of theoretical methods for the study of phase transformations and structural stability. The current accuracy of ab initio methods for the study of very subtle structural changes shows that useful insight for the mechanisms of transformations can be obtained. In the last ten years, the techniques have developed significantly so that one can now confidently obtain structural and dynamical information with experimental accuracy in many cases. We are rapidly reaching the point where experimental studies and theoretical studies will have similar accuracy and the two complementary approaches to material science will be indispensable. The rapid development of new theoretical methods for the determination of physical properties such as optical absorption, electronic and mechanical properties means that there are exciting possibilities for the rational design of new materials of technological importance. High-pressure methods will undoubtedly play a major role in this area.

7. References

[1] Anderson H.C. (1980) Molecular dynamics simulations at constant pressure and/or temperature, *J. Chem. Phys.* **72**, 2384-2393.

[2] Parrinello, M and Rahman (1980) A. Crystal structure and pair potentials: A molecular dynamics study, *Phys. Rev. Lett.* **45**, 1196-1199.

[3] Parrinello, M.and Rahman, A. (1981) Polymorphic transitions in single crystals: a new molecular dynamics method, *J. Appl. Phys.* **52**, 7182-7190.

[4] Yin, M.T and Cohen, M.L. (1982) Theory of static structural properties, crystal stability and phase transformations: Applications to Si and Ge, *Phys. Rev.* **B26**, 5668-5687.

[5] Chang, K.J. and Cohen, M.L. (1985) Solid state phase transformations and soft modes in highly condensed Si, *Phys. Rev.* **B31**, 7819-7826.

[6] Car, R. and Parrinello, M. (1985) Unified approach for molecular dynamics and density-functional theory, *Phys. Rev. Lett.*, **55**, 2471-2174.

[7] Focher, P., Chiarotti, G.L., Bernasconi M., Tossati, E., and Parrinello, M. (1994) Structural phase transformations via first-principles simulation, *Europhys. Lett.* **26**, 345-351.

[8] Bernasconi, M., Chiarotti, G.L., Focher P., Scandolo, S., Tosatti, E. and Parrinello, M. (1995) First-principles-constant pressure molecular dynamics *J. Phys. Chem. Solids*, **56**, 501-505.

[9] Thomson K.T., Wentzcovitch, R.M. and Bukowinski, M.S. (1996) *Science* **274**, 1880-1882.

[10] Alder, B.J. and Wainwright, T.E. (1957) Phase transition for a hard sphere system, *J. Chem. Phys.* **27**, 1208-1209.

[11] M. Sprik, (1996) in "Monte Carlo and Molecular Dynamics of Condensed Matter Systems", edited by K. Binder and G. Ciccotti, SIF, Bologna.

[12] M.P. Allen and D.J. Tildesley, "Computer Simulation of Liquids", Oxford Press, Oxford, 1987.

[13] Tse, J.S. and Klug, D.D. (1991) The structure and dynamics of silica polymorphs using a two-body effective potential model, *J. Chem. Phys.*, **95**, 9176-9185.

[14] Sprik, M.R., Impey, R.W., and Klein, M.L. (1984) Second-order elastic constants for the Lennard-Jones solid, *Phys. Rev.* **B29**, 4368-4374.

[15] M.Born and K. Huang, "Dynamical theory of Crystal Lattices, Clarendon, Oxford, 1954.

[16] M.L. Klein, in "Molecular Dynamics Simulation of Statistical Mechanical Systems", edited by G. Ciccotti and W.G. Hoover, North Holland, Amsterdam, 1986, p. 426.

[17] van Hove, L. (1954) Correlations in space and time and Born approximation

scattering in systems of interacting particles, *Phys. Rev.* **95**, 249-262.

[18] Car, R. and Parrinello, M. (1988), in "Simple Molecular Systems at Very High Density", Plenum, 1988, 455-476.

[19] C. Galli and M. Parrinello (1991), in *Computer Simulation in Material Science*, edited by M. Meyer and V. Pontikis, Kluwer Academic Publishers, Dortrecht, 283-304.

[20] Remler, D.A. and Madden, P.A. (1990)Molecular dynamics without effective potentials via the Car-Parrinello approach, *Mol. Phys.* **70**, 921-966.

[21] Payne, M.C., Teter, M.P., Allan, D.C. (1992) Iterative minimization techniques for ab initio total-energy calculations: molecular dynamics and conjugate gradients, *Rev. Mod. Phys.* **64**, 1045-1097.

[22] Kresse, G. and Furthmüller (1996), Efficiency of ab initio total energy calculations for metals and semiconductors using a plane-wave basis set, *Comput. Mat. Scie.*,15-50.

[23] Scandolo, S., Bernasconi, M. Chiarotti, G.L., Focher, P., and Tosatti, E. (1995) Pressure-induced transformation path of graphite to diamond, *Phys. Rev. Lett.* **74**, 4015-4018.

[24] Benoit, M., Bernasconi, M., Focher, P. and Parrinello, M. (1996) New high-pressure phase of ice, *Phys. Rev. Lett.* **76**, 2934-2936.

[25] Kresse G. and Hafner, J. (1993) Ab initio molecular dynamics for open-shell transition metals, *Phys. Rev.* **B48**, 13115-13118 ;(1994) Ab initio molecular-dynamics simulation of the liquid-metal-amorphous-semiconductor transition in germanium, *ibid*, **B49**, 14251-14269 ; Kresse, G. and Furthmuller, J. (1996) Efficient iterative schemes for *ab initio* total-energy calculations using a plane-wave basis set, *Phys. Rev.* **B54**, 11169-11186.

[26] Tse, J.S. and Klug, D.D. (1991) Mechanical Instability of α-quartz: a molecular dynamics study, *Phys. Rev. Lett.* **67**, 3559-3562.

[27] Tsuneyuki, S., Tsukada, M., Aoki, H., Matsui, Y. (1988) First-principles interatomic potential of silica applied to molecular dynamics, *Phys. Rev. Lett.* **61**, 869-872.

[28] van Beest, B.W.H., Kramer, G.J., van Santen, R.A. (1990) Force fields for silicas and aluminophosphates based on ab initio calculations, *Phys. Rev. Lett.* **64**, 1955-1958.

[29] Tse, J.S., and Klug, D.D. (1991) The structure and dynamics of silica polymorphs using a two-body effective potential model, *J. Chem. Phys.* **95**, 9176-9185.

[30] Tse, J.S. Klug, D.D., and Le Page, Y. (1992) Novel high pressure phase of silica, Phys. Rev. Lett. **69**, 3647-3649.

[31] Tse, J.S., Klug, D.D., and Le Page, Y. (1992) High pressure densification of amorphous silica, Phys. Rev. B46, 5933-5938.

[32] Tse, J.S., Klug, D.D., Ripmeester, J.A., Desgreniers, S., and K. Lagarec (1994) The role of non-deformable units in pressure-induced reversible amorphization of clathrasils, *Nature* **369**, 724-727.

[33] Tse, J.S. and Klug, D.D., (1992) Structural memory in pressure-amorphized $AlPO_4$,

Science **255**, 1559-1561.

[34] Mishima, O. Calvert, L.D., and Whalley, E. (1984) 'Melting ice' at 77 K and 10 kbar: A new method of making amorphous solids. *Nature* **310**, 393-395.

[35] Floriano, M.A., Whalley, E. Svensson, E.C., and Sears, V.F. (1986) Structure of high-density amorphous ice by neutron diffraction, *Phys. Rev. Lett.* **57**, 3062-3064.

[36] Wolf, D., Okamoto, S., Yip, S., Lutsko, J.F., and Kluge J. (1990) Thermodynamic parallels between solid-state amorphization and melting, J.Mater. Sci. **5**, 286-301.

[37] Hemley, R.J., Jephcoat, A.P., Mao, H-K. Ming, L.C., Manghnani, M.H. (1988) Pressure induced amorphization of crystalline silica, *Nature* **334**, 52-54.

[38] Ewald, P. (1921) Die Berechnung optischer und elektrostatischer Gitterpotentiale *Ann. Phys.* **64**, 253-287.

[39] Binggeli, N. and Chelikowsky, J.R. (1992) Elastic instability in α-quartz under pressure, *Phys. Rev. Lett.* **69**, 2220-2223.

[40] Tse, J.S. (1992) Mechanical Instability in ice Ih. A mechanism for pressure-induced amorphization, *J.Chem. Phys.* 5482-5487.

[41] Kingma, K.J., Hemley, R.J., Mao, H-K., Veblen, D.R. (1993) New high pressure transformation in α-quartz, *Phys. Rev. Lett.* **70**, 3927-3930.

[42] Somayazulu, M.S., Sharma, S.M., and Sikka, S.K. (1994) Structure of a new high pressure phase in α-quartz determined by molecular dynamics studies, *Phys. Rev. Lett.* **73**, 98-101.

[43] Wentzcovitch, R.M., da Silva, C., Chelikowsky, J.R. (1998) A new phase and pressure induced amorphization in silica, *Phys. Rev. Lett.* **80**, 2149-2152.

[44] Tse, J.S., Klug, D.D., and LePage, Y. (1997) High pressure four-coordinated structure of SiO_2, *Phys.Rev.* **B56**, 10878-10881.

[45] Teter, D.M. and Hemley, R.J. (1998) High-pressure polymorphism in silica, *Phys. Rev. Lett.* **80**, 2145-2148.

[46] Le Page, Klug, D.D., and Tse, J.S. (1996) Derivation of conventional crystallographic descriptions of new phases from results of ab initio inorganic structure modelling, *J. Appl. Cryst.* **29**, 503-508.

[47] McNeil, L.E., and Grimsditch, M. (1992) Pressure amorphized SiO_2 α-quartz: An anisotropic amorphous solid, *Phys. Rev. Lett.* **68**, 83-85.

[48] Tse, J.S. and Klug, D.D. (1993) Anisotropy in the structure of pressure-induced disordered solids, Phys. Rev. Lett. **70**, 174-177.

[49] Meade, C. Hemley, R.J., and Mao, H-K. (1992) High pressure x-ray diffraction of SiO_2 glass, Phys. Rev. Lett. **69**, 1387-1390.

[50] Park, K.T., Terakura, K., and Matsui, Y. (1988)Theoretical evidence for a new ultar-high-pressure phase of SiO_2, *Nature* **336**, 670-672.

[51] J.S. Tse, Klug, D.D., and Allan, D.C. (1995) The structure and stability of several

high pressure phases of silica, *Phys. Rev.* **B51**, 16392-16395.

[52] Kruger, M.B. and Jeanloz, R. (1990) Memory glass: An amorphous material formed from $AlPO_4$, *Science* **249**, 647-649.

[53] Handa, Y.P., Tse, J.S., Klug, D.D., and Whalley, E. (1991) Pressure-induced phase transitions in clathrate hydrates, *J.Chem. Phys.* **94**, 623-627.

[54] Gillet, P., Badro, J., Varrel, B., and McMillan, P.F. (1995) High-pressure behavior in $AlPO_4$: Amorphization and the memory-glass effect, *Phys. Rev.* **B51**, 11262-11268.

[55] Wigner, E. and Huntington, H.B. (1935) On the possibility of a metallic modification of hydrogen, *J. Chem. Phys.* **3**, 764-770.

[56] Narayana, C., Luo, H., Orloff, J., and Ruoff, A.L. (1998) Solid hydrogen at 342 Gpa: no evidence for an alkali metal, *Nature* **393**, 46-48.

[57] Mao, H-K., and Hemley, R.J. (1994) Ultrahigh-pressure transitions in solid hydrogen, *Rev. Mod. Phys.* **66**, 671-691.

[58] R.J. Hemley, Z.G. Soos, M. Hanfland, and H-K. Mao (1994) Charge-transfer states in dense hydrogen, *Nature* **369**, 384-387.

[59] Hemley, R.J. and Mao, H-K. (1988) Phase transition in solid hydrogen at ultrahigh pressure, *Phys. Rev. Lett.* **61**, 857-860.

[60] Tse, J.S. and Klug, D.D. (1995) Evidence from molecular dynamics simulations for non-metallic behavior of solid hydrogen above 160 GPa, *Nature* **378**, 595-597.

[61] Edwards, B. and Ashcroft, N.W. (1998) Spontaneous polarization in dense hydrogen, *Nature* **388**, 652-655.

[62] Tse, J.S. and Klug, D.D. (1998) Anomalous Isostructural Transformation in ice VIII, *Phys. Rev. Lett.* **81**, 2466-2469.

[63] Kuhs, W.F., Finney, J.L., Vettier, C. and Bliss, D.V. (1984) Structure and hydrogen ordering in ices VI, VII, and VIII by neutron powder diffraction, *J. Chem. Phys.* **81**, 3612-3623).

[64] Besson, J.M., Klotz, S., Hamel, G., Marshall, W.G., Nelmes, R.J., and Loveday, J.S. (1997) Structural instability in ice VIII under pressure, *Phys. Rev. Lett.*, **78**, 3141-3143.

[65] Hirsch, K.R. and Holzapfel, W.B. (1986) Effect of high pressure on the Raman spectra of ice VIII and evidence for ice X, *J. Chem. Phys.* 84, 2771-2775.

[66] Wong, P.T.T. and Whalley, E. (1976) Raman spectrum of ice VIII, *J. Chem. Phys.* **64**, 2359 (1976).

[67] Perdew, J.P. and Wang, Y. (1992) Accurate and simple analytic representation of the electron-gas correlation energy, *Phys. Rev.* **B45**, 13244-13249.

[68] Vanderbilt, D. (1990) Soft self-consistent pseudopotentials in a generalized eigenvalue formalism, *Phys. Rev.* **B41**, 7892-7895 (1990).

[69] Monkhorst, H.J. and Pack, J.D. (1976) Special points for Brillouin-zone integrations *Phys. Rev.* **B50**, 5188-5899.

[70] M.Y. Yin and M.L. Cohen, *Phys. Rev.* **B26**, 3259 (1980).

[71] Bruce, A.D., and Cowley, R.A. (1981) Structural phase transitions, Taylor and Frances Ltd. London.

EXPERIMENTAL TECHNIQUES IN THE DIAMOND ANVIL CELL

D.J. DUNSTAN
Physics Department, Queen Mary and Westfield College,
University of London,
London E1 4NS,
England.

Abstract. The operating principles of diamond-anvil high pressure cells are reviewed, with particular attention to the implications for design and construction. The diamond culets and gasket generate the pressure, and their behaviour dictates the requirements for the rest of the cell. The axial alignment mechanism is crucial, while tilt alignment is less important. The implication for piston-cylinder designs is that the clearance of the piston in the bore is critical, while the length of the piston is not. Good practice in the design of drive mechanisms is discussed. Finally, we consider alternatives to the standard piston-cylinder mechanism. Flexure movements, and their basic design rules are presented.

1. Introduction

The diamond-anvil cell (DAC) is now a well-established tool in condensed-matter research. It provides a means of studying solids and liquids under static hydrostatic pressure, in total safety and in most contexts essentially without limit of pressure. Only in research into the metallisation of hydrogen, and into the states of matter in the centres of the planets, is the DAC pushed to its intrinsic limits. While traditional high-pressure equipment was operated close to the engineering limit and was very dangerous if taken even a small fraction beyond its rating, a DAC of almost any design can be operated far beyond what its designer may have intended provided its principles of operation are properly respected. The purpose of this article is therefore to survey the various styles of design and to explore what these principles are.

At its centre, the DAC maintains a high-pressure volume around one or more samples. Hydrostaticity in this volume is a major concern. Working outwards from the high-pressure volume, there are a wide variety of components, of functionality, and of operating principles to be respected. But the basic questions arise in the generation and containment of the high pressure. In most areas of engineering, high pressures are contained within a strong enclosure, such as a gas cylinder or a steam boiler, characterised by a tensile strength vastly in excess of the high pressure to be confined. Thus steel has a tensile strength in the range of 5 to 20 kbar, and can more or less safely enclose pressures rather less than these. On the other hand, it is well known in mechanical engineering that steel surfaces in contact under load can generate much higher pressures, so that an effective extreme-pressure lubricant has to be able to separate steel surfaces exerting pressures up to 100 kbar on each other. If the lubricant breaks down, the surfaces pressure-weld together. No surfaces are completely flat, so that asperities meet and support the load on a small area. If the asperities are low-angle (Fig.1), the pressure at the point of contact can be many times the strength of the steel.

The DAC descends from pressure-generation equipment based on this principle, the Bridgman anvil system, in which a pair of opposed tungsten carbide anvils squeeze a gasket and a sample space between them. Bridgman observed that such a system could

R. Winter and J. Jonas (eds.), High Pressure Molecular Science, 87–101.

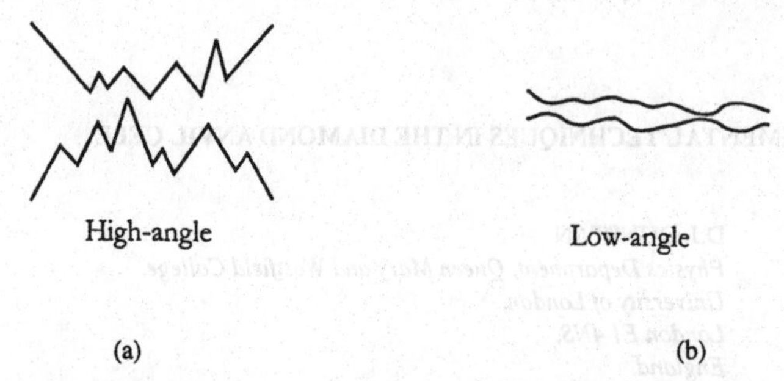

Figure 1. **In (a) two rough pieces of steel come into contact. Because the asperities are high-angle, the contact pressure will be little more than the compressive strength of the steel. In (b) the surfaces are much smoother. Because of the low angle of the asperities, the contact pressure is many times the compressive strength of the steel.**

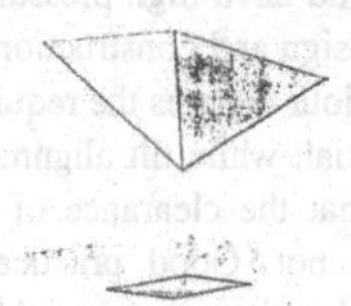

Figure 2. **The indenter has been pressed into a surface and left an indent. The hardness is defined as the force that was applied over the area of the indent**

generate pressures far beyond the normal tensile strength of the gaskets, and also well above the normal compressive strength of the anvils. The latter point was not wholly unexpected. Hardness testing relies on loading an indenter and seeing how far it presses into the material under test (Fig.2). Most engineering hardness scales (Rockwell, etc) are expressed in units of pressure — the load divided by the area of the indent — and generally the hardness is about three times the compressive strength of the material. The reason for this is what Bridgman called the massive support given by the surrounding material to the portion under load. So Bridgman anvils have a very shallow angle, to give maximum massive support to the area under pressure (Fig.3a) and get to pressures above the compressive strength of the anvil materials. Except for multi-megabar work, the pressures in a DAC are much less than the compressive strength of diamond, so shallow bevels are not necessary (Fig.3b).

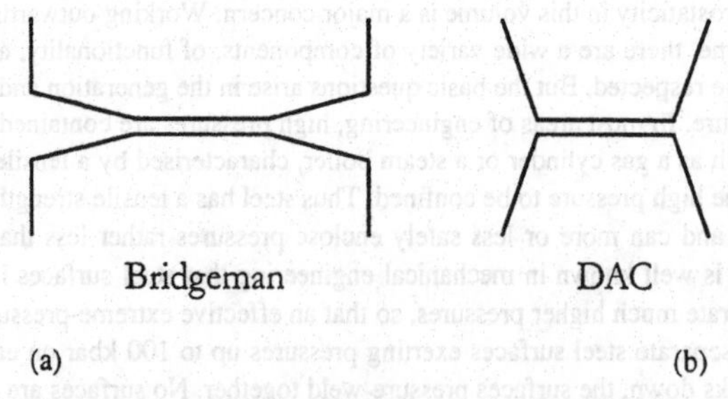

Figure 3. **Bridgman anvils are shown schematically in (a), with their shallow bevels to give massive support. Diamonds are cut without bevels (b), except for the highest pressures.**

The role of the gasket appears not to have been properly considered until recently (see Eremets [1], and references therein), except in the very different context of manufacturing or minting coins. When these were made by compressing a blank between two dies under a sledgehammer, quantitative results were hard to come by. However, when modern techniques began to be used in mints, when hydraulic presses began to replace men with sledgehammers, it was found that much larger forces are required than were expected. Rather than requiring a force equal to the compressive strength of the metal times its area, forces many times this were required. Schroeder and Webster [2] provided an explanation, which I took over later to explain the gasket in a Bridgman anvil or diamond anvil cell. Essentially, the gasket is retained by friction against the diamond faces. The hydrostatic pressure in the gasket rises linearly from the edge, at a rate proportional to the aspect ratio (diameter over thickness). Extrusion of the metal occurs down the pressure gradient. We shall see that it is best to have the gasket thin enough so that the pressure rises to a maximum higher than the experimental pressure and then drops again towards the gasket hole. This insures that the hole is pressurised by extrusion of the metal radially inwards [3].

Moving outward again from the components which generate the pressure in a DAC, one must consider the problems of alignment and force generation. Here we may note that most DAC designs are influenced by two fundamental myths or misconceptions. One is that stiffness and strength are achieved by making components massive. The other is that precision in operation can only be achieved by precision engineering, by precision in the manufacture. Both are very natural. However, the first is partly due to a natural confusion between stiffness and strength which arises partly through ignoring stress concentrations and partly through overlooking the analysis by Euler [4] of the buckling of components under compression. The second myth may be linked to the old philosophical position that the imperfect can only create the imperfect, and that perfection can only be achieved by the perfect creator. This myth was exploded by Henry Maudsley (1771-1831), the father of the machine tool. He showed that successive generations of machine tools could each be made to manufacture a more accurate successor. The myth is also exploded by the concept of the flexure movement, in which components made with little precision make an assembly which provides great precision of movement.

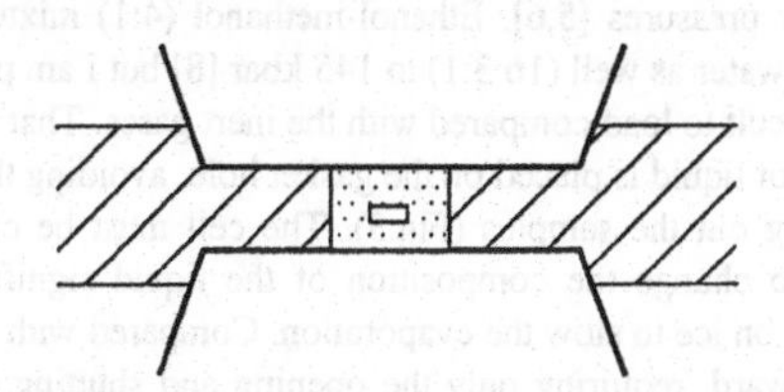

Figure 4. The pressure volume is shown schematically. The gasket is indented both sides by the diamonds, and in the centre of the indent is the pressure volume. The sample moves freely in the pressure fluid.

2. The Pressure Volume

2.1 THE FLUID

The pressure volume in a DAC is bounded by the diamond faces above and below, and by the gasket around it. It is normally cylindrical, with an aspect ratio (diameter to height) of about ten. It is filled with a fluid, to maintain hydrostatic conditions, and the sample or samples float freely in the fluid (Fig.4). Needless to say, at pressure of 100 kbar or a megabar, no fluids are truly that. The hydrostatic pressure medium is a weak solid, and there

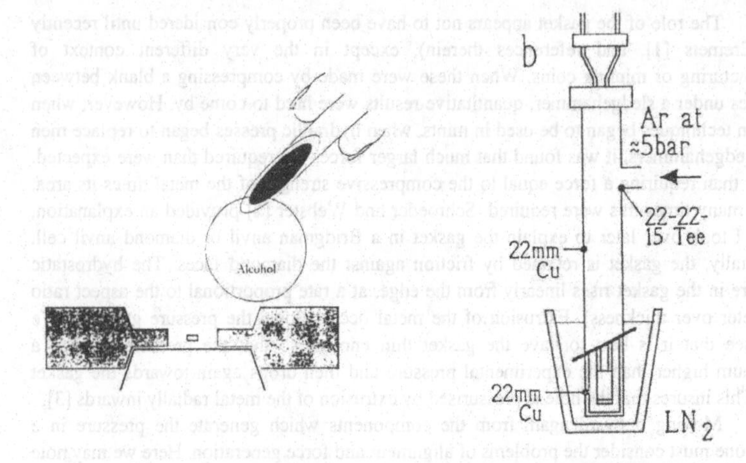

Figure 5. Alcohol loading, with a drop of fluid about to drop from a hypodermic needle onto the gasket and sample. Not to scale: the sample and sample hole are much smaller than any feasible drop of fluid than is shown here.

Figure 6. A system for loading argon. Ar gas is admitted at a pressure of a few bar, and then liquefied with the bucket of liquid nitrogen. The cell is closed when the argon is liquid or solid, and may then be brought up to room temperature.

has been immense controversy over the choice of the solid. The inert gases are favoured, although NaCl — because of its similarity to solid argon — has never fallen completely out of use. Argon is very easy to load, since it is liquid at liquid nitrogen temperatures; helium is also easy to load in a helium cryostat, and it is assumed to be a weaker solid than argon. All the inert gases require either a DAC adapted for pressure changes at cryogenic temperatures (see below) or some more-or-less complicated arrangements to supply them at room temperature and kbar pressures [5,6]. Ethanol-methanol (4:1) mixtures are fluid to about 100 kbar [7], or with water as well (16:3:1) to 145 kbar [8] but I am prejudiced against them because they are difficult to load compared with the inert gases. That is, they require manual dexterity as the drop of liquid is placed on the gasket hole, avoiding the inclusion of bubbles and avoiding washing out the samples (Fig.5). The cell must be clamped before enough alcohol evaporates to change the composition of the liquid significantly. Often, alcohol loading is carried out on ice to slow the evaporation. Compared with this, cryogenic loading (Fig.6) is straightforward, requiring only the opening and shutting of valves and filling a bucket with liquid nitrogen [9]. Enough manual dexterity and patience is required of the DAC operator without adding unnecessary extra difficulties.

The sample may of course be its own hydrostatic medium, as in studies of the viscosity of liquids under pressure in the DAC [10, 11, 12].

2.2 THICK AND THIN GASKETS

Whatever the choice of hydrostatic pressure medium, there is a clear-cut difference between compressing it by the advance of the diamonds, with the gasket hole enlarging, and compressing it by the gasket hole shrinking (Fig.7). In the former case, the hydrostatic fluid is necessarily sheared considerably, and if it is not as weak a solid as one would wish, large shear stresses are set up. In the latter case, the diamonds advance to make the pressure

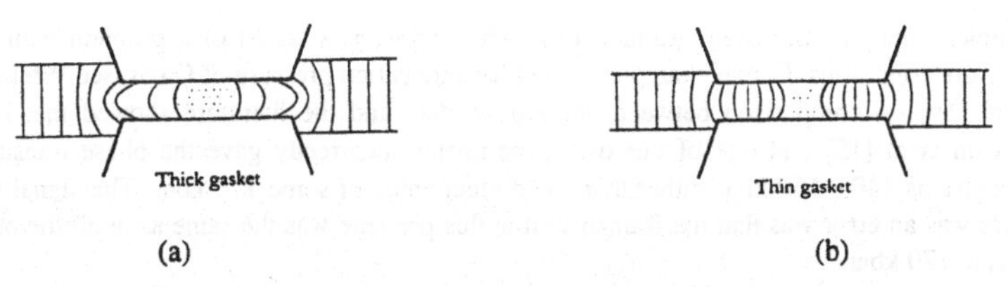

Figure 7. Thick (a) and thin (b) gaskets are contrasted. The thin lines show the movement of the steel during pressurisation. In (a), the steel moves only outwardds and the gasket hole becomes barrel-shaped. In (b), the gasket is pressurised by inward movement.

volume thinner *and* the gasket shrinks to make it smaller in diameter. Now any shearing of the pressure volume is only accidental, not intrinsic, and shear stresses are as small as they may be. To obtain a gasket hole that shrinks as the pressure is raised, requires only a gasket material that is less compressible than the hydrostatic fluid, and a gasket which is initially sufficiently thin [1, 3].

To maintain reproducibility, DAC high-pressure runs should only be carried out while increasing pressure (the upstroke). If hydrostatic fluids were perfect fluids, satisfactory data could be obtained on reducing pressure, too (the downstroke). But in practice, the sample volume can only lose pressure by leaking, and on reduction of force on the diamonds it is highly unlikely that leaking would occur at all point on the circumference of the pressure volume simultaneously. On the contrary, leaking at a single point on the circumference must occur, and must set up much severer shear strains in the fluid than those which occur on the upstroke. Any data obtained under these conditions must accordingly be treated sceptically.

The pressure volume need not be cylindrical. Eremets [1] has reviewed extensive work on Bridgman systems with hollowed-out pressure volumes which are more nearly spherical than cylindrical. This could be feasible in a DAC. The implications for the stresses on the anvils will be considered below; here we note only that pressurising a spherical gasket hole requires more movement of the gasket than the pressurisation of a cylindrical hole of the same diameter, and greater risks of non-hydrostatic conditions.

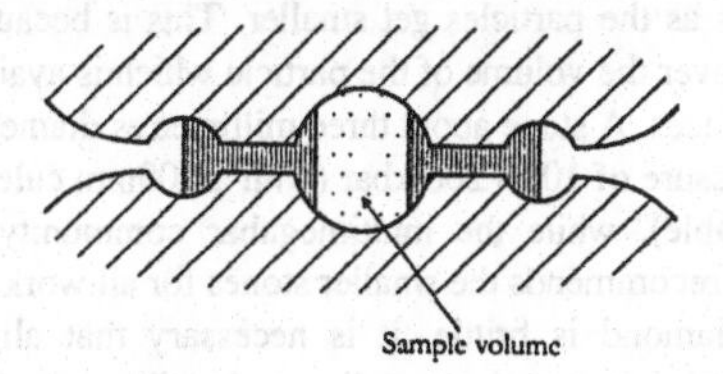

Figure 8. A large roughly spherical sample volume in a Bridgman anvil system. (After Eremets [1]).

2.3 CALIBRATION AND CONTROL

Accurate calibration of the pressure within the sample volume is far from trivial. Ruby chips may be scattered around the sample volume and the pressure they are under may be established by photoluminescence and the ruby calibration of the U.S. National Bureau of Standards. This does not guarantee that the sample is at the pressure recorded by the ruby. It is a tremendous advantage to be able to measure some standard property of the sample as well as the property which is the subject of research under pressure, in order to be able to detect anomalies. This topic is discussed elsewhere [13]. We have begun to use Raman in this way, since the Raman from a semiconductor crystal is little affected by the details of the sample. Then an anomalous Raman shift with pressure is a signal of a problem with the pressure conditions. For an example on the phase transition pressure of GaAs, see Whitaker et al [14] — comparison between our Raman data and the literature showed that both Besson et al [15] and one of our own experiments incorrectly gave the phase transition pressure as 140 - 150 kbar rather than the correct value of some 170 kbar. The signal that there was an error was that the Raman shift at this pressure was the same as in all the other data at 170 kbar.

3. The Diamonds

The diamonds have to sustain the high pressure on a small flat, the culet, and transmit the force to a much larger flat, the table, where the pressure is reduced in proportion to the areas to a value that some much weaker material such as steel can support. Early workers used brilliant-cut diamonds, with a 90° included angle at the culet and with a table much smaller than the greatest diameter of the diamond. Typically, the culet might be 500mm diameter and the table 3mm, giving a pressure multiplication from the compressive strength (or rather, indentation hardness) of the support material of times 36. Drukker introduced a better anvil cut, with a larger included angle and a larger table. Recently, Adams [16] has concluded from finite element analysis that the undercut of the brilliant cut and the Drukker cut is not necessary, and the table can be as large as the maximum diameter.

Whatever the cut, it is important to appreciate that diamonds are brittle. This means that they are strong in pure compression, and fail (shatter) unpredictably under small tensile or shear stresses. Yet the diamond must be in tension, for there is no way to transfer the normal force on the culet to a normal force on the table without quite large tangential tensile stresses. Hollowing out the culet as in the Russian Bridgman anvil work reviewed by Eremets [1], so as to give a more nearly spherical pressure volume, plainly makes little difference to the tensile stresses, since they arise fundamentally from the difference in culet and table diameter.

Given that the diamond is under tensile stress, it is important that it be small. As artists discovered many centuries ago when grinding pigments, the force required to shatter brittle particles increases as the particles get smaller. This is because it is the total elastic strain energy integrated over the volume of the particle which is available for the creation of a crack and two new surfaces. A stone about three millimetres diameter (a third of a carat) is normal for moderate pressure of 100 - 200 kbar (with a 500mm culet this gives less than 10 kbar pressure on the table), while the multimegabar community rightly prefer smaller stones. Adams [16] now recommends the smaller stones for all work.

Because the diamond is brittle, it is necessary that alignment be exact. Any asymmetry in the approach between the two diamonds will set up fairly unpredictable off-axis tensile and shear stresses and lead to fracture of the diamonds. It is normal to align the diamonds in the radial direction (*x-y*) to a few microns, and in tilt, to a couple of fringes of visible light between the optically-flat culets, to an accuracy of about 1 milliradian. However, much of the literature has probably exaggerated the need for perfect tilt alignment. Certainly, the late Ian Spain concluded that diamonds for pressures up to 100 - 200 kbar need no tilt alignment. Ordinary engineering accuracy in the machining of the table supports, together with the accuracy with which the diamond anvil manufacturers achieve parallelism of the table and culet was, he considered, adequate to eliminate the need for any tilt alignment mechanism in cells for these low pressures. All that matters is that the high points of the culets should not meet (Fig.9) and that the gasket hole should not squirt out sideways. The former criterion requires parallelism to only a tenth of a radian or so, and machining is far more accurate than that. The second requires only a gasket hole that shrinks as the pressure is raised, or in the terminology of Ref.3, a "thin" gasket: if the gasket is too

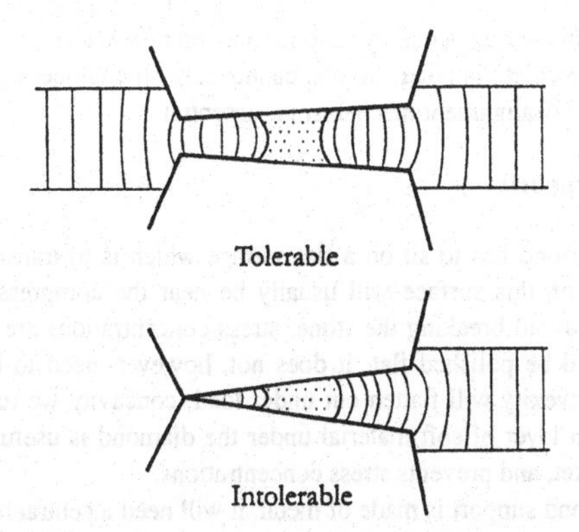

Figure 9. The diamonds can tolerate errors in tilt, as in the upper picture, provided they do not contact at the high point and as long as the gasket hole does not move sideways, as in the lower picture.

thick the hole will certainly enlarge asymmetrically towards the weakest (thickest) point of the gasket. The argument scales, of course, and implies that multi-megabar cells are in as little need of a tilt alignment mechanism as the low-pressure cells we use for semiconductor spectroscopy. Nevertheless, it would be a brave DAC designer who omitted all tilt adjustment from his cells.

Errors in *x-y* alignment of the diamonds are much more important. It is evident that they set up radial forces, which are, however, very difficult to assess quantitatively. From Fig.10, the radial force can be approximated as proportional to the axial displacement a:

$$F_r = \frac{2hF}{d}a \tag{1}$$

where F is the load on the diamonds.

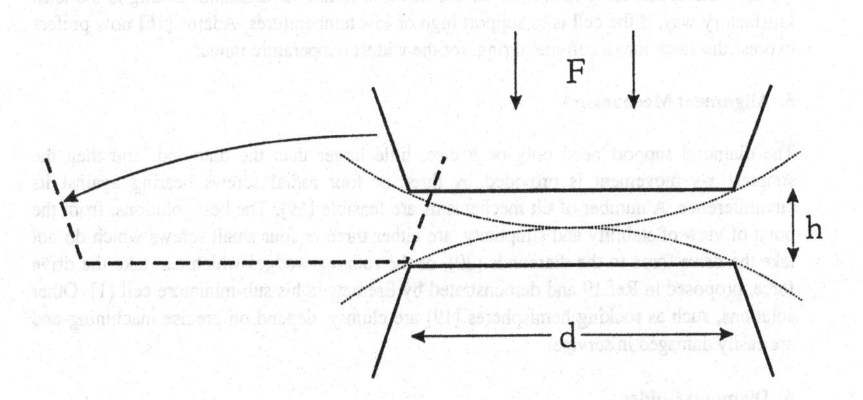

Figure 10. The culets and gaskets are approximated as spherical surfaces which slide over each other. Then as the aspect ratio d/h is large, the radial force is given by Eqn.1.

Ordinary engineering accuracy is about one thou (0.001 inch), or 25μm, and this is a large fraction of the culet diameter. So one cannot rely on engineering accuracy to centre the diamonds and an *x-y* alignment mechanism is essential.

4. The Diamond Supports

The table of the diamond has to sit on a flat surface which is to transmit the force to the stone. The pressure on this surface will usually be near the compressive strength of the material. Clearly, to avoid breaking the stone, stress concentrations are to be avoided, and so this surface should be polished flat. It does not, however, need to be a high-precision surface. A minor convexity will flatten out under load; concavity we return to below. The introduction of a thin layer of soft material under the diamond is useful. It acts in just the same way as the gasket, and prevents stress concentrations.

If the diamond support is made of metal, it will need a central hole to allow optical studies through the diamond. The diamond bridges this hole as a beam, and so shear and tensile stresses are set up in the diamond in this region. Usually the hole is tapered as little as possible, so as not to weaken the support unduly. Adams and Shaw [17] obtained the surprising result, in their finite-element analysis, that the stress patterns are improved if the support is weakened by a heavily tapered hole, but few designers have taken advantage of this.

In transferring a normal load from the table to the culet, the diamond must suffer tensile stresses. However, if the support can provide an inward radial load on the diamond, these tensile stresses are cancelled. One way to do this would be to mount the diamond in a conical hole, rather than on a flat, bearing against the bevel around the table. I think that Adams [17] achieved the same thing with the heavily tapered hole. The diamond support flexes inwards and downwards under load, and if there is no sliding between the diamond and the support, this similarly puts an inward force on the diamond.

The diamond can be secured to the support in a variety of ways. It may be metallised and then soldered on [18], or it may be pressed into a soft metal ring which is itself soldered or bolted down. As in jewellery, it may be placed in a cylindrical hole, the top of which is staked or bent inwards and down to retain the diamond. Gluing is the least satisfactory way, if the cell is to support high or low temperatures. Adams [16] now prefers to press the stone into a soft metal ring, for the widest temperature range.

5. Alignment Mechanisms

The diamond support need only be a disc, little larger than the diamond, and then the simplest *x-y* movement is provided by three or four radial screws bearing against its circumference. A number of tilt mechanisms are feasible [19]. The best solutions, from the point of view of stability and simplicity, are either three or four small screws which do not take the drive force to the diamonds [20], or the rotating wedges which can take the drive force, proposed in Ref.19 and demonstrated by Eremets in his sub-miniature cell [1]. Other solutions, such as rocking hemispheres [19] are clumsy, depend on precise machining and are easily damaged in service.

6. Diamond Guides

As we have seen, pressure is generated by advancing the diamonds under a force of the order of a tonne, while keeping *x-y* alignment accurate preferably to microns rather than tens of microns. Tilt, too, must be controlled but as we have seen above this is less of a problem. The motion has been provided, in every cell reported up to now, by a sliding component such as a piston in a cylinder. The *x-y* alignment is evidently maintained by a close fit of the piston within the bore, while the tilt alignment will be the better maintained the longer the piston is. Many megabar cells have been designed with pistons several times longer than their diameter.

6.1 PISTON CLEARANCE

To have a piston slide freely down a bore without any play, or lateral movement, is not an easy thing to achieve. It might be thought that all that is required is to make the piston the same diameter as the bore. Far from it. Traditionally, if a powerful press is required to move the piston in the bore, we have what is known as an interference fit. In machining components for a press fit, typically the inserted piece will be a thou per inch, or 10 mm per cm, or 0.1% larger than the bore. However, there is a wide tolerance on this, and any interference from zero upwards will normally require a press, or a temperature difference, to put the piston in the bore. At the other extreme, if a component is to move freely in a bore, it will normally be made about a thou per inch undersized — and some play will be detectable. In between a thou clearance and zero, we have the very ill-defined region of the hand push fit, where a component is tight, but can be pushed in by hand against a moderate frictional force. This is what is required for a DAC, and it is achievable only by very skilled craftsmanship, lapping a couple of microns off the piston or the bore until the desired fit is obtained, or by a very well set up computer controlled machine tool.

6.2 TILT CONTROL

If a reasonable fit is achieved to control *x-y* motion to a few microns, what is the tilt accuracy? Early steam engines used pistons which were very short relative to their diameter, but they were guided by being rigidly attached to a piston rod which was itself guided by an external crosshead, several diameters away. When designers of internal combustion engines began to use pistons with internal gudgeon pins and no rigid piston rod, they naturally made the piston longer than its diameter (undersquare). It seemed obvious that a piston shorter than its diameter (oversquare) would be prone to tilting and jamming in the bore. Many DAC designs have followed this principle. Yet it is a remarkable fact that the angle a piston can tilt to in a bore is not at all dependent on its diameter: it depends solely on its length and its clearance. In fact, if the length is L and the clearance is $\Delta d \ll L$, the angle it can go to is $\Delta d/L$ radians. For a clearance of a few microns and a length of a few mm, this angle is about a milliradian, which from our discussion earlier about tilt alignment should not be a problem. Whether the piston is oversquare or undersquare is irrelevant, and some of the very long pistons in some DAC designs are certainly quite unnecessary.

Just the same argument applies to the Merrill-Bassett cell design and a miniature wide-aperture cell described by Eremets [1]. In the Merrill-Bassett cell [21], instead of having a piston and a cylinder concentric with the axis of the diamonds, the diamonds are supported on two plates which are pulled together by three screws. In the Eremets cell, two similar plates are pressed together by diaphragms, and four rods or dowels pass through both plates to guide them. In both cases, precision guidance is assured by the fit of the screws or dowels in their holes. This is often assumed to lead to a rough-and-ready cell design which cannot hope to reach the pressures of a properly designed undersquare concentric piston-cylinder cell. Just because there are three or four dowels or screws at rather large distances from the centre compared with their own diameter does not mean that their precision in controlling tilt is at all compromised. Again, all that matters is the clearance of the dowels in their holes against the length of the holes. Where one might criticise these cell designs is in the demands they place on the mechanical workshop: not only must each hole fit its dowel to a few microns accuracy, but the holes and the dowels must be placed with the same precision on the plates.

It is apparent that the DAC with sliding components must be machined to great precision, and thereby places great demands on the mechanical workshops. It demands great care in use, too, for a hand push fit is easily jammed by the slightest particle of dirt, or at cryogenic temperatures by the least condensation and ice-formation of an impurity gas. To ease the demands on both the mechanical workshops and on the use what is required is a means of providing precision of motion without precision of machining and without great care in service. Any textbook on precision instruments will show that this requirement is best met by flexures rather than by sliding motions [22, 23]. We discuss this below, in the final section of the paper on future developments of the DAC.

7. Drive mechanisms

7.1 SCREW DRIVES

The generation and transmission of a force up to about one tonne, or perhaps three tonnes for some multimegabar cells, presents no engineering difficulties. Indeed, many DACs are grossly over-engineered in this area. Perhaps this is just as well, since many are also badly engineered in this area. For example, it is generally bad practice to machine a screw thread into a component if it is then to be screwed back and forth repeatedly under load. This is because the screw thread will become worn and damaged, and then the entire component has to be replaced. Much better is to use a stud and nut (Fig.11). The stud is screwed permanently into the component, and it is then only the stud and nut — cheap commercially available objects — which need be replaced.

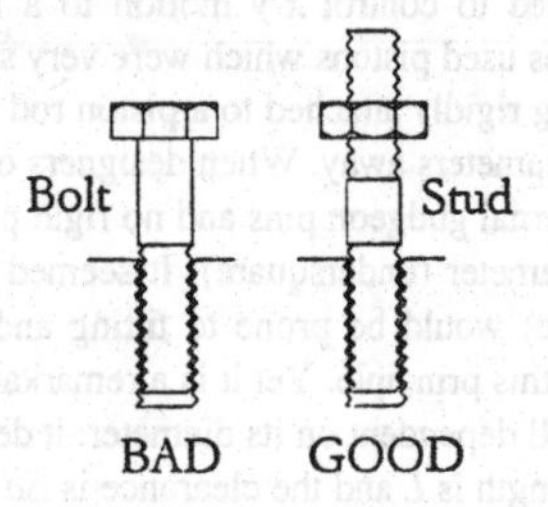

Figure 11. To have a bolt repeatedly screwed under load into a tapped hole in a complex component is bad practice. It is better to replace it with a stud and a nut, so that wear and damage occur to these two easily replaceable components rather than to the complx component.

A cell to be operated at room temperature often incorporates a screw above the piston to provide the force. This screw is often as much as 25 mm diameter, with buttress threads to improve the friction and wear characteristics. Yet a simple 5 mm or 6 mm commercial high-tensile or stainless steel screw is well able to produce a tonne force, with less torque due to friction because the diameter is so much less, and with a great deal less trouble for the mechanical workshops both in the original construction and when replacement is necessary. On my own design of DAC [20], two retaining screws are provided to enable the cell to be clamped at high pressure and removed from its drive; these screws are not intended to drive the cell, only to clamp it, and so I put the threads for them in the lower plate of the cell. Some of the cells are now over ten years old, and although these two bolts are only 3mm diameter, none have worn out yet. If the threads in the lower plates do wear out, rather than replace the plates, I will just have them drilled through to clear the bolts, and use nuts underneath — which of course is how I should have designed them in the first place.

The main problem with a simple screw mechanism in the DAC is that it restricts the choice of hydrostatic media to those which can be loaded at room temperature. Alternatively, complex mechanisms are need to allow the cell to be actuated while immersed in liquid argon or helium [1].

More elaborate mechanisms based on screw threads have been built. The Stuttgart cell used "bottle-cap" mechanisms to amplify the force from a screw. While this reduces the torque required on the screw, and permits a lightweight mechanism to actuate the screw while the cell is immersed in liquid helium, it demands a lot of the mechanical workshops, to no great advantage.

7.2 HYDRAULIC AND PNEUMATIC DRIVES

Instead of screws in the cell, many authors have used helium gas bellows [1]. This has the great advantage that remote actuation is easy, and no torques or forces need to be transmitted mechanically to the cell. If the cell has to be precisely aligned with respect to something external, such as an X-ray beam or optics, the gas bellows is probably the best drive to use to avoid any risk of upsetting the alignment during pressure changes. However, the gas bellows cannot be used to the lowest temperatures.

7.3 REMOTE DRIVES

In the author's view, the most versatile drive mechanism consists of remote force generation at room temperature, and mechanical transmission of the force to the cell (Fig.12). The force generator may be a screw mechanism, a hydraulic ram or pneumatic. Then the cell

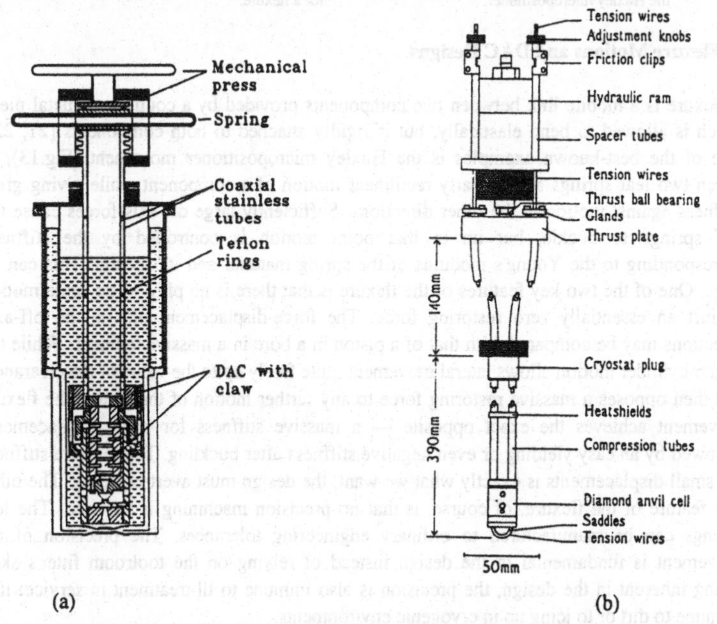

Figure 12. Schematic of DACs driven by remote drives, based on Eremets' concentric tubes (a), after [1], and my tube and wire system (b) after [20].

may be in a cryostat or furnace; it may be in a gas loading rig or out on the bench. Transmission of forces up to a tonne or so can be done safely and easily in a lightweight mechanism. Eremets et al [6] used concentric 30mm diameter 0.3 mm wall thickness stainless steel tube (Fig.12a). We use four 2mm stainless steel wires under tension, within four 3/16 inch (approx 4.5mm) stainless steel tubes under compression (Fig.12b), so that, like bicycle brake cable, the assembly is flexible and imposes no loads on any fixings [20]. This assembly is suitable for liquid helium use as the thermal load it imposes on a cryostat is minimal. Its main disadvantage in the form shown in the Figure is friction in the curves, but Adams [16] has recently modified it to use straight tubes so that the only source of friction is the O-ring vacuum seals on the wires. DACs with our standard drive mechanism also have a tendency to creep up in pressure after a pressure change, by anything from a tenth of a kbar to 2 or 3 kbar, and over a few minutes or as much as an hour, according to the conditions. We have not systematised these observations, and have no explanation for the behaviour. It would be very interesting to know if other drive mechanisms sometimes do this.

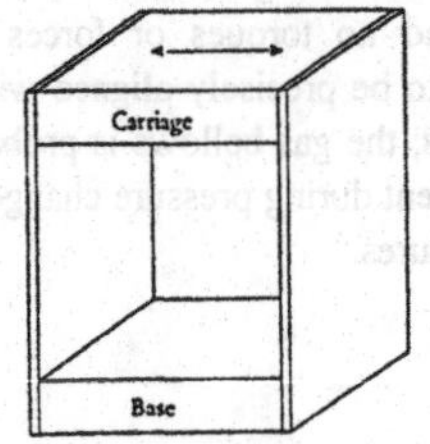

Figure 13. A flexure system giving rigid nearly rectilinear motion, used in the Huxley micropositioner.

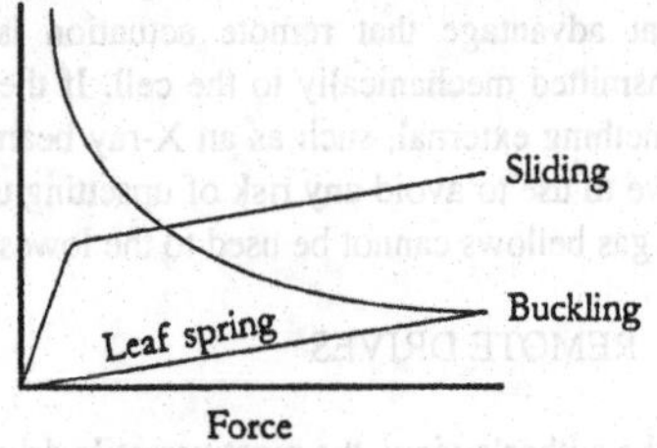

Figure 14. Schematic force-displacement plots for a sliding movement and for a flexure.

8. Flexure Motions and DAC Designs

A flexure is a mobile link between two components provided by a continuous metal piece which is allowed to bend elastically, but is rigidly attached to both components [21, 22]. One of the best-known examples is the Huxley micropositioner movement (Fig.13), in which two leaf springs allow nearly rectilinear motion of a component while giving great stiffness against motion in the other directions. Sufficiently large off-axis forces cause the leaf springs to buckle, but up to that point motion is controlled by the stiffness corresponding to the Young's modulus of the spring material and its section. This can be large. One of the two key features of the flexure is that there is no play — no finite motion against an essentially zero restoring force. The force-displacement plot in the off-axis directions may be compared with that of a piston in a bore in a massive cylinder. While the piston-cylinder motion allows lateral movement quite freely up to the value of the clearance, and then opposes a massive restoring force to any further motion of the piston, the flexure movement achieves the exact opposite — a massive stiffness for small displacements followed by an easy yielding or even negative stiffness after buckling. The massive stiffness for small displacements is exactly what we want; the design must avert buckling. The other key feature of the flexure, of course, is that no precision machining is required. The leaf springs can be manufactured to ordinary engineering tolerances. The precision of the movement is fundamental in the design instead of relying on the toolroom fitter's skill. Being inherent in the design, the precision is also immune to ill-treatment in service: it is immune to dirt or to icing up in cryogenic environments.

The simplest flexure movement for a DAC would be the Huxley micropositioner movement. The diamonds move only a short distance, and it is not necessary that their movement should be strictly rectilinear. However, it might be feared that off-axis forces due to small errors in alignment of the diamonds would create such large transverse forces as to cause the leaf springs to buckle. Analysis of this is straightforward, and due to Euler [4]. The transverse force is initially quadratic with the displacement. If this rises faster with displacement than the restoring force of the buckled springs, which is also quadratic with displacement, buckling will occur; otherwise not. I am not aware of any analysis of the transverse forces due to misaligment of the diamond culets; a plausible approximation is to treat them as frictionless rigid spherical surfaces with a radius of curvature such that while centred they are in contact, and when misaligned by half the culet diameter they advance a distance equal to the thickness of the gasket. If buckling does not occur, then the stiffness of the flexures with respect to lateral displacement is just their Young's modulus times their section.

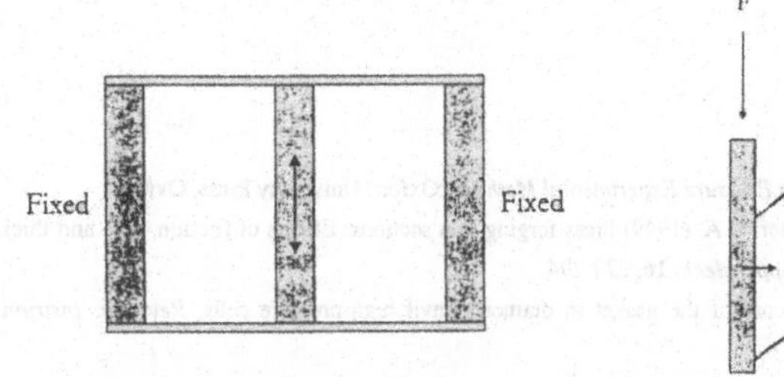

Figure 15. A double-leaf flexure system which gives less movement than the system of Fig.13, but cannot buckle and is precisely rectilinear.

Figure 16. A flexure system that converts a vertical force into a horizontal force.

If a buckling analysis is felt to give insufficient security, a system of double springs may be used. Now motion is strictly rectilinear, by symmetry. Buckling of one leaf cannot occur for it would put the opposing leaf into tension. The restoring force is now the Young's modulus times the area of the springs without a buckling limit, and it is implausible that the system would fail. However, the longitudinal motion is much more restricted that in the simple leaf-spring arrangement.

This system can be given circular (axial) symmetry, so that the moving diamond is set at the centre of one of a pair of diaphragms or membranes. When this is done, the membranes can be slotted, to increase their flexibility, at the cost of reducing their stiffness against lateral displacement. There are many patterns of slotting, including the spiral and the tangential [22]. The slotting can be done cheaply by commercial photolithography and etching. In each pattern of slotting, all that is required for stability of the cell is that the stiffness of the membrane should exceed the rate at which the radial force develops with radial or x-y displacement of the diamonds.

The two flexure systems described so far require a separate mechanism to apply the force to the moving component. They need to be made quite large (perhaps several cm), in the first case to get motion close to rectilinear, and in the second case to get sufficient longitudinal motion of the diamond to allow for gasket pre-indentation and other cell-loading operations. And the force has to be applied parallel to the optical axis of the cell, which is a disadvantage particularly for cryogenic operation. There exists another layout of flexures which overcomes both these disadvantages. The flexures themselves can be used to deliver the force to the diamonds as well as guiding them. Using the opposed flexures of Fig.15 but inclining them (Fig.16), a longitudinal force on the flexures can be converted to a transverse force on the diamonds. And the flexures can be much shorter than those of Fig.15, for they are not required to stretch to allow the diamonds to approach each other, only to bend. The resulting layout is ideal for providing transverse optical axis in a DAC which is contained in a small-bore cryostat, with the force transmitted along the axis of the cryostat.

We have done finite-element analysis on a preliminary design of such a cell. Results show the feasibility of such a layout, while showing also the need of much refinement in the design.

Acknowledgements: I am grateful to Drs J.D. Lambkin, A.D. Prins, V.A. Wilkinson, J.L. Sly, M.F. Whitaker and M.D. Frogley who have done most of the work in my laboratory. Many discussions with the late Prof. I.L. Spain and with Dr D.M. Adams have given me valuable insights into the DAC. Finite element analysis of the flexure system of Fig.16 was done by Dr S.N. Clark at Daresbury. This high-pressure work has been supported by EPSRC.

9. References

1. Eremets, M.I. (1996) *High Pressure Experimental Methods*, Oxford University Press, Oxford.
2. Schroeder, W., and Webster, D.A. (1949) Press-forging thin sections: Effects of friction, area and thickness on pressures required, *J. Appl. Mech.* **16**, 279-294.
3. Dunstan, D.J. (1989) Theory of the gasket in diamond anvil high-pressure cells, *Rev. Sci. Instrum.* **60**, 3789-3795.
4. Euler, L. (1744) *Methodus inveniendi linea curvos maximi minimive proprieatare gaudentes*, Lausanne.
5. Besson, J.M., and Pinceaux, J.P. (1979) Melting of helium at room temperature and high pressure, *Science* **206**, 1073-1075.
6. Eremets, M.I., Krasnovskij, O.A., Struzhkin, V.V., Timofeev, Yu. A., and Shirokov, A.M. (1990) Method of low-temperature optical measurements with diamond anvil cells, *High Pressure Research* **5**, 880-884.
7. Piermarini, G.J., Block, S., and Barnett, J.S. (1973) Hydrostatic limits in liquids and solids to 100 kbar, *J. Appl. Phys.* **44**, 5377-5382.
8. Fujishiro, I., Piermarini, G.J., Block, S., and Munro, R.G. (1982) Viscosities and glass transition pressures in the methanol-ethanol-water system, *Proc. 8th AIRAPT Conf.*, Uppsala, ed. C.M. Backman, T. Johannisson and L. Terner, Vol.2 pp. 608-611.
9. Spain, I.L., and Dunstan, D.J. (1989) The technology of diamond anvil high pressure cells: II. Operation and use, *J. Phys. E: Sci. Instrum.* **22**, 923-933.
10. Fujishiro, I., and Nakamura, Y. (1987) Viscosity measurements under high-pressure by diamond anvil cell, *J. Jap. Soc. Lubrication Engineers* **32**, 401-404.
11. King, H.E., Herbolzheimer, E., and Cook, R.L. (1992), The diamond-anvil cell as a high-pressure viscometer, *J. Appl. Phys.* **71**, 2071-2081.
12. Cook, R.L., Herbst, C.A., and King, H.E. (1993) High-pressure viscosity of glass-forming liquids measured by the centrifugal force diamond anvil cell viscometer, *J. Phys. Chem.* **97**, 2355-2361.
13. Frogley, M.D. (1998) this conference.
14. Whitaker, M.F., and Dunstan, D.J. (1998) Raman spectroscopy of GaAs and InGaAs under pressure, *J. Phys. Cond. Matter* (submitted).
15. Besson, J.M., Itié, J.P., Polian, A., Weill, G., Mansot J.L., and Gonzalez, J. (1991) High-pressure phase-transition and phase-diagram of gallium-arsenide, *Phys. Rev. B***44**, 4214-4234.
16. Adams, D.M. (1998) private communication.
17. Adams, D.M., and Shaw, A.C. (1982) A computer-aided design study of the behaviour of diamond anvils under stress, *J. Phys. D***15**, 1609-1635.
18. Dunstan, D.J. (1991) Soldering diamonds into the diamond anvil cell, *Rev. Sci. Instrum.* **62**, 1660-1661.
19. Dunstan, D.J., and Spain, I.L. (1989) The technology of diamond anvil high pressure cells I. Principles, design and construction, *J. Phys. E: Sci. Instrum* **22**, 913-923.
20. Dunstan, D.J., and Scherrer, W. (1988) A miniature cryogenic diamond anvil high pressure cell, *Rev. Sci. Instrum.* **59**, 627-630.

21. Merrill, L., and Bassett, W.A. (1974) Miniature diamond anvil pressure cell for single crystal x-ray diffracon studies, *Rev. Sci. Instrum.* **45**, 290-294.

22. Geary, P.J. (1961), *Flexure Devices. Pivots, Movements, Suspensions*, British Scientific Instruments Research Association, Chislehurst.

23. Trylinski, T. (1971) *Fine Mechanisms and Precision Instruments*, Pergamon Press, Oxford.

HOW ACCURATE ARE HIGH-PRESSURE EXPERIMENTS?

M. D. FROGLEY
Physics Department, Queen Mary and Westfield College,
University of London,
London E1 4NS,
England.

Abstract. A simple polynomial fit to data which is sublinear with pressure gives coefficients which vary strongly with the pressure range of the fit. Converting pressure to a variable which is linear with that measured enables direct comparison of experiments over different pressure ranges. We illustrate this for high-pressure band-gap data in GaAs. Experimental errors such as non-hydrostaticity can occur and be difficult to detect. Comparison of two spectroscopies provides a simple diagnostic. We find a linear shift of the band-gap with the Raman shift under pressure for ZnTe and propose this as a check of pressure conditions. To measure the Raman shift accurately under pressure, resonant Raman is a useful tool, but requires a tunable source. We have developed a tunable, broadband filter for high-pressure work, which rejects Rayleigh scattered light, allowing measurement of Raman shifts as small as $10cm^{-1}$.

1. Introduction

In any experiment, if the measured variable is not linear with the fixed variable, then a simple polynomial fit to the data will give coefficients which vary with the range of the fit. This effect is large and systematic if the data does not follow exactly the functional form of the fit e.g. a linear fit to parabolic data. Naturally one wishes to fit over the entire range of available data, and would report these coefficients. This has important consequences for high-pressure work when we compare pressure coefficients for data obtained over different pressure ranges. An excellent example of this is seen in the reported pressure coefficients of the band-gaps for tetrahedral semiconductors. Many such measurements have been performed over the last thirty years, and particularly since the advent of the diamond anvil cell (DAC). For most direct-gap tetrahedral semiconductors, the band-gap increases with pressure but with a clear sublinear dependence. Because of this dependence, a parabolic fit is normally used to describe the data. However, reported coefficients for a given material are widely scattered (typically by > 15%) and this has caused controversy between workers who claim much greater accuracy than this. However, these experiments were performed over a variety of pressure ranges, and the discussion above suggests that much of this scatter may be due to the use of inappropriate fitting routines. Taking GaAs as an example in section 2 we show that this is indeed the case. It would be preferable to plot the band-gap energy against a fixed variable with which there is a linear dependence, since a linear fit yields parameters independent of the range of the fit. This variable may be derived from the pressure.

R. Winter and J. Jonas (eds.), High Pressure Molecular Science, 103–108.

However, experimental errors can occur, and be difficult to detect and interpret if only one variable is measured as a function of pressure. The use of two independent measurements allows two extra checks on conditions in an experiment. Firstly, both measurements can be compared with existing data, if a reliable body of data exists, and noting the discussion above and in section 2. Secondly, the measurements can be compared directly with each other, making pressure a dummy variable and ruling out any problems in pressure calibration. For similar materials, general trends observed in such a comparison may then have predictive power, providing a general diagnostic tool. For ZnTe, we have measured both the band-gap energy and the Raman shift as a function of pressure in three high pressure experiments. In section 3 we use the comparison of these spectroscopies to discard one of the data sets.

2. Fitting Procedures for High-pressure Band-gap Data in Semiconductors

It is common to report the sublinear increase with pressure of the band-gap in terms of the parabola

$$E_g = E_0 + aP - bP^2 \tag{1}$$

Fig.1. shows reported parabolae for GaAs[1-7], plotted from 0 to 20GPa. There is over fifteen percent variation in reported values of a, even though quoted experimental errors are only a few percent. Values for b are even more scattered from 0 to 0.61μeV/GPa^2 and this may lead us to question the reliability of high-pressure methods. In Fig.2 however, we fit with Eqn.1 to a single data set [2], over increasing pressure ranges, and the returned coefficients show systematic variation. This is the reason for most of the scatter in Fig.1 and shows that the pressure dependence of the band-gap is not parabolic.

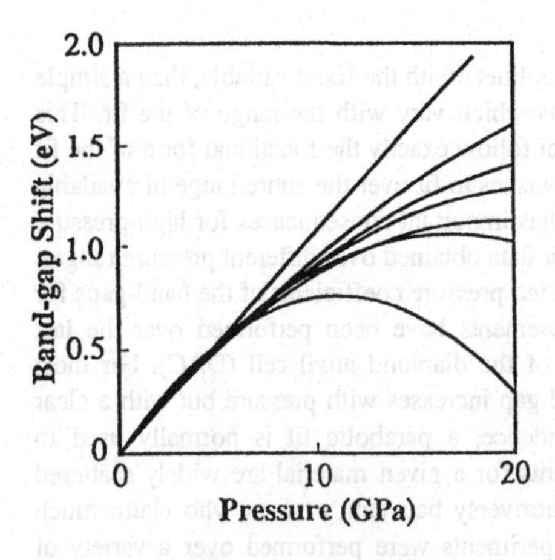

Fig.1. Reported parabolic fits to high- pressure band-gap data for GaAs. The coefficients a and b vary dramatically as do the pressure ranges used in the fits [1-7] (for curves from top to bottom in figure respectively)

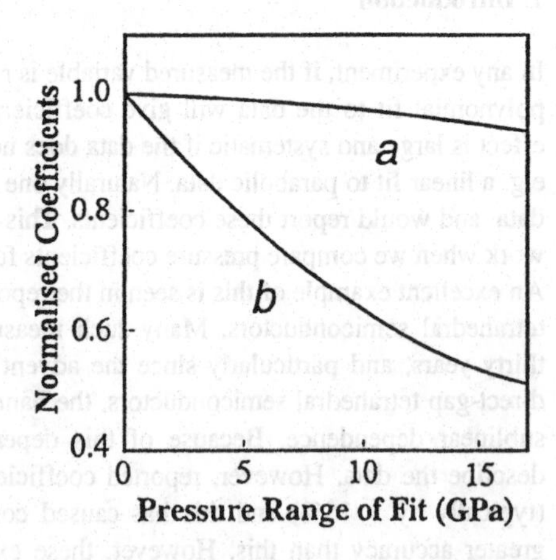

Fig.2. Variation of a and b with pressure range of a parabolic fit to a single data set [2]. The values are normalised to those at infinitesimal pressure range. This is the main reason for the difference in curves in Fig.1.

Experiments over different pressure ranges can be compared in a more useful way if a linear fit is used. In the context of tetrahedral semiconductors, Murnaghan's equation of state [8] is commonly used to relate pressure to the change in sample volume and hence lattice constant or density [2]. The band-gap of the tetrahedral semiconductors is roughly linear (within scatter) with all of these variables. On detailed analysis of the data for GaAs however, we showed elsewhere [4] that linear fits with lattice constant and volume, whilst reducing the scatter in coefficients compared with the parabolic ones, cannot fully reconcile all of the data. Fig.3 shows the seven sets of GaAs data plotted along with a linear fit against change in sample density to the lowest data in the figure [5]. All data and fits have been expanded as described below. Six of the seven data sets are in very good agreement with the fit, even though the maximum density in the experiment (and hence pressure) is very different for each. Importantly, this shows that the experiments are indeed as accurate as the authors claim. For further details of the fits of Fig.3 see Ref.4.

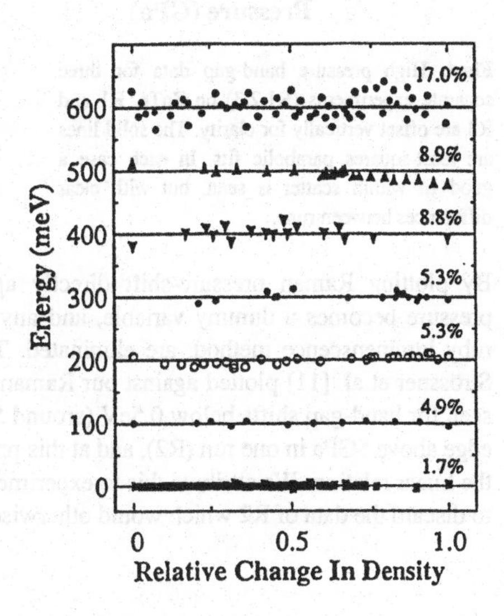

Fig.3. The seven sets of band-gap data for GaAs which were used in Fig.1 [2,3,6,7,1,4,5] (top to bottom). The horizontal axis is the relative change in sample density, which runs from 0 at ambient pressure, to 1 at the maximum change in sample density in each experiment, the value of which is given as a percentage next to each data set. Pressure was converted to change in density using Murnaghan's equation of state [8]. The solid lines are a linear fit to the lowest data set in the figure, repeated for comparison with the other data. A straight line of gradient 86.3meV/% has been subtracted from all the data so that the solid lines are horizontal. Each data set is offset on the vertical axis for clarity. All the data (except for [3] at high pressure) are in good agreement with the fit.

3. Comparison of Two Spectroscopies as a Test for High-pressure Data

Despite the high resolution possible in the measurement of material parameters such as the band-gap under pressure [5,9], experiments can go wrong. Fig.4 shows data from three absorption experiments, R1, R2 and R3, on ZnTe [10] along with parabolic fits to each. Whilst all fits are within scatter, the differences between them cannot be accounted for by the discussion above, and the problem is to know which if any of the runs are reliable. We would naturally prefer the data of R2, which is over the largest pressure range. In Fig.5, the pressure-shift of the LO phonon is plotted against the percentage change in sample density [10], as derived from the pressure using Murnaghan's equation of state. Raman and absorption measurements were made in the same experiments. All three runs are in superb agreement, suggesting that the pressure conditions in each experiment were the same. However, neither Fig.4 or Fig.5 can directly show which run(s) should be discarded.

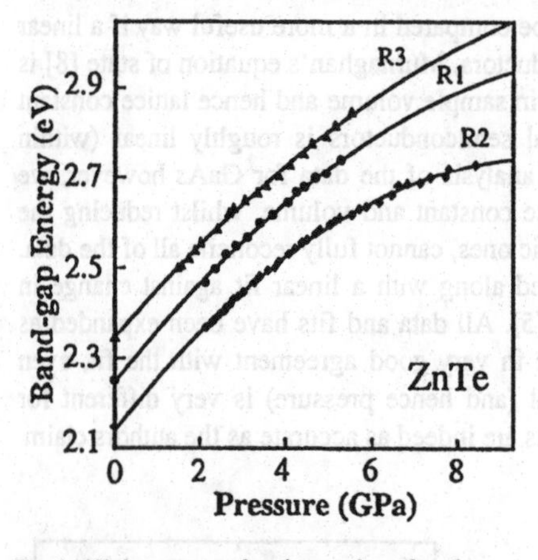

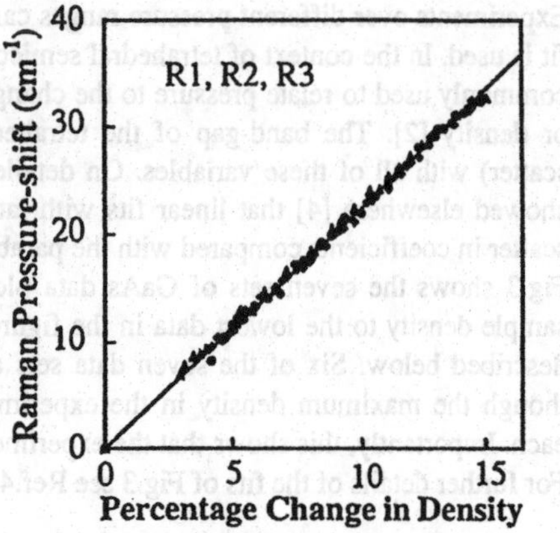

Fig.4. High pressure band-gap data for three separate experiments (R1,2,3) on ZnTe. R1 and R3 are offset vertically for clarity. The solid lines are least-squares parabolic fits. In each case a good fit within scatter is seen, but with clear differences between runs.

Fig.5. The LO phonon energy is plotted against the percentage change in sample density for R1,R2 and R3. All three runs are in good agreement, and exhibit a linear dependence of Raman shift on sample density.

By plotting Raman pressure-shift directly against the band-gap pressure-shift in Fig.6, pressure becomes a dummy variable, and any errors due to pressure calibration, using the ruby luminenscence method, are eliminated. The open symbols are the absorption data of Strössner et al. [11] plotted against our Raman data. In all three runs, a linear relationship is seen for band-gap shifts below 0.5eV (around 5GPa). We could only measure the absorption edge above 5GPa in one run (R2), and at this pressure the data shows a sharp breakaway from the linear relation. We attribute this to experimental failure, and thus the plot of Fig.6 leads us to discard the data of R2 which would otherwise have been preferred and reported.

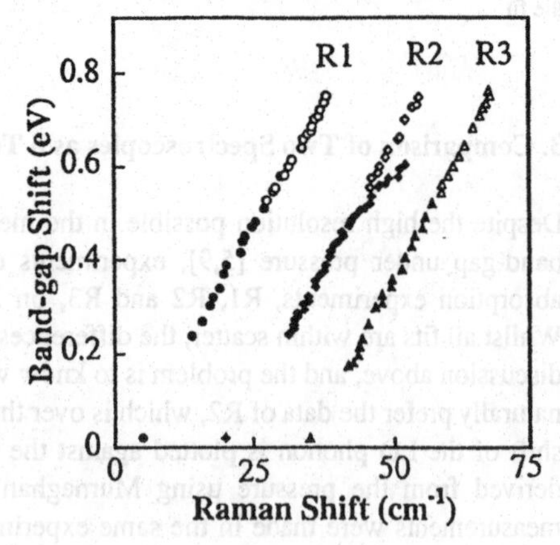

Fig.6. The pressure-shift of the band-gap is plotted against that of the Raman shift. In R1 and R3, a linear dependence is seen over the entire range. In R2 however a sharp breakaway occurs at about 0.5eV and this is indicative of experimental failure. Pressure is a dummy variable in the plot, eliminating any pressure calibration errors. The open symbols are the absorption data of Strössner et al. [11] plotted against our Raman data.

4. Resonant Raman Under Pressure

For Raman measurements on inorganic semiconductors in the DAC we use a Renishaw Raman microscope with 25cm spectrograph and CCD detection. Rayleigh scattered light is rejected by two holographic filters with a bandwidth of around 200cm^{-1}. With a few mW of He-Ne laser light incident on the sample, the Raman signal is typically so weak that collection times greater than 10 minutes are required to achieve reasonable signal-to-noise. At resonance however, the Raman intensity is enhanced by several orders of magnitude. In a high-pressure experiment using a fixed laser frequency, resonance only occurs as the band-gap moves through the lasing energy with pressure as seen in Fig.7 for a 3μm $Al_{0.3}Ga_{0.7}As$ layer and the GaAs substrate.

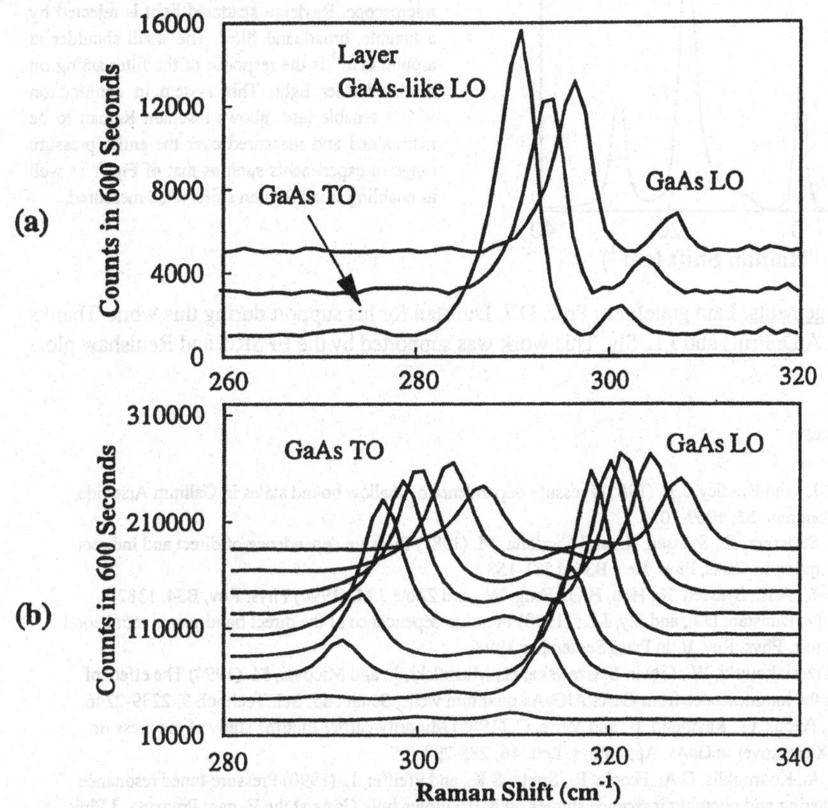

Fig.7. *Resonant Raman:* (a) The Raman spectrum of a 3μm strained $Al_{0.3}Ga_{0.7}As$ layer in a high-pressure experiment, from 2.6 to 3.9GPa, just after the layer resonance. The spectra are offset upwards for increasing pressures. (b) The same experiment from 4.7 to 10GPa, as the GaAs substrate passes through resonance. The GaAs LO and TO are then around 100 times more intense than in (a).

To maintain resonance under increasing pressure, a tunable source is required. The Raman microscope's holographic filters however, are not tunable, and a low-wavenumber filter has been developed which enables measurement of Raman shifts as small as 10cm^{-1} with broadband operation and tunability over a large frequency range. Fig.8 shows the Raman spectrum of HgI_2

recorded in 100s with the filter. The position of the filter edge is clearly visible. As well as providing quick and accurate measurements, this system allows switching between different resonances in the sample - e.g. the layer and substrate in Fig.7 - at constant pressure, and thus accurate deconvolution of the otherwise complicated spectra.

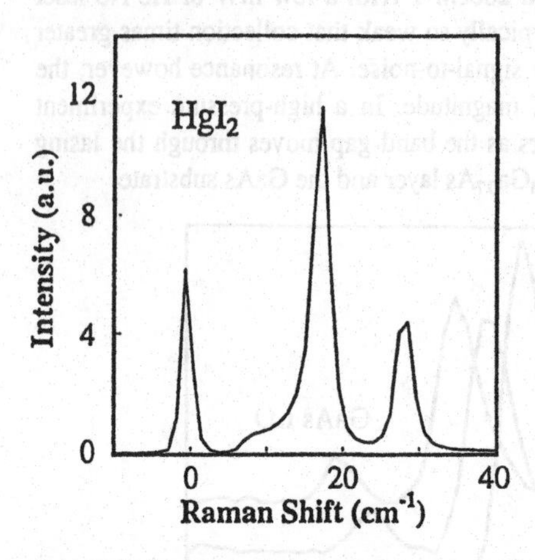

Fig.8. The Raman spectrum of HgI$_2$ near the exciting He-Ne laser line, as measured by a Raman microscope. Rayleigh scattered light is rejected by a tunable, broadband filter. The small shoulder at around 8cm^{-1} is the response of the filter acting on scattered laser light. This system in conjunction with a tunable laser allows resonant Raman to be maintained and measured over the entire pressure range in experiments such as that of Fig.7. as well as enabling small Raman shifts to be measured.

Acknowledgements: I am grateful to Prof. D.J. Dunstan for his support during this work. Thanks also to Drs. A.D. Prins and J.L. Sly. This work was supported by the EPSRC and Renishaw plc.

5. References

1. Wolford, D.J., and Bradley, J.A. (1985) Pressure dependence of shallow bound states in Gallium Arsenide, Sol. Stat. Commun. 53, 1069-1076.
2. Goñi, A.R., Strössner, K., Syassen, K., and Cardona, M. (1987) Pressure dependence of direct and indirect optical absorption in GaAs, Phys. Rev. B36, 1581-1587.
3. Li, G.H., Goñi, A.R., Syassen, K., Hou, H.Q., Feng, W., and Zhou, J.M. (1996) Phys. Rev. B54, 13820
4. Frogley, M.D., Dunstan, D.J., and Sly, J.L., (1998) Pressure dependence of the direct band-gap in tetrahedral semiconductors, Phys. Rev. B, In Press September 1998.
5. Perlin, P., Trzeciakowski, W., Litwin-Staszewska, E., Muszalski, J., and Micovic, M. (1994) The effect of pressure on the luminescence from GaAs/AlGaAs quantum wells, Semicond. Sci. Technol. 9, 2239-2246.
6. Leroux, M., Pelous, G., Raymond, F., and Verie, C. (1985) Minority-carrier lifetime study of the pressure induced Γ-X crossover in GaAs, Appl. Phys. Lett. 46, 288-290.
7. Jayaraman, A., Kourouklis, G.A., People, R., Sputz, S.K., and Pfeiffer, L. (1990) Pressure-tuned resonance Raman scattering and photoluminescence studies on MBE grown bulk GaAs at the E_0 gap, Pramana -J.Phys. 35, 167-175.
8. Murnaghan, F.D. (1944) The compressibility of media under extreme pressures, Proc. N.A.S. (USA) 30, 244-247.
9. Arnoud, G., Allègre, J., Lefebvre, P., Mathieu, H., Howard, L.K., and Dunstan, D.J. (1992) Photoreflectance and piezophotoreflectance studies of strained-layer $In_xGa_{1-x}As$-GaAs quantim wells, Phys. Rev. B46, 15290-15300.
10. Frogley, M.D., Dunstan, D.J., and Palosz, W. (1998) Combined Raman and transmission spectroscopy of ZnTe under pressure. Sol. Stat. Commun. 107, 537-541.
11. Strössner, K., Ves, S., Chul Koo Kim, and Cardona, M. (1987) Pressure dependence of the lowest direct absorption edge of ZnTe, Sol. Stat. Commun. 61, 275-278.

INVESTIGATIONS OF SEMICONDUCTOR BAND STRUCTURE USING HIGH PRESSURE.

D.J. DUNSTAN
Physics Department, Queen Mary and Westfield College,
University of London,
London E1 4NS,
England.

Abstract. Basic effects of pressure on semiconductors are the increase of direct band gap with the density of the crystal, the consequent band crossovers, and the increase in phonon frequencies. We briefly review experiments on these, with emphasis on those aspects which are not understood. These are the density-dependence of the band gap, the effects of combined pressure and strain, and the equation of state. The theoretical framework is discussed. While thermodynamics, elasticity theory, model solid theory and empirical formulae are all relevant to the experimental work, these different theoretical approaches do not converge to provide a coherent interpretation of the experimental work.

1. Introduction

High pressure is an immensely useful tool in the study of semiconductors. Quantitatively, it enables the band gap to be tuned, for the band gap of any semiconductor varies quite considerably with pressure. For example, the direct gap of InP goes from 1.3 eV at ambient pressure to 2.1 eV at 100 kbar. Qualitatively, high pressure enables one to change the ordering of the direct and indirect minima, which changes the semiconductor from direct gap to indirect gap, from GaAs-like (Γconduction-band minimum) to Si-like (X) or Ge-like (L). High pressure is also useful in the study of metals and insulators.

High pressure is not, *per se*, a meaningful physical parameter. What high pressure does is to change the interatomic separation in the sample, and all the other changes of interest follow from that. In a high-pressure experiment, then, the most fundamental thing to know is how much the interatomic separation has been changed. Then, in the interpretation of the experimental results, one should consider how this change in the interatomic separation has affected whatever physical property one has chosen to study. This paper is written from the point of view of the experimentalist who needs a theoretical framework in which to interpret experimental results. We find that such a framework is largely lacking. The author is not a theorist, and it may be that the lack of such a framework is apparent rather than real: that the required information is available in the literature. If that is so, what is certainly lacking is

R. Winter and J. Jonas (eds.), High Pressure Molecular Science, 109–120.

suitable review articles or textbooks to make this information more readily available to the experimentalist.

In this paper, we begin by presenting some typical experimental data on the high-pressure behaviour of semiconductors. Then in Section 3 we consider the theoretical framework within which the experimental data demands to be interpreted.

2. Experimental work on semiconductors

GaAs is typical of the direct-gap semiconductors. Its direct Γ_V - Γ_C gap increases from 1.4eV at room temperature and pressure, at some 10meV per kbar. At about 40 kbar, its conduction band Γ_C minimum rises above the conduction band X_C minimum, and it becomes an indirect semiconductor — Si-like. At the Γ_C - X_C crossover, the direct-gap luminescence quenches drastically; however, the direct band-gap can be followed up to the phase transition at 170 - 180 kbar by spectroscopic techniques such as absorption [1].

The band-gap is sub-linear with pressure, and is conventionally fitted with a quadratic polynomial, or with a linear fit to lattice constant through a suitable equation of state (EOS). These fits lead to considerable error, and we show elsewhere that a linear fit to density (again obtained from an EOS) leads to much better agreement between different authors' results [2].

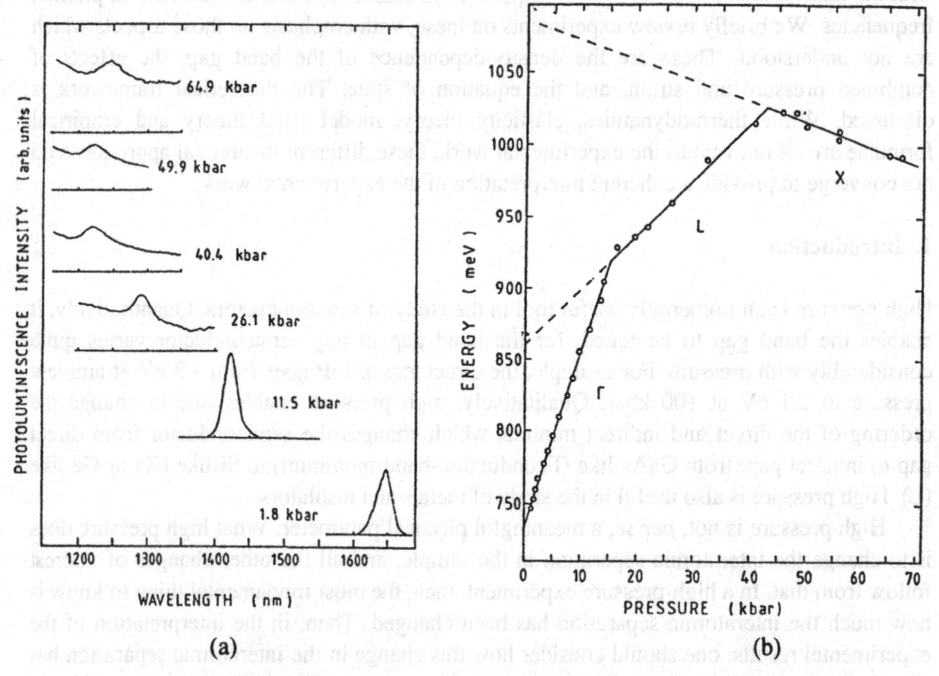

Figure 1. **In (a), the luminescence from an InGaSb quantum well in GaSb is shown moving with pressure. At 16 kbar it becomes much weaker and its pressure coefficient frops; above 43 kbar the pressure coefficient changes sign. Plotting the peak positions against pressure (b), we see the characteristic pressure dependencies of the Γ, L and X minima successively. Extrapolation of the X and L to 0 kbar gives their ambient pressure energies. (Reproduced from Ref. 3).**

When the indirect levels can be observed above crossover, as in Fig.1, they can be extrapolated back to ambient pressure to find their energies [3]. This can be important: the Gunn diode operates by the transfer of electrons from the direct to the lowest indirect minimum of GaAs, yet while devices were being manufactured there was still controversy about even the order of the indirect levels, let alone their exact energies. In Fig.1 we used a linear fit to the pressure dependence of the indirect minima to find their ambient pressure energies. We would now use a linear fit to the density of the crystal under pressure [2] and this would make a small but not negligible correction to the values. Similarly, work relying on the study of band-crossovers in heterostructures, such as determinations of the band-offset ratios [4, 5], is not strongly affected by this question, but we showed that the estimate of the valence band offset in GaAs/AlGaAs is increased a few percent [6].

In semiconductor heterostructures designed and grown for high-pressure experiments, advantage can be taken of the ability to measure differences in behaviour, between for example, quantum wells of varying thickness or composition, or between the quantum well and barrier material. In this way, we established that compressively strained quantum wells show an anomalously small band-gap pressure coefficient when they are grown in the 001 orientation [7, 8] but not in the 111 orientation [8]. The size of the anomaly is proportional to the strain (Fig.2b).

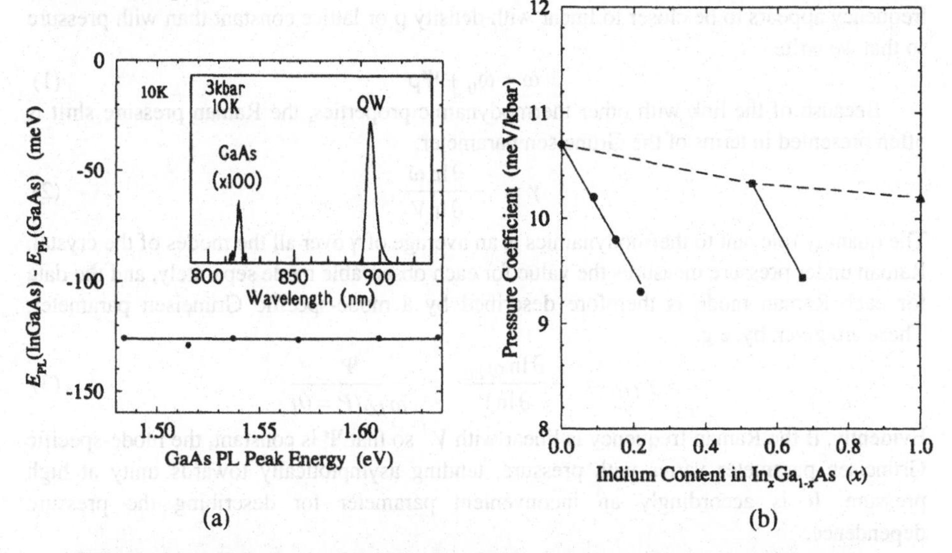

Figure 2. In (a), the spectrum of a (111) oriented InGaAs quantum well in GaAs is shown under pressure. The energy difference between the GaAs and the well emission peaks is plotted against the shift of the GaAs peak. Because the plot is horizontal we see that the pressure coefficients are very accurately the same. Similar experiments in (001) oriented structures gave the dramatic decrease in pressure coefficient shown in (b). Reproduced from Ref. 8.

Epitaxial strained layers are exemplified by InGaAs grown pseudomorphically on GaAs or Si:Ge structures grown pseudomorphically on Si or on relaxed buffer layers. They have considerable hydrostatic and shear stress built-in by the growth. In a high-pressure experiment, the first-order effect is to add the external pressure to the built-in hydrostatic pressure. In addition, smaller effects occur because of the different elastic response of the layer and its substrate. ZnSe is softer than GaAs, so that a layer of ZnSe grown on a GaAs substrate in compression shrinks faster than the substrate. The compressive strain decreases; at some pressure the layer is unstrained and at higher pressure the strain is tensile. Rockwell et al [9] observed this very nicely through the heavy-hole - light-hole splitting, which went to zero at 36 kbar and reversed at higher pressure. However, other authors reported zero strain in this system at different pressures, Cui et al [10] at 21 kbar in agreement with the linear elasticity theory they used and Tuchman et al [11] at 29 kbar in fair agreement with the non-linear theory they used.

In these experiments, highly non-linear changes in the heavy-hole - light hole splitting in the ZnSe is observed. The same effect was observed in layers of ZnTe on InAs [12]. This splitting is a good measure of the shear strain, and is expected to vary linearly with it, at least over the small range of these experiments, just as the shear strain itself is expected to be nearly linear with the pressure.

While optical spectroscopy of the band-gap gives access to electronic properties under pressure, Raman spectroscopy gives access to structural properties, through the optical phonon frequencies given by the Raman shifts. As with the band-gap, the Raman shift increases with pressure and linear or quadratic fits may be used. Again, the Raman frequency appears to be closer to linear with density ρ or lattice constant than with pressure so that we write

$$\omega = \omega_0 + \Psi\rho \tag{1}$$

Because of the link with other thermodynamic properties, the Raman pressure shift is often presented in terms of the Grüneisen parameter,

$$\gamma = -\frac{\partial \ln \omega}{\partial \ln V} \tag{2}$$

The quantity relevant to thermodynamics is an average of γ over all the modes of the crystal. Raman under pressure measures the value for each observable mode separately, and the data for each Raman mode is therefore described by a mode-specific Grüneisen parameter. These are given by, e.g.

$$\gamma_{LO} = -\frac{\partial \ln \omega_{LO}}{\partial \ln V} = \frac{\Psi}{\omega_{LO}(P=0)} \tag{3}$$

Evidently, if the Raman frequency is linear with V^1 so that Ψ is constant, the mode-specific Grüneisen parameter varies with pressure, tending asymptotically towards unity at high pressure. It is accordingly an inconvenient parameter for describing the pressure dependence.

To summarise the experimental results that we would like to understand, and the difficulties in our way, firstly, under pure hydrostatic pressure the band gap is linear with density with a suitable EOS. But neither *k.p* theory nor tight-binding calculations predict this, and in any case there are many equations of state that we could use. Secondly, the band-gap is smaller under the combined effects of hydrostatic pressure and compressive strain than expected. Thirdly, the heavy-hole - light hole splitting is highly non-linear with strain, which is not expected. It is not known whether these two unexpected effects are related. Finally, we would expect the pressure dependence of the Raman to be closely related to the EOS and so we would expect Raman to be useful in understanding experiments on the band structure under pressure.

3. Theoretical Frameworks

There are four fundamental phenomena entering into the behaviour of a solid under the effects of pressure and coherency strain. The first is thermodynamics, which reveals quite unexpected relationships between quite different physical properties. Secondly, there is linear or Hooke's Law elasticity. The third is anharmonicity, or departures from Hooke's Law arising because the atomic bonds are not ideal springs. The fourth is finite deformation. We shall summarise the thermodynamics first, and then go on to consider elasticity.

3.1 THERMODYNAMIC EQUATION OF STATE

Thermodynamics enables a maximum of conclusions to be derived from a minimum of input. Let us see what connection it has with high-pressure work. Classical models of the solid state could not account for such properties as the temperature dependence of the specific heat, but early quantum mechanics provided a better insight. In 1907, Einstein treated a crystal as a set of oscillators, with a frequency corresponding to the Einstein temperature,

$$h\nu = kT_E \tag{4}$$

He was able to predict the dramatic decrease of the specific heats of real solids at low temperature [13], but his prediction fell even faster than experiment. To get better agreement at low temperature, Debye [14] improved the model by allowing a range of frequencies from 0 to the frequency ν_D corresponding to the Debye temperature,

$$h\nu_D = kT_D \tag{5}$$

Debye's maximum frequency was determined by the minimum wavelength λ_{min} at which atoms are moving in antiphase, so that

$$\nu_D = \frac{kT_D}{h} = \frac{v}{\lambda_{\min}} = \frac{v}{a_0} \tag{6}$$

where v is the velocity of sound in the crystal and a_0 is the interatomic spacing. It bears accordingly a close relationship to the frequency of the optical phonons measured by Raman spectroscopy. Grüneisen [15] assumed that that the Debye temperature would vary as a simple power law with volume, as

$$T_D \propto V^{-\gamma} \tag{7}$$

From thermodynamics the pressure is just given by the volume dependence of the energy, so that

$$P = -\left.\frac{\partial E}{\partial V}\right|_T = -\left.\frac{\partial E_0}{\partial V}\right|_T + \frac{\gamma E_v}{V} \tag{8}$$

where E_0 accounts for other contributions to the crystal energy and E_v is the energy of the atomic oscillators. Neglecting the other contributions and identifying the energy of the oscillators with the atomic contribution to the thermal energy gives at once the Mie-

Grüneisen equation of state, relating pressure and volume in terms of the Grüneisen parameter γ

$$P_{T_a} = \frac{\gamma}{V} E_{T_a} \tag{9}$$

Then from the thermodynamic identity

$$\frac{\alpha}{\beta} = \left.\frac{\partial P}{\partial T}\right|_V \tag{10}$$

where α is the volume expansion coefficient and κ is the compressibility, and from

$$\left.\frac{\partial P}{\partial T}\right|_V = \frac{\gamma C_V}{V} \tag{11}$$

Grüneisen obtained

$$\gamma = \frac{\alpha V}{\kappa C_V} \tag{12}$$

so that the Grüneisen parameter may be calculated from experimentally measurable quantities.

In terms of the Debye model, the energy E_{T_a} is identified with the Debye energy E_D, and Kittel [16] showed that in this case the Grüneisen parameter is

$$\gamma = -\frac{\partial \ln T_D}{\partial \ln V} \tag{13}$$

It is assumed that every mode changes frequency with volume in the same way, so that

$$\nu_j = \frac{c_j}{V^\gamma} \tag{14}$$

and then

$$\gamma = \frac{\partial \ln \nu}{\partial \ln V} \tag{15}$$

Slater and Landau derived the Grüneisen parameter in a different way. To present their argument in a simplified form, they proposed that the Debye frequency would depend, in just the same way as low frequency modes, on the volume and the speed of sound u, as

$$\nu_D \propto V^{-1/3} u = \frac{V^{-1/3}}{\kappa \rho} \tag{16}$$

where κ is the compressibility of the solid and ρ its density. This gives again the fundamental relationship,

$$\gamma = \frac{\alpha V}{\kappa C_v} \tag{17}$$

For our purposes, it is more relevant that Slater and Landau recast Eqn.16 as

$$\nu_D \propto V^{-1/6} \kappa^{-1/2} \tag{18}$$

so that

$$\gamma = -\tfrac{1}{6} + \tfrac{1}{2}\frac{\partial \ln \kappa}{\partial \ln V} \tag{19}$$

They suggested that V could be expanded as a polynomial in P, whose terms could be measured experimentally at low pressures. They then had an expression for γ, implying that

$$\frac{\gamma}{V} = \frac{\gamma_0}{V_0} \tag{20}$$

and with γ for all solids tending to $\frac{2}{3}$ at infinite pressure. This gives an equation of state which has been used for shock wave experiments to very high pressures. The reader may well feel that this equation of state has emerged out of nothing! For a fuller account of this thermodynamic equation of state, see Refs 17 and 18.

3.2 ELASTICITY THEORY

The simplest approach is standard linear or second-order elasticity theory. Hooke's Law states that, as the force, so the distortion, or

$$\sigma_{ij} = c_{ijkl}\varepsilon_{kl} \quad \{i, j, k, l = 1 \text{ to } 3\} \tag{21}$$

or in the reduced Voigt notation,

$$\sigma_I = c_{IJ}\varepsilon_J \quad \{I, J = 1 \text{ to } 6\} \tag{22}$$

Note that the first form is readily rotated for different crystal orientations, while the Voigt notation is not, because in the form of Eqn.22 σ, c and ε are all tensors while in the Voigt notation they are not. Now, in the form of Eqn.3, it is clear how to handle non-linear effects. This equation can be written to as high an order as may be required,

$$\sigma_{ij} = c_{ijkl}\varepsilon_{kl} + c_{ijklmn}\varepsilon_{kl}\varepsilon_{mn} + \dots \tag{23}$$

In the Voigt notation and in cubic symmetry, the non-linear terms, the third-order terms, are reduced by symmetry to a few parameters, c_{111}, c_{112} etc, which are listed in the data books. Equations 22 and 23 are useful for small pressures. Hydrostatic pressure on a solid of cubic or higher symmetry is readily described by setting $\sigma_1 = \sigma_2 = \sigma_3 = P$, $\varepsilon_1 = \varepsilon_2 = \varepsilon_3 = 1/3\ \Delta V/V$; then Eqn.22 gives

$$P = \frac{1}{3}(c_{11} + 2c_{12})\frac{\Delta V}{V} = B_0\frac{\Delta V}{V} \tag{24}$$

defining the bulk modulus B_0, which can be found for most solids in any data book and is of order 750 kbar (GaAs). However, even at modest pressures of 10 kbar or less, one cannot treat B as a constant. Using Eqn.23 gives an expression for B under pressure which is incorrect, for levels of approximation have not been kept consistent. The finite extent of deformation must be included as well as the anharmonic terms.

3.3 FINITE DEFORMATION IN ELASTICITY THEORY

Finite deformation of a solid leads to non-linearity even if the interatomic forces obey Hooke's Law. This is because, for example, hydrostatic pressure on a specimen makes it smaller. The force on a crystal facet, the pressure times the area, increases sub-linearly with pressure because the area becomes smaller, and so the force on the Hooke's Law interatomic springs is itself sub-linear with pressure. A large part of the deviation of real crystals from Hooke's Law is due to finite deformation rather than anharmonicity.

The most fundamental parameter one can wish to know about a sample under high pressure is the distance between its atoms. This translates to knowing the equation of state, the $P(V)$ relationship for the solid in question. In order to know this, thermodynamics is not sufficient. The force law between atoms as a function of separation is required. This is not generally available, and so empirical equations of state are much used. There are many equations of state for solids under pressure. Many of them are both empirical and non-predictive. For example, polynomial fits to the experimental $P(V)$ relationship are usually good in the pressure range within which they are fitted, and useless outside. A Hooke's Law solid, without taking finite deformation into account, would reduce linearly to zero volume at a pressure equal to its bulk modulus. Plainly, this does not happen. To take non-linearity into account, Murnaghan [17] proposed that the bulk modulus increases with pressure, and he chose a linear relationship,

$$B(P) = B_0 + B'P \tag{25}$$

Writing the definition of the bulk modulus,

$$B = V\frac{dP}{dV} \tag{26}$$

and integrating, one obtains Murnaghan's well-known equation of state,

$$-\frac{\Delta a_0}{a_0} = 1 - \left[1 + \frac{B'P}{B_0}\right]^{-\frac{1}{3B'}} \tag{27}$$

This equation of state has been widely used, and agrees reasonably well with experiment. In particular, in semiconductors, it transforms sub-linear dependence of, say, the band-gap on pressure into linear dependence on a crystal parameter such as lattice constant or density. Yet it is most surprising that, while these other crystal parameters vary linearly with each other (e.g. band-gap with density), the bulk modulus itself should vary linearly with, not a crystal parameter, but the external parameter of pressure. Alternative assumptions that Murnaghan might have used include B linear with a_0, or with V or ρ, rather than with P. However, I am not aware of any discussion in the literature of alternative equations of state resulting from these assumptions. If B, which is a combination of terms of c_{IJ}, varies linearly with pressure, it would not be unreasonable to suppose that c_{IJ} do so too, so that in analogy with Eqn.25 one might write

$$c_{IJ}(P) = c_{IJ}(0) + c'_{IJ}P \tag{28}$$

The most accurate measurements of the second-order elastic constants, c_{IJ}, come from acoustic measurements, measurements of the speed of sound in a solid. For this depends on nothing but the crystal density (very accurately known) and the elastic moduli. Longitudinal acoustic waves yield the (adiabatic) Young's modulus and transverse waves yield the shear modulus. The small differences between adiabatic and isothermal moduli are within experimental error and need not concern us here. The acoustic wave can be propagated in different directions, and this gives access to both shear moduli, c_{44} and ½(c_{11} - c_{12}). Acoustic measurements are readily carried out under external pressure or uniaxial stress, and this gives access to the third-order elastic constants c_{IJK}.

Because of the accuracy of these acoustic experiments, elasticity theory has been most fully developed in this context [20, 21, 22]. Thus Sklar et al [23] give effective elastic constants for a crystal under finite normal strains, e_1, e_2 and e_3,

$$\begin{aligned} c'_{11} &= c_{11}(1-\Delta+4e_1)+c_{111}e_1+c_{112}(e_2+e_3)=c_{11}+e_P(c_{11}+c_{111}+2c_{112}) \\ c'_{12} &= c_{12}(1-\Delta+2(e_1+e_2))+c_{112}(e_1+e_2)+c_{123}e_3=c_{12}+e_P(-c_{12}+2c_{112}+c_{123}) \quad (29) \\ c'_{44} &= c_{44}(1-\Delta+2(e_2+e_3))+c_{144}e_1+c_{155}(e_2+e_3)=c_{44}+e_P(-c_{44}+c_{144}+c_{155}) \end{aligned}$$

where $\Delta = e_1 + e_2 + e_3$, and the third expression on each line gives the result for hydrostatic pressure with normal strains equal to e_P. For non-hydrostatic strain the other components of c_{ij} such as c_{22} may be obtained by rotating suffices. The contrast between these expression and Eqn.8 is glaring. The bulk modulus at ambient pressure is $B_0 = 1/3\ (c_{11} + 2c_{12})$, so for hydrostatic pressure Eqn.29 gives

$$B(P) = B_0 + \tfrac{1}{3}(c_{11} - c_{12} + c_{111} + 4c_{112} + c_{123})e_P \quad (30)$$

Thus in contrast to Murnaghan's assumption of Eqn.25, B increases linearly with deformation, not with pressure. As remarked above, it would be interesting to know what equation of state this would yield and whether it would be preferable for fitting data. For small pressures,

$$e_P = -\frac{P}{3B_0} \quad (31)$$

and substituting the GaAs values of c_{11} = 119 GPa, c_{12} = 53.8 GPa, c_{111} = -675 GPa, c_{112} = -402 GPa and c_{123} = -4 GPa, we have $B(P) = B_0 + 3\frac{1}{4}\,P$. That is, B' comes out to 3¼ instead of the accepted value of 4½ ± ½. Perhaps the book values of the third-order elastic constants — particularly c_{112} — are seriously in error, or perhaps the levels of approximation are not consistent.

The equations given by Sklar et al [23] are also applicable to the case of an 001 strained layer under pressure. There are significant mathematical difficulties in converting these equations to an equation of state, but preliminary numerical calculations we have performed show that they lead to greater stiffening in the (001) direction, which may, at least in part, account for the pressure coefficient anomalies of 001 strained layers referred to above. However, because Sklar et al include only normal strains, it is not possible to rotate their equations to the 111 orientation to test for consistency with the observed absence of anomaly in that orientation. For that, a complete set of equations requires the inclusion of shear strains. What is plainly lacking is a comprehensive tensor equation of state, expressing not the scalar volume as a function of scalar (hydrostatic) pressure, but the tensor strain as a function of tensor stress, taking into account finite deformation. This would bear the same relationship to $\sigma_{ij} = c_{ijkl}\,\varepsilon_{kl}$ as Murnaghan's equation does to the linear equation of state for infinitesimal pressures, $dP = B_0 d\ln V$.

In discussing the work on the strained ZnTe layers on GaAs, Tuchman et al [11] show the inadequacy of the linear c_{ij} approximation. Instead, they calculate the equilibrium lattice constant of each material under pressure using the Murnaghan equation, Eqn.27, and then calculate the strain in the pseudomorphic layer from the difference in the lattice constants. This procedure gives the in-plane strain (and the pressure at which this vanishes) quite correctly, but it says nothing about the strain in the growth direction and hence about the magnitude of the shear strain, and the expected heavy-hole - light-hole splitting.

Equivalently, one may say that without knowledge of the pressure dependence of the Poisson's ratio, one can take the problem no further. The nearest any authors have come to producing a tensor version of Murnaghan's equation is in the paper of Bertho et al [24], in which they state the problem in terms of pressure derivatives of the elastic constants, as in Eqn. 28. However, they state that the analytic solution is too lengthy to print, and they give only numerical solutions, which do show good agreement with the data of Rockwell et al [9].

We turn now to the interpretation of Raman experiments under pressure. In a Hooke's Law model of the lattice, the frequency of a zone-centre optical phonon is just k/m. This is not affected by pressure, contrary to some authors who have suggested that the increase of the density of the crystal will affect the optical phonon frequencies. This seems to be a misconception. While the acoustic phonon dispersion curve is undoubtedly affected by density, together with the details of the optic branch away from the zone centre, the optic branch at the zone centre surely varies only with anharmonicity — and must be closely linked to the anharmonic contribution to the changes in c_{ij}. Yet the theoretical literature is unhelpful. Most of it is expressed, not in terms of experimental observables such as B', but in terms of the behaviour of model interatomic potentials such as the Keating potential [25, 26]. It is shown, for example, that a sixth-power law potential results in a Gruneisen parameter identically equal to unity. While interesting, this leaves the experimentalist in something of a quandary how to interpret experimental Gruneisen parameters which are not equal to unity. Yet the pressure shift of the Raman frequency must be intimately linked to the anharmonic part of the pressure dependence of c_{ij} — and it is much easier to measure at high pressure.

It may be that *ab initio* methods have more to contribute here than classical analysis. Certainly, *ab initio* calculations can give directly the pressure dependence of the Raman shift. Rücker and Methfessel [26] test their Keating model against *ab initio* calculations, and what is interesting to us here is that they also obtain excellent agreement with experiment for the Raman shifts with alloy composition and with strain in Si:Ge. Such calculations could give very straightforwardly a tensor equation of state, even if it is true that it is harder to gain insight into the underlying physics than with analytic techniques.

4. Conclusions

Fig.3 shows the situation schematically. We have several theoretical approaches to help us understand the experimental behaviour of semciconductors under pressure. Without any theoretical underpinning, we have Murnaghan's empirical equation of state, Eqn.27, which does fit experiment well, as far as it has been tested in a variety of solids. This is fine simply for describing the pressure dependence of quantites such as the band-gap, although a link to electronic band structure theory would be desirable. If we try to relate Murnaghan's equation to standard elasticity theory, we find that the inconsistency between Eqn. 27 and Eqn.s 23, 28 and 29 has not been addressed. Nor has the Murnaghan approach been generalised to tensor strain, except by Bertho et al [24]. So the elasticity theory developed for acoustic work, Eqn.29, does not converge with our empirical EOS. Raman, which should be intimately linked with the pressure dependence of the elasticity, is instead

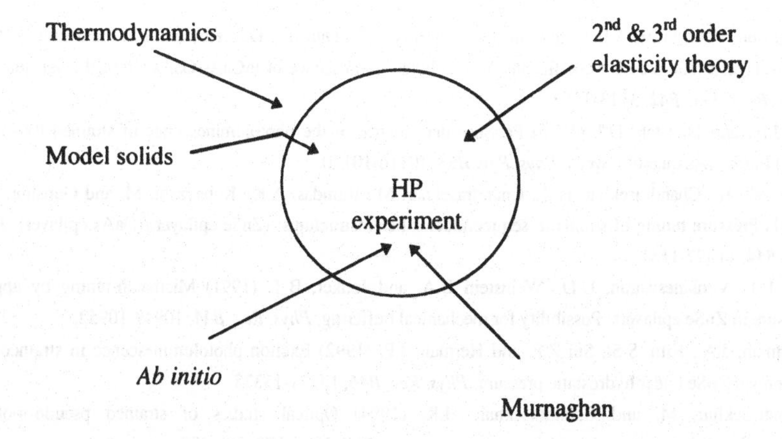

Figure 3. Experimental high-pressure work on semiconductors is represented by the circle. Different theoretical approaches are shown as arrows, which come more or less into the domain of experiment, but which do not converge on a point.

described by a parameter, γ, which comes out of thermodynamics and is interpreted in terms of model solid interatomic potentials. Again, there is no contact with the experimental band structure work.

High pressure experimentation, at present, is capable of more accuracy in the measurement than in its interpretation. As we have shown elsewhere, much of the apparent experimental error in high pressure work has been due to inappropriate fitting to inadequate analytic functions: the experimental data in the literature is better than it looked. I do not claim that the theory presented above is state-of-the-art theory — but it is all that is currently available to the experimentalist who is not also an expert theoretician. And if this is the situation with semiconductors, which are clean, simple, single crystals that have been intensively studied for some forty years, what of other materials?

References

1. Goñi, A.R., Strossner, K., Syassen K., and Cardona M. (1987) Pressure-dependence of direct and indirect optical-absorption in GaAs, *Phys. Rev. B*36, 1581-1587.
2. Frogley, M.D. (1998) How accurate is high-pressure experimentation?, this conference.
3. Warburton, R.J., Nicholas, R.J., Mason, N.J., Walker, P.J., Prins, A.D., and Dunstan, D.J. (1991) High-pressure investigation of GaSb and $Ga_{1-x}In_xSb$/GaSb quantum wells, *Phys. Rev. B*43, 4994-5000.
4. Wolford, D.J., Kuech, T.F., Bradley, J., Gell, M.A., Ninno, D., and Jaros, M. (1986) Pressure-dependence of gaas/alxga1-xas quantum-well bound-states - the determination of valence-band offsets, *J. Vac. Sci. Technol. B*4, 1043-1050.
5. Whitaker, M.F., Dunstan, D.J., Missous, M., and Gonzalez, L. (1996) A general approach to measurement of band offsets of near-GaAs alloys, *phys. stat. sol. (b)* **198**, 349-353
6. Frogley, M.D., Sly, J.L., and Dunstan, D.J. (1998) *Phys. Rev. B* in press.

7. Wilkinson, V.A., Prins, A.D., Lambkin, J.D., O'Reilly, E.P., Dunstan, D.J., Howard, L.K., and Emeny, M.T. (1990) Hydrostatic pressure coefficients of the photolumin-escence of InGaAs/GaAs strained-layer quantum wells, *Phys. Rev. B*42, 3113-3119.

8. Sly, J.L., and Dunstan, D.J. (1996) Pressure dependence of the photoluminescence of strained (001) and (111) $In_xGa_{1-x}As$ quantum wells, *Phys. Rev. B*53, 10116-10120.

9. Rockwell, B., Chandrasekhar, H.R., Chandrasekhar, M., Ramdas, A.K., Kobayashi, M. and Gunshor, R.L. (1991) Pressure tuning of strains in semiconductor heterostructures: (ZnSe epilayer)/(GaAs epilayer), *Phys. Rev. B*44, 11307-11314.

10. Cui, L.J., Venkateswaran, U.D., Weinstein B.A. and Jonker, B.T. (1991) Mismatch-tuning by applied pressure in ZnSe epilayers: Possibility for mechanical buffering, *Phys. Rev. B*44, 10949-10952.

11. Tuchman, J.A., Kim, S.S., Sui Z.F. and Herman, J.P. (1992) Exciton photoluminescence in strained and unstrained ZnSe under hydrostatic pressure, *Phys. Rev. B*46, 13371-13378.

12. Chandrasekhar, M. and Chandrasekhar, H.R. (1994) Optical studies of strained pseudo-morphic semiconductor heterostructures under external pressure, *Phil. Mag. B*70, 369-380.

13. Einstein, A. (1907) *Ann. Physik* 22, 180.

14. Debye, P. (1912) *Ann. Physik* 39, 789.

15. Grüneisen, E. (1926) in *Handbuch der Physik Vol. 10*, Springer, Berlin, p.1.

16. Slater, J.C. (1939) *Introduction to Chemical Physics*, McGraw-Hill, New York.

17. Kittel, C.,

18. Eliezer, S., Ghatak, A., and Hora, H. (1986) *An introduction to equations of state: Theory and applications*, Cambridge University Press, Cambridge.

19. Murnaghan, F.D. (1944) The compressibility of media under extreme pressures, *Proc. Nat. Acad. Sci.* 30, 244-247.

20. Thurston, R.N. (1967) Calculation of lattice parameter changes with hydrostatic pressure from third-order elastic constants, *J. Acoust. Soc. Amer.* 41, 1093-1111.

21. Brugger, K. (1964) Thermodynamic definition of higher order elastic constants, *Phys. Rev.* 133, A1611-A1612.

22. Birch, F. (1947) Finite elastic strain of cubic crystals, *Phys. Rev.* 71, 809-824.

23. Sklar, Z., Mutti, P., Stoodley, N.C. and Briggs, G.A.D. (1995) Measuring the elastic properties of stressed materials by quantitative acoustic microscopy, *Adv. Acoustic Microscopy* 1, 209-247.

24. Bertho, D., Jancu, J.M. and Jouanin, C. (1993) Equations of state and a tight-binding model for strained layers - Application to a ZnSe-GaAs epilayer, *Phys. Rev. B*48, 2452-2459.

25. Ganesan, S., Maradudin, A.A., and Oitmaa, J. (1970) A lattice theory of morphic effects in crystals of the diamond structure, *Ann. Phys.* 56, 556-594.

26. Rücker, H., and Methfessel, M. (1995) Anharmonic Keating model for group-IV semiconductors with application to the lattice dynamics in alloys of Si, Ge, and C, *Phys. Rev. B*52, 11059-11065.

CRITICAL- AND WETTING-PHENOMENA NEAR THE LIQUID-VAPOUR CRITICAL POINT OF METALS

F. HENSEL
Institute of Physical Chemistry, Nuclear Chemistry and Makromolecular Chemistry, FB Chemie, Philipps-University of Marburg, Hans-Meerwein-Straße, D-35032 Marburg

1. Introduction

Our understanding of the liquid-vapour equilibrium in metallic systems has increased enormously in the past two decades. Much of this has been stimulated by a series of pioneering papers of Mott [1] on the metal-non-metal transition which shows up when a liquid metal is heated to the region of the liquid-vapour critical point. The existence of this transition implies that the liquid-vapour phase transition of fluid metals is distinct from that of normal insulating fluids such as inert gases. An inert-gas atom retains its identity in the condensed phase and the pair potential which determines the properties of the dilute vapour phase is supplemented to a limited degree by many-body interactions, in the total potential energy of the dense phase [2]. In contrast, the electronic structures of the two coexisting phases, liquid and vapour, of fluid metals may be vastly different. The essntial point is that the metallic state is a collective phenomenon existing only when the density of atoms is sufficiently large. Unlike inert gases the electronic stucture in the high-density liquid is very different from that of an atom in the dilute vapour.

For example, liquid mercury and caesium near their triple points are considered as normal liquid metals with properties typical of the condensed state. The small changes of basic properties such as the electrical conductivity or magnetic susceptibility on melting [3] show that the electronic structure of the liquid is quite similar to that of the crystalline solid and both liquid and solid can be reasonably well described by the nearly free-electron approximation [3]. This behaviour is usually explained by the fact that the short-range atomic correlations and the atomic density in the liquid near melting are closely similar to those of the crystals.

R. Winter and J. Jonas (eds.), High Pressure Molecular Science, 121–149.

Consequently, the liquid metallic phase is usually treated as a monoatomic state which typifies the solid structure. It is regarded as built-up of single screened ions each diffusively uncoupled from every other.

In contrast to melting, however, there are substantial changes in the nature of bonding upon evaporation. At sufficiently low densities, in the vapour phase, the valence electrons occupy spatially localized atomic or molecular orbitals. In this state many caesium atoms form chemically bound dimers with a dissociation energy of 0.45 eV, whereas in the vapour phase of mercury, chemically bonded dimers do not form because the ground state of the mercury atom arises from a closed shell electronic configuration which by itself cannot form an appreciable bond owing to the inert character of the $6s^2$ shell. The interaction potential between two mercury atoms is thus generally regarded as acting between highly polarizable closed-shell systems involving very little electronic density migration from the partners, in a similar way to noble gas atoms. In this sense, mercury vapour has been denoted „pseudo-helium" while the vapour of the alkali metal caesium resembles hydrogen.

However, this description applies only to conditions near the triple point and whether the liquid-vapour phase change is always and necessarily accompanied by a metal-non-metal transition, even up to the liquid vapour critical point, is one of the points at issue [4].

During the past two decades, a considerable amount of effort has been centered on the experimental investigation of the metal-non-metal transition and its relation to the thermodynamic phase behaviour. Most of this effort has focused on mercury [5], caesium [6], rubidium [7], and potassium [7] because these metals have critical points that lie within the current limits of static temperatures and pressures available in the laboratory (see table 1).

Metal	T_{crit}(°C)	p_{crit}(bar)	ρ_{crit}(g cm^{-3})
mercury	1478	1673.0	5.80
caesium	1651	92.5	0.38
rubidium	1744	124.5	0.29
potassium	1905	148	0.18

Table 1 Critical temperatures, pressures, and densities of selected fluid metals.

These relatively favourable conditions have permitted very accurate measurements of the electrical, magnetic, optical, structural, interfacial and thermophysical properties with optimal temperature control to determine the asymptotic behaviour of these properties as the liquid-vapour coexistence curve, including the critical point, is approached.

2. Liquid-Vapour and Metal-Non-Metal Transition

The most significant experiments for the exploration of the relationship between the liquid-vapour and metal-non-metal transitions in fluid metals are measurements of the electrical, optical and magnetic properties over the whole liquid-vapour region along the coexistence curve. The latter is shown for mercury in Figure 1, along with the mean of the densities ρ_L and ρ_v of coexisting liquid and vapour phases, the so-called diameter.

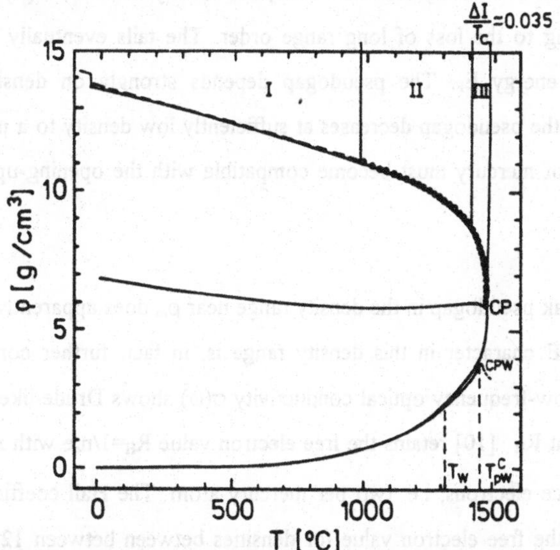

Figure 1 The bulk coexistence curve of mercury, along with the mean of the densities ρ_L and ρ_v, the so-called diameter. The prewetting line terminates at high temperatures at the prewetting critical temperature T^c_{pw} and at low temperatures at the wetting temperature T_w. The dashed extension of the prewetting line is a line of maximum two-dimensional compressibility in the prewetting supercritical one-phase region.

Three ranges, denoted I, II, III, can be distinguished. In the region I between the melting temperature T_m = 38°C and T = 980°C, the density of the coexisting liquid decreases from ρ_m = 13.65 g/cm^3 to 11 g/cm^3. The conductivity σ at ρ_m is about 10^4ohm^{-1}cm^{-1} and the nearly free electron (NFE) theory gives an electron mean free path of about 7Å which exceeds only slightly the mean interatomic distance. Application of the NFE-model leads to the conclusion that σ can be satisfactorily explained if the effect of the atomic d-states is included in the pseudopotential. Mercury is thus essentially a NFE metal at ρ_m, despite the comparatively small mean free path.

There is, however, evidence from photoelectron spectroscopy experiments [8] for a slight minimum in the density of states at the Fermi energy, i.e. a weak pseudogap. As is well known, a low density divalent metal such as mercury is predicted by the Bloch-Wilson band model, to undergo a metal-semiconductor transition when the s-(valence) and p-(conduction) bands no longer overlap. In a crystal, a real gap appears and widens further as the density decreases. Mott has proposed that the main features of the crystalline model survive in the liquid state with the band edges smeared out by disorder. Thus, the density of states N(E) is expected to tail into the gap owing to the loss of long range order. The tails eventually overlap in the region of the Fermi energy E_F. The pseudogap depends strongly on density. When the magnitude of N(E) in the pseudogap decreases at sufficiently low density to a negligibly small value, the properties of mercury must become compatible with the opening-up of an energy gap.

The existence of a weak pseudogap in the density range near ρ_m does apparently not affect the conductivity. The NFE character in this density range is, in fact, further confirmed by the observation that the low-frequency optical conductivity $\sigma(\omega)$ shows Drude-like behaviour [9] and the Hall coefficient R_H [10] retains the free electron value $R_H=1/n_e e$ with n_e given by the total number of valence electrons, i.e. two per mercury atom. The Hall coefficient begins to deviate slightly from the free electron value for densities between between 12 and 11 g/cm^3 where also changes in the shapes of the $\sigma(\omega)$ curves indicate a gradual diminution of metallic properties.

Whereas the deviation of R_H and σ from NFE behaviour is relatively small in the density region $11 \leq \rho \leq 13.65$ g/cm^3, i.e. in the region I, further expansion of mercury to a density of about 9 g/cm^3 in region II reduces σ to a value of about 450 ohm^{-1}cm^{-1} and R_H increases by about a factor of three above the free electron value. It is often assumed that a liquid conductor at high temperatures has undergone a change from metallic to nonmetallic behaviour if its σ-value falls below a few hundred $ohm^{-1}cm^{-1}$, the range of the „minimum metallic conductivity" [1]. But the magnitude of σ alone is not sufficient evidence and the fact that for densities smaller than 9 g/cm^3 the shape of the $\sigma(\omega)$ curves is that of a substance with a real energy gap [9] is a much more convincing sign that a range of energy around the Fermi energy exists that is so thinly populated with states that the contribution to the optical properties is negligible small. This view is strongly supported by the sharp drop of the Knight shift [11] for densities smaller than 9 g/cm^3 and by calculations of the total and partial densities of states at densities along the coexistence curve employing ab initio density-functional molecular dynamics techniques [12]. The central results of these calculations is that in fluid mercury decreasing density leads to a narrowing of the s- and p-bands and finally to metal-non-metal transition of the band crossing type.

The interrelation of the metal-non-metal transition and the liquid-vapour phase transition in fluid alkali metals can be seen in much the same way that one observes it in fluid mercury - by measuring the equation of state simultaneously with electrical [6], [7], optical [13], or magnetic [14], [15] properties. However, the transitions in these two kinds of systems, are clearly distinct. Hints at the differences show up at a glance from a comparison of the shape of the coexistence curve (see Figures 1 and 2). On the whole the coxistence curve of mercury is much more symmetric than that of caesium. Also the electrical properties bear a quantitative difference. Taken as a group, the alkali metals exhibit very similar electrical conductivities. At the critical point, the conductivity is about 250 $ohm^{-1}cm^{-1}$ for each of the three alkali metals (potassium, rubidium, and caesium) that have been studied experimentally in the critical region [7], while that of mercury is roughly two orders of magnitude lower. [5]

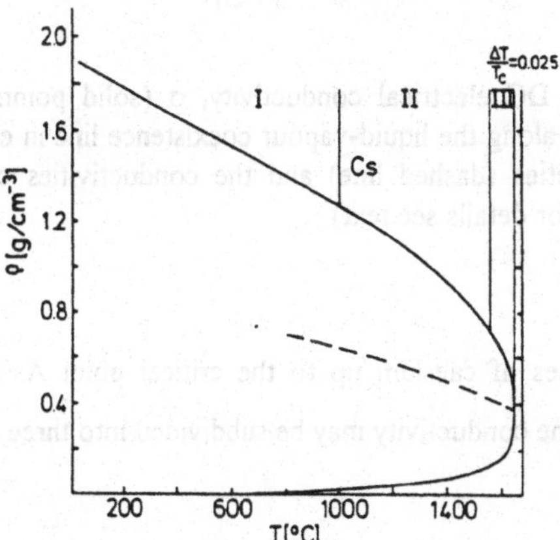

Figure 2 The bulk coexistence curve of caesium along with the mean of the densities ρ_L and ρ_v, the so-called diameter.

Such differences between divalent mercury and monovalent alkali metals might have been expected from the simplest considerations based the Bloch-Wilson model. Indeed, without electron electron interaction the latter would lead to the prediction that the alkali metals are always metallic, regardless of the density. This prediction has stimulated very early experimental attempts to observe metallic conductivity in low density alkali metal vapours but thanks to the work of Mott [16] we know now that because of electron correlations, this cannot be true. In addition, there is now considerable experimental evidence [17] from magnetic susceptibilty- and optical absorption measurements that in alkali vapours the valence electrons occupy spatially localized atomic and molecular orbitals.

The physical consequences of the substantial changes in the electronic and molecular structures upon evaporation are vividly illustrated by data such as those displayed in Figures 2 and 3 which show the most accurate density ρ and DC electrical conductivity σ data of the coexisting

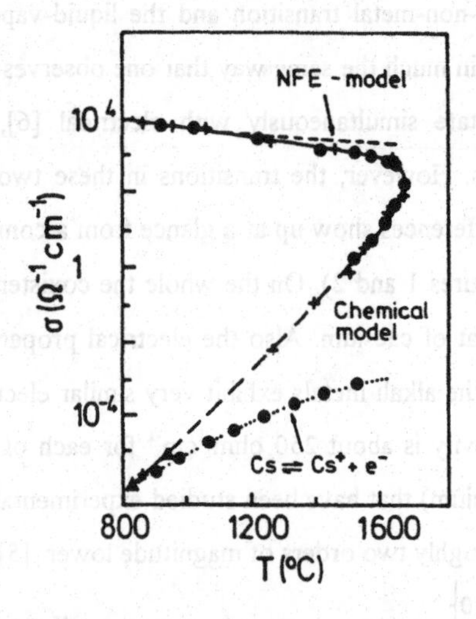

Figure 3 Experimental DC electrical conductivity, σ (solid points), of fluid caesium as a function of temperature along the liquid-vapour coexistence line in comparison with the nearly free-electron conductivities (dashed line) and the conductivities calculated using the Saha equation (dotted line) (for details see text).

liquid and vapour phases of caesium up to the critical point As for divalent mercury, the density dependence of the conductivity may be subdivided into three domains (see Fig. 2).

Between the melting temperature T_m = 30°C and T = 1100°C the density of the liquid decreases from ρ_m = 1.84 g/cm^3 to ρ = 1.3 g/cm^3. Throughout this range of initial expansion, the magnetic, optical and electrical properties are consistent with the NFE theory. Conductivities, calculated using this theory [18], match the measured values for temperatures lower than 1100°C (Figure 3). At higher temperatures, and hence lower densities, σ is lower than that predicited by the NFE-model. The gradual failure of the model at densities lower than 1.3 g/cm^3 is probably not due to the breakdown of the NFE condition, that the electron mean free path λ must exceed the mean interatomic distance. Estimates of λ from the conductivity [18] indicate that this condition is only reached in the density range $\rho \approx 0.8$ g/cm^{-3}. Rather , the NFE-breakdown is more likely to reflect the increased importance of electron correlations below this density. This view is based on magnetic susceptibility [14], NMR [15] and optical reflectivity [13] studies of low - density liquid caesium which have yielded evidence of strong electron-electron-correlations in the form of susceptibility enhancements and antiferromagnetic spin fluctuations. The susceptibility, NMR and optical data show a relative sharp onset of a correlation enhancement of the effective electron mass at a density of 1.3 g/cm^3. The combined analysis of the magnetic and NMR data shows that correlation enhancements of the effective mass $m_{eff} = 6m_0$, develop at a density of 0.8 g/cm^3, i.e. at about twice the critical density.

Alternatively, the strong decrease of σ for densities below about 1.3 g/cm^3 can also be interpreted in terms of partial ionization due to the formation of species such as Cs^0 - atoms, Cs_2^0- dimers, and Cs_2^+ - dimers which leads to a reduction of the number of free charge carriers. This approach views the low-density liquid as the high -density limit of the compressed metal vapour. It is immediately evident from a glance at Figure 3 that the formation of such species affects strongly the conductivity of the dense vapour phase of caesium. Here the measured conductivity is compared with values calculated from the thermal equilibrium fraction of ionized caesium monomers. For this calculation , the vacuum ionization potential (3.89 eV) was employed in the Saha equation. Agreement between experimental and calculated values is satisfactory only at very low temperatures and densities where the overwhelmingly dominant species in the vapour is the atomic monomer. At higher vapour densities, species such as the dimer and even larger clusters appear. The ionization potentials of these clusters are much lower than that of the free atom and we should expect the degree of ionization to be much higher than given simply by the Saha equation.

The concentrations of the various species in the coexisting vapour and liquid phases have been calculated for caesium [19], employing a quantum statistical equation of state originally derived for partially ionized plasmas which takes into account the interaction corrections between the various species in a systematic way. An approximate self-consistent solution for the system of coupled mass action laws describes the formation of atoms Cs^0 dimers Cs_2^0 and molecular ions Cs_2^+ out of the elementary particles, electrons e^- and simple ions. The numerical results obtained for the concentrations of neutral atoms, Cs^0, neutral dimers, Cs_2^0, singly ionized dimers, Cs_2^+ and thermally ionized free electrons are shown as a function of density in Figure 4.

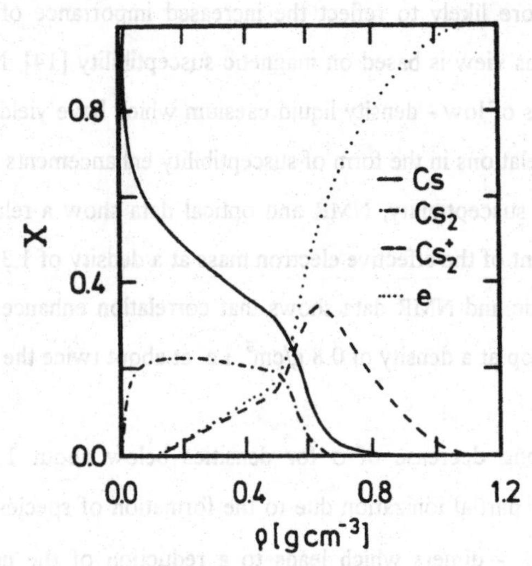

Figure 4 Relative concentrations of the free conduction electrons (dotted line) and electrons localized on neutral Cs atoms (solid line), ionized Cs_2^+ dimers (dashed line) and neutral Cs_2dimers (dashed line) as a function of density (19).

The results suggest that at about twice the critical density, the combined concentration of Cs_2^0 and Cs_2^+ dimers is of the order of 40%. Similar results have been obtained for fluid rubidium [19]. The crosses in Figure 3 show the conductivity calculated with the electron concentration in Figure 4 [20]. The agreement is surprisingly good.

3. Bulk Critical Phenomena

It is evident from the foregoing that a fundamental distinction between fluid metals and insulating fluids such as inert gases lies in the character of the interparticle interaction. Far below the critical temperature, the vapour phase is nonmetallic with weak interatomic or intermolecular interactions. In contrast, the coexisting liquid is highly conducting and the interaction is dominated by screened Coulomb potentials. However, the presence of strong state-dependent interactions in metals does not imply that the critical behaviour of the thermodynamic properties cannot be described with the same exponents as that of molecular fluids. The speculation that critical points of metals could belong to a different universality class than molecular fluids, which was mainly based on the assumption that the Hamiltonian of metallic systems consists of a sum of long-range Coulomb interactions has been clearly ruled out by accurate measurements of the shapes of the coexistence curves of mercury [5], cesium [21], and rubidium [21]. There is no doubt that they display the same critical exponents as molecular fluids, i.e. values close to those found for the three dimensional Ising-model.

The implications of the state dependence of the interatomic interactions for the liquid-vapour phase equilibrium of metals are mainly evident from the behaviour of the diameter $\rho_d = \rho_L + \rho_v / 2\rho_c$ (see Figures 1 and 2) in the transition range where metallic properties evolve continuously into those characteristic of nonmetals. Indeed far from the critical point in range I of Fig. 1 ρ_d of mercury has a positive slope as is typical of molecular fluid and as is also observed for the alkali metals. In this region the coexisting liquid is metallic ($\rho \geq 11$ g/cm^3) while the vapour consists of atoms interacting through weak forces. At higher temperatures, i.e. in range II, where the liquid is in the electronic transition range, ρ_d slopes toward higher densities, opposite to the behaviour seen in molecular fluids and the alkali metals. Thus, one observes at a density of about 11 g/cm^3 a clear correlation between an anomaly in the diameter slope and a change in the electronic structure demonstrating the interplay between the latter and the liquid-vapour phase behaviour.

Within the critical region of mercury (range III in Fig. I), there does indeed appear the theoretically expected (1-α) singularity. Theory predicts that the temperature derivative of the diameter $d\rho_d/dT$, diverges at least as fast as the constant-volume specific heat c_v. That is, as the reduced temperature $\frac{\Delta T}{T_c}$ goes to zero, the diameter varies as

$$\rho_d = 1 + D_0\left(\frac{\Delta T}{T_c}\right)^{1-\alpha} + D_1\left(\frac{\Delta T}{T_c}\right)$$

The linear term is the usual background rectilinear diameter observed in all fluids. The term $D_0\left(\frac{\Delta T}{T_c}\right)^{1-\alpha}$, where $\alpha = 0.11$ is the same exponent as that which characterizes the power-law divergence of the specific heat along the critical isochore.

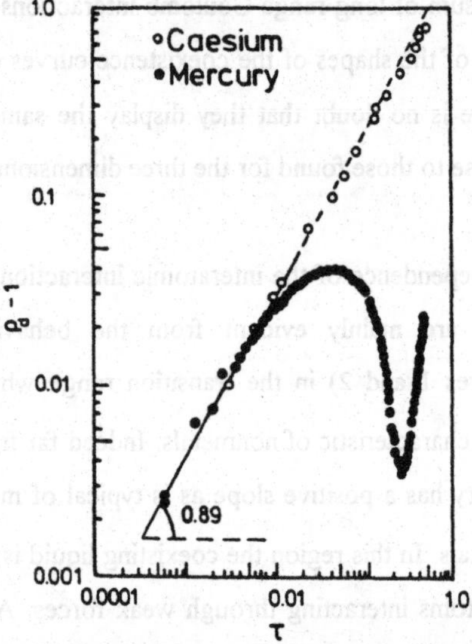

Figure 5 Power-law analysis of the diameters of mercury and caesium.

It is evident from a glance at Figure 5 that the critical diameter anomalies for the alkali metals and mercury are characterized by exponents (1-α). For the alkali metals the anomalies are so strong that a nearly pure power-law behaviour is seen over several decades in the reduced temperature $\Delta T/T_c$. The competing variations of the electronic structure of liquid mercury with density and temperature cause a pronounced wiggle at intermediate $\Delta T/T_c$ - values.

4. Wetting Phenomena Near The Bulk Critical Point Of Mercury

Another interesting example of the intimate interplay between changes in interparticle forces and changes in the electronic structure associated with the liquid-vapour critical point phase transition in mercury is provided by recent experimental work [22], [23] on the wetting behaviour of mercury films on rigid solids.

As first noted by Cahn [24], a substrate in contact with a vapour near vapour-liquid coexistence which is incompletely wet at low temperatures is expected to become completely wet as the critical temperature T_c of the bulk fluid system is approached. The change from one condition to the other is denoted the wetting transition and the corresponding temperature the wetting temperature T_w. This transition is defined only at bulk coexistence. The wetting transition can be either continuous or first order. If the transition is first order, with a discontinuous jump in the film thickness on the coexistence curve, thermodynamics requires that it is accompanied by a line of prewetting transitions [25]. These are transitions from a thin to a thick liquid layer induced, for example, by increasing p, the pressure of the vapour, towards its value on bulk saturation p_{sat} (T) at a fixed temperature T above the wetting temperature T_w. The line of first-order prewetting transitions extends smoothly from the point of the first order wetting transition (T_w, p_{sat}) into the region of unsaturated films with $p < p_{sat}$. It is expected to terminate at a prewetting critical point (T^c_{pw}, p^c_{pw}) the location of which must depend upon the properties of the substrate as well as on the interparticle interaction of the fluid system.

Since the initial prediction of wetting transitions in 1977 [24] a large amount of theoretical and experimental work has been done in an effort to determine the conditions under which wetting transitions take place, the order of the transitions, and associated phase diagrams and critical properties [25]. In spite of this effort, convincing experimental evidence for prewetting transitions at solid wall-vapour interfaces had to wait until 1992, when they were finally seen in experiments with helium on caesium [26] and hydrogen on rubidium [27]. The observation of the wetting transition in the quantum fluids helium and hydrogen was soon followed by evidence from fluid mercury on sapphire [22]. The latter system was an obvious candidate for a wetting transition because as a result of their low surface free energies, most nonmetallic solids are not wetted by nonreactive liquid metals like mercury. Indeed the closest example in common experience of a nonwetting substance is liquid mercury on glass. On the other hand, the subsequent report of the existence of a wetting transition in mercury [23] on the walls of molybdenum and niobium was a sursprise because most pure liquid metals (including mercury) wet completely a perfectly clean solid metal surface. One must be extremely careful, however, when considering the apparent nonwetting of metallic solids by liquid metals. Thin layers of oxide or sulfide are sufficient to prevent wetting by most liquid metals even when, as is usually the case, the layer is extremely thin.

Be that as it may, as mentioned before an obvious candidate for a wetting transition is mercury on a nonmetallic substrate when it is brought by heating near its liquid-vapour critical point. Unfortunately, making such measurements is technically very demanding because high pressures and high temperatures are needed to reach the critical point of mercury. A major experimental problem encountered in working with mercury at high temperatures and pressures is its containment in uncontaminated form. Materials found to be chemically inert in contact with the fluid sample are pure molybdenum, the sintered nonmetal aluminium oxide, and sapphire single crystals. However, cells from these materials cannot be constructed in such a way that they can withstand a high internal pressure. At conditions near the critical point of mercury, an arrangement must therefore be used in which the sample cell, together with the necessary furnaces, are placed in larger high pressure autoclaves filled with the pressure transmitting argon gas under the same pressure as the mercury sample inside the cell. This avoids any mechanical stress on the cell, but at the necessary high pressures severe problems of temperature measurement and control arise.

It is therefore difficult to determine directly the wetting behaviour of mercury by measurement of the wetting film thicknesses. Rather, the wetting phase diagram for mercury on sapphire has been established indirectly by reflecting light from a vertical sapphire mercury vapour interface. Successful measurements with optimal elimination of temperature gradients were achieved with a long optical reflectivity cell (Figure 6).

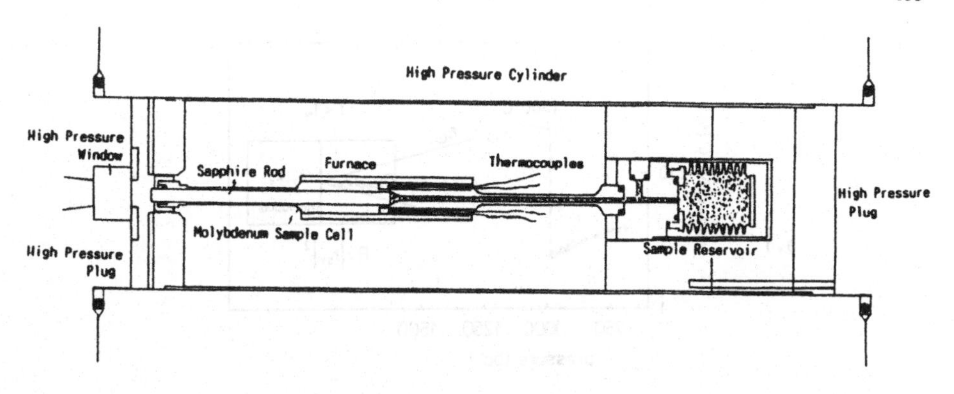

Figure 6 Autoclave with optical access for reflectivity measurements on fluid mercury.

consisting of a molybdenum cylinder, with an inner diameter of 5 mm which is closed at one end by an 80 mm long synthetic sapphire window with optically polished upper and lower surfaces. The cell was mounted in a specially constructed autoclave with high-pressure windows and an internal electric resistance furnace which is thermally insulated from the steel walls. The furnace consisted of two independently controlled heating elements made of molybdenum wire. The pressure medium was argon. The cell was connected to a reservoir maintained at the same pressures as the autoclave. The sample temperature was measured with two W-Re thermocouples in close contact with the hot part of the cell where the mercury sample was located. The sample temperature was calibrated by employing the most accurate vapour pressure data. For temperatures lower than the wetting temperature T_w the vaporization temperature at a given pressure is found as an abrupt change in the reflectivity when the mercury sample is vaporized (Figure 7a). The reproducibility of this procedure was found to be better than ± 1°C. Pressures were measured to ± 1 bar with calibrated commercial Bourdon-type manometers. A standard optical set-up was employed to measure the reflectivity. The optical system consisted of a pulse tunable dye laser pumped by an excimer laser and a commercial optical detection system.

In order to locate the prewetting line, reflectivity measurements at various wavelengths were performed at constant temperature T by increasing p, the pressure of the vapour, towards its value at bulk saturation p_{sat} or alternatively at constant p increasing T towards T_{sat}.

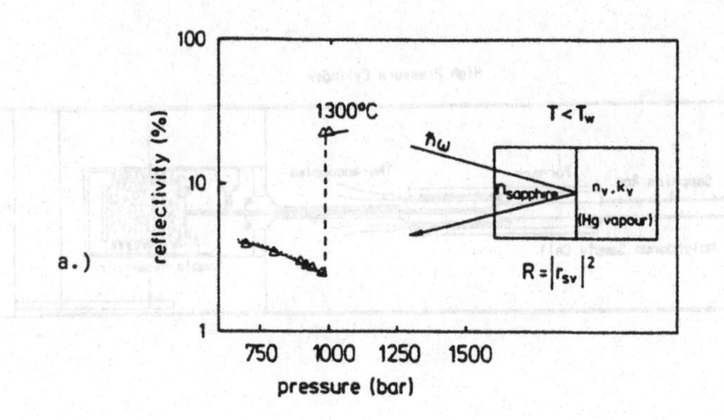

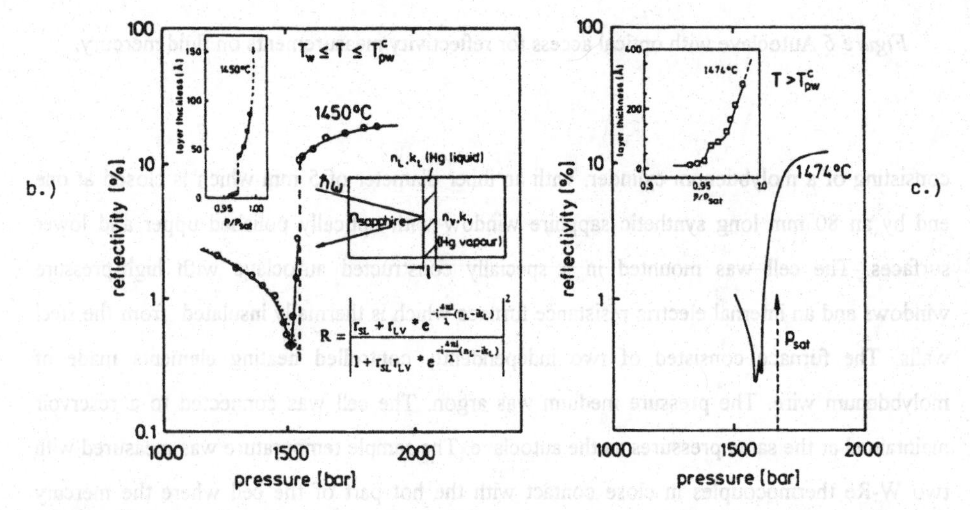

Figure 7 Typical experimentally observed reflectivity curves of mercury vapour against an optically transparent sapphire substrate at constant temperature T. *Figure 7a* displays an example for the pattern observed for $T<T_w$. The sudden increase in reflectivity indicates condensation to the liquid state

$$r_{\alpha\beta} = \frac{(n_\alpha - ik_\alpha) - (n_\beta - ik_\beta)}{(n_\alpha - ik_\alpha) + (n_\beta - ik_\beta)}$$

where n and k are usual optical constants. *Figure 7b* shows an example for the pattern observed for $T_w < T < T^c_{pw}$. The wetting layer thickness estimated with the slab model illustrated in the right part of the Figure is displayed in the inset of Figure 7b. The prewetting step at p/p_{sat} = 0.977 represents a coexistence region in which the thick phase is growing at the expense of the thin phase. *Figure 7c* shows the isotherm at 1474°C. It is a typical example for the pattern observed for $T > T^c_{pw}$. The inset shows the estimated thickness of the wetting film for $T > T^c_{pw}$ as a function of p/p_{sat}.

Figures 7a, 7b, 7c display three representative isotherms. The shape of the curve at 1300°C in Figure 7a is characteristic for isotherms below T_w. The measured reflectivity R for $p < p_{sat}$ (solid curve, and experimental points) coincides with that calculated (dotted curve) employing Fresnel's formula for normal incidence at the sapphire(s)-mercury-vapour(v) interface. There is no indication of a precursor of wetting. The sudden increase in reflectivity (broken curve) indicates bulk phase separation - that is condensation to the liquid state, i.e. R is then for normal incidence at the sapphire (s)-mercury-liquid (L) interface.

Unlike the isotherms at $T < T_w$, the measured reflectivity curve at 1450°C (Figure 7b) deviates markedly from the calculated dotted curve. The slope of the curve of reflectivity changes sign discontinuously within a certain temperature range. The pressure is below that needed for bulk condensation. The latter is again readily detected by the abrupt increase in the reflectivity (broken curve). To attribute the anomalous change in reflectivity to bulk phase behaviour would require an unphysical, discontinuous change in the refractive index of the unsaturated mercury vapour. Much more likely is that the reflectivity anomaly is due to abrupt formation of a wetting film.

On the assumption that the film formed on the sapphire window has the density and optical properties known for the coexisting liquid, one can employ the theory of the reflectance of thin absorbing slabs [28] to estimate the layer thicknesses from the reflectivity data (see Figure 7b). The density dependence of the refractive index n_v of mercury vapour was determined from reflectivity data at temperatures far above T_c, or below T_w. The wetting layer thickness for the 1450°C isotherm estimated in this way is displayed in the inset of Figure 7b. The film thickens rapidly as p approaches coexistence, i.e. at $p/p_{sat} = 1$. In principle, for complete wetting the film thickness should become infinitely thick at $p/p_{sat} = 1$. However the latter applies only to the hypothetical case of zero gravity. In terrestical experiments gravity limits the film thickness to a few hundred Angström at saturation.

In addition there is a jump in the film thickness at intermediate values of p/p_{sat}. The location of these jumps for different isotherms can be mapped by the reduced pressure at the reflectivity anomaly, p_{jump}/p_{sat}. The positions of the reflectivity anomalies measured in a series of isotherms such as that in Figure 7b define the solid curve in the p-T plane, shown in Figure 8.

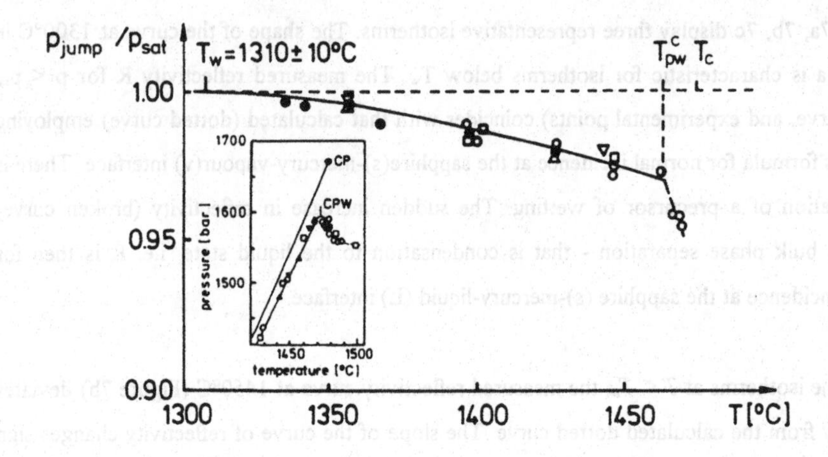

Figure 8 **p_{jump}/p_{sat} as a function of temperature T. The tangential approach of this curve to the line p/p_{sat} 0 1, i.e. the coexistence curve, locates the wetting temperature T_w. The critical temperature T^c_{pw} marks the end of the two-phase thick-thin-film coexistence. The inset shows the prewetting line in the *p-T*-plane near the prewetting critical point.**

This is the line of first-order phase transitions between thin and thick films called the prewetting line. The wetting transition temperature T_w which is the temperature at which the prewetting line meets the bulk liquid-vapour coexistence curve tangentially, is determined by extrapolating the values of p_{jump}/p_{sat} shown in Figure 8 to the coexistence curve, represented by the horizontal line p/p_{sat} = 1. The extrapolation is suggestive of a wetting temperature T_w of about 1310°C.

The prewetting critical point at which the distinction between thin and thick films disappears must lie on the p_{jump}/p_{sat} versus T curve in Figure 8. To find the prewetting critical temperature T^c_{pw} the anomalies in the reflectivity isotherms at higher temperatures must be examined.

Above T^c_{pw}, there should be no coexistence region, and the wetting film should thicken over a range of pressure, which should result in a continuous (but probably very steep) increase in an isotherm for $T \geq T^c_{pw}$. It is indeed evident from a glance at Figure 7c that the reflectivity isotherms slightly above T^c_{pw} as shown for the curve at T = 1474°C, are qualitatively different from those at lower temperatures. Two, now slightly rounded minima occur in the reflectivity

versus pressure curve. The changes in sign of the slopes seem to be no longer discontinuous. The film, calculated from the reflectivity curve in Figure 7c (see inset) thickens indeed over a range of pressure which results in a nonvertical step in the isotherm. The rounded step occurs far below p_{sat}. The position of the rounded steps in a series of isotherms such as that in Figure 7c define the dashed curve in Figure 8. It is obvious that the location of T_{pw}^{c} divides the curve in Figure 8 into two parts: for $T_w < T < T_{pw}^{c}$ the curve defines the prewetting thick-thin coexistence region, while for $T > T_{pw}^{c}$ the curve marks the maximum in the 2D compressibility of the prewetting supercritical one-phase region.

Quantitative analysis of reflectivity anomalies for isotherms above T_{pw}^{c} show that steps although progressively rounded, persists in isotherms far above the bulk critical temperature T_c (see inset in Figure 8).

In summary, the features observed in the pressure and temperature dependence of the reflectivity at the mercury-sapphire interface display all of the characteristics of a first-order wetting transition: a wetting temperature T_w, a prewetting line terminating in a prewetting critical point at T_{pw}^{c} as well as maxima in the compressibility in the prewetting supercritical single-phase region. A combination of the data presented in Figure 8 with the known p-ρ-T-data for fluid mercury allows one to construct the complete prewetting phase-diagram in the density ρ-temperature T plane as illustrated in Figure 1. The prewetting line, lies close to the bulk liquid-vapour coexistence curve. The prolongation of the prewetting phase line beyond the prewetting critical point CPW is the line of maximum two-dimensional compressibility in the prewetting supercritical single-phase region. This interpretation is supported by a recent Monte Carlo simulation investigation of the prewetting supercritical phase [29].

5. Interatomic Interaction And Structure

The problem that throughout the fluid range of metals the interparticle interaction changes significantly has been a primary motivation for measurements of the static structure factor S(Q) of mercury, caesium and rubidium over the whole liquid-vapour density range. Typical results of the Fourir transform of structure factors S(Q) of mercury measured with the energy-dispersive x-ray diffraction method [30], [31] are displayed in Figure 9.

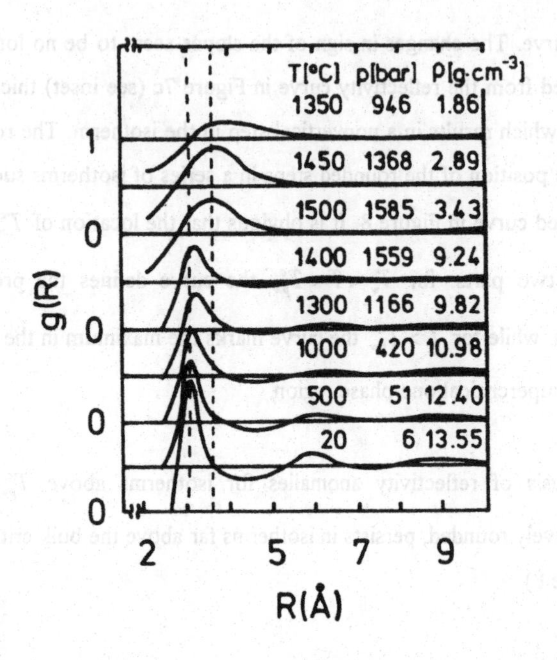

Figure 9 Pair corelation function of mercury at differnt temperatures and densities (30,31)

The Fourier transform of the structure factors S(Q) yield the radial distribution functions g(R). The data extend from a liquid density (13.55 gcm^{-3}) close to the melting point value to a vapour phase density of 1.86 gcm^{-3}.

The liquid structure just above the melting point (e.g. at 20°C and 13·55 g cm^{-3}) can be regarded as built up of single screened ions with interactions described by effective density-dependent, spherically symmetric pairwise potentials. This view is prompted by the experimental observation that relatively little change in the local atomic arrangement occurs during melting. For mercury, the molar volume increases by only about 3·6%, and the average near -neighbour distance, given by the position of the first peak of the radical distribution function g(R) in Fig. 9, is almost identical with the average near-neighbour distance R_{sol} in the crystalline structure close to melting. Thus the essential monoatomic character of the crystal seems to be preserved through the melting transition.

Monoatomic character dominates also in the coexisting vapour phase. Chemically bound mercury dimers do not form because the ground state of the mercury atom is a closed shell 6s² electronic configuration. Nevertheless, there is still a weak attraction between mercury atoms, similar to that between rare-gas atoms. Indeed, at $\rho = 1.89$ g/cm^3 and T = 1350°C (Fig. 9), the first peak in g(R) falls at the equilibrium distance $R_{Hg_2} = 3.63$ Å of the mercury dimer potential evaluated from spectroscopic data of the isolated Hg_2-species formed in a molecular beam [32]. This value is much greater than the next neighbour distance in the dense coexisting liquid, indicating that the effective interatomic interaction does indeed change with density.

It is immediately evident from a glance at Figures 9 and 10 that the structural implications of this density dependence of the effective interaction are particularly evident in the range where the gradual diminution of metallic properties in expanded fluid mercury occurs. With decreasing density or increasing temperature a number of changes in g(R) are noteworthy. First, the intensities of the main peak of g(R) are strongly reduced and broadened. The pair correlation function g(R) is related to the radial distribution function $n(R) = 4\pi R^2 n g(R)$ which determines the number of $n(R)dR$ of neighbouring atoms in a spherical shell of radius R and thickness dR centered on a particular atom of interest. The average coordination number N_1 is determined by the area under the first peak in n(R), whereas the average nearest neighbour-distance R_1 is given by the position of the first peak of g(R). An analysis of data such as those displayed in Figure 9 for mercury is shown in Figure 10 in form of N_1 and R_1 as a function of density.

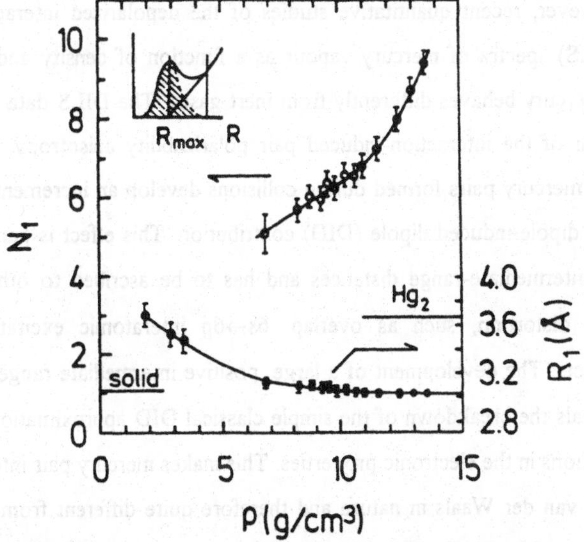

Figure 10 Average number, N_1, of nearest neighbours for expanded mercury together with the average next neighbour distance, R_1, of mercury as a function of density, ρ (30, 31).

The most significant aspect of the structural data in Figure 10 is that, within the range of NFE behaviour between 13,55 and 12 g/cm^3, thermal expansion proceeds by a structural evolution in which the average coordination number decreases roughly in proportion to the density while the average near-neighbour distance remains nearly constant. Such changes in shape and position of the pair correlation function are usually observed for fluids for which the nature of the effective pair interaction does not markedly change with liquid density. However, it is immediately evident from a glance at Figure 10 that this structural trend changes in a noteworthy way for densities smaller than 12 g cm^{-3} that is where the Hall coefficient R_H and the shape of $\sigma(\omega)$ indicate that the diminution of metallic properties sets in. The decrease in the average coordination number becomes distinctly smaller whereas the average near-neighbour distance gradually increases. This observation clearly demonstrates the interplay between changes in the electronic structure and the thermodynamic and structural properties of expanded liquid mercury.

Finally, in the vapour phase, R_1 approaches the equilibrium distance R_{Hg_2} of the weakly attractive potential of the mercury dimer. This species has been the subject of numerous experimental studies from which different pair potential models for its ground state have been derived. It was concluded from such investigations that it is a weakly bound van der Waals molecule. However, recent quantitative studies of the depolarized interaction-induced light scattering (DILS) spectra of mercury vapour as a function of density and temperature [33] indicate that mercury behaves differently from inert gases. The DILS data have been used to derive the form of the interaction-induced pair polarizability anisotropy. The result of this analysis is that mercury pairs formed during collisions develop an incremental polarizability in addition to the dipole-induced dipole (DID) contribution. This effect is particularly important at short- and intermediate-range distances and has to be ascribed to other mechanisms of electron cloud distortion, such as overlap, 6s→6p interatomic excitation, and electron correlation effects. The development of a large, positive intermediate-range pair polarizability anisotropy reveals the breakdown of the simple classical DID approximation and the onset of specific interactions in the electronic properties. This makes mercury pair interaction properties not completely van der Waals in nature and therefore quite different from noble gases. This view is consistent with an ab initio study of the individual interaction energy components in the ground state of the mercury dimer [34]. These calculations show that induction effects play an important role in the stabilization of Hg_2, unlike the situation commonly encountered for the inert gas dimers. There the assumption of unperturbed monomer electron densities is normally justified. Since the short-range induction effects lead to a significant energy lowering, the mercury dimer may be regarded as an intermediate case between a weakly bound van der Waals molecule and a chemically bonded species. Induction contributions to the bonding energy are a first indication that mercury clusters undergo a transition from weak to covalent and finally to metallic bonding [35].

As is well known, the structural properties of fluid alkali metals are fundamentally distinct from those of fluid mercury. Monoatomic character dominates in the dilute vapour phase of mercury, whereas in the vapour phase of alkali metals many alkali atoms form chemically bound dimers. The question to emerge therefore is up to what density and temperature the pairing mechanism present in the vapour can survive the condensation to the liquid state. This question is based on the view that the mechanism for the nonmetal-to metal transition which occurs in diatomic molecules with increasing density, may begin with metallization of the diatomic system by an overlap of the valence and conduction bands which at higher density is followed by a gradual dissociation into a monoatomic state.

Renewed interest in this question is motivated by the exciting progress in the search for metallic hydrogen which has come from the recent shock wave experiments of Weir et al [36]. They found a highly conducting state of hydrogen (and also deuterium) at high temperatures and and in a pressure range above 1.4 Mbar. Their measurements of the electrical conductivity at these conditions showed values of about $2000\Omega^{-1}cm^{-1}$ comparable with those for expanded liquid caesium and rubidium. Weir et al [36] interpreted their findings in terms of the existence of H_2 molecules within the highly conducting fluid phase. It has so far, however, not been possible to investigate the strucutural properties of fluid hydrogen at the extraordinarily high temperatures and pressures prevailing in the shock wave experiment in order to prove whether the pairing character persists in the highly conducting state. This has, therefore, stimulated a number of theoretical studies which suggest that significant concentrations of neutral dimers and dimer ions exist under these conditions. Given the close similarity in their atomic structure, an alternative experimental approach is to investigate the structural properties at high temperatures of the experimentally accessible fluid alkali metals, which may serve as models for the metallization of shock compressed hydrogen.

Interesting insight into the question whether remnants of the diatomic unit, present in the vapour phase of alkali metals, can survive the condensation to the liquid state can be obtained by examining recent coherent inelastic neutron scattering spectra for liquid rubidium [37], [38] which extend nearly to the critical region. From the early work of Copley and Rowe [39] it is known that rubidium near its melting point at 47°C, unlike dense Lennard-Jones (inert gas) systems, exhibits distinct longitudinal collective excitations up to relatively high Q-values. The measured dispersion relation resembles very much that of longitudinal phonons in crystalline rubidium. The more recent measurements of $S(Q,\omega)$ for expanded liquid rubidium [37], [38] extending nearly to the critical point show that these collective excitations can be observed over a very wide temperature range up to 1400°C.

With respect to changes in interatomic interactions associated with the metal-nonmetal transition, it is most significant that high-Q collective excitations are experimentally observed in liquid alkali metals, whereas they have not been detected in the insulating inert gas liquids. This distinction is reproduced well by the respective molecular dynamics calculations for these liquids. The potentials used for simulations of the inert gas liquids are the relatively short-range Lennard-Jones potentials whereas those used for rubidium are long-range oscillatory potentials typical for metals. This suggests strongly that it is the longer range of the screened coulomb potential that causes several ions to move coherently, even if for only a short time. As a result strongly damped, but observable collective excitations are present in the metals while they are absent in Lennard-Jones liquids. [38].

The results of molecular dynamics calculations are in quantitative agreement with experiment for rubidium up to a temperature of 1400°C where the density is about three times the critical density. Liquid rubidium is still basically monoatomic and the screened forces typical of metallic binding still control the liquid dynamics under these conditions. This is no longer true when the temperature is increased still further and evidence of intramolecular vibrations appears. This is immediately evident from a glance at Figure 11 which displays the longitudinal current correlation function $J_l(Q,\omega) = (\omega^2/Q^2). S(Q,\omega)$ for $Q = 1.3$ Å^{-1} at 1600°C. The liquid density at this point is 0.61 g/cm^3 about twice the critical density. Three peaks can be clearly identified as resulting from excitations at 3.2 meV with higher harmonics at 6.4 and 9.6 meV. This feature was assigned to excitation of local harmonic oscillations of pairs of rubidium atoms. These are transient dimers executing harmonic oscillations in the environment of their neighbouring atoms in the liquid.

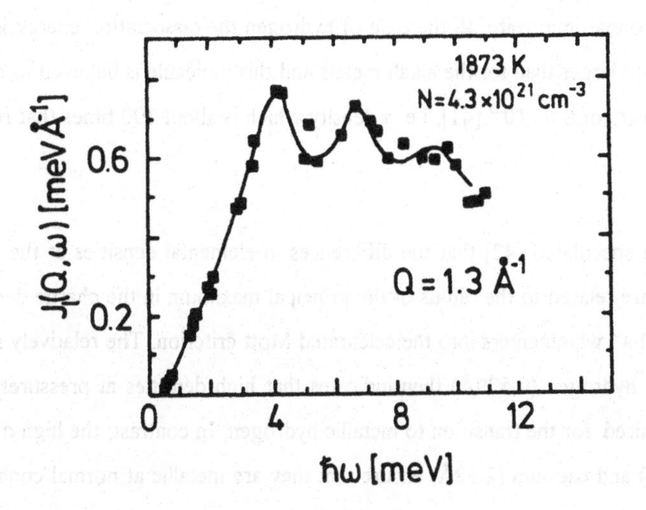

Figure 11 Experimentally determined current correlation function for liquid rubidium $J_1(Q,\omega) = (\omega^2/Q^2)S(Q,\omega)$ at $Q = 1.3$Å^{-1} at T = 1600 °C, and $\rho = 0.61$ g/cm^3. The solid line is just a guide for eye.

The measured excitation energy of 3.2 meV is surprisingly close to that calculated for Rb_2 molecules on a simple cubic lattice [38]. These are calculations for the total energy of expanded lattices of monoatomic Rb and Rb_2 dimers using density functional theory in the local density approximation. They were made for a system of Rb atoms in a body-centred lattice (bcc) and for diatomic molecules in a simple cubic lattice (sc). To create the diatomic solid the two atoms in the bcc unit cell were moved towards one another forming a simple cubic lattice (sc). The dimer bond length in the molecular lattice was varied to minimize the total energy. Below a density of 0.9 g/cm^3, which is about three times the critical density ρ_c, the sc-Rb_2 lattice has the lowest energy.

The vibron energy and dissociation energy were obtained by displacing the bond length from its equilibrium value R_0 and calculating the change in energy $E(R-R_0)$. These data were then fitted to a Morse-Potential. The calculations predict that the vibron and dissociation energy decrease with increasing density from their gas phase value to zero near 0.9 g/cm^3, the corresponding number density is $6.3 \cdot 10^{21}$cm^{-3}. Such a decrease in the vibron energy with decreasing density is a well established feature found in optical studies on solid molecular hydrogen [40]. This decrease is believed to be associated with the transfer of electron charge towards neighbouring molecules leading to a softening of the molecular bond and the formation of a monoatomic state. In the case of hydrogen the dissociation energy is nearly an order of magnitude larger than for the alkali metals and this molecule is believed to metallize in the solid at a density of $6.6 \cdot 10^{23}$ [41], i.e. a density which is about 100 times that required for rubidium.

It has been often speculated [42] that the differences in elemental densities at the metal-non-metal transition are related to the radius of the principal maximum in the charge density of the ns valence orbital a* which enters into the celebrated Mott criterion. The relatively small value of a^* for atomic hydrogen (0.529Å) then indicates that high densities at pressures of several megabar are required for the transition to metallic hydrogen. In contrast, the high a^*-values of rubidium (2.29Å) and caesium (2.52Å) ensure that they are metallic at normal conditions, and that the dimerization occurs in the expanded state at a low pressure which is stabilized in the fluid at high temperatures. Obviously, at high temperatures there will be a considerable degree of dissociation and the fluid is a partially dissociated mixture of monomers and dimers [19], [38]. Interestingly the recent experimental results for dynamically compressed fluid hydrogen [36] show that high temperature provides a path for metallization of fluid hydrogen at the relatively low atomic density of $3.7 \cdot 10^{23}\mathrm{cm}^{-3}$.

The essence of this experiment is shock compression of a thin layer (~0.5 mm) of liquid hydrogen confined between two single-crystal sapphire anvils. Shock pressures up to 1.8 Mbar are generated by the high-velocity impact (7 km s^{-1}) of a metal plate on the surface of a ductile aluminium sample cell. The reverberating shock in the liquid generates a temperature between 2500 and 4000 K, sufficiently low for the data to be directly compared with the conditions under which fluid rubidium, caesium, and potassium have been studied experimentally.

Weir et al. [36] measured the electrical conductivity of their compressed fluid hydrogen sample *in situ* using electrodes inserted through the walls of the cell, flush with the hydrogen-sapphire interface. The conductivity rises continuously over five orders-of-magnitude with pressure increasing from 1.4 Mbar to 1.8 Mbar. Figure 12 shows the conductivity versus molar atom density, m, curve of hydrogen with those obtained for the alkali metals, caesium and rubidium at supercritical temperatures.

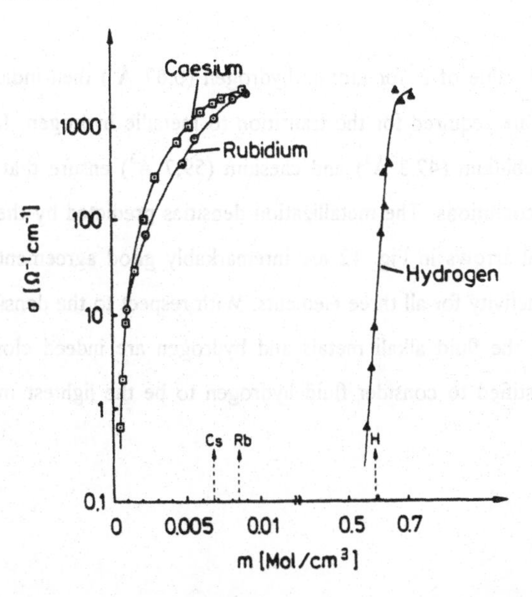

Figure 12 The electrical conductivity of fluid caesium, rubidium and hydrogen as a function of the molar atom density *m* at a temperature of kT≈0.15 eV. The arrows indicate the predict metallization densities for each element based on the Herzfeld model (see text).

It is clear that all three elements undergo a density-induced continuous transition from a low conductivity to a highly conducting state with a limiting value of the conductivity, σ~2000 $ohm^{-1}cm^{-1}$. However, the atom density at which this occurs for fluid hydrogen is about 100 times that required for rubidium and caesium. It is thus self evident that the intrinsic differences in the metal-nonmetal transition densities for these elements are related to characteristic properties, as noted over 70 years ago by Herzfeld [42]. One such property, for example is the above mentioned radius of the principle maximum in the charge density of the ns valence orbital, a^* which enters into the Mott criterion. A related property is the static polarizability α of the isolated atom. The polarizability formed the basis of the discussion of the metal-nonmetal transition by Herzfeld [42]. He pointed out that electrons localized around atomic nuclei constitute polarizable objects and their internal dynamics in dense assemblies leads to local corrections to the polarizing tendency of any external field impressed on the system. For an isotropic material, the correction factor has the form $[1-(4\pi/3N\alpha]^{-1}$ where N is the number of atoms per unit volume. If α is taken to remain roughly constant as N is increased, then the dielectric response will eventually diverge at a density $N = 3/4\pi\alpha$.

The relatively small value of α for atomic hydrogen (0.67 Å^3) then indicates that very high elemental densities are required for the transition to metallic hydrogen. In contrast, the high polarizabilities of rubidium (47.3 Å^3) and caesium (59.7 Å^3) ensure that these elements are metallic at normal conditions. The metallization densities predicted by the Herzfeld criterion, indicated by vertical arrows in Fig. 12 are inremarkably good agreement with the observed onset of high conductivity for all three elements. With respect to the density induced tansition to a metallic state, the fluid alkali metals and hydrogen are indeed closely similar. In this respect it seems justified to consider fluid hydrogen to be the lightest member of the alkali metal group [43].

References:

[1] Mott, N.F. (1974), Metal-Insulator Transitions (London: Taylor & Francis), and reterences therein

[2] Barker, J.A. (1976) Rare Gas Solids, Vol. I, edited by H.L. Klein and J.A. Venables (New York: Academic Press), p. 587

[3] Shimoji, M (1977) Liquid Metals (London: Academic)

[4] Laudau, L., Zeldovitch, G. (1943) ActaPhys.-chim. URSS, 18, 194 „On the relation between the liquid and the gaseous state of metals" and „translation of the Collected Papers of Landau, L.D., ed. D. ter Haar, Oxford, (1965), 380. „On the rélation between the liquid and the gaseous state of metals."

[5] Hensel, F. (1988) The Liquid-vapour phase transition in fluid mercury, Advances in Physics 44, 3-19

[6] Hensel, F., Stolz, M., Hohl, G. Winter, R., Götzlaff, W., (1991) Critical phenomena and the metal-nonmetal transition in liquid metals,, J.Phys.Colloq., C5-supplement 1, 191-205

[7] Freyland, W., (1981), Metal-Nonmetal transition in expanded fluid alkali metals, Comments Solid State Physics, 10, 1-10

[8] Oelhafen, P. Indlekofer, G., Günterodt, H.-J., (1988), Valence Electron Structure of Heavy Polyvalent Liquid Metals from Mercury to Bismuth, Z.Phys.Chem. N.F., 157, 483-488

[9] Hefner, W., Schmutzler, R.W. and Hensel, F., (1980) Optical Reflectivity Measurement of Fluid Mercury, J.Phys., Paris 41, C8-62-65

[10] Even, U. and Jortner, J., (1972) Evidence for the Foundation of a Pseudogap in a Divalent Metal, Phys.Rev.Lett. 28, 31-34

[11] Warren, W.W. and Hensel, F. (1982) Knight shift and dielectric anomaly in fluid mercury, Phys.Rev.B. 26, 5980-5982

[12] Kresse, G. and Hafner, J. (1997) Ab initio simulation of the metal/nonmetal transition in expanded fluid mercury, Phys.Rev.B. 55, 7539 - 7548

[13] Knuth, B., Hensel, F. and Warren, Jr., W.W. (1997) Optical Reflectivity And Electron Mass Enhancement in Expanded Liquid Cesium, J.Phys:Condens. Matter 9, 2693-2698

[14] Freyland, W. (1979) Magnetic susceptibility of metallic and nonmetallic expanded fluid cesium, Phys.Rev.B., 20, 5104-5110

[15] Warren, Jr., W.W., Brennert, G.F., El-Hannany, U., (1989) NMR investigation of the electronic structure of expanded liquid cesium, Phys.Rev.B., 39, 4038-4050

[16] Mott, N.F. (1978) Continuous and discontinuous metal-insulator transition, Phil.Mag. B37, 377-386

[17] Hensel, F. and Hohl, G.F. (1994), Expanded Fluid Alkali Metals, The Review of High Pressure Science and Technology, Vol. 3, No. 2 165-179

[18] Winter, R., Bodensteiner, T., Gläser, W., Hensel, F. (1987) The Static Structure Factor of Cesium over the whole Liquid Range up to the Critical Point, Ber. Bunseges. Phys.Chem., 91, 1327-1330

[19] Redmer, R. and Warren Jr. W.W. (1993) Magnetic susceptibility of Cs and Rb from the vapour to the liquid phase, Phys.Rev.B. 48, 14892 - 14906

[20] Redmer, R., Reinholz, H., Röpke, G., Winter, R., Noll, F., Hensel, F.(1992) The electrical conductivity of expanded liquid caesium, J.Phys.: Condens.Matter, 4, 1659 - 1669

[21] Jüngst, S., Knuth, B., Hensel, F.(1985) Observation of singular diameters in the coexistence curve of Metals, Phys.Rev.Lett. 55, 2160-2163

[22] Yao, M. and Hensel, F. (1996) Wetting of mercury on sapphire, J.Phys.Conden: Matter 8, 9547-9551

[23] Kozhevnikov, V.F., Arnold, D.I., Naurzakov, S.P. and Fisher, M.E. (1997) Prewetting Transitions in a Near Critical Metallic Vapor, Phys.Rev.Lett. 78, 1735 - 1738

[24] Cahn, J.W. (1977) Critical Point Wetting J.Chem.Phys. 66, 3667-3672

[25] Dietrich, S. (1988) Wetting Phenomena, in Phase Transitions and Critical Phenomena, Vol. 12,1 ed. By C. Domb and J.L. Lebowitz, Academic, London

[26] Rutledge, J.E. and Taborek, P. (1992) Prewetting Phase Diagram of ^{4}He on Cesium, Phys.Rev.Lett. 69, 937-940

[27] Cheng, E., Mistura, G., Lee, H.C., Chan, M.H.W., Cole, M.W., Carraro, C., Saam, W.F. and Toigo, F. (1993) Wetting Transitions of Liquid Hydrogen Films, Phys.Rev.Lett. 70, 1854 - 1857

[28] Berning, H.P. (1963) Physical Properties of Thin Films vol 1, ed. G. Hass New York: Academic

[29] Omata, K. and Yonezowa, F. (1998) Monte Carlo study of the prewetting supercritical phase, Vol. 10, 11599 - 11602

[30] Tamura, K., Inui, M. Nakaso, I., Oh´ishi, K. Funakoshi, K. and Utsumi, W. (1998) X-ray diffraction studies of expanded fluid mercury using synchrotron radiation, J.Phys.: Condens. Matter 10, 11405-11417

[31] Tamura, K. and Hosokawa, S. (1994) J.Phys.: Condens. Matter 6, A241

[32] Koperski, J., Atkinson, J.B. and Krause, L. (1997) The $GO_u^+(6^1P_1) - XO_g^+$ Excitation and Fluorescence Spectra of Hg_2 Excited in a Supersonic Jet, J. of Mol. Spectr. 184, 300-308

[33] Barocchi, F. Hensel, F. and Sampoli, M. (1995) Induced Pair Polarizability Ansotropy in Mercury from Depolarized Interaction Induced Light Scattering Spectra, Chem.Phys.Lett. 232, 445-450

[34] Kunz, C.F., Hättig, C. and Hess, B.A. (1996) Ab initio study of the individual interaction energy components in the ground state of the mercury dimer, Mol.Phys. 89, 139-157

[35] Rademann, K., Kaiser B, Even, U.and Hensel, F. (1987) Size Dependence of the Gradual Transition to Metallic Properties In Isolated Mercury Clusters, Phys.Rev.Lett. 59, 2319-2321

[36] Weir, S.T., Mitchell, A.C., and Nellis, W.J. (1996) Metallization of Fluid Hydrogen at 140Gpa (1.4 Mbar), Phys.Rev.Lett. 76, 1860 - 1863

[37] Pilgrim, W.-C., Winter, R., Hensel, F., Morkel, C. and Gläser, W. (1991) The Dynamic Structure Factor of Expanded Rubidium, Ber. Bunsenges.Phys.Chem., 95, 1133-1136

[38] Pilgrim, W.-C., Ross, M., Yang, L.H. and Hensel, F., (1997) The Monoatomic-Molecular Transition in Expanded Rubidium, Phys.Rev.Lett. 78, 3685-3688

[39] Copley, J.R.D. and Rowe, J.M., (1974), Short-Wavelength Collective Excitations in Liquid Rubidium Observed by Coherent Neutron Scattering, Phys.Rev.Lett. 32, 49 - 52

[40] Mao, H.K. and Hemly, R.J. (1994) Ultrahigh-pressure transitions in solid hydrogen, Rev.Mod.Phys.. 66, 671-691

[41] Ashcroft, N.W. (1994) The Dense Hydrogen Plasma: Translational Orientational And Electronic Structure, in: „Elementary Processes in Dense Plasma, ed. By S. Ichimaru and S. Ogata, Addison-Wesley, Reading, 251-270

[42] Herzfeld, K.F. (1927) On Atomic Properties Which Make an Element a Metal, Phys.Rev. 29, 701-705

[43] Hensel, F. and Edwards, P.P. (1996) Hydrogen, The First Alkali Metal, Chem.Eur.J. 2, 1201-1203

NEUTRON DIFFRACTION STUDIES OF LIQUID ALLOYS UP TO HIGH TEMPERATURES AND PRESSURES

R. Winter and K. Hochgesand
University of Dortmund, Department of Chemistry, Physical Chemistry I, Otto-Hahn-Strasse 6, D-44227 Dortmund, Germany

1. Introduction

Liquid alloys have been the subject of considerable interest over many years and have attracted the attention of scientists from diverse fields ranging from metallurgy, inorganic chemistry to condensed matter and theoretical physics [1-3]. Much of this work has been motivated by the strong interest in the electronic, thermodynamic and structural properties of systems which are at the borderline between liquid metals, liquid semiconductors, and ionic conductors. In this respect, the chemical bonding and chemical short-range order of liquid alkali-polyvalent alloys are topics of continuing interest. In extreme cases, a metal-insulator or metal-semiconductor transition occurs during alloying, e.g., in Cs-Au [2] and K-Pb [4] at the stoichiometric 1:1 composition. The crystalline binary alloys obtained by combining an alkali metal (e.g. Na, Cs) with polyvalent metals like Sn, Pb or Sb form classical examples of ionic alloy systems, exhibiting anion clustering. In general, if the composition is such that the number of electrons transferred is sufficient to complete the octet shell of the electronegative element, bonding and structure are salt-like, and if the electronic shell remains incomplete, a saturated chemcial bond can be formed only if valence electrons are shared among the electronegative atoms. This results in the formation of polyanionic clusters stabilized by strong covalent bonds and immersed in a matrix of the electropositive species. Zintl and Brauer proposed, that as a consequence of the large difference in electronegativity, an electron is transferred from the alkali atoms to Sn, Pb or Sb [5], and the polyanionic clusters are isostructural to the molecules or crystal structures formed by the isoelectronic neutral species. Examples are the tetrahedral Sn_4^{4-} or Pb_4^{4-} polyanions formed in equiatomic alkali-tin or alkali-lead alloys or the chainlike Sb_∞^- clusters in alkali-antimony alloys, which are isoelectronic and isomorphic to P_4 molecules or to the helical chains formed by the crystalline chalcogen elements, respectively. The role of the alkali ions is merely to separate the anions.
We will first focus on the alkali-tetravalent metal compounds. Tetrahedra have been found in the equiatomic alloys of Na, K, Rb and Cs with Pb and Sn (Fig. 1) [6-9]. The same kind of tetrahedra are also found in crystalline intermetallic compounds made of alkali metals and Si or Ge.

R. Winter and J. Jonas (eds.), High Pressure Molecular Science, 151–185.

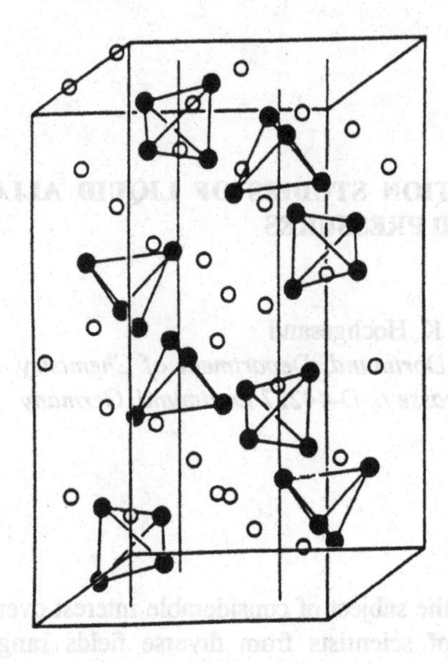

Figure 1. **Crystallographic unit cell of Na-Sn, which is isomorphous to K-Pb and Cs-Pb (adopted from Ref. [4]).**

Price, Saboungi et al. [10-15] recently found that the two Zintl compounds Cs-Pb and Na-Sn exhibit a two-stage melting process, with high-temperature solid phases characterized by rapid dynamic disorder. In Cs-Pb, this disorder is clearly associated with rapid reorientations of polyanions with the cations participating in the dynamic disorder on the same time-scale. In Na-Sn the disorder is associated with fast reorientations of the polyanions closely coupled to a slower migration of the cations. These discoveries stimulated a new research direction within the field of Zintl compounds.

Considerable evidence exists for the survival of these polyanions upon melting, and a great deal of effort has been devoted to the experimental and theoretical investigation of these liquid Zintl-alloys [16-43]. The pioneering studies by van der Lugt and his coworkers have shown that semimetallic behaviour occurs in liquid alkali - group IV metals [4]. The electrical dc-resistivity, ρ, has been used as an important physical criterion for the purpose of identifying liquid intermetallic compounds [4,18-20]. Plotted as a function of composition, its character is metallic for the pure metals, but tends to semiconducting for compositions involving compound formation due to the opening of a band gap or the formation of a pseudogap. The liquid intermetallic compounds often fall in the regime of strong scattering of the electrons with resistivities in the range between 300 and 3000 $\mu\Omega$cm (Fig. 2). In general, the temperature derivative of the resistivity, $d\rho/dT$, is negative in the region of compound formation and has a sharp minimum at the same composition as ρ peaks. Consider, for example, the system Li-Pb. Its resistivity exhibits a peak at about 20 at% Pb, and $d\rho/dT$ has a minimum at the same composition (Fig. 2). An octet compound, Li_4Pb, is formed which is ionically ordered with partial charge transfer and screening occuring. In Fig. 2 the resistivity of liquid K-Pb, Rb-Pb and Cs-Pb is also plotted as a function of composition. The position of the resistivity peak has shifted to 50 at% Pb. Such a shift in stoichiometry is considered to be an indication of a marked change in the chemical bonding, in this case for a transition from a simple octet compound to a clustered compound. The comparison with the phase diagrams, which exhibit strong congruently melting equiatomic compounds, makes this a plausible proposition. With increasing size of the alkali ion the distance between the tetrahedral anion units becomes larger. Therefore the bands of the group IV electronic states become narrower and the gap between the occupied bonding and unoccupied antibonding states becomes broader. This is in agreement with the fact that the resistivity strongly increases from K-Pb to Cs-Pb. The thermodynamic properties (e.g., the Darken stability function, mixing enthalpy and heat capacity) show drastic deviations from ideal behavior as well [21-26].

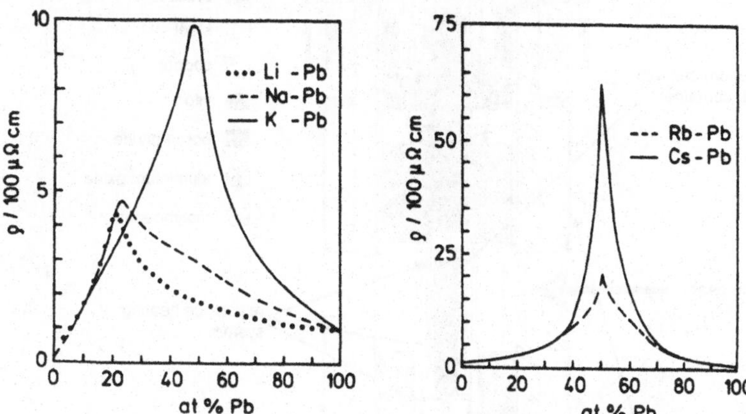

Figure 2. Electrical dc-resistivity ρ of a) liquid Li-Pb, Na-Pb, K-Pb and b) liquid Rb-Pb, CsPb as a function of composition in at% Pb (adopted from Ref. [4]).

Neutron diffraction studies of the liquid alloys at temperatures near the melting point validated this interpretation. The structure factors show a distinct peak at a momentum transfer of about 1 $Å^{-1}$, which is usually understood as indicative of intermediate-range order [28-34], due to correlations between the polyvalent metal clusters.

We investigated the structure of liquid binary alloys, such as K-Pb, Na-Sn, Cs-Pb, K-Sb and Cs-Tl, as a function of temperature, pressure and density, in order to follow changes in the structure and arrangement of the polyanions as a function of the thermodynamic variables p and T. In order to gain more detailed information on the microscopic structure of the expanded liquid alloys, we have also performed computer modelling studies based on a random packing of structural units model (RPSU) and the reverse Monte Carlo technique (RMC). The results are also compared with recent ab-initio molecular-dynamics (MD) calculations.

2. Experimental Technique and Modelling of the Data

The diffraction experiments were carried out on the LAD instrument at the ISIS pulsed spallation source (RAL, U.K.) and the SLAD instrument at the Studsvik Neutron Research Laboratory (Studsvik, Sweden). In Appendix A, a brief introduction in the neutron scattering theory of liquids is given. In order to reach the desired temperatures and pressures needed for these experiments, we used a home-built high-temperature high-pressure autoclave, which is shown in Fig. 3 and described in detail elsewhere [44].

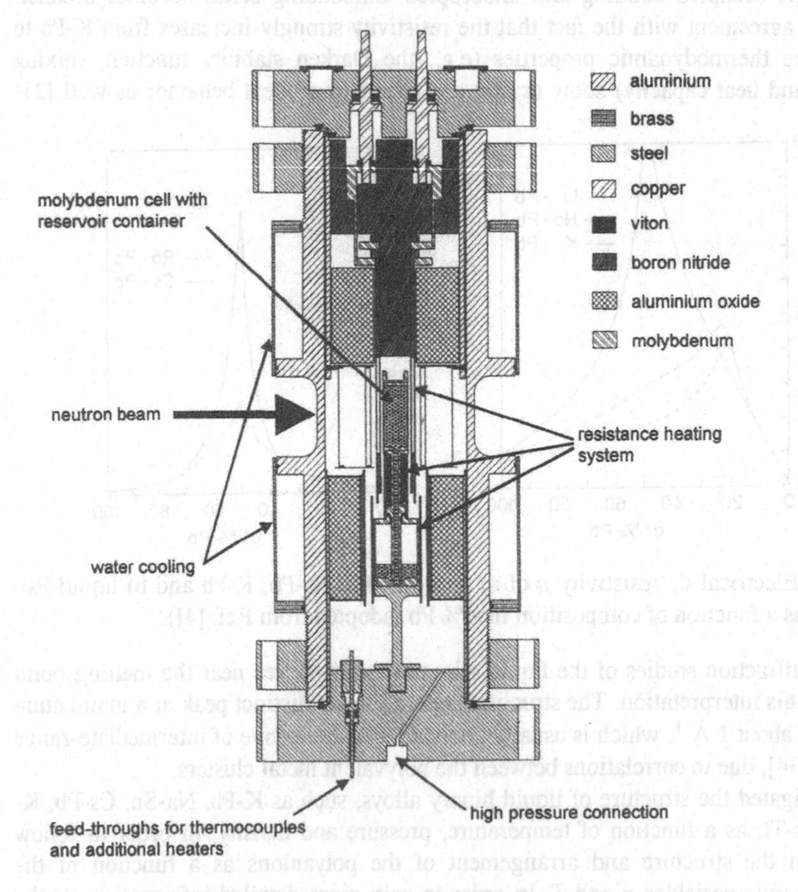

Figure 3. High-temperature high-pressure neutron scattering autoclave.

Briefly, the alloys were prepared under inert gas atmosphere and contained in a thin-walled cylindrical molybdenum cell (inner diameter: 15.6 mm, wall thickness: 0.3 mm) which was sealed with a lead (or tin) O-ring against a reservoir container. A hole in the reservoir container allows pressure equilibration between the inside and outside of the sample cell. Molybdenum was used as container because this material can withstand the high temperatures and the corrosive nature of the samples. The sample cell is mounted inside of an internally heated AlMgSi1 pressure vessel. The main heater consists of two cylinders, which are made from 0.05 mm molybdenum foils and surround the sample cell concentrically. After closing the pressure vessel, the interior was first evacuated before adding argon as pressurizing medium. The sample was then heated to the desired temperature. In order to correct for the scattering from the containment materials, three different measurements were performed: one of the sample inside the cell and pressure vessel, one of the empty cell inside the pressure vessel, and one of the empty vessel. The special experimental equipment used allowed us to collect the first two data sets in one set-up by filling the cell using pressure after the empty cell measurement has been performed under vacuum. The data were normalised and corrected for absorption, multiple scattering and inelasticity. To yield the static structure factor, $S(Q)$, the corrected sample scattering was compared with that of a vanadium cell of the same dimensions, filled with V-flakes. The density of flakes was adjusted so that its scattering power was comparable to that of the sample. For $Q > 2.7$ Å^{-1}, $S(Q)$ data were interpolated in the Q-regions where Bragg reflections due to the container materials occur, as they cannot be corrected at high temperatures with sufficient accuracy. The values for the densities were taken from the literature [4,16,17] or determined by measuring the neutron transmission of the samples.

In the random packing of structural units (RPSU) model, the structural units are distributed according to a random packing of hard spheres. The total structure factor is expressed as $S(Q) = f_1(Q) + f_2(Q)[(S_c(Q) - 1)]$, where $f_1(Q)$ is the form factor of the structural units, $f_2(Q)$ is a form factor taking into account the orientational correlation between the units, and $S_c(Q)$ is the Percus-Yevick solution for the random packing of hard spheres [45-48] (see also Appendix A). In our studies, the orientations of the structural units were taken to be random and uncorrelated. In analogy to the solid structure, the structural unit of e.g. K-Pb was defined to be K_4Pb_4, which consists of a Pb_4-tetrahedron surrounded by a larger K_4-tetrahedron. The distances within the unit were chosen to be slightly larger than in the solid state. Only two parameters are involved in the calculation of the total structure factor: the hard-sphere diameter and the packing fraction $\eta = \pi \sigma_{HS}{}^3 \rho_0 / 6m$, where m is the number of atoms in the structural unit. A Debye-Waller factor, $\exp(-Q^2 \langle u^2 \rangle / 2)$, where $\langle u^2 \rangle$ is the mean-square displacement in the bond distances within the structural unit, is applied to yield the correct damping of oscillations in $S(Q)$ at high Q values.

The reverse Monte Carlo (RMC) method has been descibed in detail elsewhere [49,50]. Briefly, RMC is a variation of the standard Metropolis Monte Carlo method in which a sequence of configurations is systematically compared with experimental structure factor data until convergence is reached. A closest distance of approach between pairs of particles was applied as a constraint in our RMC calculations. As starting configurations randomly distributed ensembles of 5200 atoms were used. A second calculation for the Zintl-alloys included the additional constraint, that the coordination found in the corresponding solid-state structure is preserved. We stress that the RMC solution is not unique, but the method has the advantage that the set of simulated functions is self consistent, it corresponds to a possible distribution of particles, and RMC calculations allow for testing suggested structural models.

3. Results and Discussion

As a first example, Fig. 4 depicts the measured structure factors $S(Q)$ of liquid CsPb up to 1900 K. These are Faber-Ziman averaged structure factors, where the partial structure factors $S_{ij}(Q)$ of the atoms i and j are weighted according to their atomic fraction c_i and mean neutron coherent scattering length (see Appendix A). The main contributions to the total $S(Q)$ derive from $S_{PbPb}(Q)$ and $S_{CsPb}(Q)$ as the weighting factor of $S_{CsCs}(Q)$ is comparably small. The structure factor of liquid Cs-Pb shows a prepeak throughout the whole temperature range investigated. Such prepeaks signify some form of intermediate range ordering, which could be consistent with an intermolecular correlation between polyanion clusters. With increasing temperature, its maximum shifts to smaller Q-values and the peak broadens. The increase of small-angle scattering has a similar extent as that observed for the expanded liquid alloys K-Pb and Na-Sn [52,53]. Similar to the systems K-Pb and Na-Sn, the existence of Pb-clusters can be assumed throughout the whole temperature range covered. At 1900 K, the density has decreased 19 %.

The structure factors are Fourier-transformed to pair correlation functions $g(r)$. Figure 5 shows the radial distribution functions $n(r) = 4\pi r^2 \rho_0 g(r)$. They exhibit a broad peak with pronounced triplet-structure. The first maximum occurs at about 3.1 Å. After passing a minimum at 3.6 Å the main maximum lies at about 4.0 Å, and a shoulder is visible around 4.7 Å. The well defined peak structure of $n(r)$ at lower temperatures flattens with increasing temperature and develops to a steep rise for $r > 4$ Å at the higher temperatures. The first and second maximum can be found throughout the whole investigated temperature range at nearly unchanged positions. The comparison of the pair correlation function with particle distances obtained from X-ray data in the crystalline solid phase [54] suggests the maximum at 3.1 Å being due to Pb-Pb-correlations. The position of the second maximum corresponds to Cs-Pb and Cs-Cs distances of the solid. The shoulder in $n(r)$ at 4.7 Å might be due to Pb-Pb correlations in the second coordination sphere. The first Pb-Pb intercluster distance in the solid occurs at 4.691 Å.

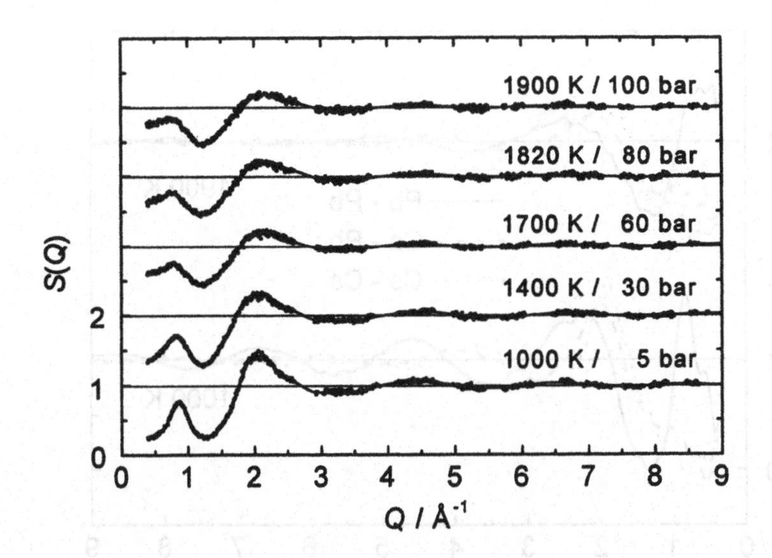

Figure 4. Static structure factors $S(Q)$ of liquid Cs-Pb.

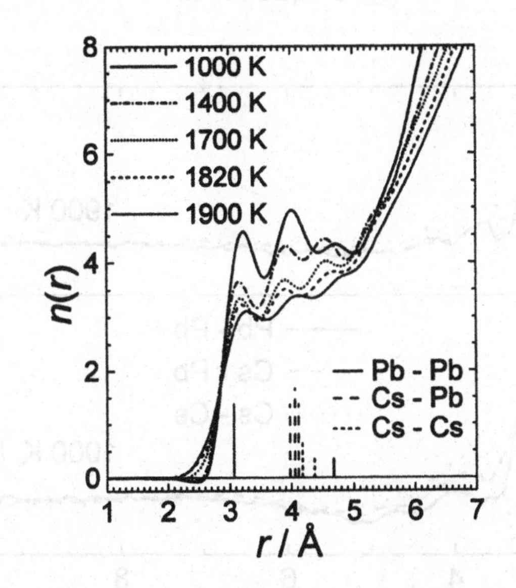

Figure 5. Radial distribution functions $n(r)$ of liquid Cs-Pb.

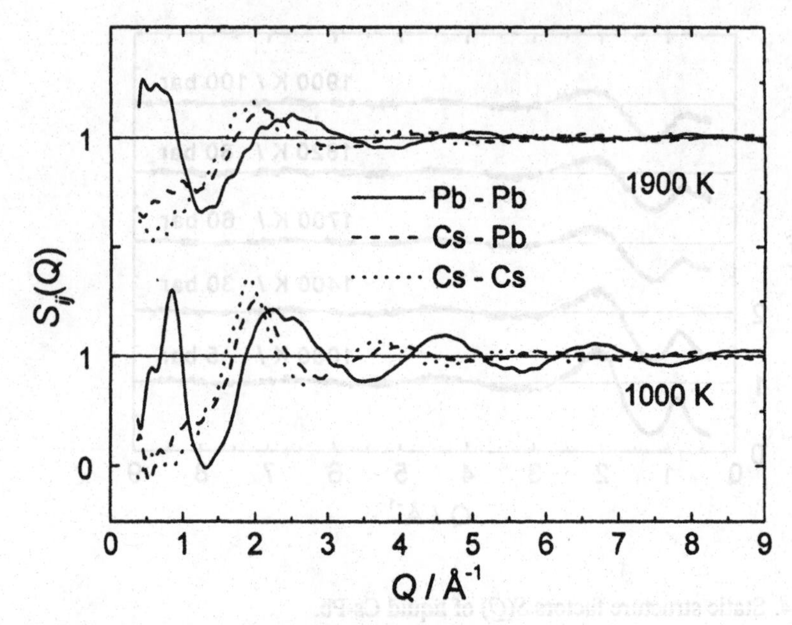

Figure 6. Partial structure factors $S_{ij}(Q)$ of liquid Cs-Pb.

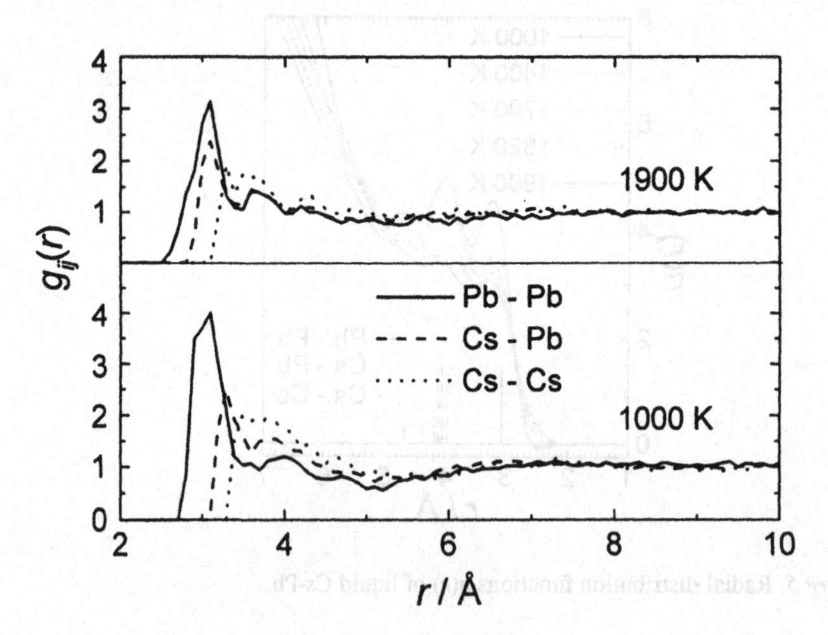

Figure 7. Partial pair correlation functions $g_{ij}(r)$ of liquid Cs-Pb.

According to eq. (35) (Appendix A), the average coordination number N_{PbPb} can be obtained by integration of $g(r)$. N_{PbPb} is found to decrease slightly in the temperature range from 1000 K up to 1900 K, with values ranging from 3.1 ± 0.2 to 2.6 ± 0.2. The values at the lower temperatures are similar to the coordination number of 3 of perfect Pb_4^{4-} tetrahedra in the crystalline solid. These results clearly show the tendency towards formation of tetrahedral structural features also in the liquid state even up to temperatures of 1900 K.
To obtain also information on the partial structural features in the alloy, RMC-simulations were carried out. From the presentation of the partial structure factors in Fig. 6 it is clearly seen that the prepeak of the total $S(Q)$ is mainly determined by Pb-Pb correlations. This striking feature of $S_{PbPb}(Q)$ is still recognizable at 1900 K. So we consider the intermediate range order in the melt to be due to pronounced Pb-Pb-interactions over the whole temperature and density range covered. Support for this conclusion arises from the plot of the partial pair correlation functions $g_{CsCs}(r)$, $g_{CsPb}(r)$ and $g_{PbPb}(r)$ (Fig. 7). $g_{PbPb}(r)$ at T = 1000 K exhibits a distinct maximum at 3.1 Å, which is sharper and much higher than the maxima of the two other partials. Even at T = 1900 K the position of g_{PbPb} remains, but its height has decreased and the difference to the other partials thus diminished. Hence, it can be inferred that the formation of Pb-clusters in the liquid alloy is favoured above a random contribution of Pb-atoms also at high temperatures in the expanded liquid state. To visualize the structural features in liquid Cs-Pb, Fig. 8 shows 40x40x40 Å^3 - sized sections of the equilibrium configuration at T = 1000 K and T = 1900 K, respectively. For easy survey only the Pb-atoms are shown. At T = 1000 K the figure shows a variety of clusters of different sizes and shapes, also tetrahedral ones. At the higher temperature the 0-, 1- and 2-fold coordinated Pb-atoms dominate, leading to a structure with smaller clusters, and also a large number of non-clustered Pb-atoms is found. To summarize, from the persistence of a distinct prepeak in the structure factors we conclude the existence of Pb clusters up to the highest temperatures measured. As the temperature increases and the density decreases in the expanded liquid state, the variety of local structural arrangements increases as the polyanions may be considered to break up.

As a second example, we focus on the system K-Sb. On melting M_3Sb or MSb, conductivities in the order of 100 $\Omega^{-1}cm^{-1}$ are observed. Because of the large difference in the electronegativities the alkali atoms transfer their valence electron to the more electronegative polyvalent element. At a composition where the charge transfer completes the octet shell of Sb, this completion leads to the formation of the ionic compound K_3Sb in the crystalline state. In K-Sb anions are found which are isoelectronic to the chalcogen atoms (e.g., Se), forming covalently bonded long helical chains in the solid state [55].
Measurements of the electrical conductivity of K-Sb and Cs-Sb show broad minima between the compound forming compositions M_3Sb and MSb. Compared to Cs-Sb, a striking concentration dependence of the alkali metal activity occurs in liquid K-Sb

near the K_3Sb composition, and even larger maxima in the excess stability function near the compositions KSb and K_3Sb are observed [56].

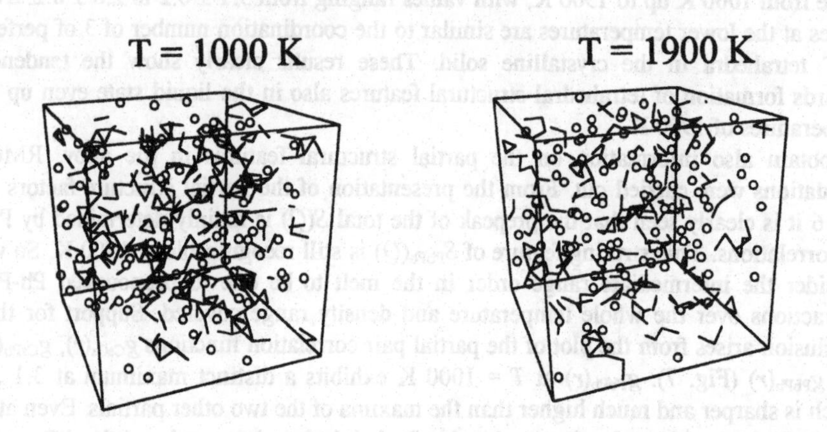

Figure 8. Configurations of Pb atoms obtained by RMC simulation of liquid Cs-Pb a) at $T = 1000$ K, and b) at $T = 1900$ K. Pb atoms within a distance of 3 Å are connected to visualize some kind of bonding (box size: 40 x 40 x 40 Å³).

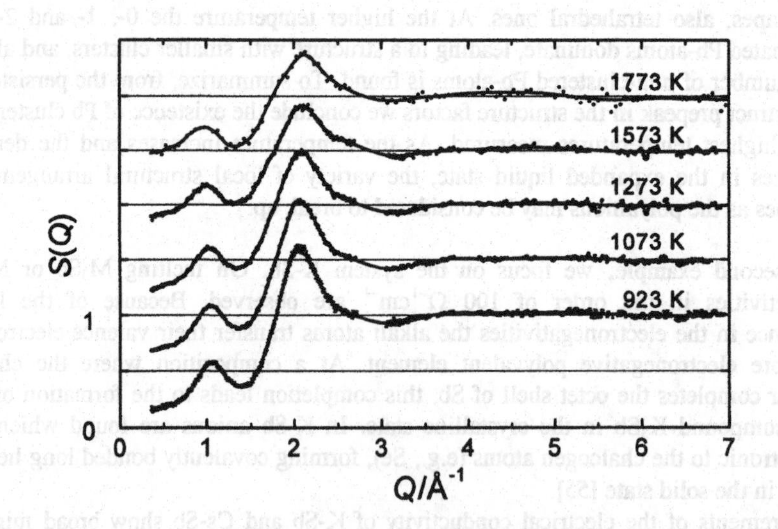

Figure 9. Temperature dependence of the experimental structure factor $S(Q)$ of liquid K-Sb.

Figure 9 shows the structure factors of liquid K-Sb for five temperatures. The line is the best fit through the experimental data points. We find a prepeak again at $Q \approx 1$ Å^{-1} indicative of intermediate-range order, and a main peak at $Q \approx 2.1$ Å^{-1}. The features of the structure factor remain, with nearly unchanged positions in Q at increased temperatures.
We performed RMC simulations to obtain the partial $S(Q)$'s and $g(r)$'s. From the three $S_{ij}(Q)$ (not displayed) only $S_{SbSb}(Q)$ has a peak at about 1 Å^{-1}, but all three partials contribute to the second peak of $S(Q)$. In the partial $g(r)$'s (Fig. 10) the overall shape of the total $g(r)$ with three main peaks can be recognized, but the first peak in $g_{SbSb}(r)$ is split into a peak at $r = 2.85$ Å, which is due to intramolecular Sb-Sb bonds like in Cs-Sb [57], and a broader peak around $r = 3.6$ Å, which is due to intermolecular distances. This splitting of the first peak becomes more pronounced when the simulation is performed with the constraint, that Sb atoms should prefer two neighbouring Sb atoms at a distance of not more than 3.0 Å, thus leading to distinct intra-chain (covalently bonded Sb atoms) and inter-chain distances. The agreement of the experimental $S(Q)$ with the calculated one from RMC is good up to a fraction $c_{SbSb} \approx 43$ % of Sb atoms satisfying this constraint; at larger fractions the agreement declines rapidly. So this configuration seems to represent the structure with the longest/most Sb chain fragments consistent with the experimental data. From c_{SbSb} and the number of Sb atoms with one neighbour only (24 %), we calculate an average Sb chain length of about 6 atoms for the lowest temperature. Note that we did not take into account the Sb atoms with no neighbours within the cutoff distance (33 %). The contribution of atoms with more than two neighbours is negligibly small. The partial $g(r)$'s and c_{SbSb} show no significant changes with increasing temperature.

Compared to the binary compounds in alkali metal - group 14 alloys, rather complex character of chemical bonding is observed in the alkali - group 13 alloys, such as K-Tl and Cs-Tl, which will be discussed as a third category of alloys now. While in the alkali metal - group 14 alloys the size of the cation determines whether polyanions occur, the cation size in alkali metal - Tl alloys delineates the shape of the polyanion clusters. The compounds of the smaller alkali metals Li and Na with Tl crystallize in the network-like diamond structure (B32), while in the equiatomic alloys of Tl with K, Rb and Cs distinct large polyanions are found. The recently resolved crystal structures of compounds in the systems K-Tl and Cs-Tl show a variety of compact polyanion clusters, such as the icosahedrally Tl_{11}^{11-} (in K_8Tl_{11} and Cs_8Tl_{11}) or the octahedrally Tl_6^{6-} (in KTl and CsTl) [58]. The equiatomic compound Cs-Tl is the Cs-richest phase in the binary system CsTl with a distinct orthorhombic crystal structure. The crystal contains Tl_6^{6-} octahedra, which are compressed with respect to the crystallographic b-axis. The bond lengths in the Tl_6^{6-} octahedron range from 3.026 to 3.434 Å. The non-bonding intra-cluster separations are 3.741 Å for the axial and 4.836 Å for the equatorial distances. The Tl-Tl-separation between the Tl_6^{6-} polyanions along the b-axis is 11.40 Å and 5.83 Å along the c axis, respectively. To create the charge balance, the Tl_6^{6-} polyanions are sheathed by 20 Cs cations.

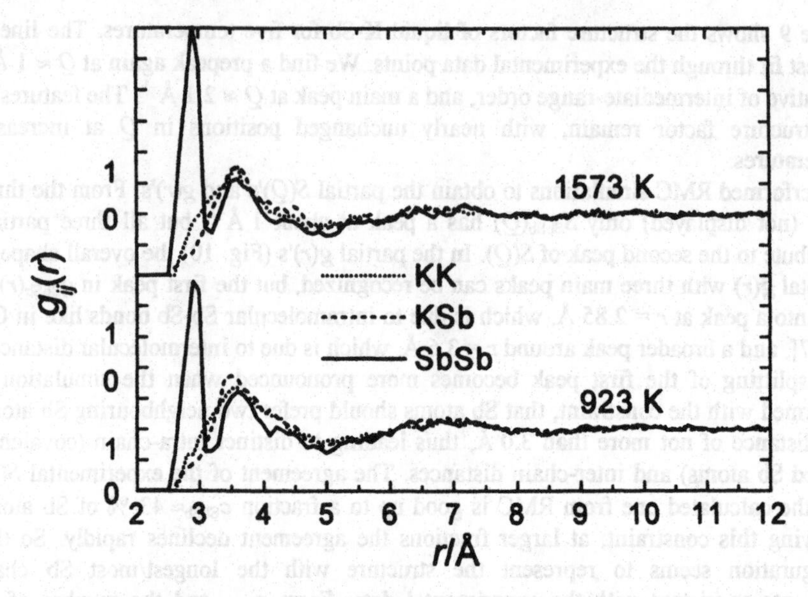

Figure 10. **Partial pair correlation functions of liquid K-Sb obtained by the RMC-simulations for T = 923 K and T = 1573 K.**

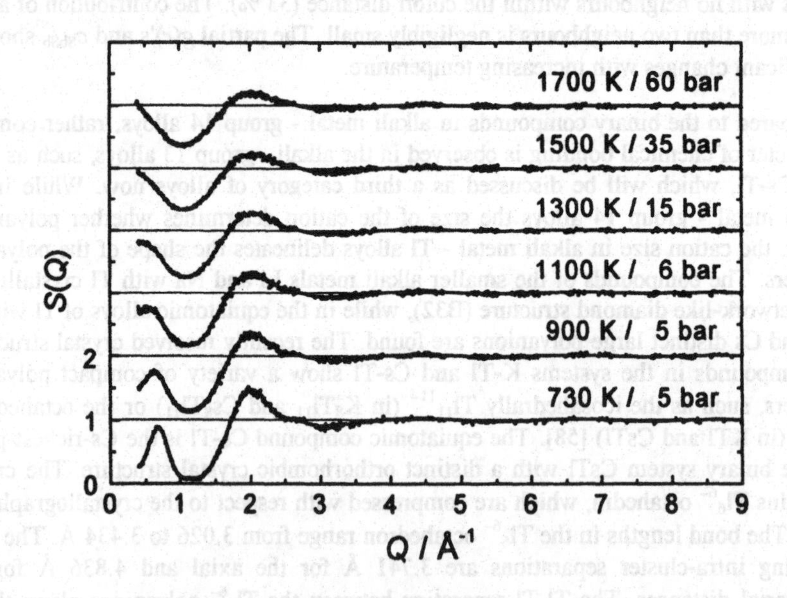

Figure 11. **The static structure factor $S(Q)$ of liquid Cs-Tl at different temperatures and pressures.**

Measurements of the density, optical and and magnetic properties of liquid Cs-Tl provide evidence for compound formation at the 50 at% composition also in the liquid state [59-61]. The electrical resistivity has a maximum (900 μΩcm) at this composition [62]. In contrast to the alkali metal - group 14 systems with resistivities up to 7000 μΩ cm (Cs-Pb) this value is still within the NFE-regime. First neutron diffraction measurements on liquid equiatomic Cs-Tl near the melting point yielded a structure factor $S(Q)$ exhibiting a prepeak at a rather small momentum transfer of Q = 0.7 $Å^{-1}$ [63,64], which corresponds to a real space distance of about 10-11 Å. This remarkably long-ranged intermediate-range structure, which is exceptional in these liquid alloys, is indicative of large Tl clusters in the melt. The measured temperature dependence of the static structure factor $S(Q)$ of liquid Cs-Tl up to 1700 K is shown in Fig. 11. Most conspicuous in the experimental data is the prepeak at an unusually small momentum transfer of Q_1 = 0.72 $Å^{-1}$ at 730 K. With increasing temperature the prepeak broadens and shifts to smaller Q-values (Q_1(1100 K) = 0.54 $Å^{-1}$). Between 1300 K and 1500 K the prepeak broadens and a drastic increase of small-angle scattering is observed, indicating a loss of coherence between Tl clusters and a tendency towards microphase separation. This marked increase of small-angle scattering has not been observed for other expanded Zintl-alloys, such as K-Pb.

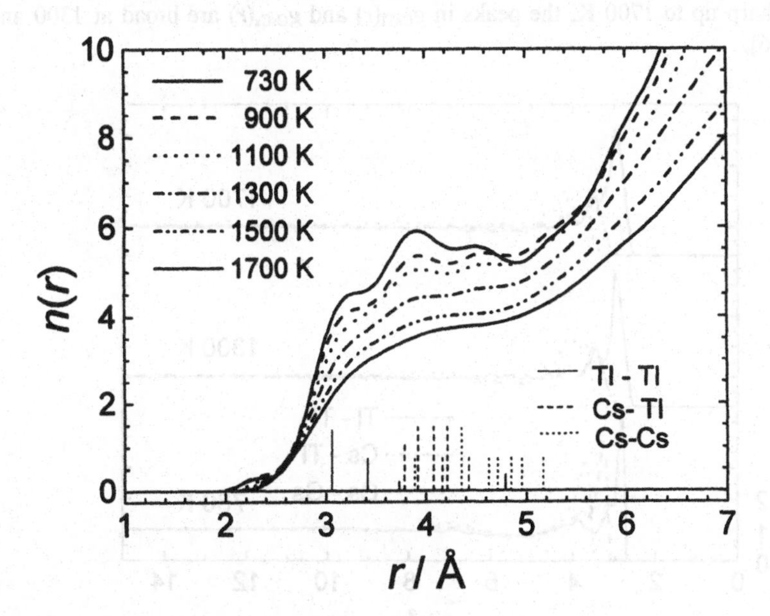

Figure 12. Radial distribution functions $n(r)$ of liquid Cs-Tl.

The radial distribution function $n(r)$ (Fig. 12) exhibits a broad peak at about 4 Å, which represents the first coordination sphere. The peak is composed of two shoulders at the lower temperatures, one centered at 3.2 Å, the other at about 4.6 Å. Comparing the interatomic separations obtained with those from the crystal structure data, the left shoulder of $n(r)$ can be assigned to Tl-Tl intracluster distances. The Cs-Tl distances contribute mainly to the peak maximum at about 4 Å, and the shoulder at the high r-side of the broad peak can be assigned essentially to Cs-Cs distances. These results emphasize that the first coordination shell is still rather well defined for the lower temperatures of the melt. With increasing temperature the maximum of the peak in $n(r)$ broadens and the two shoulders vanish. The persistence of the shoulder at 3.2 Å up to the highest temperature measured is indicative for the existence of Tl-clusters even in the expanded liquid state. Inspection of the partial structure factors reveals that the prepeak as well as the drastic increase of the small-angle scattering in the experimental structure factors at high temperatures is represented by the partial structure factor $S_{TlTl}(Q)$. This finding has been confirmed in an isotope substitution experiment [65]. Also the three partial pair correlation functions (Fig. 13) support the existence of strong Tl-Tl correlations in the liquid alloy over the whole temperature range covered, pointing to the fact that some Tl-clusters remain stable up to 1700 K. At $T = 730$ K the first maximum in $g_{CsCs}(r)$ at about 4 Å is as well pronounced as the first maximum in $g_{TlTl}(r)$ at about 3 Å, indicating a rather well defined first coordination sphere at this temperature. Whereas the first maximum in $g_{TlTl}(r)$ at 3 Å remains sharp up to 1700 K, the peaks in $g_{CsTl}(r)$ and $g_{CsCs}(r)$ are broad at 1300 and 1700 K [66].

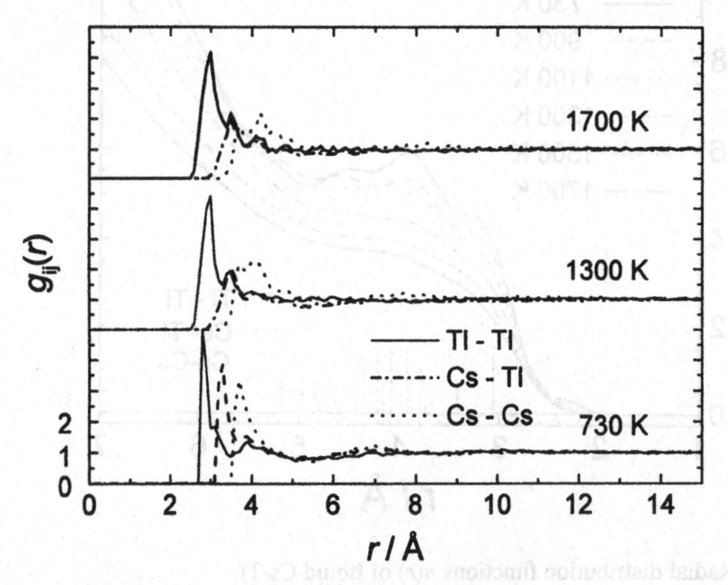

Figure 13. The partial pair correlation functions $g_{CsCs}(r)$ (- - -), $g_{CsTl}(r)$ (– – –) and $g_{TlTl}(r)$ (—) of liquid Cs-Tl at $T = 730$ K, $T = 1300$ K and $T = 1700$ K as obtained from the RMC simulation.

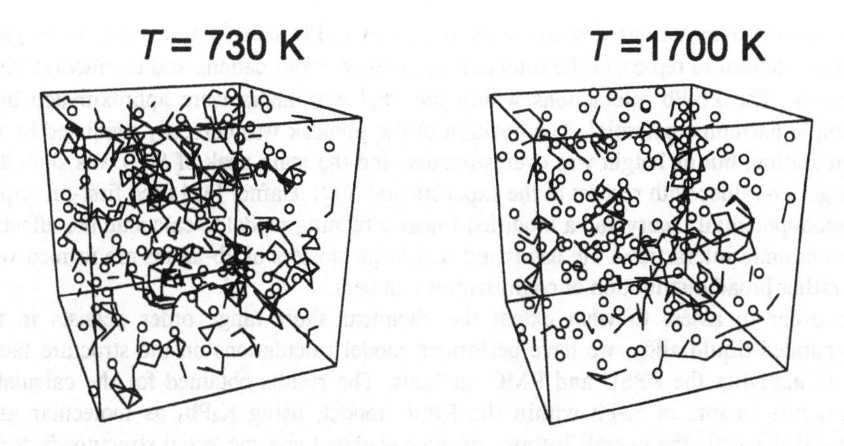

Figure 14. **Configurations of Tl atoms obtained by RMC simulation of a) liquid Cs-Tl at T = 730 K, b) liquid Cs-Tl at T = 1700 K. Tl atoms within a distance of 3.4 Å are connected to visualize some kind of bonding (box size: 40 x 40 x 40 Å^3).**

Figure 14 shows three-dimensional configurations of Tl atoms after the RMC simulations at T = 730 K and T = 1700 K, respectively. In these plots the Tl atoms with distances smaller than 3.4 Å are connected to visualize some kind of bonding. Clearly, besides some fragments of Tl clusters with octahedral symmetry, also larger clusters occur and the Tl atoms appear to be arranged in a more network-like configuration at 730 K. This network opens up at high temperatures and mainly leaves strings of Tl atoms.

4. Theoretical Modelling of the Data

A quantitative description of the microscopic liquid structure requires computer simulation or modelling studies. Reijers et al. [31,32] calculated the structure factor of liquid K-Pb at 870 K based on Pb_4^{4-}-tetrahedra using both the reference interaction site model (RISM) and a model based on random packing of structural units (RPSU). These models reproduced the gross features of the experimental data. In detail, however, deviations are observed in the first peak of the structure factor and at higher momentum transfers, where the models tended to overestimate the strength of the oscillations. Oscillations at high Q-values are not observed experimentally indicating that the structural units are not stable molecules. Reijers et al. [36,37] have also performed a molecular dynamics simulation of K-Pb using a Born-Mayer-Huggins-type potential to represent the interactions between alkali cations and tetrahedral Pb_4^{4-} anions. The Pb-Pb interactions within the Pb_4^{4-}-tetrahedra were approximated by a simple harmonic potential. The position of the prepeak was correctly predicted by the simulation, but its height was overestimated, and the main peak of $S(Q)$ was shifted to larger Q-values with respect to the experimental $S(Q)$. Hafner [38] used first-principles pseudopotential theory and a modified linear screening model to calculate the effective interatomic interactions. He has found that large clusters of Pb-atoms are formed with a rather broad distribution of coordination numbers.

In order to assess to what extent the chemical short range order persists in the expanded liquid alloy, we have performed model calculations on the structure factor data applying the RPSU and RMC methods. The results obtained for the calculated structure factors of K-Pb within the RPSU model, using K_4Pb_4 as molecular unit, reveal that only the overall features of the calculated and measured structure factor of liquid K-Pb near the melting point at 873 K are in reasonable agreement. The agreement worsens with increasing temperature, thus showing that the picture of rigid K_4Pb_4 units is probably too simple to describe the nature of the chemical short range order in the expanded liquid state.

To understand the deficiencies of the model, RMC calculations have been performed, which also yield information on higher order correlation functions [52,53], as we have seen in the last paragraph. The configurations of lead atoms obtained in the RMC model for liquid K-Pb or Cs-Pb (and of tin atoms for liquid Na-Sn) are similar to those found for molten Na-Sn, K-Si and Cs-Pb by recent ab-initio molecular dynamics simulations [40-43]. First-principles molecular dynamics as introduced by Car and Parrinello was used, with interatomic forces calculated directly from the electronic ground state. This ground state is calculated using density functional theory within the local density approximation. The ab-initio MD simulation of liquid Na-Sn [40] revealed a clear tendency to network formation of threefold and fourfold coordinated Sn atoms in the liquid. Remnants of Sn_4-complexes were found in the melt, but always connected with other Sn atoms or groups of Sn atoms. Similar results were obtained for a related system, molten K-Si, also using the ab-initio MD scheme [41]. The theoretical results exhibit all the features normally associated with Zintl-forming alloys, such as the prepeak. However, the underlying structure appears to be more complex than that predicted by the Zintl-model valid for the crystalline state. The liquid contains extended networks of Si atoms, with two-, three-, and fourfold coordinated sites, the average coordination of Si being three. The subsystem of K atoms in the melt is much less structured.

Figure 15 exhibits a snapshot of the configurations of liquid Cs-Pb at 1050 K during the ab-initio MD simulation of de Wijs et al. [42,43]. The average coordination of Pb around Pb has been found to be 3.3. Only fragments of tetrahedra rather than perfect tetrahedra are found in the liquid, and some of the fragments are linked. Like the RMC results this shows that perfect tetrahedra are not a necessary condition for the occurence of prepeaks in $S(Q)$.

Recently, an extensive ab-initio local-density functional molecular dynamics study has been performed by Hafner et al. [67-69] to examine the Zintl-picture for the formation of polyanionic clusters in solid and liquid alloys of K with Sn, Sb and Te. The calculations have been performed using the Vienna ab-initio simulation program VASP, which performs an iterative solution of the Kohn-Sham equations of local density functional theory in a plane-wave basis, using ultrasoft pseudopotentials to describe the electron-ion interaction (simulation box with 124 atoms, simulation time about 10 ps). The results demonstrate that in Na-Sn polyanions exist in the crystalline phase, but are largely destroyed on melting, as the intra- and intercluster interactions are of comparable strength. In Li-Sn a polyanionic structure is unstable, which is due to the small size of the Li-ion; strong interactions between the Sn-ions destabilize the polyanionic order. On the contrary, in K-Sn, a snapshot of a characteristic instantaneous configuation shows that at least 50 % of the Sn-atoms form almost regular Sn_4^{4-} tetrahedra, the remaining Sn atoms form fragments of polyanions (triangles, dimers). Thus the stability of tetrahedral Sn_4^{4-} polyanions in (K, Na, Li)-Sn alloys shows that their stability decreases rapidly with the decreasing size of the alkali atoms. Furthermore, it has been shown that the Zintl picture is essentially correct if charge-transfer is interpreted not in terms of an integration of the electron density over atomic cells, but in terms of the character of the occupied valence state. If all valence states are determined by the strong attractive potential of the polyvalent species and may be described in terms of the molecular eigenstates of the polyanions, then the term charge-transfer compound is justified.

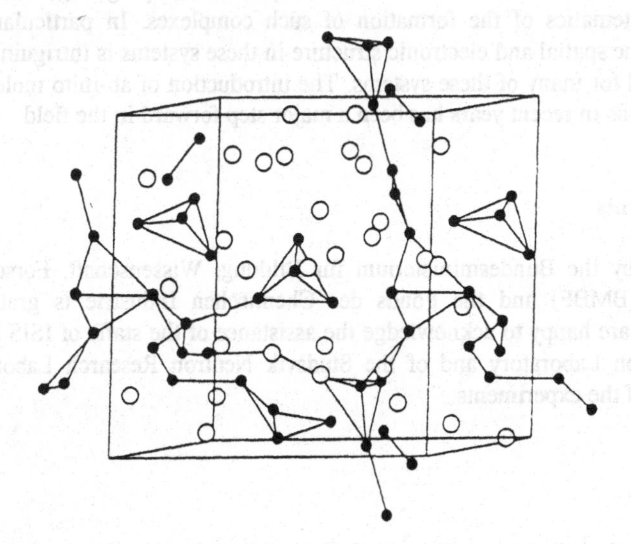

Figure 15. Snapshot of an ab-initio molecular dynamics simulation of liquid Cs-Pb (full circles: Pb atoms, open circles, Cs atoms; adopted from [42,43]).

5. Conclusions

In conclusion we can say, that polyanions of Pb atoms in K-Pb, Cs atoms in Cs-Pb and Sn atoms in Na-Sn are present at higher temperatures, in a way that is - at least partially - related to the Zintl ions in the corresponding solid compounds. With increasing temperature polyanions do not simply break up and distribute randomly. Part of the polyvalent metal atoms might be arranged in a larger loose three-dimensional network, built up by decomposition as well as by linking-up of fragments of ideal Zintl ions. The metal bonds fluctuate, as may be inferred from the absence of distinct oscillations at high values of $S(Q)$ or distinct vibrational bands in the inelastic neutron scattering spectrum. We assume that these polyanion clusters are only transient, probably with a life-time of the order of several ps. In the expanded liquid state, there is a continuous evolution from a structure with a strong tendency to form specific local clusters to a structure with a more general tendency to microphase separation.
In the crystalline alloys of thallium with K and Cs different kinds of polyanions (e.g., Tl_6 or Tl_{11} units) than the simple Zintl ions are formed. Anion clustering is also important in the K-Tl and Cs-Tl melt. The polyanions in liquid Cs-Tl disintegrate and rearrange at higher temperatures and correspondingly lower densities. A similar picture evolves for the structure of expanded liquid K-Sb.
It remains for the experimentalists to refine the experimental techniques used to study these liquids where clustering and polyanion formation is likely at the partical structure factor level, and for the theorists to identify the underlying reasons and to understand the systematics of the formation of such complexes. In particular, the interplay between the spatial and electronic structure in these systems is intriguing and not well understood for many of these systems. The introduction of ab-inito molecular dynamics calculations in recent years has been a major step forward in the field.

6. Acknowledgements

Financial support by the Bundesministerium für Bildung, Wissenschaft, Forschung und Technologie (BMBF) and the Fonds der Chemischen Industrie is gratefully acknowledged. We are happy to acknowledge the assistance of the staffs of ISIS at the Rutherford Appleton Laboratory and of the Studsvik Neutron Research Laboratory during the course of the experiments.

7. References

1. J.E. Enderby and A.C. Barnes,"Liquid Semiconductors", Rep. Prog. Phys., 53 (1990) 85-179.
2. F. Hensel, "Liquid Ionic Alloys", Z. Phys. Chem. N.F., 154 (1987) 201-219.

3. M.-L. Saboungi, W. Geertsma, and D.L. Price, "Ordering in Liquid Alloys", Ann. Rev. Phys. Chem., 41 (1990) 207-244.
4. W. van der Lugt, "Zintl Ions as Structural Units in Liquid Alloys", Physica Scripta, T39 (1991) 372-377.
5. Zintl and G. Brauer, "Über die Valenzelektronenregel und die Atomradien unedler Metalle in Legierungen", Z. Phys. Chem. B, 20 (1933) 245-271.
6. R.E. Marsh and D.P. Shoemaker, "The Crystal Structure of NaPb", Acta Crystallogr., 6 (1953) 197-205.
7. I.F. Hewaidy, E. Busmann, and W. Klemm, "Die Struktur der AB-Verbindungen der schweren Alkalimetalle mit Zinn und Blei", Z. Anorg. Allg. Chem., 328 (1964) 283-293.
8. W. Müller and K. Volk, "Die Struktur des β-NaSn", Z. Naturforsch., 32b (1977) 709-710.
9. R. Pis Diez, M.P. Iñiguez, J.A. Alonso, and J.A. Aramburu,"Charge Transfer Within Zintl Ions in Liquid Metallic Alloys", A Cluster Theory", J. Molecular Structure (Theochem), 330 (1995) 267-272.
10. D.L. Price, M.-L. Saboungi, R. Reijers, G. Kearley, and R. White, "Two-Stage Melting in Cesium Lead Alloys", Phys. Rev. Lett., 66 (1991) 1894-1897.
11. M.-L. Saboungi, J. Fortner, W.S. Howells, and D.L. Price, "Dynamic Enhancement of Cation Migration in a Zintl Alloy by Polyanion Rotation", Nature, 365 (1993) 237-239.
12. R.D. Stoddard, M.S. Conradi, A.F. McDowell, M.-L. Saboungi, and D.L. Price, "Atomic Motions in an Unusual Molecular Semiconductor: NaSn", Phys. Rev. B, 52 (1995) 13998-14005.
13. D.L. Price, M.-L. Saboungi, and W.S. Howells, "Rotor Phases in Compound Semiconductors", Physica B, 213 & 214 (1995) 547-551.
14. J. Fortner, M.-L. Saboungi, and J.E. Enderby, "Carrier Density Enhancement in Semiconducting NaSn and CsPb", Phys. Rev. Lett., 74 (1995) 1415-1418.
15. D.L. Price, M.-L. Saboungi, and W.S. Howells, "Orientational and Translational Disorder in Semiconducting Zintl compounds", Phys. Rev. B., 51 (1995) 14923-14929.
16. J. Saar and H. Ruppersberg, "Specific Heat of Liquid K/Pb Alloys Calculated from $(\partial \rho/\partial T)_S$ and $\rho(T)$ Data", Z. Phys. Chem. NF, 156 (1988) 587- 591.
17. V.D. Bushmanov and S.P. Yatsenko, "Immiscibility in Binary Systems of Cesium with Aluminium, Gallium, Indium and Thallium", Russ. Matallurgy (Metally), 5 (1981) 157-160.
18. W. van der Lugt and W. Geertsma, "Electron Transport in Liquid Metals and Alloys", Can. J. Phys., 65 (1987) 326-347.
19. J.A. Meijer, G.J.B. Vinke, and W. van der Lugt, "Resistivity of Liquid Rb-Pb and Cs-Pb Alloys", J. Phys. F: Met. Phys., 16 (1986) 845-851.
20. C. van der Marel, A. B. van Oosten, W. Geertsma, and W van der Lugt, "The Electrical Resistivity of Liquid Li-Sn, Na-Sn and Na-Pb Alloys: Strong Effects of Chemical Interactions", J. Phys. F: Met. Phys., 12 (1982) 2349-2361.

21. M.-L. Saboungi, S.R. Leonard, and J. Ellefson, "Anomalous Behaviour of Liquid K-Pb Alloys: Excess Stability, Entropy, and Heat Capacity", J. Chem. Phys., 85 (1986) 6072-6081.
22. M.-L. Saboungi, H.T.J. Reijers, M. Blander, and G.K. Johnson, "Heat Capacity of Some Liquid Zintl Compounds: Equiatomic Alkali-Lead Alloys", J. Chem. Phys., 89, (1988) 5869-5875.
23. G. K. Johnson and M.-L. Saboungi, "Heat Capacity of Liquid Equiatomic Potassium-Lead Alloy: Anomalous Temperature Dependence", J. Chem. Phys., 86 (1987) 6376-6381.
24. S. Tamaki, T. Ishiguro and S. Takeda, "Thermodynamic Properties of Liquid Na-Sn Alloys", J. Phys. F: Met. Phys., 12 (1982) 1613-1624.
25. R. Alqasmi and J.J. Egan, "Thermodynamics of Liquid Na-Sn Alloys Using CaF_2 Solid Electrolytes", Ber. Bunsenges. Phys. Chem., 87 (1983) 815-817.
26. W. Geertsma and M.-L. Saboungi, "Large Schottky-type Heat Capacity Anomalies in Liquid Alkali Group IV Alloys", J. Phys. Condens. Matter, 7 (1995) 4803-4820.
27. C. van der Marel, P.C. Stein and W. van der Lugt, "^{23}Na Knight Shift of Liquid Na-Sn Alloys", Physics Letters, 95A (1983) 451-453.
28. H. Ruppersberg and H. Egger, "Short-range Order in Liquid Li-Pb Alloys", J. Chem. Phys., 63 (1975) 4095-4103.
29. H. Ruppersberg and H. Reiter, "Chemical Short-range Order in Liquid LiPb Alloys", J. Phys. F: Met. Phys., 12 (1982) 1311-1325.
30. B.P. Alblas, W. van der Lugt, J. Dijkstra, W. Geertsma, and C. van Dijk, "Structure of Liquid Na-Sn Alloys", J Phys. F: Met. Phys., 13 (1983) 2465-2477.
31. H.T.J. Reijers, W. van der Lugt, C. van Dijk, and M.-L. Saboungi, "Structure of Liquid K-Pb Alloys", J. Phys.: Condens. Matter, 1 (1989) 5229-5241.
32. H.T.J. Reijers, M.-L. Saboungi, D.L. Price, J.W. Richardson Jr., J. Volin, and W. van der Lugt, "Structural Properties of Liquid Alkali-Metal - Lead Alloys: NaPb, KPb, RbPb, and CsPb", Phys. Rev. B, 40 (1989) 6018-6029.
33. H.T.J. Reijers, M.-L. Saboungi, and D.L. Price, "Structure of Liquid Equiatomic KSn and CsSn", Phys. Rev. B, 41 (1990) 5661-5666.
34. D.L. Price, M.-L. Saboungi, G.A. de Wijs and W. van der Lugt, "Structure of Liquid Caesium-Lead Alloys", J. Non-Cryst. Solids, 156-158 (1993) 34-37.
35. W. Geertsma, J. Dijkstra, and W. van der Lugt, "Electronic Structure and Charge transfer-induced Cluster Formation in Alkali-group-IV Alloys", J. Phys. F: Met. Phys., 14 (1984) 1833-1845.
36. H.T.J. Reijers, W. van der Lugt and M.-L. Saboungi, "Molecular-Dynamics Study of Liquid NaPb, KPb, RbPb, and CsPb Alloys", Phys. Rev. B, 42 (1990) 3395-3405.
37. K. Toukan, H.T.J. Reijers, C.-K. Loong, D.L. Price, and M.-L. Saboungi, "Atomic Motions in Liquid KPb: A Molecular-Dynamics Investigation", Phys. Rev. B., 41 (1990) 11739-11742.
38. J. Hafner, "Formation of Polyanionic Clusters in Liquid Potassium-Lead Alloys: a Molecular-Dynamics Study", J. Phys.: Condens. Matter, 1 (1989) 1133-1140.

39. R.L. McGreevy and M.A. Howe, "The Structure of Molten K-Pb, Rb-Pb and Cs-Pb Alloys", J. Phys.: Condens. Matter, 3 (1991) 577-591.
40. G. Seifert, G. Pastore, and R. Car, "Ab Initio Molecular Dynamics Simulation of Liquid NaSn Alloy", J. Phys.: Condens. Matter, 4 (1992) L179-L183.
41. G. Galli and M. Parrinello, "Theoretical Study of Molten KSi", J. Chem. Phys., 95 (1991) 7504-7512.
42. G. A. de Wijs, G. Pastore, A. Selloni, and W. van der Lugt, "Poly-Anions in Liquid CsPb: An Ab Initio Molecular-Dynamics Simulation", Europhys. Lett., 27 (1994) 667-672.
43. G.A. de Wijs, G. Pastore, A. Selloni, and W. van der Lugt, "First-principles Molecular-Dynamics Simulation of Liquid CsPb", J. Chem. Phys.,103 (1995) 5031-5039.
44. R. Winter and T. Bodensteiner, "Neutron Scattering Experiments at High Temperatures and Pressures", High Press. Res. 1 (1988) 23-37.
45. P.A. Egelstaff, An Introduction to the Liquid State (Oxford: Clarendon Press, 1992).
46. J.G. Powles, "The Structure of Molecular Liquids by Neutron Scattering", Adv. Phys., 22 (1973) 1-56.
47. J.P. Hansen and I.R. McDonald, Theory of Simple Liquids (New York, NY: Academic Press, 1990).
48. M. Shimoji, Liquid Metals - An Introduction to the Physics and Chemistry of Metals in the Liquid State (New York, NY: Academic Press, New York, 1977).
49. R.L. McGreevy and L. Pusztai, "Reverse Monte Carlo Simulation: A New Technique for the Determination of Disordered Structures", Mol. Simul., 1 (1988) 359-367.
50. R.L. McGreevy, M.A. Howe, D.A. Keen, and K.N. Clausen.,"Reverse Monte Carlo (RMC) Simulations: Modelling Structural Disorder in Crystals, Glasses and Liquids from Diffraction Data" (Inst. Phys. Conf. Ser. No. 107, 1990), 165-184.
51. M. Stolz, O. Leichtweiß, R. Winter, M.-L. Saboungi, J. Fortner, and W.S. Howells, "Survival of Polyanions in Expanded Liquid Alloys", Europhys. Lett., 27 (1994) 221-226.
52. R. Winter, O. Leichtweiß, C. Biermann, M.-L. Saboungi, R.L. McGreevy, and W.S. Howells, "Survival of Polyanions in Expanded Liquid Zintl Alloys", J. Non-Cryst. Solids, 205-207 (1996) 66-70.
53. M. Stolz, R. Winter, W.S. Howells, and R.L. McGreevy, "Computer Modelling Studies of Expanded Liquid KPb", J. Phys.: Condens. Mater, 7 (1995) 5733-5743.
54. I.F. Hewaidy, E. Busmann and W. Klemm, "Die Struktur der AB-Verbindungen der schweren Alkalimetalle mit Zinn und Blei", Z. anorg. allgem. Chem., 328 (1964) 283-293.
55. H.G. von Schnering, W. Hönle and G. Krogull, "Die Monoantimonide RbSb und CsSb", Z. Naturforsch., 34 b (1979) 1678-1682.

56. J. Bernard and W. Freyland, "Metal-semiconductor transition in liquid M_xSb_{1-x} alloys (M = Na, K, Cs): variation of electrical conductivity and thermodynamic properties", J. Non-Cryst. Solids, 205-207 (1996) 62-65.
57. P. Lamparter, W. Martin, S. Steeb, and W. Freyland, "Neutron Diffraction Study on the Structure of Liquid Cs-Sb Alloys", Z. Naturforsch., 38 a (1983) 329-335.
58. Z.-C. Dong and J.D. Corbett, "CsTl: A New Example of Tetragonally Compressed Tl_6^{6-} Octahedra. Electronic Effects and Packing Requirements in the Diverse Structures of ATl (A= Li, Na, K, Cs)", Inorg. Chem., 35 (1996) 2301-2306.
59. T. Itami, T. Sasada, N. Iwaoka, and M. Shimoji, "The Magnetic Susceptibility of Liquid Alkali-Thallium Alloys", J. Non-Cryst. Solids, 117/118 (1990) 347-350.
60. R. Fainchtein, U. Even, C.E. Krohn, and J.C. Thompson, "Reflectance of Molten Cs-Tl Alloys", J. Phys. F: Met. Phys., 12 (1982) 633-639.
61. M. Kitajima and M. Shimoji, Liquid Metals, ed. R. Evans and D.A. Greenwood (Bristol: Institute of Physics Conference Series, Vol. 30, Bristol, 1976), 226.
62. T. Itami, K. Izumi, and N. Iwaoka, "Electronic Properties of Liquid Alkali-Thallium Alloys", J. Non-Cryst. Solids, 156-158 (1993) 285-288.
63. R. Xu, P. Verkerk, W.S. Howells, G.A. de Wijs, F. van der Horst, and W. van der Lugt, "Nanometer Superstructure in Liquid Alkali-Thallium Alloys", J. Phys.: Condens. Matter, 5 (1993) 9253-9260.
64. P. Verkerk, R. Xu, W.S. Howells, G.A. de Wijs, and W. van der Lugt, "Clusters in Liquid K-Tl and Cs-Tl Alloys", J. Phys.: Condens. Matter, 6 (1994) A255-A260.
65. S.A. van der Aart, P. Verkerk, A.C. Barnes, P.S. Salmon, R. Winter, H. Fisher, L.A de Graaf, and W. van der Lugt, "Structure in liquid KTl investigated by means of neutron diffraction using ^{205}Tl isotope substitution", Physica B, 241-243 (1998) 961-963.
66. O. Leichtweiß, K. Hochgesand, C. Biermann, and R. Winter, "Neutron diffraction and computer modeling studies of expanded liquid Cs-Tl", J. Chem. Phys., 110 (1998) 498-500.
67. K. Seifert-Lorenz, J. Hafner, "Ab initio studies of local order and electronic properties in molten Zintl-alloys: K-Sb as a case study", J. Non-Cryst. Solids, 232-234 (1998) 198-204.
68. J. Hafner, K. Seifert-Lorenz, and O. Genser, "Ab-initio studies of polyanionic clustering in liquid alloys", J. Non-Cryst. Solids, in press.
69. O. Genser and J. Hafner, "Structural and electronic properties of liquid alkali-tin alloys", J. Non-Cryst. Solids, in press.

Appendix A

Elements of Neutron Diffraction Theory

a) The Structure Factor for Monoatomic Systems

In the following we show in a schematic way how the structure factor of a fluid can be measured by the neutron scattering method. A similar formalism applies to X-ray scattering experiments (Table 1 compares the two scattering probes). Figure 1 sketches a generalized picture of the elastic scattering experiment.

TABLE 1. Characteristics of neutron (*n*) and X-ray scattering probes.

Neutron scattering:

- *n* interact with nuclei
- thermal neutrons, $\lambda \approx 1$ Å
- *b* independent of *Q*
- *b* not systematically dependent on atomic number *Z*
- *b* depends on spin and isotope (isotope substitution method)
- e.g., V incoherent scatterer (sample container material)
- path length ≈ cm (*n* penetrate matter easily)
- dipole moment (magnetic scattering)
- dynamic distribution functions can easily be studied
- neutron sources are few and relatively weak

X-ray scattering:

- form factor *f* is *Q*-dependent (scattering amplitude varies with scattering angle)
- *f* strongly depends on atomic number (problem: H)
- *f* independent of isotope
- path lengths << cm (esp. high *Z*; but: high energy X-rays)
- no equivalence to V
- X-ray sources are many and strong (in particular using synchrotron radiation)
- small sample sizes can be measured (e.g. in diamond anvil pressure cells)
- anomalous scattering

In a scattering experiment, a proportion of the incident neutrons is scattered and the remaining fraction is transmitted through the sample. The intensity of the scattered neutrons can be measured as a function of the scattering angle 2θ and the energy. The energy gained by the sample (and lost by the neutron) - or vice versa - is given by

$$\Delta E = \frac{m_{\rm n}}{2}(v^2 - v_0^2) = \frac{\hbar^2}{2m_{\rm n}}(k^2 - k_0^2) = \hbar\omega \tag{1}$$

where $\vec{k}_0$ and $\vec{k}$ are the initial and final wave vectors ($k = 2\pi/\lambda$), λ_0 the wavelength, m_n the mass, and v_0 the velocity of the incident neutron; λ and v are the corresponding values of the scattered neutron. The momentum transfer is

$$\hbar\vec{Q} = \hbar(\vec{k} - \vec{k}_0)$$
$$\hbar Q = \hbar\left(k^2 + k_0^2 - 2kk_0\cos(2\theta)\right)^{1/2} \tag{2}$$

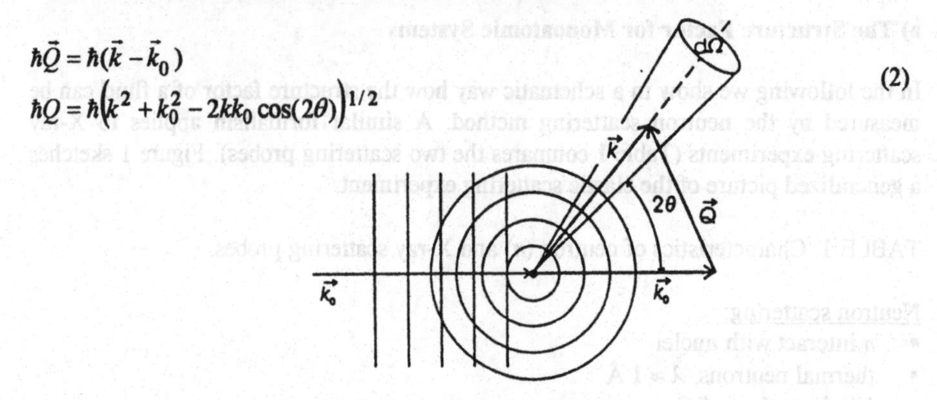

Figure 1. Schematic representation of the scattering geometry. For elastic scattering the scattering triangle will be isosceles since $\{v_1, \lambda_1, E_1\} = \{v_0, \lambda_0, E_0\}$

The ability of energy analysis is also conventionally a unique feature of the neutron probe for studying the dynamic structure factor and thus single-particle and collective dynamic properties of the sample. Here we will focus on static structural properties only.

If no energy change takes place ($\Delta E = 0$, $\lambda = \lambda_0$), the scattering is termed elastic (Fig. 1) and

$$Q = \frac{4\pi}{\lambda}\sin\theta \tag{3}$$

The wavelength of the neutrons is of the order of 0.1 nm and thus much larger than the size of the scattering nucleus, which can be viewed as a point object. The interaction between the neutron and nucleus cannot be calculated exactly, but it is known to be short-ranged compared to the wavelength of the neutron. As a consequence, the scattering is isotropic for slow neutrons and there is no angle-dependent form factor as in the case of X-rays. An incident plane wave of neutrons is described by a wave function of the type

$$\psi_0 = \frac{1}{\sqrt{v_0}} e^{-i\vec{k}_0\vec{x}} \tag{4}$$

($\vec{x}$ direction of incident neutrons). The scattered wave from a fixed nucleus is spherically symmetrical and of the form

$$\psi = -\frac{1}{\sqrt{v_0}} \frac{b}{r} e^{-i\vec{k}\vec{r}} \tag{5}$$

The quantity b has the dimension of length and is called the scattering length. In principle this is a complex quantity, but the imaginary part is important only for nuclei with a high absorption coefficient (e.g., Cd, B), and may be neglected here. For the incident energy range used for structure measurements, b can be considered to be a constant for a given nucleus (isotope).
In the experiment an incident neutron beam is scattered by an assembly of sample nuclei into solid angle dΩ. The differential scattering cross section for unit volume of sample, dσ/dΩ, is defined as the number of neutrons scattered per second into the solid angle, divided by the incident neuton flux (neutrons per second per unit area). For a single nucleus we obtain

$$\frac{d\sigma}{d\Omega} = \frac{\text{flux scattered into solid angle } d\Omega \text{ at angle } 2\Theta}{\text{incident flux density} \cdot d\Omega} = r^2 \frac{|\psi|^2}{|\psi_0|^2} = |b|^2 \tag{6}$$

If not only the direction but also the energy of the scattered neutrons is measured, the double differential scattering cross section $d^2\sigma/(d\Omega dE)$ has to be used, defined as the fraction of neutrons scattered into an element of solid angle dΩ with an energy between the values E and E+dE. In a diffraction experiment the detector integrates over all energies; hence the resulting so-called elastic of static structure factor is not strictly derived from a purely elastic scattering experiment.

The nuclear cross sections vary in an apparently random number from atom to atom and are generally treated as measured quantities rather than calculated from first principles. It is typically of the order of 10^{-15} m (1 fm). This gives rise to the usual unit for a cross section, called barn (10^{-24} cm^2). b varies from isotope to isotope and from nucleus to nucleus of the same isotope if it has nonzero spin. Because the neutron has spin 1/2, it can interact with a nucleus of spin I to form one of two compound nuclei with spins $I^+ = I + 1/2$ or $I^- = I - 1/2$, each of which has a different scattering length, b^+ or b^-, which is associated with the spin-up or spin-down states. Taking into account the degeneracy of states, the probabilities p^+ and p^- of formation of the compound nuclei with spin I^+ and I^- can be calculated. The average coherent scattering length for an isotope is then given by

$$b^{\mathrm{coh}} = \langle b \rangle = p^{+}b^{+} + p^{-}b^{-} = \frac{I+1}{2I+1}b^{+} + \frac{I}{2I+1}b^{-} \qquad (7)$$

where the brackets <...> represent an average over the spin-state populations. If there is no external field present and no magnetic sample in combination with a polarized neutron beam, the $I \pm 1/2$ spin states are distributed randomly.
The isotopes are also distributed in a random way over the chemically identical particles.

Since there is no correlation between the values of b_n and b_m, where n and m refer to different sites, it is possible to write $\langle b_n b_m \rangle = \langle b_n \rangle \langle b_m \rangle = \langle b \rangle^2$. But if $n = m$, $\langle b_n b_m \rangle = \langle b^2 \rangle$, so that in general

$$\langle b_n b_m \rangle = \langle b \rangle^2 + \delta_{nm}\left(\langle b^2 \rangle - \langle b \rangle^2\right) \qquad (8)$$

Here the brackets <...> indicate an average over isotopes and spin states.

A coherent scattering cross section can be defined for an assembly of atoms by

$$\sigma_{\mathrm{coh}} = 4\pi \langle b \rangle^2 \qquad (9)$$

with the mean coherent scattering length, averaged over the different isotopes of the element, given by

$$\langle b \rangle = \sum_j c_j b_j^{\mathrm{coh}} \qquad (10)$$

$c_j = N_j/N$ are the relative concentrations of isotopes j ($\sum_j c_j = 1$).
The total scattering cross section is defined as

$$\sigma_{\mathrm{tot}} = 4\pi \langle b^2 \rangle \qquad (11)$$

where the mean squared total scattering length, averaged over the different isotopes, is given by

$$\langle b^2 \rangle = \sum_j c_j \left(b_j^{\mathrm{coh}}\right)^2 + \sum_j c_j \left(b_j^{\mathrm{inc}}\right)^2 \qquad (12)$$

The difference between the two is the incoherent scattering cross section σ_{inc}:

$$\sigma_{inc} = \sigma_{tot} - \sigma_{coh} = 4\pi\left(\left\langle b^2\right\rangle - \left\langle b\right\rangle^2\right) \tag{13}$$

If the isotope has no spin, then there is no spin incoherent scattering. Only the coherent scattering cross section contains information on interference effects arising from spatial correlations of the nuclei in the system. The incoherent cross section forms an isotropic, flat scattering background which must be substracted in structural investigations. It does, however, contain information on the motion of single atoms, which may be investigated via energy analysis of the scattered beam. Examples of scattering cross sections are given in Table 1. It can be seen from the table, that here can be large differences in the scattering lengths of isotopes, e.g., between deuterium and hydrogen, and that the latter value is actually negative. This arises from a change in phase of the scattered wave.

TABLE 2. Some neutron scattering cross-sections (V.F. Sears, "Neutron Scattering Lengths and Cross Sections", Neutron News, Vol. 3, No. 3, 1992).

Element or isotope	Total scattering cross section /10^{-24} cm^2
^{1}H	82.03
^{2}H (D)	7.64
^{6}Li	0.97
^{7}Li	1.40
Cl	16.8
^{35}Cl	21.8
^{37}Cl	1.19
Tl	9.89
^{205}Tl	11.4

For a system of N atoms, the differential scattering cross-section $d\sigma/d\Omega$, which is be determined from the experimental data, is given by:

$$\frac{d\sigma}{d\Omega} = \left\langle \sum_{n,m}^{N} b_n b_m e^{-i\vec{Q}\cdot(\vec{r}_n - \vec{r}_m)} \right\rangle \tag{14}$$

where <...> denotes an ensemble average over all possible configurations in phase space, and the $\vec{r}_n$ and b_n refer to nuclear positions and scattering lengths respectively. The formula is generally valid, also for multi-component systems.

It is convenient to split the cross section into a coherent and incoherent part:

$$\frac{\mathrm{d}\sigma}{\mathrm{d}\Omega}=\left(\frac{\mathrm{d}\sigma}{\mathrm{d}\Omega}\right)_{\mathrm{coh}}+\left(\frac{\mathrm{d}\sigma}{\mathrm{d}\Omega}\right)_{\mathrm{inc}} \tag{15}$$

$$=\langle b\rangle^2\left\langle\sum_{n,m}^{N}\mathrm{e}^{-\mathrm{i}\vec{Q}\cdot(\vec{r}_n-\vec{r}_m)}\right\rangle+N\left(\left\langle b^2\right\rangle-\langle b\rangle^2\right) \tag{16}$$

The first term on the right hand side depends on the mean scattering length, <b>, and is called the coherent scattering since it represents the interference effects. The second term represents the incoherent scattering, which depends on the mean square deviation of the individial scattering lengths b_n. It does not contain a phase factor as the isotopes and spins are randomly distributed.
We can write equation (16) as

$$\frac{1}{N}\frac{\mathrm{d}\sigma}{\mathrm{d}\Omega}=\langle b\rangle^2 S(\vec{Q})+\left(\left\langle b^2\right\rangle-\langle b\rangle^2\right)=\frac{\sigma_{\mathrm{coh}}}{4\pi}S(\vec{Q})+\frac{\sigma_{\mathrm{inc}}}{4\pi} \tag{17}$$

where the static structure factor $S(\vec{Q})$ has been defined according to

$$S(\vec{Q})=\frac{1}{N}\left\langle\sum_{n,m}^{N}\mathrm{e}^{-\mathrm{i}\vec{Q}\cdot(\vec{r}_n-\vec{r}_m)}\right\rangle$$
$$=1+\frac{1}{N}\left\langle\sum_{n\neq m}^{N}\mathrm{e}^{-\mathrm{i}\vec{Q}\cdot(\vec{r}_n-\vec{r}_m)}\right\rangle \tag{18}$$

The distribution of atoms in real space can be described using the pair correlation function $g(\vec{r})$:

$$g(\vec{r})=\frac{V}{N^2}\left\langle\sum_{n\neq m}^{N}\delta(\vec{r}-(\vec{r}_n-\vec{r}_m))\right\rangle \tag{19}$$

The brackets denote the equilibrium ensemble statistical average. It is defined such that $\rho_0\, g(\vec{r})\,\mathrm{d}\vec{r}$ (with $\rho_0 = N/V$ number density) is the probability of finding an atom between positions $\vec{r}$ and $\vec{r}+\mathrm{d}\vec{r}$ with respect to another atom at the origin. The definition implies that for small values of $\vec{r}$, $g(\vec{r})$ must be zero since atoms cannot interpenetrate. For large values of $\vec{r}$, the correlation between atom pairs is lost and $g(\vec{r})$ tends to 1. The relation between $S(\vec{Q})$ and $g(\vec{r})$ is given by Fourier transforms:

$$S(\vec{Q})-1=\rho_0\int_0^{\infty}(g(\vec{r})-1)e^{-i\vec{Q}\vec{r}}\,d\vec{r} \tag{20}$$

$$g(\vec{r})-1=\frac{1}{(2\pi)^3\rho_0}\int_0^{\infty}(S(\vec{Q})-1)e^{i\vec{Q}\vec{r}}\,d\vec{Q} \tag{21}$$

where $\vec{r}=\vec{r}_n-\vec{r}_m$. The definitions are valid for all types of single-component systems, whether they are crystalline or not.

Making the reasonable assumption that in a liquid or other disorderd material there are no preferred directions for the vector $\vec{r}$, i.e., the material is isotropic and only the magnitude of $\vec{r}$ has significance, the integral can be performed directly over the angular part in polar coordinates. Our expressions above then simplifies to

$$S(Q)=1+4\pi\rho_0\int_0^{\infty}r^2[g(r)-1]\frac{\sin(Qr)}{Qr}\,dr \tag{22}$$

and

$$g(r)=1+\frac{1}{2\pi^2\rho_0}\int_0^{\infty}Q^2[S(Q)-1]\frac{\sin(Qr)}{Qr}\,dQ \tag{23}$$

The mean coordination number of particles between distances r_1 and r_2 is given by the integral:

$$N(r)=4\pi\rho_0\int_{r_1}^{r_2}r^2g(r)\,dr \tag{24}$$

We have, then, been able to relate the microscopic distribution of atoms in a sample to a scattering function which can be measured experimentally.

The structure factor can be shown to have certain properties that assist us in data analysis by providing a check on experimentally derived values. At Q = 0, $S(Q)$ is related through thermodynamics to the bulk isothermal compressibility χ_T,

$$S(0)=\rho_0 k_B T\chi_T \tag{25}$$

where k_B is the Boltzmann constant and T the absolute temperature. The asymptotic limit as Q tends to infinity is simply $S(\infty) = 1$.

Finally, given that $\rho(r) = 0$ as r tends to zero, we may derive a sum rule of the form:

$$\int_0^\infty Q^2[1 - S(Q)]\mathrm{d}Q = 2\pi^2 \rho_0 . \tag{26}$$

b) The Structure Factor for Multicomponent Systems

Moving on from the monoatomic systems considered so far to the case where there is more than one chemical element composing the sample requires that we generalise equations (22) and (23) still further. We first introduce the partial pair correlation function, $g_{\alpha\beta}(r)$, which represents the average distribution of atoms of type β around an atom of type α at the origin (see Fig. 2). For a system containing y atomic species there will be $y(y+1)/2$ distinct partial correlation functions, since α and β can assume any label from 1 to y. We now define the partial structure factor $S_{\alpha\beta}(Q)$ (or $A_{\alpha\beta}(Q)$) as:

$$S_{\alpha\beta}(Q) = 1 + 4\pi\rho_0 \int_0^\infty r^2 \left[g_{\alpha\beta}(r) - 1\right] \frac{\sin(Qr)}{Qr} \mathrm{d}r \tag{27}$$

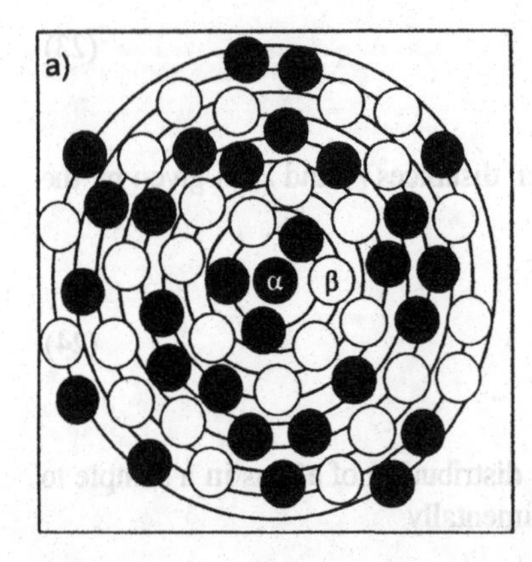

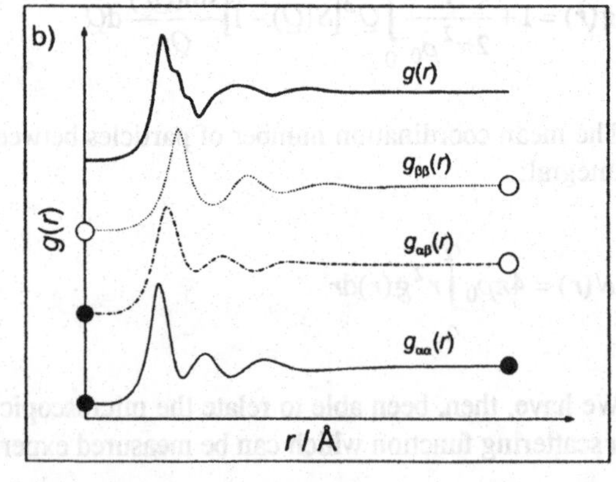

Figure 2. Schematic picture of the total and partial pair correlation functions of a binary liquid system with elements α and β (a) distribution of atoms around a central atom α up to the 9th coordination shell; b) the partial pair correlation functions which make up the total $g(r)$).

There are also other definitions for the partial structure factor, this is the Faber-Ziman form. By analogy with equation (23) we also have:

$$g_{\alpha\beta}(r) = 1 + \frac{1}{2\pi^2 \rho_0} \int_0^{\infty} Q^2 \left[S_{\alpha\beta}(Q) - 1\right] \frac{\sin(Qr)}{Qr} \mathrm{d}Q \tag{28}$$

In addition to summing over all the different isotopes and spin states we sum over atom species with their specific scattering lengths. Equation (16) can thus be formulated for each pair of atom types (α,β) as a sum weighted for concentration and scattering so that

$$\begin{aligned} \frac{1}{N}\frac{\mathrm{d}\sigma}{\mathrm{d}\Omega} &= \sum_{\alpha,\beta} c_\alpha \langle b_\alpha \rangle c_\beta \langle b_\beta \rangle \left(S_{\alpha\beta}(Q) - 1\right) + \sum_{\alpha,\beta} c_\alpha \langle b_\alpha^2 \rangle \\ &= F(Q) + \langle b^2 \rangle \end{aligned} \tag{29}$$

where $<b_\alpha>$ is the average coherent scattering length of element α, and $c_\alpha = N_\alpha/N$ is its atomic fraction. When there is no subscript to indicate the element type, the brackets <...> indicate averaging over the different elements present in the sample as well as over the isotopes and spins of the individual elements. The total structure factor $F(Q)$ is, then, a linear combination of pair-wise contributions from different atom types weighted by the relevant scattering lengths and concentrations:

$$F(Q) = \sum_{\alpha,\beta} c_\alpha \langle b_\alpha \rangle c_\beta \langle b_\beta \rangle \left(S_{\alpha\beta}(Q) - 1\right) \tag{30}$$

Alternatively, often the Faber-Ziman total structure factor is used, defined as

$$S^{\mathrm{FZ}}(Q) - 1 = \frac{F(Q)}{\langle b \rangle^2} \tag{31}$$

($<b> = c_\alpha <b_\alpha> + c_\beta <b_\beta>$). By analogy with the arguments for the monatomic case, we can write down an expression for the total pair correlation function and the Faber-Ziman pair correlation function by Fourier transformation:

$$g(r) = \sum_{\alpha,\beta} c_\alpha \langle b_\alpha \rangle c_\beta \langle b_\beta \rangle g_{\alpha\beta}(r) \tag{32}$$

$$g^{\mathrm{FZ}}(r)-1=\frac{1}{\langle b\rangle^2}\sum_{\alpha,\beta}c_\alpha\langle b_\alpha\rangle c_\beta\langle b_\beta\rangle\left(g_{\alpha\beta}(r)-1\right) \tag{33}$$

For a *binary system* with chemical constituents α and β, the total Faber-Ziman structure factor may be written as:

$$S^{\mathrm{FZ}}(Q)=\frac{1}{\langle b\rangle^2}\left(c_\alpha^2\langle b_\alpha\rangle^2 S_{\alpha\alpha}(Q)+c_\beta^2\langle b_\beta\rangle^2 S_{\beta\beta}(Q)+2c_\alpha c_\beta\langle b_\alpha\rangle\langle b_\beta\rangle S_{\alpha\beta}(Q)\right) \tag{34}$$

(with $S_{\alpha\beta}=S_{\beta\alpha}$). The Faber-Ziman total pair correlation function likewise becomes:

$$g^{\mathrm{FZ}}(r)=\frac{1}{\langle b\rangle^2}\left(c_\alpha^2\langle b_\alpha\rangle^2 g_{\alpha\alpha}(r)+c_\beta^2\langle b_\beta\rangle^2 g_{\beta\beta}(r)+2c_\alpha b_\beta\langle b_\alpha\rangle\langle b_\beta\rangle g_{\alpha\beta}(r)\right) \tag{35}$$

The corresponding partial coordination numbers (see also Fig. 2) are given, e.g. for particles β, by:

$$N_{\alpha\beta}=\frac{c_\beta}{w_{\alpha\beta}}4\pi\rho_0\int_{r_1}^{r_2}r^2 g(r)\mathrm{d}r,\quad \text{with } w_{\alpha\beta}=\frac{c_\alpha\langle b_\alpha\rangle c_\beta\langle b_\beta\rangle}{\langle b\rangle^2} \tag{36}$$

if $g(r)$ is determined by $g_{\alpha\beta}(r)$ between r_1 and r_2. Whilst it is possible to extract a great deal of useful structural information on some systems using $g(r)$ or $g^{\mathrm{FZ}}(r)$, it is often desirable to be able to derive the individual $S_{\alpha\beta}(Q)$ and hence the $g_{\alpha\beta}(r)$.
The technique of isotopic substitution has become a powerful tool in the study of multicomponent systems because it provides one route to this information. We recall that the coherent scattering length is isotope dependent. By enriching an element in a sample with a particular isotope, the scattering intensity may be changed without altering the basic chemistry or physics of the system and thus the partials.
Clearly if $F(Q)$ can be changed systematically in this way (i.e. by changing the coefficients in equation (34), enough data can be acquired through independent diffraction experiments to enable some or all of the $S_{\alpha\beta}(Q)$ to be extracted. It is a matter of solving the $y(y+1)/2$ simultaneous equations.
In common with the single component case, there are further general constraints that may be applied to the analysis of binary system data:

$$\int_0^\infty Q^2\left[1-S_{\alpha\beta}(Q)\right]\mathrm{d}Q=2\pi^2\rho_0 \tag{37}$$

The thermodynamic limit (at $Q = 0$) is now expressed as:

$$S_{\alpha\alpha}(Q) = \rho_0 k_B T \chi_T - c_\beta / c_\alpha$$
$$S_{\beta\beta}(Q) = \rho_0 k_B T \chi_T - c_\alpha / c_\beta \quad (38)$$
$$S_{\alpha\beta}(Q) = \rho_0 k_B T \chi_T + 1$$

Furthermore, because $d\sigma/d\Omega$ must be positive, we have:

$$c_\alpha (S_{\alpha\alpha}(Q) - 1) \geq 1$$
$$c_\beta (S_{\beta\beta}(Q) - 1) \geq 1 \quad (39)$$
$$(c_\alpha (S_{\alpha\alpha}(Q) - 1) + 1) \cdot (c_\beta (S_{\beta\beta}(Q) + 1) + 1) \geq c_\alpha c_\beta (S_{\alpha\beta}(Q) - 1)^2$$

We should mention that there are other schemes for separating into partial structure factors which can be illuminating. The Bhatia-Thornton forms, for instance, are derived from the mean square fluctuations in number density and concentration, together with a cross-term: $S_{NN}(Q)$, $S_{CC}(Q)$ and $S_{NC}(Q)$. $S_{NN}(Q)$ and $S_{CC}(Q)$ oscillate around, and tend towards, unity and $c_\alpha c_\beta$ respectively. $S_{NC}(Q)$ oscillates around zero and has the property that for a completely random mixture (i.e., an alloy for which the short-range number density of $\alpha+\beta$ atoms around an α atom is equal to that around a β atom) it is identically zero. Another, analogous set of partial structure factors, which is sometimes useful in the study of ionic melts, is based on number (or mass) density and charge-density correlations. For a detailed description of these partials and the relationships between different sets, the reader is directed to the references below.

c) Molecular Liquids

Although the formalism developed so far offers a valid description for all multicomponent systems, it is sometimes convenient to adopt an alternative approach for liquids in which well-defined molecular units exist (e.g., in liquid CH_4, P_4). In this approach the total structure factor is now dependent on both intra- and intermolecular correlations, i.e. on the conformation of the molecular unit itself and on the separation and relative orientation of pairs of molecules.

Scattering from an isolated, rigid molecule will exhibit interference dependent on the relative positions and the scattering lengths of its component nuclei. In the liquid phase there are additional interference terms arising from the correlation between neighbouring molecular units. We rewrite equation (14) in terms of the position vectors of nuclei within a molecule from an origin at the molecular centre and the positions of the molecular centres themselves: we replace $\vec{r}_{\alpha\beta}$ by $\vec{r}_{c'\beta} - \vec{r}_{c\alpha} + \vec{r}_{cc'}$; $\vec{r}_{c\alpha}$

denotes the distance of the αth atom from the molecular center at $\vec{r}_c$, and $\vec{r}_{cc'} = \vec{r}_{c'} - \vec{r}_c$ the distance between the molecular centers c and c'. Separating into intra- and intermolecular terms (i.e., into $c = c'$ and $c \neq c'$ terms), yields an expression of the form:

$$\left(\frac{d\sigma}{d\Omega}\right)_{coh} = N_{mol}\left\langle \sum_{\alpha(c),\beta(c)}^{n} \langle b_\alpha \rangle \langle b_\beta \rangle e^{-i\vec{Q}\cdot(\vec{r}_{c\beta}-\vec{r}_{c\alpha})} \right\rangle + \left\langle \sum_{c\neq c'}^{N_{mol}} e^{-i\vec{Q}\cdot\vec{r}_{cc'}} \sum_{\alpha(c),\beta(c')}^{n} \langle b_\alpha \rangle \langle b_\beta \rangle e^{-i\vec{Q}\cdot(\vec{r}_{c'\beta}-\vec{r}_{c\alpha})} \right\rangle \quad (40)$$

N_{mol} is the number of molecular units, each containing n atoms having average scattering lengths $\langle b \rangle$. The first term will give a normalised structure factor relating to the conformation of an individual molecule, the so-called molecular form factor:

$$f_1(Q) = \frac{1}{\left(\sum_{\alpha}^{n} \langle b_\alpha \rangle\right)^2} \sum_{\alpha,\beta}^{n} \langle b_\alpha \rangle \langle b_\beta \rangle \frac{\sin(Qr_{\alpha\beta})}{Qr_{\alpha\beta}} e^{-\langle u_{\alpha\beta}^2 \rangle Q^2/2} \quad (41)$$

where $r_{\alpha\beta}$ is the distance between atoms α and β in the molecule. We have assumed here that there are no preferred molecular orientations and performed the ensemble averaging. The additional Debye-Waller term at the end of equation (40) is usually included to take account of the thermal vibrations within the molecule; $\langle u_{\alpha\beta}^2 \rangle/2$ is the mean square amplitude of vibration between a pair of atoms α and β. The remainder of equation (40) represents the interference terms due to neighbouring molecules and depends on the distribution of molecular centres, in $\vec{r}_{cc'}$, and the relative orientation contained in $(\vec{r}_{c'\beta} - \vec{r}_{c\alpha})$. A structure factor derived from this intermolecular interference function provides information on the liquid structure as a whole, rather than on the molecules. The total structure factor of the molecular liquid is therefore often expressed as:

$$S_M(Q) = f_1(Q) + D_M(Q) \quad (42)$$

where $D_M(Q)$ is the ensemble average of the molecular correlations. For a molecular liquid the real-space correlations giving rise to $D_M(Q)$ are attenuated with increasing Q much more rapidly than is $f_1(Q)$ and are usually negligible beyond 8-10 Å^{-1}. This is not so for amorphous molecular solids, as one might expect, where $f_1(Q)$ dominates only at relatively large values of Q (pulsed sources are particularly useful for these studies, as they reach Q-values up to 50 Å^{-1}). With some foreknowledge of the molecular bond lengths and angles from gas phase (e.g., electron diffraction) measurements it is possible to fit an $f_1(Q)$ to the high Q data to give quite precise condensed phase values. We may then subtract the fit-adjusted function from the total

$S_M(Q)$ and obtain $D_M(Q)$, the first peak of which primarily depends on the distribution of molecular conters and is largely independent of the choice of model for orientational correlations (attempts have also been made to study the molecular orientation form factor, called $f_2(Q)$, which corresponds to the last term of the second part of the right hand side of eq. (40), which can also be written in the form $S(Q) = f_1(Q) + [S_c(Q)-1]\cdot f_2(Q)$, but we will not embark on this subject here).

Fourier transformation of $D_M(Q)$ provides the real-space information on the correlations between distinct molecules:

$$d_L(r) = g_L(r) - 1 = \frac{1}{2\pi^2 \rho_M} \int_0^\infty Q^2 D_M(Q) \frac{\sin(Qr)}{Qr} dQ \qquad (43)$$

For a multicomponent system, the corresponding partial structure factors could be defined. Many of the molecular fluids of most interest contain light nuclei, such as N, C and H. It follows then that corrections to the data for the effects of inelasticity will be important. However, for a strongly bonded and coupled system this calculation is far from being straightforward and the inclusion of data on the system's vibrational spectrum might be necessary.

References

1. S.W. Lovesey, Theory of Neutron Scattering from Condensed Matter, Oxford Science Publications, Oxford (1984)
2. G.L. Squires, Introduction to the Theory of Thermal Neutron Scattering, Cambridge University Press, Cambridge (1978)
3. N.E. Cusack, The Physics of Structurally Disordered Matter, Adam Hilger, Bristol (1978)
4. P. Lamparter and S. Steeb, "Structure of Amorphous and Molten Alloys", in: Materials Sciene and Technology - A Comprehensive Treatment, R.W. Cahn, P. Haasen, E.J. Kramer (eds), Vol. 1, Structure of Solids, V. Gerold (ed.), pp. 217-288 (1993).
5. A.J. Barnes, W.J. Orville-Thomas and J. Yarwood (eds.), Molecular Liquids - Dynamics and Interactions, D. Reidel Publishing Company, Dordrecht (1984).
6. Y. Waseda, The Structure of Non-Crystalline Materials: Liquids and Amorphous Solids, McGraw-Hill, New York (1980).
7. J.G. Powels, "The Structure of Molecular Liquids by Neutron Scattering", Adv. in Phys., 22, 1-56 (1973).

MUTUAL SOLUBILITY IN THE SOLID PHASE OF SIMPLE MOLECULAR SYSTEMS AT HIGH PRESSURE

J.A.SCHOUTEN AND M.E.KOOI
Van der Waals-Zeeman Institute
University of Amsterdam, Valckenierstraat 65
1018 XE Amsterdam, The Netherlands.

1. Introduction

Considerable effort has been put in the development of theories for the description of the behaviour of gas-liquid and liquid-liquid phase equilibria. Literature about the mutual solubility of simple molecular systems in the solid phase is rather limited. More work has been done on metallic systems. It is generally believed that solubility at high density, in particular in solid systems, is mainly governed by geometrical effects. For instance, the well known Hume-Rotary rule, often used by metallurgists, states that a binary mixed solid is only obtained if the diameters of the molecules differ by less than 15 %, otherwise the mixture will separate into the pure solids. We will investigate whether such a geometrical rule is also valuable for molecular systems. Nowadays, the geometrical effects can be accurately calculated by means of computer simulations as well as analytical theories such as density functional theory. We will first consider in section 2 the case of binary hard sphere mixtures, since it is to be expected that the behaviour of those mixtures will give some insight into the effects of the difference in molecular size on the phase behaviour of real binary mixtures of simple molecular systems at high density. Moreover, we will deal with the phase behaviour of colloidal mixtures which is often in close agreement with that of hard sphere mixtures. In section 3 the experimental results on disordered solid solutions are examined. In section 4 we discuss a new class of compounds, the van der Waalscompounds, characterised by only weak intermolecular forces. In section 5 the differences between the experimental behaviour of molecular and hard spheres systems will be analysed in more detail and some conclusions will be drawn. Finally in section 6 a summary will be given.

2. Hard sphere systems

The phase diagram depends strongly upon α, the ratio of the diameters of the molecules. Kranendonk and Frenkel [1,2] have made an extensive study of

R. Winter and J. Jonas (eds.), High Pressure Molecular Science, 187–204.

substitutionally disordered binary hard sphere systems (solid solutions) using both Monte Carlo and molecular dynamics simulations in the range $0.85<\alpha<1.0$. Similar simulations with comparable results have been performed by Kofke [3]. It turned out that three different regions for α could be distinguished. For these regions the p-x diagrams (x is the mole fraction of the small molecules) of binary systems are depicted schematically in fig.1.

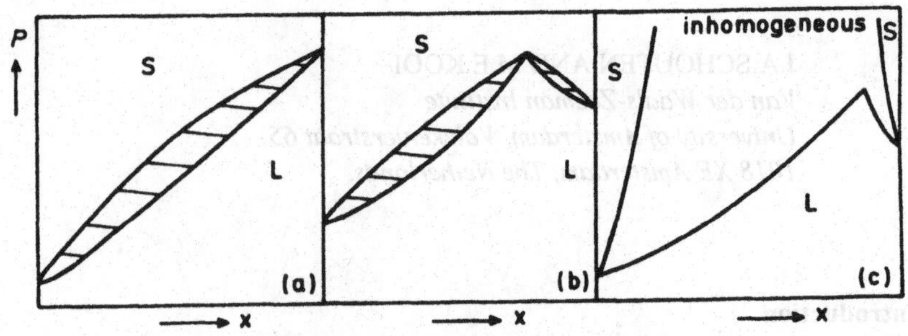

Figure 1. Phase diagram of hard sphere mixtures obtained from computer simulations (schematically): a) $\alpha > 0.94$; b) $0.87 < \alpha < 0.94$; c) $\alpha < 0.87$.

For $\alpha>0.94$ (fig.1a) both the solidus as well as the liquidus move monotonically from $x=0$ to $x=1$. The phase diagram is of the spindle type. The inhomogeneous region is very small and the mutual solubility is excellent. At high enough pressure the mixed solid can exist for all values of the mole fraction. For $0.87<\alpha<0.94$ the situation is only slightly different because the mixture shows an azeotropic behaviour (fig.1b). More interesting is the case $\alpha<0.87$ since the phase diagram changes dramatically (fig.1c). The liquidus still shows a pressure maximum, the so-called eutectic point, but the solidus is separated into two distinct branches resulting in a large miscibility gap in the solid region. The solubility of large spheres in the solid formed by small spheres is almost zero due to deformation of the structure, but also the solubility of small spheres in the solid formed by the large spheres is rather limited. In both cases the solubility initially increases with pressure, but reaches a maximum at higher pressures.

It is interesting to note that these authors found solid-solid phase separation in systems with a phase diagram showing azeotropy. The two coexisting solids have the same structure and the critical pressure decreases with decreasing value of α. Also in this case the solubility of large spheres in the matrix of small spheres is smaller than the other way round.

In addition there is an extensive literature on (modified, weighted) density functional theory of the melting behaviour of disordered solid solutions of binary hard spheres systems [4-8]. These calculations are in reasonable agreement with the results of computer simulations.

Another approach is followed by Cottin and Monson [9] who extended the well known cell theory of Lennard-Jones and Devonshire to mixtures by evaluating simultaneously all possible cell partition functions taking into account changes in composition and nearest neighbour environment. It is assumed that the distribution of the species among the lattice sites is according to the Bragg-Williams or mean field approximation. It turns out that the excess volume on mixing is positive, in contrast to hard sphere fluid mixtures. In general the predictions of this cell theory are in better quantitative agreement with the simulation results than those of density functional theory. It should be noted that the cell theory does not allow for lattice distortions, which results in a lower solubility of the small spheres in the solid of large spheres.

So far we have discussed the formation of substitutionally disordered mixed solids where the diluent molecules are not regularly arranged i.e. there is a continuous range of concentrations of the second component. The large spheres are substitutionally replaced by small spheres and *vice versa.* For smaller values of α unusual superstructures of ordered solid solutions with stoichiometric composition may be formed as shown in experiments of Bartlett *et al.*[10,11] on binary mixtures of colloidal particles. Extensive computer simulations in combination with free energy calculations have been performed by Eldridge *et al.* [12-14] in a broad range of α values. A comparison between computer simulations and experiment is given in ref. 15. Section I of that paper also gives a very good overview of the subject. The formation of these superlattices is entropy-driven [12]. In the ordered system the available free volume is larger than in the disordered system at the same density. An important quantity in the discussion is the packing or volume fraction: the fraction of the total volume occupied by the spheres. The maximum packing fraction of a closed-packed crystal of hard spheres is η=0.7405, the coexisting fluid-solid densities for a monodisperse hard sphere system are η=0.494 and η=0.545 respectively, while the maximum value for an amorphous assembly of hard spheres is $\eta \cong 0.64$. According to the assumption of Murray and Sanders [16] superlattice structures would be formed if the maximum packing fraction is larger than that of the pure solids, otherwise phase separation would occur in both pure components. This results for $0.3<\alpha<1$ in stable AB and AB_2 compounds (A is the component with the large molecules) while AB_{13} almost satisfies the criterion. The packing fraction of AB_2 is greater than 0.74 for $0.482<\alpha<0.624$ with a maximum value for α=0.577. For AB_{13} the maximum value of η occurs for α=0.558. The AB compound (NaCl structure) has a maximum value of η for α=0.414.

Eldridge *et al.* [13] found that over a range of densities and diameter ratios, AB_{13} is thermodynamically stable both with respect to the fluid mixture and pure A and B solids. Their results are in agreement with the findings of Murray and Sanders [16] except that the AB_{13} structure was not stable in the simulations for α=0.303. The stability region for AB_{13} is $0.48<\alpha<0.62$ although the lower limit is less certain because they used the Mansoori equation of state for the liquid phase, which is less valid for small diameter ratios. For AB_2 they find [14] a stability range $0.425<\alpha<0.62$. This is consistent with the experimental findings and with the results of Murray and Sanders, except that the lower limit is slightly lower. A *p-x* phase diagram of a binary mixture of hard spheres with a diameter ratio α=0.58 is presented in ref.12 (fig.2).

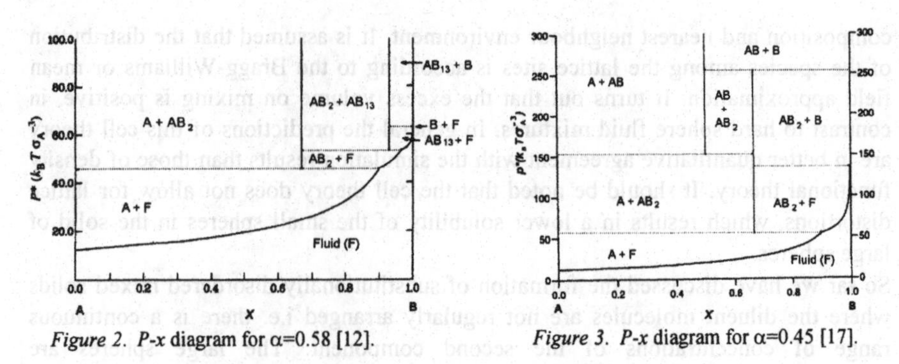

Figure 2. *P-x* diagram for α=0.58 [12].   *Figure 3.* *P-x* diagram for α=0.45 [17].

The AB phase has not been found experimentally but on the basis of the packing principle it could occur for α<0.45. It has been shown that the CsCl structure is unstable[15] but Trizac *et al.*[17] reported computer simulations of the NaCl structure and found that AB(NaCl) is stable for α=0.45 and 0.414. They presented *p-x* diagrams for these values of α. Figure 3 shows that for α=0.45 both AB and AB_2 are present and that AB_{13} is not stable For α=0.414 AB_2 has also disappeared and AB occurs already at much lower pressures.

Xu and Baus [18] have used density functional theory to determine the stability of stoichiometric phases and found that both AB_2 and AB_{13} can be formed. On the basis of the principle of maximum packing fraction AB_{13} would not be formed, but they show that entropy of mixing is responsible for the stability of the AB_{13} crystal.

Cottin and Monson also used their cell theory for the calculation of the stability of substitutionally ordered solid solutions [19]. With respect to the thermodynamic properties the agreement with Monte Carlo simulation results is again quite good. The cell theory slightly overestimates the pressure, but less than in the case of substitutionally disordered solid solutions, reflecting the fact that lattice distortions are less important for stoichiometric solids. With respect to the phase diagrams the agreement with the simulations [12] is very good, except for the range of the coexistence region of AB_{13} + fluid. They also investigated the phase diagram for α=0.732, where the CsCl structure has its maximum packing fraction, but this structure was not found to be stable. Although structures of AB_3, AB_4, and AB_5 should be stable for α<0.482 according to the space-filling principle, these structure were not found to be stable. Moreover, they noticed that all the ordered solutions form when the mole fraction of large spheres in the fluid is less than 15 %. The stability regions for the various compounds, calculated from cell theory, are the following: AB (NaCl structure): 0.2≤α≤0.42, AB_2: 0.42 ≤α≤0.59, and AB_{13}: 0.54≤α≤0.61. The stability domains of the cell theory are slightly smaller than those of the simulations, possibly because in the theory all possibilities for multiphase coexistence have been investigated. An overview of the results obtained by both cell theory and computer simulations is given in fig.4.

Eldridge *et al.* [15] investigated the difference between experiments with colloidal mixtures and simulations in more detail and concluded that these differences can be attributed to kinetic effects. They also reported that AB_{13} is easily formed from the fluid phase but that crystals of pure A and AB_2 do not crystallize readily from fluid mixtures which differ considerably in composition. It is worth noting that in the simulations as well as in the calculations compounds which melt congruently are absent [20].

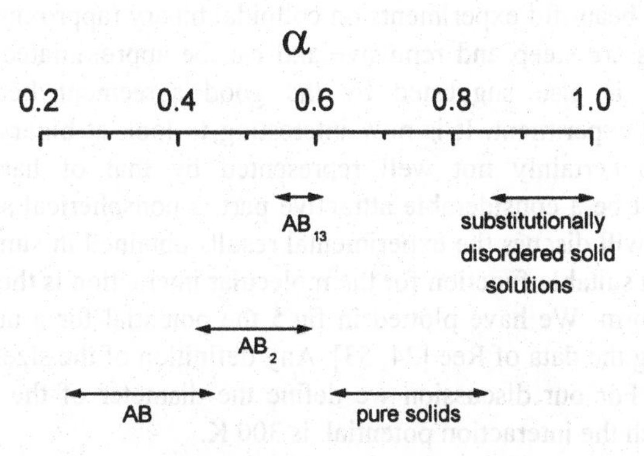

Figure 4. Stability domains in terms of α [19].

Young [21] applied a one-fluid-one-solid model to describe the phase behaviour of hard sphere mixtures. For the liquid phase the one-fluid approximation is used where the hard sphere diameter of the mixture is assumed to be composition dependent. In the solid the particles are placed at random on an fcc lattice and a modified one-solid equation of state has been used. The one-solid hard sphere diameter depends not only on composition but is also a function of α. This function has the effect of increasing the diameter of the solid above that of the liquid and is introduced because for smaller values of α the solid demixes but the liquid is still homogeneous. There is only qualitative agreement between computer simulations and this model.

Recently Cottin and Monson [22] applied classical thermodynamics to calculate solid-fluid equilibrium in hard sphere mixtures in two extreme cases where the solid solutions are either ideal (α close to unity) or where there is complete immiscibility in the solid phase ($\alpha<0.85$). The assumption of ideal fluid behaviour has hardly any influence on the results but disregarding the nonideality in the solid phase is very important. The theory is not applicable to systems exhibiting azeotropy or to substitutionally ordered solutions. For $\alpha=0.97$ and $\alpha=0.73$ the agreement is good if compressibility effects are taken into account.

Finally, Coussaert and Baus [23] calculated the phase diagram for $\alpha=0.15$ and $\alpha=0.3$ using for the liquid phase a virial approach with a negative fifth virial coefficient. They found partial freezing or sublattice melting. This solid alloy (of unknown structure) probably consists of a crystal of large spheres permeated by a fluid of small spheres.

In summary, substitutionally disordered solutions can be expected for $\alpha>0.85$; the compounds AB, AB_2, and AB_{13} are the only ones likely to form in binary hard sphere mixtures.

3. Disordered solutions.

It is generally assumed that disordered solutions can only be formed if $\alpha > 0.85$, in agreement with the hard sphere results. As mentioned before Bartlett *et al.* [11,12] have performed beautiful experiments on colloidal binary (approximately) mixtures. The interactions are steep and repulsive and can be approximated by hard sphere interactions, as is also suggested by the good agreement between computer simulations and experiment. It is now interesting to look at binary systems whose interactions are certainly not well represented by that of hard spheres. The difference might be a considerable attractive part, a nonspherical shape or both. In this section we will discuss the experimental results obtained in simple mixtures. At high pressures a suitable function for the molecular interaction is the three parameter exponential-6 form. We have plotted in fig.5 the potential for a number of simple substances using the data of Ree [24, 53]. Any definition of the size of a molecule is quite arbitrary. For our discussion we define the diameter of the molecule as the distance at which the interaction potential is 300 K.

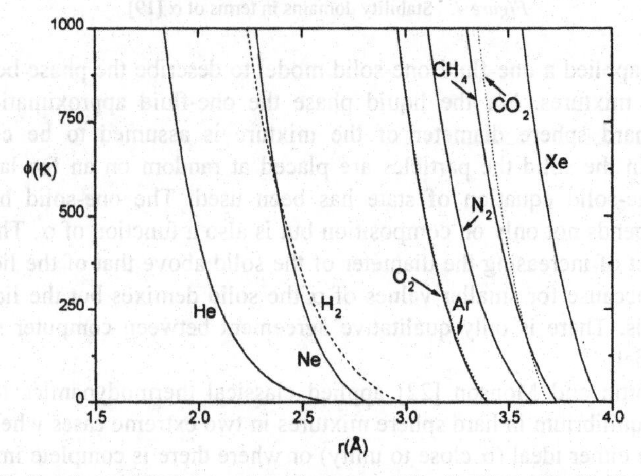

Figure 5. Exp-6 potentials for a number of substances: He, Ne, H_2, Ar, Xe, N_2, O_2, CH_4, CO_2.

Nitrogen systems

The solid mixture nitrogen-argon (α=0.95) has been studied extensively at ambient pressure because the system exhibits a number of interesting phenomena such as glass formation. At high pressure the studies have been limited to the nitrogen-rich side. Westerhoff and Feile [25] studied the system at room temperature, pressure up to about 30 GPa, and mole fraction argon $0<x<0.4$. Kooi and Schouten [26] performed measurements in a range of both temperatures and pressures, and mole fraction argon $0<x<0.25$. The solubility of argon in the solid phases β and δ of nitrogen is very high. The β-δ phase transition line shifts towards lower temperatures or higher pressure with increasing mole fraction of argon while the effect of argon on the melting line is just the reverse, thus extending the β phase

region. The δ-ε phase transition line has not been found experimentally in the mixed system. It probably shifts towards lower temperatures or higher pressure by addition of argon as shown by computer simulations of van Klaveren *et al.*[27]. It is interesting to note that from the Raman spectra it can be inferred how in δ nitrogen the argon atoms are spread over the two different sites. More details about this very interesting system are given in another contribution to this meeting.
The nitrogen-methane (α=0.95) mixture has been investigated at a few compositions and thus only limited information is available [28]. It is reported that at 1 GPa at least 25% nitrogen dissolves in methane but that at 2 GPa the solubility is much less. There seems also to be some solubility of methane in nitrogen at 2 GPa. A Raman investigation [29] of an equimolar mixture of nitrogen and carbon dioxide (α=0.95) at high pressure revealed that the mutual solubility, if any at all, is very small since no extra Raman peaks could be detected in the solid-solid coexistence region.
Recent measurements on nitrogen-neon (α=0.74) reveal that there is clearly solubility of neon in nitrogen [30], despite the small value of α. No solubility of nitrogen in neon could be detected by Raman measurements. An investigation of the system nitrogen-xenon (α=0.89) showed solubility of nitrogen in solid xenon [30]; it is limited to about 8%. The influence of xenon on the Raman spectra of δ-nitrogen turned out to be large. This suggests a considerable solubility of xenon in nitrogen, more than *vice versa.*
In particular in a comparison with nitrogen-argon the system nitrogen-oxygen (α=0.95) is very interesting, since these systems have the same diameter ratio. This mixture has been investigated at room temperature by Baer and Nicol up to 12 GPa [31] and at 18 K by Damde and Jodl up to 25 GPa [32]. At room temperature there is an enormous solubility of oxygen in δ-N_2 (up to 75%) and of nitrogen in β-O_2 (up to 60%). The solubility of oxygen in β-N_2 and of nitrogen in δ-O_2 is comparatively small, while there is no solubility of nitrogen in ε-O_2. The β-δ phase transition line of nitrogen shifts towards lower pressures with increasing mole fraction of oxygen. The pressure was not high enough to investigate the solubility in ε-N_2 but at 11 GPa a new phase appeared in equilibrium with δ-N_2. At low temperature the investigations were restricted to systems with high nitrogen concentration. Again, the solubility of oxygen in the various nitrogen structures is high. In contrast with argon, oxygen is easily soluble in ε-N_2; the ε-γ phase transition line even shifts to lower pressure with increasing oxygen concentration. The new phase detected at room temperature is also present at 18 K at about 15 GPa.

Hydrogen systems
Information about the mutual solubility in the solid phase of hydrogen systems is rather limited. As in previous cases, the solubility of hydrogen in the other component can be detected with Raman spectroscopy relatively easy and, therefore, statements about this solubility are reliable. The other way round is more difficult if the second component is not Raman active and, in that case, Raman measurements give only an indication.

In ref.33 the authors report that in the system hydrogen-argon (α=0.80) both solid H_2 and solid Ar, which are in equilibrium with the compound, are almost pure, without further specification. In ref. 34 it is stated that the solubility of H_2 in solid Ar is of the order of 0.5%. It is also suggested that the H_2 molecules have a tendency to aggregate. There is no solubility of H_2 in CH_4 [35] (α=0.72) and also not in O_2 [36] (α=0.80). Solubility of O_2 in H_2 could not be detected either [36].
At least 10% of H_2 can be dissolved in solid neon (α=0.98) at about 5 GPa, just above the crystallization surface [34]. The solubility decreases as a function of pressure and is about 5% at 15 GPa. In an earlier publication [37] the same author presented a somewhat peculiar p-x diagram. Near the crystallization pressure of hydrogen the solubility of neon in hydrogen is considerably higher than *vice versa*, while it is suggested that at high pressure this solubility is lower.

Helium systems
Considering the diameter ratio the mutual solubility in the system helium-neon (α=0.83) is substantial. The solubility of helium in solid neon at room temperature is about 6% at 5 GPa and increases to about 12% at 16 GPa [38]. About 3% Ne dissolves in solid He at 12 GPa increasing to 4% at 16 GPa.
Compared to the previous system, the mutual solubility in the helium-hydrogen (α=0.82) system is very small. The solubility of H_2 in solid He is less than 0.3% and, moreover, the hydrogen molecules form clusters [34]. The amount of He in H_2 is less than 3% at about 6 GPa and less than 1% at 12 GPa [39] but it must be noted that there is no evidence of any solubility.
The behaviour of the system helium-nitrogen (α=0.62) is rather complicated [40-43] and not yet fully understood. There is no solubility of He in the β phase of N_2. At low temperature a considerable solubility of up to about 10% of helium in nitrogen has been reported [40] while at room temperature it seems to be absent. This solubility has been confirmed by Raman measurements [44].

Argon-oxygen
This mixture (α=1.00) shows an unusual phase behaviour. Up to about 4 GPa a completely disorderly mixed solid is in equilibrium with the fluid phase and up to 5 GPa with a stoichiometric compound [45]. Between 5 and 8.5 GPa the situation is rather complex and should not be discussed here. Above 8.5 GPa there seems to be only a very limited solubility of oxygen in argon and no solubility of argon in oxygen.

4. Van der Waals compounds.

In their study of simple binary systems Vos and Schouten [40] found a disordered solid solution at high pressure and low temperature but in determining the structure of this solid at room temperature Vos *et al.* [41] detected the existence of a stoichiometric compound. It was the first time that the existence of such weakly bound compound had been proved. Because the interaction between the components is of the van der Waals type, this solid is called a van der Waals compound. Since then various other compounds have been found in binary systems composed of simple molecules. We will discuss these compounds and compare the results with the phase diagrams of hard sphere systems of the same diameter ratio.

Helium-nitrogen

The system helium-nitrogen (α=0.62) has been extensively studied from 100 to about 450 K and pressures up to 12.5 GPa. The first indications for a complicated phase diagram were given by two cusps in the three-phase line fluid-fluid-solid, representing quadruple points [42]. These quadruple points do not match with corresponding triple points in the phase diagram of nitrogen. A description of the phase behaviour of the system helium-nitrogen has been presented in ref.43 and the p-x cross-section at room temperature in ref.46. Fig. 6 shows that at room temperature the mixed solid is stable at pressures in excess of 7.7 GPa. The structure observed from x-ray experiments is hexagonal which is different from the structure of pure nitrogen under these circumstances. Therefore, the mixed solid is not just a disordered solid solution obtained by continuous admixture of helium. Moreover, in an experiment with 5 mole% helium a phase separation was observed between the mixed phase and a phase rich in nitrogen since the main vibrons of both the N_2 phases and the van der Waals compound (shifted with respect to that of N_2) could be obtained with Raman spectroscopy. Measurements with 10, 20, and 42 mole% helium gave consistent results. From the Raman spectra it could be concluded that the composition of the compound was between 5 and 10 mole% helium. It must be stoichiometric because the unit volume of the compound is independent of the initial bulk composition of the sample. On the basis of known thermodynamic data the Gibbs free enthalpy difference between the vdW compound and the coexisting phases was calculated. From this analysis it was concluded that the unit cell contains 22 N_2 molecules and 2 He atoms. The crystal structure of $He(N_2)_{11}$ has not (yet) been determined. It should be noted that the stoichiometric ratio is very unusual and that, in contrast with the hard compounds, in this case the species with the larger molecules is the major component (in the notation of the HS section $A_{11}B$).

Helium-neon

Mixtures of rare gases are ideal systems to investigate the influence of attractive forces on solid solutions since the molecules are spherical. The *p-x* diagram of the system helium-neon (α=0.83) has been determined at room temperature and pressures up to 15 GPa by Loubeyre *et al.* [38]. The components are completely miscible in the fluid phase (fig.7), although there might be some doubts about a small region around 13 GPa and x=0.66. At high helium concentrations helium-neon shows the usual eutectic point. A second eutectic point appears at a mole fraction of about 0.5. It is to be expected that such a point is related to fluid-fluid equilibrium but the authors reported only solid-fluid and solid-solid equilibria, and the existence of fluid-fluid equilibrium was explicitly excluded. The most interesting aspect of this apparently simple system is the existence of a homogeneous solid mixture at a helium mole fraction of 2/3 up to at least 25 GPa. Mixtures with other compositions resulted in a separation into two solid phases, one of these having a helium mole fraction of 2/3. X-ray diffraction experiments have been performed on a single crystal of this stoichiometric solid $Ne(He)_2$. The unit cell is hexagonal with an almost ideal c/a ratio. From the equation of state of helium and neon the number of atoms per unit cell was estimated to be 12. The structure is, therefore, assumed to be similar to that of the $MgZn_2$ Laves phase. Given the fact that fluid-fluid equilibrium is absent, the stoichiometeric solid melts congruently, in contrast with the results for hard sphere compounds.

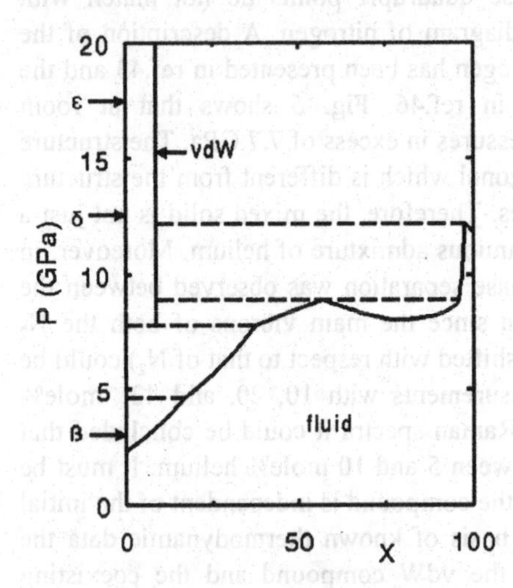

Figure 6. The p-x diagram of helium-nitrogen at room temperature [41,43]. x is mole% helium. The van der Waals compound is indicated by vdW, while β, δ, and ε are solid nitrogen polymorphs

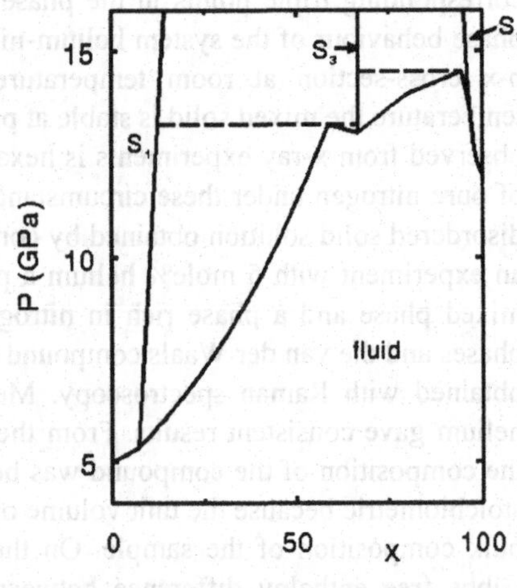

Figure 7. P-x diagram of helium-neon at room temperature [38]. x is mole% helium; S_3 is the $Ne(He)_2$ compound.

Hydrogen-argon

Similar to the previous system also in a high pressure hydrogen-argon mixture (α=0.80) a stoichiometric solid $Ar(H_2)_2$ was found [33]. At 300 K the compound was already formed at 4.3 GPa. For other compositions a solid-solid equilibrium was observed between the new compound and almost pure solid hydrogen or almost pure solid argon. As before, the unit cell is hexagonal and contains 12 atoms. At 100 K the compound was stable up to 175 GPa where a first order phase transition was detected by measuring the intramolecular hydrogen vibron as a function of pressure. The vibron suddenly disappeared and the sample, which was observed under the microscope, looked more absorbent. Similar experiments on pure hydrogen up to 220 GPa did not show these aspects. It is speculated that these phenomena are related to the metalization of hydrogen.

The authors reported that at 4.3 GPa (and room temperature) the compound is in equilibrium with a fluid mixture of the same composition. This means that also this compound melts congruently. At higher pressures the coexisting solids were almost pure components but earlier work also showed solubility of hydrogen in the argon matrix [34]. On the basis of these results the (schematic) *p-x* diagram can be constructed (fig.8). The authors obtained similar results for the $Ar(D_2)_2$ compound.

Hydrogen-oxygen

Loubeyre and LeToullec [36] investigated the hydrogen-oxygen mixture (α=0.80) at room temperature and pressures up to 10 GPa. They found a mixed solid at about 7.5 GPa with about 42 mole% oxygen. The authors are not sure whether it is a stoichiometric compound or a disordered solid but in the first case the composition would be $(O_2)_3(H_2)_4$. The melting would be incongruent, in contrast with helium-neon and argon-hydrogen. At the dilute solid sides no mutual solubility of the components has been detected.

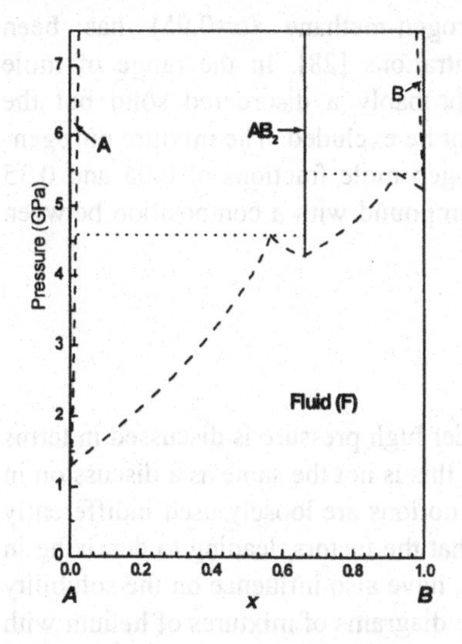

Figure 8. P-x diagram of argon-hydrogen. A is argon.

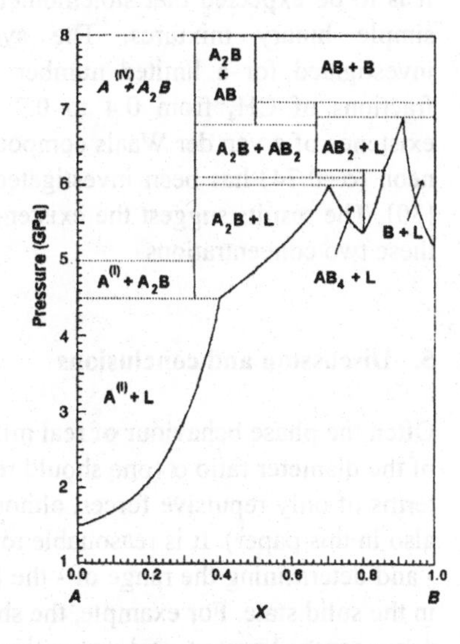

Figure 9. P-x diagram of methane-hydrogen at 298 [35]. A is methane.

Hydrogen-methane

This system (α=0.72) has been investigated by Somayazulu *et al.* [35] at room temperature and pressures up to about 8 GPa. The phase diagram is rather complicated due to the presence of four new compounds: A_2B, AB, AB_2, and AB_4 (A is methane and B is hydrogen) formed at 4.5, 6.7, 6.1, and 5.4 GPa, respectively (fig.9). AB_2 and AB_4 exist only in a very limited pressure range; at the low pressure side only AB_4 melts congruently. As in the previous examples the AB_2 compound

has the structure of $MgZn_2$, the Laves phase. The excess volume of mixing, calculated from the equation of state of hydrogen and methane, is about 1%. It is interesting to note that the compound with 1:1 stoichiometry has an amazingly large volume, 11% larger than the corresponding amounts of pure solids, but still this compound seems to be stable to at least 30 GPa.
It is suggested that there is no solubility of hydrogen in methane since they report the measurement of pure methane in coexistence with A_2B.

Argon-oxygen
Recently Loubeyre *et al.* [45] have investigated this mixture (α=1.00) at room temperature and found a rather complicated phase behaviour. Considering the value for the diameter ratio it is amazing that this system forms two compounds: $Ar(O_2)_3$ above 4 GPa and $Ar(O_2)$ above 6 GPa. Both compounds are stable up to 8.5 GPa.

Other systems
It is to be expected that stoichiometric compounds can be formed in several other simple binary mixtures. The system nitrogen-methane (α=0.95) has been investigated for a limited number of concentrations [28]. In the range of mole fractions of CH_4 from 0.4 to 0.7 there is probably a disordered solid but the existence of a van der Waals compound can not be excluded. The mixture nitrogen-neon (α=0.74) has been investigated for nitrogen mole fractions of 0.05 and 0.35 [30]. The results suggest the existence of a compound with a composition between these two concentrations.

5. Discussion and conclusions

Often the phase behaviour of real mixtures under high pressure is discussed in terms of the diameter ratio α (one should realize that this is not the same as a discussion in terms of only repulsive forces, although these notions are loosely used indifferently also in this paper). It is reasonable to assume that the factors, leading to demixing in - and determining the range of - the fluid state, have also influence on the solubility in the solid state. For example, the shape of the diagrams of mixtures of helium with neon, argon, krypton, and xenon, in particular the existence and range of fluid-fluid demixing, was explained [37] by the difference in the repulsive part of the potential. However, also the attractive part changes considerably from neon to xenon, and Kamerlingh Onnes and Keesom [47] have shown already in the beginning of this century that, given the size ratio in terms of the parameters of the Van der Waals equation, fluid-fluid equilibrium only occurs for small ratios of the attractive part of the potential. Even if the diameter ratio α=1 fluid-fluid equilibrium occurs if the ratio of the attractive depths is below 0.05, and in that case the phase diagram is certainly not of the spindle type. Note that in hard sphere systems fluid-fluid demixing does not occur (possibly with the exception of very small diameter ratios not relevant for the simple systems discussed here) but it does occur in purely repulsive systems [48]. Another important point is that systems should be compared at the same *reduced* temperature and pressure and not at the same absolute value of p and T as in ref.37.

Loubeyre [37] stated that the behaviour of H_2-rare gas mixtures at high pressure is abnormal compared to other rare gas mixtures if one takes into account the diameter ratios. To explain these results the author proposes to introduce an extra attractive interaction term due to charge transfer. We will show that there are also other possibilities.

Obviously the shape of a molecule is of less importance for the properties of the fluid phase than for those of the solid phase, where the orientational behaviour plays an important role. In simple diatomic systems such as nitrogen the shape, in combination with the accompanying quadrupole forces, leads to a complicated phase diagram. The molecular potentials (fig.5) of argon and oxygen are almost identical (α=1). However, the solubility of argon in the various high pressure phases of nitrogen is quite different from that of oxygen in nitrogen. In the case of argon the solubility in ε-N_2 is very small (if present at all [25-27]), but the solubility of oxygen in ε-N_2 at 18 K is considerable [32]. The explanation could be that the orientational degrees of freedom of N_2 and Ar are quite different, leading to a shift of the (orientational) order-disorder transition, while they are nearly the same for N_2 and O_2. There is a very high solubility of argon in β-N_2 and the β-region is extended at higher argon mole fractions at the cost of the δ-region [25,26] but the solubility of oxygen is rather limited and at increasing O_2 mole fraction the β-region shrinks in favor of the δ-region [31]. The directional forces (quadrupole or octupole moment) may be responsible for this difference in behavior. These explanations are supported by computer simulations of pure nitrogen where it has been shown that the various phases can only be obtained if both anisotropy and quadrupole moment are taken into account [49]. Moreover, the free energy differences between solid phases are very small so that small changes in the Van der Waals, anisotropy, and Coulomb contributions to the interaction shift the calculated phase transitions substantially [50]. This holds even more for mixtures where the unlike forces usually have to be estimated from the pure interactions.

It is well known that the interaction between two hydrogen molecules, obtained from *ab initio* calculations, can not be adequately described by a site-site potential. Therefore, the anisotropy might also be responsible for at least part of the 'abnormal' phase behaviour. Moreover, it is not very clear what is meant by 'abnormal' as long as we have no quantitative explanation of the behaviour. For example, the experimental data of the fluid-fluid equilibrium of the system hydrogen-helium can be well explained by e.g. statistical mechanical perturbation theory [51] but all theories and also computer simulations fail in explaining quantitatively the experimental conditions of the fluid-fluid demixing in nitrogen-helium [52].

There is at least one aspect, shown by hard sphere mixtures, which seems to hold also for most of the other systems: the solubility of the large component in the solid of the small component is much less than *vice versa*. This has been observed in He-

Ne, possibly in $He-H_2$, $He-N_2$ and N_2-Ne. However, in the system N_2-Xe the mutual solubility in the δ phase of nitrogen is considerably higher than in Xe.

For smaller values of α van der Waals compounds have been found experimentally with composition $A_{11}B(N_2\text{-}He)$, $A_2B(CH_4\text{-}H_2)$, $AB(CH_4\text{-}H_2)$, $A_3B_4(O_2\text{-}H_2)$, AB_2(Ne-He, Ar-H_2, CH_4-H_2), and $AB_4(CH_4\text{-}H_2)$. As mentioned before, hard sphere systems only exhibit the compounds AB, AB_2, and AB_{13} where B represents the smaller molecule. Thus it is clear that the occurrence is not only dictated by the diameter ratio and that the formation of these compounds is not only entropy driven.
The first calculations on real systems have been performed by Barrat and Vos [53]. They calculated the free energy of the AB_2 and AB_{13} compounds of helium-xenon relative to that of the pure solids with the local harmonic and variational Gaussian approximation schemes of LeSar *et al.* [54]. AB_{13} was found to be unstable but AB_2 (isomorphous with AlB_2) was stable with respect to the pure solids. The authors conclude that their calculations are in agreement with size ratio considerations. We remark that the size ratio for helium-xenon (α=0.55) is within the range established for hard spheres but that the compound has not yet been found experimentally. The compound in He-Ne is of the $MgZn_2$ type (Laves structure) which does not occur in hard sphere mixtures. Free energy calculations, in the local harmonic approximation, with realistic potentials show [38] that the compound is stable. Elaborate calculations have been performed for this system to investigate the driving forces for the stabilization of this phase [46]. It is pointed out that in the Laves phase the unlike interaction between the particles is less important than the like interaction. The calculations show that the volume of the compound is smaller than that of the combination of the pure substances suggesting that the compound is stabilized by an efficient packing. However, it also turns out that both the energy and entropy of the compound are lower than those of the combination of pure substances. Therefore, the formation of this compound is completely driven by the energy. In the mixtures H_2-Ar and H_2-CH_4 the AB_2 compounds have also the Laves structure.
The AB compound found in the system H_2-CH_4 does not obey the condition imposed on hard sphere systems (α<0.45) since α=0.72. Therefore, the conclusion is that for none of the van der Waals compounds the diameter ratio is the only determining factor for the formation.
As mentioned before, the behaviour of helium-nitrogen is rather complicated since at room temperature a Van der Waals compound can be formed while at lower temperatures (also) a disordered solution exists. Recent Raman measurements at room temperature up to 40 GPa [55] reveal strong similarities between the spectrum of the compound and of ε-N_2. The authors suggest a close relationship between the structures of both solids. It is also suggested that the N_2-molecules, which are substituted by helium atoms, originate from the *a* positions in ε-N_2. Therefore, it is possible that the low temperature disordered solution is substitutionally, partial disordered. One would not expect a substitutionally disordered solution of systems with such a small diameter ratio (about 0.62).

Recently Cottin and Monson [56] have applied cell theory to binary Lennard-Jones systems to examine the effect of attractive interactions on the solid-fluid behavior in disordered systems. The parameters were chosen to describe the mixtures of argon, methane, and krypton at ambient pressure. The conclusion is that the effect of a variation of the well depth is mainly a shift in temperature of the phase diagram and a change in the size of the coexistence region, while a change in size ratio has primarily influence on the overall shape of the phase diagram. However, we believe that differentiation is rather artificial since a variation of one of the potential parameters has influence on both the attractive and repulsive part of the potential [51] and the parameters can never be determined in a unique way. Unfortunately the Cottin and Monson did not present calculations in which only one type of potential parameters has been varied. Moreover, the size of the coexistence region can have a large influence on the shape of the phase diagram.
Finally, Yantsevich *et al.* [57] discussed the limited solubility in cryocrystals. Based on the data for nine binary systems of the maximum solubility at 5 K in an atomic cryomatrix, they made the statement that the difference of molecular diameters of the components is not the main parameter for the stability of a solid solution. Their conclusion is that a positive value of the mutual exchange energy is the physical reason for phase separation at low temperature in binary systems of cryocrystals.

6. Summary

For hard sphere mixtures the predictions of the theory are in close agreement with computer simulation results. The phase behaviour depends on the ratio of the diameters. Only three compounds exist: AB, AB_2, and AB_{13}. Disordered solid solutions occur for diameter ratios larger than 0.85 and the solubility of the large molecules in the solid of the smaller component is always less than *vice versa*. The experimental data of colloidal solutions compare quite well with the hard sphere results.
The ratio of the size of the molecules is only a very rough first indication for the behaviour of molecular systems. The vanderWaals interaction, the anisotropy, and the Coulomb forces play an important role both in the formation of compounds and in the solubility in the disordered solution. Subtle changes in these contributions may result in different solid phases. With good potentials it might be possible to predict whether there is any solubility in the solid phase but it will be very difficult to predict the shift of the solid-solid phase transition lines as a function of composition.

7. References:

1. Kranendonk, G.T. and Frenkel, D. (1989) Computer simulations of solid-liquid coexistence in binary hard-sphere mixtures, J.Phys.Condens.Matter 1, 7735.
2. Kranendonk, G.T. and Frenkel, D. (1991) Computer simulations of solid-liquid coexistence in binary hard-sphere mixtures, Mol.Phys. 72, 679
3. Kofke, D. (1991) Solid-fluid coexistence in binary hard sphere mixtures by semigrand Monte Carlo simulation, Molec.Simul. 7, 285
4. Lutsko, F. and Baus, B. (1990) Can the thermodynamic properties of a solid be mapped onto those of a liquid?, Phys.Rev.Lett. 64, 761.
5. Barrat, J.L., Baus, M., and Hansen, J.P. (1987) Freezing of binary hard-sphere mixtures into disordered crystals:a density functional approach, J.Phys.C 20, 1413.
6. Smithline, S.J. and haymet, A.D. (1998) Density functional theory for the freezing of 1:1 hard sphere mixtures, J.Chem.Phys. 86, 6486 (Erratum in J.Chem.Phys. (1989) 88, 4104).
7. Zeng, X.C. and Oxtoby, D.W. (1990) Density functional theory for freezing of a binary hard sphere liquid, J.Chem.Phys. 93, 4357.
8. Denton, A.R. and Ashcroft, N.W. (1990) Weighted density functional theory of nonuniform fluid mixtures: Application to freezing of binary hard sphere mixtures.
9. Cottin, X. and Monson, P.A. (1993) A cell theory for solid solutions: Application to hard sphere mixtures, J.Chem.Phys. 99, 8914.
10. Bartlett, P., Ottewill, R.H., and Pusey, P.N. (1990) Freezing of binary mixtures of colloidal hard spheres , J.Chem.Phys.93, 1299.
11. Bartlett, P., Ottewill, R.H., and Pusey, P.N. (1992) Superlattice formation in binary mixtures of hard sphere colloids , Phys.Rev.Lett. 68, 3801.
12. Eldridge, M.D., Madden, P.A., and Frenkel, D. (1993) Entropy-driven formation of a superlattice in a hard-sphere binary mixture, Nature 365, 35.
13. Eldridge, M.D., Madden, P.A., and Frenkel, D. (1993) The stability of the AB_{13} crystal in a binary hard sphere system, Molec.Phys. 79, 105.
14. Eldridge, M.D., Madden, P.A., and Frenkel, D. (1993) A computer simulation investigation into the stability of the AB_2 superlattice in a binary hard sphere system, Molec.Phys. 80, 987.
15. Eldridge, M.D., Madden, P.A, Pusey, P.H., and Bartlett, P. (1995) Binary hard-sphere mixtures: a comparison between computer simulation and experiment, Molec.Phys. 84, 395.
16. Murray, M.J. and Sanders, J.V. (1980) Close-packed structures of spheres of two different sizes II. The packing densities of likely arrangements, Phil.Mag. A 42, 721.
17. Trizac, E., Eldridge, M.D., and Madden, P.E. (1997) Stability of the AB crystal for asymmetric binary hard sphere mixtures, Molec.Phys. 90, 675.
18. Xu, H. and Baus, M. (1992) A density functional study of superlattice formation in binary hard-sphere mixtures, J.Phys.:Condens.Matter 4, L663.
19. Cottin, X. and Monson, P.A. (1995) Substitutionally ordered solid solutions of hard spheres, J.Chem.Phys. 102, 3354.
20. Cottin, X., Paras, E.P.A., Vega, C., and Monson, P.A. (1996) Solid-fluid equilibrium: new perspectives from molecular theory, Fluid Phase Equil. 117, 114.
21. Young, D.A. (1993) van der Waals theory of two-component melting, J.Chem.Phys. 98, 9819.
22. Cottin, X. and Monson, P.A. (1997) An application of classical thermodynamics to solid-fluid equilibrium in hard sphere mixtures, J.Chem.Phys. 107, 6855.
23. Coussaert, T. and Baus, M. (1997) Virial approach to hard-sphere demixing, Phys.Rev.Lett. 79, 1881.
24. Ree, F.H (1989) A high-density, high-temperature mixture model, in A.Polian, P.Loubeyre, and N.Boccara (eds.), Simple molecular systems at very high density, Plenum Press, New York, pp.153-180.
25. Westerhoff, T. and Feile, R. (1996) High-pressure Raman study of the N_2 stretching vibration in argon-nitrogen mixtures at room temperature, Phys.Rev.B 54, 913.
26. Kooi, M.E. and Schouten, J.A. (1998) Raman spectra and phase behavior of the mixed solidN_2-Ar at high pressure, Phys.Rev. B 57, 10407.

27. Van Klaveren, E.P., Michels, J.P.J., and Schouten J.A. (1998) A Monte Carlo study of disorder in the high pressure mixed crystal N_2-Ar, J. Low Temp.Phys. 111, 413.
28. Van Hinsberg, M.E. (1994) Phase behavior of the systems N_2-H_2O, N_2-CH_4 and N_2-He under extreme conditions, thesis, university of Amsterdam, pp.79-92
29. Kooi, M.E., Schouten, J.A., van den Kerkhof, A.M., Istrate, G., and Althaus, E. (1998) The system CO_2-N_2 at high pressure and applications to fluid inclusions, Geochim. & Cosmochim. Acta .
30. Kooi M.E. and Schouten, J.A. to be published.
31. Baer, B.J. and Nicol, M. (1990) High-pressure binary phase diagram of nitrogen-oxygen at 295 K determined by Raman spectroscopy, J.Phys.Chem. 94, 1073.
32. Damde, K. and Jodl, H-J. (1998) Mixtures of $(N_2)_{1-x}$: $(O_2)_x$ at high pressures and low temperatures, J.Low Temp.Phys. 111, 327.
33. Loubeyre, L., LeToullec, R., and Pinceaux, J-P. (1994) Compression of $Ar(H_2)_2$ up to 175 GPa: A new path for the dissociation of molecular hydrogen?, Phys.Rev.Lett. 72, 1360.
34. Loubeyre, P., LeToullec, R., and Pinceaux, J.P. (1992) Raman measurements of the vibrational properties of H_2 as a guest molecule in dense helium, neon, argon, and deuterium systems up to 40 GPa, Phys.Rev. B 45, 12844.
35. Somayazulu, M.S., Finger, L.W., Hemley, R.J., and Mao, H.K. (1996) High-pressure compounds in methane-hydrogen mixtures, Science 271,1400.
36. Loubeyre, P. and LeToullec, R. (1995) Stability of O_2/H_2 mixtures at high pressure, Nature 378, 44.
37. Loubeyre, P. (1991) The properties of simple molecular binary mixtures at high density: a qualitative understanding, in R.Pucci and G.Picitto (eds.), Molecular systems under high pressure, Elsevier Science Publishers, Amsterdam; pp. 245-262.
38. Loubeyre, P., Jean-Louis, M., LeToullec, R., and Charon-Gérard, L. (1993) High pressure measurements of the He-Ne binary phase diagram at 296 K: Evidence for the stability of a stoichiometric $Ne(He)_2$ solid, Phys.Rev.Lett. 70, 178.
39. Loubeyre, P. (1991) A new determination of the binary phase diagram of H_2-He mixtures at 296 K, J.Phys.: Condens. Matter 3, 3183.
40. Vos, W.L. and Schouten, J.A. (1990) Solubility of fluid helium in solid nitrogen at high pressure, Phys.Rev.Lett. 64, 898.
41. Vos, W.L., Finger, L.W., Hemley, R.J., Hu, J.Z., Mao, H.K., and Schouten, J.A. (1992) A high-pressure van der Waals compound in solid nitrogen-helium mixtures, Nature 358, 46.
42. Van den Bergh, L.C. and Schouten, J.A. (1988) The critical line and the three-phase line in a molecular helium-nitrogen mixture up to 100 kbar, Chem.Phys.Lett. 145, 471
43. Vos, W.L. and Schouten, J.A. (1992) The phase diagram of the binary mixture nitrogen-helium at high pressure, Physica A 182, 365.
44. Scheerboom, M.I.M. and Schouten, J.A. (1991) Detection of theε-δ phase transition in N_2 and a N_2-He mixture by Raman spectroscopy: new evidence for the solubility of fluid He in solid N_2, J.Phys.: Condens. Matter 3, 8305.
45. Loubeyre, P.,Jean-Louis, M., LeToullec, R., Hanfland, M., and Häusermann (1998) Structural studies and recrystallization of Ar/O_2 solid mixtures at high pressure, submitted to PRL
46. Vos, W.L. and Schouten, J.A. (1993) The stability of van der Waals compounds at high pressure, Sovj.J. Low Temp.Phys. 19, 481
47. Kamerlingh Onnes, H. and Keesom, W.H. (1907/1908) Contributions to the knowledge of theψ-surface of Van der Waals.XV. The case that one component is a gas without cohesion with molecules that have extension. Limited miscibility of two gases, Proc.Roy.Acad.Sci.Amst. 9, 786; continuation: 10, 231
48. Schouten, J.A., Sun, T.F., and De Kuijper, A. (1990) Gas-gas phase separation in binary inverse-12 systems, Physica A 169, 17.
49. Mulder, A., Michels, J.P.J., and Schouten, J.A. (1996) The importance of the anisotropic energy term for the structure of the solid phases of nitrogen, J.Chem. Phys. 105, 3235.
50. Meijer, E.J. (1998) Location of fluid-β and β-δ coexistence lines of nitrogen by computer simulation, J.Chem.Phys. 108, 5898.
51. Van den Bergh, L.C. and Schouten, J.A. (1988) Prediction of fluid-fluid and fluid-solid equilibria in the molecular system helium-hydrogen up to 1 Mbar, J.Chem. Phys. 89, 2336.

52. Schouten, J.A., van Hinsberg, M.G.E., Scheerboom, M.I.M., and Michels, J.P.J. (1994) Peculiarities in the high-pressure behaviour of binary mixtures of nitrogen with methane, helium and water, J.Phys.: Condens.Matter 6, A187.
53. Barrat, J-L. and Vos, W.L. (1992) Stability of van der Waals compounds and investigation of the intermolecular potential in helium-xenon mixtures, J.Chem.Phys. 97, 5707.
54. LeSar, R., Najafabadi, R., and Srolovitz, D.J. (1989) Finite-temperature defect properties from free-energy minimization, Phys.Rev.Lett. 63, 624.
55. Olijnyk, H. and Jephcoat, A.P. (1997) High-pressure Raman studies of a nitrogen-helium mixture up to 40 GPa, J.Phys.: Condens.Matter 9, 11219.
56. Cottin, X. and Monson, P.A. (1996) Solid-fluid phase equilibrium for single component and binary Lennard-jones systems: A cell theory approach, J.Chem.Phys. 105, 10022.
57. Yantsevich, L.D., Prokhvatilov, A.I., and Brodyanskii, A.P. (1998) On stability of solid solutions in binary systems of cryocrystals, J.Low.Temp.Phys. 111, 429.

HIGH PRESSURE BEHAVIOUR OF THE VIBRATIONAL SPECTRA OF MIXTURES IN THE FLUID PHASE AND AT THE FLUID-SOLID TRANSITION.

J.A.SCHOUTEN AND J.P.J.MICHELS
Van der Waals-Zeeman Institute
University of Amsterdam, Valckenierstraat 65
1018 XE Amsterdam, The Netherlands.

1. Introduction

Vibrational Raman spectroscopy is one of the most useful tools to study the behaviour of systems at high pressure. Phase transitions can be detected and information about the dynamics of the transition may be obtained from the spectra. Information about the microscopic behaviour of the molecules, orientational as well as translational, is revealed by both the peak position and the line width. In mixtures Raman spectroscopy can be used as a tool for determining the concentration or the concentration fluctuations.

In this paper we will show the possibilities by considering high-pressure data and simulation results obtained about the various aspects in simple systems. First we will discuss the pressure and temperature dependence of the Raman spectra of homogeneous binary fluid mixtures. A comparison of experimental data with results of molecular dynamics simulations will give some insight in the effects of the surrounding medium on the spectra. In the third section results are presented for the change in the vibrational frequency at the solid-fluid transition and the origin of this change will be investigated. In section 4 the effect of critical fluctuations on the line width will be discussed and in section 5 the influence of concentration fluctuations outside the critical region. Finally in section 6 the results will be summarized.

2. Homogeneous binary fluid mixtures

The vibrational frequency of an isolated diatomic molecule is determined by the *intra*molecular forces. Due to the anharmonicity of the intramolecular potential, the *inter*molecular forces cause a change in the frequency of the molecular vibration in a solid or a dense fluid. The fluctuations of the density in space and time in the immediate surroundings of a molecule are reflected in the profile of the spectrum, in particular in the line width.

R. Winter and J. Jonas (eds.), High Pressure Molecular Science, 205–218.

The addition of second component not only changes the intermolecular forces but also the character of the fluctuations may change considerably. In a pure substance the local environment varies due to fluctuations in the density only. Away from the critical point these fluctuations have a small wavelength and a high frequency, resulting in motional narrowing of the spectral line. In a mixture, fluctuations in the local concentration also play a role. In this section we will restrict ourselves to a discussion of the effects of the intermolecular forces and of the density fluctuations.
In pure nitrogen at ambient temperature[1-4] the experimental vibrational frequency first decreases as a function of pressure and than increases, via a minimum at 0.14 GPa, at still higher pressures (fig.1). Above 1 GPa the red shift turns into a blue shift. Sometimes, this behaviour is loosely explained as follows: at low pressure the attractive forces between the molecules are dominant leading to a larger bondlength, due to the anharmonicity of the potential, resulting in a lower frequency, while at higher pressure the repulsive forces are the most important. Such an explanation is misleading as will be shown.

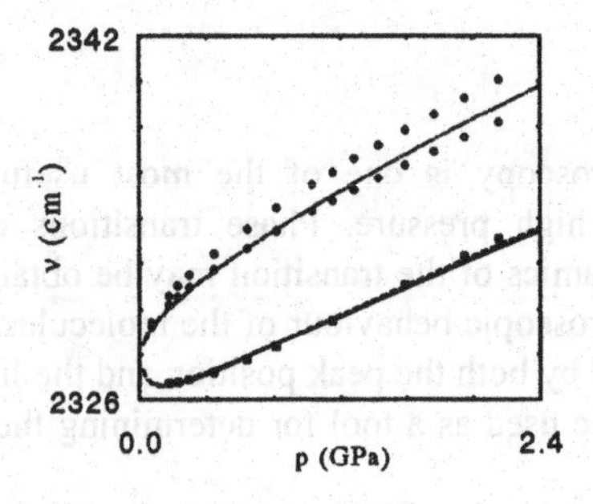

Figure 1. The shift of the Q-branch vibration line of N_2 at 296 K as a function of pressure[4]. Solid line, experimental data [2,4]; MD results without bondlength dependent contribution are represented by dots; with (full dots) and without (open dots) second order effect. The squares are for the MD calculation with bondlength dependent contribution. The HF calculation with second order effect is represented by the dashed dotted line. The full HF calculation with attractive contribution is represented by the dashed line.

Various authors have calculated the line shift by analytical methods and computer simulations but it turned out to be difficult to obtain a good quantitative agreement [3,5-9]. Quite generally, the influence of the intermolecular interaction on the frequency of a molecule is determined by the derivatives of the intermolecular potential energy with respect to the molecular bondlength. Schweizer and Chandler [6] calculated the density dependence of the frequency shift by modelling the nitrogen molecule as a hard dumbbell in a surroundings of hard spheres ("hard fluid" model). An attractive term was added to take into account the bondlength dependence of the interaction. This term was taken linear in density and the proportionality constant as an adjustable parameter. This method was modified by Ben-Amotz and co-workers [3] and resulted in a good agreement for nitrogen. They made the important statement that "the constant (adjustable parameter) depends on the differential attractive solvation energy of the solute in the ground state and excited vibrational states". However, for other systems the agreement was less satisfactory and a quadratic density dependence had to be assumed for the attractive term [10]. Levesque *et al.* [5] performed MD simulations on liquid nitrogen but the calculated shift was found to be too negative.

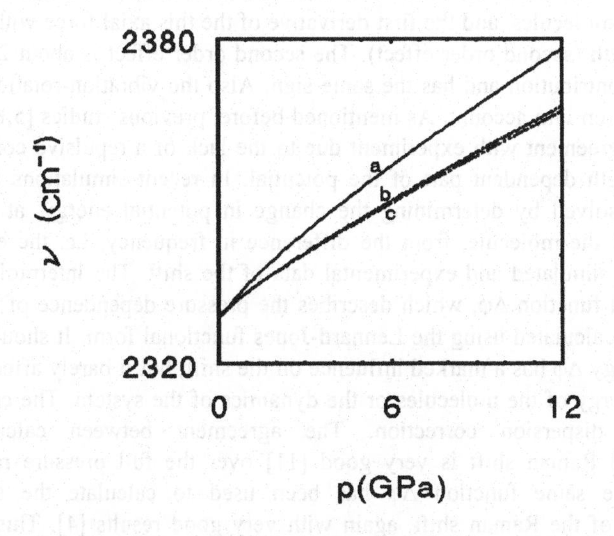

Figure 2. **Experimental Raman shift of N_2 diluted in helium as a function of pressure at 296 K (a). Results from the MD simulation without bondlength dependent contribution (b). HF results without attractive contribution (c).**

The reason for this discrepancy is most easily demonstrated by the frequency behaviour of a dilute mixture of nitrogen in helium. In figure 2 the experimental results are presented for this mixture at room temperature up to 12 GPa [4]. Also shown are the hard fluid (HF) calculations for this system. The most important feature is that, in contrast with pure nitrogen, in the mixture the frequencies calculated with the HF model are clearly below the experimental values. Therefore, it is impossible to get agreement with experiment by adding the attractive correction, which is always negative, to the HF values. Part of the discrepancy might be due to the approximations made in the HF model. Therefore, computer simulations have been performed for pure nitrogen [11] to test the validity of the approximations (the method will be explained later). Figure 1 shows that the agreement between simulations and the HF model is good. In general, it turns out that the details of the intermolecular potential hardly influence that part of the contribution to the frequency shift that is bondlength independent. Thus it is to be expected that the HF model gives good results for the dilute mixture too, as is shown in fig.2. The conclusion is, therefore, that also in the *fluid* phase the bondlength dependent part of the potential must have a repulsive component, as was already shown 20 years ago for the *solid* phases of nitrogen by Thiéry and Chandrasekharan [12]

Computer simulations provide the opportunity to investigate the relative importance of the various contributions to the frequency shift and to study the sensitivity of the results for variations in the intermolecular potential. The calculation of the so-called repulsive contribution is straightforward [8,9,13] and implies the calculation of the axial force on the molecule under consideration (first order effect), exerted by the surrounding molecules, and the first derivative of the this axial force with respect to the bondlength (second order effect). The second order effect is about 20 % of the first order contribution and has the same sign. Also the vibration-rotation coupling should be taken into account. As mentioned before, previous studies [5,8,9] did not give good agreement with experiment due to the lack of a repulsive component in the bondlength dependent part of the potential. In recent simulations [11,13] the problem is solved by determining the change in potential energy, at vibrational excitation of the molecule, from the difference in frequency, i.e. the energy gap, between the simulated and experimental data of the shift. The intermolecular site-site potential function $\Delta\varphi$, which describes the pressure dependence of this energy gap, is then calculated using the Lennard-Jones functional form. It should be noted that the energy $\Delta\varphi$ has a marked influence on the shift, but it barely affects the total potential energy of the molecules or the dynamics of the system. The correction is called the dispersion correction. The agreement between calculated and experimental Raman shift is very good [11] over the full pressure range up to melting. The same function $\Delta\varphi$ has been used to calculate the temperature dependence of the Raman shift, again with very good results [4]. This procedure works also very well in the case of the dilute mixture of nitrogen in helium in which case the *repulsive* part of $\Delta\varphi$ is dominant and assures that the bondlength dependent correction is positive.

For various mixtures the frequency has been determined as a function of concentration [14,15]. In the case of helium-nitrogen at constant pressure and temperature the concentration dependence is larger on the helium side than on the nitrogen side. However, if the frequencies are plotted as a function of volume fraction the dependence is linear within experimental uncertainty [14].

The details of the vibrational spectrum, e.g. the line width, can provide information about the dynamical behavior of the system. The line width (or dephasing time) has been determined experimentally for a number of pure substances and mixtures. As a function of pressure it exhibits the same behaviour as the frequency (fig.3): initially it decreases and increases again at higher pressures. Various investigators have tried to calculate the line width by analytical methods and computer simulations [17-23] but with limited success. To get more insight into the mechanisms which determine the line width, we will discuss some systems for which computer simulations have been performed recently, showing good agreement with experiment [11,13,24].

For a discussion of the shape the following quantities are important: i) the width of the momentary frequency distribution Δ, commonly called the amplitude of modulation, ii) the frequency autocorrelation function $\Omega(t)$, and iii) the correlation time τ_c.

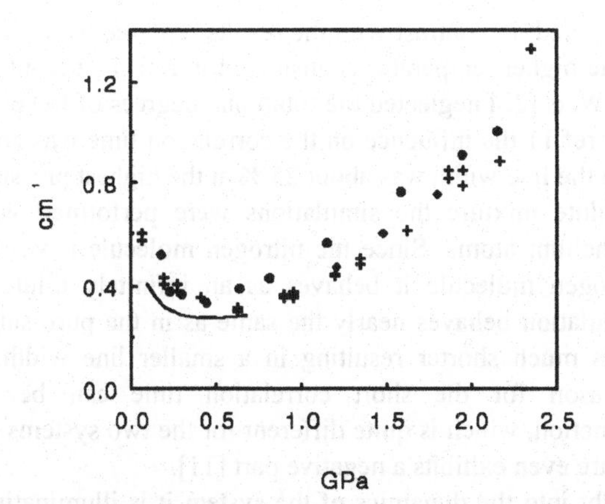

Figure 3. **The experimental linewidth (FWHM), obtained by various investigators, for the Q-branch of N_2 at 296 K as a function of pressure. Dots: MD results.**

$$\Delta = \{ \langle\omega(0)^2\rangle - \langle\omega(0)\rangle^2 \}^{1/2} \tag{1}$$

$$\Omega(t) = \frac{\langle\omega(0)\omega(t)\rangle - \langle\omega(0)\rangle^2}{\Delta^2} \tag{2}$$

$$\tau_c = \int_0^\infty \Omega(t)\,dt \tag{3}$$

The brackets in equations (1) and (2) denote the averaging over all particles throughout the entire simulation run. According to the Kubo-theory [16] the line width can be calculated in the slow and in the fast modulation regime from τ_c and Δ. In the slow modulation regime ($\Delta.\tau_c \gg 1$), which holds e.g. for critical fluctuations, the "full-width at half maximum" is FWHM=Δ and the line shape is Gaussian, while in the fast modulation regime ($\Delta.\tau_c \ll 1$), FWHM=$2\Delta^2.\tau_c$, and the line shape is Lorentzian. The line width has been calculated using simulations and determined experimentally for nitrogen and dilute mixtures of nitrogen in helium, argon, neon and xenon. In those systems the fast modulation criterion is fulfilled. Since for all these systems the agreement between experiment and simulations is good, it is interesting to look into more detail and compare the results for the extreme cases: nitrogen and nitrogen infinitely diluted in helium.
In pure nitrogen the amplitude of modulation increases monotonously as a function of pressure and has increased by a factor of 2 at the melting line. The correlation time first decreases sharply and then slowly increases again. Thus the minimum in line width is caused by the behaviour of the correlation time. The effect of a change

in bondlength is limited, in contrast with the results reported by Levesque *et al.* [5], probably due to the higher temperature considered in ref.11. It is interesting to note that Chesnoy and Weis [22] neglected the rotational degrees of freedom while in the results reported in ref.11 the influence on the correlation time was considerable and the contribution to the line width was about 25 % at the highest pressures.
In case of the dilute mixture the simulations were performed with 1 nitrogen molecule in 255 helium atoms. Since the nitrogen molecule never has a collision with another nitrogen molecule it behaves as an infinitely dilute solution. The amplitude of modulation behaves nearly the same as in the pure substance but the correlation time is much shorter resulting in a smaller line width for the dilute solution. The reason for the short correlation time can be found in the autocorrelation function, which is quite different for the two systems and in the case of the dilute mixture even exhibits a negative part [11].
To get more insight into the dynamics of the system it is illuminating to study the autocorrelation function in more detail [26]. $\Omega(t)$ is build up from the contributions of the first (F_1) and second (F_2) order effect, the dispersion correction (DC), the vibration-rotation coupling (VRC) effect and the often neglected cross correlations. The total correlation function thus consists of 4 self-correlations and 6 cross terms:

$$\Omega_{\alpha,\beta}(t) = \langle\omega_\alpha(0)\omega_\beta(t)\rangle - \langle\omega_\alpha(0)\rangle\langle\omega_\beta(0)\rangle \tag{4}$$

$$\Omega(t) = \sum_{\alpha,\beta} \Omega_{\alpha,\beta} \tag{5}$$

where α and β denote F_1, F_2, VRC or DC, and Δ is again the *total* amplitude of modulation. From these functions the separate contributions to the correlation time and the total correlation time have obtained by the integrations:

$$\tau_{\alpha,\beta} = \int_0^\infty \Omega_{\alpha,\beta}(t)\, dt \tag{6}$$

$$\tau_c = \sum_{\alpha,\beta} \tau_{\alpha,\beta} \tag{7}$$

It should be pointed out that the origin of the frequency change due to VRC is fundamentally different from the other three effects. The latter are due to intermolecular interactions but VRC is due to the velocity of rotation of the molecule under consideration. It is known from statistical mechanics that these entities are momentary independent. Thus, all the cross terms with VRC must be zero at t=0 and do not contribute to the amplitude of modulation.
The relative contributions of the other cross terms to the amplitude are very important since they sum up in nitrogen to about 35% at 2 GPa. It is shown in figure

4a that the self-correlation due to both the first order effect and VRC (and cross correlation) is significant at times above 0.1 ps and that this behavior is responsible for the long tail. The cross terms are also very important for $\Omega(t)$. Note that all the cross terms that include the VRC effect are negative.

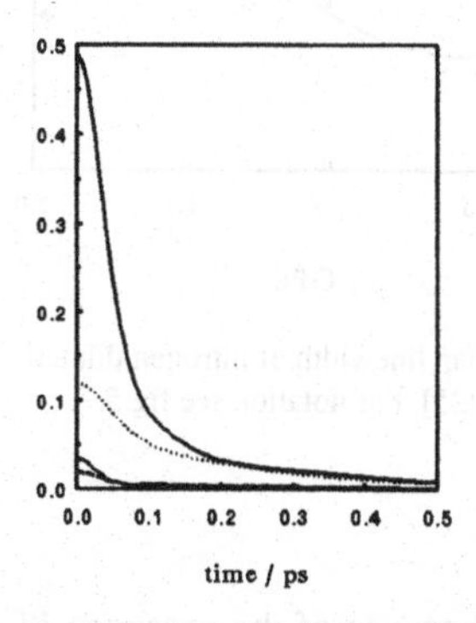

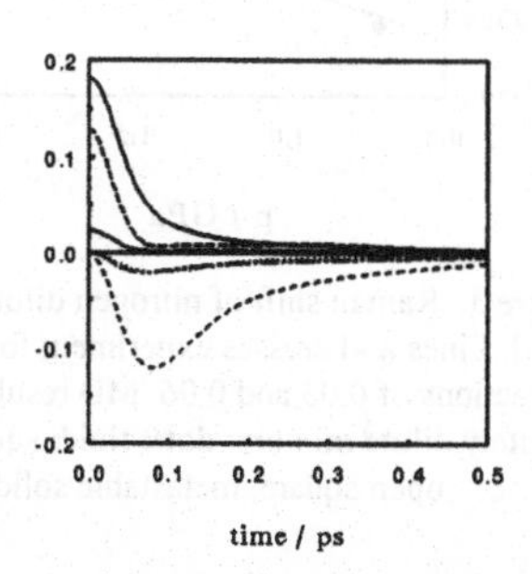

Figure 4. Contributions to the correlation function of the vibrational frequency of nitrogen at 1.1 GPa and 296 K. a) Autocorrelations due to F_1 (solid line) and F_2 (dashed), due to VRC (dotted), and to DC (dash-dotted). b) Cross-correlations with non-zero values at t=0 due to F_1-F_2 (solid), F_1-DC (dashed), F_2-DC (dash-dotted); and with zero values at t=0 due to F_1-VRC (dashed), F_2-VRC (dotted), and DC-VRC (dash-dot-dot).

For the dilute mixture of nitrogen in helium the dispersion correction can be neglected [13]. The correlation functions for the other effects are plotted in fig.4b. In comparison with pure nitrogen these functions fall off more rapidly at times less than 0.1 ps. The correlation function of the first order effect becomes negative after about 0.06 ps. This is the reason that the line width of the Raman spectrum of nitrogen diluted in helium is considerably less than that of pure nitrogen. The effect is primarily due to the intrinsic properties of helium, e.g. the mass and the mutual interaction, and not to the interaction between nitrogen and helium molecules [26].

3. Solid-fluid transition

Isobaric measurements of the vibrational frequency have shown that it is nearly indepent of temperature [4]. This suggest that it is not the density that is primarily responsible for the change in line shift but the pressure. On the other hand it is well known that generally the frequency increases discontinuously at the transition from the fluid to the solid phase. This increase is usually ascribed to the increase in density [27,28]. To solve this apparent contradiction experiments and computer simulations have been carried out on pure nitrogen [25], nitrogen diluted in argon [25], and nitrogen diluted in xenon [29] around the solid-fluid transition.

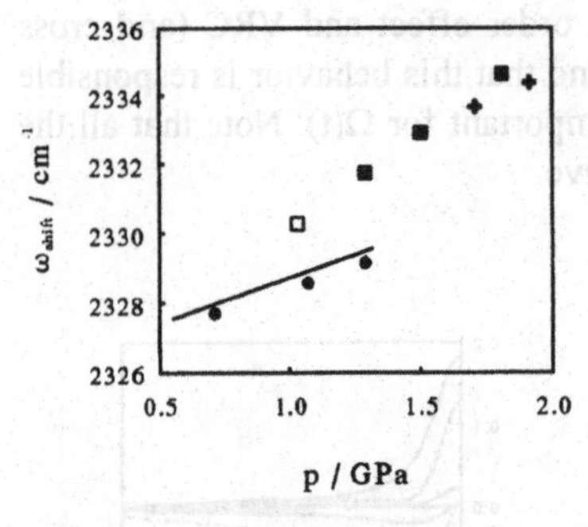

Figure 5. Raman shift of nitrogen diluted in argon [25]. Lines and crosses experiment for N_2 mole fractions of 0.03 and 0.06. MD results for an infinitely dilute mixture: dots, fluid; squares, solid; open square, metastable solid.

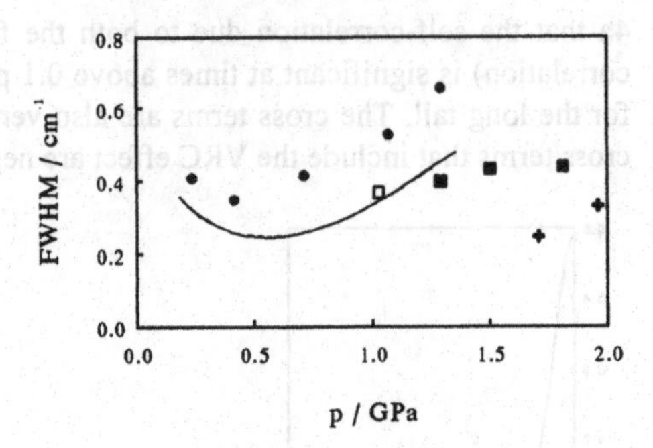

Figure 6. Raman linewidth of nitrogen diluted in argon [25]. For notation see fig.5.

This experiments should be performed at high pressures because of the presence of resonance coupling at low pressure. Figures 5 and 6 show the experimental results for the mixture argon-nitrogen at room temperature. As usual, at higher pressures both the frequency and the line width slowly increase as a function of pressure, but the frequency exhibits a jump of a few cm^{-1} at the transition to the solid phase, while the line width shows a clear drop. For pure nitrogen the behaviour is essentially the same. The figures also present the results of the simulations. The calculations are in good agreement as far as the frequency is concerned. The agreement for the line width is reasonable considering the difficulties in measuring the spectrum of nitrogen in a dilute solution. More importantly, the jump in both frequency and line width is well represented by the simulations. Thus it is meaningful to study the details of the behaviour by simulations.

The calculations have been extended into the metastable solid region in order to compare fluid and solid at the same density. Even at the same density as in the liquid the frequency is higher in the solid phase and the line width is smaller. This can be explained by considering the radial distribution functions. The radial distribution function depends on φ, which is the angle between the line connecting the center of an arbitrary molecule with the central molecule under consideration and the axis of that central molecule. Note that only the orientation of the central molecule has been taken into account. In figure 7 the radial distribution functions are presented for large angles ($|\cos\varphi| < 0.1$) and small angles ($|\cos\varphi| > 0.9$) for both the fluid and the (metastable) solid phase at a density of 48 kmol/m^3. The pressure is then 1.29 and 1.03 GPa, while the frequencies are 2329.2 and 2330.3 cm^{-1} for fluid and solid, respectively. The position of the nearest neighbour is given by the first maximum in the distribution function. In both phases this is in the repulsive region of the intermolecular interaction. This first peak for molecules with a large value of φ shifts to larger distances in going from the fluid to the solid phase, and thus the mean distance between *nearest* neighbours increases as the ordered structure is formed. Due to the resulting decrease of the repulsive forces the pressure drops in

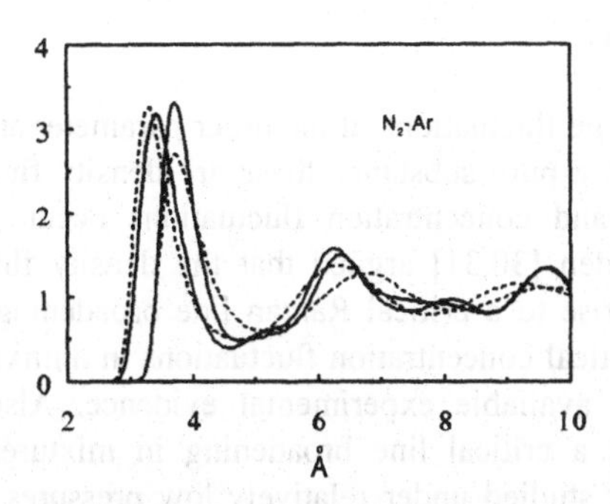

Figure 7. Pair distribution function relative to the center of mass of a nitrogen molecule for the system N_2-Ar at 48 kmol/m^3. Solid line:solid phase (metastable); dashed line: fluid phase. Left curves: perpendicular to the molecular axis; right curves: in line with the molecular axis.

the solid. However, the forces of molecules with large value of φ have only small components along the axis of the central molecule and thus hardly any influence on the vibrational frequency. In contrast, the first maximum in the distribution function of molecules with small φ values moves slightly to smaller distances at the transition from fluid to solid, and the height of the peak increases considerably. This gives rise to an increase of the compressive forces and thus to an increase of the frequency.
The collapse of the line width at the transition can also be explained. Figure 8 shows the vibration correlation function for the mixture in both the solid and the fluid phase. In the fluid phase, after the initial drop, $\Omega(t)$ has a long positive tail, while it becomes even negative in the solid phase. As a result the correlation time is reduced to about half at the transition. The amplitude of modulation hardly changes (eq.3). Therefore, the collapse of the line width is mainly due to the drop in the correlation time.

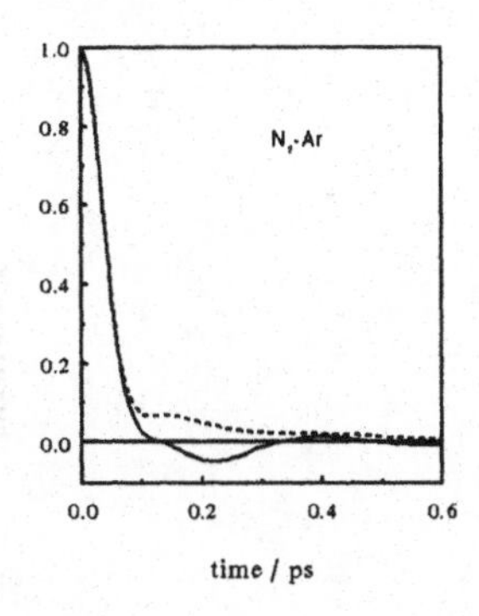

Figure 8. The vibration correlation function of nitrogen in argon at 1.29 GPa and 296 K [25]. Dashed line, fluid phase; solid line, solid phase.

4. Critical fluctuations

It is well known that large fluctuations of the order parameter arise when the critical point is approached. In a pure substance these are density fluctuations while in a mixture both density and concentration fluctuations occur. In their theoretical studies Hills and Madden [30,31] argued that the density fluctuations in a pure substance would give rise to a critical Raman line broadening, whereas no effect would occur for the critical concentration fluctuations in a mixture. This seemed to be in agreement with available experimental evidence. Also later experiments [32,33] did not reveal a critical line broadening in mixtures. Although critical phenomena are usually studied under relatively low pressures up to 100 bar, in a more recent investigation of the mixture helium-nitrogen under high pressure (up to 7 GPa) the authors reported considerably broader lines near the critical line [14]. The system helium-nitrogen is appropriate for this study because at high pressure the critical composition is nearly independent of pressure and temperature and because critical density fluctuations do not occur under these conditions. It was shown that this broadening was not due to concentrations fluctuations that already occur in non-critical mixtures. The effect of those contributions will be discussed in the next section.

The line shift as a function of composition (at constant pressure and temperature) is closely related to line broadening due to concentration fluctuations. Discarding for a moment the density fluctuations, the Raman line has a certain width since in a mixture the local composition varies in space and time and, therefore, also the vibrational frequency varies. However, if there is no shift of frequency as a function of composition, differences in local composition would not result in different frequencies, and there would be no effect of concentration fluctuations on the line width. Figure 9a shows the shift in frequency as function of volume fraction [34]. The volume fraction has been taken as order parameter in stead of the mole

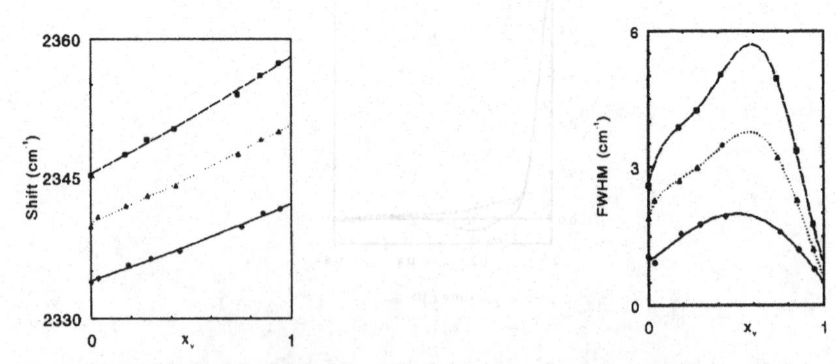

Figure 9. Raman shift (a) and FWHM (b) of N_2 as a function of the volume fraction x_v of helium at 296 K and p=2.5 GPa (circles, p=4.5 GPa (triangles) and p=6.5 GPa (squares). The lines are guides to the eye [34].

fraction because in the first case the relevant curves are more symmetric. The shift of the frequency of nitrogen by adding helium is considerable and the dependence of the shift on the volume fraction is almost linear. In particular, the behaviour is smooth near the critical volume fraction (≅0.57) which means that the *average interaction* of a particular nitrogen molecule with its environment does not show any anomaly in the critical region. Thus there is no anomalous change in the bulk density as a function of composition in the critical region and anomalous effects on the line width must be due to concentration fluctuations.

Figure 9b gives the line width as a function of volume fraction. The width clearly shows a broad maximum near the critical volume fraction. At 6.5 GPa the effect of critical broadening is much more pronounced than at 2.5 GPa because the conditions are close to the critical conditions (at room temperature the critical pressure is about 7 GPa). Results have also been obtained for other temperatures and pressures. It turns out that, if a reduced temperature is defined by $\varepsilon(T,p)=T/T_{cr}(p)-1$, where $T_{cr}(p)$ is the critical temperature of the mixture at temperature T, the broadening is nearly the same for the same values of ε and almost independent of p. Thus at a given composition one can approximately measure the distance to the critical line by regarding only the temperature. Critical enhancement in this mixture is still observable for $\varepsilon=0.8$, while in pure nitrogen it is restricted to ε values smaller than 0.4 [35].

The results for the line width can be used to make an estimate of the amplitude of the critical concentration fluctuations. The amplitude of x_v is almost half the critical value of x_v, while for pure nitrogen the amplitude of the density is about 0.3 times the critical density.

The deviation of the local composition from the bulk composition is assumed to be the main reason for an enhanced reaction rate of chemical reactions in super critical fluids [36]. If there is an increase (decrease) of the local composition of helium around the nitrogen molecule in the critical region, the frequency will show an increase (decrease) in the same region. From the regular behaviour of the shift and the symmetry of the spectrum it follows that, although the local composition might be different from the bulk composition, there is no indication of an increased *average* local composition in the critical region.

5. Concentration fluctuations

Even far from the critical region concentration fluctuations will occur with an expected maximum around x=0.5. For instance, in a very dilute mixture of helium in nitrogen a nitrogen molecule has probably at most one helium molecule as its nearest neighbour, while in an equal molar mixture the local composition can be easily 30 or 70 %. For mixtures of hard spheres of equal size this maximum indeed occurs for x=0.5, while for larger diameter ratios the maximum shifts to lower concentrations of the smaller component [37].

Knapp and Fischer [38] developed a model for calculating the contribution of non-critical fluctuations to the line width in a mixture. The main assumptions in this model are that the reference molecule is only influenced by a fixed number N of nearest neighbours, that these neighbours perturb the reference transition indepently of each other, and that neighbours of the same component have the same effect. In this model the environment of the reference molecule is then given by the composition. This results in N+1 different microscopic environmental states leading to different frequencies. It is, of course, also possible to take density fluctuations into account by varying the number N. In practice, the value for N needed to get agreement is different from the number of nearest neighbours due to the dynamics of the system [38,4]. Moreover, a few other parameters (which depend on pressure and also slightly on temperature) have to be determined from experiment, which makes an independent prediction of the line width impossible.
Recently, the computer simulations, mentioned in section 2, have been extended to the full composition range for the system neon-nitrogen [39]. Moreover, for a few compositions the experimental line width has been determined at 300 and 400 K up to high pressures. The experimental temperature and pressure are such that most probably critical fluctuations are not present or in any case negligible. Both the experiment and the simulations show a considerable line broadening for intermediate compositions. The simulations are in good agreement with experiment.

6. Conclusion

The results discussed above show that vibrational Raman spectroscopy in combination with computer simulations reveal a considerable amount of information about the microscopic behaviour of pure substances and mixtures in both the solid and liquid phase. In particular phase transition lines, the dynamics of phase transitions, solubility, disorder, density and concentration fluctuations, and orientational behaviour can be studied. Although the influence of the environment on the vibrational frequencies is due to the intermolecular forces, it is not possible to determine these forces directly from the shift of the frequency. The first and second order effect of the forces are mainly determined by the pressure and are almost independent of the potential functions, including the unlike interactions. The contribution of the vibration-rotation coupling is also independent of the interactions. An important contribution stems from the change in interaction at the vibrational excitation of the molecule. At larger distances this is due to a change in the polarizability. Under certain conditions it might be possible to obtain information on the (change in) polarizability by investigating a mixture of two Raman active components.

7. References

1. Kroon, R., Baggen, M., and Lagendijk, A. (1989) Vibrational dephasing in liquid nitrogen at high densities studied with time-resolved stimulated Raman gain spectroscopy, *J.Chem.Phys.* **91**, 74-78.
2. Lavorel, B., Oksengorn, B., Fabre, D., Saint-Loupe, R., and Berger, H. (1992) Stimulated Raman spectroscopy of the Q branch of nitrogen at high pressure: collisional narrowing and shifting in the 150-6800 bar range at room temperature, *Mol.Phys.* **75**, 397-413.
3. Devendorf, G.S. and Ben-Amotz, D. (1993) Vibrational frequency shifts of fluid nitrogen up to ultrahigh temperatures and pressures, *J.Phys.Chem.* **97**, 2307-2313.
4. Scheerboom, M.I.M., Michels, J.P.J., and Schouten, J.A. (1996) High pressure study on the Raman spectra of fluid nitrogen and nitrogen in helium, *J.Chem.Phys.* **104**, 9388-9399.
5. Levesque, D., Weis, J.-J., and Oxtoby, D.W. (1980) A molecular dynamics simulation of dephasing in liquid nitrogen. II. Effect of the pair potential on dephasing, *J.Chem.Phys.* **72**, 2744-2749.
6. Schindler,W. and Jonas, J. (1980) Solvent effects on vibrational phase relaxation and frequency shifts in isobutylene, *J.Chem.Phys.* **73**, 3547-3552.
7. Schweizer, K.S. and Chandler, D. (1982) Vibrational dephasing and frequency shifts of polyatomic molecules in solution, *J.Chem.Phys.* **76**, 2298-2314.
8. LeSar, R. (1987) Vibrational frequency shifts in dense molecular fluids,*J.Chem.Phys.* **86**, 4138-4145.
9. Belak, J., Etters, R.D., and LeSar, R. (1990) Calculated thermodynamic properties and phase transitions of solid N_2 at temperatures $0 \leq T \leq 300$ K and pressures $0 \leq P \leq 100$ GPa, *J.Chem.Phys.* **92**, 5430.
10. Ben-Amotz, D., Lee, M.R., Cho, S.Y., and List, D.J. (1992) Solvent and pressure-induced perturbations of the vibrational potential surface of acetonitrile, *J.Chem.Phys.* **96**, 8781-8792.
11. Michels, J.P.J., Scheerboom, M.I.M., and Schouten, J.A. (1995) Computer simulations of the linewidth of the Raman Q-branch in fluid nitrogen, *J.Chem.Phys.* **103**, 8338-8345.
12. Thiéry, M.M. and Chandrasekharan, V. (1977) Vibron and lattice frequency shifts in the Raman spectra of solid α-N_2 and γ-N_2 and librational force constants of diatomic molecular crystals, *J.Chem.Phys.* **67**, 3659-3675.
13. Michels, J.P.J., Scheerboom, M.I.M., and Schouten, J.A. (1996) Computer simulations of the Raman Q-branch of nitrogen in helium, *J.Chem.Phys.* **105**, 9748-9753.
14. Scheerboom, M.I.M. and Schouten, J.A. (1995) Critical broadening of the vibrational linewidth by concentration fluctuations, *Phys.Rev.* **E 51**, R2747-2750.
15. Loubeyre, P., LeToullec, R., and Pinceaux, J.P. (1992) Raman measurements of the vibrational properties of H_2 as a guest molecule in dense helium, neon, argon, and deuterium systems up to 40 GPa, *Phys.Rev.* **B 45**, 12844-12853.
16. Kubo, R. (1962) A stochastic theory of line-shape and relaxation, in D.ter Haar (ed.), Fluctuation, *Relaxation, and Resonance in Magnetic Systems*, Oliver and Boyd, Edingburgh; pp 23-68.
17. Fischer, F. and Laubereau, A. (1975) Dephasing processes of molecular vibrations in liquids, *Chem.Phys.Lett.* **35**, 6-12.
18. Rothschild, W.G. (1976) Vibrational resonance energy transfer and dephasing in liquid nitrogen near its boiling point: Molecular computations, *J.Chem.Phys.* **65**, 2958-2961.
19. Lynden-Bell, R.M. (1977) Vibrational relaxation and line widths in liquids dephasing by intermolecular forces, *Mol.Phys.* **33**, 907-921.
20. Oxtoby, D.W., Levesque, D., and Weis, J.J. (1978) A molecular dynamics simulation of dephasing in liquid nitrogen, *J.Chem.Phys.* **68**, 5528-5533.
21. Oxtoby, D.W. (1979) Hydrodynamic theory for vibrational dephasingt in liquids,*J.Chem.Phys.* **70**, 2605-2610.
22. Chesnoy, J. and Weis, J.J. (1986) Density dependence of the dephasing and energy relaxation times by computer simulations, *J.Chem.Phys.* **84**, 5378-5388.
23. Lynden-Bell, R.M. and Westland, P.-O (1987) The effects of pressure and temperature on vibrational dephasing in a simulation of liquid CH_3CN, *Mol.Phys.* **61**, 1541-1547.
24. Kooi, M.E., Ulivi, L., and Schouten J.A. (1998) Vibrational spectra of nitrogen in simple mixtures at high pressures, *Int.J.Thermophys.*, in press.

25. Michels, J.P.J., Kooi, M.E., and Schouten, J.A. (1998) The vibrational frequency of nitrogen near the fluid-solid transition in the pure substance and in mixtures, *J.Chem.Phys.* **108**, 2695-2702.
26. Michels, J.P.J. and Schouten, J.A. (1997) Time correlation functions of the vibrational frequency of nitrogen and of nitrogen in helium, *Mol.Phys.* **91**, 253-263.
27. Guissani, Y., Levesque, D., Weis, J.J., and Oxtoby, D.W. (1982) A molecular dynamics study of dephasing in solid N_2 and O_2, *J.Chem.Phys.* **77**, 2153-2158
28. Beck, R.D., Hineman, M.F., and Nibler, J. (1990) Stimulated Raman probing of supercooling and phase transitions in large N_2 clusters formed in free jet expansions, *J.Chem.Phys.* **92**, 7068-7078.
29. Michels, J.P.J., Kooi, M.E., and Schouten (1998) to be published.
30. Hills, B.P. and Madden P.A. (1979) The critical contribution to vibrational dephasing I. The gas-liquid critical point, *Mol.Phys.* **37**, 937-947.
31. Hills, B.P. (1979) The critical contribution to vibrational and rotational relaxation II. Binary liquid mixtures, *Mol.Phys.* **37**, 949-957.
32. Wood, K.A. and Stauss, H.L. (1981) Polarized Raman scattering of a binary liquid near the critical point, *J.Chem.Phys.* **74**, 6027-6036.
33. Van Elburg, H.J. and Van Voorst, J.D.W. (1987) Spontaneous polarized vibrational Raman scattering from nitrogen in fluid binary mixtures with argon, *Chem.Phys.* **113**, 463-476.
34. Schouten, J.A. and Scheerboom, M.I.M. (1995) Broadening of the vibrational linewidth in a mixture due to critical concentration fluctuations, *Int.J.Thermophys.* **16**, 585-599.
35. Chesnoy, J. and Gale, G.M. (1988) Vibrational relaxation in condensed phases, *Adv.Chem.Phys.* **70**, 297-355.
36. Roberts, C.B., Chateauneuf, J.E., and Brenneke, J.F. (1992) Unique pressure effects on the absolute kinetics of triplet benzophenone photoreduction in supercritical CO_2, *J.Am.Chem.Soc.* **114**, 8455-8463.
37. Gallego, L.J., Somoza, J.A., and Blanco, M.C. (1988) On the concentration fluctuations of a binary mixture of hard spheres, *Z.Naturforsch.* **43a**, 847-850.
38. Knapp, E.W. and Fischer, S.F. (1982) The concentration dependence of the vibrational linewidth and shift in liquid binary mixtures: An analytical model, *J.Chem.Phys.* **76**, 4730-4735.
39. Kooi, M.E., Michels, J.P.J., and Schouten, J.A. to be published.

SINGULAR SOLID-LIQUID-GAS LIKE PHASE DIAGRAM OF NEUTRAL-TO-IONIC PHASE TRANSITION

M.H. LEMÉE-CAILLEAU*, M. BURON*, E. COLLET*, H. CAILLEAU*, T. LUTY+,*
** Groupe Matière Condensée et Matériaux*
UMR 6626 CNRS - Université de Rennes 1
Bat 11A, Campus de Beaulieu 35042 RENNES Cedex FRANCE
+ Institute of Physical and Theoretical Chemistry
Technical University, 50-370 WROCLAW POLAND

For the understanding of basic mechanisms in condensed matter, organic charge-transfer (CT) complexes are of general interest because of their particular electronic tunability strongly coupled with their structural properties. Thus they can refer simultaneously to the chemical physics of organic solid state, by the way of notions like molecular multistability or solid state electron transfer reactions, and to the physics of low-dimensional systems where non-linear structurally-relaxed electronic excitations play a tremendous role, but also to the physics of structural phase transitions for the cooperativity point of view. This is particularly true for the neutral-to-ionic (N-I) transition which takes place in most of mixed-stack CT complexes [1]. These systems are formed of stacks with alternating electron donor (D) and electron acceptor (A) molecules. The transition manifests itself by a change of the degree of CT and a dimerization distortion with the formation of (D^+A^-) singlet pairs along the stacking axis in the ionic phase [2]. It is known nowadays, that the transformation can be induced by temperature [3], and/or photo-irradiation [4], but it was pressure which allowed the first observation of the N-I transition [1]. After a brief description of the mechanism of the N-I transition, considering pressure as well as temperature effects, this paper will be focussed on the thermodynamics of its recognized ambassador, the tetrathiafulvalene-*p*-chloranil (TTF-CA) on the basis of on recent experimental results [6] supported by theoretical considerations [7].

The first step for a simple description of the N-I transition is to consider a single stack in a ground state. It takes into account both electronic and structural aspects : besides the on-site Coulomb interaction, the energetic balance concerns the cost of ionization of molecules, the electrostatic Madelung energy, the transfer integral, and the electron-phonon coupling which is responsible of the dimerization process of ionic molecules. The last three factors are directly influenced by external conditions like pressure or temperature. Under pressure the cooperativity along the chain, governed by Coulomb interactions, drives the single-stack N-I transition by destabilizing the quasi-neutral state with equally-spaced molecules (N : ... $D^0A^0D^0A^0$...) towards two degenerate quasi-ionic dimerized states (I^+ : ... (D^+A^-)(D^+A^-) ... and

R. Winter and J. Jonas (eds.), High Pressure Molecular Science, 219–224.

I^- : ... $(A^-D^+)(A^-D^+)$ ...) where the partial charge transfer is due to the quantum mixing of D and A orbitals (for a sake of convinience the term "quasi" will be omitted afterwards). The two states I^+ and I^- are ferroelectrics and are deduced one from the other by the on-molecule inversion center lost at the transition. Moreover, neutral and ionic states are also very close in energy.

The effect of temperature is reflected in the external phonon frequencies (they are softer in N state than in I one [5]), but mostly by thermal mixing, i.e. by the appearance of kinks (N-I domain walls) which make the chain unhomogeneous [6-9]. Such kinks are fluctuating in time and space :

e.g. I string in N chain ... $A^0D^0A^0(D^+A^-)(D^+A^-)D^0A^0D^0$ ...
N string in I chain ... $(D^+A^-)(D^+A^-)D^0A^0D^0(A^-D^+)(A^-D^+)$ ...

At finite temperature, there is no phase transition for an isolated chain but there is a crossover between an essentially N stack at low pressure and an essentially I stack at high pressure. At the crossover, the N-I kinks are at maximum and uniformaly distributed due to the quasi-degeneracy of N and I states. As the important energy cost comes from the N-I domain walls, above the crossover point, cooperative effects give rise to long N strings in I chain and below to long I strings in N chain. In order to characterize this evolution, it is necessary to introduce two parameters C and η. C stands for the concentration of ionic dimers, analogous to the concentration of transformed species for a chemical reaction (here the transfer of electron), as for example the high spin concentration for the spin crossover transition. One can then deduce an average ionicity $<q>= Cq_I + (1-C)q_N$. This concentration shows a continuous change from 0 to 1 with a rapid change around the crossover point. With the polarization as usual image, η is the order parameter associated with the symmetry breaking and thus to the ordering process. It allows to describe locally the I^+ and I^- strings. At finite temperature, it is zero in average for the isolated stack because of the lifetime and size fluctuations of N and I strings. It would only be at zero temperature thata complete ferroelectric ordering could take place.

In three dimensions, interchain coupling may be sufficient to drive phase transitions. Similarly to the spin transition , it is possible to have a first order isostructural phase transition associated with the condensation of CT excitations, i.e. formation of ionic dimers but without any ordering between them. Moreover because of interchain dipolar interactions, of Coulomb and elastic origin, it is also possible to have a three-dimensional ordering (crystallization) of ionic strings. The simplest is a ferroelectric one but more complex ordering could also be envisaged (antiferroelectric, long-period ordering, etc ...). This is the result of the interplay of two factors : existence of quasi-degenerate states (N and I) in the chain and sufficient interchain coupling- which is at the heart of the N-I transition. Indeed this makes the difference with mixed-valence M-X chains and conducting polymers which do not exhibit any transition because there is not sufficient coupling. Therefore compounds which show the N-I transition are exceptional to study the physical properties associated with non-linear excitations (kinks, solitons, ...) because their characteristics (concentration, size, ...) can be externally controled by pressure, temperature, ...

By analogy with the models developped to describe the phase diagram of ^{3}He-^{4}He mixtures [10] or the concensation and the solidification of a simple fluid [11], the spin 1 Ising model is especially well suited in the case of the N-I transition. With assignement of states, S=0 to the neutral state and S=±1 to the ionic ones[6], the dipolar order parameter <S> can be associated with the dimerization ordering, and thus to the symmetry breaking order parameter η, while the quadrupolar order parameter $\langle S^2 \rangle$, totally symmetric, can be related to the concentration of ionic species C.

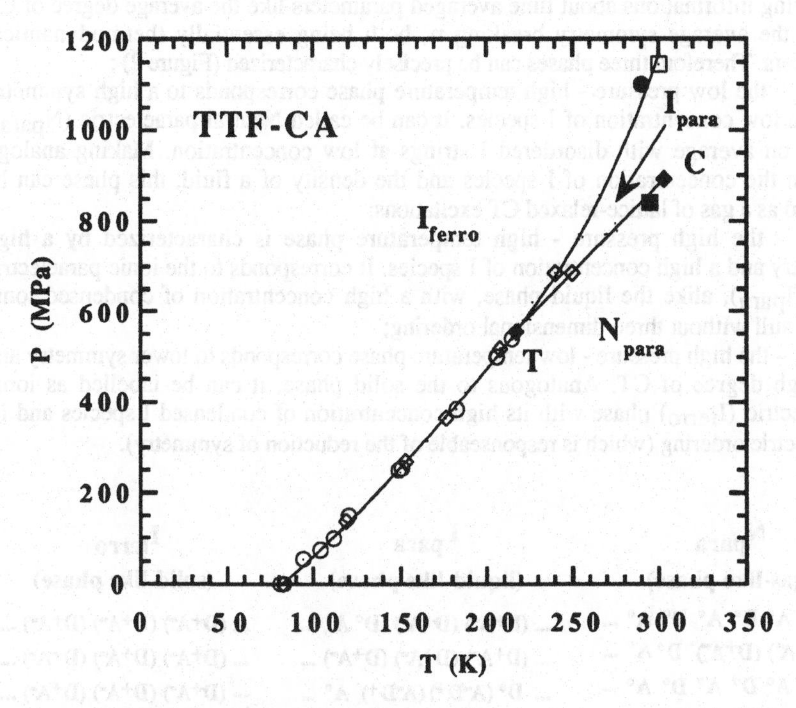

Figure 1 : Experimental phase diagram of TTF-CA
○ neutron scattering [6]
◊ ^{35}Cl NQR [6]
□ X-ray scattering [12]
■ conductivity in the high pressure part where the reliability of pressure can be judged reasonable taking into account the engaged technique [13]
● vibrational spectroscopy [14]
C and T are estimated critical and triple points respectively (◆). Lines are guides for the eyes.

This concept of N-I strings, also called lattice-relaxed CT excitations when a pair of kinks is considered, has turn out to be essential to elucidate the (P,T) phase diagram of the N-I prototype compond TTF-CA (figure 1). The phase diagram, similar to a solid-liquid-gas one, results from the compilation of results from various experimental techniques, such as diffraction [6,11], nuclear quadrupolar resonance (NQR) of ^{35}Cl [6], conductivity [13] and vibrational spectroscopies [14]. First of all, let us remind that the photoexcitation spectroscopy [4] has shown that the characteristic lifetime of lattice-relaxed CT excitations is in the range of nanosecond. This makes differences in the results provided by the different techniques mentionned above. Methods like Raman scattering or infra-red absorption, probe, in this system, "instantaneous" states of molecules with parameters like the degrees of CT, q_N and q_I, or the instantaneous local symmetry breaking. Other methods, with longer observation time, bring informations about time averaged parameters like the average degree of CT $\langle q \rangle$ or the average symmetry breaking η, both being essentially thermodynamical parameters. Therefore three phases can be precisely characterized (Figure 2) :

- the low pressure - high temperature phase corresponds to a high symmetry and to a low concentration of I species. It can be called Neutral paraelectric (N_{para}), neutral on average with disordered I strings at low concentration. Making analogy between the concentration of I species and the density of a fluid, this phase can be qualified as a gas of lattice-relaxed CT excitations;

- the high pressure - high temperature phase is characterized by a high symmetry and a high concentration of I species. It corresponds to the ionic paraelectric phase (I_{para}), alike the liquid phase, with a high concentration of condensed ionic strings, still without three-dimensional ordering;

- the high pressure - low temperature phase corresponds to lower symmetry and to a high degree of CT. Analogous to the solid phase, it can be labelled as ionic ferroelectric (I_{ferro}) phase with its high concentration of condensed I species and its ferroelectric ordering (which is responseable of the reduction of symmetry).

N_{para} (gas-like phase)	I_{para} (liquid-like phase)	I_{ferro} (solid-like phase)
$\ldots D^\circ A^\circ D^\circ A^\circ D^\circ A^\circ \ldots$	$\ldots (D^+A^-)(D^+A^-) D^\circ A^\circ \ldots$	$\ldots (D^+A^-)(D^+A^-)(D^+A^-) \ldots$
$\ldots (D^+A^-)(D^+A^-) D^\circ A^\circ \ldots$	$\ldots (D^+A^-)(D^+A^-)(D^+A^-) \ldots$	$\ldots (D^+A^-)(D^+A^-)(D^+A^-) \ldots$
$\ldots D^\circ A^\circ D^\circ A^\circ D^\circ A^\circ \ldots$	$\ldots D^\circ (A^-D^+)(A^-D^+) A^\circ \ldots$	$\ldots (D^+A^-)(D^+A^-)(D^+A^-) \ldots$
$\ldots D^\circ A^\circ D^\circ A^\circ D^\circ A^\circ \ldots$	$\ldots D^\circ (A^-D^+)(A^-D^+)(A^- \ldots$	$\ldots (D^+A^-)(D^+A^-)(D^+A^-) \ldots$
$\ldots D^\circ A^\circ D^\circ (A^-D^+) A^\circ \ldots$	$\ldots (D^+A^-)(D^+A^-) D^\circ A^\circ \ldots$	$\ldots (D^+A^-)(D^+A^-)(D^+A^-) \ldots$
$\ldots D^\circ A^\circ D^\circ A^\circ D^\circ A^\circ \ldots$	$\ldots (D^+A^-)(D^+A^-)(D^+A^-) \ldots$	$\ldots (D^+A^-)(D^+A^-)(D^+A^-) \ldots$
$0 < C < 0.5$	$0.5 < C < 1$	$C \approx 1$
$\eta = 0$	$\eta = 0$	$\eta \neq 0$

Figure 2. Schematic drawing of the three phases of TTF-CA under pressure with their order-parameter characteristics :

- N_{para} : neutral paraelectric,
- I_{para} : ionic paraelectric,
- I_{ferro} : ionic ferroelectric.

It is interesting to consider few basic arguments following from comparison of the phases and of their properties in order to explain this unique solid state phase diagram. At low pressure and low temperature, molecular ionicities q_N and q_I are substantially different and the condensation of CT excitations is instantaneously accompanied by their ordering at the ferroelectric transition. A small concentration of thermally activated CT excitations makes the transition discontinuous. The pressure not only favors the ionic phase (volume change and Madelung energy effects), but also, by the increasing of the transfer integral, makes the "compounds" N and I more and more similar ($q_N \rightarrow q_I$). Increasing the temperature stimulates the disorder between N and I strings along the chains. The coupling between chains decreases, the transition becomes less and less discontinuous and the N phase shows symptoms of disorder. Both effects, quantum mixing due to the increasing of pressure, and thermal mixing due to the increasing of temperature, favor a continuous regime for the transition. However, in the high region of the "sublimation" curve, the difference between q_N and q_I decreases and may become too small to drive the ordering of the chains. Moreover, and this is probably dominant effect, with the transition temperature increasing, the distance betweens kinks decreases, so that the dipolar interaction between smaller I strings become less effective to drive the three-dimensional ordering of I species (in an analogy to the quasi-one-dimensional Ising model). The system reachs first a triple point and then the transition of condensation of the "gas" of CT excitations to the "liquid" phase (I_{para}), where the ionic strings are disordered, dissociates from a "crystallization" transition to an ordered I phase (I_{ferro}) which takes place at lower temperature. The condensation equilibrium line ends at the critical point where there is equal concentration of neutral and ionic species. To our knowledge, it is the first time that such a (P,T) phase diagram , analogous to the solid-liquid-gas one, has been observed in the solid state.

Corresponding author : M.H. Lemée-Cailleau
email : mhlemee@univ-rennes1.fr

REFERENCES

1. J.B. Torrance,J.E. Vasquez, J.J. Mayerle, V.Y. Lee (1981) Phys. Rev. Lett. 46 253-257
2. M. Le Cointe, M.H. Lemée-Cailleau, H. Cailleau, B. Toudic, L. Toupet, G. Heger, F. Moussa, P. Schweiss, K.H. Kraft, N. Karl (1995) Phys. Rev. B 51 3374-3386 and references therein
3. J.B. Torrance, A. Girlando, J.J. Mayerle, J.I. Crowley, V.Y. Lee, P. Batail, S.J. LaPlaca (1980) Phys. Rev. Lett. 47 1747-1750
4. S. Koshihara, Y. Tokura, T. Mitani, G. Saito, T. Koda (1990) Phys. Rev. B 42 6853-6856; S. Koshihara, Y. Tokura, N. Sarukura, Y. Segawa, T. Koda, K. Takeda (1995) Synth. Met. 70 1225-1226
5. A. Moréac, A. Girard, Y. Délugeard, Y. Marqueton (1996) J. Phys. Cond. Matter 8 3553-3567

6. M.H. Lemée-Cailleau, M. Le Cointe, H. Cailleau, T. Luty, F. Moussa, J. Roos, D. Brinkmann, B. Toudic, C. Ayache, N. Karl (1997) Phys. Rev. Lett. 79 1690-1693
7. T. Luty (1997) Springer series in solid state sciences, vol. 124 "Relaxations of excited states and photo-induced structural phase transitions", Ed. K. Nasu, 142-150
8. N. Nagaosa, J.J. Takimoto (1986) J. Phys. Soc. Jpn 55 2735-2744; (1986) ibid 55 2745-2753; N. Nagaosa (1986) J. Phys. Soc. Jpn 55 2754-2764, N. Nagaosa (1986) J. Phys. Soc. Jpn 55 3488-3497
9. Y. Toyozawa (1992) Solid State Comm. 84 255-257; Y. Toyozawa (1995) Acta. Phys. Pol. A 87 47-56
10. M. Blume, V.J. Emery, R.B. Griffiths (1971) Phys. Rev. A 4 1071-1077
11. J. Lajzerowicz, J. Sivardière (1975) Phys. Rev A 4 2079-2089
12. M.H. Lemée-Cailleau, M. Le Cointe, H. Cailleau, J.P. Itié, A. Polian, unpublished
13. T. Mitani, Y. Kaneko, S. Tanuma, Y. Tokura, T. Koda, G. Saito (1987) Phys. Rev. B 35 427-429; Y. Kaneko, S. Tanuma, Y. Tokura, T. Koda, T. Mitani, G. Saito (1987) Phys. Rev. B 35 8024-8029
14. K. Takaoka, Y. Kaneko, H. Okamoto, Y. Tokura, T. Koda, T. Mitani, G. Saito (1987) Phys. Rev. B 36 3884-3887; H. Hanfland, A. Brillante, A. Girlando, K. Syassen (1988) Phys. Rev. B 38 1456_1461; H. Okamoto, T. Koda, Y. Tokura, T. Mitani, G. Saito (1989) Phys. Rev. B 39 10693-

STRUCTURE OF CARBON BLACK PARTICLES

T. W. ZERDA
Texas Christian University, Physics Department, Fort Worth, TX
A. ZERDA, Y. ZHAO, and R. B. VON DREELE
Los Alamos National Laboratory, LANSCE, Los Alamos, NM

1. Abstract

High pressure Raman and neutron scattering study of carbon blacks and HOPG graphite is reported. Carbon black particles are composed of graphitic nanocrystallites and amorphous carbon. Relative concentration of amorphous carbon decreases slightly with increased pressure. This non-reversible process differs from temperature induced transformations of amorphous carbon into ordered carbon. Post-production treatment at high temperatures results in the growth of graphitic crystallites whereas no such effect was observed when the pressure was applied to carbon black.

2. Introduction

Carbon black is used as filler to modify mechanical, electrical and optical properties of the medium in which it is dispersed. Iterfacial interactions between rubber and carbon black, occlusion of polymers in the internal voids, and filler-filler interactions depend on the structure and surface properties of carbon blacks. The surface properties of carbon blacks have been studied by gas adsorption, atomic force microscopy, TEM, X-ray diffraction and Raman spectroscopy [1-3]. In this study, we use neutron diffraction and Raman spectroscopy to characterize the internal structure of carbon black particles.

Raman spectroscopy has been successfully used to determine the average sizes of graphitic crystallites [3,4]. Temperature induced growth of nanocrystallites can be analyzed in terms of changes in position, width and relative intensity of two Raman peaks observed at 1345 cm^{-1} (the disordered, or "d" peak) and 1575 cm^{-1} (the graphitic, or "g" peak). The heat treatment induces growth of the crystallites and the peaks become narrower, their maxima move to higher frequencies and the intensity of the "g" peak systematically increases in comparison with the "d" peak [3,4]. Using the empirical formula found by Tuinstra and Koenig [5]

R. Winter and J. Jonas (eds.), High Pressure Molecular Science, 225–230.

$$L_a = 4.35\, I_g/I_d \text{ [nm]} \quad (1)$$

it is possible to evaluate L_a, the lateral size of the crystallites. The change in the band shape and frequency position of the "g" component can be explained in terms of the phonon confinement model developed by Richter et al. [6].

In addition to crystallites carbon black particles contain amorphous carbon. Amorphous carbon gives rise to a broad and asymmetric peak centered at 1530 cm^{-1}. By calculating its intensity one can evaluate the concentration of amorphous carbon. Heat treatment at temperatures below 1300 K does not change the relative concentration of amorphous carbon, but when the temperature increases above 1300 K the concentration of amorphous carbon decreases as it becomes incorporated into graphitic crystallites [4].

In this paper we concentrate on pressure induced effects on carbon-carbon bond lengths, the size of the crystallites and the relative concentration of amorphous carbon.

3. Experimental

High pressure Raman experiments were conducted using a diamond anvil cell, DAC. Two spectrometers were attached to a microscope via a swing-away mirror. Each spectrometer was set to measure a specific spectral region, a Triplemate measured the ruby fluorescence lines and its gratings were centered at 693 nm, a HoloSpec measured carbon black vibrations. Pressure was determined with precision better than 0.1 kbar.

High pressure in-situ Raman study of carbon blacks and HOPG was limited to the measurements of the "g" peak frequency position. The "d" band overlaps with a very strong peak due to the diamond. Only after the pressure was released we were able to record the whole spectra. To generate large volume of carbon black exposed to high pressure we applied a toroidal press used in the neutron diffraction experiment.

This experiment was conducted under hydrostatic compression using a Paris-Edinburgh toroidal anvil press [7]. The carbon black powder was mixed with small amount of CsCl, which served as the pressure calibrant. A powder sample of about 100 mm^3 in volume was combined with Fluorinert-70 (3M Corp.) to make a thick paste to ensure the hydrostatic compression. This sample was sealed between two WC anvils and a TiZr gasket. The high pressure neutron diffraction experiments were performed at the High Intensity Powder Diffractometer at Los Alamos Neutron Science Center [8]. The diffraction spectra were collected at $2\theta = \pm 90°$ banks in the energy-dispersive mode. The alignment of diffraction optics and pressure calibration were described by Zhao [9].

4. Results and discussion

When the external pressure did not break the HOPG crystal into smaller units, only a single peak was observed in the spectrum. The frequency of the E_{2g} peak increased

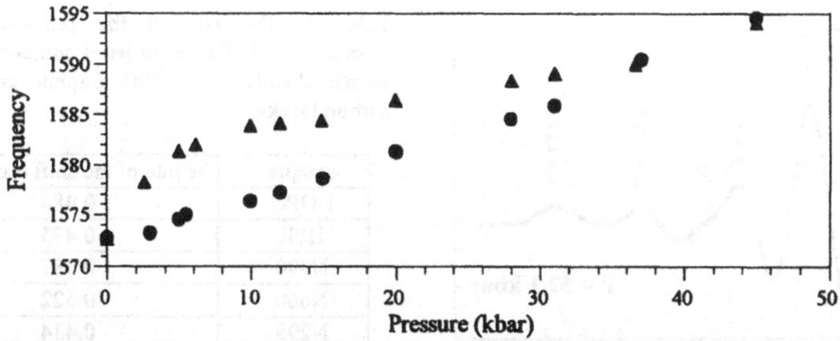

Fig. 1 Frequency position of the graphite band as a function of pressure. Circles – a single HOPG crystal, triangles – an HOPG crystal that broke into small crystallites.

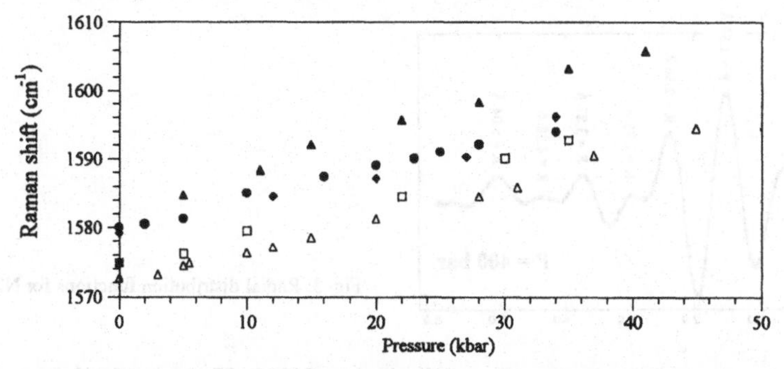

Fig. 2 Experimental frequencies of the organized band, g, as a function of pressure. Circles – N299, squares – HPR, diamonds – N660, filled triangles N990, open triangles – HOPG.

linearly with pressure, compare Fig. 1. In this figure we also depict the pressure induced shift of the E_{2g} band but for the case where the HOPG crystal broke into small crystallites of L_a=13 nm. This effect manifested itself in an abrupt appearance of the "d" component centered at 1345 cm^{-1} at 5 kbar. The intensity of the "d" component grew slightly up to about 10 kbar. This growth was accompanied by the pressure induced shift of the "g" peak. This shift can be separated into two regions of distinctly different slopes, one below and one above 10 kbar. The initial shift for P<10 kbar was primarily caused by the reduction of crystal sizes and can be explained by the phonon confinement model. The frequency shift observed at P>10 kbar was due to the changes in intermolecular potential caused by bond compression.

In Fig. 2 we show pressure induced shifts of the "g" band for various carbon blacks. The shifts are linear and the slopes obtained by fitting the experimental data to linear functions are listed in Table 1. The slopes are given with precision better than 0.005 cm^{-1}/kbar. Based on results obtained for the HOPG crystal we assumed that linear frequency shifts are attributed to pressure modified intermolecular potential and not the size reduction of the crystallites. This conclusion is supported by the neutron scattering data.

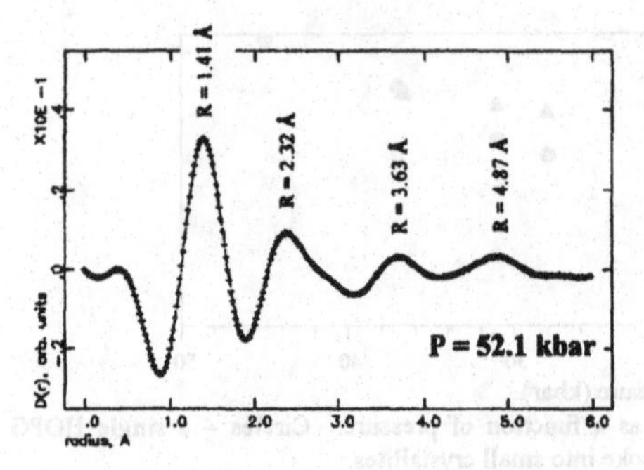

Table 1. The rate of the pressure induced frequency shift for the ordered component of the Raman doublet for HOPG graphite and various carbon blacks

sample	the rate of the shift (cm^{-1}/kbar)
HOPG	0.484
HPR	0.475
N990	0.587
N660	0.522
N299	0.434

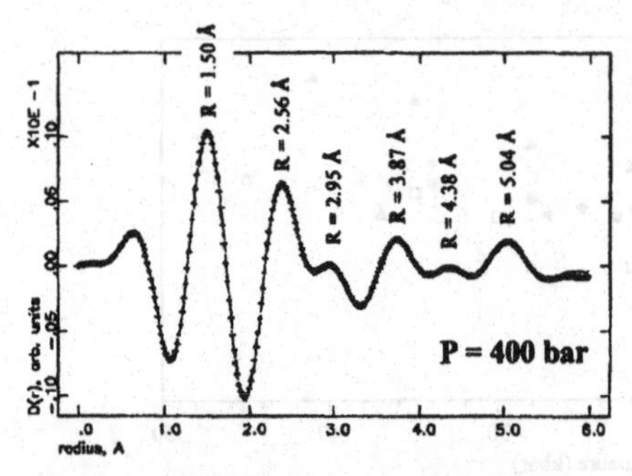

Fig. 3. Radial distribution functions for N299.

The neutron scattering experiment was limited to N299. This sample does not show apparent diffraction peaks. However, there are several diffuse "humps" in the diffraction spectrum as the results of short-range order scattering of the amorphous solids. We have incorporated a refinement routine in GSAS [10] to extract the radial distribution function (RDF) of amorphous solids from the fitting of background. The RDF function yields crucial information about important short-range order and, thereby, the nature of the chemical bonding in the amorphous solids. Plotted in Fig. 3 are RDF of the carbon black N299 at P=400 bar and P=52.1 kbar. There are six distinguishable peaks in the RDF plot for the lower pressure case and four smeared peaks clearly shown for the higher pressure case. These RDF peaks represent the carbon-carbon distances in the carbon black. It is recognized that r=1.42 Å is the closest carbon-carbon bond distance in the hexagonal ring of the graphite structure, and r=2.46 Å, r=2.84 Å, r=3.67 Å, r=4.26 Å, r=4.92 Å are the other carbon-carbon distances in the hexagonal ring-structure. The neutron diffraction experiments show that the pressure is pushing carbon-carbon bonds to form hexagonal ring structure as observed in graphite. Plotted in Fig. 4 are three RDF peak refinement results of carbon-carbon distance as a function of pressure. It shows that all these three bonds have about same linear compressibility over the pressure range of P=0-50 kbar. It shows that the most compression of the carbon bond is achieved within the pressure range of P<10 kbar, which indicates that the formation of the hexagonal ring occurs in this initial compression stage.

Table 2. Crystallite sizes, L_a, and relative content of amorphous carbon, I_{amor}/I_{total}, for N299 and N660.

	L_a [nm]		I_{amor}/I_{total}	
Pressure	1 bar	30 kbar	1 bar	30 kbar
N299	2.82	2.48	0.22	0.20
N660	2.72	2.62	0.25	0.22

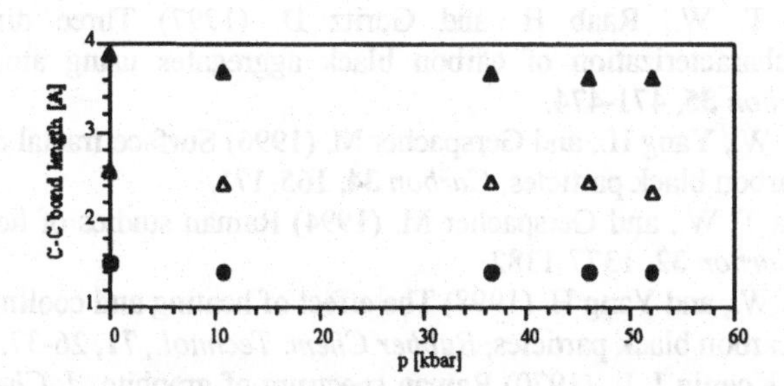

Fig. 4 Pressure dependence of C-C bond lengths of nanocrystallites in untreated carbon black N299.

Transformation of amorphous carbon into ordered structures has been also observed by Raman spectroscopy. Concentration of amorphous carbon has been evaluated and the results are given in Table 2. Heat treatment at 3000 K results in a dramatic reduction of amorphous carbon, but hydrostatic pressure has only a small effect on the concentration of amorphous carbon. Although this effect is small, it is reproducible and greater than experimental error.

Figure 4 also shows that the carbon-carbon bond lengths changed only slightly at pressures greater than 10 kbar. This is consistent with the observation in high pressure experiments on graphite and hexagonal boron nitride [11,12]. The pressure induced changes in the C-C bond lengths are continuous and support the notion that changes in force constants are responsible for the observed band shifts. This conclusion is further supported by the comparison of the L_a values (Table 2) before and after the pressure was released. It is seen that L_a decrease slightly when the pressure of 30 kbar is released, but those changes are too small to be considered significant.

5. Conclusions

Hydrostatic pressure does not change the size of the crystallites in carbon black particles. This means that the pressure induced shift of Raman band of the graphitic component is a result of increased intermolecular forces and not due to smaller crystallites. The amorphous carbon content is reduced after high pressure is applied to carbon blacks. This process observed by both Raman scattering and neutron diffraction decreases the population of five and seven member rings as well as the population of the sp^3 carbons. Since the size of the crystallites did not increase we explain the reduction in the amorphous carbon content as a result of formation of new crystallites.

6. References

1. Xu W., Zerda T. W., Raab H. and Goritz D. (1997) Three dimensional morphological characterization of carbon black aggregates using atomic force microscopy, *Carbon* **35**, 471-474.
2. Xu W., Zerda T. W., Yang H., and Gerspacher M. (1996) Surface fractal dimension of graphitized carbon black particles, *Carbon* **34**, 165-171.
3. Gruber T., Zerda T. W., and Gerspacher M. (1994) Raman studies of heat treated carbon blacks, *Carbon* **32**, 1377-1382.
4. Zerda T. W., Xu W., and Yang H. (1998) The effect of heating and cooling rates on the structure of carbon black particles, *Rubber Chem. Technol.*, **71**, 26-37.
5. Tuingstra F and Koenig J. L. (1970) Raman spectrum of graphite, *J. Chem. Phys.*, **53**, 1126-1130.
6. Richter H., Wang, Z. P., and Ley L. (1981) The one phonon Raman spectrum in microcrystalline silicon, Solid State Communications. **39**, 625-629.
7. Besson J.M., S. Klotz, G. Hamel, I. Makarenko, J.S. Loveday, R.M. Wilson, and W. G. Marshall (1995) High pressure neutron diffraction: present and future possibilities using the Paris-Edinburgh cell, *High Pressure Research*, **14**, 1-6.
8. Von Dreele (1995) High pressure neutron powder diffraction at LANSCE, *High Pressure Research*, **14**, 13-20, (1995).
9. Zhao, Y ., R. B. Von Dreele, D. J. Weidner (1998) Correction of *P-V-T* data and diffraction optics using thermoelastic equations of state of multiple phases, *High Pressure Research*, (in press).
10. Larson, A.C., and R.B. Von Dreele, (1994) GSAS manual, Report LAUR 86-748, Los Alamos National Laboratory.
11. Zhao, Y. X., and I. L. Spain (1989) X-ray diffraction data for graphite to 20 GPa, *Phys. Rev.*, **B40**, 993-997.
12. Zhao, Y., R. B. Von Dreele, D. J. Weidner, and D. Schiferl (1997) P-V-T data of hexagonal boron-nitride and determination of pressure and temperature using thermoelastic equation of state of multiple phases," *High Pressure Research*, **16**, 1-18.

HIGH-RESOLUTION NMR SPECTROSCOPY AT HIGH PRESSURE

JIRI JONAS
Beckman Institute for Advanced Science and Technology
University of Illinois at Urbana-Champaign
405 North Mathews Avenue
Urbana, IL 61801 USA

1. Abstract

The field of high-resolution nuclear magnetic resonance spectroscopy at high pressure is reviewed. The various applications of the high-pressure NMR spectroscopy to chemical systems are given out in the introduction. After a discussion of high-resolution NMR instrumentation for experiments at high pressures up to 950 MPa, the two main sections cover the recent applications of the high-resolution NMR spectroscopy at high pressure to chemical and biochemical systems. In particular, the use of 2D NMR techniques will be emphasized as the combination of advanced NMR methods with the high-pressure capability represents a new, powerful tool in studies of chemical and biochemical systems.

2. Introduction

Several recent articles reviewed the field of NMR spectroscopy at high pressures [1-5]. In earlier reviews, which have appeared in the NATO ASI series [6-8], the following applications of high-pressure NMR spectroscopy were covered:

- Relaxation and diffusion studies of the dynamics of simple liquids, including water
- Kramer's turnover for reactions in liquid solutions
- Kinetics of solid-solid phase transformations
- Polymer crystallization kinetics
- Transport and relaxation in highly viscous complex molecular liquids
- Dynamics of disordered solids
- Transport and relaxation in supercritical fluids
- Dynamics in model membranes
- Dynamics of liquids in confined geometries

R. Winter and J. Jonas (eds.), High Pressure Molecular Science, 231–259.

Advances in superconducting magnet technology have resulted in the development of superconducting magnets capable of attaining a high homogeneity of the magnetic field over the sample volume, so that even without sample spinning, high resolution can still be achieved. The ability to record high-resolution NMR spectra on dilute spin systems opened a new field of high-pressure NMR spectroscopy which deals with pressure effects on biochemical systems [2].

The fundamental reasons why it is important to carry out high-pressure experiments on chemical or biochemical systems are as follows:

- During experiments at constant pressure, a change in temperature produces both a change in density and a change in kinetic energy of the molecules. Therefore, in order to separate the effects of density (volume) and temperature on a dynamic process, one has to use both pressure and temperature as experimental variables.
- Use of pressure allows one to extend the range of measurements above the boiling point of a liquid, and also permits the study of supercritical dense fluids.
- Studies of simple liquids indicated that volume effects often determine the mechanism of a specific dynamic process, whereas temperature only changes the frequency of the motions without actually affecting the mechanism.

For studies of biochemical systems, one should point out several additional reasons for high-pressure experiments.

- Because noncovalent interactions play a primary role in the stabilization of biochemical systems, the use of pressure allows one to change, in a controlled way, the intermolecular interactions without the major perturbations produced by changes in temperature and/or chemical composition.
- Pressure affects chemical equilibria and reaction rates. The following standard equations define the reaction volume ΔV and the activation volume $\Delta V^{\neq}$

$$\Delta V = -\left[\frac{RT\partial \ln K}{\partial P}\right]_T, \quad \Delta V^{\neq} = -\left[\frac{RT\partial \ln k}{\partial P}\right]_T \tag{1}$$

  where K is the equilibrium constant, and k is the reaction rate.
- Although proteins are known to undergo pressure denaturation, few details are known about this important process or how it is related to thermal- or solvent-induced denaturation.
- According to the high-pressure phase diagram of water, even at –15°C water is still a liquid. Therefore, protein solutions can be measured at subzero temperatures to investigate their cold-denaturation behavior.
- Finally, the phase behavior and dynamics of phospholipid membranes can be explored more completely by carrying out experiments at high pressure.

The lipid-phase transitions are influenced by pressure and unique high-pressure gel phases are produced.

Pressures used to investigate chemical and biochemical systems range from 0.1 MPa to 1 GPa (01.MPa = 1 bar; 1 GPa = 10 kbar); such pressures only change intermolecular distances and affect conformations but do not change covalent bond distances or bond angles. In fact, pressures in excess of 30 GPa are required to change the electronic structure of a molecule [9].

After a discussion of high-resolution NMR instrumentation for experiments at high pressure, the two main sections cover the applications of the high-resolution NMR spectroscopy to chemical systems and biochemical systems.

3. High-Resolution NMR Instrumentation for Experiments at High Pressure

Nearly all components necessary for building high-pressure setup, such as piston hand pumps, high-pressure tubing and valves, intensifiers, and high-pressure separators, are currently available from commercial sources. The high-pressure setup consists of three main parts: (a) the pressure-generating system, (b) the pressure-measuring system, and (c) the NMR probe, which is located in a high-pressure vessel. Figure 1 shows a schematic drawing of the high-pressure-generating system used in our laboratory, illustrating the relative simplicity of the standard equipment used to generate high hydrostatic pressure. This system can produce hydrostatic pressure up to 1 GPa..

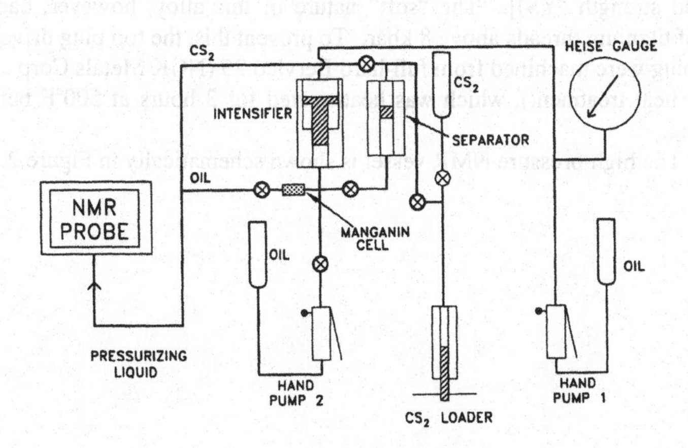

Figure 1. Schematic diagram of the high pressure-generating equipment.

3.1. SPECTROMETER AND PRESSURE-GENERATING SYSTEM

Our high-frequency NMR systems operate at a proton Larmor frequency of 300 MHz and 500 MHz. The system is composed of a General Electric GN-300 FT NMR with an Oxford Instruments, Inc., wide-bore superconducting magnet (ϕ = 89 mm, 7.04 T). The GN-300 is interfaced to a Tecmag Scorpio data acquisition system for pulse programming and experimental control, using MacNMR software. In addition, we have recently obtained access to a commercial 500 MHz system composed of an Oxford wide-bore superconducting magnet (ϕ = 89 mm, 11.7 T) and a Varian UNITY NOVA spectrometer (Varian-Oxford Instruments Center for Excellence NMR Laboratory, School of Chemical Sciences, University of Illinois).

The hydraulic pressure generation system is similar to the system described previously [8]. As with the earlier system, carbon disulfide (CS_2) is used as the pressure-transmitting fluid for proton studies. An alkane liquid mixture is used for studies of other nuclei.

3.2. HIGH-PRESSURE NMR VESSEL

3.2.1. *Materials and High-Pressure Vessel*

The material for any high-pressure NMR vessel must be nonmagnetic and of high mechanical strength. The most commonly used materials for this purpose are beryllium copper (Berylco) and titanium alloys. While our early vessels were of Berylco, the advantages of titanium have caused us to use the latter whenever possible. For this particular vessel, the body and the top plug were machined from the high-strength beta alloy of titanium, 3A1-8V-6Cr-4Zr-4Mo [RMI Titanium, 180 kpsi yield strength (YS)]. The "soft" nature of this alloy, however, can lead to galling of titanium threads above 8 kbar. To prevent this, the top plug driver and the bottom plug were machined from full-hard Berylco 25 (NGK Metals Corp., 179 kpsi YS after heat treatment), which was heat-treated for 3 hours at 500°F before final machining.

The high-pressure NMR vessel is shown schematically in Figure 2.

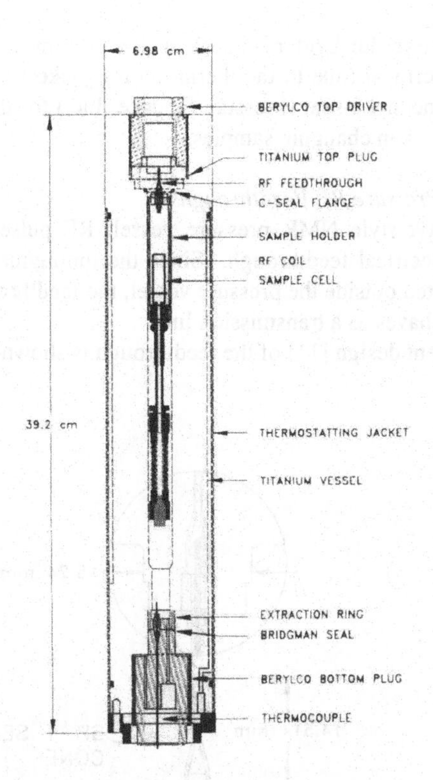

Figure 2. **Schematic drawing of titanium high-pressure vessel used in the wide-bore 7.04 T superconducting magnet. For clarity, only one of the three RF feedthroughs is shown.**

The vessel's maximum outer diameter (6.98 cm) is limited by the inner diameter of the room-temperature shim coils, while the inner diameter (1.5 cm) was chosen to accommodate an 8 mm sample tube and an RF coil. The vessel has a calculated ultimate burst pressure of 15.5 kbar at the sealing flanges and has been pressure-tested (while tuned) to 9.25 kbar. The top plug of the vessel can accommodate up to three electrical feedthroughs.

Thermal control is provided by circulating 50/50 ethylene glycol/water through etched grooves in the exterior of the vessel. Temperatures inside the vessel are measured by a copper-constantan thermocouple encased in stainless steel (Omega Engineering, Inc.). The thermocouple enters the vessel through an additional feedthrough in the bottom plug with a brazed stainless-steel cone seated inside a copper cone providing the high-pressure seal.

3.2.2. *High-Pressure Seals*

Our previous designs have utilized either Bridgman seals or c-seals to contain the pressure [5, 10]. In the current design [11], both types of seals are employed. An

Everdur-lead-Everdur Bridgman seal for the bottom plug allows use of a narrower flange at the critical (due to the thermostating jacket) bottom plug. A lead-coated, heat-treated Inconel c-seal (Pressure Science, Inc.) for the top plug allows relatively facile sealing when changing samples.

3.2.3. *High-Pressure RF Feedthroughs*

In an autoclave-style NMR pressure vessel, RF pulses enter the pressure vessel through an electrical feedthrough. Since the tuning/matching capacitor network is typically located outside the pressure vessel, the feedthrough becomes part of the RF circuit and behaves as a transmission line.

The current design [11] of the feedthrough is shown in Figure 3.

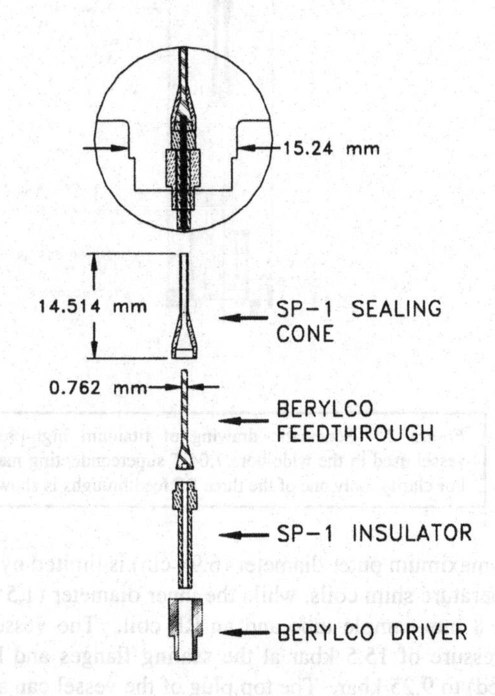

Figure 3. **Schematic drawing of the 300 MHz NMR high-pressure-probe RF feedthrough.**

This new feedthrough design has proven to be quite durable over a temperature range of –15°C to 80°C and a tested pressure range of up to 10 kbar. This feedthrough has been included in single-tuned and double-tuned RF circuits at frequencies ranging from 45 to 500 MHz.

3.2.4. *Sample Cells*

The type, as well as the shape, of the sample cell is related to the type of the experiment performed. In general, we use two types of high-pressure sample cells which differ in the way the pressure is transmitted to the sample. One is a piston-based design while the other uses a bellows system. Both sample cells are shown in Figure 4.

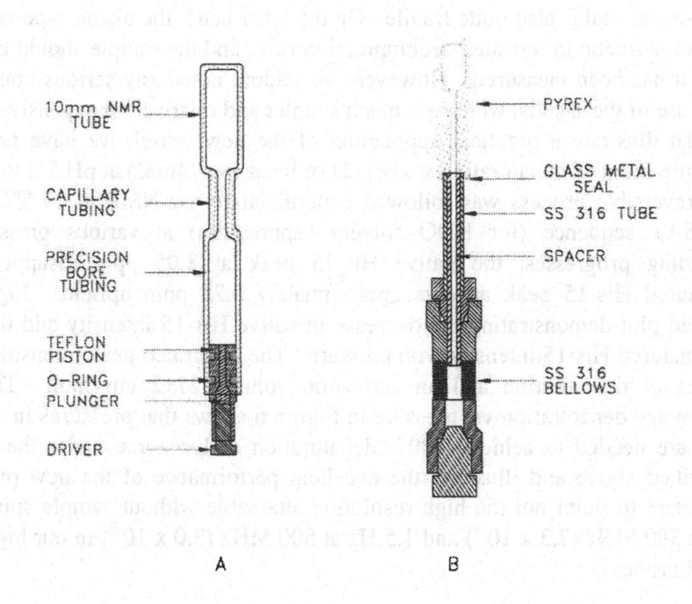

Figure 4. **Types of sample cells used in the high-pressure NMR studies: (A) the 10 mm, piston-type sample cell; and (B) the 8 mm sample cell using metallic bellows.**

In the piston design (Figure 4A), the part of the sample cell which is in the RF coil region is made of a 10 mm NMR tube which is then connected via capillary tubing to precision-bore tubing that contains the piston. The piston, which is made of Teflon, has a plunger equipped with a driver to provide tighter contact with the glass surface. Two O rings made of rubber (Parker Seal Group) are used on the piston in order to ensure no sample contamination and a good seal.

In the sample cell which uses bellows (Figure 4B), a glass (Pyrex 7740) or quartz-to-metal (SS-316) seal (Quartz Scientific, Inc., Palo Alto, California) connects the actual 10 mm glass sample chamber to the SS-316 tubing. Stainless steel bellows (Mechanized Science Seals, Inc., Los Angeles, California) accommodate the volume change of the liquid due to compression. Hermetic closure of the sample cell is attained by a copper cone brazed to an SS-316 bellow plug and seated in a flange made of stainless steel. The spacer in the sample cell serves only to reduce the total volume of the sample. Either SS-316 or high-temperature polyimide plastic (Vespel, Du Pont de Nemours & Co., Wilmington, Delaware) is used to make this spacer.

The bellows, as well as the piston-type cells, have both advantages and disadvantages. The bellows with the glass sample cell represent a hermetically closed system where the danger of contamination of the sample does not exist. The sample can be kept in such a system for a long time and remeasured later. However, the bellows themselves, as well as glass-metal seals, are quite expensive and this glass-metal seal is also quite fragile. On the other hand, the piston-type sample cells are not hermetic in repeated decompression runs, and the sample should be replaced once it has been measured. However, we seldom noted any serious contamination from use of these cells, which are much simpler and relatively inexpensive.

To illustrate a practical application of the new vessel, we have performed a pressure-induced denaturation study [11] of lysozyme (4mM) at pH 2.2 to 8.25 kbar. The reversible process was followed experimentally by NMR at 37.5°C using the PRESAT sequence (for HDO solvent suppression) at various pressures. As unfolding progresses, the native His-15 peak at 8.95 ppm disappears and a denatured His-15 peak appears approximately 0.25 ppm upfield. Figure 5 is a stacked plot demonstrating the decrease in native His-15 intensity and the increase in denatured His-15 intensity with pressure. The integrated peak intensities give the degree of denaturation and an activation volume 27±2 cm^3/mol. The plot of percentage denaturation vs. pressure in Figure 6 shows that pressures in excess of 8 kbar are needed to achieve 100% denaturation of lysozyme under the conditions described above and illustrate the excellent performance of the new probe. It is important to point out the high resolution attainable without sample spinning---0.7 Hz at 300 MHz (2.3 x 10^{-9}) and 1.5 Hz at 500 MHz (3.0 x 10^{-9}) in our high-pressure NMR probes.

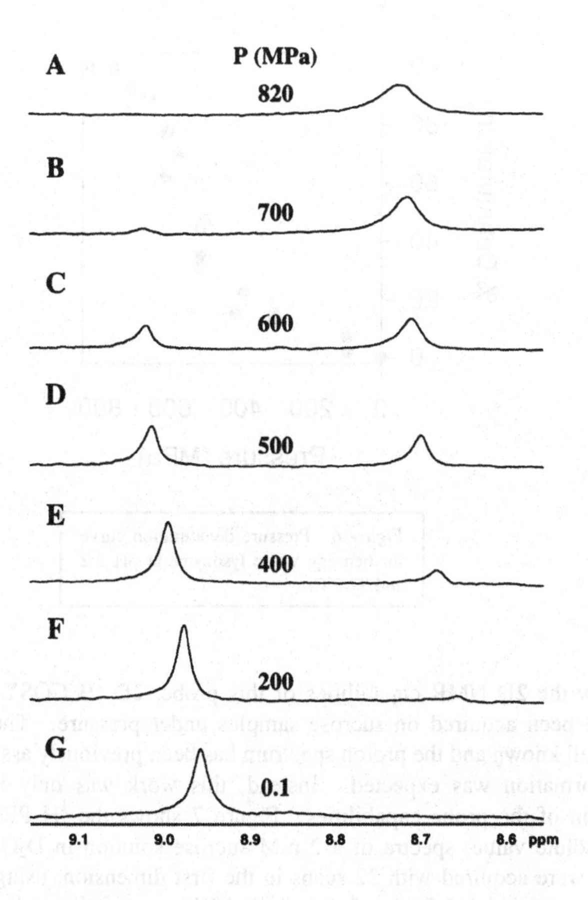

Figure 5. Stacked ^{1}H NMR plots of the lysozome (pH 2.2) histidine region at selected pressures at 37.5°C. Note the disappearance of the native His-15 residue peak (~8.95 ppm) and the appearance of the denatured His-15 residue peak (~8.70 ppm) with pressure; complete denaturation is attained between 700 and 825 MPa.

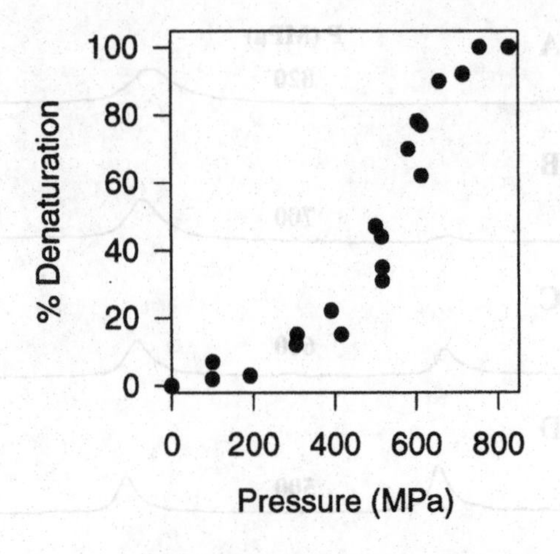

Figure 6. Pressure denaturation curve for hen egg white lysozyme at pH 2.2 and 37.5°C.

To show the 2D NMR capabilities of this probe, 2D ^{1}H COSY and NOESY spectra have been acquired on sucrose samples under pressure. The structure of sucrose is well known and the proton spectrum has been previously assigned [12], so no new information was expected. Instead, this work was only intended as a demonstration of the probe capabilities. Figure 7 shows the ^{1}H PRESAT COSY [13-15] (absolute value) spectra of a 2 mM sucrose solution in D_2O at 475 MPa. The spectra were acquired with 32 scans in the first dimension, using a 3.2s delay between scans, of which 1.5s was for an 8 dB CW presaturation pulse (for residual HDO suppression). 256 x 1K data matrices were obtained and processed (NUTS software) with zero-filling to 1K x 1K.

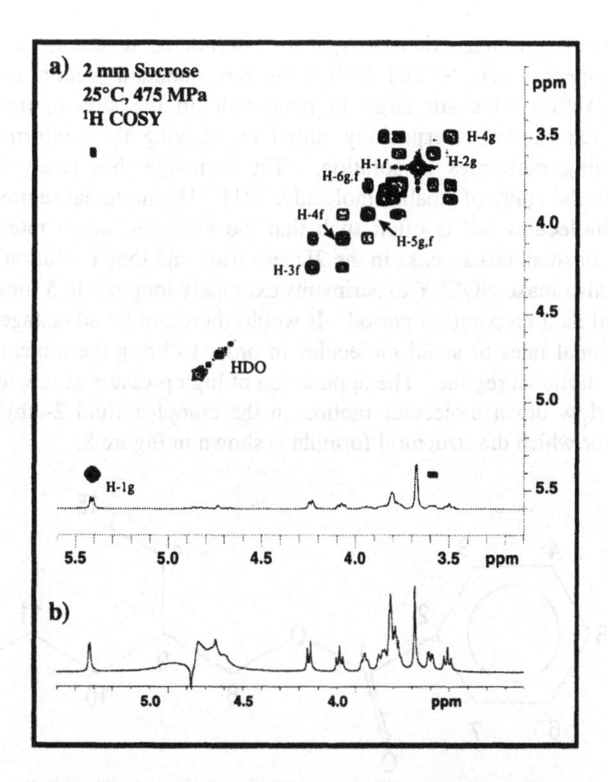

Figure 7. 2D ^{1}H PRESAT COSY spectrum of 2mM sucrose at 25°C and 475 MPa of pressure obtained at 500 MHz. (a) Contains the 2D COSY absolute value display, while (b) contains the 1D spectrum obtained using the PRESAT sequence for solvent suppression.

4. Applications of High-Resolution, High-Pressure NMR Spectroscopy to Chemical Systems

4.1. 2D NOESY STUDY OF THE COMPLEX LIQUID OF 2-ETHYLHEXYL BENZOATE AT HIGH PRESSURE

Two-dimensional nuclear Overhauser effect spectroscopy (NOESY) has been proven to be a valuable technique in determining the conformations of large polypeptides [16-19] and oligonucleotides [20]. The slow-motional regime in which such molecules lie (where $\omega\tau > 1$) causes the zero quantum transition to be extremely important in determining the rate of cross-relaxation. Cross-relaxation in the homonuclear case is described by the expression [18, 20]

$$R_{ij}^{C} = \frac{\gamma^4 \hbar^2}{10\, r_{ij}^6}\left[6J_2(2\omega) - J_0(0)\right] \tag{2}$$

where γ is the gyromagnetic ratio, r_{ij} is the internuclear distance, $J_2(2\omega)$ is the double quantum spectral density, and $J_0(0)$ is the zero quantum spectral density. Because cross-relaxation rates are large in magnitude in the slow-motional regime, the NOESY experiment is especially suited to studying the conformations of large, slow-moving molecules in solution. The technique has rarely been employed, though, in the study of smaller molecules [21]. The motional regime in which such smaller molecules fall is often such that the cross-relaxation rates are near zero, resulting in small cross-peaks in the 2D spectrum and long evolution times. Lengthy T_1's can also make NOESY experiments extremely long if 3 to 5 times the maximum T_1 is used as a preparation period. It would therefore be advantageous to decrease the rotational rates of small molecules in order to bring the molecular motion into the slow-motional regime. The application of high pressure at low temperatures was used to slow down molecular motion in the complex fluid 2-ethylhexyl benzoate (EHB), for which the structural formula is shown in Figure 8.

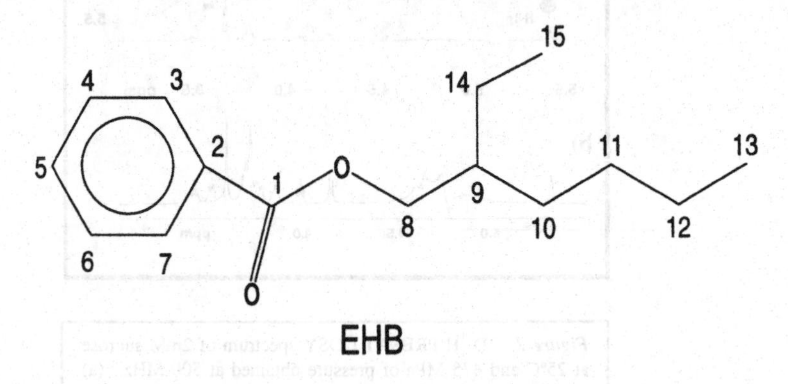

Figure 8. Structural formula of 2-ethylhexylbenzoate (EHB). Carbons are numbered according to international rules.

This fluid has been the subject of previous transport and relaxation studies carried out in our laboratory [22-24]. The high-pressure/low-temperature NMR capability allowed the 2D NOESY experiment [25] to be performed in the slow-motional regime, as indicated by the presence of positive cross-peaks in a phase-sensitive NOESY spectrum. In order to separate intramolecular and intermolecular cross-relaxation interactions, NOESY cross-peak evolution rates were measured as a function of the mole fraction of EHB in perdeuterated EHB-d_{22}. Cross-relaxation rates at infinite dilution were assumed to occur only from intramolecular interactions. We have therefore assumed that cross-relaxation between two protons occurs at a significantly greater rate than cross-relaxation occurring in a multi-step spin diffusion process. Spin diffusion is expressed in the quadratic and higher order terms in the Taylor expansion describing the mixing coefficient a_{kl} (which governs the intensity of the peaks in a NOESY spectrum) [18, 26]

$$a_{kl}(\tau_m) = \left(\delta_{kl} - R_{kl}\tau_m + 1/2\sum_j R_{kj}R_{jl}\tau_m^2 + \ldots\right)M_Z \quad (3)$$

where $\delta = 1$ when $k = 1$ and $\delta = 0$ when $k \neq 1$, and τ_m is the mixing time.

The primary objective of this study [25] was to see if pressure could be used to slow down the rotational motion of a small molecule so that its conformation could be studied in the negative cross-relaxation rate regime. To our knowledge, EHB is the smallest molecule ever studied by NOESY. The second objective was to investigate the effect of pressure on cross-relaxation in EHB. Third, it was hoped that some insight could be gained into the preferential conformation of EHB under such conditions of high density. Finally, we sought further dynamical information about EHB since the cross-relaxation rate is dependent on the rate of molecular motion.

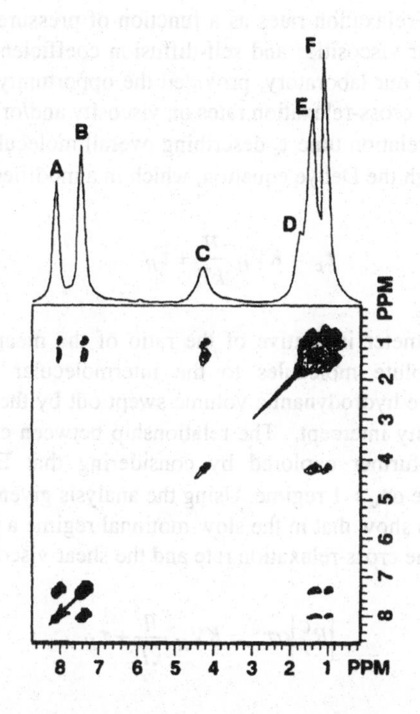

Figure 9. 1D and 2D NOESY spectra of 10% EHB in EHB-d_{22} at –20°C and 100 MPa.

Figure 9 shows a 1D ^{1}H spectrum of 10% EHB in EHB-d_{22} at -20°C and 100 MPa, along with a 2D NOESY spectrum taken under the same conditions with a mixing time of 200 ms. Of particular interest are the cross-peaks occurring between aromatic and methyl protons. We believe the cross-peaks to be the result of through-space dipolar coupling and not the result of spin diffusion propagating through the molecule. A quantitative explanation is reserved for later. Figure 9 also shows cross-peaks between protons connected to carbon 8 and the methyl protons. Additional cross-peaks exist between chain methylene protons and aromatic protons. The experimental NOESY spectra showed a drastic effect of pressure on the cross-peak evolution rate (equal to the cross-relaxation rate).

In order to separate the intermolecular and intramolecular contributions to cross-relaxation, values of cross-relaxation rate R_{ij}^{C} were plotted as a function of mole fraction of EHB in EHB-d_{22}. The experimental points were least squares fit to a straight line and the *y*-intercept was taken to be the intramolecular contribution to the total cross-relaxation rate. This analysis neglects any cross-relaxation due to deuterium.

From the experimental data we were able to determine both the intramolecular and intermolecular relaxation rates as a function of pressure and temperature. The availability of shear viscosities and self-diffusion coefficients of EHB, which were measured earlier in our laboratory, provided the opportunity to test the dependence of the experimental cross-relaxation rates on viscosity and/or diffusion of EHB. The reorientational correlation time τ_c describing overall molecular motion is coupled to the η/T term through the Debye equation, which in a modified form is [27]

$$\tau_c = KV_H \frac{\eta}{kT} + \tau_H \tag{4}$$

where K is a parameter indicative of the ratio of the mean square intermolecular torques on the solute molecules to the intermolecular forces on the solvent molecules, V_H is the hydrodynamic volume swept out by the reorienting vector, and τ_H is a zero viscosity intercept. The relationship between cross-relaxation rate and viscosity can be further explored by considering that EHB molecular motion definitely lies in the $\omega\tau_c > 1$ regime. Using the analysis given in detail in the original study [25], one can show that in the slow-motional regime a proportionality between the magnitude of the cross-relaxation rate and the shear viscosity is expected:

$$\left|R_{ij}^{C}\right| \alpha \tau_c = KV_H \frac{\eta}{kT} + \tau_H \tag{5}$$

where $\omega\tau_c > 1$.

In order to test the validity of Eq. (7) for the EHB system, we have plotted $\left|R_{ij}^{C}\right|$ as a function of η/T, an example of which is shown in Figure 10.

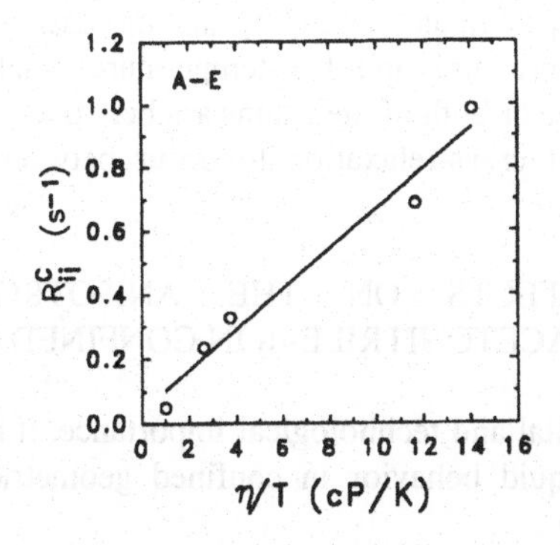

Figure 10. The linear dependence of the intramolecular cross-relaxation rate on η/T for the cross-peaks between hydrogens on ring carbons 3, 7 and hydrogens on side chain carbons 10, 11, 12, 14.

The linear behavior of the plot indicates good agreement with Eq. (7). Analogous plots for A-F; B-E; and B-F relaxation rates as η/T also showed good linearity. The linearity of the $\left|R^C_{ij}\right|$ versus η/T plots implies that overall rotation is extremely important in determining intramolecular cross-relaxation rates and that reorientation of each of the relaxation vectors is adequately described by one correlation time.

An investigation of intermolecular cross-relaxation can also be made. Intermolecular cross-relaxation rates, equal to the differences between rates in the neat EHB liquid and corresponding intramolecular rates were calculated. It is expected that the rate of intermolecular relaxation should be related to the rate of diffusion. In the pressure and temperature range covered in the study, the EHB density changes by about 7% and the translational diffusion coefficient changes by about an order of magnitude so that R^C_{inter} should be more heavily dependent on D.

The limited resolution available under high-pressure/low-temperature conditions did not permit differentiation between the methyl group on the hexyl chain and the methyl group on the ethyl chain. It can therefore not be determined whether the aromatic-methyl cross-peaks are due to the hexyl chain or the ethyl chain or both of the chains being close enough to the ring in order to have cross-relaxation. In order to gain some qualitative insight into the conformational possibility of one or both of the chains folding back to be close to the ring, possible minimum conformational energy structures of EHB were analyzed using Allinger's MM2 force field parameters [28]. The calculations give qualitative information only, being that no intermolecular interactions are assumed and though distances between atoms can be constricted, the total volume cannot.

In view of the approximate nature of the discussion above, one can only conclude that at high-pressure and low-temperature conditions, EHB exists in conformations where one or both of the chains are bent over toward the ring to allow a significant amount of cross-relaxation to occur between methyl and aromatic protons.

4.2. PRESSURE EFFECTS ON THE ANISOTROPIC ROTATIONAL DIFFUSION OF ACETONITRILE-d_3 IN CONFINED GEOMETRY

In view of its fundamental and technological importance, it is not surprising to find that the problem of liquid behavior in confined geometrics continues to attract attention.

In connection with our systematic NMR studies to improve the understanding of the dynamical behavior of molecular liquids in confined geometry [29-36], we have become interested in problems related to the confinement effects on anisotropic rotational diffusion of symmetric-top molecules at high pressures. Several results from our previous studies motivated this experiment. First, the two-state fast exchange model has been found to be valid for analyzing the spin-lattice relaxation time, T_1, data for polar liquids confined to porous silica glasses at ambient pressure and pressures up to 500 MPa [29, 32, 35]. It is important to point out that this approach allows us to investigate the effects of pressure on the reorientational dynamics of the molecules near the solid-liquid interface. Second, the results of deuteron and nitrogen spin-lattice relaxation measurements for acetonitrile-d_3 indicated that $D_{\parallel}$, the diffusion constant for rotation about the symmetry axis, is almost independent of pressure, whereas $D_{\perp}$, the diffusion constant for rotation about an axis perpendicular to the symmetric axis, shows a significant dependence on pressure [37]. The results of a temperature-dependence study of acetonitrile-d_3 indicated that E_a became higher for liquid confined to porous glasses than for the bulk liquid [35]. Therefore, it was of interest to determine the activation volumes for confined acetonitrile in order to determine whether the trend of ΔV^{*} as a function of pore size parallels the behavior of activation energies for the confined acetonitrile liquid. In a general sense, the main purpose of this study was to further examine the applicability of the two-state fast exchange model, as applied to the specific case of rotational diffusion of a symmetric-top molecule confined to porous silica glasses, under the extreme condition of high pressure. In addition, by taking advantage of the two-state fast exchange model and by assuming the validity of the classical rotational diffusion equation, we wanted to find out the pressure effect on the anisotropic rotational diffusion for liquid acetonitrile in confined geometry.

The NMR measurements were carried out using an NMR pulse spectrometer equipped with an Oxford 4.2 T superconducting magnet. The operating frequencies were 27.6 MHz for deuterium and 13.0 MHz for nitrogen-14. Sol-gel prepared porous silica glasses with a narrow pore size distribution were used as the porous media. The ^{2}H and ^{14}N spin-lattice relaxation times in bulk liquid and in liquid confined to porous glasses with 96, 60, 40 and 31 Å pore radii were measured as a function of pressure up to 500 MPa at 300± 1 K. The errors in the T_1 measurements were estimated to be about 7% for deuterium and 10% for nitrogen. The detailed experimental procedures, including the sol-gel porous silica glass preparation, sample loading, and high-pressure NMR techniques, have been described elsewhere [32-34].

According to the two-state fast exchange model [38, 39], the liquid in pores is assumed to have two distinct phases: a bulk phase and a surface-affected phase. If diffusion between the two phases is much faster than the relaxation rate, the observed T_1^{-1} of a liquid inside pores has both bulk, T_{1B}^{-1}, and surface, T_{1S}^{-1}, contributions,

$$T_1^{-1} = T_{1B}^{-1} + \frac{2\alpha}{R}(T_{1S}^{-1} - T_{1B}^{-1}), \tag{6}$$

where α is the thickness of the surface layer and R is the average pore radius.

As mentioned in our previous studies [33, 34], the NMR relaxation experiments on strongly interacting (polar) liquids, Pyridine-d_5 and nitrobenzene-d_5, confined to porous silica glasses were analyzed by the two-state model at pressures up to 500 MPa. It was important to find in the present experiment that this model also describes well the experimental relaxation data for acetonitrile-d_3. Figures 11 and 12 show the ^{2}H and ^{14}N T_1, respectively, of liquid acetonitrile-d_3 as a function of the inverse pore radius (R^{-1}) in porous silica glasses at pressures from 0.1 MPa to 500 MPa. As a result, we were able to separate the effect of pressure on T_{1S}^{-1} from the observed T_1^{-1}. Moreover, by assuming the validity of the classical rotational diffusion equation, we were able to calculate the anisotropic diffusion constants, $D_\parallel$ and $D_\perp$, for CD_3CN molecules near the liquid-solid interface.

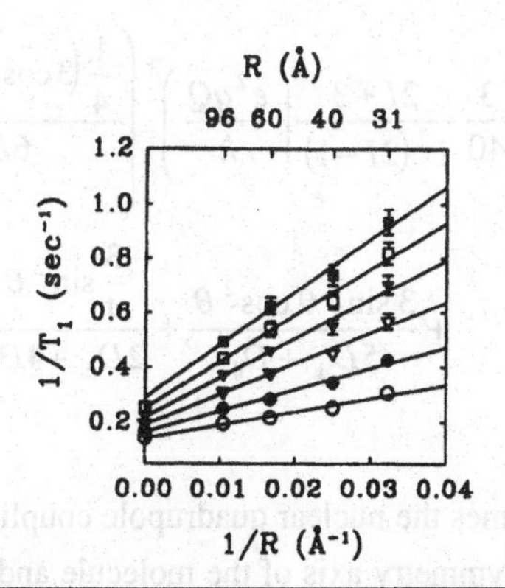

Figure 11. ^{2}H spin-lattice relaxation rate (T_1^{-1}) of acetonitrile-d_3 as a function of pore radius (R^{-1}) in porous silica glasses at 300 K. (O) 0.1 MPa; (●) 100 MPa; (▽) 200 MPa; (▾) 300 MPa; (□) 400 MPa; (■) 500 MPa.

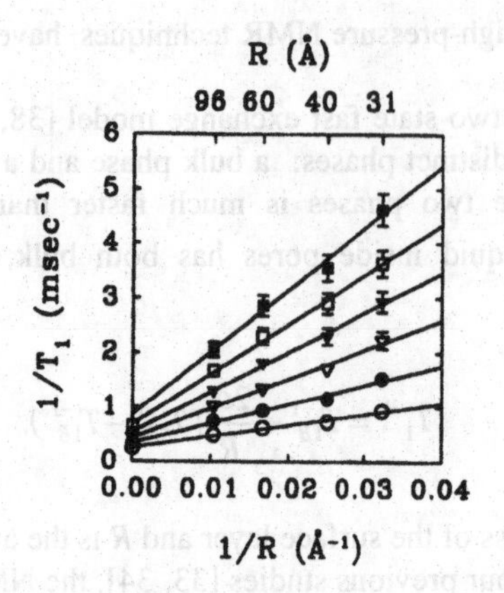

Figure 12. ^{14}N spin-lattice relaxation rate (T_1^{-1}) of acetonitrile-d_3 as a function of pore radius (R^{-1}) in porous silica glasses at 300 K. (O) 0.1 MPa; (●) 100 MPa; (▽) 200 MPa; (▼) 300 MPa; (□) 400 MPa; (■) 500 MPa.

In the limit of extreme narrowing, by neglecting the asymmetry parameter, one can relate the spin lattice relaxation rates of the nuclei ^{2}H (I=1) and ^{14}N (I=1) of acetonitrile-d_3 with diffusion constants as [40]

$$\frac{1}{T_1}=\frac{3}{40}\frac{2I+3}{I^2(2I-1)}\left(\frac{e^2qQ}{\hbar}\right)^2\left(\frac{\frac{1}{4}\left(3\cos^2\theta-1\right)^2}{6D_\perp}+\frac{3\sin^2\theta\cos^2\theta}{5D_\perp+D_\parallel}+\frac{\frac{3}{4}\sin^2\theta}{2D_\perp+4D_\parallel}\right) \tag{7}$$

where $e^2qQ/\hbar$ is 2π times the nuclear quadrupole coupling constant in Hz, and θ is the angle between the symmetry axis of the molecule and the z axis of the molecule coordinate system which diagonalizes the field gradient tensor at the nucleus. All other symbols have their usual meaning. For ^{14}N, $\theta = 0°$ and therefore $T_1^{-1} \propto D_\perp^{-1}$. Thus, the measured T_1 allows us to directly calculate $D_\perp$, the diffusion constant of the tumbling motion about the axis perpendicular to the main symmetry axis. Due to the linear dependence of T_1^{-1} on $D_\perp^{-1}$, this leads from Eq. (1) to another useful expression relating the diffusion constant $D_\perp$ with the pores size R, simply by replacing T_1^{-1} with $D_\perp^{-1}$ [35],

$$D_\perp^{-1} = D_{\perp B}^{-1} + \frac{2\alpha}{R}\left(D_{\perp S}^{-1} - D_{\perp B}^{-1}\right) \tag{8}$$

where $D_{\perp B}$ and $D_{\perp S}$ are the tumbling diffusion constants for bulk liquid and liquid in surface layer, respectively. For ^{2}H, θ=109° 33′, and we can readily determine the parallel diffusion constant $D_\parallel$, from Eq. (2) by substituting in the $D_\perp$ obtained from the ^{14}N data. The following values of the quadrupole coupling constants are used in the calculations: 4.24 MHz for ^{14}N and 172.5 kHz for ^{2}H. The thickness of the surface layer is presumed to be approximately two molecular diameters. The hard sphere radius of CD_3CN, calculated from PVT data at 298 K, is equal to 4.09 Å. In agreement with our earlier results [37], we find $D_{\perp B}$ = 1.75 x 10^{11}/s for bulk acetonitrile-d_3 at 1 bar and 300 K, and $D_{\perp S}$ = .58 x 100^{10}/s for acetonitrile in the surface layer. As expected, the tumbling mobility of acetonitrile appears to be significantly reduced due to the effect of geometrical confinement.

The overall anisotropic rotational diffusion constants, $D_\perp$ and $D_\parallel$, calculated using Eq. (2) for acetonitrile confined to glasses with different pore sizes are plotted as a function of pressure in Figure 13. In view of the inherent error of the $D_\parallel$ values due to the strong inertial effect [41] and the high anisotropic

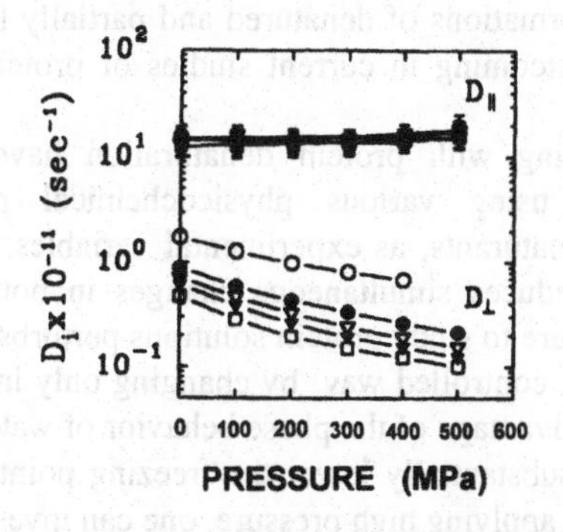

Figure 13. Rotational diffusion constants ($D_\perp$ and $D_\parallel$) of acetonitrile-d_3 as a function of pressure in porous silica glasses at 300 K, (O) bulk liquid; (●) 96 Å; (▽) 60 Å; (▾) 40 Å; (□) 31 Å.

rotational diffusion of CD_3CN molecules [42], the values of parallel diffusion constant are not as reliable as the values of tumbling diffusion constant. Therefore, a quantitative interpretation of the trend in the $D_\parallel$ values must be approached with caution. However, we can conclude that $D_\parallel$ does not change significantly with the pore size and pressure as compared to the marked decrease of $D_\perp$ with decreasing pore size or increasing pressure. The low $\Delta V^{*}(D_\parallel)$ value indicates that the intermolecular potential energy is largely independent of the angle of orientation about the main symmetry axis and as the CD_3 group behaves like a free rotor. As a support for the finding that $\Delta V^{*}(D_\parallel) \approx 0$ for acetonitrile confined to porous silica glasses, we note the results of the study of acetonitrile in the solid state where the line width measurements showed a CH_3 rotation about the C_3 axis for temperatures above 90 K [43].

5. Applications of High-Resolution, High-Pressure NMR Spectroscopy to Biochemical Systems

5.1. PRESSURE-ASSISTED COLD DENATURATION OF PROTEINS

Since Anfinsen and colleagues [44] first studied the renaturation of reduced and unfolded ribonuclease A (RNase A), much effort has been expended in attempting to understand the relationships between the amino acid sequence, the structure, and the dynamic properties of the native conformation of proteins. Recently, increasing attention has been focused on denatured and partially folded states because determination of their structure and stability may provide critical insights into the mechanisms of protein folding [45]. The native conformations of hundreds of proteins are known in great detail from structural determinations by X-ray crystallography and, more recently, NMR spectroscopy. However, detailed knowledge of the conformations of denatured and partially folded states is lacking, which is a serious shortcoming in current studies of protein stability and protein folding pathways [46].

Most studies dealing with protein denaturation have been carried out at atmospheric pressure using various physicochemical perturbations such as temperature, pH, or denaturants, as experimental variables. Compared to varying temperature, which produces simultaneous changes in both volume and thermal energy, the use of pressure to study protein solutions perturbs the environment of the protein in a continuous, controlled way by changing only intermolecular distances. In addition, by taking advantage of the phase behavior of water [47] shown in Figure 14, high pressure can substantially lower the freezing point of an aqueous protein solution. Therefore, by applying high pressure, one can investigate in detail not only pressure-denatured proteins, but also cold-denatured proteins in aqueous solution.

One major application of NMR to protein chemistry is the use of hydrogen exchange kinetics as a probe of protein structure. Our group has recently applied this method to investigate structure in proteins denatured by high pressure at various temperatures. Previous work [48-50] on pressure and cold denaturation has suggested that these methods can leave appreciable residual structure in proteins, particularly when compared to other methods such as thermal or urea denaturation. In RNase A, for example [48], the extent of residual structure measured by hydrogen exchange methods is similar to that present in molten globules [51] and other well-characterized, partially structured proteins. The cold denaturation appears to be a milder method of denaturation than the more conventional methods of heat and chemical (e.g., urea) denaturation.

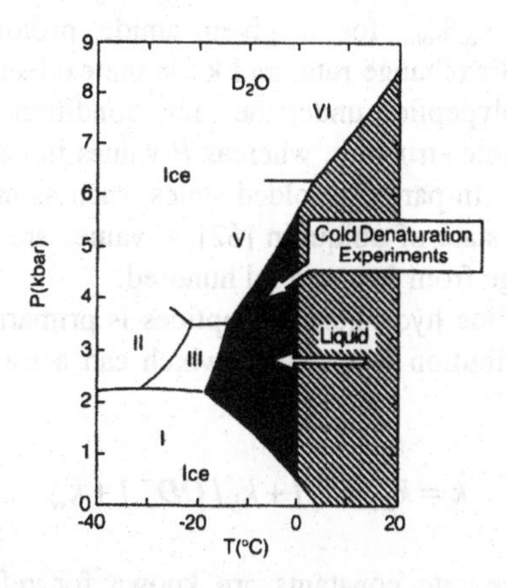

Figure 14. **High-pressure phase diagram of D_2O.**

Such results are interesting enough, but they become more important when related to modern protein-folding studies. Structures that persist in a protein upon denaturation, even relatively mild denaturation, are likely to be highly stable. If no appreciable barriers exist to the rapid formation of such structures, they should be among the first to appear in the refolding of a protein. Consequently, mildly denatured states may serve as useful models for species present early in refolding, with the advantage that such species could be studied for hours or even days, rather than fractions of a second. There is already considerable evidence that the equilibrium collapsed unfolded states of some proteins, obtained under mildly denaturing conditions, are structurally similar to the early, collapsed states occurring when the protein begins to fold. Hence, a major motivation for our earlier research [52, 53] was to characterize several proteins by cold denaturation and compare them to known species observed during folding.

The proteins discussed here were chosen for several reasons. All three – lysozyme, ribonuclease A, and ubiquitin – are small, well-characterized proteins that have been studied before. The solution NMR structures for all three are known [54-56]. Numerous studies have been done on all of these proteins in various denatured and partially folded states. For example, the folding pathways and intermediates of ribonuclease A [57, 58] and of lysozyme [59] have been described in detail. Various denatured states of ribonuclease A [46], lysozyme [51], and ubiquitin [60] are available for comparison. The cold-denatured state of ribonuclease A has been characterized before [48] and evidence for structure in the pressure-denatured state has been described. Additionally, in the case of ubiquitin [61], no evidence was found for significant protection from exchange at early stages of folding; therefore, to test the hypothesis of this parallel, we have also investigated hydrogen exchange in the pressure-assisted cold-denatured state of ubiquitin.

The experimental procedures were described in detail in the original studies [48, 52, 53] of cold-denaturation of ribonuclease A, lysozyme, and ubiquitin. Hydrogen exchange data for proteins are typically expressed in terms of the protection factor P = k_{rc}/k_{obs} for a given amide proton, where k_{obs} is the experimentally measured exchange rate, and k_{rc} is the exchange rate for a proton in an ideal unstructured polypeptide under the same conditions. *P* values close to 1 indicate lack of appreciable structure, whereas *P* values in certain regions of native proteins can exceed 10^6. In partially folded states, such as molten globules [51] or the methanol-induced A state of ubiquitin [62], *P* values are intermediate. Typical values in such states range from 1 to several hundred.

The exchange of amide hydrogens in peptides is primarily caused by acid and base, with a small contribution from water, which can act as a weak acid or base [63]:

$$k = k_a[D^+] + k_b[OD^-] + k_w \tag{9}$$

Values for the three rate constants are known for reference "random coil" materials, specifically unstructured oligopeptides and random-coil polypeptides, at 293 K and ambient pressure [63]. To obtain a value for k_{rc} for cold denaturing conditions, the reference values were corrected for temperature and pressure by using the activation energies and the activation volumes, respectively,

$$k_i(T) = k_i(T_0)exp\left(-\frac{E_a}{R}\left[\frac{1}{T}-\frac{1}{T_0}\right]\right) \tag{10}$$

$$k_i(P) = k_i(P_0)exp\left(-\frac{(P-P_0)\Delta V^{\ddagger}}{RT}\right) \tag{11}$$

where *i* denotes any of the three individual reactions and E_a and $\Delta V^{\ddagger}$ are the activation energy and activation volume. The activation volume, defined from

$$\Delta V^{\ddagger} = -RT\frac{\partial \ln k}{\partial P} \tag{12}$$

was assumed to be constant, over the pressure range of interest, to derive Eq. 3. This assumption has been shown to be valid over a wide pressure range for k_a and k_b in poly-D,L-lysine [64], and because the reaction mechanism for k_w is similar, this is a reasonable assumption there as well. The activation energies for k_a, k_b, and k_w are 14 kcal/mol, 3 kcal/mol, and 19 kcal/mol, respectively [63]. Activation volumes were obtained from data on model compounds: random coil poly-D,L lysine for k_a and k_b, and N-methylacetamide for k_w [65]. The values of $\Delta V^{\ddagger}$ are 0 ±1, +6 ± 1, and -9.0 ± 1.8 cm^3/mol for k_a, k_b, and k_w, respectively.

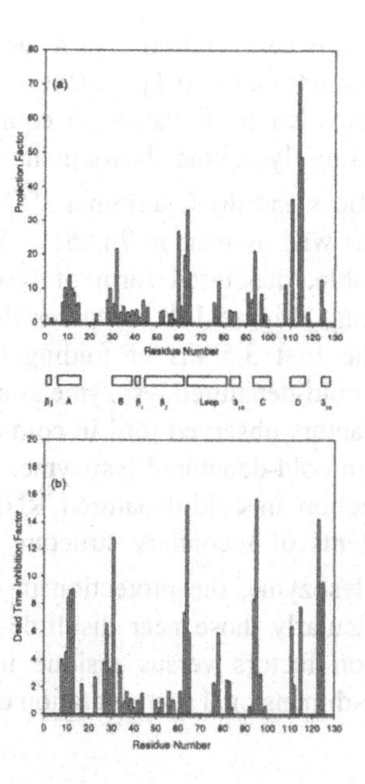

Figure 15. (a) Protection factors for the pressure-assisted cold-denatured state (p = 375 MPa, T = -13°C) of lysozyme. Secondary structural elements in the native state are indicated below. A-D denote the four α-helices; β_1 and β_2 denote the two- and three-stranded β-sheet regions, respectively, (b) dead-time inhibition factors obtained for the first 3.5 ms of lysozyme folding [61].

Figure 15a shows the protection factors in cold-denatured lysozyme (T = -13 °C, P = 3750 bar) as a function of residue number and their correspondence with secondary structural elements in the native state [52].

From the hydrogen exchange rate data, it is clear that many regions of lysozyme are markedly protected from exchange, with P values exceeding 10, in the cold-denatured state. The range of P factors runs from 1.19 for F3 to 71.0 for R114. It should be made clear that the pressure-assisted cold-denatured state is not a "pure" cold-denatured state because the pressure affects the protein directly. Nevertheless, any effect that pressure has is clearly very different from that induced by high temperature or chemical denaturants such as urea.

By using pulsed-labeling hydrogen exchange studies and confining exchange to the dead time of their instrumentation (~3.5 ms), one could observe relatively early stages in the folding of lysozyme [61]. Slowing of exchange measured during this time can be quantified by comparing the measured exchange rate to that for a random coil, as in other hydrogen exchange studies. The resulting quantity, though similar to a protection factor, is determined from a rate that changes considerably as the protein refolds; as a result, it is best referred to as a "dead-time inhibition factor" (I_D) rather than a true protection factor [61]. Nevertheless, I_D values provide information similar to that provided by P values in equilibrium-denatured states. During the first 3.5 ms of folding, lysozyme shows moderate degrees of protection ($I_D > 5$), not only in the α-helices and the C-terminal 3^{10} helix, but also in the loop region from residues 60-65, as well as residue 78 [61]. We have thus observed a strong parallel between a stable, denatured form of lysozyme and the transient species observed during folding. Figure 15b compares the inhibition of hydrogen exchange observed during the first 3.5 ms of folding to the protection factors obtained for pressure-assisted cold-denatured lysozyme as shown in Figure 15a.

The range of protection factors observed [66] in cold-denatured RNase A (2.8-78) is similar to that present in cold-denatured lysozyme. In contrast to lysozyme, however, the extent of protection in cold-denatured RNase A is less organized. Rather than having large extents of secondary structure protected, such as large portions of the α-helices in lysozyme, the protection in cold-denatured RNase A occurs in small regions, particularly those near disulfide-linked cysteine residues. Figure 16 plots the protection factors versus residue in pressure-assisted cold-denatured RNase A on a three-dimensional representation of the native structure.

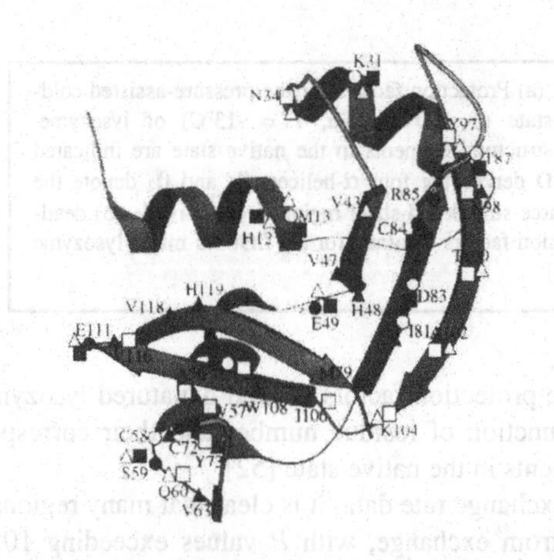

Figure 16. Qualitative comparison of protection factors for the pressure-denatured state [48] and pressure-assisted cold-denatured state [66] of RNase A compared to protection factors for early folding intermediate for RNase A [57]. Amide protons are protected in pressure-denatured state with $\rho > 10, <20$ (O); pressure-denatured state with $\rho > 20$ (●); cold-denatured state with $\rho > 10, <20$ (□); cold-denatured state with $\rho > 20$ (■); folding intermediate with "strong" protection (△); and folding intermediate with "weak", "medium", or "ill-defined" protection (▲).

The most notable such region of high protection in RNase A, 3 residues with $P > 10$, is centered in C84 in the central strand of the large β-sheet. Protection in this β-sheet is highly nonuniform. For example, the region of the first strand from residues 35 to 48 is essentially unprotected ($P < 5$ for all residues), in contrast to the region around C84.

In a qualitative comparison, one can see that the pressure-assisted cold-denatured state exhibits patterns of protection factors (relatively low) resembling the pattern of protection factors observed by Udgaonkar and Baldwin [57] and Houry and Scheraga [58] for the folding intermediate of ribonuclease A. Figure 15 also gives a qualitative comparison of protection factors for the early folding intermediate [57] and those observed of hydrogen exchange in the pressure-assisted cold- denatured state of ribonuclease A [66].

In the case of lysozyme and RNase A, the patterns of protection against hydrogen exchange are similar to those observed in early (refolding time <10 ms) folding intermediates for these proteins, leading to the idea that the cold-denatured state is structurally similar to such intermediates. To help test this idea, ubiquitin, which has folding kinetics [67] that are markedly different from those of either RNase A or lysozyme, was investigated [53]. In particular, ubiquitin shows much less evidence of early structure formation than lysozyme; it more closely resembles a random coil.

Cold-denatured ubiquitin (P=225 MPa, T = -16 °C) shows little deviation from a random coil in its hydrogen exchange kinetics, with no P values above 5 and most below 2. These values are typical of highly denatured proteins such as the urea-denatured state of lysozyme. It is important to point out that in ubiquitin, the same dead-time inhibition study [61] showed no evidence for protection from exchange, with the largest dead-time inhibition factors being ~2 (Figure 17). Moreover, there was no correlation of even these modestly elevated factors within significant elements of secondary structure as there were in lysozyme.

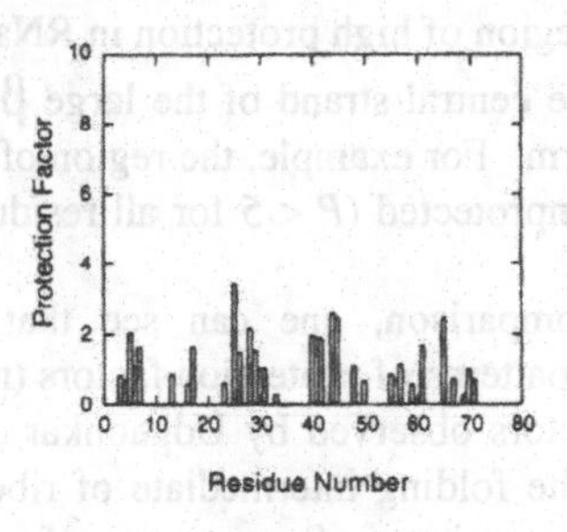

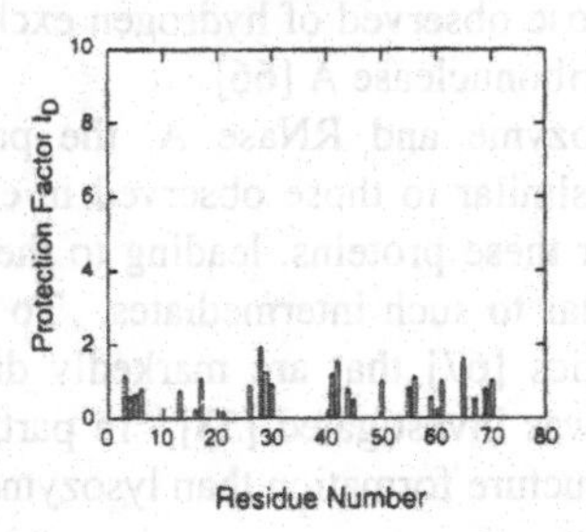

Figure 17. **Protection factors versus residue in pressure-assisted cold-denatured (*$T = -16°V$, $\rho = 225$ MPa*) ubiquitin (*A*) compared with dead-time exchange inhibition factors (*B*) (protection factor I_D obtained during the first 3.5 ms of refolding [61]).**

The similarity of cold-denatured state protection factors to the protection observed in protein refolding studies has led to the idea that the cold-denatured state is populated by species comprising elements of secondary and perhaps tertiary structure that are comparatively stable. The similarities between the cold-denatured state and the early folding states suggest that these structural elements would be expected to form first during folding. In the case of ubiquitin, the folding reaction appears to proceed in one highly cooperative step, with little of the multiphasic behavior observed in proteins like lysozyme or ribonuclease A. Although a possible intermediate for ubiquitin refolding has been characterized by other means [68], the intermediate thus identified shows no protection from hydrogen exchange. The results for the cold-denatured state of ubiquitin parallel these results: there is no partially folded state that is stable enough to be characterized by hydrogen exchange methods when ubiquitin is cold-denatured.

Work is in progress in our laboratory to investigate systematically the pressure- and pressure-assisted cold-denaturation of other proteins, such as α-lactalbumin and apomyoglobin, to obtain novel information about their folding intermediates or to simply indicate key stable structures favored preferentially in the folding process. The methods used to investigate these states include both the hydrogen exchange method outlined above and more direct NMR methods (e.g., COSY, NOESY, and TOCSY) using the improved high-pressure NMR probes and the 500 MHz NMR instrument.

6. Acknowledgments

This work was supported in part by the National Science Foundation under grant NSF CHE 95-26237 and National Institutes of Health under grant PHS 5 RO1 6M42452.

7. References

1. Ballard, L. and Jonas, J. (1997) *Annual Reports on NMR Spectroscopy* **33,** 115.

2. Jonas, J. and Jonas, A. (1994) *Ann. Rev. Biophys. Biomol. Structure* **23,** 287.

3. Lang, E.W. and Lüdemann, H.-D. (1993) *Prog. NMR Spectrosc.* **25,** 507.

4. van Eldik, R. and Merbach, A.E. (1992) *Comments Inorg. Chem.* **12,** 341.

5. Jonas, J. (1990) High-Pressure NMR, in J. Jonas (ed.), *NMR Basic Principles and Progress,* Springer-Verlag, Heidelberg, p. 85.

6. Jonas, J. (1978) *NATO ASI, Series C* **41,** 65.

7. Jonas, J. (1987) *NATO ASI, Series C* **197,** 193.

8. Jonas, J. (1993) *NATO ASI, Series C* **401,** 393.

9. Drickamer, H.G. (1987) *NATO ASI, Series C* **197,** 263.

10. Jonas, J., Koziol, P., Peng, X., Reiner, C., and Campbell, D. (1993) *J. Magn. Reson. B* **102,** 299.

11. Ballard, L., Reiner, C., and Jonas, J. (1996) *J. Magn. Reson. A* **123,** 81.

12. du Penhoat, C.H., Imberty, A., Roques, N., Michon, V., Mentech, J., Descotes, G., and Pérez, S. (1991) *J. Am. Chem. Soc.* **113,** 3720.

13. Sanders, J.K.M. and Hunter, B.K. (1993) *Modern NMR Spectroscopy,* 2nd ed., Oxford University, New York.

14. Bodenhausen, G., Freeman, R., Niedermeyer, R., and Turner, D.L. (1977) *J. Magn. Reson.* **26,** 133.

15. Bachmann, P., Aue, W.P., Muller, L., and Ernst, R.R. (1977) *J. Magn. Reson.* **28,** 29.

16. Kumar, A., Wagner, G., Ernst, R.R., and Wüthrich, K. (1981) *J. Am. Chem. Soc.* **103,** 3654.

17. Wüthrich, K., Billeter, M., and Braun, W. (1984) *J. Mol. Biol.* **180,** 715.

18. Mirau, P.A. and Bovey, F.A. (1986) *J. Am. Chem. Soc.* **108,** 5130.

19. Wright, P.E., Dyson, H.J., and Lerner, R.A. (1988) *Biochemistry* **27,** 7167.

20. Baleja, J.D., Moult, J., and Sykes, B.D. (1990) *J. Magn. Reson.* **87,** 375.

21. Anderson, N.H., Eaton, H.L., and Lai, X. (1989) *Magn. Reson. Chem.* **27,** 515.

22. Jonas, J., Adamy, S.T., Grandinetti, P.J., Masuda, Y., Morris, S.J., Campbell, D.M., and Li, Y. (1990) *J. Phys. Chem.* **94,** 1157.

23. Walker, N.A., Lamb, D.M., Adamy, S.T., Jonas, J., and Dare-Edwards, M.P. (1988) *J. Phys. Chem.* **92,** 3675.

24. Adamy, S.T., Grandinetti, P.J., Masuda, Y., Campbell, D., and Jonas, J. (1991) *J.Chem. Phys.* **94,** 3568.

25. Adamy, S.T., Kerrick, S.T., and Jonas, J. (1994) *Z. Phys. Chem. (Munich)* **184,** 185.

26. Macura, S. and Ernst, R.R. (1980) *Mol. Phys.* **41,** 95.

27. McClung, R.E. and Kivelson, D. (1968) *J. Chem. Phys.* **49,** 3380.

28. Allinger, N.L. (1977) *J. Am. Chem. Soc.* **99,** 8127.

29. Liu, G., Li, Y., and Jonas, J. (1989) *J. Chem. Phys.* **90,** 5881.

30. Mackowiak, M., Liu, G., and Jonas, J. (1990) *J. Chem. Phys.* **93,** 2154.

31. Liu, G., Mackowiak, M., Li, Y., and Jonas, J. (1991) *J. Chem. Phys.* **94,** 239.

32. Liu, G., Li, Y., and Jonas, J. (1991) *J. Chem. Phys.* **95,** 6892.

33. Xu, S., Zhang, J. and Jonas, J. (1992) *J. Chem. Phys.* **97,** 4564.

34. Jonas, J., Zhang, J., and Xu, S. (1993) *Materials Research Society Symposium of Confined Systems* **290,** 95.

35. Zhang, J. and Jonas, J. (1993) *J. Chem. Phys.* **97,** 8812.

36. Korb, J.-P., Xu, S., and Jonas, J. (1993) *J. Chem. Phys.* **98,** 2411.

37. Bull, T.E. and Jonas, J. (1970) *J. Chem. Phys.* **53,** 3315.

38. Brownstein, K.R. and Terr, C.E. (1977) *J. Magn. Res.* **26,** 17.

39. Gallegos, D.P. and Smith, D.M. (1988) *J. Colloid Interface Sci.* **122,** 143.

40. Woessner, D.E. (1962) *J. Chem. Phys.* **37,** 647.

41. Bopp, T.T. (1967) *J. Chem. Phys.* **47,** 3621.

42. Allerhand, A. (1970) *J. Chem. Phys.* **52,** 3596.

43. Gutowsky, H.S. and Pake, G.E. (1950) *J. Chem. Phys.* **18,** 162.

44. Anfisen, C.B. (1973) *Science* **181,** 223.

45. Kim, P.S. and Baldwin, R.L. (1990) *Annu. Rev. Biochem.* **59,** 631.

46. Robertson, A.D. and Baldwin, R.L. (1991) *Biochemistry* **30,** 9907.

47. Jonas, J. (1982) *Science* **216,** 1179.

48. Zhang, J., Peng, X., Jonas, A., and Jonas, J. (1995) *Biochemistry* **34,** 8631.

49. Konno, T., Kataoka, M., Kamatari, Y., Kanaori, K., Nosaka, A., and Akasaka, K. (1995) *J. Mol. Biol.* **251,** 95.

50. Wong, K.-B., Freund, S.M.V., and Fersht, A.R. (1996) *J. Mol. Biol.* **259,** 805.

51. Buck, M., Radford, S.E., and Dobson, C.M. (1994) *J. Mol. Biol.* **237,** 247.

52. Nash, D. and Jonas, J. (1997) *Biochemistry* **36,** 14375.

53. Nash, D. and Jonas, J. (1997) *Biochem. Biophys. Res. Commun.* **238,** 289.

54. Di Stefano, D.L. and Wand, A.J. (1987) *Biochemistry* **26,** 7272.

55. Redfield, C. and Dobson, C.M. (1988) *Biochemistry* **27,** 122.

56. Rico, M., Santoro, J., González, Bruix, M., Neira, J.L., Nieto, J.L., and Herranz, J. (1991) *J. Biomol. NMR* **1,** 283.

57. Udgaonkar, J.B. and Baldwin, R.L. (1990) *Proc. Natl. Acad. Sci .USA* **87,** 8197.

58. Houry, W.A. and Scheraga, H.A. (1996) *Biochemistry* **35,** 11734.

59. Radford, S.E., Dobson, C.M., and Evans, P.A. (1992) *Nature* **358,** 302.

60. Harding, M.H., Williams, D.H., and Woolfson, D.N. (1991) *Biochemistry* **30,** 3120.

61. Gladwin, S.T. and Evans, P.A. (1996) *Folding Des.* **1,** 407.

62. Pan, Y. and Briggs, M.S. (1992) *Biochemistry* **31,** 11405.

63. Bai, Y., Milne, J.S., Mayne, L., and Englander, S.W. (1993) *Proteins Struct. Funct. Genet.* **17,** 75.

64. Carter, J.V., Knox, D.G., and Rosenberg, A. (1978) *J. Biol. Chem.* **253,** 1947.

65. Mabry, S.A., Lee, B.-S., Zheng, T., and Jonas, J. (1996) *J. Am. Chem. Soc.* **118,** 8887.

66. Nash, D., Lee, B.-S., and Jonas, J. (1996) *Biochim. Biophys. Acta.* **1297,** 40.

67. Briggs, M.S. and Roder, H. (1992) *Proc. Natl. Acad. Sci. USA* **89,** 2017.

68. Khorasanizadeh, S., Peters, I.D., and Roder H. (1996) *Nature Struct. Biol.* **3,** 193.

PRESSURE-ENHANCED MOLECULAR CLUSTERING IN LIQUID DIMETHYL SULFOXIDE STUDIED BY RAMAN SPECTROSCOPY

C. CZESLIK, J. JONAS
School of Chemical Sciences and Beckman Institute for Advanced Science and Technology, University of Illinois, Urbana, IL 61801, USA

1. Abstract

Raman spectroscopy was used to study intermolecular interactions in liquid dimethyl sulfoxide. The non-coincidence of the vibrational wavenumbers of the in-phase and the out-of-phase SO stretching mode was analyzed in the temperature range of 30 - 70 °C under pressures up to 2200 bar. The observed non-linear increase of this splitting parameter as a function of density can be explained by an increase in local order leading to an enhanced coupling of neighboring oscillators. The isotropic SO stretching vibrational band consists of three components which are attributed to clusters differing in intermolecular association strength. The pressure dependence of the integrated intensities of the three SO band components is interpreted as a pressure-induced formation of clusters consisting of strongly associated molecules.

2. Introduction

The liquid phase of dimethyl sulfoxide (DMSO) has been the subject of numerous publications in which the existence of molecular clusters is controversially discussed, e.g., thermodynamic studies [1,2], X-ray and neutron diffraction experiments [3,4,5], molecular dynamics simulations [6,7], and infrared and Raman spectroscopic studies [8,9,10,11] have been performed. In most cases, the authors find indications for DMSO clusters in the liquid phase which are either cyclic dimers with an antiparallel arrangement of the two neighboring dipole vectors or linear chains in which the molecular dipoles are oriented in a parallel fashion.

In order to study the nature of molecular clustering in liquid DMSO in more detail, pressure has to be applied, because pressure affects the density of the system by reducing intermolecular distances. Pressure studies on liquid DMSO must be carried out at elevated temperatures to prevent pressure-induced solidification. If the liquid/solid coexistence curve of DMSO is calculated from thermodynamic data, a liquid/solid transition pressure of, e.g., 2200 bar at 70 °C is obtained. Since increasing temperature changes both the thermal energy and the density of the system, the density of liquid DMSO has to be known as a function of temperature and pressure in order to study the

R. Winter and J. Jonas (eds.), High Pressure Molecular Science, 261–266.

effect of temperature at constant density. Several examples exist where the effect of temperature on system properties is mainly due to a change in density [12].

In this sudy, the Raman non-coincidence effect (NCE) serves to investigate intermolecular interactions in liquid DMSO. It is defined as the difference between the wavenumbers of the anisotropic and the isotropic band of a vibrational mode:

$$\Delta\tilde{\nu} = \tilde{\nu}_{aniso} - \tilde{\nu}_{iso} \tag{1}$$

Most often, the NCE is due to a transition dipole-transition dipole coupling. According to Fini and Mirone [10], this splitting parameter can be explained by the presence of molecular clusters in which molecules are oriented in an at least partially ordered way. The isotropic band is then assigned to the in-phase vibration of all cluster molecules, whereas the anisotropic band reflects the mode in which the molecules of a given dipole direction vibrate out of phase with the molecules of the opposite direction. Logan has developed a theory which predicts the NCE of polar modes in neat liquids [13]:

$$\Delta\omega = \frac{48}{25\pi\varepsilon_0}\left(\frac{d\mu}{dQ}\right)^2 \frac{\xi}{m\omega\sigma^3} \tag{2}$$

where $\Delta\omega$ is the non-coincidence value in the dimension of angular frequency, μ is the dipole moment, Q is the normal coordinate, m is the reduced mass, ω is the vibrational angular frequency of the free molecule, σ is the hard-sphere diameter, and the parameter ξ can be calculated numerically according to the following equation:

$$\frac{\mu_0^2\rho}{3kT\varepsilon_0} = \frac{(1+4\xi)^2}{(1-2\xi)^4} - \frac{(1-2\xi)^2}{(1+\xi)^4} \tag{3}$$

where μ_0 is the permanent dipole moment and ρ is the number density. This equation Logan obtained by using the mean spherical approximation according to which the liquid is composed of sperical molecules that possess a permanent dipole moment.

3. Experimental Section

DMSO was purchased from Fisher Scientific and had a purity of 99.9 %. It was used without further purification. High-pressure density measurements were carried out using a home-built high-pressure relative density meter. At each temperature the relative densities as a function of pressure were calibrated using the corresponding absolute density at 1 bar from the literature [2]. High-pressure Raman experiments were performed using a home-built high-pressure cell. Raman spectra were recorded in the temperature interval of 30 - 70 °C under pressures up to 2200 bar within the liquid phase region of DMSO. The light source for the excitation was the 488 nm line of an argon ion laser (Spectra-Physics). The scattered Raman radiation was analyzed by a double monochromator (Spex 1403) with a slit width of 24 µm and detected by a liquid-nitrogen cooled CCD detector (Princeton Instruments). Stokes Raman spectra were collected in the interval of 1160 - 940 cm^{-1} with parallel ($I_{\parallel}$) and perpendicular ($I_{\perp}$) polarization. The Raman spectra were analyzed with Grams/32 software (Galactic Industries Corp.). Isotropic and anisotropic Raman spectra were then calculated as

$$I_{iso} = I_{\parallel} - \frac{4}{3} I_{\perp} \tag{4}$$

$$I_{aniso} = I_{\perp} \tag{5}$$

4. Results and Discussion

As an example, the isotropic and anisotropic Raman spectrum of DMSO in the SO stretching vibration region at 50 °C and 1000 bar are given in Figure 1. The isotropic SO stretching band at about 1043 cm^{-1} can be assigned to the in-phase vibration of cluster molecules [14,10], whereas the smaller peak at about 953 cm^{-1} reflects a methyl rocking mode [14]. The anisotropic SO stretching band consists of three components. The first at about 1028 cm^{-1} is assigned to a methyl rocking mode, the second reflects the depolarized part of the SO in-phase vibration and coincides with the first moment of the isotropic SO stretching band, and the third component at about 1056 cm^{-1} is assigned to the out-of-phase vibration of cluster molecules [14,10].

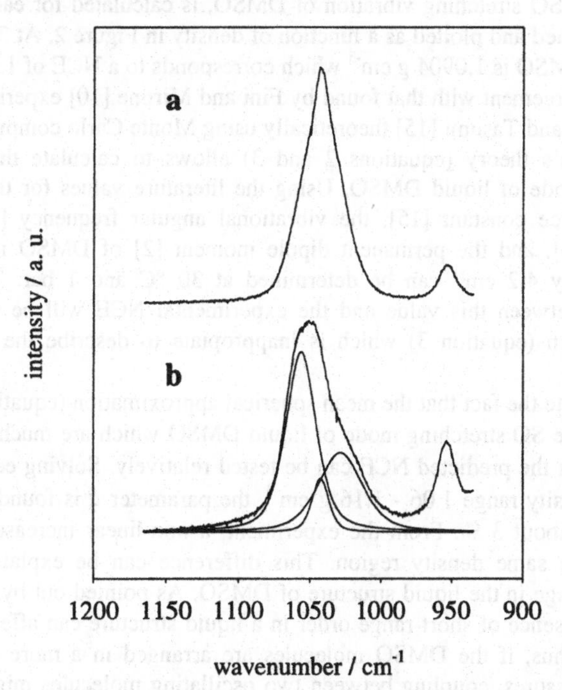

Figure 1. Isotropic (a) and anisotropic (b) Raman spectrum of DMSO at 50 °C and 1000 bar.

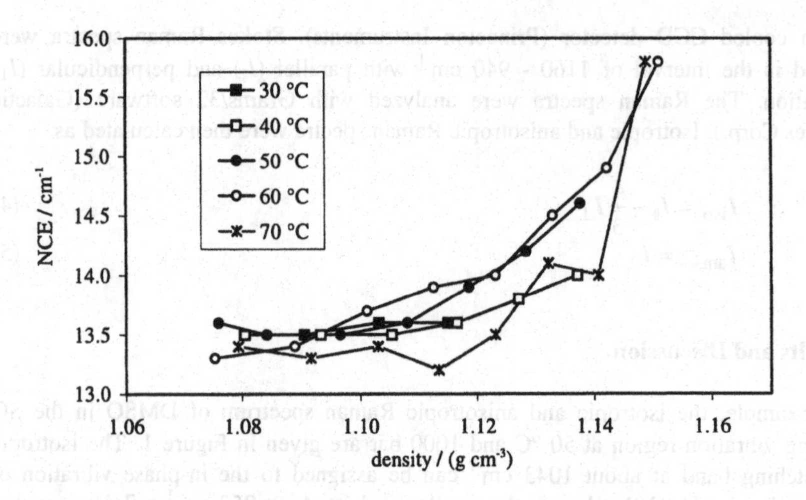

Figure 2. The non-coincidence effect (NCE) of the SO stretching mode of liquid DMSO as a function of density.

The NCE, as the difference between the wavenumbers of the out-of-phase and the in-phase SO stretching vibration of DMSO, is calculated for each temperature and pressure studied and plotted as a function of density in Figure 2. At 30 °C and 1 bar, the density of DMSO is 1.0904 g cm^{-3} which corresponds to a NCE of 13.5 cm^{-1}. This value is in good agreement with that found by Fini and Mirone [10] experimentally (14 cm^{-1}), and by Torii and Tasumi [15] theoretically using Monte Carlo computer simulations (11 cm^{-1}). Logan's theory (equations 2 and 3) allows to calculate the NCE of the SO stretching mode of liquid DMSO. Using the literature values for the transition dipole [15], the force constant [15], the vibrational angular frequency [14], the molecular diameter [16], and the permanent dipole moment [2] of DMSO molecules, an NCE value of only 4.2 cm^{-1} can be determined at 30 °C and 1 bar. The reason for the difference between this value and the experimental NCE will be the mean spherical approximation (equation 3) which is inappropiate to describe the liquid structure of DMSO.

Despite the fact that the mean spherical approximation (equation 3) leads to NCE values for the SO stretching mode of liquid DMSO which are much too low, the effect of density on the predicted NCE can be tested relatively. Solving equation 3 for 50 °C over the density range 1.06 - 1.16 g cm^{-3}, the parameter ξ is found to increase almost linearly by about 3 %. From the experiment, a non-linear increase of about 17 % is found in the same density region. This difference can be explained by a pressure-induced change in the liquid structure of DMSO. As pointed out by Wang and McHale [17], the presence of short-range order in a liquid structure can affect the magnitude of the NCE. Thus, if the DMSO molecules are arranged in a more ordered way under elevated pressures, coupling between two oscillating molecules might occur in a more efficient way leading to the observed non-linear increase of the NCE.

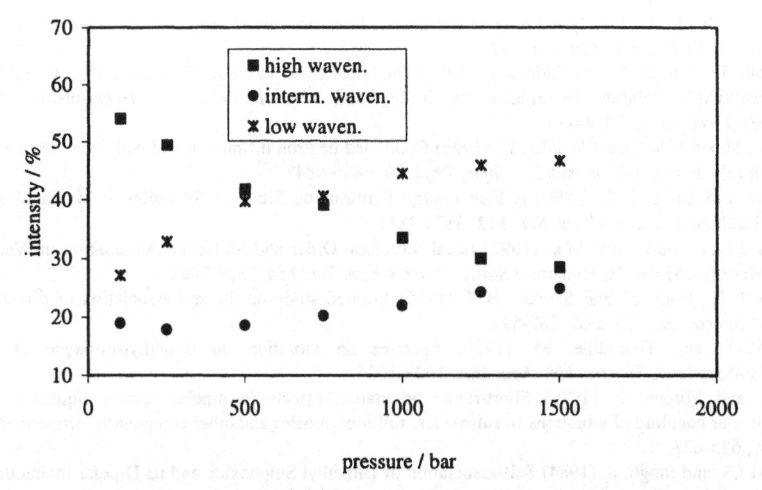

Figure 3. Integrated intensities of the three components of the isotropic SO stretching band at 50 °C as a function of pressure.

The isotropic SO stretching band of DMSO (Fig. 1) was fitted to 50:50 combinations of Gaussian and Lorentzian functions. In the temperature and pressure region studied, three components are needed to reproduce the shape of this band. In Figure 3, the integrated intensities of these three components are plotted as a function of pressure at e. g. 50 °C. The intensity of the high-wavenumber component which can be attributed to weakly associated molecules is decreasing drastically with pressure, whereas the intensity of the low-wavenumber component, assigned to strongly associated molecules, shows the opposite behavior. If the integrated intensities of the three SO band components are correlated with concentrations, a pressure-induced formation of clusters with strong intermolecular association forces is observed. Thus, from both the NCE of the SO stretching mode and the decomposition of the isotropic SO band, a pressure-induced ordering in the liquid structure of DMSO can be concluded.

5. Acknowledgment

This work was supported in part by the National Science Foundation under grant NSF CHE 95-26237.

6. References

1. Schläfer, H.L. and Schaffernicht, W. (1960) Dimethylsulfoxyd als Lösungsmittel für anorganische Verbindungen, *Angew. Chem.* **72**, 618-626.
2. Casteel, J.F. and Sears, P.G. (1974) Dielectric Constants, Viscosities, and Related Physical Properties of 10 Liquid Sulfoxides and Sulfones at Several Temperatures, *J. Chem. Eng. Data* **19**, 196-200.

3. Itoh, S. and Ohtaki, H. (1987) A Study of the Liquid Structure of Dimethyl Sulfoxide by the X-Ray Diffraction, *Z. Naturforsch.* **42a**, 858-862.
4. Bertagnolli, H., Schultz, E., and Chieux, P. (1989) The Local Order in Liquid Dimethylsulfoxide and KI-Dimethylsulfoxide Solution Determined by X-Ray and Neutron Diffraction Experiments, *Ber. Bunsenges. Phys. Chem.* **93**, 88-95.
5. Luzar, A., Soper, A.K., and Chandler, D. (1993) Combined neutron diffraction and computer simulation study of liquid dimethyl sulphoxide, *J. Chem. Phys.* **99**, 6836-6847.
6. Rao, B.G. and Singh, U.C. (1990) A Free Energy Perturbation Study of Solvation in Methanol and Dimethyl Sulfoxide, *J. Am. Chem. Soc.* **112**, 3803-3811.
7. Vaisman, I.I. and Berkowitz, M.L. (1992) Local Structural Order and Molecular Association in Water-DMSO Mixtures. Molecular Dynamics Study, *J. Am. Chem. Soc.* **114**, 7889-7896.
8. Figueroa, R.H., Roig, E., and Szmant, H.H. (1966) Infrared study on the self-association of dimethyl sulfoxide, *Spectrochim. Acta* **22**, 587-592.
9. Forel, M.-T. and Tranquille, M. (1970) Spectres de vibration du diméthylsulfoxyde et du diméthylsulfoxyde-d_6, *Spectrochim. Acta* **26A**, 1023-1034.
10. Fini, G. and Mirone, P. (1976) Short-range orientation effects in dipolar aprotic liquids - III. Intermolecular coupling of vibrations in sulfoxides, sulfones, nitriles and other compounds, *Spectrochim. Acta* **32A**, 625-629.
11. Sastry, M.I.S. and Singh, S. (1984) Self-association of Dimethyl Sulphoxide and its Dipolar Interactions with Water: Raman Spectral Studies, *J. Raman Spectrosc.* **15**, 80-85.
12. Jonas, J. and Lee, Y.T. (1991) NMR and laser Raman scattering studies of fluids at high pressure, *J. Phys.: Condens. Matter* **3**, 305-338.
13. Logan, D.E. (1986) The non-coincidence effect in the Raman spectra of polar liquids, *Chem. Phys.* **103**, 215-225.
14. Tranquille, M., Labarbe, P., Fouassier, M., and Forel, M.T. (1971) Détermination des champs de forces de valence et des modes normaux de vibration des molécules $(CH_3)_2S$ et $(CH_3)_2SO$, *J. Mol. Structure* **8**, 273-291.
15. Torii, H. and Tasumi, M. (1995) Raman Noncoincidence Effect and Intermolecular Interactions in Liquid Dimethyl Sulfoxide: Simulations Based on the Transition Dipole Coupling Mechanism and Liquid Structures Derived by the Monte Carlo Method, *Bull. Chem. Soc. Jpn.* **68**, 128-134.
16. Zeidler, M.D. (1965) Umorientierungszeiten, Sprungzeiten und Quadrupolkopplungskonstanten in einigen organischen Flüssigkeiten aus kernmagnetischen Relaxationszeitmessungen, *Ber. Bunsenges. Phys. Chem.* **60**, 659-669.
17. Wang, C.H. and McHale, J. (1980) Vibrational resonance coupling and the noncoincidence effect of the isotropic and anisotropic Raman spectral components in orientationally anisometric molecular liquids, *J. Chem. Phys.* **72**, 4039-4044.

INORGANIC AND BIOINORGANIC REACTION KINETICS UNDER HIGH PRESSURE

RUDI VAN ELDIK
Institute for Inorganic Chemistry
University of Erlangen-Nürnberg
Egerlandstr. 1
91058 Erlangen
Germany

ABSTRACT. The elucidation of the mechanisms of thermal reactions in inorganic and bioinorganic chemistry through the application of high pressure kinetic techniques is described. High pressure kinetic and thermodynamic data are used to construct volume profiles for typical reactions that include ligand substitution, activation of small molecules, and electron transfer processes. The mechanistic information obtained from such profiles is discussed in detail.

1. Introduction

It is the objective of this contribution to review the advances made in the elucidation of inorganic and bioinorganic reaction mechanisms through the application of high pressure kinetic techniques in the study of a wide range of reactions in solution. This topic has been covered in chapters of earlier proceedings in this series [1,2], as well as in a number of reviews that have appeared in recent years [3-7]. Readers are advised to consult these references for further background information. This research area has shown a consistent interest from various groups and more than 1500 data sets were published for inorganic systems over the periode 1987 to 1996 [6].

The approach adopted in this presentation first involves a short outline of the basic principles and experimental techniques involved in this work, in an effort to answer the question *Why?* and *How?* do we do these measurements. Following this, an account on the mechanisms of fundamental ligand substitution reactions of metal complexes and the mechanistic insight gained through the application of high pressure kinetic techniques will be presented. This section forms the basis for many chemical processes in inorganic and bioinorganic systems, and then allows a treatment of the activation mechanisms of small molecules by metal complexes in the subsequent section. Such activation processes usually involve a direct interaction between the small

R. Winter and J. Jonas (eds.), High Pressure Molecular Science, 267–289.

molecule and the metal center, which is in many cases a substitution controlled process. Once the small molecule is bound to the metal complex, the actual activation in many cases involves an electron transfer process. Thus, this section in a natural way leads to a treatment of the effect of pressure on electron trasfer processes that are of inorganic and bioinorganic interest. Numerous examples will be used to demonstrate the kind of mechanistic information that can be obtained through the application of high pressure kinetic techniques in the study of these reactions.

2. Fundamental Principles and Experimental Techniques

It has been our philosophy in the past to obtain as much as possible mechanistic insight from detailed kinetic investigations as a function of as many as possible experimental variables. These include the concentrations of the reactants and products, the composition of the solvent, pH, buffer composition, and ionic strength, which all form part of the chemical variables in mechanistic studies. In addition, there are two physical variables that can be applied in such studies, viz. temperature and pressure, and we have made use of these where possible. Such detailed kinetic data can provide along with all other known chemical and structural properties of the system and possible intermediates, detailed insight into the nature of the underlying reaction mechanism. In many cases our insight is restricted by the number of variables covered during such investigations. The suggested mechanism based on this information will only then approach the "real" mechanism when it is in line with all available experimental as well as theoretical data.

In this presentation we will focus on the role of the pressure variable in such mechanistic studies. Almost all chemical reactions in solution exhibit a characteristic pressure dependence over a moderate pressure range of a few hundred MPa. The pressure dependence of an equilibrium (K) or a rate constant (k) results in the reaction volume, ΔV, or the volume of activation, $\Delta V^{\#}$, via the relationships $(\delta \ln K/\delta P)_T = -\Delta V/RT$ or $(\delta \ln k/\delta P)_T = -\Delta V^{\#}/RT$, respectively. This information combined with partial molar volume data for reactant and product species can now be used to construct a volume profile for the studied reaction, which represents the partial molar volume changes associated with the process along the reaction coordinate in a kinetic and thermodynamic sence. Such volume profiles, of which many have been reported [3,6] and will be discussed in this contribution, greatly assist the assignment of the intimate mechanism on the basis of the location of the transition state in reference to that of the reactant and product species on a volume basis. Thus the reaction mechanism is interpreted in terms of specific volume changes along the reaction coordinate, which mainly consist of intrinsic and solvational contributions that result from changes in bond lengths and angles, and changes in electrostriction or dipole moment, respectively.

Much of the achieved advances result from the development and availability of instrumentation to study slow and fast reactions at pressures up to 300 MPa, including stopped-flow, T-jump. P-jump, NMR, ESR, flash-photolysis and pulse-radiolysis instrumentation [3,4,6,8,9]. Readers are advised to consult the quoted

references for more detailed information, since these present detailed accounts on the present instrumentation situation and also on the commercial availabilty of such equipment.

3. Ligand Substitution Reactions

The general understanding of ligand substitution mechanisms of square planar and octahedral complexes since the earlier concepts were developed, benefitted greatly from the numerous high pressure kinetic studies performed during the past two decades. In this respect it has especially been Merbach and coworkers that have contributed in an impressive way to resolve mechanisms of symmetrical solvent and ligand exchange processes on transition metal complexes [4,10-12]. Solvent exchange processes are the most fundamental substitution reactions that characterize the lability and reaction mechanism of a metal center within a coordination geometry. For these reactions the interpretation of the volume of activation becomes rather straight forward since these reactions do not involve major changes in solvation due to changes in dipole moments and electrostriction, which can in the case of non-symmetrical reactions significantly complicate the interpretation. The results and mechanistic information obtained from these studies are presently well accepted by the mechanistic community, but have in the past resulted in several discrepancies in the literature, especially with theoretical chemists.

The activation volumes for solvent exchange on the divalent cations of the first row of transition metals exhibit clear evidence for a mechanistic changeover along the series, i.e. from more associative for the earlier, larger cations to a more dissociative one for the later, smaller cations [10]. It has been claimed on the basis of theoretical calculations [13,14] that the interpretation of such activation volume data could be considered faulty, and arguments were presented against a mechanistic changeover along the series. Additional theoretical calculations were needed in order to investigate the validity of the mechanistic proposals that were based on calculations, in an effort to resolve the disparate conclusions.

The original studies [13,14] involved *ab initio* self consistent field (SCF) calculations of the binding energies, ligand-field effects, water exchange reactions, and exchange of hexahydrated divalent first row transition metal cations. In subsequent work, Rotzinger [15,16] succeeded in computing the structures of the transition states and the intermediates formed during the water exchange reactions of the first row transition metals with *ab initio* methods at the Hartree-Fock or CAS-(complete active space)-SCF level. It was now possible to generate A, I_a, I_d and D pathways, and to optimize the structures of the transition and intermediate states. Furthermore, the calculated bond length changes that occurred during the activation process were entirely consistent with the activation volume data, and indicated that an associative interchange mechanism can operate for some metal ions, in contrast to the previous theoretical prediction [14].

Density functional theory has also been applied successfully to describe the

solvent exchange mechanism for aquated Pd(II), Pt(II) and Zn(II) cations [17,18]. Our own work on aquated Zn(II) [18] was stimulated by our interest in the catalytic activity of such metal ions and by the absence of any solvent (water) exchange data for this cation. The optimized transition state structure clearly demonstrated the dissociative nature of the process; in no way could a seventh water molecule be forced to enter the coordination sphere without the simultaneous dissociation of one of the six coordinated water molecules.

The level of understanding reached for solvent exchange reactions on solvated metal ions has encouraged investigations of non-symmetrical complex-formation and ligand substitution reactions. Complex-formation of the divalent first row transition elements exhibited volumes of activation that are very similar to those observed for the water exchange reactions, from which it followed that very similar mechanisms must be operating [6,10]. The same was observed for an extended series of aquation reactions for complexes of the type $M(NH_3)_5L^{3+}$, M = Co(III), Cr(III) and L = neutral nucleophile [6]; the results once again agreed very well with the solvent exchange data. Along these lines, a volume profile was recently constructed for the reaction

$$Fe^{II}(CN)_5NH_3^{3-} + H_2O \rightleftharpoons Fe^{II}(NH_3)_5H_2O^{3-} + NH_3$$

The pressure dependence of the forward and reverse reactions resulted in significantly positive volumes of activation [19], and the volume profile reported in Figure 1 clearly demonstrates the dissociative nature of the substitution process.

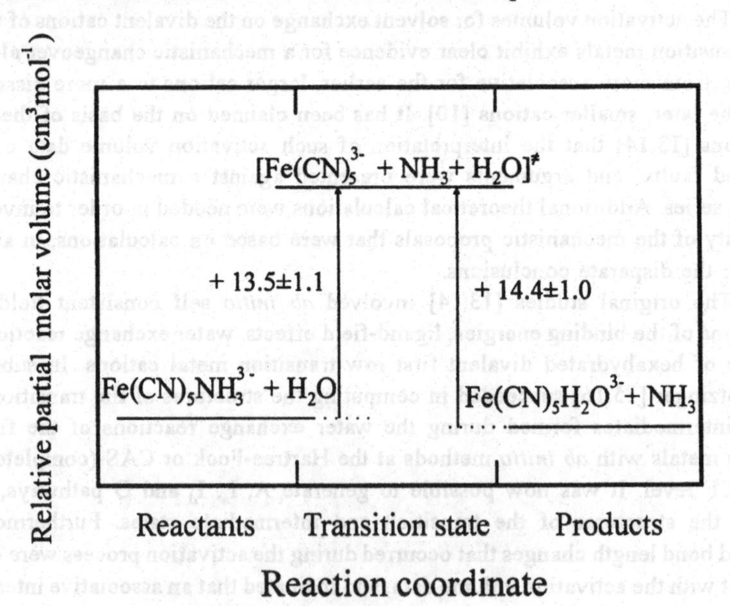

Figure 1. Volume profile for the reaction
$$Fe^{II}(CN)_5NH_3^{3-} + H_2O \rightleftharpoons Fe^{II}(NH_3)_5H_2O^{3-} + NH_3$$

On the basis of the available mechanistic information, a challenging question is to what extent can the substitution mechanism of metal complexes be tuned by structural modifications of the systems. These usually involve the tuning of steric and electronic effects. A number of such examples are treated in the following sections.

3.1. LABILIZATION BY A HYDROXO LIGAND

The mechanistic understanding of solvent exchange reactions has reached the point where specific labilization effects can be studied in a systematic way. In this respect it is appropriate to refer to the significant trans-labilization caused by the deprotonation of a coordinated water molecule. In the case of hexaaqua complexes of Fe(III), Rh(III) and Ir(III), such deprotonation can cause an increase in the water exchange rate of between 700 and 20000 times at 298 K [20]. This labilization is also accompanied by a changeover in mechanism from a more associative interchange for the hexaaqua complex ions to a more dissociative interchange mechanism for the pentaaquamonohydroxo complex ions. As a result of rapid proton exchange, labilization by coordinated hydroxide is not site specific, with the result that all five coordinated water molecules are labilized to the same extent.

We recently studied the effect of pressure on the water exchange reaction of the dihydroxo bridged Rh(III) dimer in which the hydroxo labilization must now be site specific [21]. ^{17}O NMR experiments clearly demonstrated that the coordinated water molecules located *cis* and *trans* to the hydroxo bridging ligands exhibit different chemical shifts, and exchange significantly faster with the bulk solvent than the bridging hydroxo groups. Surprisingly, however, was the finding that these water molecules exchange at a rather similar rate (ca. 5×10^{-7} s^{-1} at 298 K), which is ca. 10^2 faster than water exchange on the hexaaquarhodium(III) monomer, but ca. 10^2 slower than exchange on the pentaaquamonohydroxo monomer. The estimated volumes of activation were found to be between +9 and +10 cm^3 mol^{-1}, which is a clear indication for a dissociative exchange mechanism. The surprising similarity in the exchange rates of the *cis* and *trans* water molecules becomes quite understandable in terms of a dissociative mechanism. Dissociation of the more labile *trans* water molecules will result in the formation of a trigonal bipyramidal five-coordinate intermediate, which can via a Berry pseudo-rotation convert to a tetragonal pyramidal species, with the consequence that the entering solvent molecule will take the *cis* position. Thus dissociative solvent exchange in such complexes results in an almost equal labilization of both the *cis* and *trans* bound water molecules.

3.2. LABILIZATION BY A METAL-CARBON BOND

Other examples of large labilization effects include the introduction of metal-carbon bonds on traditionally inert metal centers such as Co(III), Rh(III) and Ir(III). For instance, the presence of Cp* (Cp* = Me_5Cp, Cp = cyclopentadienyl) in the complexes $M(Cp^*)(H_2O)_3^{2+}$ (M = Rh(III) and Ir(III)) causes an increase in the solvent exchange rate constant of 10^{14} as compared to the hexaaqua species [22]. The volumes of activation support the operation of a dissociative interchange mechanism in both cases. In the case of the $Co(NH_3)_5CH_3^{2+}$ complex, the strong trans labilization caused by the metal-carbon bond does not only show up in the ground state structure [23], but also causes this complex to become extremely labile and the substitution reaction is characterized by a very positive volume of activation, suggesting a limiting dissociative mechanism [24]. The introduction of carbon bonded alkanes on rhodoximes causes a drastic increase in the lability of the *trans* position. The nature of the organic ligand not only controls the rate of the substitution process, but also the nature of the mechanism. Based on the reported volumes of activation [25], it could be concluded that CH_3 induces a fast substitution process that follows a dissociative interchange mechanism, whereas for the weaker CH_2CF_3 donor group, the substitution reaction is significantly slower and follows an associative interchange mechanism. A typical volume profile constructed for the reversible binding of iodide in the case of the methyl complex is shown in Figure 2. Interesting enough, whether the organic group was varied or not for a given nucleophile, all reactions studied were characterized by moderately negative entropies of activation. Thus the additional potential mechanistic discrimination power of the pressure variable has been aptly exploited in *trans*-rhodoxime substitution behaviour.

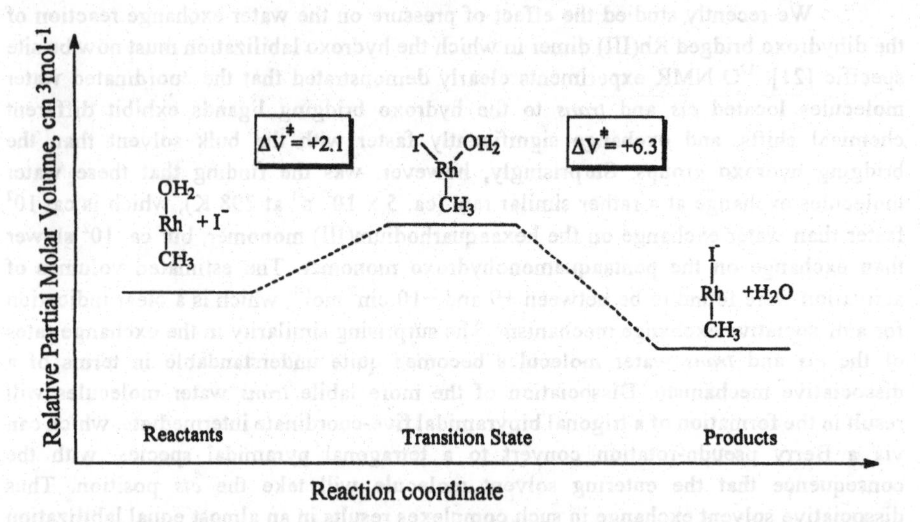

Figure 2. Volume profile for the nucleophilic substitution of *trans*-$Rh^{III}(dmg)_2(CH_3)(H_2O)$ by iodide

3.3. LABILIZATION BY CHELATE EFFECTS

Solvent exchange and ligand substitution reactions can be drastically affected by the influence of chelating ligands. For instance, solvent exchange on $Fe(H_2O)_6^{3+}$ occurs at a rate of 2×10^2 s^{-1} at 298 K and is characterized by an activation volume of -5.4 cm^3 mol^{-1}, typical for an associative interchange mechanism [26]. Introduction of hexadentate chelating ligands such as ethylenediaminetetraacetate, cyclohexyldiaminetetraacetate and phenylenediaminetetraacetate to produce seven-coordinate complexes of the type $Fe^{III}(L)H_2O^-$, result in solvent exchange rates of ca. 10^7 s^{-1} at 298 K (i.e. an acceleration of 10^5) and volumes of activation of between +3.2 and +4.6 cm^3 mol^{-1}, which are typical for a dissociative interchange process [27,28]. Thus the increase in lability is once again accompanied by a changeover in the nature of the ligand substitution mechanism from more associative to more dissociative.

Introduction of a multidentate ligand into the coordination sphere of an aquated metal ion can also cause a change in coordination geometry accompanied by a drastic decrease in lability. One such example we have dealt with involved ligand substitution reactions on Cu(II). The extremely rapid dissociative solvent exchange on $Cu(H_2O)_6^{2+}$ [29] is slowed down by three orders of magnitude for the trigonal bipyramidal $Cu(tren)H_2O^{2+}$ (tren = tris(aminoethyl)amine) ion. Water exchange and complex-formation reactions of the tren complex proceed by an associative interchange mechanism as evidenced by significantly negative volumes of activation, viz. -4.7 cm^3 mol^{-1} for water exchange and between -7.5 and -10 cm^3 mol^{-1} for ligand substitution, respectively [30]. Some typical volume profiles for such ligand substitution reactions are shown in Figure 3, from which the compact nature of the transition state can be seen. Modifications of the tren ligand by introducing methyl substituents on the tripodal arms of the chelate, i.e. Me_3tren and Me_6tren, drastically affect the coordination geometry as well as the lability of the complexes [31,32]. Substitution reactions of the Me_3tren complex are still characterized by negative volumes of activation in support of an I_a mechanism, whereas reactions on the Me_6tren complex clearly follow as a result of the steric hindrance a dissociative mechanism.

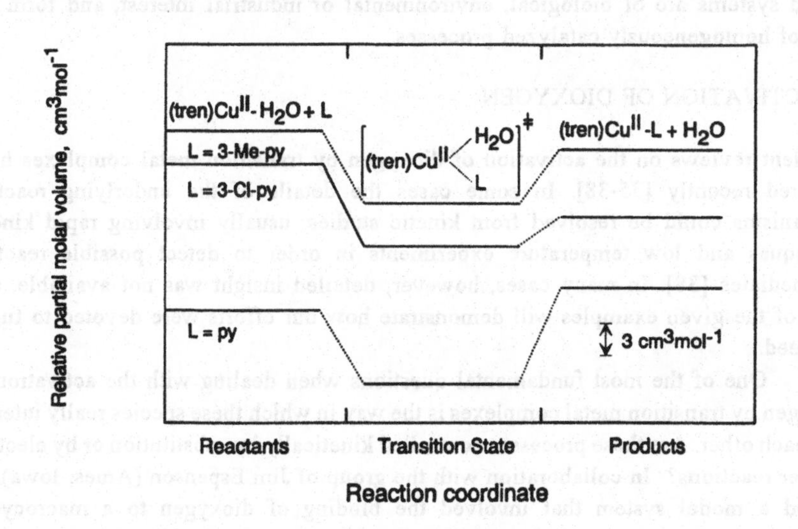

Figure 3. Volume profile for complex-formation reactions of $Cu(tren)H_2O^{2+}$ by a series of pyridine ligands

3.4. SUBSTITUTION REACTIONS IN BIOINORGANIC SYSTEMS

The chemical behaviour and catalytic function of metalloenzymes incorporating for instance Fe(II), Fe(III), Zn(II) and Co(III) metal centers, in many cases involve ligand substitution reactions very similar to those encountered in model coordination complexes. Also the successful application of Au and Pt complexes in the treatment of arthritis and tumors, respectively, depends to a large extent on the ability of these complexes to bind other biological molecules via ligand substitution reactions. In the previous proceedings [2] in this series we documented our interest in the substitution behaviour of vitamin B_{12}, a Co(III) metalloenzyme, and reported typical volume profiles for such ligand substitution processes [33]. Since then we have investigated the substitution behaviour of the coenzyme B_{12} in which a cobalt-carbon bond is present between the Co(III) center and the 5'-desoxyadenosyl moiety in the β position [34]. Surprisingly we found that the presence of the metal-carbon bond does not induce a dissociative ligand substitution mechanism. All available kinetic data, including a very negative volume of activation, support the operation of an associative substitution mechanism. Further work is presently performed on model systems in efforts to further resolve the nature of this process.

4. Activation of small molecules

In this section we will address the coordination chemistry of a wide variety of systems that all, in a direct or indirect way, involve the activation of small molecules. The treated systems are of biological, environmental or industrial interest, and form the basis of homogeneously catalyzed processes.

4.1 ACTIVATION OF DIOXYGEN

Excellent reviews on the activation of dioxygen by transition metal complexes have appeared recently [35-38]. In some cases the details of the underlying reaction mechanisms could be resolved from kinetic studies, usually involving rapid kinetic techniques and low temperature experiments in order to detect possible reaction intermediates [38]. In many cases, however, detailed insight was not available, and some of the given examples will demonstrate how our efforts were devoted to fulfill this need.

One of the most fundamental questions when dealing with the activation of dioxygen by transition metal complexes is the way in which these species really interact with each other. Are these processes controlled kinetically by substitution or by electron transfer reactions? In collaboration with the group of Jim Espenson (Ames, Iowa) we studied a model system that involved the binding of dioxygen to a macrocyclic hexamethylcyclam Co(II) complex to form the corresponding Co(III) superoxo species [39], thus modelling the first redox activation step of dioxygen.

$$Co^{II}(L)(H_2O)_2^{2+} + O_2 \rightleftharpoons Co^{III}(L)(H_2O)(O_2^-)^{2+} + H_2O$$
$$(L = Me_6cyclam)$$

The overall reaction thus involves ligand substitution and electron transfer, the question being which occurs first. From the pressure dependence of the overall equilibrium constant a reaction volume of -22 cm^3 mol^{-1} was determined, which demonstrates that the displacement of a water molecule on the Co(II) complex by dioxygen is accompanied by a significant volume collapse, probably mainly due to the oxidation of Co(II) to Co(III) during the overall reaction. The kinetics of the reaction could be studied by flash-photolysis, since the dioxygen complex can be photo-dissociated by irradiation into the CT band, and the subsequent reequilibration could be followed on the microsecond time scale. From the effect of pressure on the binding and release of dioxygen, the activation volumes for both processes could be determined. A combination of these activation volumes resulted in a reaction volume that is in excellent agreement with the value determined directly from the equilibrium measurements as a function of pressure. The volume profile for this reaction is given in Figure 4. The small volume of activation associated with the binding of dioxygen is clear evidence for a rate-limiting interchange of ligands, dioxygen for water, which is followed by an intramolecular electron-transfer reaction between Co(II) and O_2 to form Co^{III}-O_2^-, the superoxo species. It is the latter process that accounts for the large volume reduction *en route* to the reaction products. Thus during flash-photolysis, electron transfer in the reverse direction occurs due to irradiation into the CT band, which is followed by the rapid release of dioxygen. Thus our volume profile analysis has resolved the question concerning the nature of the rate-determining step during the activation of dioxygen by this model macrocyclic Co(II) complex [39].

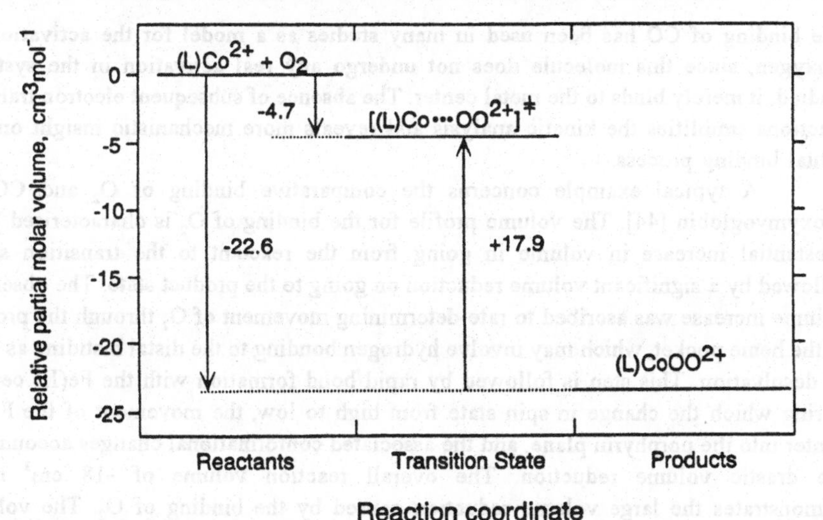

Figure 4. Volume profile for the reaction of dioxygen with the hexamethylcyclam complex of Co(II)

Following earlier work on the binding of dioxygen to myoglobin [40,41], we more recently generated a volume profile for the binding of dioxygen to hemerythrin [42]. The $\Delta V^{\#}$ values for the "on" and "off" reactions as well as the overall reaction volume are ca. twice the magnitudes of those for the corresponding myoglobin case. In the hemerythrin system two Fe(II) centers are oxidized to Fe(III) during which dioxygen is reduced and bound as hydroperoxide to one Fe(III) center. The $\Delta V^{\#}_{on}$ value can partly be accounted for in terms of desolvation of oxygen during its entrance into the protein. The value is, however, such that it suggests some form of dynamic "breathing" motion of the protein that momentarily causes an opening up of a cleft and enables oxygen to enter the protein. The significant volume decrease that occurs following the formation of the transition state can be ascribed to the oxidation of the Fe centres and the reduction of O_2 to O_2^{2-}. The fact that the overall volume collapse is almost double that observed for the oxygenation of myoglobin may indicate similar structural features in oxyhemerythrin and oxymyoglobin. This would suggest that a description of the bonding mode as Fe^{III}-O_2^- or Fe^{III}-O_2H (H from histidine E7) instead of Fe^{II}-O_2, may be more appropriate for oxymyoglobin.

A suitable model for the oxygen carrier protein hemerythrin is [Fe_2(Et-HPTB)(OBz)](BF_4)$_2$, Et-HPTB = N,N,N',N'-tetrakis[(N-ethyl-2-benzimidazolyl)methyl]-2-hydroxy-1,3-diaminopropane, OBz = benzoate; it can mimic the formation of a binuclear peroxo iron complex in the natural system [43]. In this case it was possible to follow the irreversible uptake of dioxygen. The measured value of -12.8 $cm^3\ mol^{-1}$ for the activation volume of the reaction together with the negative value of the activation entropy, confirm the highly structured nature of the transition state.

4.2. ACTIVATION OF CARBON MONOXIDE

The binding of CO has been used in many studies as a model for the activation of dioxygen, since this molecule does not undergo any real activation in the systems studied, it merely binds to the metal center. The absence of subsequent electron transfer reactions simplifies the kinetic analysis and reveals more mechanistic insight on the actual binding process.

A typical example concerns the comparative binding of O_2 and CO to deoxymyoglobin [44]. The volume profile for the binding of O_2 is characterised by a substantial increase in volume in going from the reactant to the transition state, followed by a significant volume reduction on going to the product state. The observed volume increase was ascribed to rate-determining movement of O_2 through the protein to the heme pocket, which may involve hydrogen bonding to the distal histidine as well as desolvation. This step is followed by rapid bond formation with the Fe(II) center, during which the change in spin state from high to low, the movement of the Fe(II) center into the porphyrin plane, and the associated conformational changes account for the drastic volume reduction. The overall reaction volume of -18 $cm^3\ mol^{-1}$ demonstrates the large volume reduction caused by the binding of O_2. The volume profile for the binding of CO shows a considerable volume decrease on going from the reactant to the transition state, which has been ascribed to rate-determining bond

formation. The reverse bond cleavage reaction is accompanied by a volume decrease, which may be related to the different bonding mode of CO compared with O_2. This difference in bonding mode must also account for the much smaller absolute reaction volume observed in this case.

In another study the binding of CO to lacunar Fe(II) complexes was studied in detail as a function of temperature and pressure [45,46]. In these systems the high spin Fe(II) center is five coordinate and has a vacant pocket available for the binding of CO. These systems can therefore be considered as ideal for the modelling of biological processes. A detailed kinetic analysis of the "on" and "off" reactions, as well as a thermodynamic analysis of the overall equilibrium, enabled the construction of the energy and volume profiles for the binding of CO to $[Fe^{II}(PhBzXy)](PF_6)_2$, of which the latter is shown in Figure 5 [46]. The free energy profile demonstrates the favourable thermodynamic driving force for the overall reaction, as well as the relatively low activation energy for the binding process. The entropy profile demonstrates the high degree of order in the transition state on the binding of CO. The large volume collapse associated with the forward reaction is very close to the partial molar volume of CO, which suggests that CO completely disappears within the ligand pocket cavity of the complex in the transition state during partial Fe-CO bond formation. It is also known [45] that Fe^{II}-CO bond formation is accompanied by a high-spin to low-spin conversion of the Fe(II) center. In forming the six-coordinate, low-spin Fe(II) complex, the metal moves into the plane of the equatorial nitrogen donors. Thus following the transition state for the binding of CO, there is a high-spin to low-spin change during which bond formation is completed and the metal center moves into the ligand plane. These processes account for the subsequent volume decrease observed from the transition to the product state. The overall reaction volume of -49 cm^3 mol^{-1} therefore consists of a volume decrease of ca. -37 cm^3 mol^{-1} associated with the disappearance of CO into the ligand cavity, and ca. -12 cm^3 mol^{-1} for the high-spin to low-spin transition.

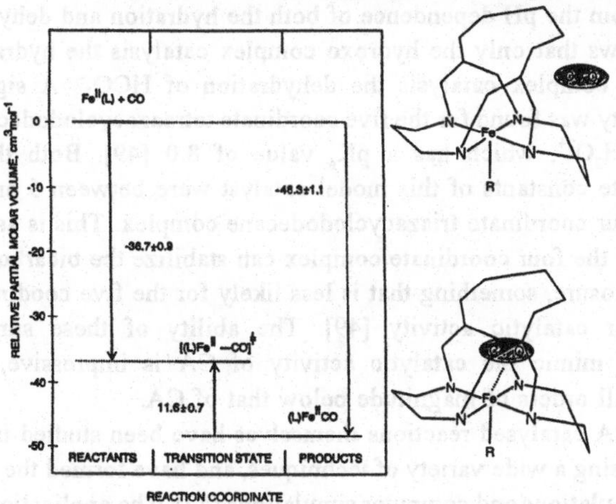

Figure 5. Volume profile for the reaction of $[Fe^{II}(PhBzXy)](PF_6)_2$ with CO in acetonitrile

4.3. ACTIVATION OF CARBON DIOXIDE

One of the most fundamental processes dealing with the activation of CO_2, involves the hydration of CO_2 to produce bicarbonate, as well as the reverse dehydration of bicarbonate to produce CO_2. These processes are of biological and environmental interest since they control the transport and equilibrium behaviour of CO_2. The spontaneous hydration of CO_2 and dehydration of HCO_3^- are processes that are too slow and must therefore be catalyzed by metal complexes in order to expedite the overall conversion rate. In biological systems, enzymes are the efficient catalysts, and we have undertaken some studies to clarify the catalytic mechanism of such systems.

The active centre of the zinc containing metalloenzyme carbonic anhydrase (CA) consists of three histidine residues and one water molecule coordinated to zinc in a slightly distorted tetrahedral geometry. Catalytic activity is integrally related to the ionisation (pK_a value ca. 7) of the coordinated water molecule, and for human CA II the mechanism is referred to as the zinc hydroxide mechanism, which has been described and modelled theoretically in considerable detail [47]. According to this mechanism it is the hydroxo form of the enzyme that can bind CO_2 to produce a bicarbonato complex, which subsequently undergoes a ligand exchange reaction with water to rapidly release HCO_3^-. During the reverse dehydration reaction, it is the aqua form of the enzyme that is the reactive species, rapidly binds HCO_3^- via a substitution of coordinated water, followed by decarboxylation to release CO_2. Thus the hydration and dehydration reactions exhibit very characteristic pH dependences. Model complexes must therefore on the one hand mimic the active site of the enzyme, and on the other hand exhibit the characteristic pH dependence observed in the catalytic activity of the enzyme.

The first model complex that could adhere to both these requirements was the triazacyclododecane complex of Zn(II), viz. $Zn([12]aneN_3)H_2O^{2+}$, which has a pK_a value of 7.3 [48]. From the pH dependence of both the hydration and dehydration reactions, it clearly follows that only the hydroxo complex catalysis the hydration of CO_2, and only the aqua complex catalysis the dehydration of HCO_3^-. A significantly higher catalytic activity was found for the five coordinate tetraazacyclododecane complex, viz. $Zn([12]aneN_4)H_2O^{2+}$, which has a pK_a value of 8.0 [49]. Both the hydration and dehydration rate constants of this model catalyst were between 5 and 6 times higher than for the four coordinate triazacyclododecane complex. This is associated with the possibility that the four coordinate complex can stabilize the bicarbonate intermediate through ring-closure, something that is less likely for the five coodinate complex, and thus its higher catalytic activity [49]. The ability of these simple coordination compounds to mimic the catalytic activity of CA is impressive, but their actual reactivity is still orders of magnitude below that of CA.

The CA catalysed reactions themselves have been studied in detail by many investigators using a wide variety of techniques, and have formed the subject of several theoretical calculations and computer simulations [47]. The application of high pressure kinetic measurements provided further mechanistic distinction than previously available [50]. A close agreement was obtained between the volume profiles for the uncatalysed

reaction obtained earlier [51] and that derived from kinetic measurements on the catalysed reaction. The first complete, detailed volume profile (see Figure 6) for an enzyme catalysed reaction could be generated. The Zn(II) bound hydroxyl moiety subjects the carbon of CO_2 to nucleophilic attack resulting in the formation of an oxygen-carbon bond, and the results are consistent with a unidentate bonding of bicarbonate. For this process the transition state lies approximately halfway between the reactant and product states (see left part of volume profile). The substitution of coordinated bicarbonate by water tends more toward a limiting D mechanism (see right part of volume profile), which may result from the influence of the environment of the active centre of the enzyme.

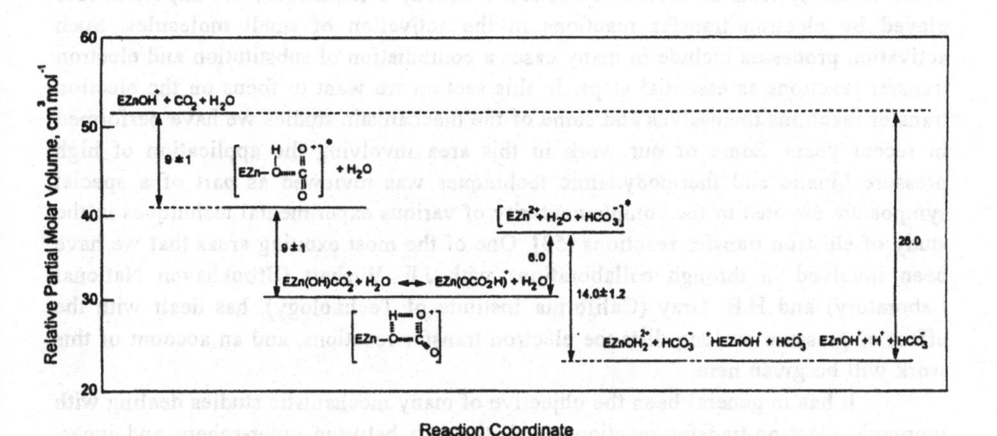

Figure 6. Volume profile for the carbonic anhydrase catalyzed hydration of CO_2 and dehydration of HCO_3^-

Fujita *et al* [52] and others have used a wide range of methods to study cobalt(I) complexes with tetraazamacrocyclic ligands as potential catalysts for the reduction of CO_2. The interaction of the low spin $[Co^IHMD]^+$ species, HMD = 5,7,7,12,14,14-hexamethyl-1,4,8,11-tetraazacyclotetradecane-4,11-diene, with CO_2 in CH_3CN leads to a five-coordinate species, $[CoHMD(CO_2)]^+$, which is in equilibrium with a six-coordinate complex ion, $[CoHMD(CO_2)(CH_3CN)]^+$, formed through addition of CH_3CN. Results from an XANES study together with other information provide a clear indication that in the six-coordinate complex cobalt is in the +3 oxidation state, meaning that the complex ion is Co^{III}-CO_2^{2-} (i.e. CO_2 is coordinated as carboxylate). Hence the initial cobalt(I) complex has reduced the bound CO_2 to carboxylate. The change in coordination number equilibrium can be studied readily by UV-vis spectrophotometry; a decrease in temperature or an increase in pressure favour the formation of the Co^{III}-CO_2^{2-} species. The thermodynamic parameters for this equilibrium are $\Delta H° = -29$ kJ mol^{-1}, $\Delta S° = -113$ J mol^{-1} K^{-1} and $\Delta V° = -17.7$ cm^3 mol^{-1}. The latter two are mutually compatible and consistent with a highly ordered and compact six-coordinate complex ion. It has been proposed that a major part of the volume decrease arises from the intramolecular electron transfer process accompanied by a shortening of the Co-CO_2 bond (as supported by XANES and EXAFS studies) and an increase in electrostriction. Only a relatively minor contribution to the large negative reaction volume is suggested to result from the intrinsic effect of CH_3CN addition [52].

5. Electron transfer reactions

Some of the systems described in Section 4 already demonstrated the important role played by electron transfer reactions in the activation of small molecules. Such activation processes include in many cases a combination of substitution and electron transfer reactions as essential steps. In this section we want to focus on the electron transfer reactions themselves and some of the mechanistic studies we have performed in recent years. Some of our work in this area involving the application of high pressure kinetic and thermodynamic techniques was reviewed as part of a special symposium devoted to the complementarity of various experimental techniques in the study of electron transfer reactions [53]. One of the most exciting areas that we have been involved in through collaborations with J.F. Wishart (Brookhaven National Laboratory) and H.B. Gray (California Institute of Technology), has dealt with the effect of pressure on long distance electron transfer reactions, and an account of this work will be given here.

It has in general been the objective of many mechanistic studies dealing with inorganic electron-transfer reactions to distinguish between outer-sphere and inner-sphere mechanisms. Along these lines high pressure kinetic methods and the construction of reaction volume profiles have also been employed to contribute towards a better understanding of the intimate mechanisms involved in such processes. The differentiation between outer-sphere and inner-sphere mechanisms depends on the nature of the precursor species, Ox//Red in the following scheme, which can either be an ion pair or encounter complex, or a bridged intermediate, respectively.

$$\mathrm{Ox + Red \rightleftharpoons Ox//Red} \quad , K$$
$$\mathrm{Ox//Red \rightarrow Ox^-//Red^+} \quad , k_{ET}$$
$$\mathrm{Ox^-//Red^+ \rightleftharpoons Ox^- + Red^+}$$

This means that the coordination sphere of the reactants remains intact in the former case and is modified by ligand substitution in the latter, which will naturally affect the associated volume changes.

We have in our earlier work mainly concentrated on the analysis of non-symmetrical electron-transfer reactions [53], i.e. reactions in which redox products are formed and for which the overall driving force and reaction volume will not be zero. Some typical examples of such studies have been discussed in more detail [53] and form an important basis for the following section dealing with "long distance" electron-transfer processes. A challenging question concerns the feasibility of the application of high pressure kinetic and thermodynamic techniques in the study of such reactions. Do "long-distance" electron transfer processes exhibit a characteristic pressure dependence, and to what extent can a volume profile analysis reveal information on the intimate mechanism?

The systems that we investigated in collaboration with others involved intermolecular and intramolecular electron-transfer reactions between ruthenium complexes and cytochrome *c*. We also studied a series of intermolecular reactions between chelated cobalt complexes and cytochrome *c*. A variety of high pressure experimental techniques, including stopped-flow, flash-photolysis, pulse-radiolysis and voltammetry, were employed in these investigations. As the following presentation will show, a remarkably good agreement was found between the volume data obtained with the aid of these different techniques, which clearly demonstrates the complementarity of these methods for the study of electron-transfer processes.

Application of pulse-radiolysis techniques revealed that the following intramolecular and intermolecular electron-transfer reactions all exhibit a significant acceleration with increasing pressure. The reported volumes of activation are -17.7 ± 0.9, -18.3 ± 0.7, and -15.6 ± 0.6 $cm^3\ mol^{-1}$, respectively, and clearly demonstrate a significant volume collapse on going from the reactant to the transition state [54].

$$(NH_3)_5Ru^{II}\text{-(His33)cyt } c^{III} \rightarrow (NH_3)_5Ru^{III}\text{-(His33)cyt } c^{II}$$
$$(NH_3)_5Ru^{II}\text{-(His39)cyt } c^{III} \rightarrow (NH_3)_5Ru^{III}\text{-(His39)cyt } c^{II}$$
$$Ru^{II}(NH_3)_6^{2+} + \text{cyt } c^{III} \rightarrow Ru^{III}(NH_3)_6^{3+} + \text{cyt } c^{II}$$

At this stage it was uncertain what the negative volumes of activation really meant since overall reaction volumes were not available. There was however data, now in the literature [55], that suggested that the oxidation of $Ru(NH_3)_6^{2+}$ to $Ru(NH_3)_6^{3+}$ is accompanied by a volume decrease of ca. 30 $cm^3\ mol^{-1}$, which would mean that the activation volumes quoted above could mainly arise from volume changes associated with the oxidation of the ruthenium redox partner.

In order to obtain further information on the magnitude of the overall reaction volume and the location of the transition state along the reaction coordinate, a series of intermolecular electron-transfer reactions of cyt *c* with pentaammineruthenium complexes were studied, where the sixth ligand on the ruthenium complex was selected in such a way that the overall driving force was low enough so that the reaction kinetics could be studied in both directions [56,57]. The selected substituents were isonicotinamide (isn), 4-ethylpyridine (etpy), pyridine (py), and 3,5-lutidine (lut). The overall reaction can be formulated as

$$Ru^{III}(NH_3)_5L^{3+} + \text{cyt } c^{II} \rightleftharpoons Ru^{II}(NH_3)_5L^{2+} + \text{cyt } c^{III}$$

For all the investigated systems, the forward reaction was significantly decelerated by pressure, whereas the reverse reaction was significantly accelerated by presssure. The absolute values of the volumes of activation for the forward and reverse processes were indeed very similar, demonstrating that a similar rearrangement occurs in order to reach the transition state. In addition, the overall reaction volume for these systems could be determined spectrophotometrically by recording the spectrum of an equilibrium mixture as a function of pressure, and electrochemically by recording cyclic and differential pulse voltammograms as a function of pressure [58]. A comparison of the ΔV data demonstrates the generally good agreement between the values obtained from the difference in the volumes of activation for the forward and reverse reactions, and those obtained thermodynamically. Furthermore, the values also clearly demonstrate that $|\Delta V^{\#}| \approx 0.5\,|\Delta V|$, i.e. the transition state lies approximately halfway between the reactant and product states on a volume basis independent of the direction of electron transfer. The typical volume profile in Figure 7 presents the overall picture, from which the location of the transition state can clearly be seen.

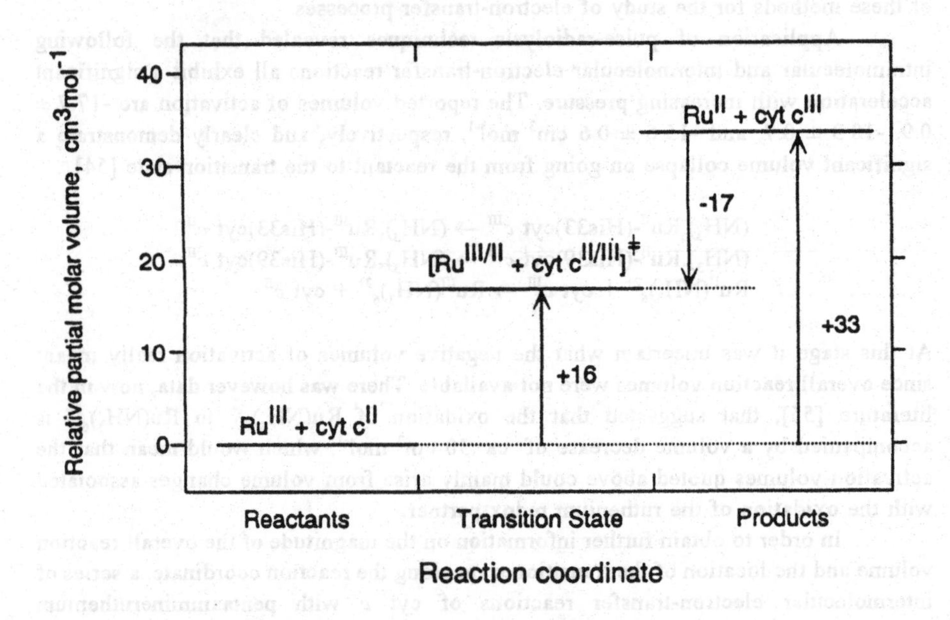

Figure 7. Volume profile for the reaction
$Ru(NH_3)_5isn^{3+} + Cyt\ c^{II} \rightleftharpoons Ru(NH_3)_5isn^{2+} + Cyt\ c^{III}$

Similar results were recently obtained for the redox reactions of a series of cobalt diimine complexes with cytochrome *c* [59,60]. In general a good agreement exists between the kinetically and thermodynamically determined parameters, and the typical volume profile in Figure 8 once again demonstrates the symmetrical location of the transition state with respect to the reactant and product states.

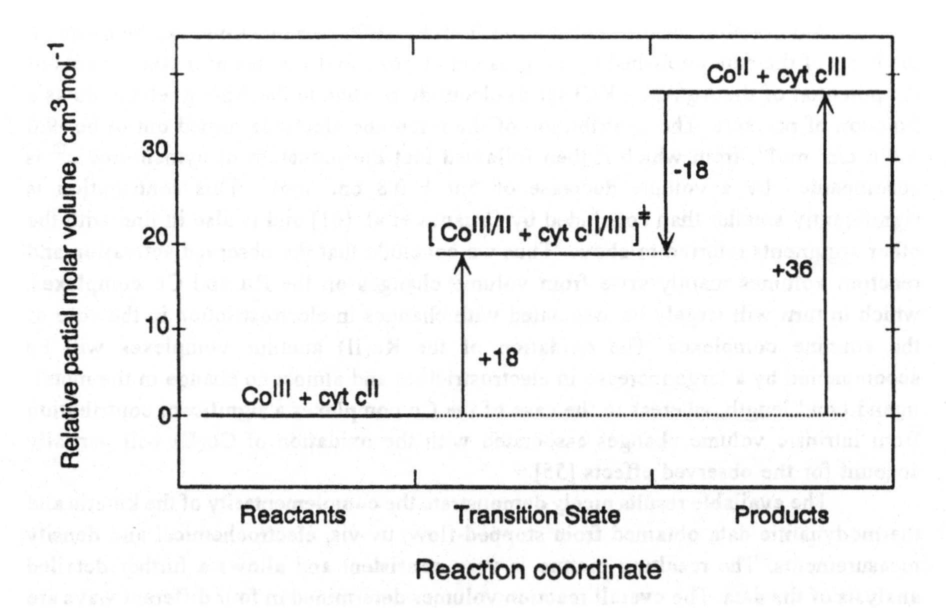

Figure 8. Volume profile for the reaction
$$\text{Co(terpy)}_2^{3+} + \text{Cyt } c^{\text{II}} \rightleftharpoons \text{Co(terpy)}_2^{2+} + \text{Cyt } c^{\text{III}}$$

At this point it is important to ask the question where these volume changes really come from? We have always argued that the major volume change arises from changes on the redox partner and not on cytochrome *c* itself. This was suggested by the fact that the change in partial molar volume associated with the oxidation of the investigated Ru(II) and Co(II) complexes as obtained from electrochemical and density measurements, almost fully accounted for the observed overall reaction volume. Thus the reduction of cytochrome *c* can only make a minor contribution towards the overall volume change.

These arguments were apparently in contradiction with electrochemical results reported by Cruanes et al. [61], according to which the reduction of cytochrome *c* is accompanied by a volume collapse of 24 cm^3 mol^{-1}. This value is so large that it almost represents all of the reaction volume found for the investigated reactions discussed above. A reinvestigation of the electrochemistry of cytochrome *c* as a function of pressure, using cyclic and differential pulse voltammetric techniques [58], revealed a reaction volume of -14.0 ± 0.5 cm^3 mol^{-1} for the reaction

$$\text{Cyt } c^{\text{III}} + \text{Ag(s)} + \text{Cl}^- \rightarrow \text{Cyt } c^{\text{II}} + \text{AgCl(s)}$$

A correction for the contribution from the reference electrode can be made on the basis of the data published by Tregloan et al. [62], and a series of measurements of the potential of the Ag/AgCl(KCl sat′d) electrode relative to the Ag/Ag^+ electrode as a function of pressure. The contribution of the reference electrode turned out to be -9.0 ± 0.6 cm^3 mol^{-1}, from which it then followed that the reduction of cytochrome c^{III} is accompanied by a volume decrease of 5.0 ± 0.8 cm^3 mol^{-1}. This contribution is significantly smaller than concluded by Cruanes et al. [61] and is also in line with the other arguments referred to above. Thus we conclude that the observed activation and reaction volumes mainly arise from volume changes on the Ru and Co complexes, which in turn will largely be associated with changes in electrostriction in the case of the ammine complexes. The oxidation of the Ru(II) ammine complexes will be accompanied by a large increase in electrostriction and almost no change in the metal-ligand bond length, whereas in the case of the Co complexes a significant contribution from intrinsic volume changes associated with the oxidation of Co(II) will partially account for the observed effects [55].

The available results nicely demonstrate the complementarity of the kinetic and thermodynamic data obtained from stopped-flow, uv-vis, electrochemical and density measurements. The resulting picture is very consistent and allows a further detailed analysis of the data. The overall reaction volumes determined in four different ways are surprisingly similar and underline the validity of the different methods employed. The volume profiles in Figures 7 and 8 demonstrate the symmetric nature of the intrinsic and solvational reorganization in order to reach the transition state of the electron-transfer process. In these systems the volume profile is controlled by effects on the redox partner of cytochrome *c*, but this does not necessarily always have to be the case. The location of the transition state on a volume basis will reveal information concerning the "early" or "late" nature of the transition state and reveal details of the actual electron-transfer route followed.

Recent investigations on a series of intramolecular electron transfer reactions, closely related to the series of intermolecular reactions described above, revealed non-symmetrical volume profiles [63]. Reactions of the type

$$(NH_3)_4(L)Ru^{III}\text{-(His33)-Cyt } c^{II} \rightleftharpoons (NH_3)_4(L)Ru^{II}\text{-(His33)-Cyt } c^{III}$$

where L = isonicotinamide, 4-ethylpyridine, 3,5-lutidine, and pyridine, all exhibited volumes of activation for the forward reaction of between +3 and +7 cm^3 mol^{-1}, compared to overall reaction volumes of between +19 and +26 cm^3 mol^{-1}. This indicates that electron transfer from Fe to Ru is characterized by an "early" transition state in terms of volume changes along the reaction coordinate (see Figure 9). The overall volume changes could be accounted for in terms of electrostriction effects centered around the ammine ligands on the ruthenium center. A number of possible explanations in terms of the effect of pressure on electronic and nuclear factors were offered to account for the asymmetrical nature of the volume profile [63].

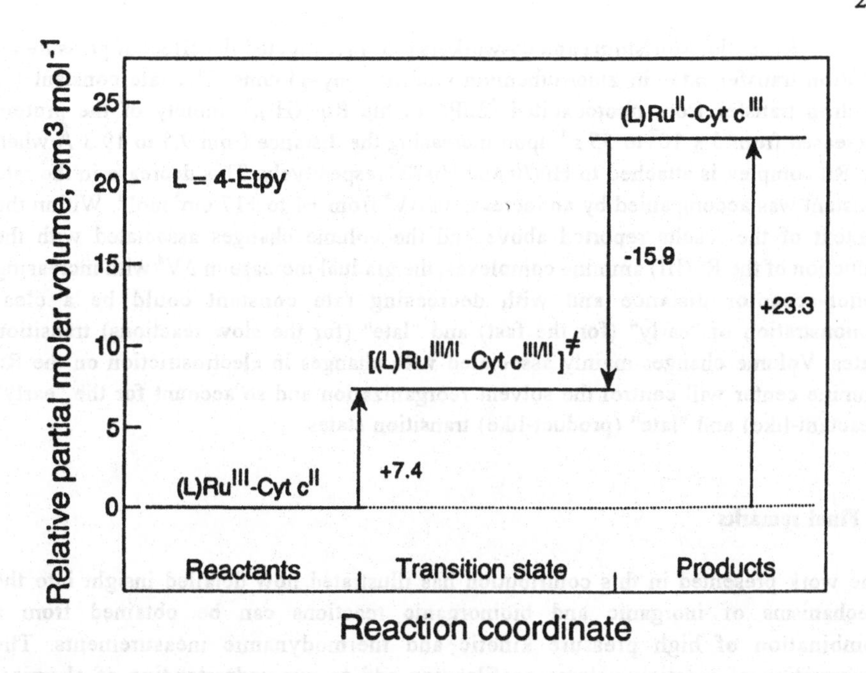

Figure 9. Volume profile for the reaction
$(NH_3)_4Ru^{III}$(4-Etpy)-Cyt c^{II} ⇌ $(NH_3)_4Ru^{II}$(4-Etpy)-Cyt c^{III}

One system was investigated where the effect of pressure on the electron-transfer rate constant revealed information on the actual electron-transfer route. We investigated the effect of pressure on distant electronic coupling in $Ru(bpy)_2(im)$-modified His33 and His72 cytochrome *c* derivatives, for which the electron transfer from Fe(II) to Ru(III) is activationless [64]. In the case of the His33-modified system the electron-transfer rate constant exhibited no dependence on pressure within experimental error limits. However, the rate constant for the His72-modified protein increased significantly with increasing pressure, corresponding to a $\Delta V^{\#}$ value of -6 ± 2 cm^3 mol^{-1}. Since this value is exactly opposite to that expected for the reduction of Ru(III), the result was interpreted as an increase in electronic coupling at elevated pressure. The application of moderate pressures will cause a slight compression of the protein that in turn shrinks the through-space gaps that are key units in the electron-tunneling pathway between the heme and His72. A decrease of 0.46 Å in the tunneling path length at a pressure of 150 MPa can account for the observed increase in rate constant. This in turn means that there is an average decrease in the space-gap of 0.1 Å. The absence of an effect for the His33-modified species is understandable since electronic coupling through covalent and hydrogen bonds will be less pressure sensitive than coupling via van der Waals gaps [64].

Recently, Morishima and co-workers [65] investigated the effect of pressure on electron transfer rates in zinc/ruthenium modified myoglobins. The rate constant for electron transfer from photoexcited $^3ZnP^*$ to the $Ru(NH_3)_5^{3+}$ moiety of the protein decreased from 5 x 10^7 to 55 s^{-1} upon increasing the distance from 9.5 to 19.3 Å when the Ru complex is attached to His70 and His83, respectively. This decrease in the rate constant was accompanied by an increase in $\Delta V^{\#}$ from +4 to +17 cm^3 mol^{-1}. Within the context of the results reported above and the volume changes associated with the reduction of the Ru(III) ammine complexes, the gradual increase in $\Delta V^{\#}$ with increasing donor-acceptor distance and with decreasing rate constant could be a clear demonstration of "early" (for the fast) and "late" (for the slow reactions) transition states. Volume changes mainly associated with changes in electrostriction on the Ru ammine center will control the solvent reorganization and so account for the "early" (reactant-like) and "late" (product-like) transition states.

6. Final remarks

The work presented in this contribution has illustrated how detailed insight into the mechanisms of inorganic and bioinorganic reactions can be obtained from a combination of high pressure kinetic and thermodynamic measurements. The construction of reaction volume profiles can add to our understanding of chemical processes and contribute to our ability to control and design the reactivity of such systems. These aspects therefore complement synthetic/structural studies and highlight the magnificant reaction possibilities as well as reactivities available in such systems. Our mechanistic insight is restricted by experimental limitations, which calls for kinetic studies to be as detailed as possible, and for suitable theoretical calculations, in order to improve our understanding. In this repect the physical variable pressure can play an important role and add a further dimension to mechanistic studies.

Acknowledgements

Sections of this review were composed by the author while being a Visiting Erskine Fellow at the Department of Chemistry, University of Canterbury, Christchurch, New Zealand. The many stimulating discussions with Don House and his colleagues during that time is very much appreciated. Financial support for our work came form the Deutsche Forschungsgemeinschaft, Bundesministerium für Bildung, Wissenschaft, Forschung und Technologie, Fonds der Chemischen Industrie, Volkswagen-Stiftung, Max-Buchner Forschungsstiftung, German-Isreali Foundation, NATO Scientific Affiars Division, DAAD and the Alexander von Humboldt Stiftung.

7. References

[1] van Eldik, R. and Jonas, J. (Eds.) (1987) *High Pressure Chemistry and Biochemistry*, D. Reidel, Dordrecht, Series C197.

[2] Winter, R. and Jonas, J. (Eds.) (1993) *High Pressure chemistry, biochemistry and Materials Science*, Kluwer Academic Publishers, Dordrecht, Series C401.

[3] van Eldik, R. and Hubbard, C.D. (Eds.) (1997) *Chemistry under Extreme or Non-Classical Conditions*, Wiley, New York, Chapter 2.

[4] Frey, U., Helm, L. and Merbach, A.E. (1995) in *Dynamics of Solutions and Fluid Mixtures by NMR*, Delpuech, J.-J. (Ed.), Wiley, Chichester, 263.

[5] van Eldik, R. and Ford, P.C. (1998) in *Advances in Photochemistry*, Vol. 24, Neckers, D.C., Volman, D.H. and von Brünau, G. (Eds.), Wiley, New York, 61.

[6] Drljaca, A., Hubbard, C.D., van Eldik, R., Asano, T., Basilevsky, M.V., le Noble, W.J. (1998) *Chem. Rev*., **98**, 2167-2289.

[7] Stochel, G. and van Eldik, R. (1998) *Coord. Chem. Rev*., in press.

[8] Hubbard, C.D. and van Eldik, R. (1995) *Instrum. Sci. Technol.*, **22**, 1.

[9] Magde, D. and van Eldik, R. (1997) in *High Pressure Techniques in Chemistry and Physics: a Practical Approach*, Holzapfel, W. and Isaacs, N. (Eds.), Oxford University Press, Oxford, Chapter 6.

[10] van Eldik, R. and Merbach, A.E. (1992) *Comments Inorg. Chem*., **12**, 341.

[11] Lincoln, S.F. and Merbach, A.E. (1995) *Adv. Inorg. Chem*., **42**, 1.

[12] Frey, U., Helm, L., Merbach, A.E. and Roulet, R. (1996) in *Advanced Applications of NMR to Organometallic Chemistry*, Gielen, M., Willem, R. and Wrackmeyer, B. (Eds.), Wiley, Chichester, 193.

[13] Akesson, R., Petterson, L.G.M., Sandström, M. and Wahlgren, U. (1994) *J. Am. Chem. Soc*., **116**, 8691.

[14] Akesson, R., Petterson, L.G.M., Sandström, M. and Wahlgren, U. (1994) *J. Am. Chem. Soc*., **116**, 8705.

[15] Rotzinger, F.P. (1996) *J. Am. Chem. Soc*., **118**, 6760.

[16] Rotzinger, F.P. (1997) *J. Am. Chem. Soc*., **119**, 5230.

[17] Deeth, R.J. and Elding, L.I. (1996) *Inorg. Chem*., **35**, 5019.

[18] Hartmann, M, Clark, T. and van Eldik, R. (1997) *J. Am. Chem. Soc*., **119**, 5867.

[19] Maciejowska, I., van Eldik, R., Stochel, G. and Stasicka, Z. (1997) *Inorg. Chem*., **36**, 5409.

[20] Cusanelli, A., Frey, U., Ritchens, D.T. and Merbach, A.E. (1996) *J. Am. Chem. Soc*., **118**, 5265, and literature cited therein.

[21] Drljaca, A., Zahl, A. and van Eldik, R. (1998) *Inorg. Chem*., **37**, 3948.

[22] Dadci, L., Elias, H., Frey, U., Hörnig, A., Koelle, U., Merbach, A.E., Paulus, H. and Schneider, J.S. (1995) *Inorg. Chem*., **34**, 306.

[23] Kofod, P., Harris, P. and Larsen, S. (1997) *Inorg. Chem*., **36**, 2258.

[24] Eckhard, C., Dücker-Benfer, C. and van Eldik, R. (1998) submitted for publication.

[25] Dücker-Benfer, C., Dreos, R. and van Eldik, R. (1995) *Angew. Chem., Int. Ed. Engl.*, **34**, 2245.

[26] Laurenczy, G., Rapaport, I., Zbinden, D. and Merbach, A.E. (1991) *Magn. Res. Chem.*, **29**, 545.

[27] Mizuno, M., Funahashi, S., Nakasuka, N. and Tanaka, M. (1991) *Inorg. Chem.*, **30**, 1550.

[28] Tregloan, P.A., Seibig, S., Zahl, A. and van Eldik, R. (1998) unpublished results.

[29] Powell, D.H., Furrer, P., Pittet, P.-A. and Merbach, A.E. (1995) *J. Phys. Chem.*, **99**, 16622.

[30] Powell, D.H., Merbach, A.E., Fabian, I., Schindler, S. and van Eldik, R. (1994) *Inorg. Chem.*, **33**, 4468.

[31] Dittler-Klingemann, A.M., Orvig, C., Hahn, F.E., Thaler, F., Hubbard, C.D., van Eldik, R., Schindler, S. and Fabian, I. (1996) *Inorg. Chem.*, **35**, 7798.

[32] Thaler, F., Hubbard, C.D., Heinemann, F.W., van Eldik, R., Schindler, S., Fabian, I., Dittler-Klingemann, A.M., Hahn, F.E. and Orvig, C. (1998) *Inorg. Chem.*, **37**, 4022.

[33] Prinsloo, F.F., Meier, M. and van Eldik, R. (1994) *Inorg. Chem.*, **33**, 900.

[34] Brasch, N.E., Hamza, M.S.A. and van Eldik, R. (1997) *Inorg. Chem.*, **36**, 3216.

[35] Valentine, A.M. and Lippard, S.J. (1997) *J. Chem. Soc. Dalton Trans.*, 3925.

[36] Que, L. (1997) *J. Chem. Soc. Dalton Trans.*, 3933.

[37] Tolman, W.B. (1997) *Acc. Chem. Res.*, **30**, 227.

[38] Karlin, K.D., Kaderli, S. and Zuberbühler, A.D. (1997) *Acc. Chem. Res.*, **30**, 139.

[39] Zhang, M., van Eldik, R., Espenson, J.H. and Bakac, A. (1994) *Inorg. Chem.*, **33**, 130.

[40] Projahn, H.-D., Dreher, C. and van Eldik, R. (1990) *J. Am. Chem. Soc.*, **112**, 17.

[41] Taube, D.J., Projahn, H.-D., van Eldik, R., Magde, D. and Traylor, T.G. (1990) *J. Am. Chem. Soc.*, **112**, 6880.

[42] Projahn, H.-D., Schindler, S., van Eldik, R., Fortier, D.G., Andrew, C.R. and Sykes, A.G. (1995) *Inorg. Chem.*, **34**, 5935.

[43] Feig, A.L., Becker, M., Schindler, S., van Eldik, R. and Lippard, S.J. (1996) *Inorg. Chem.*, **35**, 2590.

[44] Projahn, H.D. and van Eldik, R. (1992) *Inorg. Chem.*, **30**, 3288.

[45] Buchalova, M., Warburton, P.R., van Eldik, R. and Busch, D.H. (1997) *J. Am. Chem. Soc.*, **119**, 5867.

[46] Buchalova, M., Busch, D.H. and van Eldik, R. (1998) *Inorg. Chem.*, **37**, 1116..

[47] Silverman, D.N. and Lindskog, S. (1988) *Acc. Chem. Res.*, **21**, 30, and the literature survey in ref. 50.

[48] Zhang, X., van Eldik, R., Koike, T. and Kimura, E. (1993) *Inorg. Chem.*, **32**, 5749.

[49] Zhang, X. and van Eldik, R. (1995) *Inorg. Chem.*, **34**, 5606.

[50] Zhang, X., Hubbard, C.D. and van Eldik, R. (1996) *J. Phys. Chem.*, **100**, 9161.

[51] van Eldik, R. and Palmer, D.A. (1982) *J. Sol. Chem.*, **11**, 239.

[52] Fujita, E. and van Eldik, R. (1998) *Inorg. Chem.*, **37**, 360 and literature cited therein.

[53] van Eldik, R. (1998) in *Photochemisty and Radiation Chemistry: Complementary Methods in the Study of Electron Transfer*, Nocera, D. and Wishart, J.F. (Eds.), Advances in Chemistry Series **254**, American Chemical Society, Washington, Chapter 19.

[54] Wishart, J.F., van Eldik, R., Sun, J., Su, C. and Isied, S.S. (1992) *Inorg. Chem.*, **31**, 3986.

[55] Sachinidis, J.J., Shalders, R.D. and Tregloan, P.A. (1996) *Inorg. Chem.*, **35**, 2497.

[56] Bänsch, B., Meier, M., Martinez, M., van Eldik, R., Su, C., Sun, J., Isied, S.S. and Wishart, J.F. (1994) *Inorg. Chem.*, **33**, 4744.

[57] Meier, M., Sun, J., Wishart, J.F. and van Eldik, R. (1996) *Inorg. Chem.*, **35**, 1564.

[58] Sun, J., Wishart, J.F., van Eldik, R., Shalders, R.D. and Swaddle, T.W. (1995) *J. Am. Chem. Soc.*, **117**, 2600.

[59] Meier, M. and van Eldik, R. (1994) *Inorg. Chim. Acta*, **225**, 95.

[60] Meier, M. and van Eldik, R. (1997) *Chem. Eur. J.*, **3**, 33.

[61] Cruanes, M.T., Rodgers, K.K. and Sligar, S.G. (1992) *J. Am. Chem. Soc.*, **114**, 9660.

[62] Sachinidis, J.J., Shalders, R.D. and Tregloan, P.A. (1994) *Inorg. Chem.*, **33**, 6180.

[63] Sun, J., Su, C., Meier, M., Isied, S.S., Wishart, J.F. and van Eldik, R. (1998) *Inorg. Chem.*, in press.

[64] Meier, M., van Eldik, R., Chang, I.-J., Mines, G.A., Wuttke, D.S., Winkler, J.R. and Gray, H.B. (1994) *J. Am. Chem. Soc.*, **116**, 1577.

[65] Sugiyama, Y., Takahashi, S., Ishimori, K. and Morishima, I. (1997) *J. Am. Chem. Soc.*, **119**, 8592.

EFFECT OF PRESSURE ON REACTION KINETICS.
THE COMPONENTS OF THE ACTIVATION VOLUME REVISITED

G. Jenner

Laboratoire de Piézochimie Organique (UMR 7509)
Université Louis Pasteur, Strasbourg, France

1. Introduction

Volume is a perfectly tangible parameter in current life. On the macroscopic level the volume profile is a pictorial view of a chemical reaction on the basis of volume changes. When reactants are converted into products the volume change is expressed by the reaction volume based on partial molar volumes depending on the medium (Eq. 1)

$$\overline{\Delta V} = \overline{V}_P - \overline{V}_R \tag{1}$$

The quantity $\overline{\Delta V}$ is easily measurable by physical techniques such as dilatometry. However, $\overline{\Delta V}$ does not provide any mechanistic perception. The advent of the transition state theory opened a fruitful area for mechanistic delineation. This theory learns about how the reaction goes from reactants to products. In his volume version the basic relationship associates the volume of the transition state, the equilibrium or rate constant and pressure (Eq. 2)

$$\partial \operatorname{Ln} k / \partial P = - \Delta V^{\neq} / RT \tag{2}$$

While the relation looks conceptually simple, two observations must be made:

- no transition state has actually ever been observed although physicochemists do not desesperate to catch it with the help of more and more sophisticated ultrafast detection techniques. To date, the volume of the transition state $V^{\neq}$ remains a mere virtual quantity so as [1]:

$$\partial G^{\neq} / \partial P = V^{\neq} \tag{3}$$

Importantly to say, the fictive volume $V^{\neq}$ is the only kinetic property of the transition state which can be obtained from experimental data, albeit indirectly, with the help of Eq. 1 and 2.

$$V^{\neq} = \Delta V^{\neq} + \overline{V}_R \tag{4}$$

- the quantity $\Delta V^{\neq}$ itself is a reflection of all volume changes $\Delta V_i^{\neq}$ occurring during the progression of the reaction from initial state to transition state and within the transition state. Hence:

$$\Delta V^{\neq} = \Sigma \Delta V_i^{\neq} \tag{5}$$

Eq. 5 is the basic relation describing the overall reaction profile. The purpose of this chapter is to survey the various volume contributions $\Delta V_i^{\neq}$ including the most recent

R. Winter and J. Jonas (eds.), High Pressure Molecular Science, 291–311.

findings in this respect arising from volume effects associated with the formation of the transition state. Although Eq. 5 will be treated in a comprehensive way, the examples given for illustration are obviously not exhaustive.

2. The components of the activation volume

The original suggestion made by Eyring at the time he developped the transition state theory considered a merely structural contribution $\Delta V^{\neq}_S$ and an environmental term $\Delta V^{\neq}_m$ stemming from medium effects [1]:

$$\Delta V^{\neq} = \Delta V^{\neq}_S + \Delta V^{\neq}_m \tag{6}$$

- $\Delta V^{\neq}_S$ is the volume variation due to changes in the nuclear positions of the reactants during the formation of the transition state. It evidently relates to the volume effects generated by bond cleavage and new bond formation and represents roughly the geometric balance of these transformations. In numerous reactions where the medium effect is negligible, $\Delta V^{\neq}$ is a one-component expression and therefore, permits the positionning of the transition state. This is a very important point since the location of the transition state along the reaction axis has mechanistic implications (see 3.1). To this respect, the volume of the transition state $V^{\neq}$ must be compared to $\overline{V}_R$ and $\overline{V}_P$ or, expressed differently, the ratio $\theta = \Delta V^{\neq} / \overline{\Delta V}$ is the best indicator of the progression of the transition state along the reaction coordinate. This view has been controversed with arguments such as inconsistency with the calculated intrinsic molar volumes (van der Waals volumes) and conflict with the Bell-Evans-Polanyi-Hammond principle [2]. However, from an experimental point of view, significant deviations from unity of θ - values have been observed in pericyclic reactions clearly occurring via stepwise pathways whereas concerted processes showed θ - values close or around unity [3].

- $\Delta V^{\neq}_m$ is *inter alia* the volume effect arising from changes in solute-solvent interactions when the reaction reaches the transition state. Until recent years $\Delta V^{\neq}_m$ has been assimilated to the electrostriction component of Eq. 6. At variance with $\Delta V^{\neq}_S$, $\Delta V^{\neq}_m$ is usually impossible to estimate correctly. Charge build-up in the transition state means creation of charged atoms or free ions. Interactions of such species with the medium lead to a volume decrease known as electrostriction volume.

However, it became clear in the last two decades that Eq. 6 was unable to describe the volume profile in many cases so as it needed appreciable refinement. Both $\Delta V^{\neq}_S$ and $\Delta V^{\neq}_m$ may include additional volume contributions. Another argument against the validity of the $\Delta V^{\neq}$ criterion lies in the fact that, rigorously speaking, $\Delta V^{\neq}$ should accommodate dynamic effects related to pressure-induced changes of transport properties such as self-diffusion and viscosity [4]. Viscosity influences reaction rates and must be taken into account either at very high pressures for common organic solvents [5] or at lower pressures for highly viscous media [6] in such way that Eq. 7 becomes valid

$$\Delta V^{\neq} \equiv RT (\partial \ln \eta / \partial P)_T + \text{other terms} \tag{7}$$

However, thoughout this article, we shall consider that the chemistry presented here embodies "normal high pressure" reactions occurring outside such viscosity limits.

In addition, volume changes mean the volume balance between initial state and transition state, unless explicitely specified.

3. Structural volume changes ($\Delta V^{\neq}_s$)

3.1. VOLUME CHANGES RESULTING FROM BOND TRANSFORMATIONS AND CONFORMATIONAL CHANGES

This is the simplest case for the interpretation of the activation volume. We will describe mainly pericyclic reactions. These reactions designate processes involving cyclic or pseudocyclic transition states. Mechanistically, concerted or stepwise diradical pathways are both available. They encompass [m+n] cycloadditions such as Diels-Alder reactions, [2+2] cycloadditions, 1,3-dipolar cycloadditions, [$\pi^2 + \pi^2 + \pi^2$] cycloadditions (homo-Diels-Alder reactions), ene additions, sigmatropic rearrangements, electrocyclic reactions. The basic expression of $\Delta V^{\neq}$ in these processes is usually (but not in all cases) given by the simple approximation:

$$\Delta V^{\neq} \sim \Delta V^{\neq}_s$$

The assumption is vindicated by the fact that pericyclic reactions show relative solvent unsensitivity (however, [2+2] processes [7] and, even, some [4+2] cycloadditions [8] may not obey the rule).

The van der Waals activation volume $V^{\neq}_w$ of a Diels-Alder reaction is in the same order of magnitude for both mechanistic (concerted and stepwise) pathways [9]. However, the packing coefficient χ defined as the ratio $\chi = V^{\neq}_w / V^{\neq}$ is definitely larger for a cyclic transition state than for the acyclic one related to the stepwise process [10]. With such argument, the determination of θ appears to be a reliable quantity to estimate the location of the transition state and, possibly, to decide whether the reaction proceeds concertedly or stepwise.

It is not the aim of this paper at reviewing the mechanisms of various reactions investigated through pressure kinetics. For the information of the reader, there are excellent reviews and anthologies in this field [3,11]. Some examples are given hereafter for illustration.

TABLE 1. Activation volumes of Lewis acid catalyzed cycloadditions[a]

Type	Reaction	Catalyst	$\Delta V^{\neq}_c$	$\Delta V^{\neq}_{nc}$
			($cm^3.mol^{-1}$)	
$\pi^4 + \pi^2$	2-methylfuran + acrylonitrile	$LiClO_4$.ether	- 29.3	- 28.8
$\pi^4 + \pi^2$	isoprene + methyl vinyl ketone	$ZnCl_2$	- 33.3	- 38.0
$\pi^4 + \pi^2$	acroleine + ethyl vinyl ether	$Yb(fod)_3$	- 31.7	- 29.6
$\pi^4 + \pi^2$	crotonaldehyde + ethyl vinyl ether	$Eu(fod)_3$	- 31.9	- 27.7
$\pi^2 + \pi^2 + \pi^2$	norbornadiene + methyl propiolate	$AlCl_3$	- 32	- 32.0

[a] T (25.0°C except second line - 34°C-). $\Delta V^{\neq}_c$ and $\Delta V^{\neq}_{nc}$: activation volumes for the catalyzed and uncatalyzed reaction respectively

Diels-Alder reactions show large volume contractions and are notoriously promoted by pressure. For normal electron demand partners e.g. the diene bearing electron-donating substituents and the dienophile beeing substituted by electron-withdrawing groups, the activation volume is invariably close to the reaction volume ($\theta \sim 1$). Such result is consistent only with a cyclic transition state and a concerted mechanism although the formation of the two bonds may be asynchronous. The $\Delta V^{\neq}$ - data were tabulated in two reviews [11,12] and do not need any further comment. More recently, the case of Lewis acid catalyzed [$\pi^4 + \pi^2$] and [$\pi^2 + \pi^2 + \pi^2$] cycloadditions has been examined (Table 1) [13]. The results indicate that Lewis acid catalysis does not significantly affect the activation volume with respect to the $\Delta V^{\neq}$ - value determined in the corresponding uncatalyzed reaction, meaning conservation of concertedness. The catalyzed homo-Diels-Alder reaction described in Table 1 affords actually two products: the homo-Diels-Alder adduct 1 and an exo [2+2]-cycloadduct 2 (Scheme 1):

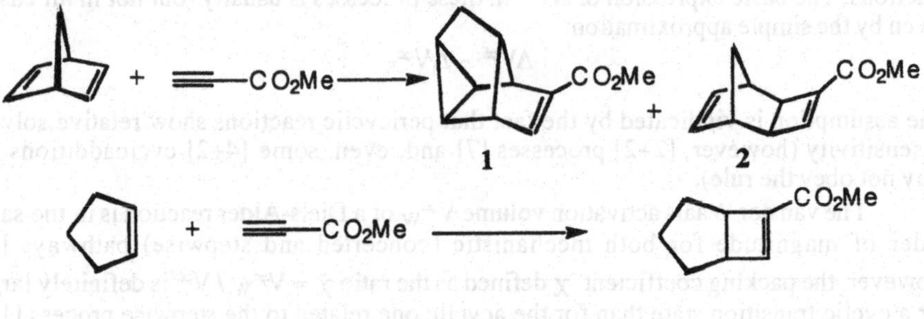

Scheme 1

The activation volumes for the two processes differ by $\Delta\Delta V^{\neq} = -12.5\ cm^3.mol^{-1}$. A precise determination of the activation volume in the chemoselective $AlCl_3$-catalyzed [2+2] addition of methyl propiolate to cyclopentene gives a value of $-20.5\ cm^3.mol^{-1}$ [14]. This low absolute value is indicative of a stepwise process for which electrostriction revealed by $\Delta V^{\neq}$ is low or inexistent, since the rate determining step (RDS) is the addition of the alkene to the complex formed between the acetylenic ester and the Lewis catalyst to yield an intermediate. This step is independent of the medium [13] (scheme 2).

$$\equiv\!-E + MX_n \xrightarrow{\text{fast}} [\text{complex}] \xrightarrow[\text{RDS}]{+\text{ alkene}} [\text{INT}] \xrightarrow{\text{collapse}} [2+2]\ \text{adduct}$$

Scheme 2

It should be pointed out that uncatalyzed [2+2] cycloadditions can be very pressure dependent with considerable electrostriction. In this case, Eq. 6 must be used (for more details see paragraph 4.1).

An interesting result was obtained in a [2+2] organometallic cycloaddition of unsaturated Fischer carbene complexes with 3,4-dihydro-2H-pyrane (scheme 3) [15]. The $\Delta V^{\neq}$ - values were found independent of solvent which does not change significantly the reaction rate. It was concluded that the reaction proceeded according to a non polar concerted mechanism.

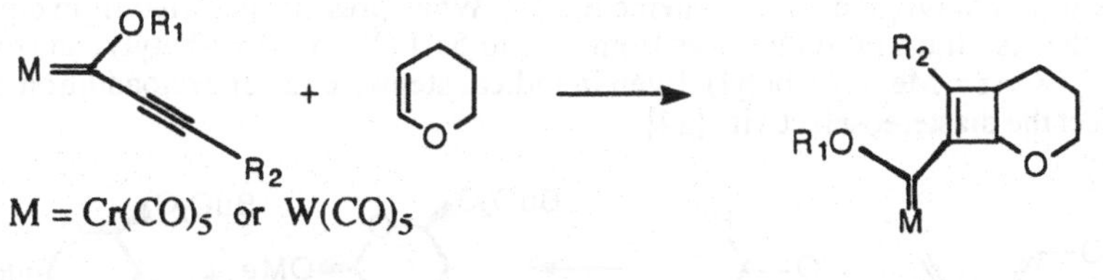

M = Cr(CO)$_5$ or W(CO)$_5$

Scheme 3

In the field of pericyclic sigmatropic rearrangements a recent paper reported the Cope reaction of hexadiene **3** yielding rearranged products E,Z (**4**) and E,E (**5**) (scheme 4) [16]. The $\Delta V^{\neq}$ - values for the formation of both products amount - 13.3 (**4**) and - 8.8 $cm^3.mol^{-1}$(**5**) respectively. The $\Delta\Delta V^{\neq}$ difference must obviously be ascribed to different volume requirements, the chair-like transition state leading to **4** involving higher steric hindrance, in accordance with comments of paragraph 3.4.

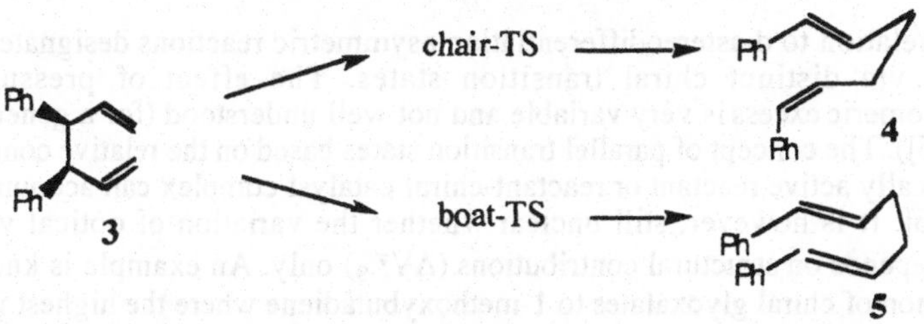

chair-TS

boat-TS

Scheme 4

The same reasoning applies in the condensation of octanal with (E)-crotyltributyltin (scheme 5) [17]. Pressure promotes the predominant formation of **8** with the (E)-isomer of **6** as reactant and of **7** when the (Z)-isomer of **6** is involved. Both pathways (E)-**6** -------> **8** and (Z)-**6** --------> **7** occur via boat-like transition states with $\Delta\Delta V^{\neq}$ = + 5.0 $cm^3.mol^{-1}$ for the (E)-isomer reaction and - 4.3 $cm^3.mol^{-1}$ for the corresponding (Z)-isomer condensation. These results confirm the conclusions of an earlier paper of the same authors concerning the aldol reaction of silyl enol ethers with aldehydes in which pressure induces a crossover from a chair-transition state to a boat-preferred transition state[18].

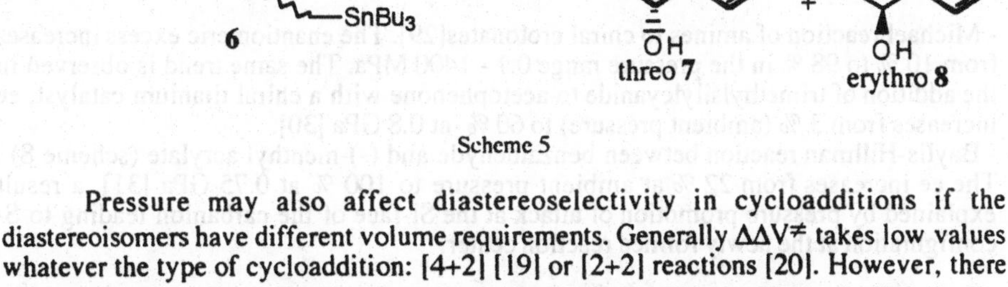

RCHO + **6** (SnBu$_3$) → threo **7** + erythro **8**

Scheme 5

Pressure may also affect diastereoselectivity in cycloadditions if the diastereoisomers have different volume requirements. Generally $\Delta\Delta V^{\neq}$ takes low values whatever the type of cycloaddition: [4+2] [19] or [2+2] reactions [20]. However, there are examples for which $\Delta\Delta V^{\neq}$ - values are appreciable [21,22] (see Table 5). This is also the case in the Diels-Alder reaction of trialkylsilyloxy substituted 1-methoxy-1,3-

butadiene and butyl glyoxalate (scheme 6) [23]. When pressure is varied from ambient to 1 GPa, the cis : trans ratio increases from 1 : 1 to 5 : 1 (X =Y=Z = t-butyl) and from 7 : 3 to 10 : 1 (X =Y= Me, Z= t-butyl). Even in radical stepwise [2+2] cycloaddition pressure can affect the diastereoselectivity [24].

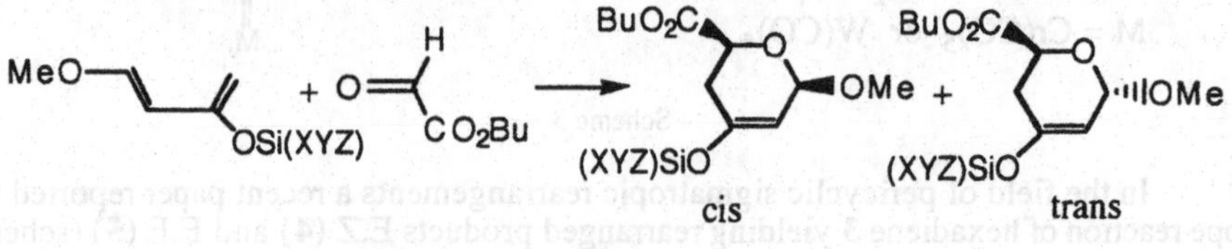

MeO... + O=C(H)CO2Bu ⟶ cis + trans

Scheme 6

3.2. STRUCTURAL VOLUME CHANGES RELATED TO CHIRAL TRANSITION STATES

In relation to diastereodifferentiation asymmetric reactions designate processes occurring via distinct chiral transition states. The effect of pressure on the diastereoisomeric excess is very variable and not well understood (for a general review, see ref. [25]). The concept of parallel transition states based on the relative conformations of the optically active reactant or reactant-chiral catalyst complex can account for some results [26]. It is however, still unclear whether the variation of optical yields with pressure depends on structural contributions ($\Delta V^{\neq}_S$) only. An example is known in the cycloaddition of chiral glyoxalates to 1-methoxybutadiene where the highest increase in optical yield with pressure is obtained with toluene as solvent [27]. The "solvation" effect of toluene on the direction of asymmetric induction is additive with the effect of pressure.

The most representative examples in this area showing a significant pressure enhancement of the enantiomeric excess (ee) can be listed as:

- reduction of prochiral ketones [28]. In the reaction shown in scheme 7, ee varies from 57 % at ambient pressure to 90 % at 0.6 GPa.

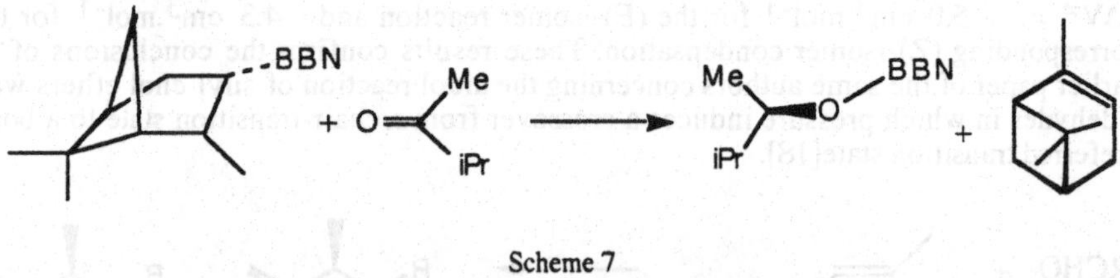

Scheme 7

- Michael reaction of amines to chiral crotonates[29]. The enantiomeric excess increases from 10 % to 98 % in the pressure range 0.1 - 1400 MPa. The same trend is observed in the addition of trimethylsilylcyanide to acetophenone with a chiral titanium catalyst, ee increases from 3 % (ambient pressure) to 60 % at 0.8 GPa [30].
- Baylis-Hillman reaction between benzaldehyde and (-)-menthyl acrylate (scheme 8). The ee increases from 22 % at ambient pressure to 100 % at 0.75 GPa [31], a result explained by pressure promotion of attack at the Si-face of the carbanion leading to S-configuration at the newly formed reaction center.

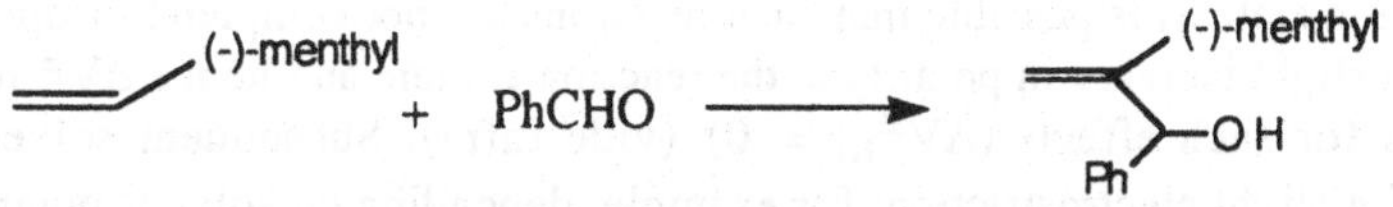

(-)-menthyl + PhCHO ⟶ (-)-menthyl, OH, Ph

Scheme 8

However, puzzling effects can be met. The intramolecular Diels-Alder reaction of a benzylidene compound proceeds in the presence of a chiral titanium catalyst [32]. The enantioselectivity increases from 4.5 % to 20.4 % when pressure is raised from ambient to 0.5 GPa. One possible interpretation would be considering that one enantiomer has a smaller activation volume. In a subsequent study reporting the intermolecular reaction of isoprene and a dienophile derived from an oxazolidin-2-one substituted by an unsaturated lateral chain and using the same chiral titanium catalyst, an opposite result was obtained [33]. There is a decrease in ee from 38 % to 21 % in the same pressure range. This could be understood when assuming that the diene attacks the dienophile-titanium complex from the less hindered side. Actually, the reason for the decrease in ee is due to an unfavorable pressure effect on the stability of the chiral titanium complex.

A considerable decrease of ee with pressure is observed in the Baylis-Hillman addition of aldehydes to methyl vinyl ketone in the presence of quinidine (scheme 9) [34]. When R is cyclohexyl, ee decreases from 46 to 13 % in the pressure range 0.1-1800 MPa. When R = nC_9H_{19}, ee first increases with pressure before reaching an extremum at 1 GPa, then decreases at higher pressures. These intriguing results are interpreted in terms of steric interactions and conformational changes with pressure.

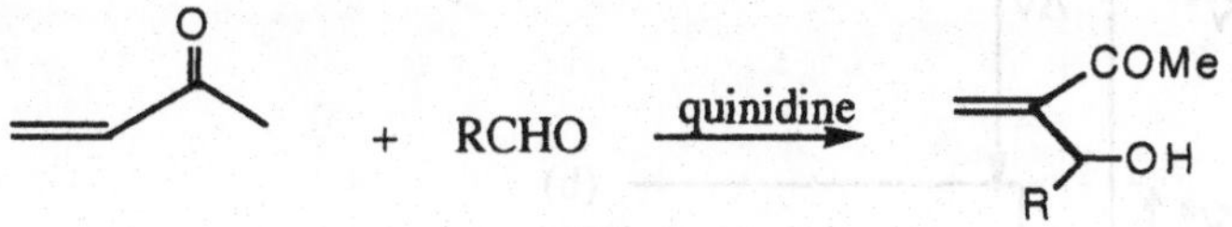

O + RCHO —quinidine→ COMe, OH, R

Scheme 9

3.3. MINIMIZATION IN THE VOLUME PROFILE

In some unhindered Diels-Alder reactions a rather strange phenomenon was observed as $|\Delta V^{\neq}| > |\overline{\Delta V}|$. Since $\Delta V^{\neq}_m$ is in this case assumed to be negligible, $\Delta V^{\neq} \sim \Delta V^{\neq}_s$. However, these reactions are concerted so that $\Delta V^{\neq}_s$ must approach $\Delta V^{\neq}$. Such result is compatible with a transition state more compact than the product and located at a minimum in the volume profile. For these cycloadditions it was suggested that the additional volume contraction within the transition state could be due to secondary orbital interactions $\Delta V^{\neq}_\mu$ (Eq. 8) [36] which would contribute to the stability of the transition state (Fig. 1). Such phenomenon was evidenced in reactions reported in ref. 3 and listed in Table 2.

$$\Delta V^{\neq} = \Delta V^{\neq}_s + \Delta V^{\neq}_\mu \qquad (8)$$

$\Delta V^{\neq}_\mu$: volume contraction within the transition state leading to a minimum in the volume profile

The reactions exhibiting minimization in volume profiles mainly involve maleic anhydride which is a rigid dienophile. There is seemingly a transient shrinkage *within* the transition state. It is possible that such reactions are not completely isopolar. There might be a slight increase in polarity at the reaction centers and hence $\Delta V^{\neq}$ must make allowance for such effects ($\Delta V^{\neq}{}_{m} \neq 0$) (vide infra). Subsequent solvent studies confirmed a slight electrostriction: for example, depending on solvent polarity, $\Delta V^{\neq}$ varied from - 37.0 to - 39.5 $cm^3.mol^{-1}$ for the cycloaddition of maleic anhydride to isoprene and from - 43.0 to - 53.6 $cm^3.mol^{-1}$ in the corresponding reaction involving 1-methoxybutadiene. However, the $\Delta V^{\neq}$ - values shown in Table 2 cannot be fully rationalized in such way. As correctly pointed out by Firestone [2], the phenomenon is not a simple artefact of calculation. The kinetic pressure effect on the retrodiene reaction involving the cycloadduct formed by the addition of acrylonitrile to 2-methylfuran is, surprisingly, negative with $\Delta V^{\neq}$ = - 2 $cm^3.mol^{-1}$ [37]. Such result is *a priori* difficult to understand and to rationalize, since pressure should retard decomposition reactions ($\overline{\Delta V}$ is positive). However, the result fits the principle of microreversibility and is in accordance with the $\Delta V^{\neq}$ - value for the forward reaction and also with $\Delta V^{\neq}{}_{\mu}$ = - 1.6 $cm^3.mol^{-1}$ (Table 2).

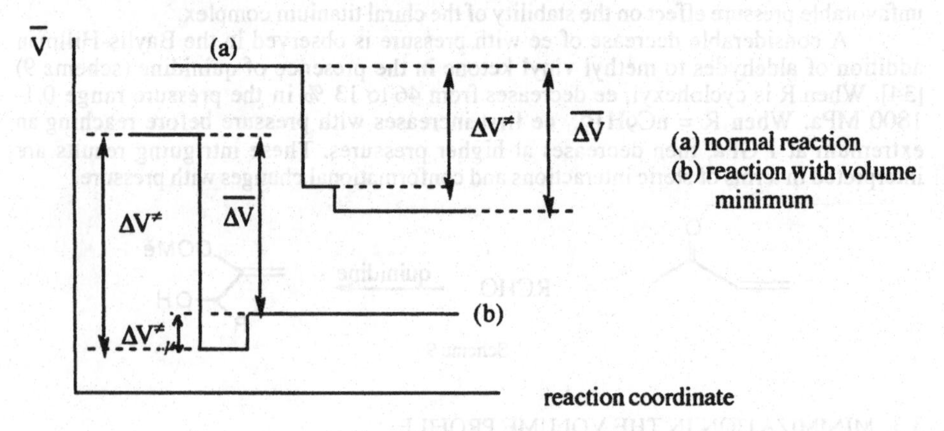

Figure 1. Volume profiles for Diels-Alder reactions

TABLE 2. Volume data for Diels-Alder reactions with possible secondary orbital interactions

Diene[a]	Dienophile	Medium	$\Delta V^{\neq}$ $cm^3.mol^{-1}$	$-\Delta V^{\neq}{}_{\mu}$ $cm^3.mol^{-1}$	θ
2-methylfuran	acrylonitrile	CH_2Cl_2	- 30.3	1.6	1.06
2-methylfuran	methyl acrylate	CH_2Cl_2	- 32.2	3.3	1.11
1,3-pentadiene	maleic anhydride	C_4H_9Cl	- 43.1	11.8	1.37
DMB	maleic anhydride	C_4H_9Cl	- 41.3	5.0	1.14
CHX	maleic anhydride	CH_2Cl_2	- 39.6	9.3	1.30
MOB	maleic anhydride	C_4H_9Cl	- 45.4	9.9	1.27

[a] DMB (2,3-dimethylbutadiene), CHX (1,3-cyclohexadiene), MOB(1-methoxybutadiene)

The extent of stabilization by these interactions is also responsible for the stereochemistry of adducts obtained from cyclic dienes or dienophiles. In an earlier study the pressure effect on the endo : exo ratio of cycloadducts showed only low $\Delta\Delta V^{\neq}$ differences meaning that the volume shrinkage in both transition states is nearly identical [38]. However, the study reported normal electron demand reactions. In fact, examples are known where $\Delta V^{\neq}_{endo} - \Delta V^{\neq}_{exo}$ have significant differences (entries 1,2 in Table 3).

TABLE 3. Effect of pressure on endo : exo ratios in cycloaddition reactions [36,38]

Entry	Diene	Dienophile	Pressure (MPa)	endo (%)	$\Delta V^{\neq}_{endo} - \Delta V^{\neq}_{exo}$ ($cm^3.mol^{-1}$)
1	norbornadiene	maleic anhydride	0.1	81	
			1000	92	- 3.4
2	1,3-cyclohexadiene	cyclohexadiene	0.1	89	
			700	96	- 6.0
3	cyclopentadiene	methyl methacrylate	0.1	35.3	
			300	36.8	- 0.8
4	cyclopentadiene	n-butyl methacrylate	0.1	30.6	
			300	32.1	- 0.8
5	cyclopentadiene	2-ethylacrylonitrile	0.1	10.0	
			300	10.6	- 1.0

It is generally agreed that the endo transition structure is promoted by application of pressure due to the enhanced rigidity of the transition state arising from geometrical factors. This is also reflected by a higher reaction volume shrinkage for the endo stereoisomer. As an example (scheme 10) [39]:

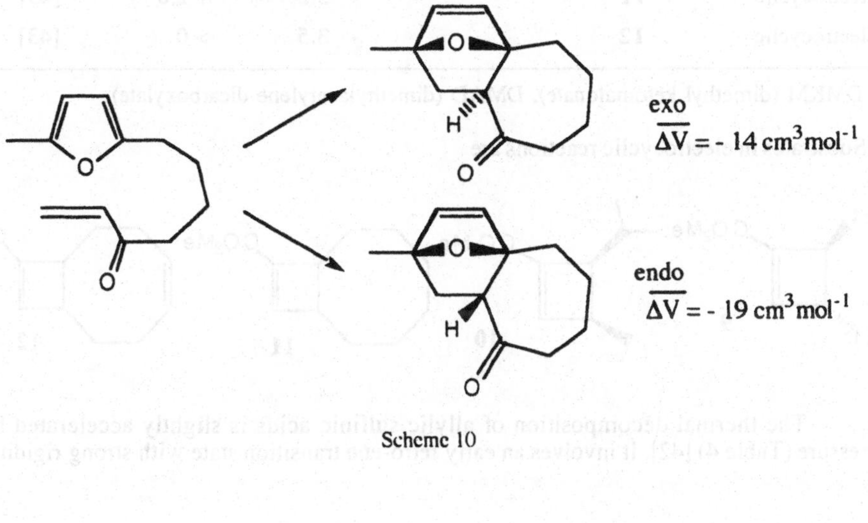

exo $\overline{\Delta V} = -14\ cm^3 mol^{-1}$

endo $\overline{\Delta V} = -19\ cm^3 mol^{-1}$

Scheme 10

Like Diels-Alder cycloadditions other reactions were found to exhibit a minimum in the volume profile. Ene reactions involving C--H--O and C--H--C hydrogen transfer show θ - values higher than unity (Table 4). In this case, the transition state is assumed to be more strained than the final state and is indicative of an angular process for hydrogen transfer (Fig. 2) [40,41].

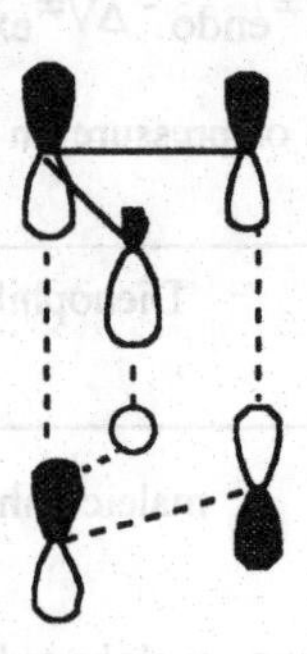

Fig. 2
Angular H-transfer
in ene reactions

TABLE 4. Reactions with minimization of volume profiles

Type	Reactants[a]	$-\Delta V^{\neq}$ ($cm^3.mol^{-1}$)	$-\overline{\Delta V}$ ($cm^3.mol^{-1}$)	Ref.
Ene	1-hexene + DMKM	- 28.4	- 27.0	[40]
Ene	2-ethyl-1-butene + DMKM	- 34.0	- 31.3	[40]
Ene	ß-pinene + DMKM	- 31.3	- 29.4	[41]
Ene	ß-pinene + DMAD	- 39.6	- 35.4	[41]
Retro-ene	allylsulfinic acid	- 5.5	+ 15	[42]
Electrocyclic	9	- 2.2	> 0	[43]
Electrocyclic	10	- 2.4	> 0	[43]
Electrocyclic	11	- 3.2	+ 2.0	[43]
Electrocyclic	12	- 3.5	> 0	[43]

[a] DMKM (dimethyl ketomalonate), DMAD (dimethylacetylene dicarboxylate)

Substrates in electrocyclic reactions are

The thermal decomposition of allylic sulfinic acids is slightly accelerated by pressure (Table 4) [42]. It involves an early retro-ene transition state with strong rigidity.

A very recent study reported the effect of pressure on electrocyclic reactions such as the conrotatory opening of cyclobutenes (scheme 11) [43]. All reactions were found to be pressure accelerated despite the cleavage of a σ-bond which should lead to a volume increase. It was suggested that such concerted processes require a strict orbital topology fixing the direction of twisting of the cyclobutane ring. This is nothing else as a process involving a very compact pericyclic transition state before opening of the σ-framework [43].

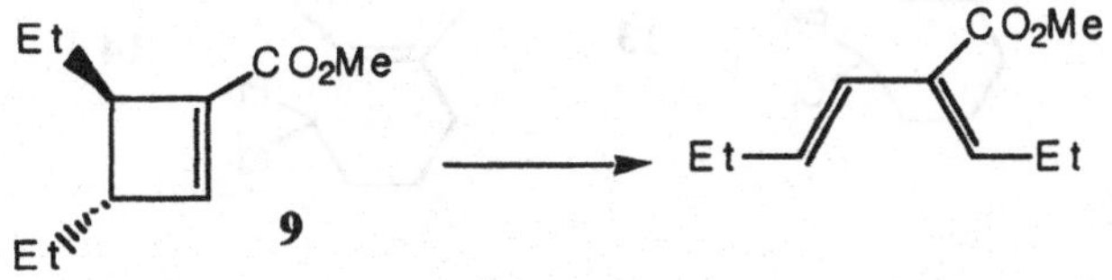

Et CO2Me Et 9 → Et CO2Me Et

Scheme 11

3.4. VOLUME CHANGES RESULTING FROM STERIC HINDRANCE AND STRAIN

While studying the effect of pressure on reactions involving molecules frozen by steric requirements, several authors observed an enhancement of the rate response to pressure with an increase of steric compression in Menshutkin reactions [44,45], copolymerizations [46], solvolyses [47]... Such an effect was interpreted as resulting from a displacement of the transition state according to the Hammond postulate [48] or from a more compact spatial arrangement of the transition state [44]. More recently, it has been proposed to write Eq. 5 as [49]:

$$\Delta V^{\neq}_{S} = \Delta V^{\neq}_{O} + \Delta V^{\neq}_{\sigma} \quad (9)$$

where $\Delta V^{\neq}_{O}$ is defined as the volume of activation of the unhindered reaction whereas $\Delta V^{\neq}_{\sigma}$ would make allowance for the volume effect of additional steric hindrance. $\Delta V^{\neq}_{\sigma}$ is defined as steric volume of activation. Although the physical meaning underlying $\Delta V^{\neq}_{\sigma}$ remains to be clarified, it is evident that the effect is real. High pressure induces a stronger promotion of the most hindered process. The additional putative volume $\Delta V^{\neq}_{\sigma}$ is a dynamic term. If it is related to the Hammond postulate, this would mean that more crowded transition states are shifted downwards toward the product along the reaction axis. Such reactions are most pressure accelerated. Number of examples highlighting such a pressure effect are known e.g. Menshutkin reactions [48], aromatization of Dewar benzenes [50], hydrogen transfer reactions [51] reduction of hindered ketones mediated by tributyltin hydride [52], hydroboration reactions [53] ...(for a review see [54]). More recent examples mostly centered on syntheses encompass the methylation of pyridines [55], the Ugi reaction applied to the synthesis of peptides [56], glycosylation reactions [57], aromatic nucleophilic substitution of halobenzenes with amines [58], synthesis of phenyl sulfides from alcohols [59], synthesis of rigid crown-shaped macrocycles [60], bisesterification of sterically hindered alcohols [61].

Although solvolysis reactions obey rather Eq. 6, these reactions are strongly accelerated by pressure when π-participation is involved [35]. The pressure dependence of the solvolysis of chlorides **13** and **14** in aqueous ethanol yields: $\Delta V^{\neq}$ = - 13.3 $cm^3.mol^{-1}$ for **13** and - 24.0 $cm^3.mol^{-1}$ for **14**. The lower activation volume of the solvolysis of **14** is ascribed to pressure-induced conformational changes required for the prealignment of double bonds necessary for extended π-participation. This introduces an additional volume contraction. However, as pointed out by le Noble, the extra volume contribution is too large to be ascribed to this effect only. We suggest that an additional steric term is operating.

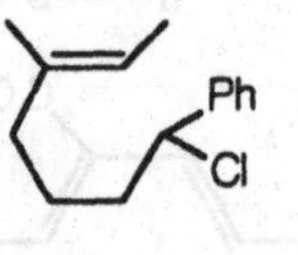

Particular volume effects are expected in pericyclic reactions producing diastereoisomers. Large $\Delta\Delta V^{\neq}$ (up to - 11 $cm^3.mol^{-1}$) may characterize the two reaction pathways leading to cis and trans diastereoisomers in intermolecular Diels-Alder reactions [62]. Table 5 portrays $\Delta\Delta V^{\neq}$ - values for cis and trans diastereoisomers formed in the hetero-Diels-Alder reaction of enaminoketones and vinyl ethers (scheme 12) [63]. Clearly, there is a relationship between the steric bulk of R and the cis : trans ratio. The steric demand of CCl_3 vs CF_3 is reflected in the more negative $\Delta\Delta V^{\neq}$ - values.

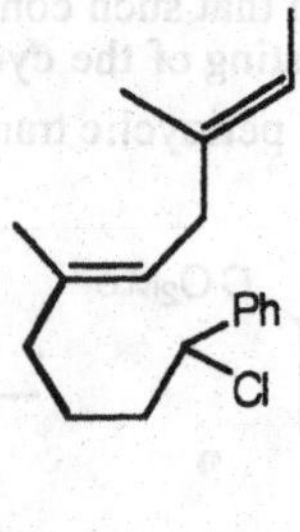

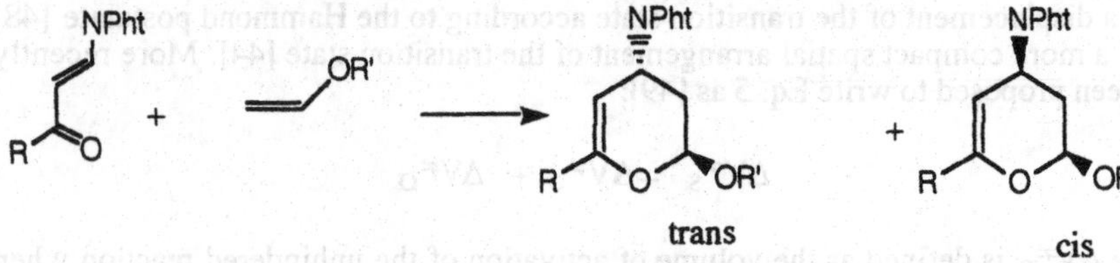

Scheme 12

Pht is

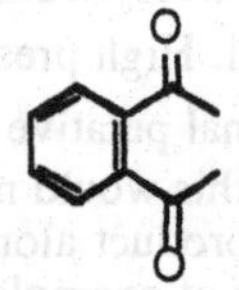

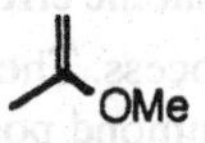

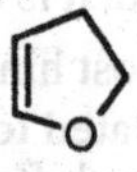

TABLE 5. Effect of substituent R on the diastereoselectivity under pressure

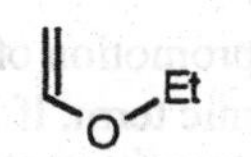

Vinyl ether	R	$\Delta V^{\neq}_{cis} - \Delta V^{\neq}_{trans}$ ($cm^3.mol^{-1}$)
15	CF_3	+ 4.1
15	CCl_3	- 0.6
16	CF_3	- 3.9
16	CCl_3	- 7.3
17	CF_3	- 6.0
17	CCl_3	- 9.2

Enhanced strain can produce a similar pressure effect as it was shown some time ago in the copolymerization of maleic anhydride with strained cycloalkenes [64]. Examples using strain as a driving force for the high pressure synthesis of strained polycyclic molecules are scarce [65]. One may mention the dimerization of dimethylsubstituted propenes to yield bicyclobutanes [66] and the Diels-Alder reaction of [2.2]-paracyclophane [67].

In a very recent paper, Buback et al. examined the pressure behavior of furans tethered by bicyclopropylidene and methylene cyclopropane moieties [68,69]. The activation volumes of the intramolecular Diels-Alder reaction were found to be dependent on the additional strain as well as steric hindrance introduced by the cyclopropane rings (Table 6). The changes in $\Delta V^{\neq}$ upon substituting the double bond in **18** by an additional cyclopropane ring as in **19** reflect a more crowded transition state.

TABLE 6. Pressure effect on the intramolecular Diels-Alder reactions of **18** and **19**.

Furan	R	T(°C)	$\Delta V^{\neq}$ ($cm^3.mol^{-1}$)
18	H	110	- 28.4
19	H	110	- 31.9
18	OMe	80	- 35.8
19	OMe	80	- 40.8

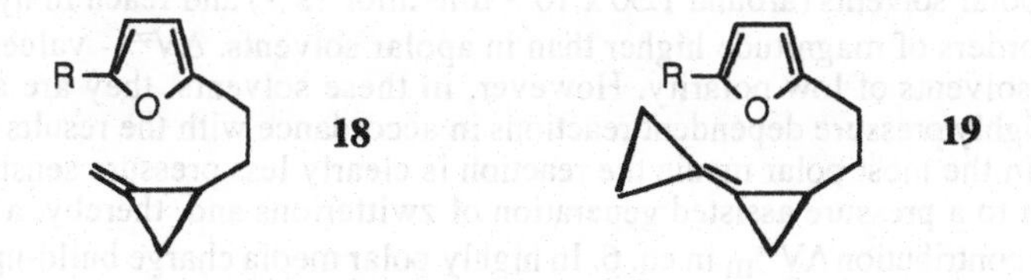

4. Environmental activation volume ($\Delta V^{\neq}_m$)

4.1. ELECTROSTRICTION

The importance of the medium on the course of organic reactions is a well established fact [70]. The effects of the medium can be various since they may result from dispersion, repulsive, electrostatic forces and specific interactions such as hydrogen bonding. The most significant contribution to $\Delta V^{\neq}_m$ originates from changes in polarization and polarity. When the polarity of the transition state and the initial state differs, it means involvement of interactions of charged species or free ions with their neighbors or with the medium. From the viewpoint of volume effects, the solvation of the transition state is a major contributor. It is related to the pressure change of polarization which increases intermolecular forces. Electrostriction results from the compression of molecules by vicinal charged species with a volume shrinkage V_e given in rough approximation by the Drude-Nernst equation:

$$V_e = -\frac{q^2}{2 r \varepsilon} \frac{\partial \operatorname{Ln} \varepsilon}{\partial P} \quad (10)$$

(q: ionic charge, r : ionic radius, ε : dielectric constant)

In terms of kinetic parameters, $\Delta V^{\neq}_m$ is related to the pressure dependence of ε. $\Delta V^{\neq}_m$ can vary in large proportions depending on the polarity of the medium. Representative examples are the solvent dependence of the activation volume in Menshutkin reactions [45] and [2+2] cycloadditions [5]. In the addition of methyl iodide to pyridine, $\Delta V^{\neq}$ varies from - 37.5 to - 21.3 $cm^3.mol^{-1}$ in carbon tetrachloride (ε = 2.24) and nitrobenzene (ε = 34.8) respectively. In the cycloaddition of tetracyanoethylene to n-butyl vinyl ether, $\Delta V^{\neq}$ increases from - 50 (carbon tetrachloride) to - 29 $cm^3.mol^{-1}$ (acetonitrile). A more recent example is provided by the Michael reaction of t-butylamine and acrylonitrile [71]. This reaction involves in the rate-determining step a nucleophilic attack on the activated double bond of acrylonitrile with complete development of a zwitterionic specie like :

tBu, H, H on N⁺ — CH₂ — C⁻ (CN, H)

It is obvious that Eq. 6 applies. The rate constants are very low at atmospheric pressure in the least polar solvents (around 1.50 x 10^{-6} $dm^3.mol^{-1}s^{-1}$) and reach in hydroxylated media two orders of magnitude higher than in apolar solvents. $\Delta V^{\neq}$ - values are fairly constant in solvents of low polarity. However, in these solvents, they are also lowest indicating highly pressure dependent reactions in accordance with the results reported in ref. 72,73. In the most polar media the reaction is clearly less pressure sensitive. These results point to a pressure assisted generation of zwitterions and, thereby, a significant electrostatic contribution $\Delta V^{\neq}_m$ in eq. 6. In highly polar media charge build-up is already accomplished at ambient pressure amply explaining the high rates and lower absolute $\Delta V^{\neq}$ - values.

It should be noted that $\Delta V^{\neq}$ can be negative in bond-breaking processes for reasons different of those presented in paragraph 3.3. The respective activation volumes in the cycloaddition of tetracyanoethylene to α-methylstyrene are - 27 $cm^3.mol^{-1}$ for the forward reaction and - 3.3 $cm^3.mol^{-1}$ for its reverse [74]. This cycloaddition has a strong dipolar character as it involves a charge-transfer complex both in the forward and the reverse reaction. In the latter case, $\Delta V^{\neq}_s$ is positive whereas $\Delta V^{\neq}_m$ is negative with $|\Delta V^{\neq}_m| > |\Delta V^{\neq}_s|$.

The isopolarity of Diels-Alder reactions is not universal. The [4+2] cycloaddition of tropone with maleic anhydride and norbornene shows slight $\Delta V^{\neq}$ - variations when the reactions are carried out in isopropylbenzene or N,N-dimethylformamide (Table 7) [75]. The interpretation is based on the large dipole moment of tropone. Contrastingly with normal Diels-Alder reactions, the transition state is less polar than the reactants.

TABLE 7. Diels-Alder reactions of tropone

Dienophile	Solvent	ΔV $cm^3.mol^{-1}$	$\Delta V^{\neq}$ $cm^3.mol^{-1}$	θ
maleic anhydride	isopropylbenzene	nd	- 21.4	nd
	DMF	- 25.5	- 16.8	0.66
norbornene	isopropylbenzene	-33.5	- 30.0	0.90
	DMF	- 32.1	- 27.8	0.87

Takeshita et al. also investigated the high pressure kinetics of the $[\pi^6 + \pi^4]$ cycloaddition of tropone with 2,3-dimethylbutadiene [76] and 1,3-cyclohexadiene [77] (scheme 15, Table 8). In such processes involving tropone, $\Delta V^{\neq}_m$ has a low value and is positive.

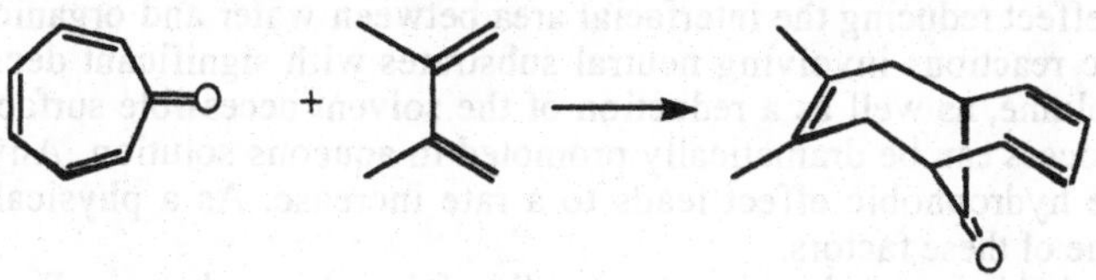

Scheme 13

TABLE 8. Solvent effect in the [6+4] cycloaddition of tropone to cyclohexadiene [77]

Solvent	ΔV $cm^3.mol^{-1}$	$\Delta V^{\neq}$ $cm^3.mol^{-1}$
cumene	- 36.1	- 37.6
dioxane	- 31.8	- 28.2
DMF	- 34.2	- 32.6

A special volume effect induced by the solvent is disclosed in the 1,3-dipolar cycloaddition of diphenyldiazomethane to acetylenic and ethylenic diesters as well as to maleic anhydride (scheme 14) [78]. 1,3-Dipolar reactions are similar to Diels-Alder reactions as they involve the same number of electrons. To this respect, transition and initial states are isopolar. However, in the above reactions there is a complex solvent dependence of $\Delta V^{\neq}$ which may contain a volume term originating only partly from electrostatic interactions. Swieton suggested that the solvation of the transition state occurs with reorientation in the solvation shell of the acetylenic diester, in line with the displacement of solvent molecules from the center of the cyclic compound during the reaction.

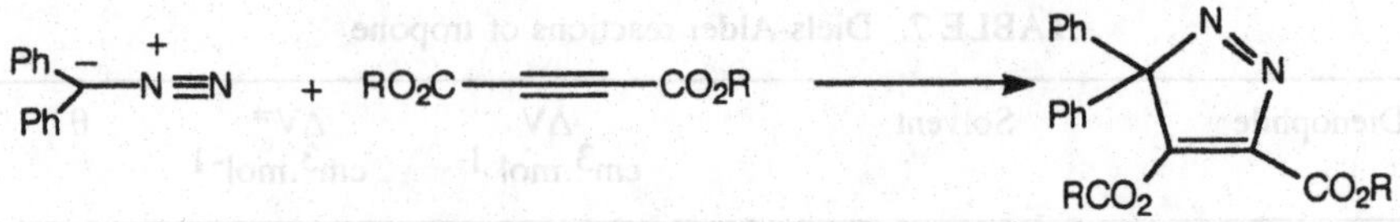

$$Ph_2C^{-}-N^{+}\equiv N + RO_2C-C\equiv C-CO_2R \longrightarrow$$

Scheme 14

4.2. SOLVOPHOBIC INTERACTIONS

At last, $\Delta V^{\neq}_m$ can make allowance for volume effects due to the medium which have no electrostatic origin. It has become a known fact that many reactions can be subjected to dramatic rate accelerations when the medium is water [79]. The origin of this kinetic change has been ascribed to several factors: high cohesive energy density of water (δ^2 = 550 cal.cm^{-3}), hydrogen bonding, polarity, enforced hydrophobic interactions [80]. It is doubtless that hydrophobic effects play the major role. These effects should be interpreted as the entropy-driven association of hydrophobic molecules in water (associative effect reducing the interfacial area between water and organic molecules) [81]. Organic reactions involving neutral substrates with significant decrease of the molecular volume, as well as a reduction of the solvent accessible surface during the activation process can be dramatically promoted in aqueous solution. Any factor that increases the hydrophobic effect leads to a rate increase. As a physical parameter, pressure is one of these factors.

Solvophobic interactions involve small to fair volume changes. Pressure kinetic studies can be valuable in their diagnosis and estimation of their magnitude [82]. It is clear that, in order to get reliable activation volume data, it is necessary to operate in homogeneous solution. However, such studies are faced with the rather poor solubility of organic molecules in water. Kinetic rate constants have been determined in the last years with highly dilute solutions [83,84]. The Diels-Alder reactions of conjugated dienes with methyl vinyl ketone show different $\Delta V^{\neq}$ - values visibly depending on the medium (Table 9) [84]. Other reactions were examined in water and CH_2Cl_2 (Table 10).

TABLE 9. Effect of solvent on $\Delta V^{\neq}$ in the Diels-Alder reaction of methyl vinyl ketone and isoprene (T = 313.7 K)

Solvent	δ^2 (cal.cm^{-3})	$\Delta V^{\neq}$ (cm^3.mol^{-1})
bulk	~ 80	- 41.0
dichloromethane	104	- 39.5
nitrobenzene	110	- 35.2
methanol	208	- 35.0
glycol	213	- 32.5
water	555	- 33.9

TABLE 10. Activation volumes in aqueous Diels-Alder reactions [84]

Reaction	T (K)	$-\Delta V^{\neq}$ ($cm^3.mol^{-1}$) CH_2Cl_2	water
Cyclohexadiene + methyl vinyl ketone	313.2	38.0	32.0
Furan + methyl vinyl ketone	303.8	32.4	28.5
Isoprene + methyl acrylate	335.3	33.3	31.6
Hexachlorocyclopentadiene + styrene	313.7	35.4	28.0

The addition of a generator of hydrogen bonds does not alter the value of the activation volume in dichloromethane, at variance with the $\Delta V^{\neq}$ change when the reaction is carried out in water. Fig. 3 portrays this behavior in the cycloaddition of methyl vinyl ketone to isoprene. Let us define α as the ratio of rate constants with (+) and without (-) p-nitrophenol (PNP) added (PNP is the hydrogen bond generator).

$$\alpha = kPNP(+) / kPNP(-)$$

This ratio is nearly constant in the organic solvent while it increases linearly with pressure in aqueous solution. This means a less negative $\Delta V^{\neq}$- value in water and, therefore, a less pressure sensitive reaction.

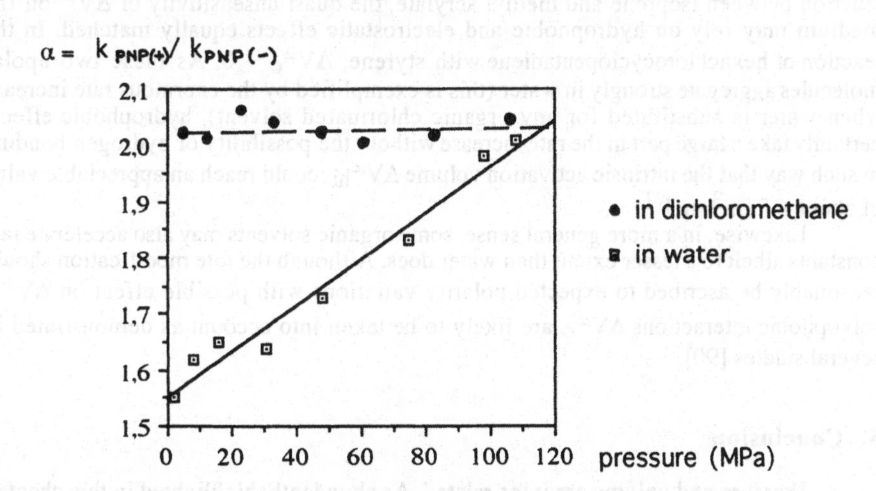

Figure 3. Effect of the addition of PNP in the cycloaddition of methyl vinyl ketone to isoprene

Methyl vinyl ketone is a polarizable molecule. The formation of the complex diene-dienophile in aqueous solution is accompanied by a decrease of the molecular volume that has to be hydrated. Electric polarization of the transition state is enhanced in water. This view is comforted by computational studies relative to the aqueous Diels-Alder reaction of cyclopentadiene and methyl vinyl ketone [85] and the Claisen rearrangement of allyl vinyl ether [86]. It is, therefore, justified that the pressure kinetic results portrayed in Fig. 3 give credit to hydrogen bonding as the cause of the increased polarization of the activated complex and, subsequently, as one of the reason leading to the rate acceleration of the aqueous pericyclic reactions.

Taking into account hydrophobic and electrostatic effects, $\Delta V^{\neq}_m$ can be expressed as eq. 11:

$$\Delta V^{\neq}_m = \Delta V^{\neq}_e + \Delta V^{\neq}_\phi \tag{11}$$

The volume $\Delta V^{\neq}_\phi$ makes allowance for specific volume changes associated with the interactions between water and the activated complex. We suggest that $\Delta V^{\neq}_\phi$ itself may be decomposed into two terms related to hydrogen bond effects $\Delta V^{\neq}_H$ and repulsive hydrophobic effects $\Delta V^{\neq}_{hi}$ respectively:

$$\Delta V^{\neq}_\phi = \Delta V^{\neq}_H + \Delta V^{\neq}_{hi} \tag{12}$$

Pressure normally enhances the miscibility of liquids [87]. This means that genuine hydrophobic effects are disfavored by pressure in such way that $\Delta V^{\neq}_{hi} > 0$. At variance, hydrogen bond formation is accompanied by negative volume changes due to shortening of the interatomic distances [88]. The scarce values of $\Delta V^{\neq}_H$ available amount 0 to - 4 $cm^3.mol^{-1}$ [89]. $\Delta V^{\neq}_{hi}$ and $\Delta V^{\neq}_\phi$ are unknown quantities at the present time. They can be roughly estimated if $\Delta V^{\neq}_e$ and $\Delta V^{\neq}_H$ can be approximated from eq. 11 and 12. Usually, only qualitative information is available. For example, in the aqueous cycloaddition of methyl vinyl ketone to isoprene $|\Delta V^{\neq}_H| > |\Delta V^{\neq}_{hi}|$. In the Diels-Alder reaction between isoprene and methyl acrylate, the quasi unsensitivity of $\Delta V^{\neq}$ on the medium may rely on hydrophobic and electrostatic effects equally matched. In the reaction of hexachlorocyclopentadiene with styrene, $\Delta V^{\neq}_e \sim 0$. As these two apolar molecules aggregate strongly in water (this is exemplified by the enormous rate increase when water is substituted for any organic chlorinated solvent), hydrophobic effects certainly take a large part in the rate increase without the possibility of hydrogen bonding in such way that the intrinsic activation volume $\Delta V^{\neq}_{hi}$ could reach an appreciable value of about + 7 $cm^3.mol^{-1}$.

Likewise, in a more general sense, some organic solvents may also accelerate rate constants albeit to a lesser extent than water does. Although the rate modification should reasonably be ascribed to expected polarity variations with possible effect on $\Delta V^{\neq}_e$, solvophobic interactions $\Delta V^{\neq}_\phi$ are likely to be taken into account as demonstrated in several studies [90].

5. Conclusion

Pressure and volume are inter-related. As abundantly highlighted in this chapter, the activation volume can be a very complex quantity resulting from structural, electrostatic and solvophobic effects. The quantification of these activation volume components depends on parameters and assumptions that are not easily manageable except in simple apolar reactions involving unhindered reactants and carried out in non associative solvents of low polarity. Even in normal processes where $\Delta V^{\neq} = \Delta V^{\neq}_s$ ($\Delta V^{\neq}_m \sim 0$), the activation volume may deviate from comforting expected values making the interpretation delicate as is the case in enzymatic reactions occurring via successive steps [91]. In spite of these difficulties, pressure should be considered as a fundamental tool in elucidating the mechanism of a vast spectrum of reactions including enzymatic processes.

Symbolism

$\overline{\Delta V}$: reaction volume based on partial molar volumes
$\Delta V^{\neq}_{s}$: structural component of the activation volume
$\Delta V^{\neq}_{m}$: environmental component of the activation volume
$\Delta V^{\neq}_{\sigma}$: steric component of the activation volume
$\Delta V^{\neq}_{\phi}$: activation volume resulting from hydrophobic effects

6. References

1. Le Noble, W.J. and Kelm, H. (1980) *Angew. Chem., Int. Ed. Engl.* **11**, 841-946.
2. Firestone, R.A. (1996) *Tetrahedron* **46**, 14459-14468.
3. Jenner, G. (1988) *Organic High Pressure Chemistry*, W.J. le Noble (ed.), Elsevier, Amsterdam, pp. 143-203.
4. Kowa, L.N., Schwarzer, D., Troe, J., and Schroeder, J. (1992) *J. Chem. Phys.*, **97**, 4827-4835.
5. Kelm, H., and Palmer, D. (1978) *High Pressure Chemistry*, H. Kelm,(ed.), Reidel Publishers, Dordrecht, pp. 281-309.
6. Asano, T., Furuta, T., and Sumi, H. (1994) *J. Am. Chem. Soc.* **116**, 5545-5550.
7. Scheeren, J.W. (1986) *Rec. Trav. Chim. Pays-Bas* **105**, 71-84.
8. Jenner, G. *New J. Chem.*, in press.
9. Yoshimura, Y., Osugi, J., and Nakahara, M. (1983) *J. Am. Chem. Soc.*, **105**, 5414-5418.
10. Klärner, F.G., Krawczyk, B., Ruster, V., and Deiter, U.K. (1994) *J. Am. Chem. Soc.* **116**, 7646-7657.
11. Van Eldik, R., Asano, T., and le Noble, W.J. (1989) *Chem. Rev.* **89**, 549-688.
12. Asano,T., and le Noble, W.J. (1978) *Chem. Rev.* **78**, 407-489.
13. Jenner, G. (1997) *New J. Chem.* **21**, 1085-1090.
14. Jenner, G., and Papadopoulos, M. (1996) *Tetrahedron Lett.* **37**, 1417-1420.
15. Pipoh, R., Van Eldik, R., Wang, S.L., and Wulff, W.D. (1992) *Organometallics* **11**, 490-492.
16. Diedrich, M.K., Hochstrate,D., Klärner, F.G., and Zimny, B. (1994) *Angew. Chem., Int Ed. Engl.* **33**, 1079-1081.
17. Yamamoto, Y., and Saito, K. (1989) *J. Chem. Soc., Chem. Comm.*, 1676-1678.
18. Yamamoto, Y., Maruyama, K., and Matsumoto, K. (1983) *J. Am. Chem. Soc.*, **105**, 6965-6966.
19. Buback, M., Abeln, J., Hübsch, T., Ott, C., and Tietze, L.F. (1995) *Liebigs Ann.* 9-11. Tietze, L.F., Hübsch,T., Ott, C., Kuchta, G., and Buback, M. ib., 1-7.
20. Buback, M., Bünger, J., and Tietze, L.F. (1992) *Chem. Ber.* **125**, 2577-2582.
21. Buback, M., Tost, W., Hübsch, T., Voss, E., and Tietze, L.F. (1989) *Chem. Ber.* **122**, 1179-1186.
22. Tietze, L.F., Hübsch, T., Oelze, J., Ott, C., Tost, W., Wörner, G., and Buback, M. (1992) *Chem. Ber.* **125**, 2249-2258.
23. Jurczak, J., Golebiowski, A., and Rahm, A. (1986) *Tetrahedron Lett.* **27**, 853-856.
24. Dolbier, W.B., and Weaver, S.L. (1990) *J. Org. Chem.* **55**, 711-715.
25. Jenner, G. (1997) *Tetrahedron* **53**, 2669-2695.
26. Golebiowski, A., Jurczak, J., and Pikul, S. (1989) *High Pressure Chemical Synthesis*, J. Jurczak and B. Baranowski (eds.), Elsevier, Amsterdam, pp. 210-254.
27. Jurczak, J., and Bauer, T. (1986) *Tetrahedron* **42**, 5045-5052.
28. Midland, M.M., and Mc Laughlin, J.L. (1984) *J. Org. Chem.* **49**, 1316-1317.

29. Dumas, F., Mezrhab, B., d'Angelo, J., Riche, C., and Chiaroni, A. (1996) *J. Org. Chem.* **61**, 2293-2304.
30. Choi, M.C., Chan, S.S., and Matsumoto, K. (1997) *Tetrahedron Lett.* **38**, 6669-6672.
31. Gilbert, A., Heritage, T.W., and Isaacs, N.S. (1991) *Tetrahedron:Asym.* **2**, 969-972.
32. Tietze, L.F., Ott, C., Gerke, K., and Buback, M. (1993) *Angew. Chem., Int. Ed. Engl.,* **32**, 1485-1486.
33. Tietze, L.F., Ott, C., and Frey, U. (1996) *Liebigs Ann.*. 63-67.
34. Marko, I.E., Giles, P.R., and Hindley, N.J. (1997) *Tetrahedron* **53**, 1015-1024.
35. Ho, N.H., and le Noble, W.J. (1989) *J. Org. Chem.* **54**, 2018-2021.
36. Jenner, G. (1991) *New J. Chem.* **15**, 897-899.
37. Jenner, G., Papadopoulos, M., and Rimmelin, J. (1983) *J. Org. Chem.* **48**, 748-749.
38. Seguchi, K., Sera, A., and Maruyama, K. (1973) *Tetrahedron Lett.* 1585-1588.
39. Harwood, L.M., and Isaacs, N.S. personal communication.
40. Jenner, G., and Papadopoulos, M. (1982) *J. Org. Chem.* **47**, 4201-4204.
41. Jenner, G., Ben Salem, R., El'yanov, B.S., and Gonikberg, E.M. (1989) *J. Chem. Soc., Perkin Trans. II* 1671-1675.
42. Hiscock, S.D., Isaacs, N.S., King, M.D., Sue, R.E., White, R.H., Young, D.J. (1995) *J. Org. Chem.* **60**, 7166-7169.
43. Jenner, G. (1998) *Tetrahedron* **54**, 2771-2776.
44. Gonikberg, M.G., Zhulin, V.M., and El'yanov, B.S. (1963) *The Physics and Chemistry of High Pressures (Soc. Chim. Ind. London)* pp. 212-218.
45. Le Noble, W.J., and Ogo, Y. (1970) *Tetrahedron* **26**, 4119-4124.
46. Jenner, G., Kellou, M. (1981) *Tetrahedron* **37**, 1153-1160.
47. Okamoto, Y., and Yano, T. (1971) *Tetrahedron Lett.*, 919-922.
48. Le Noble, W.J., and Asano, T. (1975) *J. Am. Chem. Soc.*, **97**, 1778-1782.
49. Jenner, G. (1985) *J. Chem. Soc., Faraday Trans. 1* **81**, 2437-2460.
50. Le Noble, W.J., Brower, K.R., Brower, C., and Chang, S. (1982) *J. Am. Chem. Soc.* **104**, 3150-3152.
51. Ewald, A.H. (1959) *Trans. Faraday Soc.* **55**, 792-797.
52. Degueil-Castaing, M., Rahm,A., and Dahan, N. (1986) *J. Org. Chem.* **51**, 1672-1676.
53. Hou, C.J., and Okamoto, Y. (1982) *J. Org. Chem.* **47**, 1977-1979.
54. Jenner, G. (1992) *High Pres. Res.* **11**, 21-32.
55. Cheung, C.K., Wedinger, R.S., and le Noble, W.J. (1989) *J. Org. Chem.* **54**, 570-573.
56. Yamada, T., Hanagi, Y., Omote, Y., Mizayawa, T., Kuwata, S., Sugima, M., and Matsumoto, K. (1990) *J. Chem. Soc., Chem. Comm.* 1640-1641.
57. Sasaki, M., Gama, Y., Yasumoto, M. and Ishigami, Y. (1990) *Chem. Lett.* 2099-2102.
58. Ibata, T., Isogami, Y., and Toyoda, J. (1991) *Bull. Chem. Soc. Jpn* **64**, 42-49.
59. Kotsuki, H., Matsumoto, K., and Nishizawa, H. (1991) *Tetrahedron Lett.* **32**, 4155-4158.
60. Benkhoff, J., Boese, R., Klärner, F.G., and Wigger, A.E. (1994) *Tetrahedron Lett.* **35**, 73-76.
61. Shimizu, T., Kobayashi, R., Ohmori, H., and Nakaya, T. (1995) *Synlett* 650-652.
62. Tietze, L.F., Hübsch, T., Oelze, J., Ott, C., Tost, W., Wörner, G., and Buback, M. (1992) *Chem. Ber.* **125**, 2249-2258.
63. Buback, M., Kuchta, G., Niklaus, A., Henrich, M., Rothert, I., and Tietze, L.F. (1996) *Liebigs Ann.* 1151-1158.
64. Jenner, G., Kellou, M., and Papadopoulos, M. (1982) *Angew. Chem., Int. Ed. Engl.* **21**, 929-933.
65. Jenner, G. (1991) *Molecular Systems under High Pressure*, R. Pucci and G. Piccitto (eds.), Elsevier, Amsterdam, pp. 361-380.
66. Franck-Neumann, M., Miesch, M., Barth, F., and G. Jenner (1989) *Bull. Soc. Chim. France* 661-666.
67. Matsumoto, K. (1985) *Chem. Lett.* 1681-1682.
68. Heiner, T., Michalski, S., Gerke, K., Kuchta, G., de Meijere, A., and Buback, M. (1995) *Synlett* 355-357.
69. Buback, M., Heiner, T., Hermans, B., Kowollik, C., Kozhushkov, S.I. and de Meijere, A. (1988) *Eur. J. Org. Chem.* 107-112.
70. Abraham, H. (1974) *Progr. Phys. Org. Chem.* **11**, 1-87.

71. Jenner, G. in preparation.
72. Jenner, G. (1995) *New J. Chem.* **19**, 173-178.
73. Jenner, G. (1996) *Tetrahedron* **43**, 13557-13568.
74. Tsuzuki, H., Uosaki, Y., Nakahara, M., Sasaki, M;, and Osugi, J. (1982) *Bull. Chem. Soc. Jpn* **55**, 1348-1351.
75. Takeshita, H., Sugiyama, S., and Hatsui, T. (1985) *Bull. Chem. Soc. Jpn* **58**, 2490-2493.
76. Sugiyama, S., and Takeshita, H. (1987) *Bull. Chem. Soc. Jpn* . **60**, 977-980.
77. Takeshita, H., Sugiyama, S., and Hatsui, T. (1986) *J. Chem. Soc., Perkin Trans. II* 1491-1493.
78. Swieton, G., Jouanne, J., Kelm, H. and Huisgen, R. (1983) *J. Org. Chem.* **48**, 1035-1040.
79. Breslow, R. (1991) *Acc. Chem. Res.* **24**, 159-164.
80. Blokzijl, W., and Engberts, J.B. (1993) *Angew. Chem.. Int. Ed. Engl.*, **32**, 1545-1579.
81. Blokzijl, W., Blandamer, M.J., and Engberts, J.B. (1991) *J. Am. Chem. Soc.* **113**, 4241-4246.
82. Le Noble, W.J., Srivastava, S., Breslow, R., and Trainor, G. (1983) *J. Am. Chem. Soc.*, **105**, 2745-2748.
83. Isaacs, N.S., Maksimovic, L., and Laila, A. (1994) *J. Chem. Soc., Perkin Trans. 2* 495-498.
84. Jenner, G., Ben Salem, R. (1998) *Rev. High Press. Sci. Technol.* **7**, 1265-1267.
85. Blake, J.F., Lim, D. and Jorgensen, W.L. (1994) *J. Org. Chem.* **59**, 803-805.
86. Cramer, C.J., and Truhlar, D.G. (1994) *J. Am. Chem. Soc.* **114**, 8794-8799.
87. Lüdemann, H.D., (1992) High Press. Biotechol., C. Balny, R. Hayashi, K. Heremans and P. Masson ((eds.), John Libbey Eurotext Lrd. **224**, pp. 371- 379.
88. Morild, E. (1981) *Adv. Protein Chem.* **34**, 93-166.
89. Makimoto, S., Suzuki, K., and Taniguchi, Y. (1984) *J. Phys. Chem.* **88**, 6021-6024.
90. Schneider, H.G., and Sangwan, N.K. (1986) *J. Chem. Soc., Chem. Comm.* 1787-1789.
Dunams, T., Hoekstra, W., Pentaleri, M., and Liotta, D. (1988) *Tetrahedron Lett.* **29**, 3745-3748.
Hill, J.S., and Isaacs, N.S. (1990) *J. Phys. Org. Chem.* **3**, 285-288.
91. Balny, C., Masson, P., and Travers, F. (1989) *High Press. Res.* **2**, 1-28.

ACTIVATION OF ORGANIC REACTIONS. HIGH PRESSURE VS NEW EMERGING ACTIVATION MODES

G. Jenner

Laboratoire de Piézochimie Organique (UMR CNRS 7509), Université Louis Pasteur, Strasbourg, France

1. Introduction

Six years ago, during a preceeding NATO ASI devoted to *High Pressure Chemistry, Biochemistry and Material Science*, we reported on "*The Future of High Pressure Organic Chemistry*""[1]. The accent was focused on the specific advantage of using high pressure activation vs traditional activation methods apparently competing with pressure in organic synthesis. From 1992 to date, the armory of activation methods has been enriched in such way that the access to sophisticated organic molecules is now possible.

The optimization of a chemical reaction due to specific activation is not only related to the yield, but also, and perhaps more importantly, to the selectivity. In the last decades, new activation modes have been applied to organic reactions.

The aim of this chapter is at comparing pressure (N.B. Pressure will be expressed throughout either in megapascal (MPa, 1 bar = 0.1 MPa) or gigapascal (GPa, 10000 bars = 1 GPa) with new activation modes able to stimulate the reactivity in organic reactions. From a general point of view, it presents the most recent developments in this area, from 1992, albeit some reminiscence of previous references will be given for illustration when necessary.

2. Activation modes

The rate of a bimolecular reaction depends on relative concentration of reactants and rate constant. This constant is affected by number of factors such as physical and chemical parameters. In the transition state theory, k can be related to pressure by the well known expression :

$$\partial \, \mathrm{Ln}\, k \,/\, \partial P = -\, \Delta V^{\neq} \,/\, RT \qquad \text{(Evans-Polanyi equation)}$$

$$\text{with } \Delta V^{\neq} = \partial \, \Delta G^{\neq} \,/\, \partial P$$

$\Delta G^{\neq}$ is the activation energy which in the transition state theory is a measure of the height of the potential barrier that the reaction must overcome in order to occur. The higher $\Delta G^{\neq}$, the more difficult it will be for the reaction to proceed. In order to modify $\Delta G^{\neq}$, a specific activation must be exerted on the reactional system.

To this purpose, organic synthesis benefits of the following armory of activation modes (the list is not exhaustive):

R. Winter and J. Jonas (eds.), High Pressure Molecular Science, 313–330.

a) Physical activation

° Traditional modes

- thermal activation
- photonic activation
- pressure activation

° New modes

- activation by ultrasonic waves
- activation by microwaves

b) Chemical activation

° Traditional modes

- Lewis acid catalysis

° New modes

- lanthanide catalysis
- catalysis by organic solutions of perchlorates
- catalysis by transition metals

c) Physicochemical activation (new mode)

- activation by solvophobic effects

d) Biochemical activation (new mode)

We will briefly define and review the recent activation modes.

2.1. ACTIVATION BY ULTRASOUNDS (SONOCHEMISTRY)

The propagation of ultrasonic waves through a liquid produces acoustic cavitation processes. The destruction of such cavities generate instant and local high temperatures and pressures (up to 100 MPa). For this reason, sonoactivation might be compared in first approximation to pressure activation. These effects have been exploited in organic synthesis [2], often leading to substitution of ultrasound [symbol)))] for pressure [symbol ΔP] [3] as portrayed in the following reactions (Tables 1,2).

- The aldol reaction between a silyl enol ether and benzaldehyde yields two diastereoisomeric addition products (scheme 1).

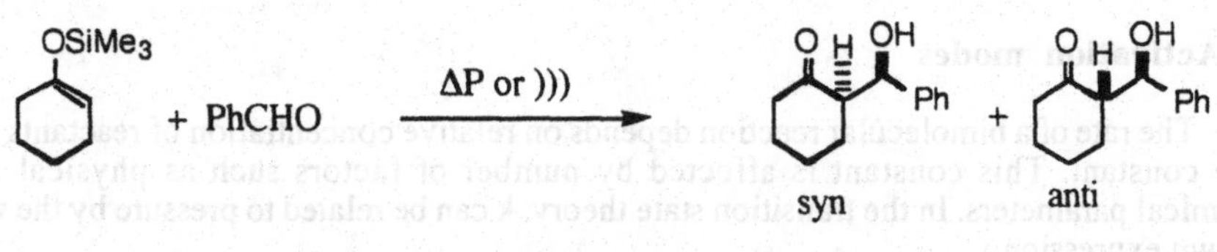

$OSiMe_3$ + PhCHO —(ΔP or)))→ syn + anti

Scheme 1

TABLE 1. Pressure and ultrasonic activation in an aldol reaction

Activation mode	Conditions	Yield	syn:anti	Ref.
ΔP	CH_2Cl_2, 1GPa, 60°C, 9d	90 %	75 : 25	[4]
)))	DME-H_2O, 0.1 MPa, 55°C, 2d[a]	68 %	73 : 23	[3]

[a] DME (1,2-dimethoxyethane)

Under ordinary conditions, the reaction is practically non existent. However, either pressure or ultrasound activates the aldol reaction leading to excellent yields (Table 1). It can be observed that these yields are nearly equivalent with an obvious advantage for the sonication method (2 days are sufficient to reach 68 % yield against 90 % obtained in 9 days under pressure). The fact that the syn : anti ratio is not modified could be an argument to support the similar nature of both activation modes.

- Another example concerns the dipolar cycloaddition of benzenesulfonyl azides to methylcyclohexenyl ethers followed by rearrangement [5] (scheme 2, Table 2). Although the experimental conditions (temperature, reaction time and medium) are not strictly identical, sonication is a valuable alternate method to promote such cycloadditions.

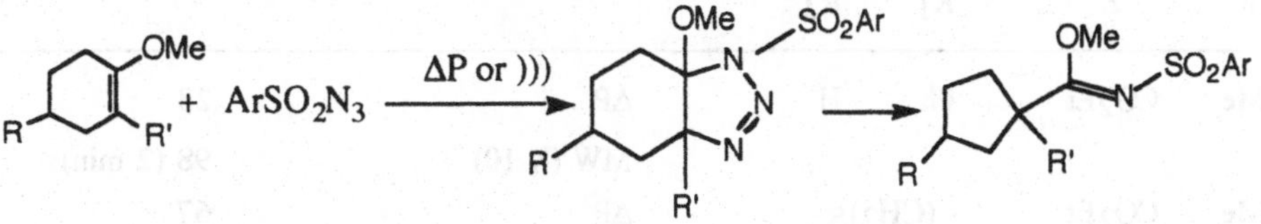

OMe; + $ArSO_2N_3$; ΔP or))); OMe; SO_2Ar; N; N; N; R; R'; OMe; SO_2Ar; N; R; R'

Scheme 2

TABLE 2. Pressure and ultrasonic activation in the addition of benzenesulfonyl azide to enol ethers

Enol ether	Activation mode	Conditions	Yield
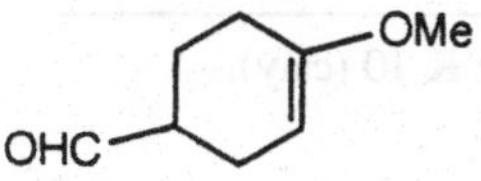 OHC–, –OMe	ΔP	MeCN, 1 GPa,25°C,18h	85 %
	)))	neat, 0.1 MPa, 35°C, 46h	78 %
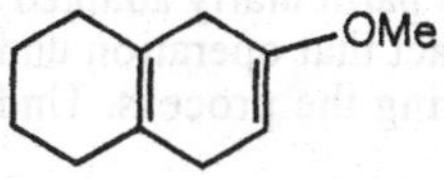 –OMe	ΔP	MeCN, 1 GPa, 25°C,18h	95 %
	)))	MeCN, 0.1 MPa,35°C,10h	83 %

2.2. ACTIVATION BY MICROWAVES

The actual nature of microwaves (symbolism MW) has an electromagnetic origin [6]. Polar molecules are subjected to dipole orientation which alternatively changes when the electric field is generated by a high frequency alternative current. As a result, the dielectric relaxation induces molecular friction leading to a rapid temperature increase depending on the dipole moment of the molecules.

The use of microwaves has been proved to be rewarding in numerous syntheses. To our knowledge there is only one case permitting comparison within some restrictions between activation by pressure [7] and microwave irradiation [8]. The synthesis of enaminoketones proceeds in the simplest way via the condensation of an amine and a 1,3-diketone or 3-ketoester (scheme 3). The reaction is facile with uncrowded primary amines. Aromatic or hindered amines react sluggishly or not at all. The formation of enamino compounds proceeds via equilibria and involves elimination of water, making microwave activation appropriate. The results are listed in Table 3.

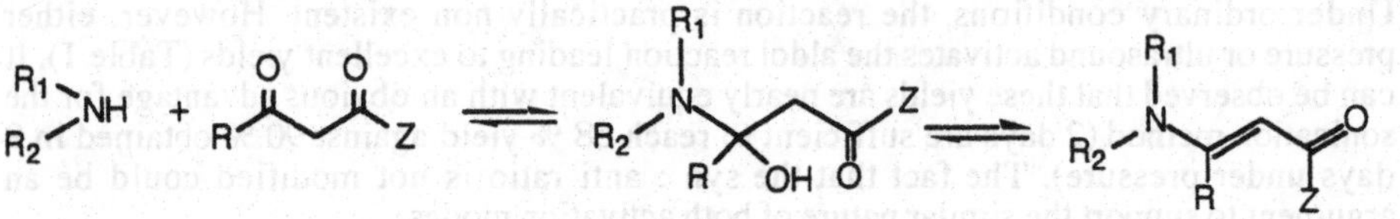

Scheme 3

TABLE 3. Synthesis of enamino compounds

Keto compound R	Z	Amine R_1	R_2	Conditions[a]	Yield (%)
Me	CO_2Et	Ph	H	ΔP	23
				MW (K 10)	98 (2 min)
Me	CO_2Et	- $(CH_2)_5$ -		ΔP	67
				MW (PTSA)	69 (2 min)
Me	CO_2Et	Me	iPr	ΔP	62
Me	CO_2Et	Me	Bu	MW (K 10)	53 (5 min)
Me	Me	- $(CH_2)_5$ -		ΔP	63
				MW (K 10)	71 (15 min)

[a] ΔP (0.3 GPa, $CHCl_3$, 20°C, 24 h), MW with supported acid catalysts: K 10 (clay), PTSA (p-toluenesulfonic acid)

Considering these results, it is clear that microwave activation is particularly adapted vs pressure which gives less efficient results probably due to the fact that operation under pressure requires closed systems retaining water produced during the process. Under MW conditions equilibria are readily shifted toward products [8].

2.3. ACTIVATION BY LANTHANIDE CATALYSIS

Lanthanide as well as scandium complexes have coordination properties which are effective for site selectivity. The preferred binding sites are those bearing carbonyl or nitrile or imine bonds. The catalytic promotion is ascribed to the oxophilicity of lanthanide cations, mainly La^{+++}, Yb^{+++} [9]. There exists several examples comparing pressure activation and lanthanide catalysis. They include hetero-Diels-Alder reactions [10] and the conjugate addition of amines to acrylic compounds [11,12] (scheme 4). The results are listed in Table 4.

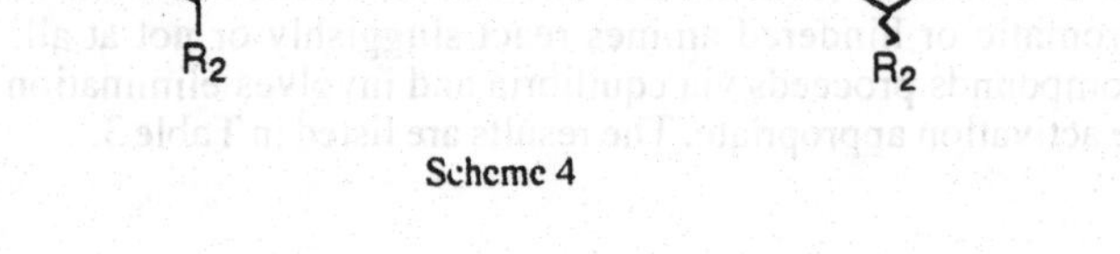

Scheme 4

(B) $R_1CH{=}C(R_2)CO_2Me$ + $RR'NH$ → $RR'N{-}CH(R_1){-}CH(R_2){-}CO_2Me$ → (- MeOH (if R' = H)) (β-lactam)

(C) $R_1CH{=}C(R_2)CN$ + $RR'NH$ → $RR'N{-}CH(R_1){-}CH(R_2){-}CN$

Scheme 4 (continued)

TABLE 4. Comparison of lanthanide catalysis with pressure activation

Entry	Type	Reaction	Conditions	Yield (%)
1	A	$R_1 = R_2 = H$	catalytic (40°C)[a]	20
			ΔP (40°C)[a]	24
2	A	R_1= Me, R_2 = H	catalytic (30°C)[a]	15
			ΔP (30°C)[a]	12
3	B	$R_1 = R_2$= H , R = R' = iPr	catalytic (50°C)[b]	17
			ΔP (50°C)[b]	13
4	B	R_1 =H R_2=Me, R=H R'=tBu	catalytic (30°C)[b]	4
			ΔP (30°C)[b]	11
5	B	R_1 =Me R_2=H, R=Me R'=iPr	catalytic (50°C)[b]	6
			ΔP(50°C)[b]	55
6	B	R_1 =Me R_2=H, R=R'=iBu	catalytic (50°C)[b]	7
			ΔP (50°C)[b]	20
7	C	R_1 =R_2=H, R = tBu R'= H	catalytic (30°C)[b]	35
			ΔP (30°C)[b]	62
8	C	R_1 =R_2=H, R = R'= iPr	catalytic (50°C)[b]	2
			ΔP (50°C)[b]	8
9	C	R_1=H R_2=Me, R+R'= $-(CH_2)_5-$	catalytic (20°C)[b]	5
			ΔP (20°C)[b]	19

[a] $CHCl_3$. 24h. Catalytic runs ($Eu(fod)_3$ 1%. 0.1 MPa), pressure runs (0.3 GPa)
[b] MeCN. 24h. Catalytic runs ($YbTf_3$ 5%. 0.1 MPa), pressure runs (0.3 GPa)

From Table 4, lanthanide catalysis is as efficient as pressure (0.3 GPa) in hetero-Diels-Alder reactions (type A). This may be ascribed to the high oxophilicity of Eu^{+++} toward carbonyl bonds in the heterodiene. In Michael-like reactions of type B, the Lewis acidity of $YbTf_3$ (ytterbium triflate) promotes the reaction, however to a lower extent than in Diels-Alder reactions of type A. When the acrylic compound is unhindered (entry 3), pressure activation is not necessary. However, simply placing a methyl group on either position of the double bond gives an advantage to pressure activation (entries 4-6). The difference in yields for both methods increases with increasing steric hindrance of the reaction centers in either the acrylate or the amine (compare entries 4,5 vs entry 3). This is discussed later (see section 3.2) and may be in relation to the general pressure effect on steric hindrance [13]. Similar comments apply to reactions of type C for which pressure is revealed as an appropriate activation parameter (compare entries 8,9 with entry 7).

2.4. ACTIVATION BY LPDE (SOLUTION OF LITHIUM PERCHLORATE IN DIETHYL ETHER)

It was shown that the accelerating effect of LPDE is due to catalysis through ionic aggregates containing the cation Li^+ complexed by ether [14]. LPDE catalysis has been found to rival pressure activation for some addition reactions listed in Table 5 [ref. 14 for entries 1-4 and ref.15 for entry 5]. Entry 5 refers to the emblematic high pressure preparation of cantharidine [16] which was the parangon of high pressure synthesis until the discovery of LPDE catalysis.

TABLE 5. Comparison between LPDE catalysis and pressure activation

Entry	Reaction	Conditions	Yield (%) LPDE cat	Yield (%) ΔP activ[a,c]
1	isoprene + 1,4-benzoquinone	20°C, 24h	69[b]	32
2	isoprene + toluquinone	20°C, 24h	89[b]	29
3	isoprene + 2,6-dimethylbenzoquinone	20°C, 24h	19[b]	8
4	1-hexene + diethylketomalonate	80°C, 24h	40[b]	14
5	furan + DHTDA[d]	20°C, 6h	70[b]	100

[a] ΔP (0.3 GPa) except in entry 6 (1.5 GPa)
[b] LPDE: 1 M (entries 1-3), 2.5 M (entry 4), 5 M (entry 5)
[c] Solvent is acetone (entries 1-3), ether (entry 5)
[d] 2,5-dihydrothiophene-3,4 dicarboxylic anhydride

However, at variance with the example shown in entry 5, there is no generality as demonstrated in the synthesis of (±)-palasonin obtained by cycloaddition of citraconic anhydride to furan followed by catalytic hydrogenation [17]. The synthesis needs high pressure assistance exclusively. In addition, LPDE may induce polymerization of sensitive unsaturated compounds [18].

2.5. ACTIVATION BY TRANSITION METAL CATALYSIS

The specificity of transition metal catalysis lies in the fact that transition metals complexed by ligands (M_xL_n) activate not only classical reactions like traditional Lewis acid catalysts, but they can also lead to products unavailable by other means by priviledging specific pathways due to the various combination possibilities of M_xL_n with

+ E≡E → 1 + 2
(E = CO_2Me)

Scheme 5

substrates. Consequently, the chemoselectivity may be completely altered when selecting pressure activation vs transition metal catalysis [19] (Table 6). The ruthenium catalyst promotes the [2+2] adduct **2** whereas pressure favours the $[\pi^2+\pi^2+\pi^2]$ cycloadduct **1**. This is particularly exemplified in the pressure catalyzed reaction (scheme 5).

TABLE 6. Chemoselectivity in the cycloaddition of dimethylacetylene dicarboxylate to norbornadiene

Conditions	ΔP (MPa)	Yield (%)	1	2
benzene, 80°C, 16 h	300	100	100 %	0 %
catalyst[a], dioxan, 80°C, 96 h	0.1	100	31 %	69 %
catalyst[a], dioxan, 80°C, 16 h	300	100	87 %	13 %

a $RuH_2CO(PPh_3)_2$

2.6. PHYSICOCHEMICAL ACTIVATION

Activation can arise from the solvent itself or added complexing molecules. Since several years numerous studies have revealed that aqueous media are capable to induce dramatic rate accelerations in some reactions [20]. The origin of the kinetic alteration consists mainly of hydrophobic packing of the reactive molecules. A reaction proceeds via the small amount of dissolved reactants in water. As a consequence of the generally limited solubility, organic synthesis in water proceeds in a heterogeneous way.

In some cases, activation through hydrophobic effects can be very efficient compared to pressure. This is illustrated by the reactions listed in Table 7 reporting three different reactions [21]: Diels-Alder additions (entries 1-3), Michael reactions (entries 4-7) and Baylis-Hillman reactions (entries 8,9).

- Entries 1-3 refer to Diels-Alder reactions all involving ketones or quinones. They are promoted in aqueous solution. The yields reached are even higher than does a 0.3 GPa pressure in organic solvents. The cause of the enhanced rate acceleration is very probably the complexing ability of carbonyl bonds to form hydrogen bonds with water contributing to the stabilization of the transition state [22].
- The Michael-like reactions involving unsaturated nitriles (entries 4-6) are considerably promoted in water. In some cases the reaction does not occur at all under pressure whereas it proceeds smoothly in aqueous solution. As the mechanism involves zwitterions, polar media such as water favour their formation with a concomitant rate

increase [12]. It should be pointed out that the modest or zero yields obtained under pressure is surprising since such process is normally very sensitive to pressure due to considerable electrostriction as highlighted by the strong negative $\Delta V^{\neq}$ - values (- 45 to - 60 $cm^3.mol^{-1}$) [23]. When acrylic esters are used in place of nitriles (entry 7), no Michael adduct at all is formed in aqueous solution at variance with the pressure assisted reaction. This contrasting behavior is easily rationalized by the fast reversibility of β-aminoesters in highly polar media whereas β-aminonitriles are quite stable in water.

TABLE 7. Comparison of the water effect vs pressure activation

Entry	Reaction	Yields (%) organic solvent[a]	Yields (%) water[b]
1	furan + methyl vinyl ketone (20°C, 16 h)	17 (ether)	27
2	isoprene + p-benzoquinone (20°C, 5h)	12 (acetone)	21
3	isoprene + 2,6-dimethylbenzoquinone (20°C,24 h)	8 (acetone)	12
4	acrylonitrile + t-butylamine (50°C, 24 h)	62 (acetonitrile)	98
5	crotononitrile + t-butylamine (50°C, 24 h)	0 (acetonitrile)	30
6	methacrylonirile + t-butylamine (50°C, 24 h)	0 (acetonitrile)	40
7	methyl crotonate + t-butylamine (50°C, 24 h)	23 (acetonitrile)	0
8	dimerization of acrylonitrile (35°C, 24 h)[c]	81 (neat)	3
9	phenyl vinyl sulfone + acetaldehyde (20°C, 24 h)[c]	100 (neat)	24

[a] Pressure is 0.3 GPa
[b] Pressure is 0.1 MPa. Concentration of reactants is 1 mmol each in 3.5 mL water. Due to their limited solubility it is clear that the final yields depend upon their concentration
[c] Catalyst is the tertiary amine DABCO (10 % molar)

- In the two Baylis-Hillman reactions examined (entries 8,9 and scheme 6) pressure proves to be an exceptional activation mode compared to reactions in aqueous solution. The Baylis-Hillman reaction is the amine-catalyzed addition of keto compounds to acrylic derivatives. Such reactions are highly pressure sensitive ($\Delta V^{\neq}$ - values range from - 60 to - 80 $cm^3.mol^{-1}$)[24]. The low yields obtained in water may be ascribed to the poor solubility of DABCO whereas the quantitative yields obtained at 0.3 GPa under neat conditions are in relation with the very large negative $\Delta V^{\neq}$ - value resulting from a considerable electrostriction volume.

CN + CN —DABCO→ CN, CN (entry 8)

SO_2Ph + RCHO —tertiary amine→ SO_2Ph, R, HO (entry 9)

Scheme 6

3. Specific pressure activation

Pressure is generally considered as a mild non-destructive activation mode. Its specific effects can be of important value for the organic synthesis.

3.1. SOLVATION EFFECT

Reactions involving solvent-solute interactions are affected by pressure independently of the normal volume change due to structural modifications when the reaction enters the transition state. Pressure magnifies the corresponding electrostatic volume change. As a result, pressure can be a valuable activation method in syntheses involving creation of charged species in the transition state. This is highlighted for example in the synthesis of hindered β-aminoesters [23]. These reactions are characterized by electrostriction volumes which may amount - 40 $cm^3.mol^{-1}$ (acetonitrile or solvents of lower polarity).

The synthesis of host compounds such as cryptands has been achieved via pressure activation (scheme 7). The remarkable efforts of Jurczak et al. in this field (see ref. 70 in our previous review [1]) have resulted recently in the synthesis of N,N'-dimethyl diazacoronands [25] and tricyclic cryptands [26]. Diaza-crown ethers can be functionalized with heterocycles via high pressure nucleophilic aromatic substitution (0.8 GPa, 100°C) [27]. The diazacrown ethers are specific binders for Ag^+.

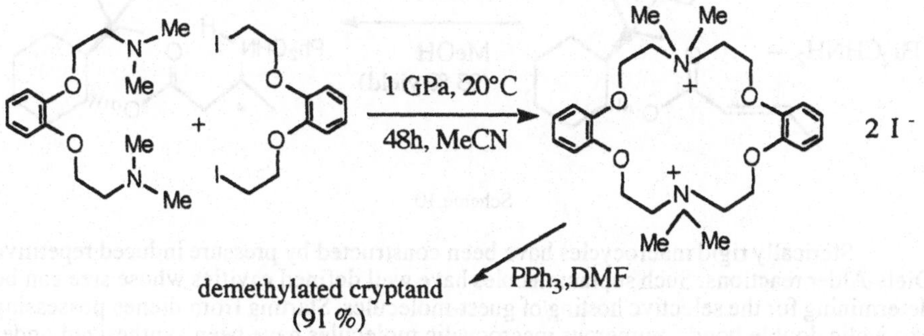

1 GPa, 20°C, 48h, MeCN; $2\ I^-$; PPh_3,DMF → demethylated cryptand (91 %)

Scheme 7

3.2. EFFECT ON STERICALLY HINDERED OR STRAINED TRANSITION STATES

It has been recognized for some time that pressure is capable to remove steric and strain inhibition in sterically hindered reactions and strained molecular systems [13,28]. Recent studies encompass Diels-Alder reactions and Michael reactions :

- cycloaddition of Danishefsky's diene with a sterically hindered dienophile angularly substituted by a trifluoromethyl group (scheme 8) [29]

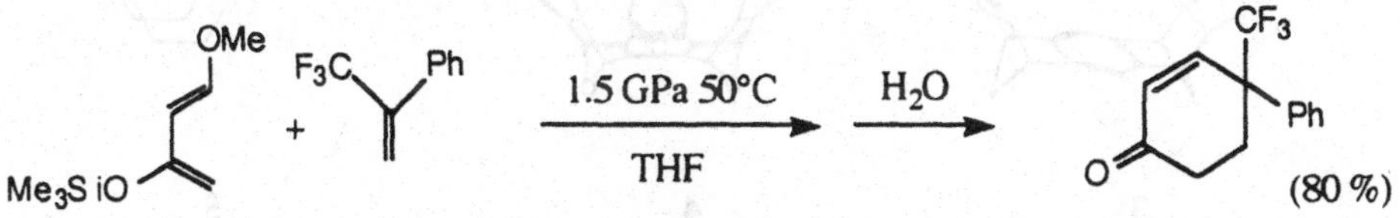

1.5 GPa 50°C, THF; H_2O (80 %)

Scheme 8

- inverse electron demand cycloaddition of enol ethers with nitroalkenes (scheme 9) [30]

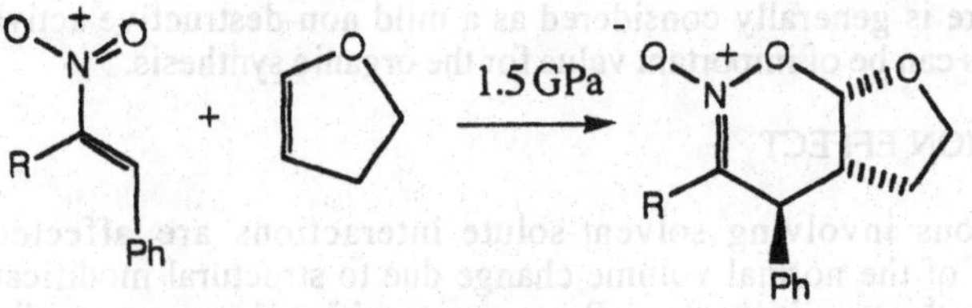

1.5 GPa

Scheme 9

- in the asymmetric conjugate Michael addition of amines to chiral crotonates (scheme 10), the alkyl moiety of the crotonate is the chiral inducer [31]. The role of high pressure is essential for efficient stereo- and enantioselectivity since the diastereoisomeric excess increases from 10 % at ambient pressure to 98 % at 1.4 GPa. The high optical efficiency is ascribed to the orientational pressure effect on the π-stacked conformation of the alkyl moiety of the crotonate.

Ph_2CHNH_2 + ... 1.4 GPa, 20°C, MeOH (93 % yield)

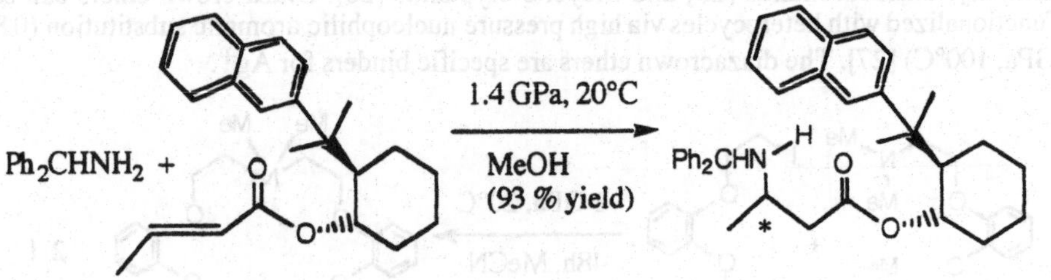

Scheme 10

Sterically rigid macrocycles have been constructed by pressure induced repetitive Diels-Alder reactions. Such supramolecules have well defined cavities whose size can be determining for the selective hosting of guest-molecules. Starting from dienes possessing exocyclic double bonds, numerous macrocyclic molecules have been synthesized under high pressure (> 1 GPa). For example, **3** [32], **4** [33], **5** [34].

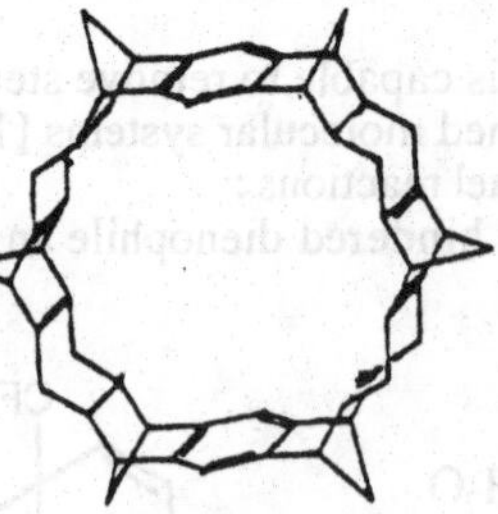

3 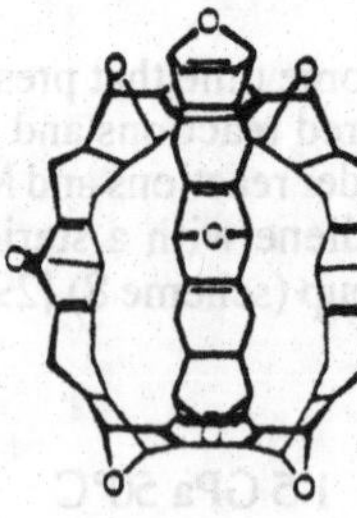 **4** 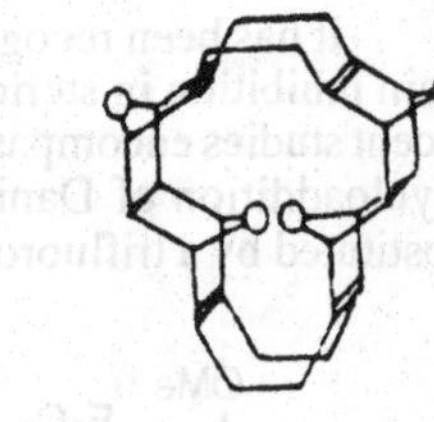 **5**

In the construction of strained polycyclic molecules retaining the methylene or the oxygen-bridged norbornane skeleton, high pressure allows use of building blocks like **6** which are unstable under thermal conditions [35].

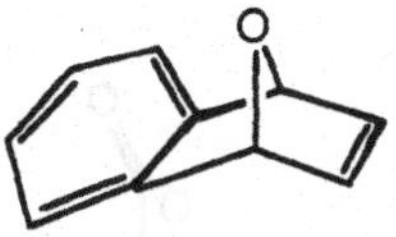

6

4. Pressure activation as an alternative for the synthesis of sensitive molecules

4.1. ALTERNATIVE FOR THE SYNTHESIS OF HEAT SENSITIVE MOLECULES

This is probably the most fruitful application field to date. At ambient pressure the rate of numerous reactions becomes observable only at elevated temperatures. Very often, in addition to chemoselectivity problems, the product stability is not secured under these conditions, either decomposing to other products or reverting to starting materials. Since our former description of the application of pressure in this area, various papers have been published. They refer to cycloadditions almost exclusively.

- A new high pressure route has been proposed to prepare β-lactam derivatives [36]. It involves the dipolar cycloaddition of nitrones and trimethylsilylacetylene to give Δ^4-isoxazolines in excellent yields which are converted subsequently to β-lactams by known procedures (scheme 11).

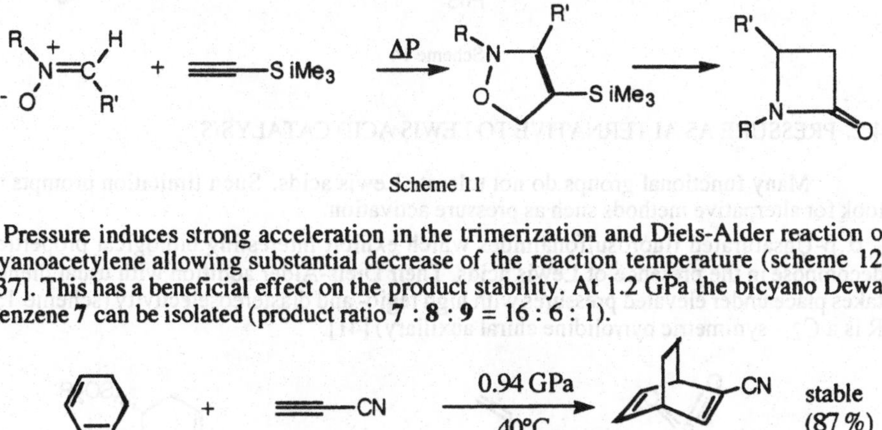

Scheme 11

- Pressure induces strong acceleration in the trimerization and Diels-Alder reaction of cyanoacetylene allowing substantial decrease of the reaction temperature (scheme 12) [37]. This has a beneficial effect on the product stability. At 1.2 GPa the bicyano Dewar benzene **7** can be isolated (product ratio **7** : **8** : **9** = 16 : 6 : 1).

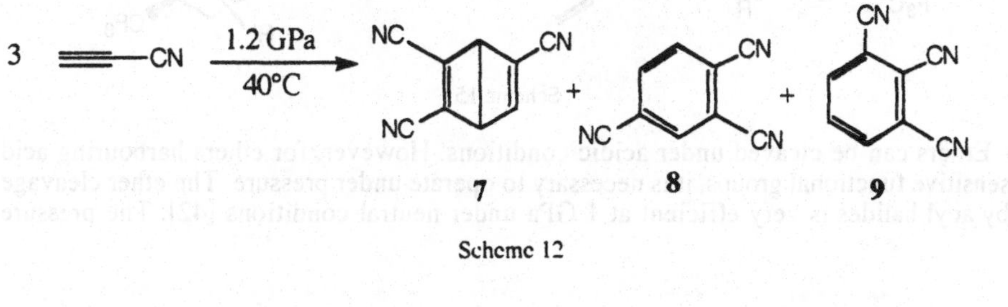

Scheme 12

- In the synthesis of (±)-sativene, the first step refers to the Diels-Alder reaction of methyl coumalate and 2,3-dimethyl-1,3-cyclohexadiene (scheme 13) [38]. In this reaction methyl coumalate (which is a pyrone derivative) reacts as a diene.

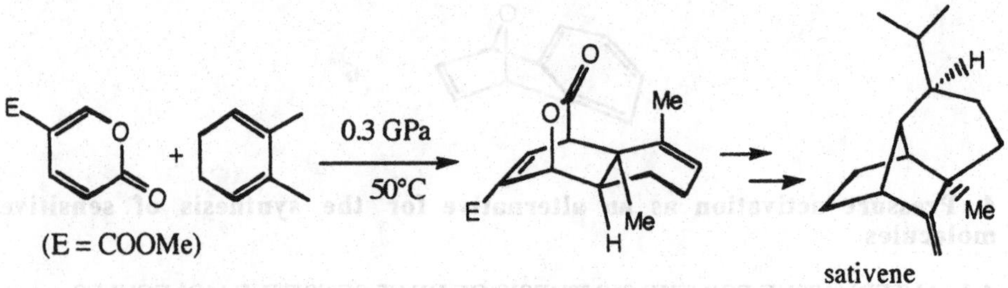

(E = COOMe) 0.3 GPa 50°C sativene

Scheme 13

- The Diels-Alder reactivity of heterodienes is almost exclusively restricted to furans. The aromaticity of pyrroles prevents them to react as dienes in [4+2] cycloadditions. With adequately N-substituted pyrroles, it is possible to obtain Diels-Alder adducts under high pressure [39]. A possible route using this activation for the synthesis of epibatidine analogues (which is a potent analgesic) has been reported (scheme 14) [40].

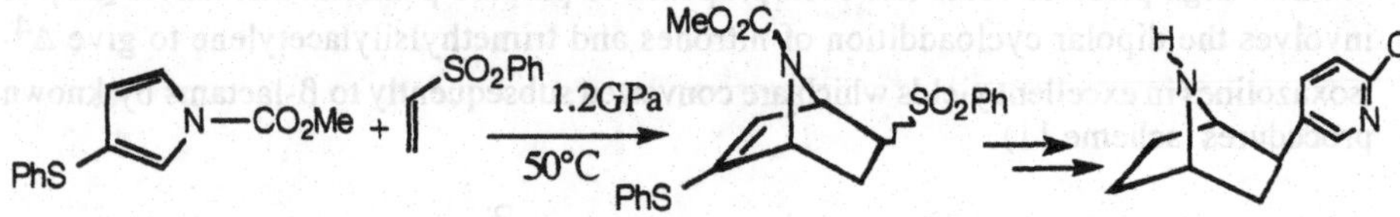

1.2GPa 50°C

Scheme 14

4.2. PRESSURE AS ALTERNATIVE TO LEWIS ACID CATALYSiS

Many functional groups do not tolerate Lewis acids. Such limitation prompts to look for alternative methods such as pressure activation.

- α,β-Unsaturated fluorosulfonamides which exhibit interesting biological properties decompose in the presence of Lewis acids. Their Diels-Alder addition with usual dienes takes place under elevated pressures with high regio- and diastereoselectivity (scheme 15, R is a C_2 - symmetric pyrrolidine chiral auxiliary) [41].

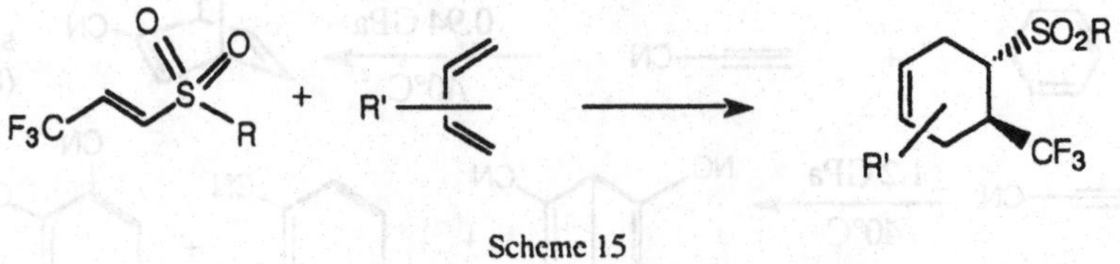

Scheme 15

- Ethers can be cleaved under acidic conditions. However, for ethers harbouring acid sensitive functional groups, it is necessary to operate under pressure. The ether cleavage by acyl halides is very efficient at 1 GPa under neutral conditions [42]. The pressure acceleration is presumed to be caused by priviledged formation of the oxonium complex followed by the nucleophilic attack of the halide ion (scheme 16).

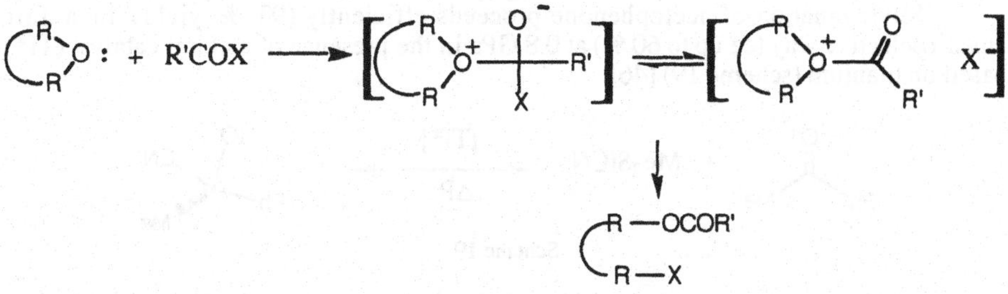

Scheme 16

5. Multiactivation based on pressure

The combination of two or more activation modes can possibly lead to convergent and, sometimes, synergistic effects. The purpose of this paragraph is to report recent high pressure syntheses using biactivation.

5.1. ACTIVATION BY PRESSURE AND TRADITIONAL LEWIS ACID CATALYSIS

Such multiactivation methodology has been applied for several years. A recent study described the quantitative effect of pressure on Lewis acid catalyzed addition reactions [43]. It was found that introduction of such catalysts in the reactional system did not modify the activation volume. This is an important result underlining the advantage of using both activation methods. Novel 4-alkoxy-β-lactams derived from aminoesters could be synthesized via $ZnCl_2$-catalyzed high pressure [2+2] cycloadditions (scheme 17) [44].

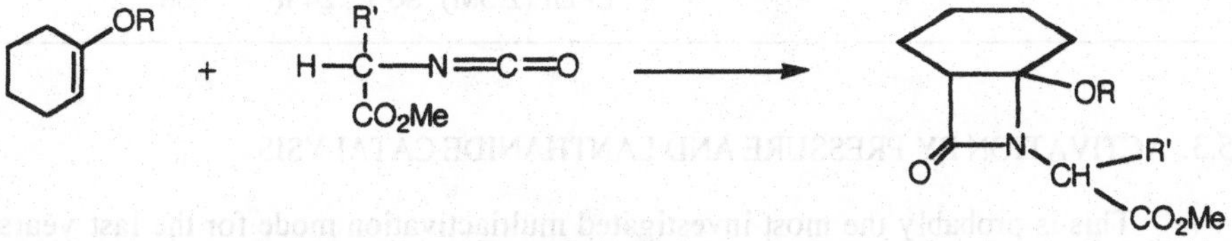

Scheme 17

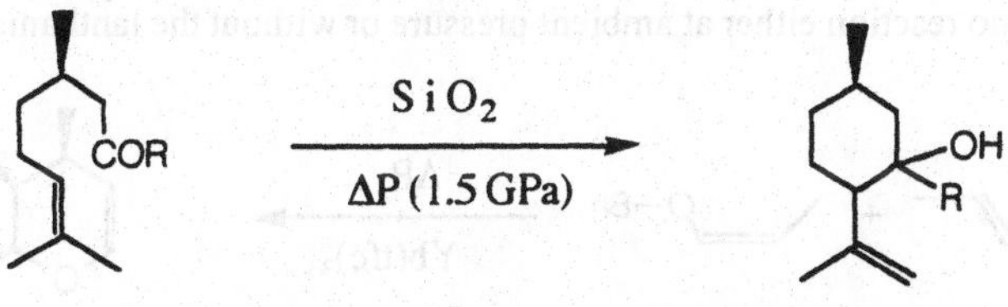

Scheme 18

The acidic properties of silicagel have been exploited in the heterogeneous high pressure ene-like cyclizations of unsaturated carbonyl compounds (scheme 18). It seems that the acidity of silicagel is enhanced at high pressure [45].

Silylcyanation of acetophenone proceeds efficiently (93 % yield) in a fair enantioselective way (ee up to 60 %) at 0.8 GPa in the presence of a chiral catalyst (Ti*) based on titanium (scheme 19) [46].

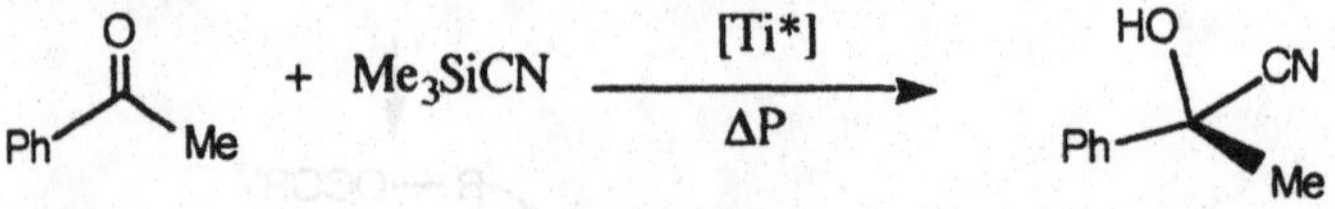

$$PhC(O)Me + Me_3SiCN \xrightarrow[\Delta P]{[Ti^*]} Ph(Me)C(OH)CN$$

Scheme 19

5.2. ACTIVATION BY PRESSURE AND LPDE CATALYSIS

There is one study to date reporting such multiactivation procedure [14] (Table 8). The results clearly demonstrate the mutual beneficial effect of LPDE and pressure in Diels-Alder as well as in ene reactions.

TABLE 8. High pressure (0.3 GPa) syntheses in LPDE solution

Reaction	Conditions	Yield (%)
isoprene + benzoquinone	acetone, 20°C, 24 h	32
	LPDE(1M), 20°C, 24 h	100
isoprene + 2,6-dimethylbenzoquinone	acetone, 20°C, 24 h	8
	LPDE(1M), 20°C, 24 h	68
furan + methyl vinyl ketone	ether, 20°C, 16 h	17
	LPDE(3M), 20°C, 16 h	83
1-hexene + diethyl ketomalonate	chloroform, 80°C, 24 h	14
	LPDE(2.5M), 80°C, 24 h	58

5.3. ACTIVATION BY PRESSURE AND LANTHANIDE CATALYSIS

This is probably the most investigated multiactivation mode for the last years. Bicyclic lactones are obtained in the cycloaddition of pyrones and vinyl ethers under the combined influence of pressure (1.2 GPa) and catalytic amounts of ytterbium derivatives [47]. There is no reaction either at ambient pressure or without the lanthanide catalyst.

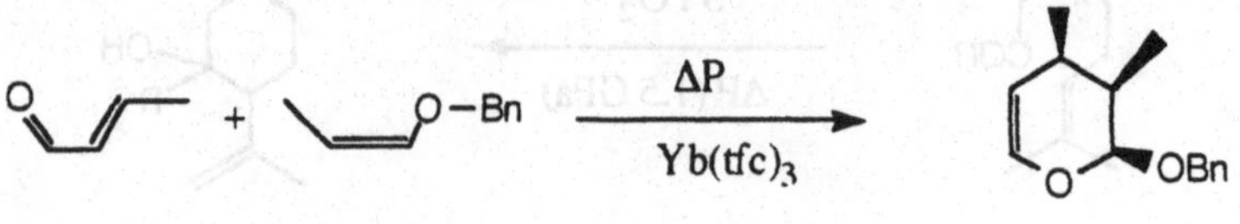

$$\text{crotonaldehyde} + \text{(Z)-propenyl O-Bn} \xrightarrow[Yb(tfc)_3]{\Delta P} \text{dihydropyran-OBn}$$

Scheme 20

The hetero-Diels-Alder reaction between crotonaldehyde and benzyl vinyl ethers occurs under 1.5 GPa with the chiral catalyst $Yb(tfc)_3$ (scheme 20) [48].

Ring opening of epoxides with amines readily takes place under pressure with ytterbium triflate [49]. Using biactivation conditions epoxides add to indoles affording tryptophol derivatives (scheme 21) [50]. A most interesting application is the enantioselective synthesis of (S,S)-diolmycin A2.

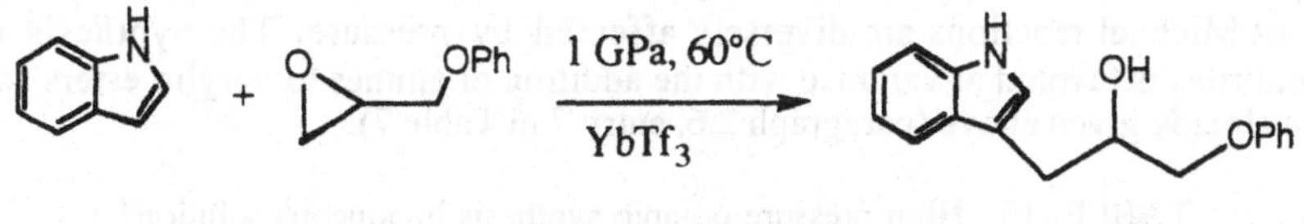

1 GPa, 60°C

$YbTf_3$

Scheme 21

Michael reactions are also excellent candidates for the application of this biactivation method. The addition of β-ketoesters to α,β-unsaturated carbonyl compounds proceeds efficiently under 0.8 GPa and $YbTf_3$ catalysis [51]. The conjugate addition of bulky amines to hindered acrylates is strongly accelerated by pressure and lanthanide catalysis [12,52]. Representative examples are given in Table 9.

TABLE 9. Catalytic high pressure synthesis of hindered β-aminoesters[a]

Methyl ester	Amine	Pressure (GPa)	Yields (%) no catalyst	Yields (%) with $YbTf_3$
methacrylate	Ph_2CHNH_2	0.3	0	0
		0.95	34	100
3,3-dimethylacrylate	iPrNH2	0.3	1	7
		0.95	45	78

[a] MeCN, 50°C, 24 h, catalyst (5% mole)

5.4. ACTIVATION BY PRESSURE AND TRANSITION METAL CATALYSIS

This multiactivation refers to specific transition metal catalysed processes promoted by pressure, such as Heck reactions and related versions. The typical Heck reaction is the Pd-catalyzed coupling of alkenes with aryl or vinyl halides. Inter- as well as intramolecular Heck reactions are known to be promoted by pressure application [53]. This is explained by the negative activation volumes refering to the formation of active zerovalent palladium species as well as the oxidative addition of these species to the aryl halide. Scheme 22 shows an example of Heck reactions occurring only under pressure.

The multiactivation method was used in the synthesis of isoquinolines and benzazepines [54] and isochromanes [55].

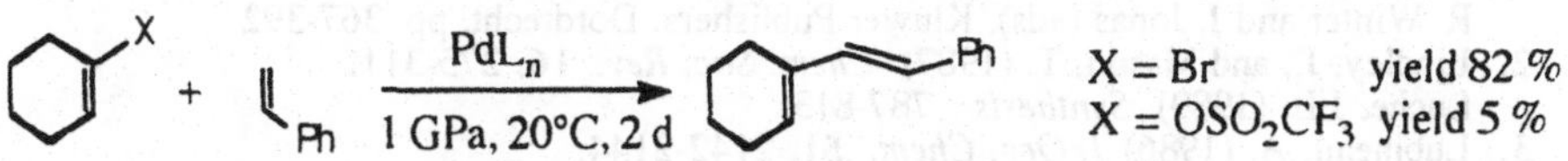

PdL_n

1 GPa, 20°C, 2 d

X = Br yield 82 %

X = OSO_2CF_3 yield 5 %

Scheme 22

5.5. ACTIVATION BY PRESSURE AND SOLVOPHOBIC EFFECTS

The effect of pressure on organic reactions in water solution has been examined only recently [21]. Some results are listed in Table 10. Pressure promotes the aqueous [4+2] cycloadditions, however to a lower extent than in organic solvents: the yields are increased by a factor of 2.5 - 4 only instead of 10 - 30 in ether or chloroform. This is in relation with less efficient hydrophobic interactions when pressure is increased [22]. The aqueous Michael reactions are diversely affected by pressure. The synthesis of β-aminonitriles is favored at variance with the addition of amines to acrylic esters for the reason already given above (paragraph 2.6, entry 7 in Table 7).

TABLE 10. High pressure organic synthesis in aqueous solution[a]

Reaction		Yields (%) 0.1 MPa	300 MPa
furan + methyl vinyl ketone	(20°C, 16 h)	27	87
isoprene + p-benzoquinone	(20°C, 5 h)	21	52
isoprene + toluquinone	(20°C, 5 h)	15	42
isoprene + 2,6-dimethylbenzoquinone	(20°C,24 h)	12	47
crotononitrile + t-butylamine	(30°C, 24 h)	6	45
methyl crotonate + t-butylamine	(50°C, 24 h)	0	0
methyl methacrylate + t-butylamine	(50°C, 24 h)	0	0

[a] See comment in footnote b (Table 10)

6. Conclusion

The diversity of activation methods has grown in recent years. Some of them may largely rival high pressure organic synthesis depending on the reaction considered. Nonetheless, the pressure parameter allows the implementation of synthetic strategies not easily conceivable with other activation modes. One further aspect of this approach is its general compatibility with acid and heat sensitive compounds as well as with sterically congested transition states. Combination of high pressure and another activation method (generally Lewis acid catalysis) represents an efficient synthetic approach.

7. References

1. Jenner, G. (1993) *High Pressure Chemistry, Biochemistry and Materials Science* in R. Winter and J. Jonas (eds), Kluwer Publishers, Dordrecht, pp. 367-392
2. Lindley, J., and Mason, T. (1987) *Chem. Soc. Rev.* **16**, 275-311.
   Luche, J.L. (1989) *Synthesis* 787-813.
3. Lubineau, A. (1986) *J. Org. Chem.* **51**, 2142-2144.

4. Yamamoto, Y., Maruyama, K., and Matsumoto, K. (1983) *J. Am. Chem.Soc.* **105**, 6963-6965.
5. Goldsmith, D., and Soria, J.J. (1991) *Tetrahedron Lett.* **32**, 2457-2460.
6. Caddick, S. (1995) *Tetrahedron* **51**, 10403-10432.
7. Jenner, G. (1996) *Tetrahedron Lett..* **37**, 3691-3694.
8. Rechsteiner, B., Texier-Boullet, F., and Hamelin, J. (1993) *Tetrahedron Lett.* **34**, 5071-5074.
9. Kobayashi, S. (1994) *Synlett* 689-701.
10. Jenner, G. (1995) *High Press. Res.* **13**, 321-326.
11. Jenner, G. (1995) *Tetrahedron Lett.* **36**, 233-236.
12. Jenner,G. (1996) *Tetrahedron* **52**, 13557-13568.
13. Jenner,G. (1985) *J. Chem. Soc., Faraday Trans. I* **81**, 2437-2460.
14. Jenner, G., and Ben Salem, R. (1997) *Tetrahedron* **53**, 4637-4648.
15. Grieco, P., Nunes, J.J., and Gaul, M.D., (1990) *J. Am. Chem. Soc.* **112**, 4595-4596
16. Dauben, W.G., Kessel, C.R., and Takemura, K.H. (1980) *J. Am. Chem. Soc.* **102**, 6893-6894.
17. Dauben, W.G., Lam, J.Y., and Guo, Z.R. (1996) *J. Org. Chem.* **61**, 4816-4819.
18. Jenner, G. (1993) *High Press. Res.* **11**, 257-262.
19. Papadopoulos, M., Ben Salem, R., and Jenner, G. (1990) *High Press. Res.* **5**, 644-646.
20. Lubineau, A., Augé, J. and Queneau, Y. (1994) *Synthesis* 741-760.
21. Ben Salem, R., and Jenner, G. (1998) *Rev. High Press. Sci. Technol.* **7**, 1268-1270.
22. Jenner, G., and Ben Salem, R. (1998) *Rev. High Press. Sci. Technol.* **7**, 1265-1267.
23. Jenner, G. (1995) *New J. Chem.* **19**, 173-178.
24. Hill, J.S., and Isaacs, N.S. (1986) *Tetrahedron Lett.* **27**, 5007-5010.
25. Jurczak, J., Ostaszewski, R., Salanski, P., and T. Stankiewicz, (1993) *Tetrahedron* **49**, 1471-1477.
26. Ostaszewski, R., Jurczak, J. (1997) *Tetrahedron* **53**, 7967-7974.
27. Tsukube, H., Minatogawa, H., Munakata, M., Toda, M. and Matsumoto, K. (1992) *J. Org. Chem.* **57**, 542-547.
28. Jenner, G. (1992) *High Press. Res.* **11**, 107-118.
29. Bégué, J.P., Bonnet-Delpon, D., Lequeux, T., d'Angelo, J., and Guingant, A. (1992) *Synlett* 146-148.
30. Uittenbogaard, R.M., Seerden, J.P., and Scheeren, H.W. (1997) *Tetrahedron* **53**, 11929-11936.
31. Dumas, F., Mezrhab, B., d'Angelo, J., Riche, C., and Chiaroni, A. (1996) *J. Org. Chem.* **61**, 2293-2304.
32. Benkhoff, J., Boese, R., Klaerner, F.G., and Wigger, A.E. (1994) *Tetrahedron Lett.* **35**, 73-76.
33. Ashton, P.R., Girreser, U., Giuffrida, D., Kohnke, F.H., Mathias, J.P., Raymo, F.M., Slawin, A.M., Stoddart, J.F., and Williams, D.J. (1993) *J. Am. Chem. Soc.* **115**, 5422-5429.
34. Pollmann, M., and Müllen, K. (1994) *J. Am. Chem. Soc.* **116**, 2318-2323.
35. Warrener, R.N., Margetic, D., Tietink, E.R., and Russell, R.A. (1997) *Synlett* 196-198.
36. Kennington, J.W., Li, W., and Deshong, P. (1992) *High Press. Res.* **11**, 163-166.
37. Breitkopf, V., Hopf, H., Klärner, F.G., Witulski, B., and Zimny, B. (1995) *Liebigs Ann.* 613-617.
38. Hatsui, T. Hashiguchi, T., and Takeshita, H. (1994) *Chem. Lett.* 1415-1416.
39. Keijsers, J., Hams, B., Kruse, C.G., and H.W. Scheeren, H.W. (1989) *Heterocycles* **29**, 79-86.
40. Aben, R.W., Keijsers, J., Hams, B., Kruse, C.G., and H.W. Scheeren, (1994) *Tetrahedron Lett.* **35**, 1299-1300.
41. Tsuge, H., Nogori, T., Okano, T., Eguchi, S., and Kimoto, H. (1996) *Synlett* 542-547.
42. Kotsuki, H., Ichikawa, Y., Iwasaki, M., Kataoka, and Nishizawa, M.I., (1992) *High Press. Res.* **11**, 107-118.
43. Jenner, G. (1997) *New J. Chem.* **21**, 1085-1090.
44. Aben, R.W., Limburg, E.P., and Scheeren, H.W. (1992) *High Press. Res.* **11**, 167-170.
45. Dauben, W.G., Hendricks, R.T. (1992)*Tetrahedron Lett.* **33**, 603-606.

46. Choi, M.C., Chen, S.S., and Matsumoto, K. (1997) *Tetrahedron Lett.* **38**, 6669-6672.
47. Posner, G.H., and Ishihara, Y. (1994) *Tetrahedron Lett.* **35**, 7545-7548.
48. Vandenput, D.A., and Scheeren, H.W. (1995) *Tetrahedron* **51**, 8383-8388.
49. Meguro, M., Asao, N., Yamamoto, Y. (1994) *J. Chem. Soc., Perkin Trans. 1* 2597-2601.
50. Kotsuki, H., Teraguchi, M., Shimomoto, N., and Ochi, M. (1996) *Tetrahedron Lett.* **37**, 3727-3730.
51. Kotsuki, H., Arimura, K. (1997) *Tetrahedron Lett.* **38**, 7583-7586.
52. Jenner, G. (1995) *Tetrahedron Lett.* **36**, 233-236.
53. Voigt, K., Schick, U., Meyer, F.E., and de Meijere, A. (1994) *Synlett* 189-190.
54. Tietze, L.F., Burkhardt, O., and Henrich, M. (1997) *Liebigs Ann.* 1407-1413.
55. Tietze, L.F., Burkhardt, O., and Henrich, M. (1997) *Liebigs Ann.* 887-891.

FREE-RADICAL POLYMERIZATION UNDER HIGH PRESSURE

Dedicated to Professor Heinz Georg Wagner on the occasion of his 70th birthday

SABINE BEUERMANN and MICHAEL BUBACK
Institut für Physikalische Chemie
Georg-August-Universität Göttingen
Tammannstraße 6
D – 37077 Göttingen
Germany

1. Introduction

Ethene homo- and copolymerizations are important technical processes. At pressures up to about 3000 bar and temperatures up to 300 °C, approximately 16 million tons LDPE (**l**ow **d**ensity **p**oly**e**thylene) have been produced worldwide in 1997. The continued interest in the LDPE process is primarily due to the enormous flexibility of this reaction which is carried out under supercritical (sc) conditions. A particular advantage of free-radical polymerization in sc fluid phase relates to the potential of widely tuning polymer properties just by continuously varying polymerization conditions. Further advantages consist in the tunability of solvent properties (which allows for choosing p and T conditions such that either homogeneity, e.g., for reaction, or inhomogeneity, e.g., for the subsequent separation step, are achieved) and in heat and mass transfer processes being very efficient under sc conditions. It is occasionally overlooked that the high-pressure ethene polymerization is the archetype of an extremely useful and successful sc fluid phase process. The situation met with the high-pressure ethene polymerization is referred to as *reactive sc fluid phase*. Ethene acts as both reactant and tunable supercritical fluid medium.

The advantages of polymerization in sc fluid phase have been considered sufficiently important to induce also considerable activities in the field of free-radical polymerization in *inert* sc fluid phase. Particularly attractive are polymerizations in scCO_2 where additional interest originates from the reduction of environmental concerns that are associated with reaction in this particular fluid [1]. It should, however, be noted that the pressures suggested for polymerization in scCO_2 are by about one order of magnitude below pressures applied in the LDPE process.

Arguments for fundamental studies into high-pressure free-radical polymerization are related to the interest in elucidating the mechanism of individual reaction steps, in particular of initiation, propagation, termination, and of chain-transfer reactions.

In Section 2 of this article, experimental procedures for spectroscopic online monitoring of polymerizations and of initiator decomposition reactions under high pressure will be illustrated, as will be the advanced laser-assisted techniques that have

R. Winter and J. Jonas (eds.), High Pressure Molecular Science, 331–367.

been used in the study of high-pressure homo- and copolymerizations. In Section 3, high-pressure propagation and termination rate coefficients of homo- and copolymerizations, mostly of (meth)acrylic ester monomers, are presented and discussed. Results for high-pressure ethene polymerization and for associated copolymerizations will be specifically addressed as will be initiator decomposition reactions in solution of *n*-heptane.

2. Experimental Methods

2.1 OPTICAL HIGH PRESSURE CELL

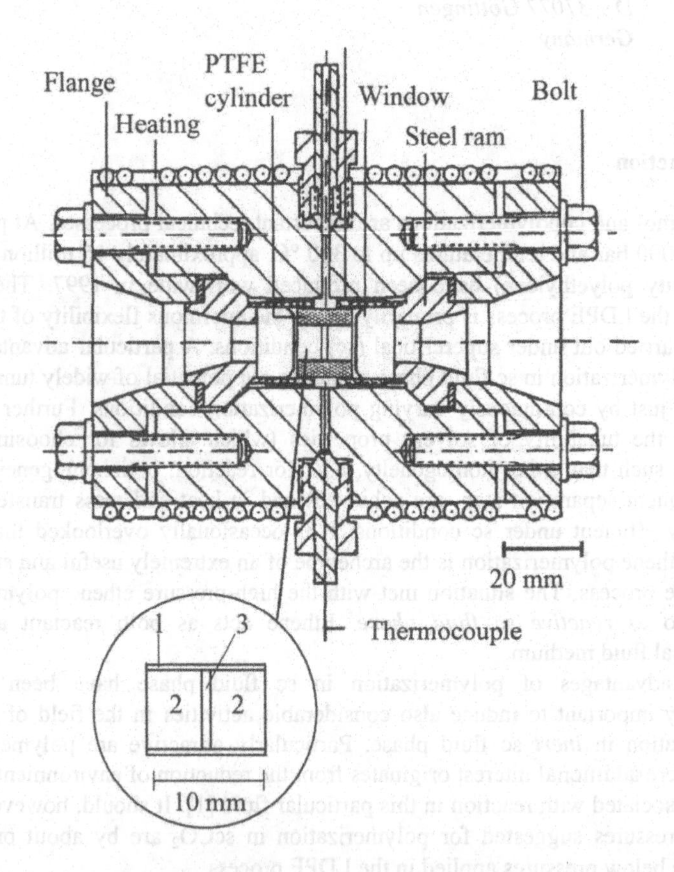

Figure 1: Optical high-pressure cell containing an internal cell: teflon® tube (1), CaF_2 windows (2), and sample volume (3).

To reduce the number of time-consuming and mostly rather difficult experiments in studying reactions under extreme conditions of p and T, it is highly desirable to apply online techniques which allow for the reliable quantitative analysis of the major reaction species during the entire course of the reaction. The application of infrared (IR) and near-infrared (NIR) spectroscopy has proven to be perfectly suited for this purpose, in particular for polymerization reactions [2]. Fourier-Transform (FT) spectroscopy allows for the fast simultaneous measurement of IR and NIR spectral ranges. The results reported in this work, with a very few exceptions, are obtained from experiments using optical high-pressure cells for operation up to 3500 bar and to temperatures of 350 °C. Cells equipped with windows made from synthetic sapphire are transparent in the spectral range from 2 000 to 50 000 cm^{-1}. In addition to being well suited for IR and NIR analysis, they allow to introduce UV laser pulses into the cell. For studies at wavenumbers below 2000 cm^{-1}, polycrystalline silicon is a suitable window material. Some reduction in maximum pressure and temperature, however, goes with the substitution of sapphire by silicon windows.

For the study of the polymerization of liquid monomers, high-pressure cells with an internal cell turned out to be very useful. The internal cell consists of a teflon® tube (with an internal diameter of 9 mm and a length of about 12 mm) which is closed at each end by a cylindrical CaF_2 window (with a diameter of 10 mm and a thickness of 5 mm). The reaction mixture is contained in the volume between the two CaF_2 windows. Figure 1 shows an autoclave with internal cell. The internal cell is positioned into the center of the high-pressure cell which enables irradiation of the entire reaction volume. In addition, by using an internal (teflon® – CaF_2) cell, metal contact of the reaction system is avoided. This is essential for carrying out reliable investigations, e.g., into decomposition kinetics of peroxides as these reactions may be influenced by the catalytic action of metal walls. A detailed description of optical high-pressure cells and procedures is given in reference [3].

2.2 KINETIC COEFFICIENTS FOR FREE-RADICAL POLYMERIZATIONS VIA PULSED LASER POLYMERIZATION AND SIZE-EXCLUSION CHROMATOGRAPHY

Among the significant number of kinetic coefficients that need to be known for a full description of free-radical polymerizations, the propagation rate coefficient, k_p, the termination rate coefficient, k_t, and the chain-transfer to monomer rate coefficient, $k_{tr,M}$, are of particular importance. Procedures of measuring these coefficients will be presented in the subsequent text. The coefficients relate to the following processes:

Initiation $$I_2 \xrightarrow{k_d} 2I\cdot \quad (1)$$

Propagation $$R_s + M \xrightarrow{k_p} R_{s+1} \quad (2)$$

Termination $$R_s + R_r \xrightarrow{k_t} P \quad (3)$$

Chain transfer to monomer $$R_s + M \xrightarrow{k_{tr,M}} P_s + R_1 \quad (4)$$

I_2 and M refer to initiator and monomer, respectively. R and P indicate radicals and polymeric molecules with degrees of polymerizations of s and r. R_1 represents a small radical of similar size as the monomer molecule. It should be noted that k_t refers to the sum of termination rate coefficients, by disproportionation and by combination.

The PLP-SEC technique, introduced by Olaj and coworkers [4, 5], combines pulsed laser polymerization (PLP) with analysis of the generated polymer by size-exclusion chromatography (SEC). The technique, which is recommended by the IUPAC Working Party "Modeling of Polymerisation Kinetics and Processes" as the method of choice for reliable k_p determination, has been extensively used to derive k_p of homopolymerizations for a wide variety of monomers, such as styrene, vinyl acetate, acrylic and methacrylic esters, up to high pressure. A compilation of "benchmark k_p values" for the two most widely studied monomers, methyl methacrylate (MMA) and styrene, is presented in two publications by this Working Party [6, 7]. An attractive feature of applying this PLP-SEC method toward measurement of k_p at high pressure is that only the laser pulsing needs to be carried out at high p whereas the SEC analysis is performed at ambient pressure. PLP-SEC may also be used to study copolymerization k_p (see chapter 3.3).

Experimental details of the PLP-SEC technique have already been reported in the literature [8]. Therefore, only a brief description will be given here: A mixture of monomer and photoinitiator is subjected to pulsing by an evenly spaced sequence of laser pulses. Due to the almost instantaneous generation of free radicals by each laser pulse, an enhanced probability of termination for radicals from the preceding pulse(s) occurs. The periodic changes in radical concentration result in a characteristic structure of the MWD. As has been shown in the literature [4, 5] propagation rate coefficients k_p may be calculated according to equation 5 from the degree of polymerization L_i at the position of points of inflection of the MWD:

$$L_i = i \cdot k_p \cdot c_M \cdot t \qquad i = 1, 2, 3, \ldots \tag{5}$$

where c_M is the monomer concentration and t the time between two subsequent laser pulses. Mostly equation 5 is applied for the first point of inflection with i = 1: $L_1 = k_p \cdot c_M \cdot t$. The position of the inflection points may be identified by calculating the first derivative of the MWD. In Figure 2, the MWD (full line) with typical PLP structure and the corresponding first derivative plot (dashed line) are given for a polymerization of glycidyl methacrylate (GMA) at 30 °C and 2000 bar. The polymerization was initiated by a laser pulse sequence with a pulse repetition rate of 15 Hz. Inflection points are derived from the maxima of the derivative plot, marked as L_1 and L_2. The occurrence of at least one higher order inflection point, L_2 at about $2 \cdot L_1$, serves as a consistency criterion which indicates suitable reaction conditions for PLP-SEC. The higher order inflection points prove that laser pulse induced termination is the main chain-stopping event. The PLP-SEC technique is restricted to low conversions. Other limitations of the technique may result from termination and transfer rates being too high [9, 10, 11].

In addition to measuring k_p, the PLP-SEC technique has been used for the determination of transfer and termination rate coefficients, too. Clay and Gilbert [12] showed that the transfer rate coefficient, k_{tr}, may be obtained from plotting the log number distribution vs. molecular weight. In contrast to the classical PLP-SEC experiment for k_p determination, PLP-SEC studies directed toward the evaluation of k_{tr}

values require chain transfer to be the dominant chain stopping event [9, 13]. This requirement may be fulfilled by applying low initiator concentrations and laser pulse repetition rates. Under these conditions, the PLP-SEC procedure provides access to coefficients for transfer to monomer [14] and to chain transfer agents [13, 15].

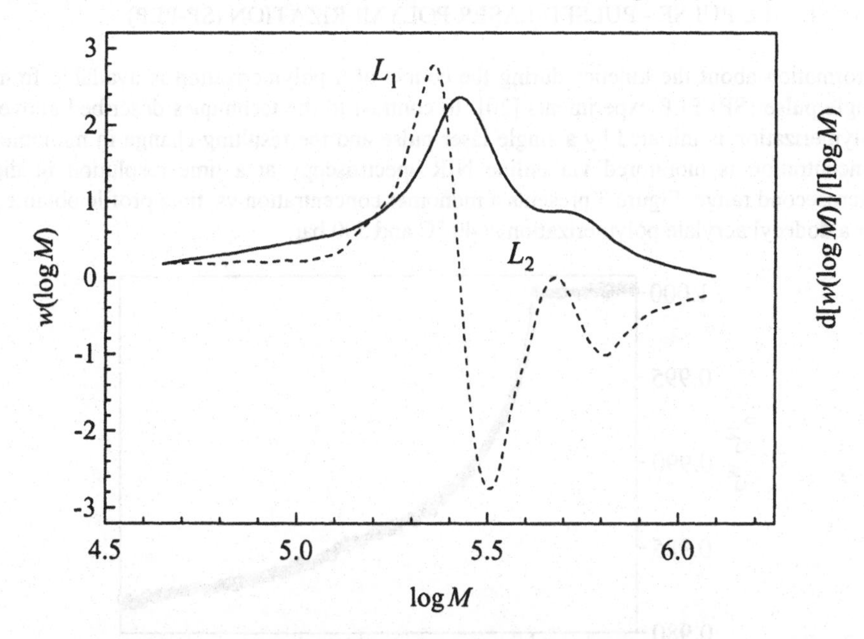

Figure 2: **Molecular weight distribution (full line) and corresponding first derivative curve (dashed line) obtained for polymer from a glycidyl methacrylate homopolymerization at 30 °C and 2000 bar at a pulse repetition rate of 15 Hz, a laser pulse energy of 7 mJ and an initiator (2,2-dimethoxy-2-phenylacetophenone) concentration of $4 \cdot 10^{-3}$ mol·L^{-1}.**

Buback and Lämmel [16] suggested an experimental procedure for the simultaneous measurement of k_p and k_{tr} from pulse-laser-induced polymerization. A narrow pulse sequence with laser pulse repetition rate being suitable for k_p evaluation is followed by an extended dark time (of up to several seconds) before a subsequent narrow pulse sequence is applied. During the narrow pulse sequence, the main chain stopping event is pulse-induced termination resulting in a MWD with a typical PLP structure. During the dark time interval chain transfer is the dominating termination event and high-molecular weight material is produced. The resulting MWD is strongly bimodal. This so-called "railroad" experiment appears to be especially interesting for high-pressure polymerization studies as the number of kinetic measurements may be reduced.

Using PLP-SEC data also for k_t determination has been advocated by Lämmel [17, 18]. The procedure uses kinetic modeling via the PREDICI program [19], to simulate both conversion and MWD for a PLP experiment. Only part of the MWD trace, the region where the distinct PLP structure occurs, is used for the k_t analysis. Actually, the ratio of areas, RA, under the number MWD curve, the area between the second and third point of inflection divided by the area between the first and second point of inflection, is considered. k_t is obtained from a fitting procedure using either the PREDICI program or a lumping scheme. The applicability of the procedure has been demonstrated for the butyl methacrylate polymerization at high pressure [18].

2.3 SINGLE PULSE - PULSED LASER POLYMERIZATION (SP-PLP)

Information about the kinetics during the course of a polymerization is available from single pulse (SP)-PLP experiments [20]. In contrast to the techniques described above, polymerization is initiated by a single laser pulse and the resulting change in monomer concentration is monitored via online NIR spectroscopy at a time resolution in the microsecond range. Figure 3 presents a monomer concentration vs. time profile obtained for a dodecyl acrylate polymerization at 40 °C and 500 bar.

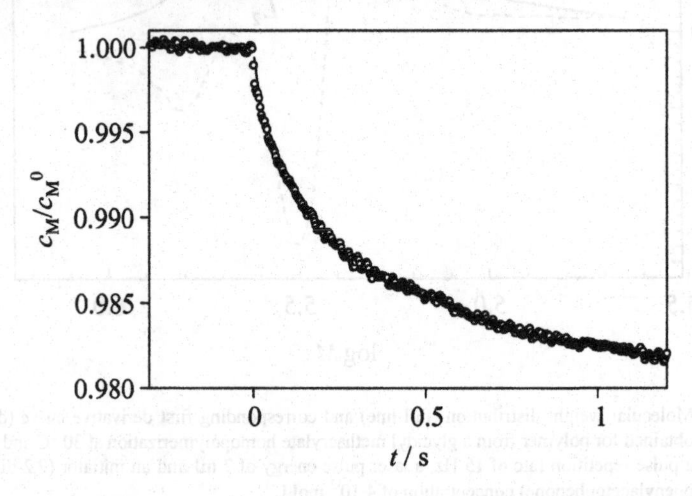

Figure 3: Variation of the relative monomer concentration with time (after application of a laser pulse) in a dodecyl acrylate polymerization at 40 °C, 500 bar, and 40 per cent initial polymer content. c_M^0: monomer concentration before firing the laser pulse at $t = 0$.

Fitting of equation 6 to the resulting conversion vs. time trace yields the coupled parameters k_t/k_p and $k_t \cdot c_R^0$.

$$\frac{c_M(t)}{c_M^0} = \left(2 \cdot k_t \cdot c_R^0 \cdot t + 1\right)^{\frac{-k_p}{2k_t}} \tag{6}$$

where c_R^0 denotes the free-radical concentration that is instantaneously generated by the single laser pulse and c_M^0 is the monomer concentration prior to irradiation with this laser pulse. Details about deriving c_R^0 are given elsewhere [21]. With k_p being known from PLP-SEC, an individual k_t value is directly available from k_t/k_p. This k_t value is a chain- length averaged quantity which refers to the narrow conversion range covered by an individual single pulse experiment. The SP-PLP method may be carried out up to high degrees of conversion, thus providing k_t as a function of monomer conversion. A detailed description of the experimental procedure is given elsewhere [22].

The SP-PLP technique is particularly well suited for high k_p monomers such as the acrylates. However, k_t values have recently also been determined for the slowly propagating methyl methacrylate monomer [22] (see also chapters 3.2 and 3.3). For monomers where k_p is very low, e.g., for styrene, an alternative laser-assisted technique (PS-PLP) has been applied which uses sequences of laser pulses differing in pulse repetition rate [21, 23].

2.4 MEASUREMENT OF INITIATOR DECOMPOSITION KINETICS

With IR spectroscopy being used to measure the decay of peroxide concentration (or the increase in the absorption of one or several products), discontinuous and continuous techniques have been applied to measure peroxide decomposition rates. The different types of experimental setups and procedures are described in some detail elsewhere [3], as are the optical high-pressure cells used in these studies. The design of these optical cells is close to or even identical to the device shown in Figure 1.

Two types of discontinuous procedures are used: At lower reaction temperatures, e.g., up to 120 °C, the peroxide solution is contained in an internal cell (see Figure 1), which is filled and assembled in a glove box under an Ar atmosphere. The internal cell allows for isolating the reacting system from steel parts of the autoclave which may induce catalytic activity. The internal cell is positioned into the optical high-pressure cell and pressurized with *n*-heptane acting as the pressure-transmitting medium. The assembly is heated to the reaction temperature and the collection of IR spectra is started.

At higher reaction temperature, decomposition rate may become too fast and a major fraction of the peroxide is decomposed before reaction conditions of constant T and p are reached. This problem may be circumvented by applying a procedure where the peroxide solution is directly fed into an autoclave manufactured from catalytically less active material (which however has the disadvantage of some reduction in maximum attainable pressure). The empty autoclave is positioned into the sample chamber of the IR spectrometer and is heated to the pre-selected decomposition temperature. As soon as this temperature is reached, the valve between the pre-pressurized peroxide solution and the optical cell is opened and the solution introduced into the cell. The time period required for reaching constant T and p conditions is typically of the order of a few seconds. The collection of IR spectra is immediately started and the reaction monitored up to completion.

Kinetic analysis under conditions where decomposition is fast, e.g. reaction half-lives are well below one minute, has been carried out using a continuous tubular reactor which essentially consists of a high-pressure capillary of 10 m (or even larger) length and 0.5 mm internal diameter [24]. IR spectroscopic analysis is performed under reaction pressure, but at lower temperature, in an optical cell (as used in the discontinuous experiments) which is positioned directly behind the tubular reactor. The continuous experiment is started by passing, at ambient T and p, the freshly prepared and air-liberated peroxide solution through the reactor until the peroxide absorbance, measured in the optical high-pressure cell maintained e.g. at 50 °C, stays constant. The system is then brought to the desired reaction pressure and an IR spectrum is taken, which serves as the reference spectrum for c_0, the initial peroxide concentration. The reactor is then heated to the desired temperature. To determine the decomposition kinetics, at each set of T and p conditions 5 to 10 solution flow rates, corresponding to different residence times, were adjusted and the corresponding spectra collected.

2.5 CONTINUOUSLY OPERATED SET-UP

In addition to batch-type experiments which are used in the laser-assisted experiments to deduce propagation and termination rate coefficients, free-radical high-pressure polymerizations need to be carried out in continuously operated devices, preferably in a continuously stirred tank reactor (CSTR) under close-to-ideal mixing conditions. The main reason for performing such experiments is that the full kinetic information is contained in the polymer properties. Product characterization thus is mandatory. Sufficient quantities of polymer prepared under well defined conditions are required for these studies. They are best provided, in particular for copolymerization of monomers which are rather dissimilar in reactivity ratio, such as the ethene – acrylate systems, by reaction in a CSTR. Moreover, in order to stay sufficiently close to application and thus to investigate the various aspects of technical polymerizations, e.g., aspects related to the optimum choice of initiators and initiation strategies, fundamental research carried out in continuously operated devices is highly important.

It is beyond the scope of this article to describe continuously operated high-pressure, high-temperature devices in any detail. A setup which has been built for operation up to 3000 bar and 300 °C at ethene mass flows up to 5 kg/h [25, 26, 27] will be briefly mentioned. The assembly contains two continuously stirred tank reactors, CSTR1 and CSTR2, aligned in series. These two autoclaves [28] are virtually identical. Both are equipped with a sapphire window of 20 mm aperture which allows for visual or spectroscopic control of homogeneity of the system and/or photochemical initiation of the polymerization. The residence time behavior of the CSTRs may be spectroscopically measured during ethene polymerization under high pressure [29]. The entire set-up contains multiple facilities for generating pressure, for injecting liquids (initiators, comonomers, chain-transfer agents, cosolvents, etc.), for measuring pressure and temperature, and for IR/NIR spectroscopic monitoring of the polymerizing system. The back pressure control valves are specially designed for expansion of the reaction mixture under conditions of huge pressure gradients and of small flow rates [25].

The device with only one CSTR being in operation is sufficient for the investigation of copolymerization kinetics within extended p and T ranges. Using both CSTRs allows to study cascade (n = 2) performance in high-pressure (co)polymerizations and, as another attractive feature, provides the possibility of producing copolymer under stationary conditions in CSTR1 and of mapping out, in CSTR2, the cloud point behavior of this particular monomer/comonomer/copolymer mixture. Within these thermodynamic studies, (co)polymerization conditions in CSTR1 and thus the properties of the resulting product mixture are kept constant over many hours, whereas p and T in CSTR2 are varied and the cloud point curve is measured. The temperature in CSTR2 is always below T of CSTR1 in order to prevent further polymerization in the second autoclave. To study the cosolvent quality of the comonomer, additional comonomer may be introduced into the mixture just before entering CSTR2. The details of such cloud point measurements are described elsewhere [30].

3. Results and Discussion

3.1 PROPAGATION RATE COEFFICIENTS IN HOMOPOLYMERIZATIONS

The PLP-SEC technique has been applied toward the study of k_p for a wide variety of monomers. Literature reflects the excellent agreement of PLP-SEC data obtained by different groups. Within this chapter a brief overview will be given of recently reported k_p data, with special emphasis on high-pressure k_p. Figure 4 illustrates the pressure dependence of k_p for methacrylic acid ester polymerizations at 30 °C. k_p of linear alkyl esters increases in going from MMA to dodecyl methacrylate (DMA) [31, 32]. For the monomers with cyclic ester groups, glycidyl methacrylate (GMA) and cyclohexyl methacrylate (CHMA), higher k_p values are observed [33]. For 2-hydroxyethyl methacrylate (HEMA), containing a polar ester group, a particularly large k_p value is found [33]. Although k_p values of MMA and HEMA differ by a factor of four, the activation volumes, $\Delta V^{\neq}$, calculated according to equation 7, are identical within experimental accuracy: $\Delta V^{\neq} = -(16.7 \pm 1.1)\ \mathrm{cm^3 \cdot mol^{-1}}$ and $\Delta V^{\neq} = -(15.8 \pm 1.1)\ \mathrm{cm^3 \cdot mol^{-1}}$, respectively.

$$\left(\frac{\partial \ln k_p}{\partial p}\right)_T = -\frac{\Delta V^{\neq}(k_p)}{\mathrm{R} \cdot T} \qquad (7)$$

Table 1 lists the individual activation volumes obtained for methacrylate polymerizations at 30 °C. The entire set of activation volumes may be represented by a single value of $-(16 \pm 2)\ \mathrm{cm^3 \cdot mol^{-1}}$. In addition, Table 1 contains values for methyl (MA) and dodecyl acrylate (DA) as well as for styrene.

Table 1: Activation volumes derived from PLP-SEC experiments at 30 °C, $\Delta V^{\neq}$ for MA was determined at -15 °C and for DA at 15 °C.

methacrylates	MMA	$-(16.7 \pm 1.1)\ \mathrm{cm^3 \cdot mol^{-1}}$	[31]
	BMA	$-(16.5 \pm 1.8)\ \mathrm{cm^3 \cdot mol^{-1}}$	[32]
	DMA	$-(16.0 \pm 3.0)\ \mathrm{cm^3 \cdot mol^{-1}}$	[32]
	GMA	$-(15.0 \pm 0.7)\ \mathrm{cm^3 \cdot mol^{-1}}$	[33]
	CHMA	$-(16.2 \pm 1.1)\ \mathrm{cm^3 \cdot mol^{-1}}$	[33]
	HEMA	$-(15.8 \pm 1.1)\ \mathrm{cm^3 \cdot mol^{-1}}$	[33]
acrylates	MA	$-(11.3 \pm 0.7)\ \mathrm{cm^3 \cdot mol^{-1}}$	[34]
	DA	$-(11.7 \pm 1.8)\ \mathrm{cm^3 \cdot mol^{-1}}$	[34]
styrene		$-(12.1 \pm 1.1)\ \mathrm{cm^3 \cdot mol^{-1}}$	[35]

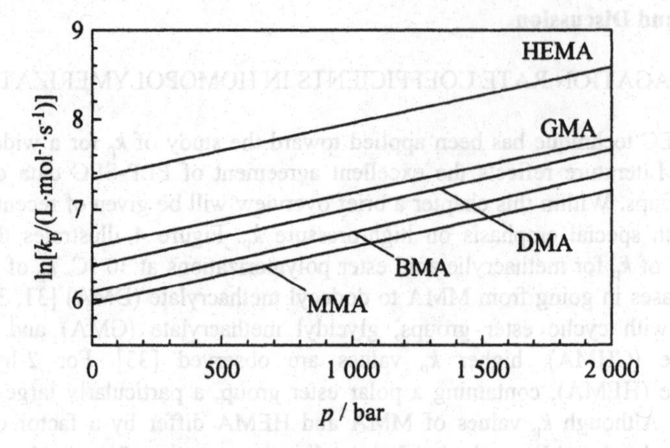

Figure 4: Pressure dependence of k_p for methacrylic acid esters at 30 °C.

An overview over k_p as a function of pressure for different monomer families is given in Figure 5. The data for styrene are from Buback and Kuchta [35] and for methyl and dodecyl acrylate from Buback et al. [34]. As with the methacrylates, k_p for the acrylates increases toward larger ester size.

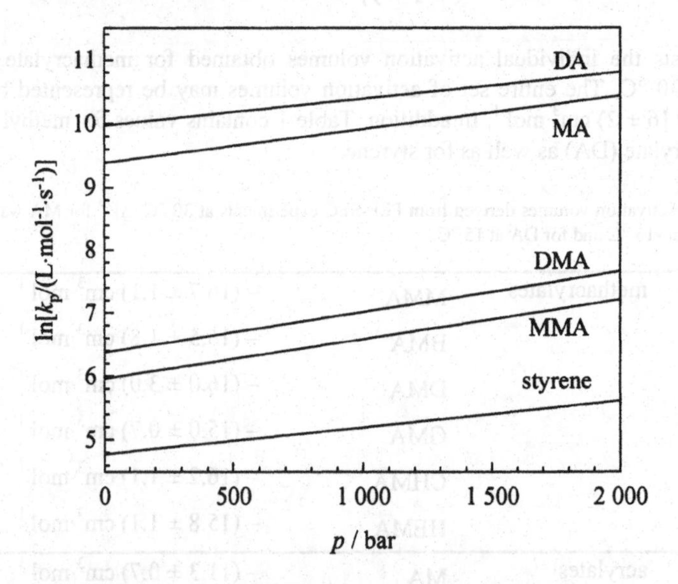

Figure 5: Pressure dependence of k_p for different monomer families at 30 °C.

The $\Delta V^{\neq}$ values for MA and DA are identical within experimental accuracy. Comparison of the different types of monomers shows that k_p increases strongly in the order styrene, methacrylates, and acrylates. k_p of dodecyl acrylate is by two orders of magnitude higher than k_p of styrene (at 30 °C and ambient pressure). In spite of the large difference in k_p, the activation volumes for styrene and the acrylates are very similar, whereas $\Delta V^{\neq}$ for the methacrylates is significantly lower. This observation indicates an influence of the different types of substitution at the double bond. Whereas styrene and the acrylates have a single substituent at the double bond, two substituents at one of the C atoms of the double bond occur with the methacrylates. This results in an enhanced steric hindrance for the transition state of methacrylate propagation going with a larger negative $\Delta V^{\neq}$.

A family-type behavior for the acrylates and for the methacrylates is also observed with respect to the temperature dependence of k_p, as is described in more detail in references [33, 36, 37]. Reliable data are available for styrene [7], vinyl acetate [11, 38], methacrylic and acrylic acid esters [9, 11, 33, 34, 36, 39]. The k_p values for vinyl acetate in absolute value are in between the data for the methacrylates and acrylates. The activation energies for the entire set of methacrylates are in the narrow range of (22 ± 2) $kJ \cdot mol^{-1}$ and for the acrylates at (17 ± 2) $kJ \cdot mol^{-1}$. In clear contrast to the situation seen with the activation volumes, $E_A(k_p)$ for styrene polymerizations (32.5 $kJ \cdot mol^{-1}$ [7]) differs significantly from $E_A(k_p)$ for acrylate polymerizations. For vinyl acetate, an activation energy of 20.6 $kJ \cdot mol^{-1}$ [11, 38] is reported, that is intermediate between the corresponding acrylate and methacrylate values.

Applications of the PLP-SEC method toward investigations into the solvent influence on k_p may be found in the original literature [40]. The solvent influence on k_p has also been measured at high pressure. Studies carried out in solution of $scCO_2$ will be presented in section 3.5.

3.2. TERMINATION RATE COEFFICIENTS IN HOMOPOLYMERIZATIONS

The termination kinetics of free-radical polymerizations have been determined for several monomers. This chapter will focus on k_t of bulk styrene high-pressure homopolymerizations. The data have been deduced from laser-induced (PS-PLP) experiments and from thermally initiated reactions with AIBN as the initiator. Details on both techniques are given in reference [23]. k_t values for ethene homopolymerizations will be given in chapter 3.5. The variation of k_t with monomer conversion will be presented for dodecyl acrylate, methyl acrylate, and for dodecyl methacrylate homopolymerizations as well as for binary copolymerizations of these monomers in chapter 3.4.

The conversion dependence of k_t for a styrene polymerization at 50 °C and 2800 bar is presented in Figure 6. The data is obtained from three independent PS-PLP experiments [23]. In the initial stage of the polymerization, up to about 20 % conversion (monomer conversion = 0.2), a constant value of k_t is observed. The initial plateau region is followed by a range of significant decrease in k_t, between 20 and 50 % monomer conversion. To even higher conversion the variation of k_t with conversion is again less pronounced. At very high conversion, e.g., above 75 % in Figure 6, kinetic parameters can no longer be determined as the reacting system becomes inhomogeneous. Significant changes of k_t with conversion, plateau - steep decrease - plateau, have also been observed for some other monomers. The characteristic k_t vs. conversion behavior is assigned to diffusion control of the termination reaction and, in particular, to different mechanisms being operative during the course of the reaction. In the initial stage of styrene polymerization, k_t is determined by the properties of the polymeric coil. k_t is assumed to be controlled by segmental diffusion [41] with contributions also from steric shielding of the free-radical chain end (see section 3.3). The overall viscosity of the system which enormously increases with monomer conversion obviously does not influence termination rate, as a constant value of k_t is observed up to about 20 % monomer conversion. At higher conversion, k_t is under translation diffusion control with the center of mass diffusion of polymeric radicals being clearly reduced toward higher bulk viscosity. At even higher conversion, above 60 %, k_t is controlled by reaction diffusion, where termination essentially proceeds via propagation steps. A detailed discussion of such conversion dependence of k_t is given in reference [42].

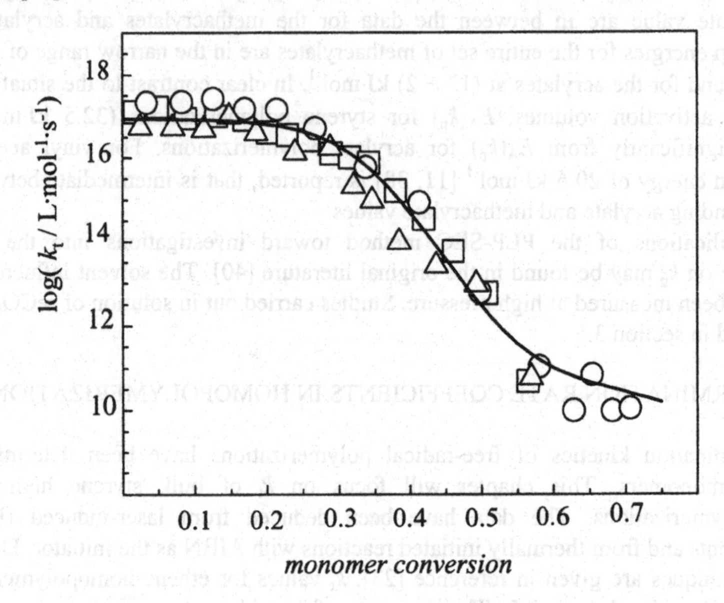

Figure 6: Conversion dependence of the termination rate coefficient k_t in pulsed-laser induced bulk polymerization of styrene at 50 °C and 2800 bar; different marker styles represent results from separate experiments. The line gives the best fit of equation 8 to the data.

Equation 8 has proven to be well suited for the representation of the conversion dependence of k_t in free-radical bulk polymerizations:

$$k_t = \left(k_{SD}^{-1} + k_{TD}^{-1}\right) + k_p^0 \; C_{RD} \cdot (1-U) \tag{8}$$

k_{SD} accounts for contributions from segmental diffusion (including steric shielding effects), k_{TD} characterizes translational diffusion, C_{RD} is the reaction diffusion constant, and $k_p{}^0$ is the propagation rate coefficient at low and moderate conversion (where propagation is not diffusion-controlled). Equation 8 and its modifications have been successfully used to model the conversion dependence of k_t for various monomers. The line in Figure 6, for example, represents a fit of equation 8 to the experimental data.

In the remainder of this chapter, the influence of p and T on termination rate will be investigated for the initial conversion range, where k_t remains constant over a fairly extended range, e.g., up to about 20 % monomer conversion in the styrene homopolymerization of Figure 6. In Figure 7, the variation of this conversion-averaged k_t with pressure is plotted for 30 °C (A) and for 90 °C (B). The lower temperature data were obtained from PS-PLP experiments and the higher temperature data from polymerizations with AIBN acting as the thermally decomposing initiator.

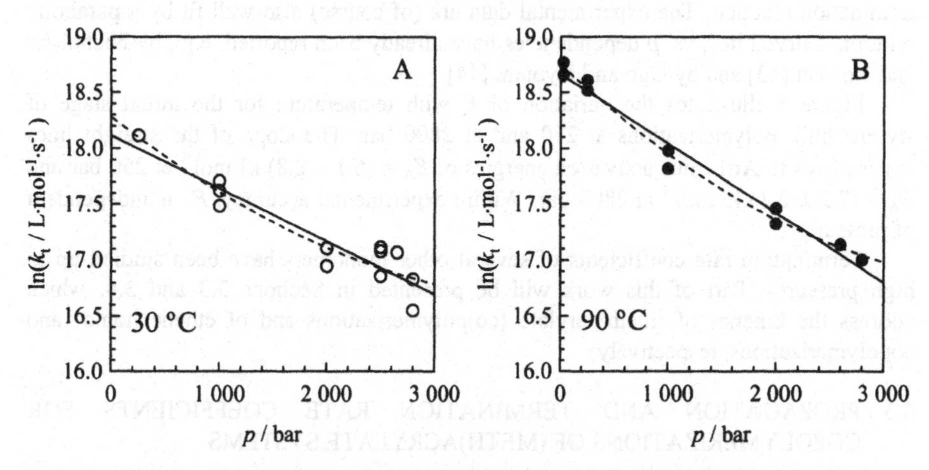

Figure 7: Variation of styrene (low-conversion) k_t with pressure at 30 °C (A) and 90 °C (B). Open markers refer to PS-PLP experiments and filled markers to chemically initiated polymerizations. The full lines are linear fits to the experimental data, the dashed curves refer to parabolic fitting.

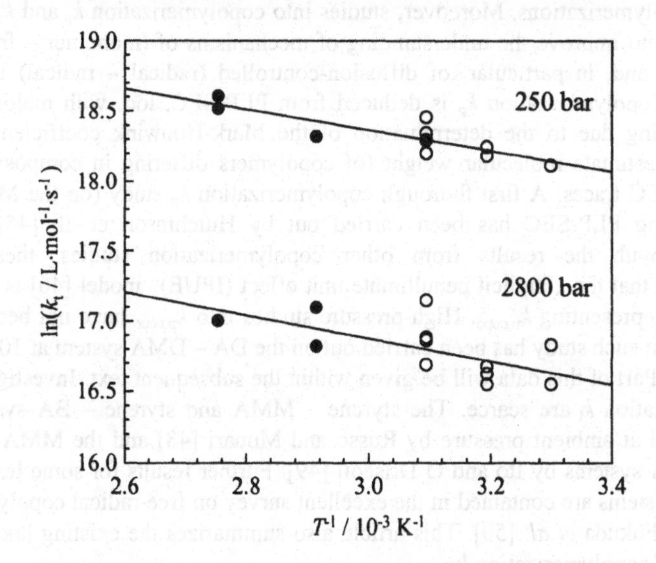

Figure 8: Variation of styrene (low conversion) k_t with temperature at 250 and at 2800 bar. Open markers refer to PS-PLP experiments and filled markers to chemically initiated polymerizations.

The full lines in Figure 7 are linear fits and the dashed lines parabolic fits of the dependence of k_t on pressure. Activation volumes of $\Delta V^{\neq}$(30 °C) = (13.6 ± 2.6) $cm^3 \cdot mol^{-1}$ and $\Delta V^{\neq}$(90 °C) = (18.9 ± 2.2) $cm^3 \cdot mol^{-1}$ are calculated from the slopes of the linear ln k_t vs. p relations. The activation volumes for 50 and 70 °C, which have also been determined, are in between the $\Delta V^{\neq}$(30 °C) and $\Delta V^{\neq}$(90 °C) values [23]. The positive numbers of activation volume are indicative of diffusion control of the termination reaction. The experimental data are (of course) also well fit by a parabolic relation. Curved ln k_t vs. p dependencies have already been reported, e.g., by Nicholson and Norrish [43] and by Ogo and Kyotani [44].

Figure 8 illustrates the variation of k_t with temperature for the initial stage of styrene bulk polymerizations at 250 and at 2800 bar. The slope of the straight lines corresponds to Arrhenius activation energies of E_A = (6.1 ± 2.8) $kJ \cdot mol^{-1}$ at 250 bar and E_A = (7.2 ± 2.1) $kJ \cdot mol^{-1}$ at 2800 bar. Within experimental accuracy, E_A is independent of pressure.

Termination rate coefficients of several other monomers have been studied up to high pressures. Part of this work will be presented in Sections 3.3 and 3.4, which address the kinetics of (meth)acrylate (co)polymerizations and of ethene homo- and copolymerizations, respectively.

3.3 PROPAGATION AND TERMINATION RATE COEFFICIENTS FOR COPOLYMERIZATIONS OF (METH)ACRYLATE SYSTEMS

Investigations into free-radical copolymerization kinetics are of considerable technical importance as systems containing more than one monomer are frequently met in industrial polymerizations. Moreover, studies into copolymerization k_p and k_t should be very helpful to improve the understanding of mechanisms of (monomer – free-radical) propagation and, in particular, of diffusion-controlled (radical – radical) termination processes. Copolymerization k_p is deduced from PLP-SEC, too, with major problems however being due to the determination of the Mark-Houwink coefficients that are required to estimate molecular weight (of copolymers differing in composition) from measured SEC traces. A first thorough copolymerization k_p study (on the MMA – BA system) using PLP-SEC has been carried out by Hutchinson et al. [45]. In good agreement with the results from other copolymerization studies, these authors demonstrate that the „implicit penultimate unit effect (IPUE)“ model [46] is capable of adequately representing $k_{p,copo}$. High-pressure studies into $k_{p,copo}$ have not been reported so far. A first such study has been carried out on the DA – DMA system at 1000 bar and 40 °C [47]. Part of this data will be given within the subsequent text. Investigations into copolymerization k_t are scarce. The styrene – MMA and styrene – BA systems have been studied at ambient pressure by Russo and Munari [48] and the MMA-BMA and MMA-DMA systems by Ito and O'Driscoll [49]. Further results for some less common monomer systems are contained in the excellent survey on free-radical copolymerization kinetics by Fukuda *et al.* [50]. This article also summarizes the existing knowledge on modeling of copolymerization k_t.

The study by Buback and Kowollik [51] into high-pressure $k_{t,copo}$ of DA – DMA and of DA – MA bulk free-radical copolymerizations is the first one that is exclusively based on data from pulsed-laser techniques. The interest in these particular systems arises from the fact that the homopolymerization k_p data of MA and DA differ only by a factor of 1.4 whereas the associated k_t values are significantly apart, by a factor of 55 (with these numbers referring to 40 °C and 1000 bar). Just the opposite is true for the DA – DMA system: The numbers for k_t are almost identical whereas the k_p values differ by a factor of 29. A further argument for selecting just these two DA-containing systems relates to the excellent signal-to-noise quality of SP-PLP experiments carried out on monomers (such as DA) which are low in k_t and high in k_p. The strategy of deriving $k_{t,copo}$ consists of combining $(k_t/k_p)_{copo}$ values from SP-PLP with $k_{p,copo}$ from separate PLP-SEC experiments.

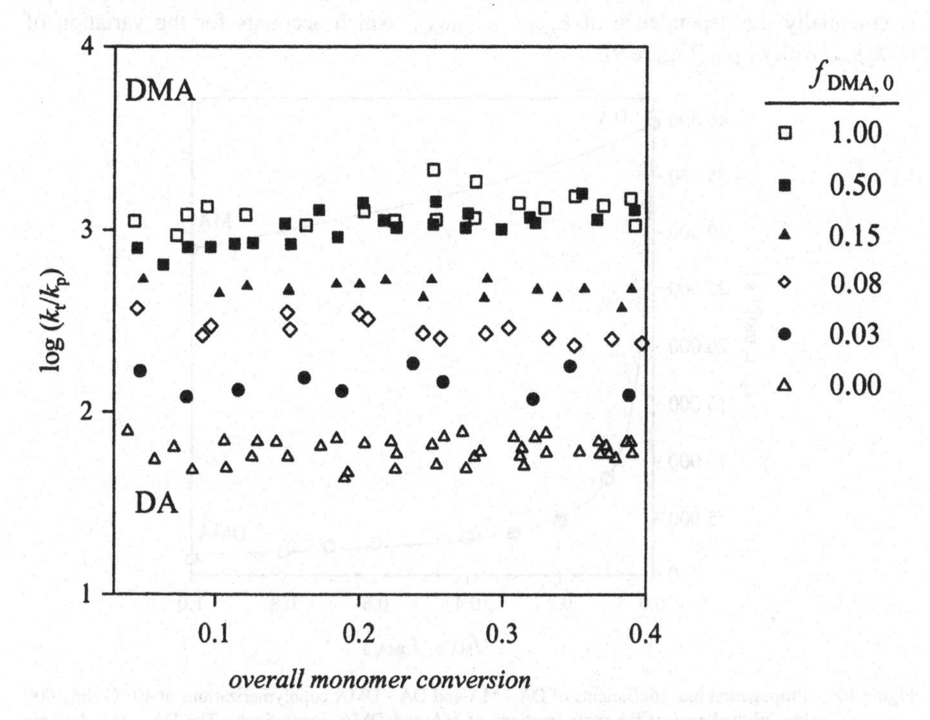

Figure 9: k_t/k_p of DA - DMA bulk copolymerizatons at 40 °C and 1000 bar plotted against overall monomer conversion; $f_{DMA,0}$: mole fraction of DMA in the (initial) comonomer mixture.

During the course of each DA – DMA and DA – MA bulk copolymerization at different initial DA monomer feed concentrations, indicated by $f_{DMA,0}$ or by $f_{MA,0}$, several SP-PLP experiments have been carried out. Figure 9 shows the measured $(k_t/k_p)_{copo}$ data of the DA – DMA system for copolymerizations at initial DMA mole fractions, $f_{DMA,0}$, of 0.03, 0.08, 0.15, and 0.50. Inspection of the measured reactivity ratios [47]: r_{DA} = 0.93 and r_{DMA} = 2.81, tells that the monomer mixture composition stays very close to the initial composition $f_{DMA,0}$ at conversions up to 0.4 (40 % overall monomer conversion). The k_t/k_p values measured for the homopolymerizations of DA [52] and DMA [53] are also contained in Figure 9. No significant variation of k_t/k_p with the degree of (overall) monomer conversion occurs for either the homo- or copolymerizations. Thus, arithmetic mean values of log k_t/k_p taken over the conversion range up to 40 % appear to adequately represent k_t/k_p (in this conversion range) over the entire range of monomer feed compositions extending from $f_{DMA,0} = 0$ to $f_{DMA,0} = 1$. k_t/k_p stays independent of the composition of the monomer mixture between $f_{DMA,0} = 1.0$ and $f_{DMA,0} = 0.5$, but strongly varies toward lower DMA contents, in particular at DA mole fractions above 0.92 ($f_{DMA,0}$ below 0.08). In Figure 10, $k_{p,copo}$ and the associated homo-propagation rate coefficients are given for the entire range between $f_{DMA,0} = 0$ and 1. The characteristic feature of the DA – DMA k_p data in Figure 10 is the pronounced decrease in k_p upon the addition of small amounts of DMA to DA and the extended region with only a slight reduction in k_p at DMA monomer mole fractions above 0.2. As will be shown below, it is essentially the dependence of $k_{p,copo}$ on $f_{DMA,0}$ which accounts for the variation of $(k_t/k_p)_{copo}$ with $f_{DMA,0}$ (Figure 9).

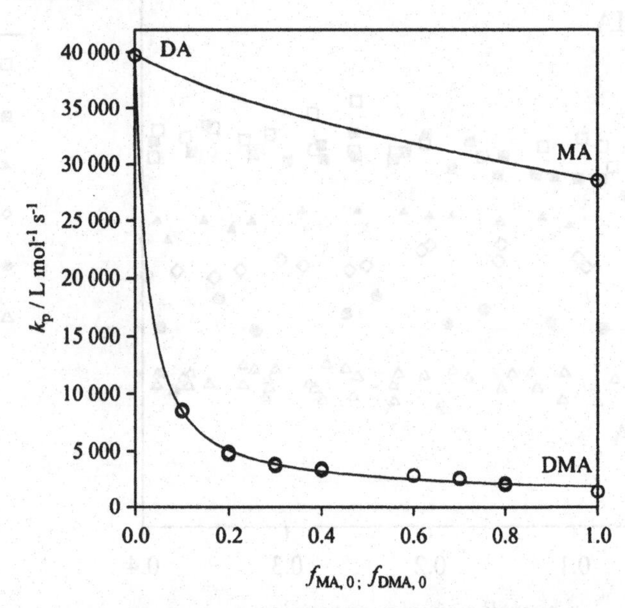

Figure 10: Propagation rate coefficients of DA - MA and DA - DMA copolymerizations at 40 °C and 1000 bar, plotted against the mole fractions of MA and DMA, respectively. The DA - MA data are obtained via the IPUE model [51]. The DA - DMA data are measured by PLP-SEC (see text).

The log (k_t/k_p) vs. (overall) monomer conversion data for DA – MA copolymerizations at initial MA contents of $f_{MA,0} = 0.25$, 0.50, and 0.80 are shown together with the corresponding homopolymerization data for DA and MA [52] in Figure 11. Contrary to what has been observed for the DA – DMA system, in MA homopolymerization (and in copolymerizations which are high in MA content), k_t/k_p clearly depends on conversion. In MA homopolymerization, a plateau region of invariant k_t/k_p extends only up to about 10 to 15 % overall monomer conversion and a pronounced decrease in k_t/k_p occurs upon further polymerization (see the styrene data in Figure 6 for comparison). The MA homopolymerization could not be followed to conversions above 40 %, as the system then becomes inhomogeneous. The clear reduction in k_t/k_p at MA conversions of 15 % can be described by equation 8. The subsequent discussion will be restricted to the initial plateau region of constant k_t/k_p for which, as with the DA – DMA system (and with styrene homopolymerization, see above), arithmetic mean values of k_t/k_p have been calculated. Using $k_{p,copo}$ data for DA – MA that have been estimated via an IPUE procedure [51] yields $k_{t,copo}$ for the DA – MA system.

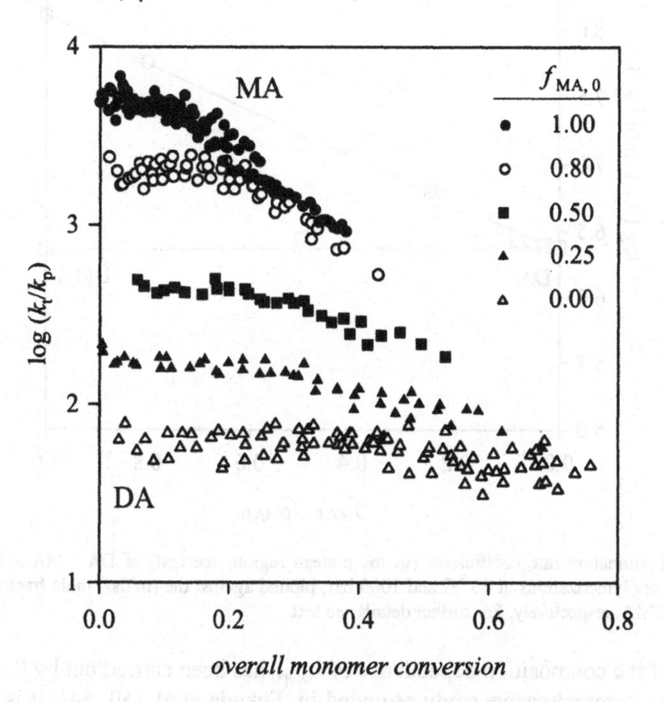

Figure 11: k_t/k_p of DA - MA bulk copolymerizatons at 40 °C and 1000 bar plotted against overall monomer conversion; $f_{MA,0}$: mole fraction of MA in the (initial) comonomer mixture.

The $k_{t,copo}$ values for 40 °C and 1000 bar of the two copolymerization systems, DA – DMA and DA – MA are plotted together with the corresponding homopolymerization values in Figure 12. The k_t values refer to the (initial) plateau region, where no variation of k_t with conversion is seen. The very remarkable observation from Figure 12 is that the data for each system closely fit to a straight line in the log $k_{t,copo}$ vs. mole fraction ($f_{MA,0}$ or $f_{DMA,0}$) representation. The $k_{t,copo}$ values for DA – DMA demonstrate that the very pronounced variation of $(k_t/k_p)_{copo}$ with $f_{DMA,0}$ at low DMA contents (Figure 9) reflects the associated variation in $k_{p,copo}$ (Figure 10). For the special situation of the two homo-termination rate coefficients, $k_{t,DA}$ and $k_{t,DMA}$, being almost identical, $k_{t,copo}$ is close to these values and independent of the monomer mixture composition. That an approximately linear log $k_{t,copo}$ vs. monomer feed relation is also found for the DA - MA system where the homopolymerization k_t data differ by almost two orders of magnitude could not be expected. The $k_{t,copo}$ behavior of the DA – MA system is very attractive with respect to predicting copolymerization k_t.

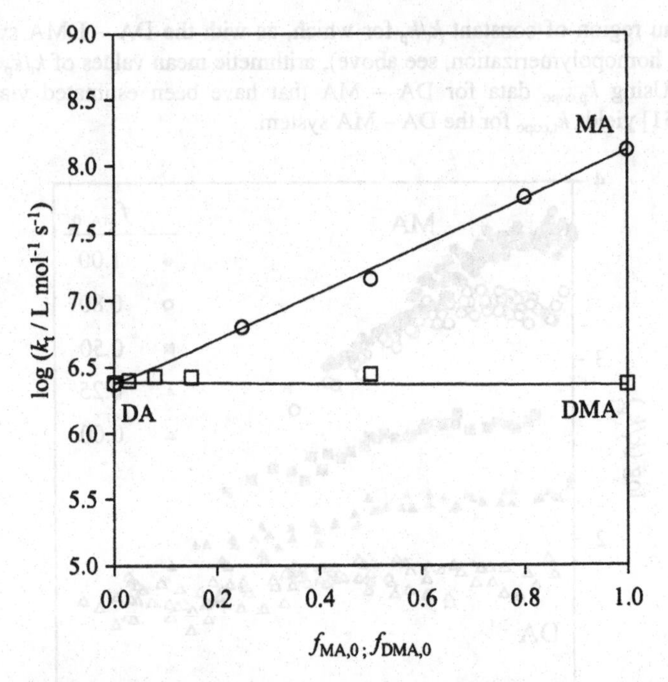

Figure 12: Termination rate coefficients (in the plateau region, see text) of DA - MA and DA - DMA copolymerizations at 40 °C and 1000 bar, plotted against the (initial) mole fraction of MA and DMA, respectively. For further details see text.

Modeling of the composition dependence of $k_{t,copo}$ has been carried out by Kowollik [51] based on the comprehensive study provided by Fukuda et al. [50, 54]. It is beyond the scope of this article to present the results in any detail. The essential message of this investigation should however be given: The penultimate unit k_t model (equation 9) is perfectly suited for representation of $k_{t,copo}$.

$$k_{t,copo}{}^{0.5} = k_{t11,11}{}^{0.5} \cdot P_{11} + k_{t21,21}{}^{0.5} \cdot P_{21} + k_{t22,22}{}^{0.5} \cdot P_{22} + k_{t12,12}{}^{0.5} \cdot P_{12} \tag{9}$$

where the coefficients $k_{t\,11,11}$, $k_{t\,21,21}$, $k_{t\,22,22}$, and $k_{t\,12,12}$ characterize the rate coefficients of termination between "like" free radicals terminating in the units 11, 21, 22, and 12, respectively. P_{ij} and P_{kk} in equation 9 indicate the relative populations of the four types of "penultimate" free radicals. It goes without saying that the four termination rate coefficients in equation 9 can not account for all kinds of "penultimate" termination processes. Actually, within a binary copolymerization, a full "penultimate k_t" treatment requires ten coefficients to be considered, as may be seen from equation 10:

$$k_{t,copo} = \sum_{l=1}^{2}\sum_{k=1}^{2}\sum_{i=1}^{2}\sum_{j=1}^{2} P_{ij} \cdot P_{kl} \cdot k_{tij,kl} \tag{10}$$

According to a suggestion first been made by Russo and Munari [48], the termination rate coefficients for reaction between "unlike" free radicals are eliminated via the so-called geometric mean approximation (equation 11):

$$k_{\mathrm{ti,kl}} = \left(k_{\mathrm{tij,ij}} \cdot k_{\mathrm{tkl,kl}}\right)^{0.5} \tag{11}$$

Among the models that have been tested to fit the DA – MA data and also $k_{t,copo}$ of a few other (meth)acrylate systems [51], the penultimate unit model (equation 9), in conjunction with applying the geometric mean approximation (equation 11) toward estimating cross-termination from rate coefficients for the termination of "like" species, is by far best suited to fit experimental $k_{t,copo}$. As has been shown by Fukuda et al. [54], equations 9 and 11 can also be used to fit curved log $k_{t,copo}$ vs. monomer feed composition dependencies, including situations with maxima or minima in between the homo-termination rate coefficients. Thus the almost linear change of $k_{t,copo}$ with monomer feed composition that is seen for the (meth)acrylate systems (Figure 12) is particularly noteworthy. Such linear correlation may be very helpful for estimating copolymerization k_t. This aspect is especially important for systems under high pressure where collecting experimental data may be very difficult. The rather simple $k_{t,copo}$ behavior of the (meth)acrylate systems is certainly associated with an identical mode of diffusion control being operative in the initial plateau region of the (meth)acrylate copolymerizations. The modeling via equations 9 and 11 may perhaps be even further simplified, as it turns out that the coefficients $k_{t12,12}$ and $k_{t21,21}$ are very close to each other in the (meth)acrylate systems [51].

Although the excellent representation of $k_{t,copo}$ via equations 9 and 11 does not necessarily mean that the underlying mechanistic concept is truly operative, it appears worthwhile to consider why a penultimate unit effect (PUE) model for k_t is so successful. The major argument that is seen in favor of a k_t PUE model assumes steric factors to be important. In particular, a shielding of the free-radical chain end is considered to play a significant role. Substituents on the terminal and on the penultimate unit will contribute to such steric effects. The shielding hypothesis would also explain the pronounced difference in homo-termination rate coefficients of methyl acrylate and dodecyl acrylate (Figure 12). That steric effects may significantly contribute to k_t is also expected from the very low values of k_t that have been reported for highly substituted monomers such as itaconates [55] and fumarates [56].

It goes without saying that at higher degrees of monomer conversion where either translational diffusion or reactive diffusion [42] may become termination rate controlling, the influence of steric arguments will be reduced or will even be lost. As has been illustrated in Section 3.2, the plateau region of initial k_t vs. monomer conversion behavior has also been assigned to diffusion control by segmental mobility. A clear separation of segmental and steric influences on k_t is certainly difficult. Thus Russo and Munari [48] suggested that the rotational motion of a very few units at the chain end determines free-radical diffusion rate via their conformational behavior to a stronger extent than do motions of larger segments. It appears, however, that the remarkable success of the PUE k_t model (with additionally using the geometric mean approximation, equation 11) to fit the copolymerization data is somewhat easier understood in terms of steric hindrance than of segmental mobility.

Further studies need to be carried out to elucidate the range of monomers and of polymerization conditions where modeling via equations 9 and 11 is successful.

3.4 ETHENE HOMO-AND COPOLYMERIZATIONS

Although ethene high-pressure polymerizations are carried out on a large industrial scale since more than 60 years, detailed kinetic investigations into this reaction, preferably being carried out in conjunction with simulation studies, are still an active field of research. The essential reason behind this continued interest is the enormous tunability of the high-pressure process. By varying reaction pressure an almost unlimited variety of polyethylenes may be produced. As has been mentioned in the Introduction, high-pressure ethene polymerization is the archetype of a supercritical fluid phase process. Copolymerization allows for a further significant enhancement in flexibility of providing widely differing types of polymeric products by appropriate design of reaction parameters. In order to handle the multitude of potential process conditions, simulation of kinetics and polymer properties is particularly important. These simulations need to be based on accurately measured kinetic data. It is this requirement which stimulates detailed experimental studies into rate coefficients of ethene high-pressure poly–merization and of copolymerizations with ethene being one of the monomers. Only a few selected results from these investigations can be briefly presented here.

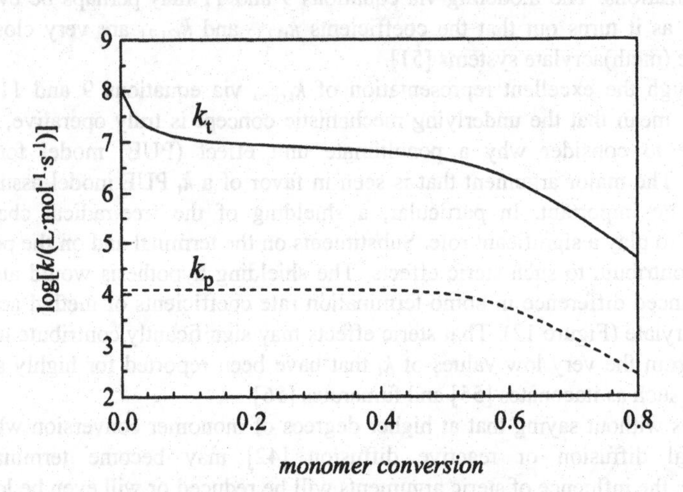

Figure 13: Conversion dependence of the termination and propagation rate coefficients from SP-PLP experiments at 230 °C and 2550 bar. The lines are calculated according to equations 12 and 13. Details are given in reference [58].

Among the laser techniques that have been described in Sections 2.2 and 2.3, the SP-PLP method has been successfully applied to study ethene homopolymerization. Schweer [57] performed an extended SP-PLP study at temperatures between 190 and 230 °C and pressures between 1950 and 2900 bar. The discussion of the primary experimental parameters is presented in reference [57, 58]. In Figure 13 is shown the measured dependence on monomer conversion of k_t and of k_p at 230 °C and 2550 bar.

An enormous change in k_t, by about three (decadic) orders of magnitude is seen. A clear decrease of k_t in the initial conversion range is followed by a rather weak decay in the 20 to 50 % conversion region. Toward still higher polymer content, k_t is again significantly reduced upon further polymerization. According to the interpretation underlying equation 8, the sequence of these characteristic three types of k_t vs. conversion behavior indicates three modes of diffusion control of k_t becoming successively operative: translation diffusion, reactive diffusion without diffusion control of k_p, and reactive diffusion with diffusion control of k_p [57, 58]. This assignment of modes of diffusion control is supported by the dependence of k_p on monomer conversion: k_p stays constant at conversion up to about 50 % and is significantly lowered toward higher conversion (Figure 13).

Schweer [57] derived the following expressions for the temperature, pressure, monomer conversion (U), and viscosity (η) dependence of termination and propagation rate coefficients of ethene homopolymerization:

$$k_t(T,p,U,\eta) = \left(0.832 \cdot \frac{1}{\eta_r} + 8.04 \cdot 10^{-6} \cdot (1-U) \cdot \frac{k_p^0}{1 + \frac{k_p^0}{1.13 \cdot 10^{10}} \cdot \eta_r} \right) \cdot k_t^0 \qquad (12)$$

$$k_p(T,p,U,\eta,) = \frac{k_p^0}{1 + \frac{k_p^0}{1.13 \cdot 10^{10}} \cdot \eta_r} \qquad (13)$$

where η_r indicates relative bulk viscosity, $\eta_r = \eta/\eta_0$, with η_0 referring to the pure monomer viscosity at identical p and T. k_t^o and k_p^o in equations 12 and 13 are the zero conversion limiting values of termination and propagation rate coefficients, respectively. These two quantities are deduced from SP-PLP experiments, too:

$$k_t^0 \,/\, \mathrm{L \cdot mol^{-1} \cdot s^{-1}} = 8.11 \cdot 10^8 \cdot \exp\left(\frac{-553 - 0.190 \cdot p/\mathrm{bar}}{T/\mathrm{K}} \right) \qquad (14)$$

$$k_p^0 \,/\, \mathrm{L \cdot mol^{-1} \cdot s^{-1}} = 1.88 \cdot 10^7 \cdot \exp\left(\frac{-4126 + 0.33 \cdot p/\mathrm{bar}}{T/\mathrm{K}} \right) \qquad (15)$$

A detailed discussion of these kinetic results including the procedure of estimating relative bulk viscosity is given in reference [57].

Whereas conversion vs. reaction time behavior is almost entirely determined by propagation, termination, and initiation processes, the size (distribution) and architecture of polymeric material essentially depends on chain-transfer processes (in which the free-radical activity of a growing chain is transferred to a monomer, a polymer, or to a solvent molecule), to back-biting processes (which are very important in ethene high-pressure polymerization), and to β-scission processes where a C – C bond in β-position to the free-radical site is broken. Rate coefficients of all these reactions should be known for the (range of) pressures and temperatures of a particular polymerization process in order to allow for the modeling of product properties [59]. These rate coefficients need to be extracted from careful analysis of molecular weight and molecular structure of polyethylenes prepared under well defined reaction conditions at quite different polymerization pressures and temperatures. In order to give just one example of such studies, kinetic data for the chain-transfer-to-monomer step in ethene polymerization will be presented. This reaction proceeds according to equation 16 with the l.h.s. of this equation being identical to the l.h.s. expression for the propagation step (compare equations 2 and 4).

$$R_s + E \xrightarrow{k_{tr,M}} P_s + R_1 \tag{16}$$

R_s and P_s refer to a free radical and a polymer molecule of degree of polymerization s, respectively; E is ethene and R_1 is a small free radical of similar size as is ethene.

According to standard practice, the transfer to monomer kinetics is expressed by $C_{tr,M} = k_{tr,M}/k_p$, the chain-transfer-to-monomer constant. This quantity has been deduced from ethene homopolymerization experiments carried out at several pressures and temperatures under conditions of very low initiation rate, by so-called thermal (or spontaneous) initiation [60, 61]. The Arrhenius expression for $C_{tr,M}$ reads:

$$\ln C_{tr,M} = 2.90 - (T/\mathrm{K})^{-1} \cdot \left(5524 + 0.257 \cdot ((P/\mathrm{bar}) - 2000)\right) \tag{17}$$

Being aware of the difficulties encountered in the detailed kinetic analysis of high-pressure ethene homopolymerization, it comes as no surprise that the kinetics of high-pressure copolymerizations (of ethene with polar comonomers) are far from being fully understood. Whereas suitable treatments of copolymerization k_p and k_t have now become available, general agreement about suitable kinetic schemes to be used for the adequate representation of chain-transfer steps has not yet been reached. The set of kinetic equations that needs to be considered in copolymerization modeling is considerably enhanced with respect to homopolymerization. Moreover, structural information on copolymer samples that is required for deducing chain-transfer rate coefficients, is not easily obtained. NMR spectroscopy appears to be best suited for measuring the amount and the type of branching in ethene – acrylate copolymers [62, 63]. It is beyond the scope of this article to touch these aspects. A brief account of copolymerization kinetics in high-pressure ethene polymerizations will be given by presenting some recent work on reactivity ratios.

Reactivity ratios r_i are defined as $r_i = k_{ii}/k_{ij}$, the ratio of homo-propagation, k_{ii}, to cross-propagation, k_{ij}, rate coefficients where k_{ij} refers to the addition of monomer j to a free-radical chain-end terminating in species i. Under ideal polymerization conditions, the mole fraction of monomer units i within the copolymer, F_i, is given by equation 18, which holds for both the terminal and the implicit penultimate model (see section 3.3) [54].

$$F_i = \frac{r_i \cdot f_i^2 + f_i \cdot (1-f_i)}{r_i \cdot f_i^2 + 2 \cdot f_i \cdot (1-f_i) + r_j \cdot (1-f_i)^2} \tag{18}$$

where f_i is the mole fraction of monomer i in the monomer mixture under reaction conditions.

Equation 18 is used to derive r_i and also $r_j = k_{jj}/k_{ji}$ from the measured compositions of the monomer mixture and of the copolymer, f_i and F_i, respectively.

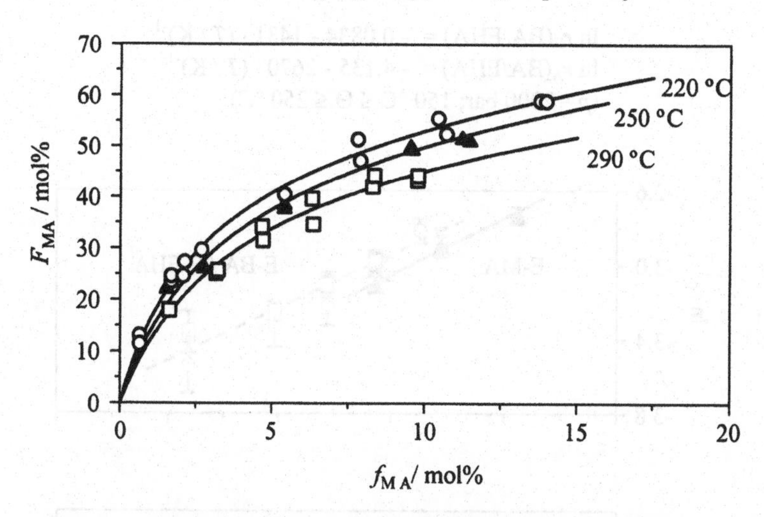

Figure 14: MA content of the copolymer, F_{MA} (in mol %), plotted against the MA content of the monomer mixture, f_{MA}, for E-MA copolymerizations at 2000 bar and temperatures of 220, 250, and 290 °C. The curves give the fits of equation 18 to the experimental data points [65].

Plotted in Figure 14 is the copolymer mole fraction, F_{MA}, vs. the monomer mole fraction, f_{MA}, for E – MA copolymerizations carried out in a continuously operated reactor at 2000 bar and temperatures of 220, 250, and 290 °C. By visual and/or spectroscopic monitoring of the copolymerizing system it is ensured that the reaction takes place in homogeneous phase. Total monomer conversion has been kept small in these reactions, mostly below 1 %. As can be seen from Figure 14, copolymerization is associated with a significant increase in MA content (in going from the monomer mixture to the copolymer). From pairs of F_{MA} and f_{MA} values measured at identical pressure and temperature, the two reactivity ratios, r_E and r_{MA}, are obtained via equation 18. Actually, these two reactivity ratios are strongly correlated which may be taken into account by calculating 95 % confidence intervals for these quantities according to the procedure suggested by van Herk [64].

Arrhenius plots of the ethene and acrylate reactivity ratios, r_E and r_A, for E – MA, E – BA, and E – EHA copolymerizations at 2000 bar are shown in Figure 15. The r_E values of the three systems are rather similar whereas r_A of E – MA is slightly, but significantly above the corresponding values for the E – BA and E – EHA systems. The fitted lines in Figure 15 are given by the following Arrhenius-type relations (referring to 2000 bar):

E – MA system [65]

$$\ln r_E(\mathrm{MA}) = -0.0202 - 1516 \cdot (T/\mathrm{K})^{-1} \quad (19)$$
$$\ln r_A(\mathrm{MA}) = -3.170 - 2406 \cdot (T/\mathrm{K})^{-1} \quad (20)$$
$$(p = 2000\ \text{bar};\ 220\ ^\circ\text{C} \le \Theta \le 290\ ^\circ\text{C})$$

E – BA and E – EHA systems [27]

$$\ln r_E(\mathrm{BA/EHA}) = -0.0834 - 1431 \cdot (T/\mathrm{K})^{-1} \quad (21)$$
$$\ln r_A(\mathrm{BA/EHA}) = -4.135 - 2670 \cdot (T/\mathrm{K})^{-1} \quad (22)$$
$$(p = 2000\ \text{bar};\ 150\ ^\circ\text{C} \le \Theta \le 250\ ^\circ\text{C})$$

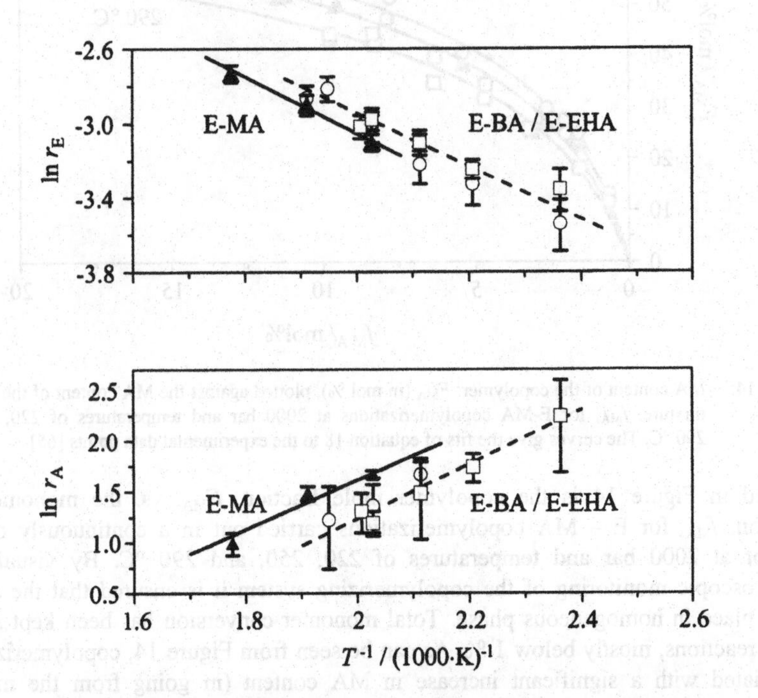

Figure 15: Variation of r_E and r_A with temperature for copolymerizations of E-MA (triangles), E-BA (squares), and E-EHA (circles) at 2000 bar. The dashed lines represent the fit to the combined data set of E-BA and E-EHA [27].

These equations allow to estimate reactivity ratios and thus to correlate monomer feed concentration and copolymer composition in these ethene – acrylate copolymerizations. Moreover, from r_i values in conjunction with the associated homo-propagation rate coefficients, k_{ii}, cross-propagation rate coefficients k_{ij} may be estimated (via the terminal model approximation: $r_i = k_{ii}/k_{ij}$.). In reference [65], cross-propagation rate coefficient k_{XE} and k_{EX} data are presented. It turns out that the activation energies, $E_A(k_{XE})$, are rather similar, which suggests the major influence of the monomeric species (ethene) on the activation energy. The pre-exponential factor, $A(k_{XE})$, on the other hand, varies largely and thus indicates control of this quantity by the terminal unit (comonomer X) of the free radical. Consistent with this latter finding is the observation that $A(k_{EX})$, the pre-exponential of cross-propagation rate coefficients of (all these) free radicals terminating in an ethylene unit, varies comparatively little upon changing the type of monomer species.

The variation of reactivity ratios with pressure is small which poses problems toward reliable measurement of activation volume data, in particular of $\Delta V^{\neq}(r_A)$. The pressure range of ethene – acrylate copolymerizations is not easily extended as pressure is limited toward high p by the equipment and toward low p by inhomogeneity of the reaction mixture. A few experiments, in which pressure was changed during the course of continuous copolymerizations under otherwise constant reaction conditions, have been carried out to estimate $\Delta V^{\neq}(r_E)$ [65]. The numbers obtained for the E – MA and E – EHA systems are: $\Delta V^{\neq}(r_E) = -(7.4 \pm 4.1)$ $cm^3 \cdot mol^{-1}$ and $\Delta V^{\neq}(r_E) = -(8.2 \pm 3.5)$ $cm^3 \cdot mol^{-1}$, respectively. A recently reported value of $\Delta V^{\neq}(r_E)$ for the ethene – butyl methacrylate system is $-(6.0 \pm 3.9)$ $cm^3 \cdot mol^{-1}$ [66]. These negative $\Delta V^{\neq}(r_E)$ values indicate that r_E, which is well below unity for each of the three systems, increases with pressure.

Also by using the terminal model, an estimate of $\Delta V^{\neq}(k_{EX})$, the activation volume associated with the cross-propagation rate coefficient k_{EX}, based on $\Delta V^{\neq}(r_E)$ and on the activation volume of the ethene homopolymerization, $\Delta V^{\neq}(k_{EE})$, can be made:

$$\Delta V^{\neq}(k_{EX}) = \Delta V^{\neq}(k_{EE}) - \Delta V^{\neq}(r_E) \qquad (23)$$

Analysis via equation 23 [65] suggests that the resulting value of $\Delta V^{\neq}(k_{EX})$ is close to the arithmetic mean of the associated homo-propagation activation volumes, $\Delta V^{\neq}(k_{EE})$ and $\Delta V^{\neq}(k_{XX})$. It is highly interesting to try to find out, whether such a simple estimate may also apply to $\Delta V^{\neq}(k_{XE})$.

3.5 FREE-RADICAL POLYMERIZATION IN SUPERCRITICAL CO_2

This chapter presents experimental results on polymerizations in homogeneous phase of supercritical (sc) CO_2. The experiments cover a wide pressure range, from 200 to 2000 bar.

Because of the poor solubility of conventional polymers such as polystyrene, polyacrylates, and polymethacrylates in $scCO_2$, knowledge of the phase behavior for the monomer/polymer/CO_2 system is essential to identify suitable reaction conditions. Investigations into the phase behavior of the butyl acrylate (BA)/poly(butyl acrylate)/ CO_2 system [67] showed that polymerizations may be carried out in a homogeneous phase up to considerable degrees of monomer conversion as long as sufficient amounts of monomer (which acts as a cosolvent) are available.

It was expected that the presence of CO_2 will influence the termination rate coefficient, whereas no variation of k_p should occur. A few PLP-SEC experiments were however performed to check whether k_p is indeed independent of CO_2 concentration and to measure the pressure and temperature dependence of k_p for BA polymerizations in $scCO_2$. Because the PLP-SEC technique is not easily applied to acrylate monomers [9, 11, 34, 39], experiments were also performed for methyl methacrylate. Figure 16 illustrates the variation of k_p with pressure (A) at 0 °C and an Arrhenius diagram for 1000 bar (B) for BA polymerizations at a CO_2 content of 42 wt. %. For comparison, the corresponding dependencies measured for bulk polymerizations are also given in Figure 16. The slope of the lines, and thus E_A and $\Delta V^{\neq}$, are obviously identical within experimental accuracy. The absolute values of k_p are, however, clearly different for polymerizations in bulk and in CO_2.

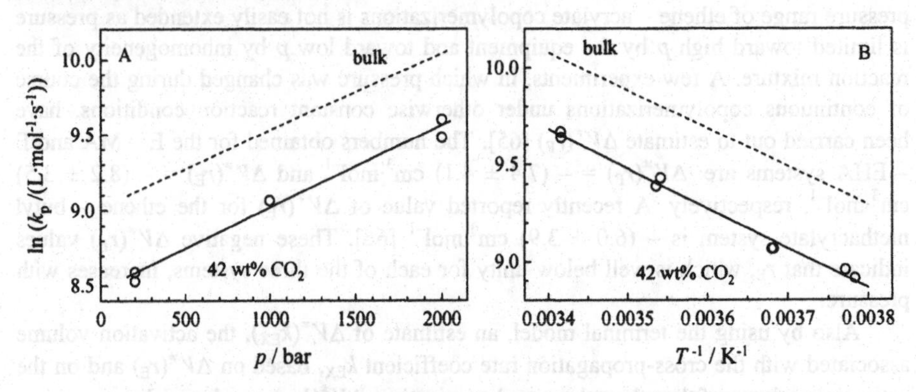

Figure 16: Variation of k_p with pressure at 0 °C (A) and with temperature at 1000 bar (B) for butyl acrylate polymerizations in CO_2 and in bulk (dashed line).

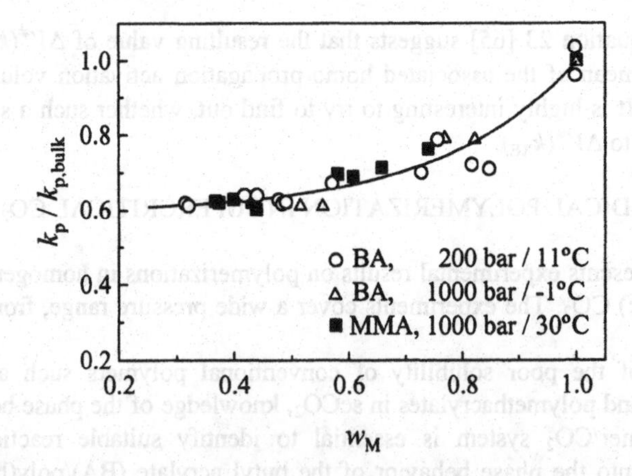

Figure 17: Variation of relative k_p for butyl acrylate and methyl methacrylate polymerizations as a function of relative initial monomer concentration, w_M.

The variation of k_p with CO_2 concentration was determined for three systems [68]: (i) for BA polymerizations at 200 bar and 11 °C, (ii) for BA polymerizations at 1000 bar and −1 °C, and (iii) for MMA polymerizations at 1000 bar and 30 °C. Figure 17 presents k_p as a function of the (initial) monomer weight fraction, w_M, relative to the bulk monomer concentration at identical p and T. The k_p values are also given relative to the corresponding bulk value, $k_{p,bulk}$. As can be seen from Figure 17, k_p decreases significantly toward lower w_M. Surprisingly, almost identical dependencies of k_p on w_M are observed for the three systems. Below $w_M = 0.6$ further reduction in relative monomer concentration is not associated with a further lowering of k_p. For w_M between 0.35 and 0.6, a constant value of $k_p/k_{p,bulk}$ is observed.

That E_A and $\Delta V^{\neq}$ are invariant toward changing CO_2 content provides some indication that the variation of k_p with CO_2 concentration is not primarily due to kinetic reasons, but originates from thermodynamic influences. As $scCO_2$ is a poor solvent for PBA and PMMA, solvent quality arguments and the resulting effects on polymer coil dimensions [69] need to be considered, as has first been suggested by O'Driscoll [70]. It is to be expected that polymer coils in $scCO_2$ are contracted with respect to the situation in bulk polymerization. As a consequence, a higher segment density within the polymer coil and an associated reduction (with respect to overall monomer concentration) of local monomer concentration, $c_{M,loc}$, may occur. Equation 5 shows that it is the product $k_p \cdot c_M$ which is directly obtained from the experimental quantities L_1 and t. The observed changes (Figure 17) thus may reflect changes in local monomer concentration (at the site of the propagating radical) rather than changes in the kinetic coefficient. A detailed discussion on these effects is given elsewhere [71]. It should further be noted that for very low molecular weights no influence of CO_2 on k_p is seen for either MMA or styrene polymerization [72].

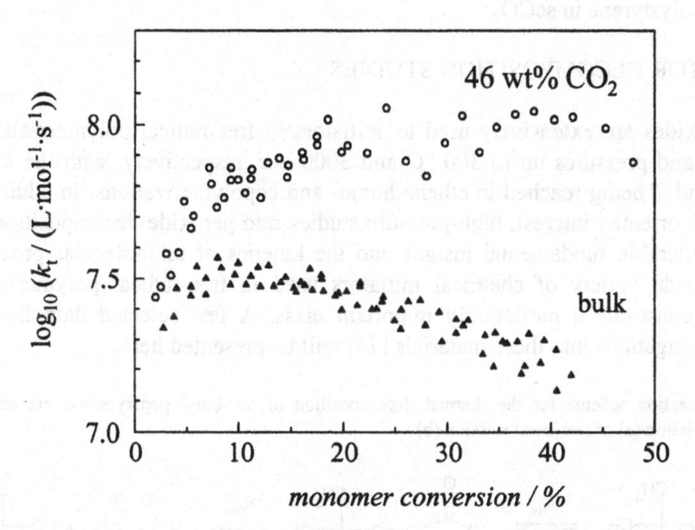

Figure 18: Conversion dependence of k_t for butyl acrylate polymerizations at 40 °C and 1000 bar in bulk and in solution of CO_2.

Performing SP-PLP experiments provides k_t as a function of monomer conversion [20]. As is to be expected, a strong influence of CO_2 on k_t is observed. As an example, k_t values for BA polymerizations at 40 °C and 1000 bar are given in Figure 18. The closed triangles represent data obtained for bulk polymerization, the open circles for polymerization in CO_2 (46 wt. %). In both systems, after an initial increase of k_t with conversion a constant k_t value is found. The plateau region observed for the bulk polymerizations is followed by a moderate decrease of k_t toward higher conversion. For the BA polymerization in CO_2 no decrease of k_t toward high degrees of monomer conversion is seen. At about 40 % conversion, k_t for polymerization in CO_2 is by a factor of 3.5 above bulk k_t.

These findings may again be understood in terms of solvent quality arguments. In both systems monomer conversion results in a lowering of solvent quality. This effect is however much larger in the presence of CO_2. A decrease in the size of polymer coils is associated with enhanced diffusion rate of the coils and thus with a faster termination rate. In BA bulk polymerization at moderately high conversion, this effect is more than compensated by other contributions, e.g., resulting from viscosity being largely enhanced with monomer conversion. In the presence of CO_2, the changes in solvent quality and thus in coil dimension are expected to be more pronounced. This is indeed what is observed. A more elaborate discussion of the influence of CO_2 on k_t will be given in reference [73].

The concept of solvent quality being responsible for the observed effects is supported by investigations into styrene homopolymerizations in $scCO_2$ [74]. Again, higher k_t values are obtained for polymerization in CO_2 as compared to polymerization in bulk. The effect is even more pronounced than with BA and amounts to about one order of magnitude. The stronger effect on styrene k_t is probably due to the poorer solubility of polystyrene in $scCO_2$.

3.6 INITIATOR DECOMPOSITION STUDIES

Organic peroxides are extensively used as initiators in free-radical polymerizations at temperatures and pressures up to 300 °C and 3000 bar, respectively, with the highest values of p and T being reached in ethene homo- and copolymerizations. In addition to an application-oriented interest, high-pressure studies into peroxide decomposition may provide considerable fundamental insight into the kinetics of unimolecular processes. Among the wide variety of chemical initiators used in free-radical polymerization, peroxyesters constitute a particularly important class. A few selected data from the extended investigations into these materials [75] will be presented here.

Scheme 1: Reaction scheme for the thermal decomposition of *tert*-butyl peroxyesters: via one-bond scission (a) or two-bond scission (b).

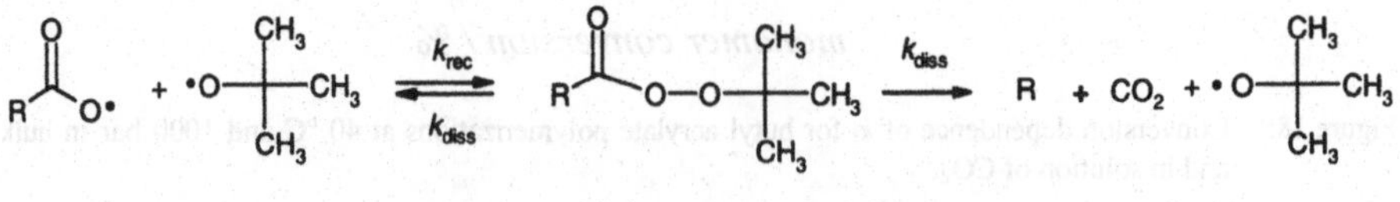

$$\mathrm{RC(=O)O\bullet} + \bullet\mathrm{O{-}C(CH_3)_3} \underset{k_{diss}}{\overset{k_{rec}}{\rightleftharpoons}} \mathrm{RC(=O)O{-}O{-}C(CH_3)_3} \xrightarrow{k_{diss}} \mathrm{R} + \mathrm{CO_2} + \bullet\mathrm{O{-}C(CH_3)_3}$$

(a) (b)

In addition to knowing the rate of initiator decay, the mechanism of the decomposition process should be clear. Scheme 1 illustrates two decomposition pathways, (a) and (b), for *tert*-butyl peroxyesters. By one-bond homolysis, mechanism (a), an acyloxy and an alkoxy radical are produced. Cage recombination of these two species restores the peroxyester. Concerted two-bond scission, mechanism (b), directly yields CO_2, the *tert*-butoxy radical, and an alkyl radical. If mechanism (b) applies, cage recombination after the primary decomposition event does not restore the peroxyester, but produces an ether which, under typical polymerization conditions, is not capable of producing free radicals by a subsequent dissociation step. As a consequence, mechanisms (a) and (b) should be associated with quite different efficiencies of free-radical production. Even if cage recombination plays no major role, knowing whether mechanisms (a) or (b) apply, is important as different types of free radicals, oxygen-centered or carbon-centered, are produced in the primary step. They differ in chain-transfer activity which may result in structural differences of polymeric products. Another point: If mechanism (a) is operative, the observed decomposition rate depends on primary bond scission (k_1) and on the competition between the subsequent processes of recombination and out-of-cage diffusion. In case of concerted two-bond scission, where peroxide decomposition occurs irreversibly, the measured decay rate is fully described by k_1 (see Scheme 1).

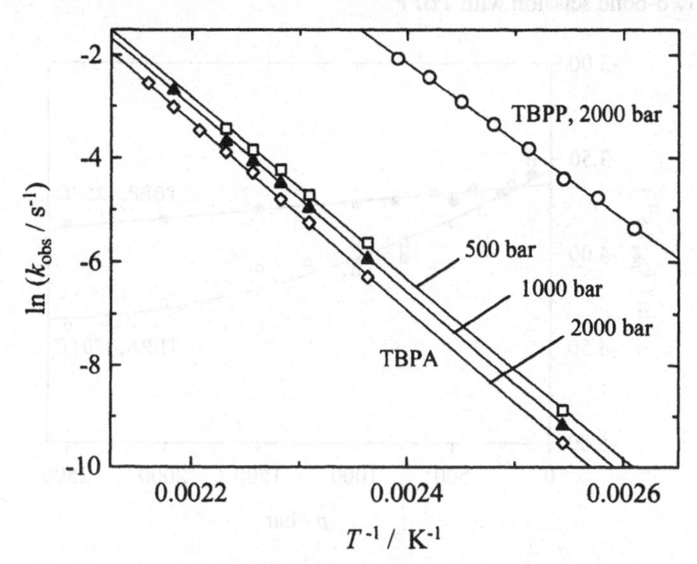

Figure 19: Temperature dependence of k_{obs} for TBPP at 2000 bar and for TBPA at 500, 1000, and 2000 bar [75].

Evidence from literature strongly suggests that *tert*-butyl peroxyacetate (TBPA) and *tert*-butyl peroxypivalate (TBPP) may be regarded as the archetypes of decomposition mechanisms (a) and (b), respectively. It thus appeared desirable to accurately measure decomposition rate coefficients for these two peroxides (in dilute solution of *n*-heptane) within extended ranges of T and p [75]. According to equation 24, overall first-order rate coefficients, k_{obs}, have been derived from the experimental peroxyester concentration (c) vs. time profiles:

$$\frac{\mathrm{d}\,c}{\mathrm{d}\,t} = -\,k_{\mathrm{obs}} \cdot c \qquad (24)$$

This first-order rate law is found to adequately describe the decomposition kinetics of both peroxides (in 1.0×10^{-2} M solution of *n*-heptane) over several half-lives. The peroxide samples were kindly provided by AKZO NOBEL in high purity at active oxygen contents of 94.6 and 99.0 % for TBPA and TBPP, respectively, and were used as obtained. The peroxide concentration is sufficiently low to eliminate induced decomposition, but is high enough to allow for quantitative IR spectroscopic detection of peroxide concentration via the carbonyl fundamental mode.

In Figure 19, the temperature dependence of k_{obs} is plotted at pressures of 500, 1000, and 2000 bar for TBPA and at 2000 bar for TBPP. The decomposition of TBPP is much faster. As can be seen from the TBPA data, pressure lowers decomposition rate. The activation energy, E_{A}, for TBPP decomposition ((125.4 ± 3.9) $\mathrm{kJ \cdot mol^{-1}}$, at 2000 bar) is by about 25 $\mathrm{kJ \cdot mol^{-1}}$ below the corresponding number for TBPA. The observed E_{A} values fully meet the expectation of one-bond scission taking place with TBPA and of concerted two-bond scission with TBPP.

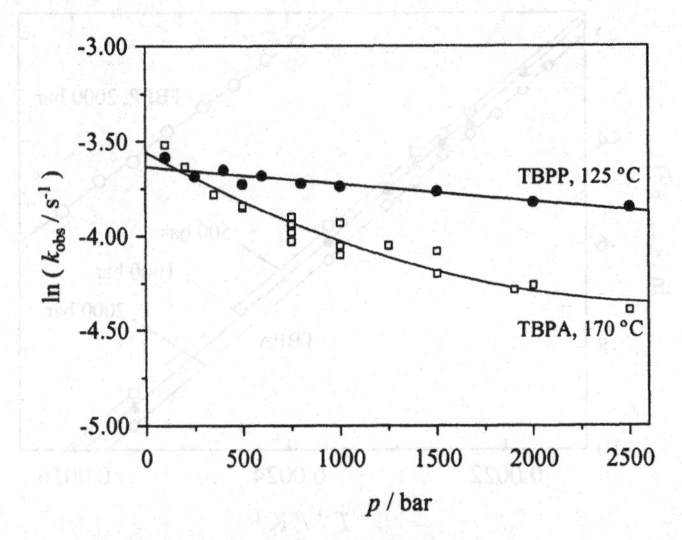

Figure 20: Pressure dependence of k_{obs} for the thermal decomposition of TBPP at 125 °C (filled circles) and of TBPA at 170 °C (open squares) in *n*-heptane solution [75].

A clear difference between TBPA and TBPP decomposition kinetics is also seen with respect to their pressure dependence. In Figure 20 is plotted ln k_{obs} vs. p for the decomposition of TBPP at 125 °C and of TBPA at 170 °C. The pressure dependence of k_{obs} is much weaker with TBPP. Moreover, the ln k_{obs} vs. p data for TBPP closely fit to a straight line. The corresponding fit for TBPA shows a clear curvature. In Figure 21 the entire set of decomposition rate coefficients for TBPA from experiments between 120 and 185 °C is shown [75]. Activation volumes, $\Delta V_{\mathrm{obs}}^{\neq}$, are determined according to equation 7.

$\Delta V_{\mathrm{obs}}^{\neq}$ for TBPP may be directly obtained from the linear ln k_{obs} vs. p fit. The pressure-dependent activation volume for TBPA decomposition is derived from the first derivative of ln k_{obs} vs. p data fitted to a quadratic polynomial. The activation volume for TBPP is (3.0 ± 0.9) $cm^3 \cdot mol^{-1}$. From the pressure-dependent $\Delta V_{obs}^{\neq}$ data for TBPA, a value of (17 ± 2) $cm^3 \cdot mol^{-1}$ is found for 1000 bar (at temperatures between 120 and 170 °C). The small activation volume observed for TBPP and the much larger value found for $\Delta V_{obs}^{\neq}$ of TBPA are consistent with the expectations of Neuman and Behar [76] for concerted two-bond and for single-bond scission, respectively.

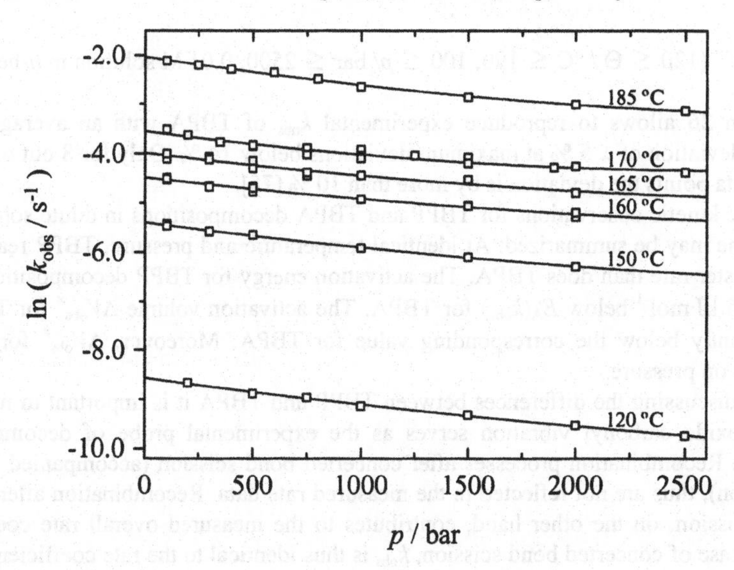

Figure 21: Variation of the decomposition rate coefficient k_{obs} with pressure for TBPA at 120, 150, 160, 165, 170, and 185 °C in n-heptane solution [75].

The TBPP data in Figure 19 and 20 are consistent with previously reported [77] TBPP decomposition rate coefficients that have (also) been measured in n-heptane solution, but at lower temperatures. Common fitting of the two data sets yields equation 25, from which k_{obs} may be estimated for TBPP at temperatures between 70 and 145 °C and at pressures between 100 and 2500 bar.

$$k_{obs}(p,T)/s^{-1} = 1.04 \cdot 10^{15} \cdot \exp - \left(\frac{15227 + 0.0370 \cdot p/\text{bar}}{T/\text{K}} \right) \quad (25)$$

<u>*TBPP*</u> (70 ≤ Θ / °C ≤ 145; 100 ≤ p/ bar ≤ 2500; 0.01 M solution in n-heptane)

Equation 25 is based on 38 data points which are represented with a mean square deviation of 4 % (at a maximum deviation below 11 %).

In the representation of the TBPP k_{obs} data via equation 25 it is assumed that activation energy is independent of pressure and that $\Delta V_{obs}^{\neq}$ is independent of temperature. Both assumptions are assumed to be approximately valid for k_{obs} of TBPA, too. A pressure dependent activation volume, however, needs to be taken into account in fitting the measured TBPA decomposition rate coefficients. The resulting expression reads:

$$k_{obs}(p, T)/s^{-1} = 6.78 \cdot 10^{15} \cdot \exp - \left(\frac{17714 + (0.2471 - 3.336 \cdot 10^{-5} \cdot p/\text{bar}) \cdot p/\text{bar}}{T/\text{K}} \right) \quad (26)$$

TBPA (120 ≤ Θ / °C ≤ 190; 100 ≤ p/ bar ≤ 2500; 0.01 M solution in n-heptane)

Equation 26 allows to reproduce experimental k_{obs} of TBPA with an average mean square deviation of ± 5 % at maximum deviations below 14 %. Only for 8 out of a total of 81 data points the deviation is by more than 10 % [75].

The kinetic observations for TBPP and TBPA decompositions in dilute solution of n-heptane may be summarized: At identical temperature and pressure, TBPP reacts at a much faster rate than does TBPA. The activation energy for TBPP decomposition is by about 25 kJ·mol^{-1} below $E_A(k_{obs})$ for TBPA. The activation volume $\Delta V_{obs}^{\neq}$ for TBPP is significantly below the corresponding value for TBPA. Moreover, $\Delta V_{obs}^{\neq}$ for TBPA depends on pressure.

In discussing the differences between TBPP and TBPA it is important to note that the peroxide carbonyl vibration serves as the experimental probe of decomposition kinetics. Recombination processes after concerted bond scission (accompanied by CO_2 formation), thus are not reflected in the measured rate data. Recombination after single-bond scission, on the other hand, contributes to the measured overall rate coefficient k_{obs}. In case of concerted bond scission, k_{obs} is thus identical to the rate coefficient of the primary dissociation step whereas equation 27 has to be used in situations where primary recombination inside the solvent cage and out-of-cage diffusion occur as competing processes after one-bond homolysis:

$$k_{obs} = k_{diss} \cdot \frac{k_{diff}}{k_{rec} + k_{diff}} \quad (27)$$

k_{diss} and k_{diff} are the first-order rate coefficients for primary one-bond homolysis and for out-of-cage diffusion, respectively, and k_{rec} is the (quasi-first-order) rate coefficient for cage recombination of the free radicals produced within the primary bond scission step. Modeling of k_{obs} for single-bond decompositions is difficult as (at least) the three rate coefficients k_{diss}, k_{diff}, and k_{rec} of equation 27 need to be known in absolute value and also in their pressure and temperature dependence. As is demonstrated in reference [75], the experimental observations for the pressure dependence of k_{obs} for TBPA are consistent with the view that TBPA decomposes via single-bond scission. The subsequent competition of recombination and out-of-cage diffusion gives rise to the enhanced pressure dependence of $\Delta V_{obs}^{\neq}$. The curvature of the ln k_{obs} vs. p relations is primarily assigned to diffusivity (and thus viscosity) contributions.

The linear correlation observed for the ln k_{obs} vs. p behavior of TBPP decomposition rate indicates that the overall rate coefficient is essentially determined by

k_{diss} as is to be expected for concerted two-bond scission. The associated activation volume of about 3 $cm^3 \cdot mol^{-1}$ appears to be rather small for a process that goes with breaking of two bonds. The small value may be understood as being due to the structure of the CO_2 moiety being almost linear in the transition state. Under such conditions, the volume of the transition state must not be significantly above the volume of the TBPP molecule. Investigations into the decomposition kinetics of peroxyesters with substitution at the α-C atom that is intermediate between the situations in peroxyacetates and peroxypivalates appear to be rewarding. Interesting candidates for such studies are *tert*-butyl peroxy-*n*-butyrate, *tert*-butyl peroxy-*iso*-butyrate, *tert*-butyl peroxyethyl-hexanoate, and *tert*-butyl peroxy-*neo*-decanoate.

Another important aspect of kinetic studies into peroxyester decomposition deals with the influence that substituents on the "alcohol side" of the peroxyester have on decomposition rate and on selectivity. Finally, it remains to be studied within polymerization processes, to which extent the many detailed aspects of peroxide decomposition, that are now well understood for reaction in *n*-heptane, translate into overall polymerization kinetics and into the properties of polymeric products.

4. Conclusions

The advent of laser-assisted techniques has enormously improved the quality of measuring propagation and termination rate coefficients of high-pressure polymerizations. Chain-transfer coefficients are available from careful MWD analysis of polymer produced at quite different reaction conditions. Via online quantitative IR spectroscopy, also the decomposition kinetics of peroxides have been measured up to high pressure. The availability of precise rate coefficients for these individual reactions allows to model kinetics and polymer properties in free-radical high-pressure polymerizations, with particular interest in ethene homo- and copolymerizations. The experimental procedures are also well suited to measure kinetic parameters for free-radical polymerization in supercritical CO_2.

5. Acknowledgments

The authors are grateful to the "*Deutsche Forschungsgemeinschaft*" for support of several of the Ph.D. projects which provided a major part of the results underlying this article. Additional support by the "*Fonds der Chemischen Industrie*" is gratefully acknowledged, as is the cooperation with *BASF AG* in the field of high-pressure ethene polymerization and the support by *AKZO NOBEL* in the peroxide decomposition work.

6. References

[1] DeSimone, J. M., Guan, Z., Elsbernd, C. S. (1992) Synthesis of Fluoropolymers in Supercritical Carbon Dioxide, *Science* **257**, 945-947

[2] Buback, M. (1994) Absorption Spectroscopy in Fluid Phases, in E. Kiran and J. Levelt-Sengers (eds.), *Supercritical Fluids - Fundamentals for Application*, Kluwer Academic Publishers, Dordrecht, Boston,

London, pp. 499-526

[3] Buback, M. and Hinton, C. (1997) Vibrational Spectroscopy in Dense Fluid Phases, in W. B. Holzapfel and N. S. Isaacs (eds.), *High-pressure techniques in chemistry and physics: a practical approach*, Oxford University Press, pp. 151 - 186

[4] Olaj, O. F., Bitai, I., Hinkelmann, F. (1987) The Laser-Flash-Initiated Polymerization as a Tool of Evaluating (Individual) Kinetic Constants of Free-Radical Polymerization: 2. The Direct Determination of the Rate Constant of Chain Propagation, *Makromol. Chem.* **188**, 1689-1702

[5] Olaj, O. F., and Schnöll-Bitai, I. (1989) Laser Flash-Initiated Polymerization as a Tool of Evaluating (Individual) Kinetic Constants of Free-Radical Polymerization: 5. Complete Analysis by Means of a Single Experiment, *Eur. Polym. J.* **25**, 635-641

[6] Beuermann, S., Buback, M., Davis, T. P., Gilbert, R. G., Hutchinson, R. A., Olaj, O. F., Russell, G. T., Schweer, J., van Herk, A. M. (1997) Critically evaluated rate coefficients for free-radical polymerization, 2 Propagation rate coefficients for methyl methacrylate, *Macromol. Chem. Phys.* **198**, 1545-1560

[7] Buback, M., Gilbert, R. G., Hutchinson, R. A., Klumperman, B., Kuchta, F.-D., Manders, B. G., O'Driscoll, K. F., Russell, G. T., Schweer, J. (1995) Critically evaluated rate coefficients for free-radical polymerization, 1 Propagation rate coefficients for styrene, *Macromol. Chem. Phys.* **196**, 3267-3280

[8] Hutchinson, R. A., Aronson, M. T., Richards, J. R. (1993) Analysis of Pulsed-Laser-Generated Molecular Weight Distributions for the Determination of Propagation Rate Coefficients, *Macromolecules* **26**, 6410-6415

[9] Beuermann, S., Paquet, D. A., Jr., McMinn, J. H., Hutchinson, R. A. (1996) Determination of Free-Radical Propagation Rate Coefficients of Butyl, 2-Ethylhexyl and Dodecyl Acrylates by Pulsed-Laser Polymerization, *Macromolecules* **29**, 4206-4215

[10] Busch, M. and Wahl, A. (1998) The significance of transfer reactions in pulsed laser polymerization experiments, *Macromol. Theory Simul.* **7**, 217-224

[11] Manders, L. G. (1997) *Pulsed Initiation Polymerization*, Universiteitsdrukkerij, Eindhoven

[12] Clay, P. A. and Gilbert, R. G. (1995) Molecular Weight Distribution in Free-Radical Polymerizations. 1. Model Development and Implications for Data Interpretation, *Macromolecules* **28**, 552-569

[13] Hutchinson, R. A., Paquet, Jr., D. A., McMinn, J. H. (1995) Determination of Free-Radical Chain-Transfer Rate Coefficients by Pulsed-Laser Polymerization, *Macromolecules* **28**, 5655-5663

[14] Kukulj, D., Davis, T. P., Gilbert, R. G. (1998) Chain Transfer to Monomer in the Free-Radical Polymerizations of Methyl Methacrylate and α-Methylstyrene, *Macromolecules* **31**, 994-999

[15] Christie, D. I. and Gilbert, R. G. (1996) Transfer constants from complete molecular weight distributions, *Macromol. Chem. Phys.* **197**, 403-412

[16] Buback, M. and Laemmel, R. A. (1997) Simultaneous determination of free-radical propagation and transfer rates from a novel type of PLP-SEC experiment, *Macromol. Theory Simul.* **6**, 145-150

[17] Lämmel R. A. (1996), *Ph.D. Thesis*, Göttingen

[18] Buback, M. and Lämmel, R. A. (1998) Termination rate coefficients derived from PLP-SEC data, *Macromol. Theory Simul.* **7**, 197-202

[19] Wulkow, M. (1996) The simulation of molecular weight distributions in polyreaction kinetics by discrete Galerkin methods, *Macromol. Theory Simul.* **5**, 393-416

[20] Buback, M., Hippler, H., Schweer, J., Vögele, H.-P. (1986) Time-resolved study of laser-induced high-pressure ethylene polymerization, *Makromol. Chem., Rapid Commun.* **7**, 261-265

[21] Beuermann, S., Buback, M., Russell, G. T. (1995) Kinetics of free-radical solution polymerization of methyl methacrylate over an extended conversion range, *Macromol. Chem. Phys.* **196**, 2493-2516

[22] Buback, M. and Kowollik, C. (1998) Termination Kinetics of Methyl Methacrylate Free-Radical Polymerization Studied by Time-Resolved Pulsed Laser Experiments, *Macromolecules* **31**, 3211-3215

[23] Buback, M. and Kuchta, F.-D. (1997) Termination kinetics of free-radical polymerization of styrene over an extended temperature and pressure range, *Macromol. Chem. Phys.* **198**, 1455-1480

[24] Buback, M. and Klingbeil, S. (1995) Hochdruck-Strömungsrohrreaktor zur FT-IR-spektroskopischen Analyse von Initiatorzerfallsreaktionen, *Chem. Ing. Tech.* **67**, 493-495

[25] Buback, M., Busch, M., Lovis, K., Mähling, F.-O. (1995) Mini -Technikumsanlage für Hochdruck-Polymerisationen bei kontinuierlicher Reaktionsführung, *Chem.-Ing.-Tech.* **67**, 1652-1655

[26] Buback, M., Busch, M., Lovis, K., Mähling, F.-O. (1996) High-pressure free-radical copolymerization of ethene and butyl acrylate, *Macromol. Chem. Phys.* **197**, 303-313

[27] Buback, M., Dröge, T., v. Herk, A., Mähling, F.-O. (1996) High-Pressure Free-Radical Copolymerization of Ethene and 2-Ethylhexyl acrylate, *Macromol. Chem. Phys.* **197**, 4119-4134

[28] Buback, M., Busch, M., Lovis, K., Mähling, F.-O. (1994) Entwicklung eines kontinuierlich betriebenen Hochdruck-Hochtemperatur-Rührkessels mit Lichteinkopplung, *Chem.-Ing.-Tech.* **66** , 510-513

[29] Buback, M., Busch, M., Panten, K., Vögele, H.-P. (1992) Direkte spektroskopische Bestimmung von Verweilzeitverteilungen bei Fluidphasenreaktionen bis zu hohem Druck, *Chem.-Ing.-Tech.* **64**, 352-354

[30] Buback, M., Busch, M., Dietzsch, H., Dröge, T., Lovis, K. (1996) Cloud-Point Curves in Ethylene-Acrylate Poly(ethylene-co-acrylate) Systems, in R. v. Rohr and Ch. Trepp (eds), *Process Technology Proceedings 12, "High Pressure Chemical Engineering"*, Elsevier, Amsterdam, pp. 175-168

[31] Beuermann, S., Buback, M., Russell, G. T. (1994) Variation with pressure of the propagation rate coefficient in free-radical polymerization of methyl methacrylate, *Macromol. Rapid Commun.* **15**, 351-355

[32] Buback, M., Geers, U., Kurz, C. H. (1997) Propagation rate coefficients in free-radical homopolymerizations of butyl methacrylate and dodecyl methacrylate, *Macromol. Chem. Phys.* **198**, 3451-3464

[33] Buback, M. and Kurz, C. H. (1998) Free-radical propagation rate coefficients for cyclohexyl methacrylate, glycidyl methacrylate and 2-hydroxyethyl methacrylate homopolymerization, *Macromol. Chem. Phys.* **199**, 2301-2310

[34] Buback, M., Kurz, C. H., Schmaltz, C. (1998) Pressure dependence of propagation rate coefficients in free-radical homopolymerizations of methyl acrylate and dodecyl acrylate, *Macromol. Chem. Phys.* **199**, 1721-1727

[35] Buback, M. and Kuchta, F.-D. (1995) Variation with pressure and temperature of the propagation rate coefficient in free-radical bulk polymerization of styrene, *Macromol. Chem. Phys.* **196**, 1887-1898

[36] Hutchinson, R. A., Beuermann, S., Paquet, D. A., Jr., McMinn, J. H. (1997) The Determination of Free-Radical Propagation Rate Coefficients for Alkyl Methacrylates by Pulsed-Laser Polymerization, *Macromolecules* **30**, 3490-3493
Hutchinson, R. A., Beuermann, S., Paquet, D. A., Jr., McMinn, J. H., Jackson, C. (1998) The Determination of Free-Radical Propagation Rate Coefficients for Cycloalkyl and Functional Methacrylates by Pulsed-Laser Polymerization, *Macromolecules* **31**, 1542-1547
Zammit, M. D., Coote, M. L., Davis, T. P., Willett, G. D. (1998) Effect of the Ester Side-Chain on the Propagation Kinetics of Alkyl Methacrylates-An Entropic or Enthalpic Effect ?, *Macromolecules* **31**, 955-963

[37] Beuermann, S. and Buback, M. (1998) Critically-evaluated propagation rate coefficients in free radical polymerizations - II. Alkyl methacrylates, *Pure & Appl. Chem.* **70**, 1415-1416

[38] Hutchinson, R. A., Richards, J. R., Aronson, M. T. (1994) Determination of Propagation Rate Coefficients by Pulsed-Laser Polymerization for Systems with Rapid Chain Growth: Vinyl Acetate, *Macromolecules* **27**, 4530-4537

[39] Lyons, R. A., Hutovic, J., Piton, M. C., Christie, D. I., Clay, P. A., Manders, B. G., Kable, S. H., Gilbert, R. G. (1996) Pulsed-Laser Polymerization Measurements of the Propagation Rate Coefficient for Butyl Acrylate, *Macromolecules* **29**, 1918-1927

[40] Zammit, M. D., Davis, T. P., Willett, G. D., O'Driscoll, K. F. (1997) The Effect of Solvent on the Homo-Propagation Rate Coefficients of Styrene and Methyl Methacrylate, *J. Polym. Sci., Polym. Chem. Ed.* **35**, 2311-2321
O'Driscoll, K. F., Monteiro, M. J., Klumperman, B. (1997) The Effect of Benzyl Alcohol on Pulsed Laser Polymerization of Styrene and Methylmethacrylate, *J. Polym. Sci., Polym. Chem. Ed.* **35**, 515-520
Morrison, B. R., Piton, M. C., Winnik, M. A., Gilbert, R. G., Napper, D. H. (1993) Solvent Effects on the Propagation Rate Coefficient for Free Radical Polymerization, *Macromolecules* **26**, 4368-4372
Beuermann, S., Buback, M., Russell, G. T. (1994) Rate of propagation in free radical polymerization of methyl methacrylate in solution, *Macromol. Rapid Commun.* **15**, 647-653

[41] Benson, S. W. and North, A. M. (1962) The Kinetics of Free Radical Polymerization under Conditions of Diffusion-controlled Termination, *J. Am. Chem. Soc.* **84**, 935-940

[42] Buback, M. (1990) Free-Radical polymerization to high conversion. A general kinetic treatment, *Makromol. Chem.* **191**, 1575-1587

[43] Nicholson, A. E. and Norrish, R. G. W. (1956) Polymerization of styrene at high pressures using the

sector technique, *Disc. Faraday Soc.* **22**, 104-113

[44] Ogo, Y. and Kyotani, T. (1978) Effect of Pressure on the Termination Rate Constant in Free-Radical Polymerization. Correlation between Rate Constant and Monomer Viscosity, *Makromol. Chem.* **179**, 2407-2417

[45] Hutchinson, R. A., McMinn, J. H., Paquet, D. A., Jr., Beuermann, S., Jackson, C. (1997) A Pulsed-Laser Study of Penultimate Copolymerization Propagation Kinetics for Methyl Methacrylate/n-Butyl Acrylate, *Ind. Eng. Chem. Res.* **36**, 1103-1113

[46] Fukuda, T., Ma, Y.-D., Kubo, K., Inagaki, H. (1991) Penultimate-Unit Effects in Free-Radical Copolymerization, *Macromolecules* **24**, 370-375

[47] Buback, M., Isemer, C., Kowollik, C., Lacík, I. (1998), *Macromol. Chem. Phys.*, in preparation

[48] Russo, S. and Munari, S. (1968) A Model for the termination stage of some Radical Copolymerizations, *J. Macromol. Sci. Chem.* **2**, 1321-1332

[49] Ito, K. and O'Driscoll, K. F. (1979) The termination reaction in Free-Radical Copolymerization. I. Methyl methacrylate and Butyl - or Dodecyl Methacrylate, *J. Polym. Sci., Polym. Chem. Ed.* **17**, 3913-3921

[50] Fukuda, T., Ide, N., Ma, Y.-D. (1996) Propagation and termination processes in free radical copolymerization, *Macromol. Symp.* **111**, 305-316

[51] Buback, M. and Kowollik, C. (1998) Termination Kinetics in Free-Radical Bulk Copolymerization - The Systems Dodecyl Acrylate - Dodecyl Methacrylate and Dodecyl Acrylate - Methyl Acrylate, *Macromolecules*, submitted

[52] Kurz, C. H. (1995) *Laserinduzierte radikalische Polymerisation von Methylacrylat und Dodecylacrylat in einem weiten Zustandsbereich,* Verlag Graphikum, Göttingen

[53] Bergert, U. (1994) *Laserinduzierte radikalische Polymerisation von Butylmethacrylat und Dodecylmethacrylat in einem weiten Zustandsbereich*, Cuvillier Verlag, Göttingen

[54] Fukuda, T., Kubo, K., Ma, Y.-D. (1992) Kinetics Of Free Radical Copolymerization, *Prog. Polym. Sci.* **17**, 875-916

[55] Matsumoto, A. and Otsu, T. (1995) Detailed mechanism of radical high polymerization of sterically hindered Dialkyl fumarates, *Macromol. Symp.* **98**, 139-152

[56] Otsu, T., Yamagishi, K., Matsumoto, A., Yoshioka, M., Watanabe, H. (1993) Effect of α- and β-Ester Alkyl Groups on the Propagation and Termination Rate Constants for Radical Polymerization of Dialkyl Itaconates, *Macromolecules* **26**, 3026-3029

[57] Schweer, J. (1988), *Ph.D. Thesis*, Göttingen

[58] Buback, M. and Schweer, J. (1989) Conversion and Chain-Length Dependence of Rate Coefficients in Free-Radical Polymerization, *Z. Phys. Chem. N. F.* **161**, 153-165

[59] Beuermann, S., Buback, M., Busch, M. (1998) Free-Radical Polymerization in Reactive Supercritical Fluids, in W. Leitner (ed.), "Chemical Synthesis using Supercritical Fluids" VCH, Weinheim, in press

[60] Buback, M. Choe, C.-R., Franck, E. U. (1984) Thermisch und UV-photochemisch initiierte Hochdruckpolymerisation des Ethylens, 1, Molmassen-Mittelwerte, *Makromol. Chem.* **185**, 1685-1697

[61] Buback, M. Choe, C.-R., Franck, E. U. (1984) Thermisch und UV-photochemisch initiierte Hochdruckpolymerisation des Ethylens, 2, Molmassen-Verteilung, *Makromol. Chem.* **185**, 1699-1717

[62] McCord, E. F., Shaw, W. H., Jr., Hutchinson, R. A. (1997) Short-Chain Branching Structures in Ethylene Copolymers Prepared by High-Pressure Free-Radical Polymerization: An NMR Analysis, *Macromolecules* **30**, 246-256

[63] v. Boxtel, H. C. M., Buback, M., Busch, M., Lehmann, S. (1998) Long- and Short-Chain Branching in Ethene-Methyl Acrylate Copolymers Studied by Quantitative ^{13}C-NMR Spectroscopy, *Macromol. Chem. Phys.*, in preparation

[64] v. Herk, A. M. (1995) Least-Squares Fitting by Visualization of the Sum of Squares Space, *J. Chem. Educ.* **72**, 138-140

[65] Buback, M. and Dröge, T. (1997) High-pressure free-radical copolymerization of ethene and methyl acrylate, *Macromol. Chem. Phys.* **198**, 3627-3638

[66] Buback, M. and Dröge, T. (1998) High-pressure free-radical copolymerization of ethene and butyl methacrylate, *Macromol. Chem. Phys.*, in press

[67] McHugh, M. A., Rindfleisch, F., Kuntz, P. T., Schmaltz, C., Buback, M. (1998) Cosolvent effect of alkyl acrylates on the phase behavior of poly(alkyl acrylates)-supercritical CO_2 mixtures, *Polymer* **39**, 6049-6052

[68] Beuermann, S., Buback, M., Kuchta, F.-D., Schmaltz, C. (1998) Determination of Free-Radical Propagation Rate Coefficients for Methyl Methacrylate and *n*-Butyl Acrylate Homopolymerization in CO_2, *Macromol. Chem. Phys.* **199**, 1209-1216
[69] Dionisio, J., Mahabadi, H. K., O'Driscoll, K. F., Abuin, E., Lissi, A. (1979) High-Conversion Polymerization. IV. A Definition of the Onset of the Gel Effect, *J. Polym. Sci., Polym. Chem. Ed.* **17**, 1891-1900
[70] O'Driscoll, K. F. (1977) Pressure dependence of the Termination Rate Constant in Free Radical Polymerization *Makromol. Chem.* **178**, 899-903
[71] Beuermann, S., Buback, M., Schmaltz, C. (1998) Pressure and Temperature Dependence of Butyl Acrylate Propagation Rate Coefficients in Fluid CO_2, *Macromolecules* **31**, 8069-8074
[72] v. Herk, A. M., Manders, B. G., Canelas, D. A., Quadir, M. A., DeSimone, J. M. (1997) Propagation Rate Coefficients of Styrene and Methyl Methacrylate in Supercritical CO_2, *Macromolecules* **30**, 4780-4782
[73] Beuermann, S., Buback, M., Schmaltz, C. (1998) *Ind. Eng. Chem. Res.*, in preparation
[74] Beuermann, S., Buback, M., Isemer, C., Wahl, A. (1998), *Macromol. Chem. Phys.*, in preparation
[75] Buback, M., Klingbeil, S., Sandmann, J., Sderra, M.-B., Vögele, H.-P., Wackerbarth, H., Wittkowski, L. (1998) Pressure and temperature dependence of the decomposition rate of *tert*-butyl peroxypivalate, *Z. Phys. Chem.*, in press
[76] Neuman, R. C. and Behar, J. V. (1969) Cage Effects and Activation Volumes for Homolytic Scission Reactions, *J. Am. Chem. Soc.* **91**, 6024-6031
[77] Buback, M. and Lendle, H. (1982) Die Kinetik des Zerfalls von *tert*-Butylperoxypivalat bei hohen Drücken und Temperaturen, *Z. Naturforsch.* **36a**, 1371-1377

PRESSURE EFFECTS ON LYOTROPIC LIPID MESOPHASES AND MODEL MEMBRANE SYSTEMS - EFFECTS ON THE STRUCTURE, PHASE BEHAVIOUR AND KINETICS OF PHASE TRANSFORMATIONS

R. Winter, A. Gabke, J. Erbes, and C. Czeslik
University of Dortmund, Department of Chemistry, Physical Chemistry I, Otto-Hahn-Straße 6, D-44227 Dortmund, Germany

1. Introduction

Amphiphilic lipid molecules, which provide valuable model systems for lyotropic mesophases and biomembranes, display a variety of polymorphic phases, depending on their molecular structure and environmental conditions, such as the water content, pH, ionic strength, temperature and pressure [1-5]. The basic structural element of biological membranes consists of a lamellar phospholipid bilayer matrix. Due to the large hydrophobic effect, most phospholipid bilayers associate in water already at very low concentrations ($< 10^{-12}$ mol·L^{-1}). Saturated phospholipids often exhibit two thermotropic lamellar phase transitions, a gel to gel ($L_{\beta'}/P_{\beta'}$) pretransition and a gel to liquid-crystalline ($P_{\beta'}/L_{\alpha}$) main transition at a higher temperature T_m (see Fig. 1). In the fluid-like L_{α}-phase, the acyl chains of the lipid bilayers are conformationally disordered, whereas in the gel phases, the chains are more extended and ordered. In addition to these thermotropic phase transitions, also pressure-induced phase transformations have been observed (see, e.g., [6-20]). Upon compression, the lipids adopt to volume restriction by changing their conformation and packing system. When such adjustments are no longer possible, any further increase in pressure results mainly in a reduction of bond lengths, which affects the frequency of the stretching vibrations in the IR spectra.
It is now well-known that many biological lipid molecules also form non-lamellar liquid-crystalline phases (see Fig. 1) [1-5,21-23]. Lipids, which can adopt the hexagonal phase are present at substantial levels in biological membranes, usually at least 30 mol% of total lipids. For the double-chain lipids found in membranes, the polar/apolar interface curves towards the water (such phases are called inverted or type II). Some lipid extracts, such as those from archaebacteria (*S. solfataricus*), exhibit a cubic liquid-crystalline phase [24,25].
It is generally assumed that the non-lamellar lipid structures, such as the inverted hexagonal (H_{II}) and cubic lipid phases, are of signifiant biological relevance. They probably play an important functional role in some cell processes [25-30]. Fundamental cell processes, such as endo- and exocytosis, membrane recycling, protein trafficing, fat digestion, membrane budding, fusion and enveloped virus infection, involve a rearrangement of biological membranes where non-lamellar lipid phases are probably involved. A recent reanalysis of a large number of published electron micrographs of cell membranes has shown that static cubic structures (cubic membranes) might also occur in biological cells [25,31]. Furthermore, the cubic phases can be used as controlled-release drug carriers and

R. Winter and J. Jonas (eds.), High Pressure Molecular Science, 369–403.

crystallization media for proteins [32-36].

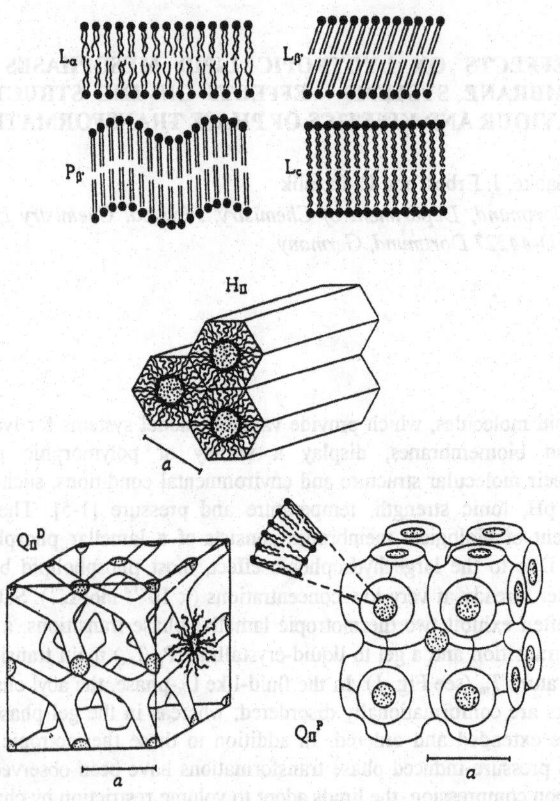

Figure 1. Schematic drawing of lamellar gel (L_c, $L_{\beta'}$ and $P_{\beta'}$) and liquid-crystalline L_α, inverted hexagonal (H_{II}) and two bicontinuous cubic (Q_{II}) lipid phases.

To date, seven cubic lipid phases are known. At a macroscopic level, they are very viscous and optically isotropic. The lipid cubic phases can be sorted in two main classes: bicontinuous and micellar [1,2,21,22,37-40]. The bicontinuous cubic phases of type II (Q_{II}) can be visualized in terms of a highly convoluted lipid bilayer, which subdivides three-dimensional space into two disjointed polar labyrinths separated by an apolar septum. The structure of three of these phases Q^{230}, Q^{224}, and Q^{229}, are closely related to the Schoen Gyroid (G), the Schwarz D and the Schwarz P infinite periodic miminal surfaces (IPMS). An IPMS is an intersection-free surface periodic in three dimensions with a mean curvature that is everywhere zero. The surface, that sits at the lipid bilayer midplane, separates two interpenetrating but not connected water networks.

A prerequisite for the formation of the H_{II} or inverse cubic phases is that the opposing monolayers wish to bend towards the water region. This desire arises because of differential lateral pressures which are present through the monolayer films (see below). It increases, for instance, if the lateral chain pressure increases due to increase of trans/gauche isomerisations of the acyl chains at high temperatures or if the level of head group hydration decreases (e.g., due to Ca^{2+} adsorption at the polar/apolar interface).

Generally, the molecular organization within a membrane lipid aggregate can be understood in terms of a balance of attractive and repulsive forces acting at the level of the lipid polar headgroups and non-polar acyl chains (Fig. 2). Within the headgroup (lipid/water) region there is an effectively attractive force $F_{l/w}$, which arises from the unfavourable contact of the hydrocarbon chains with water (the hydrophobic effect); attractive contributions from hydrogen bonding, as in the case of phosphatidylethanolamines, may also be present. The repulsive headgroup pressure F_{head} is the result of hydrational, steric, and - for charged head groups - electrostatic contributions. For the acyl chain region the attractive van der Waals interactions among the CH-groups are compensated by the repulsive lateral pressure, F_{chain}, due to thermally activated dihedral angle isomerizations. In a membrane bilayer at equilibrium, the various lateral pressures are balanced.

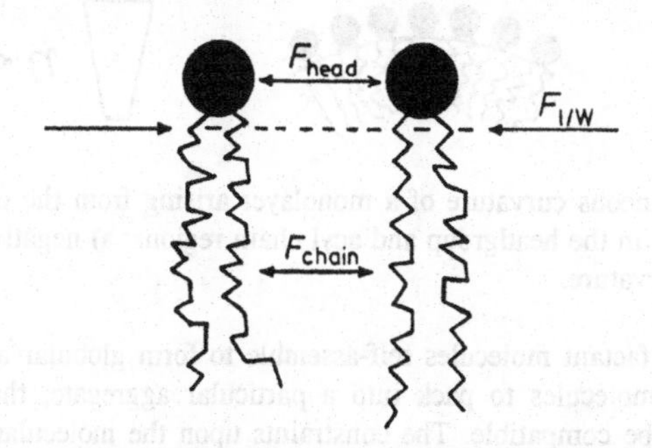

Figure 2. Illustration of the balance of lateral forces across a lipid monolayer.

An imbalance of attractive and repulsive forces at the level of the headgroup and acyl chains within a given lipid monolayer yields a spontaneous curvature. Clearly the two monolayers cannot be simultaneously at a free energy minimum with regards to their intrinsic or spontaneous curvature. It follows that a symmetrical bilayer composed of such non-lamellar forming lipids is under a condition of curvature elastic stress. This curvature free energy associated with the membrane lipid-water interface then leads to formation of non-lamellar (cubic and/or H_{II}) phases, e.g. as the temperature is increased.

Figure 3 provides an illustrative representation of these concepts. Figure 3a shows that for headgroups, that are small or favour hydrogen bonding (e.g. PE headgroups), a relatively small equilibrium separation is favoured. If the chains favour a larger separation (e.g., long, bulky polyunsaturated of phytanoyl chains), then a significant imbalance of the forces across the layer yields a curvature stress, producing a negative spontaneous curvature. It follows that a H_{II} phase is favoured as the temperature is increased. Figure 3b shows, that, if the forces within the headgroup region favour an equilibrium interfacial area, that is similar to the chain cross-sectional area (e.g., in the case of PC's), then the spontaneous curvature is zero - the attractive and repulsive forces within the headgroup and acyl chain regions are balanced. Such lipids tend to form lamellar phases. Finally, Fig. 3c indicates, that if the forces within the headgroup region lead to larger separations than in the acyl chains (e.g., in the case of gangliosides and detergents) then there results a positive spontaneous curvature. Consequently micelles and the normal hexagonal H_I phase is favoured.

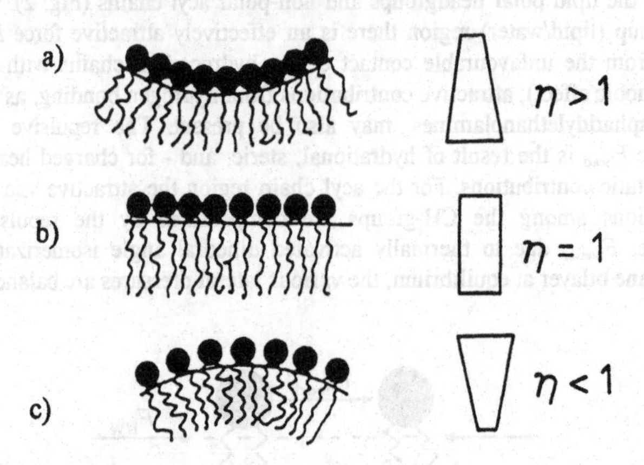

Figure 3. Spontaneous curvature of a monolayer arising from the distribution of lateral forces within the headgroup and acyl chain regions: a) negative, b) zero, c) positive mean curvature.

Amphiphilic, surfactant molecules self-assemble to form globular aggregates. In order for these molecules to pack into a particular aggregate, their molecular dimension must be compatible. The constraints upon the molecular geometry of the amphiphilic molecules, which are determined by the balance between the repulsive and attractive forces across the lipid bilayer, can be quantified by the so-called surfactant parameter η. This parameter describes the dynamically averaged shape of the molecule in terms of the volume of the hydrophobic region (the chain volume), V, the preferred length of the acyl chains, l, and the polar head group cross-sectional area, A. The magnitude of the dimensionless surfactant parameter is given by

$$\eta = V/(A \cdot l) \qquad (1)$$

Its parameters depend on the temperature, pressure, water content, steric effects, pH and ionic strength of the solvent. η defines the preferred direction of (mean) curvature of the aggregate. If $\eta = 1$, the hydrophobic chain region of each molecule has, on average, a cylindrical shape. If $\eta < 1$, the molecule is wedge-shaped, tapered towards the free chain ends. If $\eta > 1$, it is tapered in the opposite sense. Close packing of these molecules leads to curved interfaces, and the direction of the curvature towards the polar or hydrophobic parts depends upon the value of η, as indicated in Fig. 3 (we adopt the convention that the mean curvature is positive if the interface is curved towards the polar regions, and negative

otherwise). In general, knowledge of η is not sufficient to determine uniquely the interfacial geometry, as it is not directly related to the curvatures of the interface. Both the mean and Gaussian curvature can be varied cooperatively without altering the value of η. Nevertheless, the surfactant parameter does furnish a local constraint upon the curvatures of the interface.
However, the molecular shape is not the sole determinant of the structure of an aggregate. If the surfactant/water mixture is to from a single phase, such as in bicontinuous cubic phases, the surface and volume requirements set by the composition of the mixture must also be satisfied. Typically, the surfactant parameter varies between about 1.05 and 1.5 for cubic phases.

So far, no full theoretical description of lyotropic lipid phase behaviour exists, though some progress has been made in recent years [40-47]. Often a concept is used that can be explained in terms of a small set of parameters, irrespective of the precise chemical nature of the lipid molecule. Helfrich described the surface curvature energy contribution associated with amphiphile films in terms of three curvature elastic parameters: the spontaneous mean curvature H_s, the mean curvature modulus κ_m, and κ_G, the Gaussian curvature modulus [48]. Applying differential geometry, the surface energy per unit area for small curvatures is given by

$$G_{bend}/S = g_{bend} = 2\kappa_m\langle(H-H_s)^2\rangle+\kappa_G\langle K\rangle \quad (2)$$

$H = (C_1+C_2)/2$ is the mean interfacial curvature, which is equal to half the sum of the principal curvatures $C_1 = 1/R_1$ and $C_2 = 1/R_2$ at the interface (Fig. 4), and $K = C_1C_2$ is the Gaussian curvature at the interface, given by the product of the principal curvatures C_1 and C_2 of the interface (e.g., $C_1 = C_2 = 0$ for a planar bilayer; $C_1 < 0$, $C_2 = -C_1 > 0$, i.e., $H = 0$, $K < 0$, for a hyperbolic saddle surface). The spontaneous mean curvature H_s is that mean curvature the lipid aggregate would wish to adopt in the absence of external constraints (Fig. 3), and κ_m tells us what energetic cost there would be for deviations away from this. Besides the curvature energetic contribution, there will be other energetic contributions. Due to the desire to fill all the hydrophobic volume by the amphiphile chains (due to the hydrophobic effect), there will be also a contribution quantifying an eventual packing frustration. A further contribution might be due to interlamellar interactions (e.g., van der Waals interaction, hydration repulsion, undulation forces). The curvature elastic energy is believed to be the crucial term governing the stability of non-lamellar phases and the ability of lipid membranes to bend, in particular at high levels of hydration. To probe the concept of any energetic description and the resultant set of parameters necessary to provide a general explanation of universal lyotropic phase behaviour, one needs to scan the appropriate parameter space experimentally.
Most of the experimental work so far has relied on temperature and sample composition as the tools to attack this problem. A further important thermodynamic variable is pressure. Besides the general physico-chemical interest in using high pressure as a tool for understanding phase behaviour, structure and energetics of amphiphilic molecules, high pressure is also of considerable physiological and biotechnological (e.g., deep sea diving, high pressure food

processing) interest [49-53]. Hydrostatic pressure significantly influences the structural properties and thus functional characteristics of cell membranes, yet this has not prevented the invasion of high pressure and cold habitats by deep-sea organisms (up to about 10000 m depths - corresponding to 1 kbar of pressure - at 2-3 °C). 70 % of the surface of the earth is covered by oceans, and the average pressure on the ocean floor is about 380 bar. Without any compensatory adjustments the membranes of deep-sea organisms should be highly ordered. It is now well established that deep-sea organisms display a variety of adaptations to high hydrostatic pressure and low temperature at the molecular level of organization to keep their membranes in a fluid-like state which is a prerequisite for optimal physiological function [54].

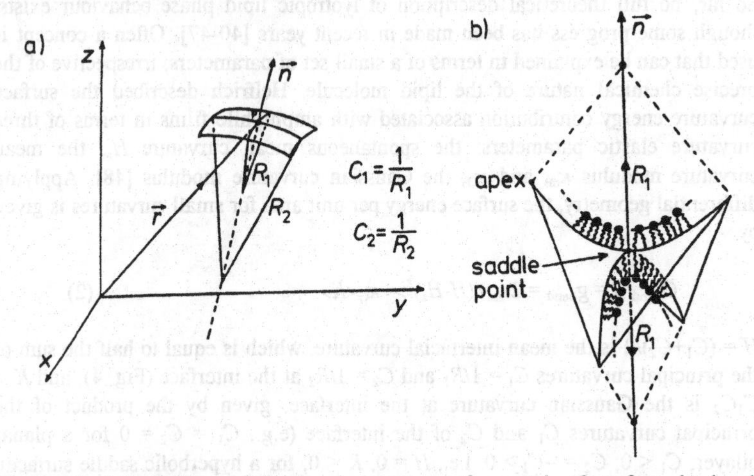

Figure 4. a) Definition of the local curvature of a patch of surface; b) local curvature of a saddle surface with negative Gaussian curvature ($\vec{n}$ normal vector to the surface at point *P*).

In this review, we discuss experimental techniques for studying the high pressure structure of lyotropic lipid mesophases, and we present data on the temperature and pressure dependent phase behaviour, energetics and stability of single- and two-component lipid dispersions. Then, we discuss the pressure-jump relaxation technique for studying the kinetics of phase transformations between different lyotropic mesophases.

2. Experimental Techniques

Most of the experimental results presented are based on small-angle X-ray (SAXS) and Fourier-transform infrared spectroscopy (FT-IR) studies, which have been designed to allow also for pressure dependent studies. The techniques are briefly discussed in the following section.

X-ray diffraction. The small- and wide-angle X-ray diffraction experiments were performed at Synchrotron X-ray beam lines, e.g. at beam line X13 of the EMBL outstation at DESY [55]. With the installed germanium monochromator, the X-ray wavelength was fixed to 1.5 Å. Sets of guard slits were used to adjust the beam size at the sample (0.5 mm in height and 3 mm in width) and to reduce parasitic scattering. One-dimensional diffraction patterns in the small- and wide-angle regime were recorded simultaneously using two sealed linear detectors with delay line readout connected in series. To avoid radiation damage, a small selenoid-driven lead shutter protected the samples from excess radiation within the periods where no data were recorded. Additional information, like temperature, pressure, time, and X-ray flux measured with an ionisation chamber in front of the sample, was stored in the local memory.

The reciprocal spacings $s = 1/d_{hkl} = (2/\lambda)\sin\theta$ (d_{hkl} lattice spacing, 2θ scattering angle, λ wavelength of radiation, see also below) were calibrated in the small-angle region by the diffraction pattern of rat-tail collagen. The calibration is accurate to within 1.7 %. This means that in a typical measurement of a d-spacing of 60 Å, which is typical for the repeat distance of a lamellar lipid structure, we might expect d with an accuracy of about ± 1 Å. The relative accuracy is better than 0.1 Å, however.

For the investigation of the structure and phase behaviour of the lipid system at elevated pressures as well as for studies of the kinetics of lipid phase transitions using the pressure-jump technique, we built a high pressure X-ray cell suitable for studies up to pressures of 8 kbar at temperatures ranging up to 140 °C. Figure 5 shows the cross-sectional view of the essential parts of the cell. The pressure vessel is made from stainless steel of high tensile strength, it contains two high pressure connections perpendicular to each other and a bore for a thermocouple. Temperature control is achieved by circulating water from a thermostat through the outside jacket of the vessel. The temperature is controlled to within 0.2 °C. The sample of 40 μL volume is hold in a PTFE-ring that is closed with mylar foils glued on both sides of the ring to separate the sample from the pressurizing medium (distilled water). Beryllium discs of 1.0 to 2.0 mm thickness are used as X-ray windows up to pressures of 2 kbar [56]. For higher pressures, flat diamond windows of 0.5-1 mm thickness are used. The window holders are sealed with Viton-O-rings and are kept by closure nuts. The maximum scattering angle observable with this configuration is 25°. The pressures were measured using a Heisse Bourdon gauge with an accuracy of 2 bar.

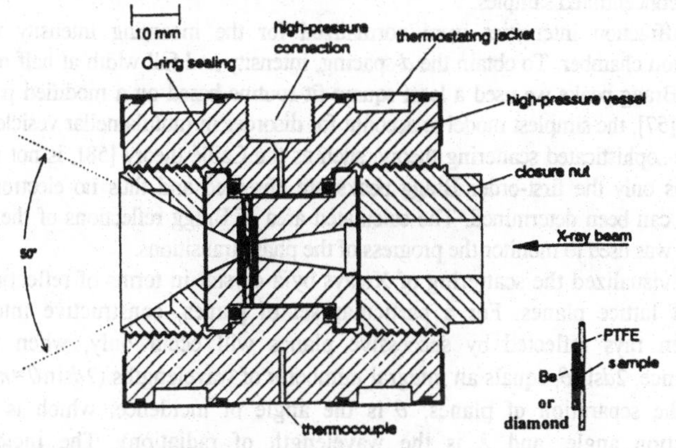

Figure 5. Schematic drawing of the high-pressure X-ray cell.

The pressure-jumps are performed by a computer controlled opening of an air operated valve between the high pressure cell and a liquid reservoir container. With the pressure-jump apparatus rapid (< 5 ms) and variable amplitude pressure-jumps are possible. To minimize the effect of an adiabatic temperature change in the course of the pressure-jump, the high pressure cell was constructed to hold only a very small volume of the pressurizing medium.

In relaxation kinetic measurements, the pressure-jump technique has been shown to offer several advantages over the temperature jump approach: 1) pressure propagates rapidly so that sample homogeneity is less of a problem, 2) pressure-jumps can be performed bidirectional, i.e. in the pressurization and depressurization direction, and 3) the amplitude of the pressure-jump can be easily and repeatedly varied to a level of high accuracy, thus also allowing to average sets of diffraction data to improve the counting statistics in the case of fully reversible phase transitions.

For the pressure range up to 20 kbar, a diamond anvil cell (DAC) was used [55]. In these experiments, no bidirectional pressure runs can be performed. The sample was sealed in a gasket of 0.25 mm thickness with a hole of 0.37 mm diameter. Pressure was calibrated by adding a small ruby crystal to the sample and measuring the shift of the R1 ruby fluorescence line. The light source for the excitation of the ruby was a tungsten lamp in combination with a 650 nm edge filter. The fluorescence spectra were analysed with a grating monochromator equipped with a diode array detector from AMKO (Uetersen, Germany). A focal length of 200 mm and a grating with 1800 lines/mm give a reciprocal linear dispersion of 2.7 nm/mm. A diode width of 25 μm corresponds to a wavelength shift of 0.068 nm. Using the relation $\Delta\lambda/\Delta p$ = 0.0365 nm/kbar, the wavelength of the R1 line is converted to pressure. When fitting the ruby spectra, pressure can be determined to within 300 - 500 bar.

The X-ray transmission and thus the scattering background drastically depends on the density of the solvent (water) and the path length of the pressure cell, both being pressure-dependent. To perform an accurate background correction, one therefore has to measure both the sample and reference solution under the same cell conditions at each pressure. This condition requires high reproducibility of the path-length change by pressure. The DAC is therefore not suited for measuring diffuse small-angle scattering of low concentrated samples.

The diffraction intensities were normalized for the incoming intensity using an ionisation chamber. To obtain the *d*-spacing, intensity and full width at half maximum of the Bragg peaks we used a least square fit routine based on a modified paracrystal model [57], the simplest model to account for disorder in multilamellar vesicles. Use of a more sophisticated scattering theory, such as the Caillé-theory [58], is not necessary here, as only the first-order Bragg peaks are observed and thus no electron density profile can been determined. The calculated area of Bragg reflections of the different phases was used to monitor the progress of the phase transitions.

Bragg visualized the scattering of X-rays by a crystal in terms of reflections from sets of lattice planes. For a particular set of planes, constructive interference between rays reflected by successive planes will occur only, when the path difference, $2d\sin\theta$, equals an integral number n of wavelengths ($2d\sin\theta = n\lambda$, where d is the separation of planes, θ is the angle of incidence, which is half the diffraction angle, and λ is the wavelength of radiation). The incident and diffracted beams are specified by wavevectors $\bar{k}$ and $\bar{k}_0$, whose moduli are $2\pi/\lambda$.

The diffracted intensity is often plotted as a function of the scattering vector $\vec{Q}$, where $\vec{Q} = \vec{k} - \vec{k}_0$. For elastic scattering its modulus is given by $Q = 4\pi\sin\theta/\lambda$. In X-ray scattering also the modulus of the scattering vector $s = |\vec{s}| = 2\sin\theta/\lambda = Q/(2\pi)$ is in use. An equivalent statement of Bragg's law is then $Q_n = n(2\pi/d)$. For a set of equally spaced planes, the scattering intensity $I(\vec{Q})$ is everywhere zero except where Bragg's law is satisfied, and the diffraction pattern consists of a set of equally-spaced Bragg peaks, $(2\pi/d)$ apart, along a direction normal to the planes.
Powder samples - which are of relevance here only - are composed of very many randomly distributed microdomains, so that the Bragg condition is automatically satisfied for all values of n, and all of the possible diffraction peaks are simultaneously recorded. The positions of the diffraction peaks are reciprocally related to the separations betwen molecules (or group of molecules) within the lyotropic mesophase. The sharpness of peaks is related to the extent to which the separations extend periodically over large distances. For phases with Bragg peaks, the ratio of the peak positions reveal the long-range organisation of the phase, otherwise diffusive small-angle scattering is observed only. Low-angle peak ratios of 1, 2, 3, ... indicate a lamellar phase (1-dimensional stack of layers); ratios of 1, $\sqrt{3}$, 2, $\sqrt{7}$, 3, ... are indicative of a hexagonal phase (2-dimensional lattice of densely packed rods), and ratios of e.g. 1, $\sqrt{2}$, $\sqrt{3}$, 2, $\sqrt{5}$, $\sqrt{6}$, $\sqrt{8}$, 3, ... indicate a cubic phase. The wide-angle peaks relate to the translational order over short distances, e.g., to the lateral in-plane packing of the lipids in the mesophase.
Bragg's law does not say anything about the intensities of the various diffraction peaks. The observed scattering intensity $I(\vec{Q})$ in a particular direction is the modulus-quared of the scattering amplitude $F(\vec{Q})$, which is given by the Fourier transform of the electron density distribution $\rho(\vec{r})$. For a periodic structure, $F(\vec{Q})$ is only non-zero for those discrete diffraction directions, denoted by the reciprocal lattice vectors $\vec{Q}_{hkl}$, for which the Bragg condition is satisfied. The set of integers (hkl) are the Miller indices of the various lattice planes of the crystal.

FT-IR spectroscopy. Fourier-transform infrared spectroscopy has been shown to be a valuable tool for studying the phase behaviour and the conformation of lipid membrane systems. Applying pressure in the 15 kbar regime only affects the intermolecular distances and, unlike temperature, does not impart thermal energy to the system. Therefore changes in the infrared spectrum of the lipid system can be directly related to variations in the intermolecular interactions, that can be caused by bond strengthening or weakening, conformational changes or hydrational effects at the polar/apolar interface. In general, higher pressure decreases the intermolecular distances and increases the repulsion forces between the atoms, leading to a linear increase of the stretching vibrations. Deviations from this linear pressure dependence can be interpreted in terms of phase transformations in connection with structural or conformational changes of the lipid bilayer system. Many infrared spectral parameters, particularly the frequencies, widths, intensities, shapes, and splittings of the IR bands, are very sensitive to the structural and dynamical properties of membrane lipid molecules [11, 59]. In particular cases, such as by analyzing the CH_2 wagging vibrations, even quantitative conformational information on conformer population can be obtained.

For the pressure dependent infrared studies, small amounts of the sample were placed, along with powdered α-quartz which was used as an internal pressure calibrant [60, 15], in a 0.5 mm diameter hole of a 0.05 mm thick stainless steel gasket. The pressure is accurate to within about 200 bar by this method. A diamond anvil cell (DAC) with type IIa diamonds was used for applying pressure. Temperature was controlled to within 0.1 °C by means of water flowing from a thermostat through the thermostating jacket of the DAC and was monitored via a thermocouple placed in the cell body close to the diamond window. Infrared spectra were collected on a Nicolet Magna 550 Fourier-transform infrared spectrometer. A beam condensing system focussed the light on the center of the diamond windows. For each spectrum 512 scans were coadded at a spectral resolution of 2 cm^{-1}, and apodized with a Happ-Genzel function. This set-up leads to a total measuring time per spectrum of about 8 minutes. The sample chamber was purged with dry and carbon dioxide-free air during the data collection to avoid spectral contributions from atmospheric gases.

3. Results and Discussion

3.1 PHASE TRANSITIONS OF SINGLE-COMPONENT PHOSPHOLIPID BILAYER DISPERSIONS

Lamellar Phase Transitions. The increase in entropy with chain rotational disorder, together with the increase in intermolecular entropy and the increased headgroup hydration are the driving forces for the $L_{\beta'}(P_{\beta'})/L_\alpha$ chain melting (gel/fluid) transition of membranes. Opposing the chain melting is the increase in internal energy due to increased rotational isomerism, the decreased van der Waals attraction between chains, and the increased hydrophobic exposure at the polar/apolar interface that results from the lateral expansion of the bilayer accompanying the increase of chain isomerism. The balance of these opposing contributions to the bilayer free energy, which dependens on the geometry of the lipid molecule, determines the chain melting transition temperature T_m. Generally, the lamellar gel phases prevail at high pressure and low temperature. They give way to the lamellar liquid-crystalline L_α phase as pressure is lowered and temperature is raised. A common value for the L_α/gel transition slope of about 22 °C/kbar has been observed for the saturated phosphatidylcholines DMPC, DPPC and DSPC [6-13] (see Fig. 6; for abbreviations see below). The positive slope can be explained by the endothermic enthalpy change ΔH_m and the volume increase ΔV_m at the gel to L_α transition through the Clapeyron relation $dT_m/dp = T_m\Delta V_m/\Delta H_m$. Similar transition slopes have been found for the mono-cis-unsaturated POPC, the phosphatidylserine DMPS, and for the phosphatidylethanolamine DPPE. Only those of cis-unsaturated DOPC and DOPE have been found to be markedly smaller [6,11]. The transition slope does not significantly depend on the hydrocarbon chain length or the type of headgroup, they affect the transition temperature, mainly. The existence of cis double-bonds in the chain region drastically affects the transition slope, however. The introduction of cis double-bonds leads to the lowest transition temperatures and smallest transition slopes, presumably as they impose a kink in the linearity of the acyl chains, thus creating significant free volume in the bilayer, which reduces the ordering effect of high pressure. Also further pressure-induced gel phases have been observed in single-component phospholipid dispersions, such as an interdigitated high pressure gel phase $L_{\beta i}$ in DPPC and DSPC [8,7]. These studies clearly demonstrate that, by regulating the lipid composition of the cell membrane through changes in lipid chain length, degree of unsaturation and headgroup structure, biological organisms are already provided with a mechanism for efficiently modulating the physical state of their membranes in response to changes in the external environment ("homeoviscous adaptation"), such as high hydrostatic pressure in the deep sea [6,54]. However, nature has also further means to regulate membrane fluidity, such by changes in concentration of cholesterol and protein in the lipid bilayer [6,15,61].

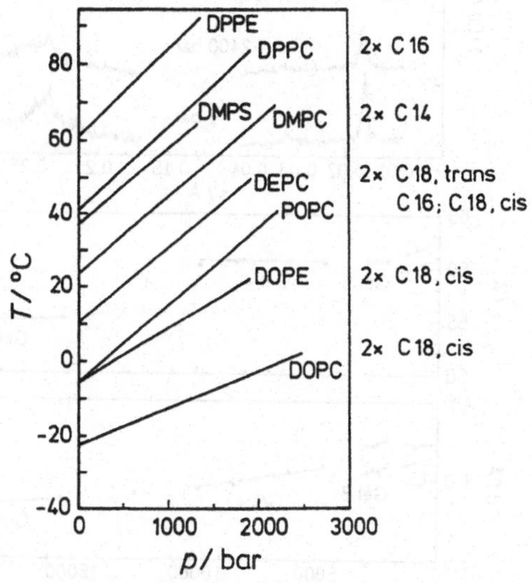

Figure 6. T,p-phase diagram for the main (chain-melting) transition of different phospholipid bilayer systems. The L_α-phase is observed at the low-pressure high-temperature corner of the phase diagram.

At higher pressures, further pressure-induced gel phases are generally observed. The temperature-pressure phase diagram of DPPC/water dispersions has been explored by performing X-ray diffraction measurements up to 20000 bar [62]. In the small-angle X-ray scattering (SAXS) region, Bragg reflections of first order occur ($s_{001} = 1/d_{001} = 1/a$) indicating a lamellar packing of the lipid bilayers in form of multilamellar vesicles. The repeat distance of the lamellar lattice constant a is the sum of the lipid bilayer thickness and the thickness of one adjacent interlamellar water layer. In the wide-angle X-ray scattering (WAXS) region the packing of the lipid acyl chains is detected, i.e., the chain-chain lattice spacings s_{hk}. In the gel phase, the chain lattice is normally either of an undistorted or a distorted hexagonal type. Assuming a hexagonal packing of the lipid acyl chains, reflections occur at $s_{11}=s_{20}=1/d_{hex}=2/(\sqrt{3}\ a_{hex})$ where a_{hex} is the distance between two acyl chains. A distorted hexagonal or orthorhombic lattice of the lipid acyl chains has two reflections in the wide-angle region. A sharp reflection at $(1/d_{20})$ and a usually broad one at $(1/d_{11})$. The orthorhombic lattice parameters a_{orth} and b_{orth} of the chain packing can be calculated using $a_{orth}=2d_{20}$ and $b_{orth}=d_{11}/[1-(d_{11}/2d_{20})^2]^{1/2}$. The area per chain perpendicular to the chain axis is given by $A=a_{orth}b_{orth}/2$. In the fluid, liquid-crystalline phase, only a diffuse broad high-angle reflection is observed centered around 1/(4.5 Å), which is indicative of disordered chains.

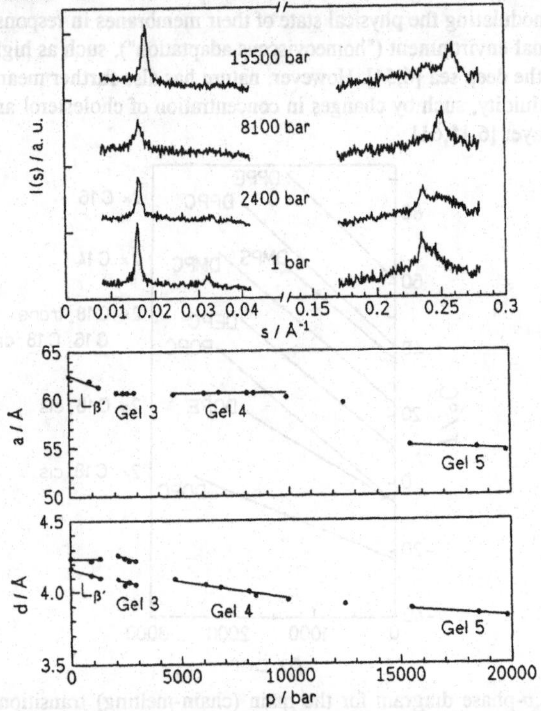

Figure 7. a) X-ray small- and wide-angle diffraction patterns of a 30 wt-% DPPC dispersion at 23 °C and selected pressures; b) lamellar lattice constants *a* and *d*-spacings as a function of pressure.

As an example of the quality of X-ray diffraction patterns one can get even in a diamond anvil cell, Fig. 7 shows selected SAXS and WAXS patterns of DPPC in excess water for T = 23 °C at several pressures. It is clearly seen that peak positions and shapes are changing with pressure indicating different phases. The corresponding lamellar lattice constant a is plotted for this temperature as a function of pressure. Two phase transitions can be observed: The $L_{\beta'}$(Gel 2)/Gel 3 transition at about 1700 bar and the Gel 4/Gel 5 transition at about 12500 bar. No significant change in small-angle diffraction pattern is observed in passing the Gel 3/Gel 4 transition which has been found by FT-IR measurements and Raman studies [11,62]. Whereas the lamellar spacing a decreases with increasing pressure in the $L_{\beta'}$-phase, $a(p)$ seems to be almost constant in the Gel 3 and Gel 4 phase. A drastic drop to about a = 55 Å is found when entering the Gel 5 phase. From WAXS, the d-spacing is obtained, which is also plotted as a function of pressure. In the $L_{\beta'}$ and in the Gel 3 phase, the wide-angle reflection is asymmetric, having a shoulder on the high s side. By fitting the diffraction data this asymmetric peak was resolved into two. In contrast to the Gel 3 phase (between 1.7 and 3.8 kbar), the pressure dependence of these two peaks is different in the $L_{\beta'}$-gel phase. Similar diffraction patterns have been obtained for other temperatures as a function of pressure, thus allowing us to construct a phase diagram of the system, which is shown in Fig. 8. Data points of phase transitions obtained by FT-IR and X-ray diffraction measurements are given. At ambient conditions, the structure is the $L_{\beta'}$(Gel 2)-phase which is well known. The lipid acyl chains, which are packed in an orthorhombic lattice, are tilted with respect to the bilayer normal with an angle of $\theta \approx 31°$. The direction of the tilt angle is towards nearest neighbours, i.e., in the direction of the b_{orth}-lattice vector of the orthorhombic chain lattice. With increasing pressure, we observe almost no change in the d_{20}-spacing, but a reduction of the d_{11}-spacing: At 23 °C and 1 bar, we obtain a chain cross sectional area of A = 20.3 Å^2, and at 1300 bar a reduced area of A = 19.8 Å^2. Since only the b_{orth}-lattice constant is changed, it is likely that the tilt angle is reduced by increase of pressure, which leads to a smaller head group area and a lower level of hydration at the lipid/water interface. The water layer thickness decreases upon pressurization leading to a reduction of the lamellar lattice constant a. At lower temperatures, we observe two wide-angle peaks (one asymmetric peak) in the pressure region of the Gel 3 phase. In contrast to the $L_{\beta'}$-phase, the pressure dependence of the peak position is the same for both peaks. The two-dimensional lattice is uniformly compressed by about 1.2 %/kbar. We can speculate that within the Gel 3 phase the tilt angle of the lipid chains remains constant leaving the hydration of the lipid headgroups essentially unchanged. The interlamellar water layer thickness does not change with pressure and thus a pressure independent lamellar lattice constant a is observed. At higher temperatures (above 40 °C), an gel phase $L_{\beta i}$ appears, where the acyl chains are partially interdigitated. In the pressure region of the Gel 4 phase only one sharp symmetric reflection is found in the wide-angle region below 36 °C. From this, we can conclude that the lipid acyl chains are packed in a hexagonal lattice and that the tilt angle is nearly zero. At 23 °C and 4700 bar, the lateral spacing is d_{hex} = 4.09 Å, which corresponds to a chain distance of a_{hex} = 4.72 Å and a molecular cross sectional area of 38.6 Å^2. At 12500 bar, the lateral spacing is reduced to 3.91 Å leading to a_{hex} = 4.51 Å and a molecular area of 35.3 Å^2. The higher density of the lipid headgroup packing at this pressure largely prevents the penetration of water molecules into the lipid bilayer and thus a hydration of the carbonyl groups. There is no difference in the wide-angle diffraction patterns of the Gel 4 and Gel 5 phases, so that the packing of the lipid chains is the same in both phases, hexagonal and not tilted. The Gel 5 phase is characterized by a very small lattice constant a which is about 55 Å at 23 °C. A DPPC lipid bilayer with untilted chains is nearly as thick as this lattice constant. This means that the lipid molecules have lost essentially all the hydration water, which now exists as bulk water frozen as ice, probably ice VI.

Information about hydrational properties of the polar/apolar interface can be obtained by analysis of the C=O stretching vibration of DPPC using FT-IR spectroscopy. At ambient pressure and temperature it is located around 1737 cm^{-1}. The contour and position of the carbonyl band is sensitive to conformational changes in the interfacial region and to the level of hydration of the lipid headgroup. Analysis of the band by Fourier self-deconvolution gives rise to three overlapping bands centred at about 1744 cm^{-1}, 1725 cm^{-1} and 1696 cm^{-1}, that can be assigned to stretching modes of free-, mono- and higher hydrated carbonyl groups. The integral intensities of these sub-components were determined by fitting the experimental C=O stretching band with three Gaussian-Lorentzian functions. The integral intensity ratio of the band at 1744 cm^{-1} with respect to the two lower frequency components is a measure for the degree of hydration of the interfacial region, assuming that the transition dipole moments of the three spezies are similar. Figure 9 shows the relative contribution of non-hydrated C=O groups ($n_{C=O,free}/n_{C=O,total}$) at 20 °C as a function of pressure. Whereas at pressures below 2 kbar, i.e. within the $L_{\beta'}$ gel phase, the level of hydration decreases drastically with increasing pressure, the amount of hydrated carbonyl groups does not depend significantly on the applied pressure within the Gel 3 phase. Above about 4 kbar the Gel 4 phase is formed and increasing pressure again leads to a diminishing of the level of hydration of the C=O groups.

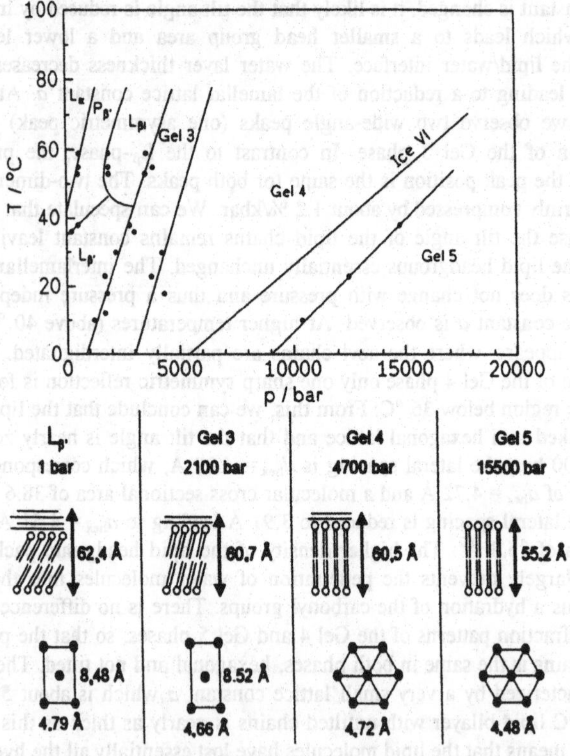

Figure 8. *T,p*-phase diagram of DPPC in excess water and schematic drawing of the lamellar repeat unit and lipid packing in the bilayer plane of some of these phases.

The use of pressure to study the phase behaviour of DPPC has allowed us to observe phase changes that are not seen at ambient temperature. The structural differences observed for the different phases are essentially due to changes in chain packing area, lipid length, chain tilt angle and level of hydration.

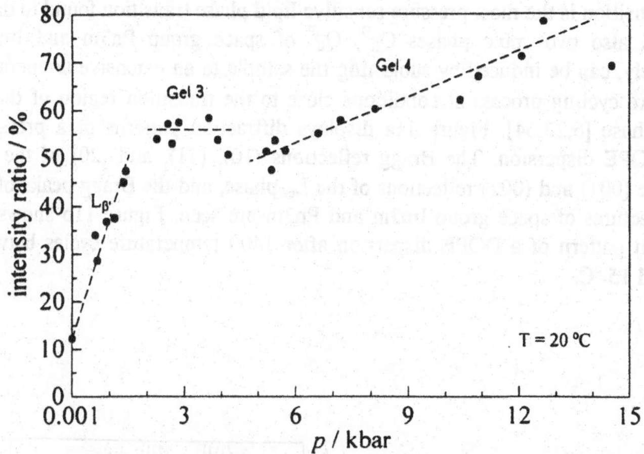

Figure 9. Relative concentration of non-hydrated carbonyl groups of DPPC at 20 °C as a function of pressure.

Non-lamellar phase transitions. For a series of lipid molecules, also non-lamellar phases occur as thermodynamically stable phases or they can often be induced as long-lived metastable states. Here we discuss examples, taken from different groups of amphiphilic molecules. Contrary to DOPC, the corresponding cis-unsaturated phospholipid with ethanolamine as headgroup, DOPE, in addition to the lamellar L_β/L_α transition, exhibits a lamellar L_α to inverted hexagonal (H_{II}) transition. As pressure forces a closer packing of the chains, which results in a reduction of the number of gauche bonds and kinks within the chains, both transition temperatures increase with increasing pressure.

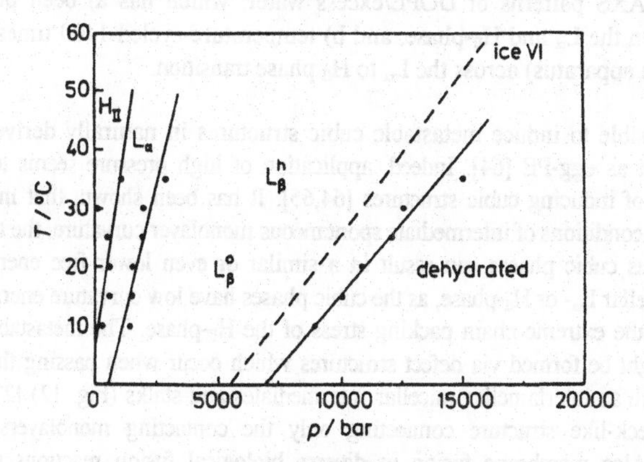

Figure 10. T,p-phase diagram of DOPE in excess water.

Figure 10 displays the corresponding T,p-phase diagram obtained. The transition slope of the H_{II}/L_{α} transition ($dT_h/dp \approx 40$ °C/kbar) is almost three times steeper as the slope of the lamellar chain-melting transition ($dT_m/dp \approx 14$ °C/kbar). A similar steep slope for the H_{II}/L_{α}-transition has also been observed for egg-PE [6,63]. The H_{II}/L_{α}-transition is the most pressure-sensitive lipid phase transition found to date. In DOPE also two cubic phases Q_{II}^D, Q_{II}^P of space group Pn3m and Im3m, respectively, can be induced by subjecting the sample to an extensive temperature or pressure cycling process at conditions close to the transition region of the L_{α} and H_{II}-phase [6,23,64]. Figure 11a displays diffraction patterns of a pressure-cycled DOPE dispersion. The Bragg reflections (10), (11), and (20) of the H_{II}-phase, the (001) and (002) reflections of the L_{α}-phase, and the Bragg peaks of the cubic structures of space group Im3m and Pn3m are seen. Figure 11b shows the diffraction pattern of a DOPE dispersion after 1400 temperature cycles between −5 °C and 15 °C.

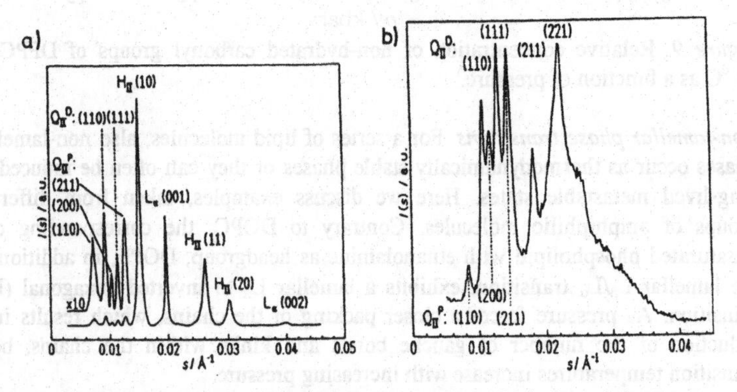

Figure 11. SAXS patterns of DOPE/excess water, which has a) been pressure-cycled between the L_{α} and H_{II}-phase, and b) temperature-cycled 1400 times (using a roboter-type apparatus) across the L_{α} to H_{II} phase transition.

It is also possible to induce metastable cubic structures in naturally derived lipid systems, such as egg-PE [64]. Indeed, application of high pressure seems to be an effective way of inducing cubic structures [64,65]. It has been shown, that in certain situations, for conditions of intermediate spontaneous monolayer curvature, the topology of bicontinuous cubic phases can result in a similar or even lower free energy than either the lamellar L_{α}- or H_{II}-phase, as the cubic phases have low curvature energies and do not suffer the extreme chain packing stress of the H_{II}-phase. The metastable cubic structures might be formed via defect structures which occur when passing the L_{α}/H_{II} transition, such as interlamellar micellar intermediates and stalks (Fig. 12) [27,66]. A stalk is a neck-like structure connecting only the contacting monolayers of the membranes. Also membrane fusion in diverse biological fusion reactions probably involves formation of some specific intermediates, such as stalks and pores, and the energy of these intermediates, and, consequently, the rate and extent of fusion, depend on the propensity of the corresponding monolayers of membranes to bend. Lipid composition may be altered by enzymatic (fusion proteins) attack. From a membrane point of view, to be best suited for fusion, membranes should be asymmetrical, with the contracting monolayers containing H_{II}-phase-promoting lipids (e.g., cholesterol, PE's with negative H_s), and inner (distal) monolayers micelle-forming lipids (e.g., lysolipids). The concentration of various lipid species on the cytosolic and plasma or lumenal monolayers of biological membranes is highly regulated. Among the possible functional roles this transbilayer lipid asymmetry may have is thus the regulation of membrane fusion through the influence of spontaneous monolayer curvature.

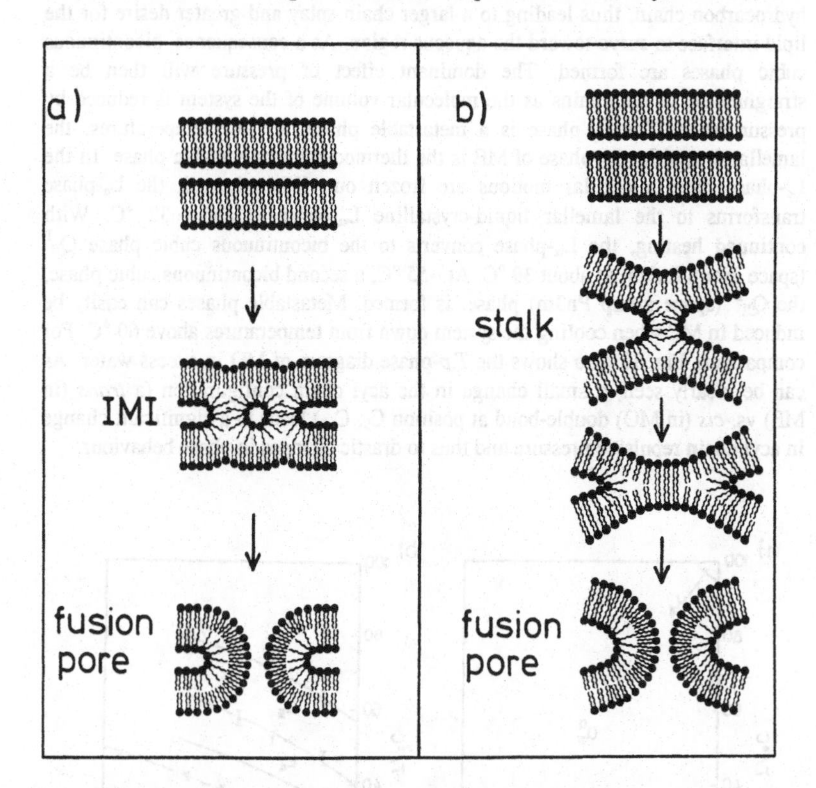

Figure 12. Possible mechanisms of membrane fusion indicating specific intermediates structures; a) two opposed membranes form an interlamellar micelle intermediate (IMI), which leads to the formation of an interlamellar attachment or fusion pore; b) the monolayers of opposed membranes mix to form a stalk intermediate that expands, leading to rupture as a result of curvature and interstitial stress and then to a fusion pore (adopted from [66]).

It has been shown for a wide variety of lipid compositions exhibiting fusion that the formation of a stalk intermediate is energetically favoured. However, it is probable that fusion proteins affect the energies of these proposed structures or, possibly, give rise to different intermediate lipid structures in the fusion of biological membranes. As a result, it remains difficult to definitely assign the non-bilayer lipid structures that may be involved in membrane fusion.

Fourier-transform infrared spectroscopy has been used to characterize differences in conformation and hydration between different lamellar and non-lamellar phases of the single-chain lipid monoelaidin (ME) in excess water [59] and, in combination with synchrotron X-ray diffraction results [16], to establish the *T,p*-phase diagram of the system over an extended temperature and pressure range (Fig. 13b). Increasing the temperature will introduce greater disorder in the hydrocarbon chain, thus leading to a larger chain splay and greater desire for the lipid interface to curve toward the aqueous region. As a consequence, bicontinuous cubic phases are formed. The dominant effect of pressure will then be a straightening of the chains as the molecular volume of the system is reduced by pressurization. The L_β phase is a metastable phase. At low temperatures, the lamellar crystalline L_c-phase of ME is the thermodynamically stable phase. In the L_c-phase most molecular motions are frozen out. When heated, the L_c-phase transforms to the lamellar liquid-crystalline L_α-phase at about 32 °C. With continued heating, the L_α-phase converts to the bicontinuous cubic phase Q_{II}^P (space group Im3m) at about 39 °C. At ~55 °C, a second bicontinuous cubic phase, the Q_{II}^D (space group Pn3m) phase, is formed. Metastable phases can easily be induced in ME when cooling the system down from temperatures above 60 °C. For comparison, Fig. 13 also shows the *T,p*-phase diagram of MO in excess water. As can be clearly seen, a small change in the acyl chain configuration (a *trans* (in ME) vs. *cis* (in MO) double-bond at position C_9, C_{10}) leads to a significant change in acyl chain repulsive pressure and thus to drastic changes in phase behaviour.

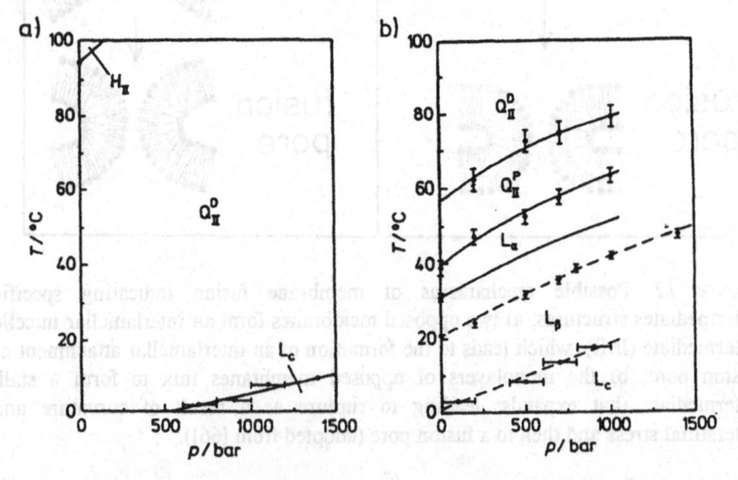

Figure 13. T,p-phase diagram of a) monoolein (MO) and b) monoelaidin (ME) in excess water. The lamellar L_β-phase of ME is probably metastable (dashed phase lines).

Modelling of Phase Behaviour. It is clear, that an understanding of the form of the T,p-phase diagram of lipid systems such as ME/water would require a detailed consideration of all the complex interactions involved, such as interfacial, hydration and van der Waals forces, steric repulsion, hydrogen bonding, as well as the geometry of the lipid molecule as a function of the thermodynamic parameters temperature and pressure. At present, modelling of these contributions is still in its infancy, and one is not yet able to predict lyotropic phase stability and phase sequence as a function of temperature and pressure. Only selected contributions can be modelled so far. We used X-ray diffraction measurements of the variation of lattice parameters a with hydration and pressure at 50 °C of hydrated monoelaidin to calculate the curvature elastic energy contribution for the lipid Q_{II}^{G} (Ia3d) cubic mesophase, which is observed under limited hydration conditions [65]. To calculate the curvature energy, the monolayer radius of curvature has been defined at the so-called neutral interface, which represents a surface which does not register a molecular cross-sectional area change upon bending by hydration. The neutral area surface is displaced from the minimal surface by a distance ξ. The cross-sectional area of a lipid molecular at a distance x from the minimal surface is given by [41,67]

$$s(x) = \frac{2S(x)v_L}{\phi_L a^3}. \tag{3}$$

with $S(x) = \sigma a^2 + 2\pi\chi x^2$ the total surface area of the monolayer, χ the Euler characteristic of the surface (−8 for the G-surface), σ = area of minimal surface/(unit cell volume)$^{2/3}$ ($\sigma = 3.091$ for the G-surface), v_L and ϕ_L are the molecular volume and volume fraction of the lipid, respectively. v_L can be determined by $v_L = Mv_{sp,L}/N_A$ with $v_{sp,L}$ the lipid specific volume, and M the molar mass of the lipid. Knowing ϕ_L and a from measurement we can calculate the lipid length l for a given phase [67,68]:

$$\phi_L = (1-\phi_W) = 2\sigma\left(\frac{l}{a}\right) + \frac{4\pi\chi}{3}\left(\frac{l}{a}\right)^3 \tag{4}$$

The volume fractions of lipid (ϕ_L) and water (ϕ_w) have been determined from the mass fraction and densities of water and ME, respectively, and the lattice constant a has been obtained from X-ray diffraction measurements. Using Eq. 4, a value of l = 15.1 Å is found for ϕ_L values between 0.4 and 0.5. In measurements on MO/water, a value of l = 17.3 Å has been found for T = 25 °C [43,45]. An appropriate neutral surface analysis involves calculating $s(x)$ at different positions along the length of the lipid molecule as a function of hydration. With increasing level of hydration, the cross-sectional area of the lipid molecule increases at the headgroup and decreases at the chain ends. The required neutral surface is located at that value of x in the cross-sectional area profile where $s(x)$ remains constant, i.e., where the profiles cross in the hydration series. At ξ = 9.3 Å for 50 °C and 1 bar, no change of $s(x)$ is observed. Here the neutral surface area is situated with a cross-sectional area per lipid molecule of 41.5 Å^2. Using the analysis of Templer

[42,43], ξ can be obtained in a more rigorous way. Referring the free bending energy per surface area for the two monolayers, which can be characterized by two neutral surfaces, a distance $\pm\xi$ apart from the IMPS, one obtains

$$G_{\mathrm{bend}} / S_0 = g_{\mathrm{bend}} = \kappa_{G,b}\langle K_0 \rangle + \kappa_{G,b}\langle K_0^2 \rangle \quad (5)$$

where $\kappa_{G,b}$ and $\kappa_{G,b}'$ are the first and second order Gaussian curvature moduli for the bilayer (index b), respectively, with $\kappa_{G,b} = 2\kappa_G - 8\kappa_m H_s \xi$ and $\kappa_{G,b}' = 4\kappa_m \xi^2 - (2\kappa_G - 8\kappa_m H_s \xi)\xi^2 / f$ and $f = 1.2187$. From the experimental determination of a and l as a function of hydration, ξ and the ratio of the bending constants (κ_G/κ_m) [41-43]

$$\frac{\kappa_G}{\kappa_m} = -\frac{8\pi\chi f\xi^2}{\sigma a^2 - 4\pi\chi\xi^2} + 4H_s\xi \quad (6)$$

can be calculated. It should be pointed out that the neutral surface is assumed to be parallel to the minimal surface, which is likely to be the case at high levels of hydration. A similar value of ξ of 9.6 to 9.7 Å can be determined using this method. This places the neutral surface at the upper part of the lipid chain which is generally believed to be the most incompressible region. Assuming that the spontaneous curvature H_s of the ME monolayers at 50 °C is similar to the mean curvature of the neutral surface, $\langle H_\xi \rangle = -5.32 \cdot 10^{-3}$ Å^{-1}, a value of $(\kappa_G/\kappa_m) = 0.012$ is obtained for $\phi_w = 0.49$, $\xi = 9.5$ Å, and $a = 174.6$ Å. Alternatively, the bending constants can be measured using osmotic pressure [46]. A similar value of $(\kappa_G/\kappa_m) = 0.05$ has been obtained for MO/water at 25 °C using $\xi = 12.6$ Å. As $\langle K_\xi \rangle < 0$, a positive value of (κ_G/κ_m) says that the free energy of the lipid system is indeed lowered upon formation of the bicontinuous cubic phase. Modifications of the free energy expression to include additional terms are being tested currently ([44]).

Energetic models based on first-order bending energy arguments predict that the different inverse bicontinuous cubic phases Q_{II}^G, Q_{II}^D and Q_{II}^P are energetically degenerate. In fact, this energetic degeneracy is not observed in the experimental results for establishing the *T,p*-phase diagram of this and other lipid systems. Only in excess water dispersions coexisting cubic phases can be observed for longer times, because here the two interfaces are almost parallel to the IPMS. Recently, it has been shown by Templer et al. [44] that the energetic degeneracy of the bicontinuous cubic phases can indeed be broken by changes in the geometry of the interface. Analysis of 2:1 fatty acid/phospholipid mixtures (see below) suggests that the destabilisation of Q_{II}^P with respect to Q_{II}^D as one increases the temperature may be understood in terms of the requirements for the interface to alter geometry as it goes from one of almost constant mean curvature to one which approaches a constant thickness monolayer.

3.2 PHASE TRANSITIONS OF BINARY LIPID MIXTURES

Fatty Acids/Phosphatidylcholines. The addition of fatty acids drastically changes the phase behaviour of aqueous phospholipid (lecithin) dispersions. Dispersions of pure phosphatidylcholines merely exhibit lamellar phases. Non-lamellar inverted hexagonal (H_{II}) and/or inverted bicontinuous cubic phases can be induced by adding fatty acids, such as lauric acid (LA), myristic acid (MA), palmitic acid (PA), or stearic acid (SA) [56,69,70]. Fatty acids probably influence also the fusogenicity of biological membranes because they relieve the formation of non-lamellar intermediate structures which have to occur in the process of membran fusion.

The gel to fluid phase transition temperature of e.g. the DMPC/MA (1:2) mixture is 25 °C above the main transition temperature T_m of pure DMPC dispersions (T_m = 23.9 °C). The rationale for this observation is that the fatty acid molecules act as spacers between the lecithin molecules and reduce the crowding of the relatively bulky phosphatidylcholine headgroups, and their ability to form hydrogen bonds, even rather stable phospholipid/fatty acid 1:2 complexes in the gel phase. This change in the steric balance of the bilayer results in the non-lamellar phases being energetically favoured over the fluid lamellar L_α-phase, immediately that the chain melting transition occurs. The lamellar L_α phase can be observed under non-equilibrium conditions after pressure jumps over the L_β to H_{II}/Q_{II}^P transition, however (see below).

In liquid-crystalline phases of DMPC/MA, DPPC/PA and DSPC/SA 1:2 mixtures, the incorporated fatty acid molecules increase the spontaneous curvature of the lipid monolayers, so that one does not observe the lamellar liquid-crystalline phase (L_α) in equilibrium measurements; the systems show a direct transition from the lamellar L_β to non-lamellar phases. Figure 14 shows the T,p-phase diagrams for the 1:2 mixtures of DLPC/LA, DMPC/MA, DPPC/PA and DSPC/SA containing excess water [56,70]. The latter three systems exhibit a phase separation into two crystalline phases at low temperatures in the T-range shown here, one of these crystalline phases consisting of the pure crystalline fatty acid (L_c) and the other of a mixture of the lecithin and the fatty acid (L_c^{com}). When chain melting occurs, the inverted hexagonal phase is found to be the stable liquid-crystalline phase in DPPC/PA (1:2) and DSPC/SA (1:2) dispersions, whereas in DLPC/LA (1:2) and DMPC/MA (1:2) dispersions the Q_{II}^{P}-phase is observed, eventually in coexistence with the H_{II}-phase.

In the systems DPPC/PA and DSPC/SA, neither the L_α nor cubic phases are found. The large splay of their fluid acyl chains leads to a large spontaneous curvature that can only be adopted by the H_{II} structure. Whereas the packing constraints at gentle curvatures still favour cubic phases in DLPC/LA and DMPC/MA, the higher degrees of curvatures in DPPC/PA and DSPC/SA seem to kill off cubic phases. Recently it has also been shown that the effects of reducing the propensity for interfacial curvature by reducing the PC/FA composition is far more dramatic than by changing pressure [70].

The marked differences between the effects of pressure and PC/FA composition reflect the fact that compositional variations cause large differential changes in the lateral pressure between amphiphiles, whereas the hydrostatic pressure does not. This means that pressure provides an extremely fine resolution parameter for probing the stability and geometry of lyotropic mesophases.

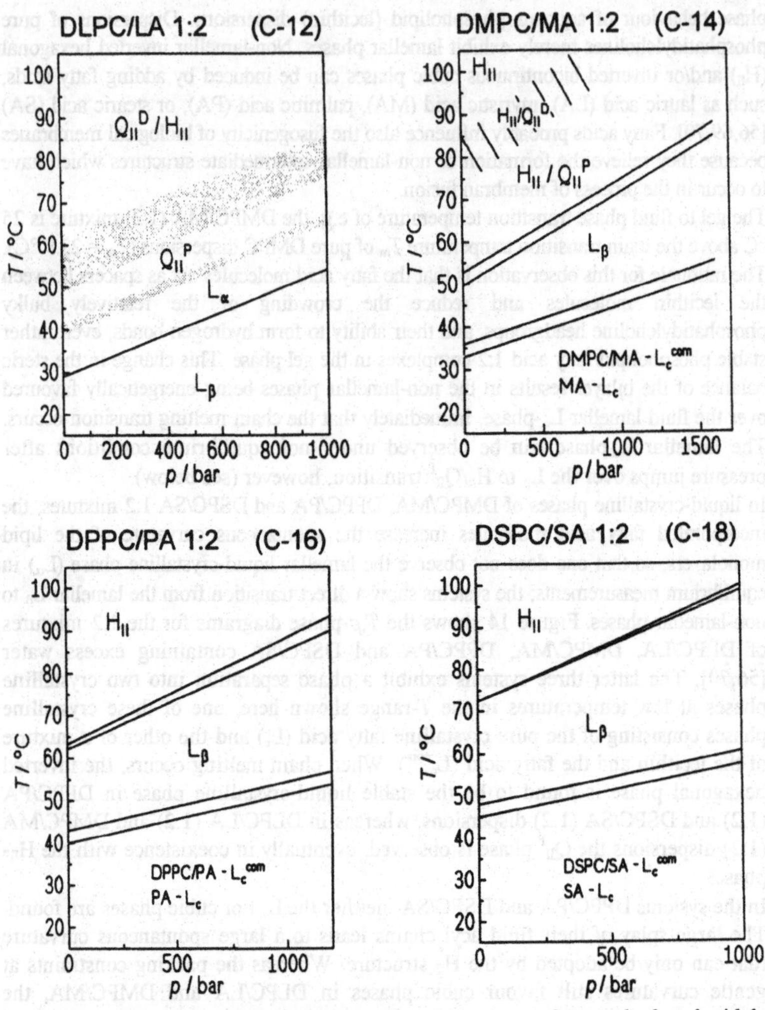

Figure 14. T,p-phase diagrams of aqueous dispersions of several phosphatidylcholine/fatty acid (1:2) mixtures in excess water.

Phase-separated phospholipid mixtures. There is still considerable controversy regarding the lateral organization of lipid mixtures [71-75]. To understand the diverse effects of the presence of several components, studies of membranes composed of two components have been carried out. To address this problem, we used small-angle neutron scattering (SANS) experiments in combination with the H/D contrast variation technique [75]. We investigated mixtures where both components are saturated phospholipids of different acyl chain-lengths, such as DMPC(di-C_{14})/DSPC(di-C_{18}) dispersions. The scattering cross section density of the D_2O/H_2O solvent mixture was adjusted to that of an equimolar DMPC(d_{54})/DSPC mixture, assuming a homogeneous distribution of the two components. Under these so-called matching conditions, the scattering is determined by large-scale concentration fluctuations, only. The differential coherent scattering cross section of the sample can be written as

$$\frac{d\Sigma}{d\Omega} = n_p \Delta\rho^2 V_p^2 P(Q) S(Q) \tag{7}$$

where n_p denotes the number density of lipid molecules, V_p their volume, $\Delta\rho = \rho_p - \rho_s$ the contrast, i.e., the difference in mean scattering-length density of the particles (ρ_p) and the solvent (ρ_s) respectively. $P(Q) = \langle |F(Q)|^2 \rangle$, with the form factor $F(Q)$ of the particles, and $S(Q)$ is the structure factor describing the spatial distribution of the particles [75].

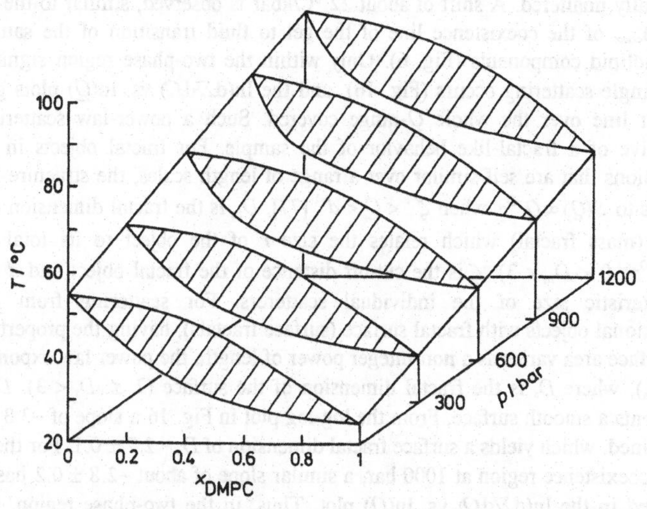

Figure 15. *T,x*-phase diagram of the binary lipid mixture DMPC/DSPC (1:1) in excess water as a function of pressure.

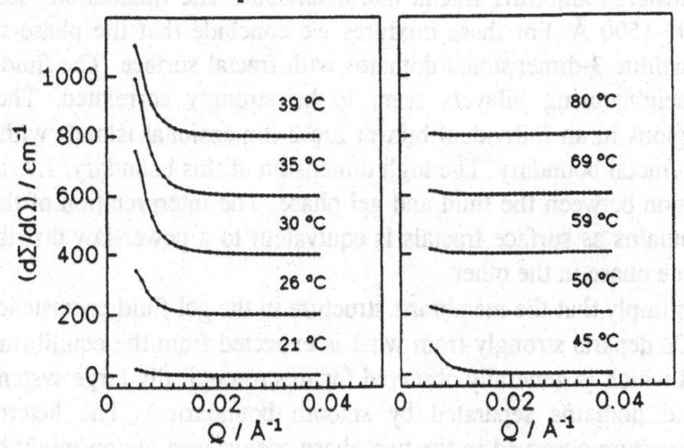

Figure 16. SANS curves of a contrast matched DMPC(d_{54})/DPPC (1:1) dispersion at selected temperatures (p = 1 bar).

The temperature-pressure-concentration phase diagram of the DMPC/DSPC aqueous lipid dispersion has been determined by differential thermal (DTA) analysis [76]. The T,p-projection of the phase diagram of the 1:1 (mol:mol) mixture is depicted in Fig. 15. With increasing temperature, the gel, two-phase coexistence and fluid region of the mixture is shown. Deviations from ideal mixing behaviour of the two lipid components are considerable, as can already be inferred from the broad gel-fluid coexistence region. The system is close to a gel-gel immiscibility. The transition temperatures T_m of the pure lipid components at ambient pressure are 55 °C for DSPC-h_{88} and 24 °C for DMPC-h_{72}, respectively. Deuteration lowers the phase-transition temperature of DMPC-d_{54} by about 4 °C and thus shifts the solidus-line of the two-phase region of the mixture to slightly lower temperatures. With increasing pressure, the two-phase coexistence region is shifted towards higher temperatures, and the shape of the phase diagram is essentially unaltered. A shift of about 22 °C/kbar is observed, similar to the slope $(dT/dp)_{coex}$ of the coexistence line of the gel to fluid transition of the saturated phospholipid components (Fig. 6). Only within the two-phase region significant small-angle scattering occurs (Fig. 16), and the $\ln(d\Sigma/d\Omega)$ vs. $\ln(Q)$ plots give a straight line over the whole Q-range covered. Such a power-law scattering is indicative of a fractal-like behavior of the sample. For fractal objects in three dimensions that are self-similar over a range of length scales, the structure factor reduces to $S(Q) \propto Q^{-D_m}$ when $\zeta^{-1} < Q < d^{-1}$ [77]. D_m is the fractal dimension of the object (mass fractal) which relates the size r of the object to its total mass ($m \propto r^{D_m}$, $0 < D_m < 3$); ζ is the cut-off distance of the fractal object and d is the characteristic size of the individual scatterers. For scattering from three-dimensional objects with fractal surface (surface fractals), having the property that the surface area varies as a non-integer power of length, the power law exponent is $-(6-D_s)$, where D_s is the fractal dimension of the surface ($2 \le D_s < 3$). $D_s = 2$ represents a smooth surface. From the log-log plot in Fig. 16 a slope of -3.3 ± 0.1 is obtained, which yields a surface fractal dimension of $D_s = 2.7 \pm 0.1$. For the two-phase coexistence region at 1000 bar, a similar slope of about -2.8 ± 0.2 has been observed in the $\ln(d\Sigma/d\Omega)$ vs. $\ln(Q)$ plot. Thus, in the two-phase region, large-scale fluctuations abound, which can be characterized as non-uniform system of coexisting clusters exhibiting fractal-like behaviour. The fluctuations seen lie in the range 200-1500 Å. For these mixtures we conclude that the phase-separated domains constitute 3-dimensional domains with fractal surface. The fluid and gel islands in neighbouring bilayers seem to be strongly correlated. The phase-separated regions in an individual bilayer are 2-dimensional islands with (D_s-1)-dimensional fractal boundary. The high dimension of this boundary, 1.7, implies a low line tension between the fluid and gel phase. The interpretation of the phase-separated domains as surface fractals is equivalent to a power-law distribution of droplets of one phase in the other.

These results imply that the membrane structure in the gel-fluid coexistence region of binary PC's departs strongly from what is expected from the equilibrium phase diagram and which is generally observed for macroscopically large systems (large gel and fluid domains separated by smooth boundaries). The heterogeneous membrane structure observed in the two-phase coexistence region might be due to interfacial wetting effects such as those suggested recently from computer simulations [78,79]. In these calculations, the interface between coexisting gel and fluid phases domains is found to be enriched by one of the lipid species, leading to a decrease of the interfacial tension and hence to a stabilization of non-equilibrium lipid domains. These and further results on similar binary lipid systems suggest that such heterogeneous and fractal-like domain morphologies might be a common phenomenon. Depending on the acyl chain mismatch of the lipid components, the clusters scatter like surface or mass fractals, implying that gel and fluid domains are correlated across many bilayers, and that segregation into a minority and majority phase occurs, respectively [80]. These results will certainly have also drastic consequences for the transport properties of membrane components.

3.3 KINETICS OF LIPID MESOPHASE TRANSFORMATIONS

Although the static structure and phase behaviour of many lipid systems is rather well established, considerable lack of knowledge exists regarding the understanding of the kinetics and mechanisms of lipid phase transformations. We used the synchrotron X-ray diffraction technique to record the temporal evolution of the structural changes after induction of the phase transition by a pressure-jump across the phase boundary [56]. In the following we will discuss a few representative examples.

DEPC. First we present pressure-jump experiments carried out in DEPC/water dispersions to study its L_β/L_α transition. Selected SAXS diffraction patterns at 18 °C after a pressure-jump from 200 to 370 bar are depicted in Fig. 17. An intermediate structure is clearly observable here. The first order Bragg reflection of the initial L_α (a = 66 Å) phase vanishes in the course of the pressure-jump (5 ms). The first diffraction pattern collected after the pressure-jump exhibits a Bragg reflection of a new lamellar phase $L_?$ with a larger a-value, which increases with time. The lattice constant of the L_β-phase formed is 78 Å. The transition is complete after about 15 s. In equilibrium measurements, no such intermediate lamellar structure is detectable (see Fig. 15).

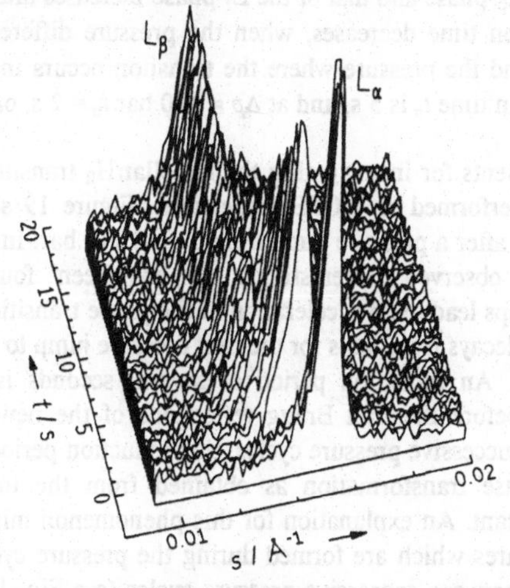

Figure 17. Diffraction curves of DEPC/excess water after a p-jump from 200 bar to 370 bar at T = 18 °C.

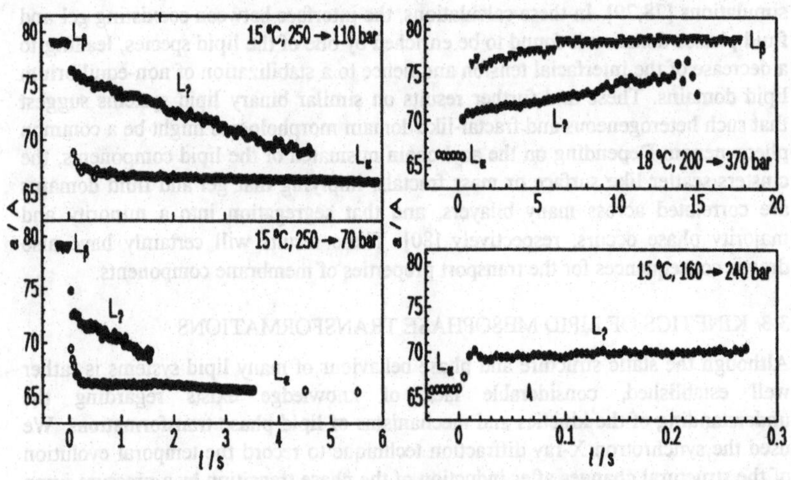

Figure 18. Lamellar lattice constants of DEPC/excess water after pressure jumps a) into the L_α-phase, and b) into the L_β-phase using different pressure-jump amplitudes.

Figure 18 exhibits the lattice constants of the L_β- and L_α-phase of DEPC/water after pressure-jumps in both directions. The transition to the intermediate lamellar phase $L_?$ takes place within about 10 ms. The lattice constant first shifts a few Å, afterwards the lattice constant of the lamellar intermediate phase increases or decreases linearly with time, due to the uptake or loss of interlamellar water. It can also be seen from the figure, where kinetics curves with different end pressure are seen, that the final pressure has a significant influence on the transition time. With decreasing pressure from the transition line, the difference between the lattice constant of the L_β-phase and that of the $L_?$-phase increases after the pressure jump, and the transition time decreases, when the pressure difference Δp between the final pressure and the pressure where the transiton occurs increases. At $\Delta p \approx 70$ bar, the transition time t_{tr} is 5 s, and at $\Delta p \approx 110$ bar $t_{tr} = 2$ s, only.

DOPE. Experiments for investigating the lamellar/H_{II} transition kinetics have for example been performed on DOPE dispersions. Figure 19 shows the diffraction pattern at 20 °C after a pressure jump from 300 to 110 bar. In this case, a two-step mechanism is observed. Interestingly, it has been found that successive temperature jumps lead to an acceleration of the phase transition kinetics. The half transition time decays from 8.5 s for the first pressure jump to 5.3 s after the fourth jump (Fig. 20). An induction period of several seconds is observed after the pressure-jump before the first Bragg reflections of the newly formed H_{II}-phase appears. Upon successive pressure cycles, this induction period decreases, whereas the rate of phase transformation as obtained from the intensity curves stays essentially constant. An explanation for this phenomenon might be the formation of defect structures which are formed during the pressure cycles and which have not healed out between successive pressure cycles (see Fig. 12). This observation also shows that the history of sample preparation plays an essential role in these kind of studies.

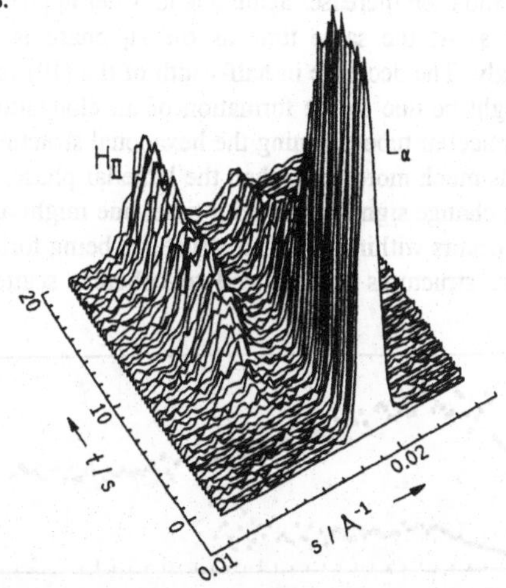

Figure 19. Selected diffraction patterns of DOPE in excess water after a pressure-jump from 300 to 110 bar at 20 °C.

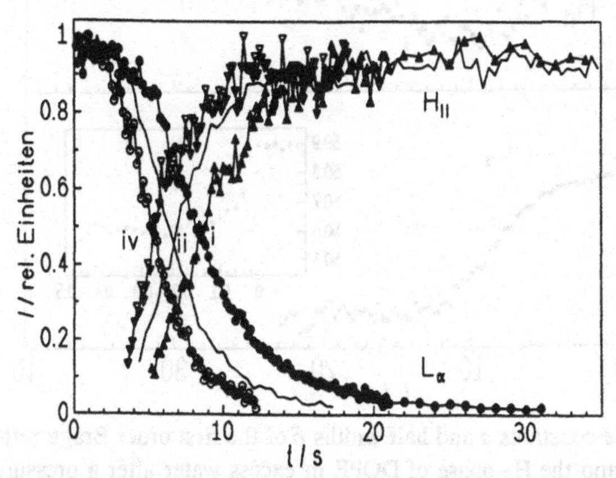

Figure 20. Intensities of Bragg reflections of the L_α- and H_{II}-phase after the first (i), second (ii) and forth (iv) *p*-jump from 300 to 120 bar at $T = 20$ °C.

With increasing pressure-jump amplitude, the induction period decreases and the rate of phase transfomation increases (e.g., 10%/0.6 s phase change for a 300→110 bar jump (induction period 5 s) and 10%/0.15 s phase change for a 300→1 bar pressure jump (induction period 0.5 s)).

Figure 21 shows the lattice constants and half-widths of the Bragg reflections of the L_α- and H_{II}-phase at 20 °C after a pressure jump from 300→110 bar. After 20 ms the lattice constant of the L_α-phase has decreased by 0.2 Å, due to fast conformational changes (probably on the ns time scale) of the lipid acyl chains. After that fast "burst phase" $a(L_\alpha)$ decreases slowly to 50.6 Å after 250 ms.

Following an induction period, the Bragg reflection of the H_{II}-phase appears. $a(H_{II})$ first decreases slightly, and then increases again due to water uptake by 0.5 Å up to 73.9 Å after about 25 s. At the same time as the H_{II}-phase is formed, $a(L_\alpha)$ decreases correspondingly. The decrease in half-width of the (10) reflection of the H_{II}-phase with time might be due to the formation of an elongation and a more densely packing of the micellar tubes forming the hexagonal structure. As the fully hydrated H_{II}-phase needs much more water than the lamellar phase, and the lattice constant $a(H_{II})$ does not change significantly with time, one might assume that the necessary water uptake occurs within the defect structures being formed during the induction period. These structures do not lead to coherent scattering patterns, however.

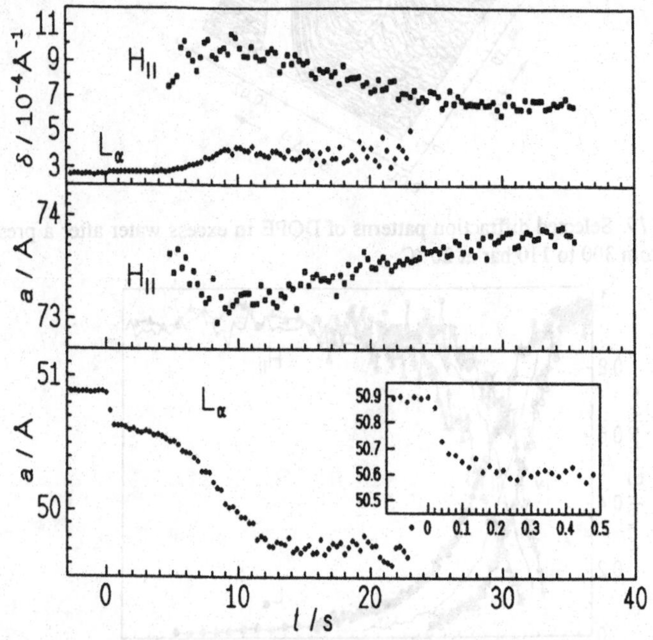

Figure 21. Lattice constants a and half widths δ of the first order Bragg reflections of the L_α-phase and the H_{II}-phase of DOPE in excess water after a pressure-jump from 300 to 110 bar at 20 °C.

DMPC/DSPC. As a further example we show measurements of the kinetics when jumping into the two-phase fluid/gel coexistence region of a DMPC/DSPC (1:1) dispersion. At ambient pressure, the gel/fluid coexistence region is observed between 30 and 45 °C, and the phase transition lines are shifted with about 22 °C/kbar to higher pressures (Fig. 15). Figure 22 shows the lattice constants and intensities of (001) Bragg reflections after a pressure-jump from 400 bar (gel-phase) to 180 bar (into the two-phase region) at 40 °C. The lattice constant of the gel phase increases from 73 to 75 Å after the pressure-jump, possibly due to the increase of DSPC content in the gel lipid mixture. After a fast conformational change of the lipid molecular system also in this binary system, a slower component (probably a diffusion process) takes place, and the whole transition is completed after about 30 s.

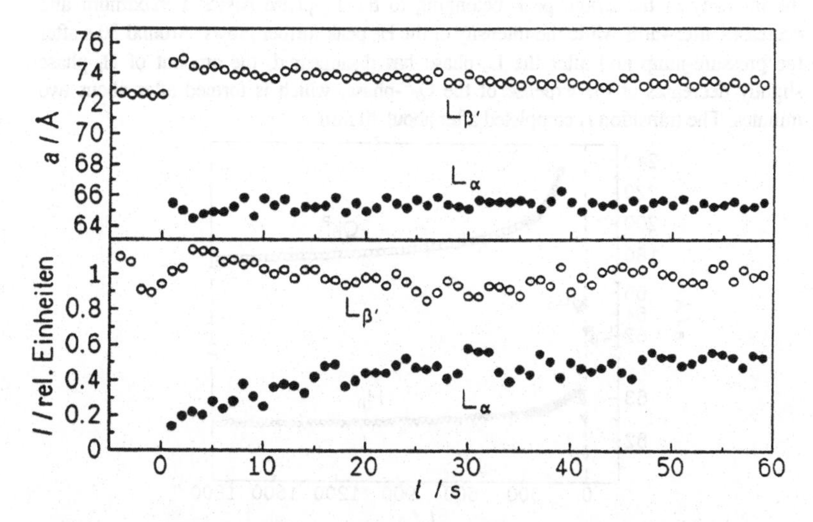

Figure 22. Lattice constants *a* and intensities of the first order Bragg reflections of the L_α- and $L_{\beta'}$-gel of DMPC/DSPC (40 wt%) in excess water after a pressure-jump from 400 to 180 bar at 40 °C.

DMPC/MA (1:2). The pressure-jump experiments across the L_β to H_{II}/Q_{II}^P chain melting transition reveals a rather complex transition mechanism. Figure 23 shows the lattice constants $a(t)$ of the structures of a 75 wt% H_2O DMPC/MA (1:2) dispersion as a function of time after a pressure-jump from 400 to 110 bar, carried out at 58 °C. The lamellar L_β-phase transforms to the H_{II}- and Q_{II}^P-phases, and there appears an intermediate lamellar L_α-phase. The lattice constant of the intermediate L_α structure is 64.5 Å, which is about 2 Å larger than that of the L_β structure, indicating that this phase is more hydrated than the L_β-phase. The line width of the (001) Bragg reflection of the disappearing L_β-phase increases about 30 %, whereas that of the initially formed hexagonal phase decreases by about 30 % until completion of the phase transformation. The single Bragg reflection of the lamellar lattice vanishes after 50 s, that is about 50 s before the Bragg reflections of the cubic lattice Q_{II}^P appear. Before well-defined Bragg peaks of the cubic lattice occur, a broad area of increased diffuse scattering intensity is visible in the small-angle scattering region.

The Bragg reflections of the H_{II}-phase appear immediately after the pressure-jump. It is remarkable that the lattice constant of the hexagonal structure first increases by about 1.5 Å before it decreases again to reach an equilibrium value of about 62.6 Å, and the lattice constant of the cubic phase decreases from an initial high value of about 233 Å slowly to its equilibrium value of 176 Å within 30 minutes. The rates of both decay processes are similar. The unusual pressure-dependence of the H_{II} lattice constant might be due to the coexistence of phases with different levels of hydration or a partial dissociation of the 1:2 lecithin/fatty acid complex. A detailed analysis of the intensities of the Bragg peaks as a function of time indicates that the amount of the H_{II}-phase runs through a broad maximum. While the intensity of the L_β peak decreases, first the H_{II} peak appears and begins to increase before the L_α peak sets in. As the L_β-phase vanishes the intensity of the Bragg peak belonging to the L_α-phase passes a maximum and decreases afterwards while the intensity of the H_{II} peak further grows. Around 50 s after the pressure-jump and after the L_α-phase has disappeared, the amount of H_{II}-phase slightly decreases at the expense of the Q_{II}^D-phase, which is formed after about two minutes. The transition is completed after about 30 min.

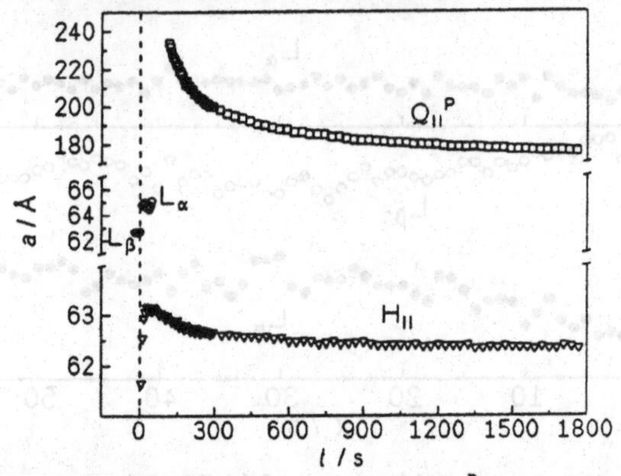

Figure 23. Lattice constants of the L_β-, L_α-, H_{II}- and Q_{II}^P-phase of DMPC/MA (1:2, 25 wt%) after a *p*-jump from 400 bar to 100 bar at $T = 58$ °C.

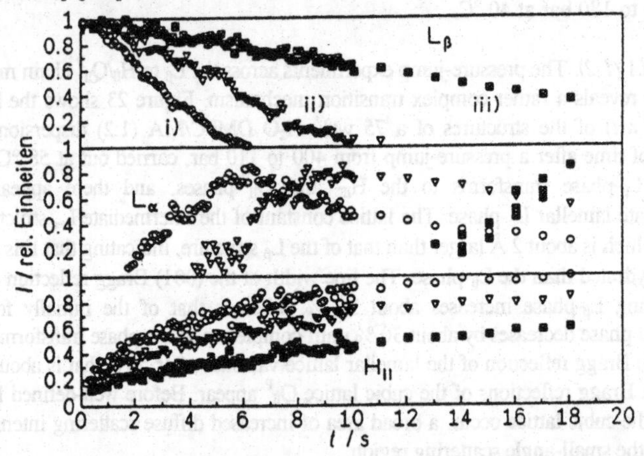

Figure 24. Intensities of Bragg reflections of the L_β-, L_α-, and H_{II}-phase of DMPC/MA (1:2, 25 wt%) after a *p*-jump from 660 bar to (i) 300 bar, (ii) 365 bar, and (iii) 430 bar at $T = 61$ °C.

Figures 24 gives the SAXS intensities of the DMPC/MA (1:2) mixture at 61 °C after pressure-jumps Δp starting at 660 bar in the L_β-phase and ending at a final pressure of (i) 300 bar, (ii) 365 bar, and (iii) 430 bar, respectively. As can be clearly seen, the lifetime of the L_β-phase significantly increases with decreasing pressure-jump amplitude Δp, and the transient L_α-phase starts to develop later. The decay of the intensity of the

L_β lattice is linear in time within the accuracy of the measurements. The intensity has decayed to zero after 6.5 s for Δp = 360 bar, after 10 s for Δp = 295 bar, and after 30 s for Δp = 230 bar. The time of the first observation of the L_α-phase increases correspondingly with decreasing pressure amplitude, from 1.5 s at Δp = 360 bar and 4.5 s at Δp = 295 bar to about 12 s at Δp = 230 bar. The intensity of the L_α peak increases as long as the L_β-phase coexists, and begins to decay then. Zero intensity is reached after about 28 s and 34 s, respectively, for the first two *p*-jump amplitudes. The H_{II} intensity reaches its plateau value after about 30-40 s, and decays then again after about 1-2 minutes when the cubic phase Q_{II}^P starts to build up.

The pressure-jump amplitude has been found to have a significant influence on the kinetics of the phase transformation, the mechanism staying the same. Increase of the pressure-jump amplitude leads to an increase in chemical potential difference between phases and thus a faster decay of the L_β-phase and the more rapid formation of the H_{II}-phase. Indeed, it seems that the rate of the L_β/H_{II}-phase transformation depends on the distance of the final pressure p_{end} from the pressure of the phase transition boundary p_{tr}. For the differences $p_{end} - p_{tr}$ of 80 bar, 145 bar and 210 bar, the initial formation of the L_α-phase is observed about 13 s, 5 s and 2 s after the pressure jump.

In the initial step the L_β-phase probably transforms to the H_{II}-phase ($L_\beta \rightarrow H_{II}$) without forming a long-lived intermediate phase, because the water contents of phases is low. When there is enough water diffused into the lamellar structure, another transformation pathway probably opens, the transformation to the H_{II}-phase with the L_α-phase as an intermediate state ($L_\beta \rightarrow L_\alpha \rightarrow H_{II}$). First Bragg peaks of the cubic structure Q_{II}^P are observed about 1-2 min after the lamellar phases have disappeared. The formation of the cubic structure thus does not seem to be involved in the primary phase transformation process. Precursors of the cubic structure might be formed at an earlier stage, however. Interestingly, also in *T*-jump experiments on aqueous dispersions of particular phospholipids structural intermediates at the lamellar to H_{II} transition have been found under non-equilibrium conditions [81,82].

Generally, as has also been found in studies of pressure- and temperature-jump induced phase transitions of other systems [56,81-83], the results show that the relaxation behaviour and the kinetics of lipid phase transformations drastically depend on the topology and symmetry of the lipid phases, as well as on the applied jump amplitude Δp. In most cases the rate of the transition is probably limited by the transport and redistribution of water into and in the new phase, rather than being controlled by the required time for a rearrangement of the lipid molecules. This can be inferred from lattice relaxation experiments in the lipid one-phase regions. For example, in the H_{II}-phase of DOPE, lattice relaxation times are on the time scale 10-20 s, depending on Δp and *T*, which are about an order of magnitude slower than those in its lamellar phases. The transition times in cubic phase also drastically depend on Δp and *T*; they vary for example in the Q_{II}^D-phase of MO after *p*-jumps from 330 to 110 bar from 10 s to 1800 s at 60 °C and 45 °C, respectively. The turtuosity factor of the different structures, especially in cases where non-lamellar (hexagonal and cubic) phases are involved, is thus likely to control the different kinetic components. In addition, nucleation phenomena and domain size growth of the structures evolving might also play a significant role.

We conclude that pressure work on model membrane and lipid systems can yield a wealth of enlightening new information on the structure, energetics and phase behaviour of these systems and on the transition kinetics between lipid mesophases. In addition, these results are of biological and biotechnological relevance.

4. Acknowledgements

Financial support from the Deutsche Forschungsgemeinschaft (DFG), the Fonds der Chemischen Industrie and the DAAD (ARC) is gratefully acknowledged. We thank Drs. J. Seddon and R. Templer for many valuable discussions.

5. Abbreviations

PC: phosphatidylcholines, PE: phosphatidylethanolamines, PS: phosphatidylserines, FA: fatty acid, SA: stearic acid, LA: lauric acid, MA: myristic acid, PA: palmitic acid, MO: monoolein, ME: monoelaidin, DMPC: 1,2-dimyristoyl-sn-glycero-3-phosphatidylcholine (di-$C_{14:0}$), DMPS: 1,2-dimyristoyl-sn-glycero-3-phosphatidylserin (di-$C_{14:0}$), DPPC: 1,2-dipalmitoyl-sn-glycero-3-phosphatidylcholine (di-C16:0), DSPC: 1,2-distearoyl-sn-glycero-3-phosphatidylcholine (di-C18:0), DOPC: 1,2-dioleoyl-sn-glycero-3-phosphatidylcholine (di-C18:1,cis), DOPE: 1,2-dioleoyl-sn-glycero-3-phosphatidylethanolamine (di-C18:1,cis), POPC: 1-palmitoyl-2-oleoyl-sn-glycero-3-phosphatidylcholine, (C16:0,C18:1,cis), DEPC: 1,2-dielaidoyl-sn-glycero-3-phosphatidylcholine (di-C18:1,trans), egg-PE: egg-yolk phosphatidylethanolamine.

6. References

[1] Seddon J. M. 1990 Biochim. Biophys. Acta **1031** 1

[2] Seddon J M, Templer R H 1993 Phil. Trans. R. Soc. Lond. A **344** 377

[3] Cevc G, Marsh D 1987 *Phospholipid Bilayers* (New York: John Wiley & Sons); Cevc G (Ed.) 1993 *Phospholipids Handbook* (New York: Marcel Dekker)

[4] Epand R M (Ed.) 1997 *Lipid Polymorphism and Membrane Properties, Current Topics in Membranes.* Vol. 44 (San Diego: Academic Press)

[5] Lipowski R, Sackmann E (Eds.) 1995 *Structure and Dynamics of Membranes*, Vol. 1A, 1B (Amsterdam: Elsevier)

[6] Winter R, Landwehr A, Brauns Th, Erbes J, Czeslik C, Reis, O 1996 In *High Pressure Effects in Molecular Biophysics and Enzymology* Markley J L, Northrop D B, Royer C A (Eds.) (Oxford: Oxford University Press) p. 274

[7] Winter R, Pilgrim, W C 1989 Ber. Bunsenges. Phys. Chem. **93** 708

[8] Braganza L F, Worcester D L 1986 Biochemistry **25** 2591

[9] Böttner M, Ceh D, Jacobs U, Winter R 1994 Z. Phys. Chem. **184** 205

[10] Chong P-L G, Weber G 1983 Biochemistry **22** 5544

[11] Wong P T T, Siminovitch D J, Mantsch H H 1988 Biochim. Biophys. Acta **947** 139; Carrier D, Wong P T T 1997 Eur. J. Solid State Inorg. Chem. **34**, 733

[12] Driscoll D A, Jonas J, Jonas A 1991 Chem. Phys. Lipids **58** 97
[13] Landwehr A, Winter R 1994 Ber. Bunsenges. Phys. Chem. **98** 214
[14] So P T C, Gruner S M, Shyamsunder E S 1993 Phys. Rev. Lett. **70** 3455
[15] Reis O, Winter R, Zerda, T W 1996 Biochim. Biophys. Acta **1279** 5
[16] Czeslik C, Winter R, Rapp G, Bartels K 1995 Biophys. J. **68** 1423
[17] Cheng A, Mencke A, Caffrey M 1996 J. Phys. Chem. **100** 299
[18] Duesing P M, Seddon J M, Templer R H, Mannok D A 1997 Langmuir **13** 655
[19] Bonev B B, Morrow M R 1995 Biophys. J. **69** 518
[20] Czeslik C, Reis O, Winter R, Rapp G 1998 Chem. Phys. Lip. **91** 135
[21] Lindblom G, Rilfors L 1989 Biochim. Biophys. Acta **988** 221
[22] Tate M W, Eikenberry E F, Turner D C, Shyamsunder E, Gruner S M 1991 Chem. Phys. Lipids **57** 147
[23] Erbes J, Czeslik C, Hahn W, Rappolt M, Rapp G, Winter R 1994 Ber. Bunsenges. Phys. Chem. **98** 1287
[24] Luzzati V, Vargas R, Mariani P, Gulik A, Delacroix H 1988 J. Mol. Biol. **229** 540
[25] Luzzati V 1995 J. Phys. II France **5** 1649
[26] Gruner S M 1987 In *Liposomes, from Biophysics to Therapeutics* Ostro M J (Ed.) (New York: Marcel Dekker) p. 1
[27] Chernomordik L, Kozlov M M, Zimmerberg J 1995 J. Membrane Biol. **146** 1
[28] Bouligand Y 1990 Colloq. Phys. **C7** 35
[29] Mariani P, Luzzati V, Delacroix H 1988 J. Mol. Biol. **204** 165
[30] Luzzati V 1997 Curr. Opin. Struct. Biol. **7** 661
[31] Landh T 1995 FEBS Lett. **369** 13
[32] Luzzati V, Delacroix H, Gulik A, Gulik-Krzywicki T, Mariani P, Vargas R 1997 In *Current Topics in Membranes*, Vol. 44, pp 3-24 (San Diego: Academic Press)
[33] Engström S, Lindahl L, Wallin R, Engblom J 1992 Int. J. Pharm. **86**, 137
[34] Puvvada S, Qadri S B, Naciri J, Ratna B R 1993 J. Phys. Chem. **97**, 11103
[35] Pebay-Peyroula E, Rummel G, Rosenbusch J P, Landau E M 1997 Science **277**, 1676
[36] Landau E M, Rosenbusch J P 1996 Proc. Natl. Acad. Sci. U.S.A. **93**, 14532
[37] Charvolin J, Sadoc J-F 1996 Phil. Trans. R. Soc. London A **354** 2173
[38] Hyde S T 1989 J. Phys. Chem. **93** 1458
[39] Riste T, Sherrington D (Eds.) 1989 *Phase Transitions in Soft Condensed Matter* (New York: Plenum Press)
[40] Hyde S, Andersson S, Larsson K, Blum Z, Landth T, Lidin S, Ninham B W 1997 *The language of shape. The role of curvature in condensed matter: physics, chemistry and biology* (Amsterdam: Elsevier).
[41] Templer R H, Seddon J M, Warrender N A 1994 Biophysical Chemistry **49** 1
[42] Templer R H, Turner D C, Harper P, Seddon J M 1995 J. Phys. II France **5** 1053
[43] Templer R H 1995 Langmuir **11** 334
[44] Templer R H, Seddon J M, Duesing P M, Winter R, Erbes J 1998 J. Phys. Chem. **102**, 7262; Templer R H , Seddon J M, Warrender N A, Syrykh A, Huang Z, Winter R, and Erbes J 1998 J. Phys. Chem. **102**, 7251
[45] Chung H, Caffrey M 1994 Biophys. J. **66** 377
[46] Chung H, Caffrey M 1994 Nature **368** 224

[47] Mariani P, Paci B, Bösecke P, Ferrero C, Lorenzen M, Caciuffo R 1996 Phys. Rev. E **54** 5840.
[48] Helfrich W 1990 In *Liquids at interfaces* (Les Houches summer school session XLVIII) Charvolin J, Joanny J-F, Zinn-Justin J (Eds.) (Amsterdam: North-Holland)
[49] Rostain J C, Martinez E, Lemaire C (Eds.) 1989 *High pressure nervous syndrome - 20 years later* 1989 (Marseille: ARAS-SNHP Publications)
[50] Balny C, Hayashi R, Heremans K, Masson P (Eds.) 1992 *High Pressure and Biotechnology* (Montrouge, France: Colloque Inseram, Vol. 224, John Libbey Eurotext)
[51] Winter R, Jonas J (Eds.) 1993 *High Pressure Chemistry, Biochemistry and Materials Science* (NATO ASI C 401) (Dordrecht, The Netherlands: Kluwer Academic Publishers)
[52] Macdonald A G 1986 In *Topic in Lipid Research* Klein R, Schmitz B (Eds.) (London: The Royal Society of Chemistry) p. 319
[53] Macdonald A G 1984 Phil. Trans. R. Soc. Lond. B **304** 47
[54] Behan M K, Macdonald A G, Jones G R, Cossins A R 1992 Biochim. Biophys. Acta **1103** 317
[55] Czeslik C, Malessa R, Winter R, Rapp G 1996 Nuclear Instruments and Methods in Physics Research A **368** 647
[56] Erbes J, Winter R, Rapp G 1996 Ber. Bunsenges. Phys. Chem. **100** 1713
[57] Kotlarchyk M, Ritzau S M 1991 J. Appl. Cryst. **24**, 753
[58] Zhang R, Tristram-Nagle S, Sun W, Headrick R L, Irving T C, Suter R M, Nagle J F 1996 Biophys. J. **70**, 349
[59] Reis O, Winter R 1998 Langmuir **14** 2903
[60] Wong, P T T, Moffatt D J, Baudais F L 1985, Appl. Spectrosc. **39** 733
[61] Bernsdorff S, Wolf A, Winter R, Gratton E 1997 Biophys. J. **72** 1264
[62] Czeslik C, Reis O, Winter R, Rapp G 1998 **91** 135
[63] Chang E L, Yager P 1983 Mol. Cryst. Liq. Cryst. **98** 125
[64] Shyamsunder E, Gruner S M , Tate M W, Turner D C, So P T C, Tilcock C P S 1988 Biochemistry **27** 2332
[65] Winter R, Erbes J, Czeslik C, Gabke A 1998 J. Phys.: Condens. Matter **10** 11499
[66] Siegel D P 1993 Biophys. J. **65** 2124
[67] Turner D C, Wang Z-G, Gruner S M, Mannock D A, McElhaney R N 1992 J. Phys. II France **2** 2039
[68] Anderson D M, Gruner S M, Leibler S 1988 Proc. Natl. Acad. Sci. U.S.A. **85** 5364
[69] Koynova R D, Tenchov B G, Quinn P J, Laggner P 1988 Chem. Phys. Lipids **48** 205; Koynova R D , Boyanov A I, Tenchov B G 1987 Biochim. Biophys. Acta **903** 186; Koynova R D, Tenchov B G, Rapp G 1997 Chem. Phys. Lipids **88** 45
[70] Winter R, Erbes J, Templer R H, Seddon, J M, Syrykh, A, Warrender N A, Rapp G 1999 PCCP, in press
[71] Almeida P F F, Vaz W L C, Thompson T E 1992 Biochemistry **31** 7198
[72] Morrow M R, Srinivasan R, Grandal N 1991 Chem. Phys. Lipids **58** 63
[73] Schmidt G, Knoll W 1985 Ber. Bunsenges. Phys. Chem. **89** 36
[74] Sankaram M B, Thompson T E 1992 Biochemistry **31** 8258
[75] Czeslik C, Erbes J, Winter R 1997 Europhys. Lett. **37** 577

[76] Landwehr A, Winter R 1994 Ber. Bunsenges. Phys. Chem. **98** 1585
[77] Pfeifer P, Obert M 1989 In *The Fractal Approach to Heterogeneous Chemistry - Surface, Colloids, Polymers* Avnir D (Ed.) (Chichester: John Wiley & Sons) p. 11
[78] Jørgensen K, Sperotto M M, Mouritsen O G, Ipsen J H, Zuckermann M J 1993 Biochim. Biophys. Acta **1152** 135
[79] Jørgensen K, Mouritsen O G 1995 Biophys. J. **95** 942
[80] Winter R, Gabke A, Czeslik C, and Pfeifer P, submitted
[81] Laggner P, Kriechbaum M, Rapp G 1991 J. Appl. Cryst. **24** 836
[82] Rapp G, Rapport M, Laggner P 1993 Progr. Colloid Polym. Sci. **93** 25
[83] Caffrey M 1987 Biochemistry **26** 6349

OPTICAL SPECTROSCOPIC TECHNIQUES IN HIGH PRESSURE BIOSCIENCE

Claude BALNY and Reinhard LANGE
INSERM U 128, IFR 24
1919, route de Mende
34293 MONTPELLIER Cedex 5, France
Email: balny@crbm.cnrs-mop.fr

Abstract

High hydrostatic pressure induces changes in protein conformation, solvation and enzyme activities. To have access to these structural modifications, various biophysical techniques have been specially designed for high pressure experiments. Among them, spectroscopic methods are rather simple and easily adaptable both to high pressure and controlled temperatures. In the present paper, different improvements are presented together with some applications.

1. Introduction

There is growing evidence that the key to understanding enzyme function may be found in the conformational flexibility of proteins. Hydrostatic pressure is an elegant way to perturb the conformational equilibrium since it does not change the chemical composition of the solutions or its energy [1-4]. To have access to these phenomena, one main aim is the adaptation of biophysical methods and to attribute the pressure effects to the different structural aspects of proteins, i.e. secondary, tertiary and quaternary structures [1].

When biochemical compounds have chromophores, the best way to study their reactions is to use optical methods which can be easily adapted for high pressure measurements [5].

In the present paper we want review some of them which have been developed recently in our laboratory. They work both at high pressure and in a large temperature range, and are applied to the study of protein conformation or enzyme reactions.

2. Materials and methods

2.1. UV/VIS SPECTROSCOPY

It is relatively easy to characterize conformational equilibrium of proteins and kinetic reaction steps of enzymes when chromophores absorbing in the visible or ultraviolet

R. Winter and J. Jonas (eds.), High Pressure Molecular Science, 405–422.

regions are present in a rather high concentration. The literature provides many descriptions of high-pressure devices for optical spectroscopic studies [5,6]. Several commercial optical bombs are also available. However, the systems described are generally designed to be used only at ambient temperature. This limits our understanding of the thermodynamics of the reaction wich can be obtained only by the association of temperature and pressure [7]. On the other hand, in carrying out their biological functions, proteins go through a number of subtle conformational changes that are related to their dynamic structural flexibility. These processes are usually very rapid and therefore difficult to study. One way of reducing this rapidity is to carry out experiments at low temperatures [8]. This necessitates a special optical device that combines both high pressure and low temperatures.

The design developed in our laboratory is an improvement of devices already described for room temperature experiments at high pressure. In addition to the classical characteristics of a device working at room temperature, a high pressure optical bomb for biochemical sub-zero temperature measurements should permit accurate measurements in absorbency and fluorescence mode, with, in addition, the possibility of carrying out photochemistry, in a temperature range from - 50 to 90 ° C. The design should include a device that will avoid any ice formation at the windows or at the sample cell. This high pressure optical bomb should also allow the sample cell to be precooled at a set temperature, have a special system to quickly close the sample cell, operate at a pressure of several hundred MPa without leakage, even at sub-zero temperatures, have a glass or plastic sample cell to avoid problems of metal ion contamination, and be easily interfaced to conventional spectrophotometers.

The actual high-pressure cell that we have recently constructed is similar to the device that we described 15 years ago [9]. The cell core, built by S.O.F.O.P. (Rodez, France), was made of Marval X 12 special steel allowing experiments up to 700 MPa in absorbency or in fluorescence mode. Around this core an envelope was added in which the thermostating alcohol circulated. The geometry of the assembly was miniaturized so that the cell could be placed in the sample compartment of a Cary 3 E spectrophotometer (for absorbency) or a Bowman 2 spectrofluorimeter (SLM). The windows were made of synthetic sapphire (diameter 10 mm, h = 8 mm) from RSA (Jarrie, France) which has a transmittance at 250 nm of 50 % for a light path of 8 mm. The windows were placed directly upon the optical opturators. The tightness was assured by the reversible deformation of a polyurethane seal, which was placed between the external face of the obturator and metallic counterpart. Pressure was generated by a manual pump (700 MPa) and controlled by a metallic gauge type PR 851 C from Top Industries (Dammarie les Lys, France). The working pressure was measured by an Infinity Indicator, Model Infcp from Newport Electronique (Trappes, France). The pressure vector, water, was mediated through stainless steel capillaries (outer diameter 1/16 inch, inner diameter 0.2 mm). Protein samples were measured in a polyethylene stretch film maintained by a rubber O-ring. The stability of the temperature was ± 0.2 ° C. At 700 MPa the stability of the pressure was ± 1 MPa.

Figure 1 shows the diagram of the high pressure bomb.

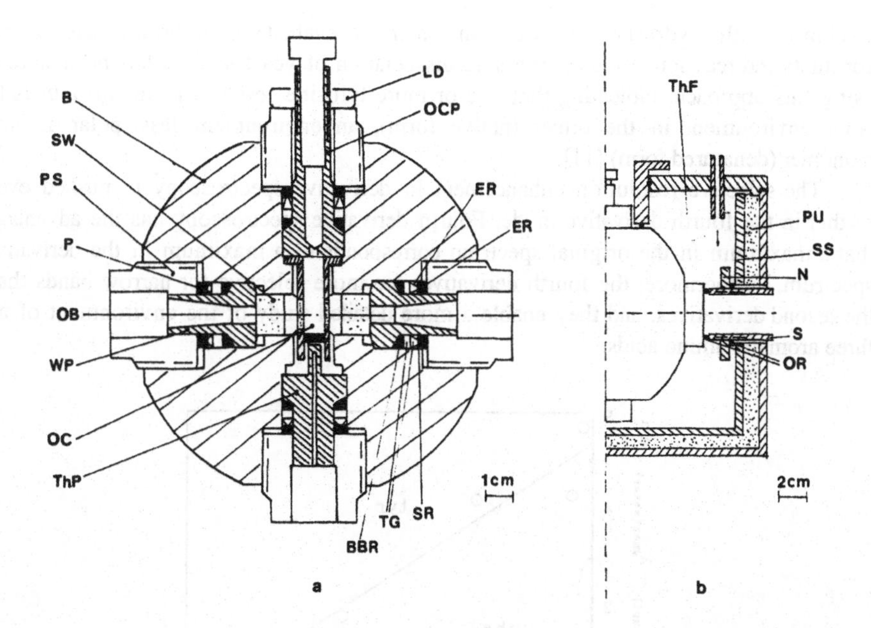

Figure 1 : (a) Side-view diagram of the high pressure bomb. B, body of the bomb; SW, sapphire windows; PS, packing screws; P; packing, Bridgman type; OB, optical beam; WP, window plugs; OC, observation cell; ThP, thermocoulpe plug; LD, locking device; OC, optical cell plug; ER, Elco bronze rings; BBR, bronze beryllium ring; TG, Teflon gaskets; SR, support ring.
(b) Half view of the thermostatization device for the high pressure bomb. ThF, thermostated fluid; PU, polyurethane isolation; SS, stainless steel wall; N, nut in polyvinyl chloride; S, sleeve; OR, O-rings.
(Adapted from ref. 9)

2.2. FOURTH DERIVATIVE UV-SPECTROSCOPY

It is difficult to obtain protein structural information from bands absorbing in the ultraviolet region, since many overlapping individual electronic transitions broadened by their vibronic structure are found here. Not only the three aromatic amino acids phenylalanine, tyrosine and tryptophan, but also cysteine, histidine and UV components of visible chromophores such as the heme group contribute to the ultraviolet absorbency of proteins, resulting in rather structureless spectra. Yet, a better resolution can be achieved, by calculating the second or the fourth derivative of the spectra [10]. Second derivative spectroscopy is the more widely used method which has been recently adapted to high pressure conditions [11]. This method has been used in the ultraviolet region to study the polarity of the regions surrounding tyrosine residues showing that it can also be a tool to study the degree to which proteins associate and that it can be effectively combined with hydrostatic pressure in order to evaluate equilibrium dissociation constants and reaction volumes. Pressure dissociation of yeast enolase has been studied using this approach, indicating that one or more tyrosine residues is in an unusually polar environment in the dimer (native form), an environment less polar in the monomer (denatured form) [11].

The selective resolution enhancement in derivative spectroscopy is pushed even further in the fourth derivative mode. Fourth derivative spectroscopy has the advantage that a maximum in the original spectrum corresponds to a maximum in the derivative spectrum. Furthermore, the fourth derivatives are more selective for narrow bands than the second derivatives, and they enable a more detailed study of the environment of all three aromatic amino acids.

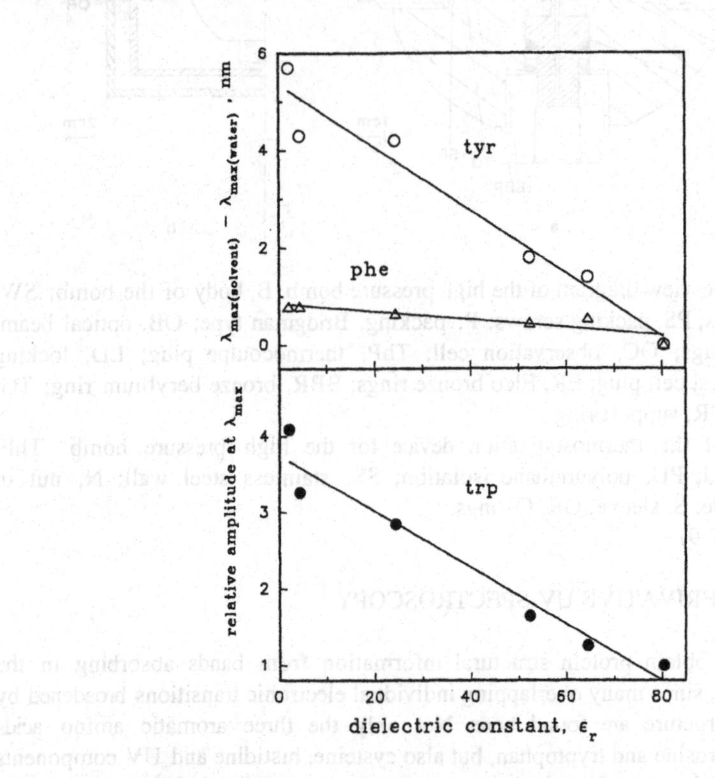

Figure 2 : Effect of the dielectric constant ε_T on the fourth derivative spectral properties of the aromatic amino acids. The solvents were : water ε_T = 80.4; 50% ethylene glycol ε_T = 64.5; 50 % ethanol ε_T = 52.4; ethanol ε_T = 24.3; diethyl ether ε_T = 4.34; cyclohexane ε_T = 2.02 (Adapted from ref. 12).

A detailed description of the method adapted to high pressure has been published elsewhere [12]. The derivatives are computed by shifting a spectrum for a given wavelength difference (the derivative window) and its subsequent subtraction from the original spectrum. The choice of the derivative window is very important since its choice allows one to study selectively certain electronic transitions, according to their spectral band widths. In contrast to many commercial spectrophotometers which impose fixed values for these parameters, we used a tunable method, allowing the optimization of the necessary compromise between spectral amplitude and resolution for the aromatic amino acids. For this, we used a Cary 3 E (Varian) spectrophotometer which has a high wavelength reproducibility (SD < 0.02 nm) where baseline corrected UV spectra were recorded in the double beam mode. For phenylalanine the bandwidth was 0.5 nm, the data acquisition time 1.5 s per data point and the data interval was 0.05 nm. For tyrosine and for tryptophan the bandwidth was 1.0 nm, the data acquisition time 1.0 s and the data interval was 0.1 nm. The fourth derivatives were computed by the spectral shift with a Sigma-plot based program using the spectra saved in ASCII-format. (see ref. 12).

As in the case of second derivative spectroscopy, the amplitude and the position of the derivative spectral bands of the aromatic amino acids are related to the polarity of the medium. As shown in Figure 2, for a variety of solvents ranging from water to cyclohexane, simple and adequate relationships between the solvent effect on the fourth derivative spectra of the aromatic amino acids and the dielectric constant can be found. The strong red shift of wavelength of tyrosine and the relatively weak shift of phenylalanine depend linearly on the dielectric constant. In the case of tryptophan, it is mainly the amplitude of the fourth derivative spectrum that depends on the dielectric constant.

For high pressure experiments, the cell described above was used.

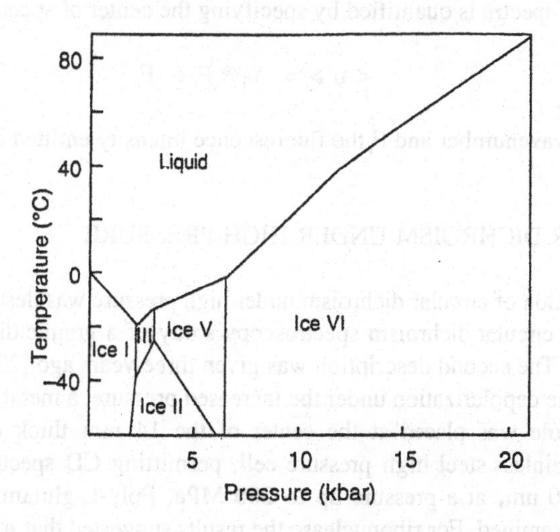

Figure 3. Solid-liquid phase diagram of water where solid lines show phase boundaries.

The method has been applied to study the pressure reversible structural changes of various proteins to probe their important mechanistic aspects : adrenodoxin, ribonucleases from various sources, thermolysin, yeast enolase (dimeric protein) and methanol dehydrogenase (tetrameric protein) [13-17]. It has been shown that the local environmental changes can either reflect the cooperative denaturation of the protein or the molten globule predenaturation state. Particular attention has been given to heat and cold denaturation showing the complexity of the denaturation process. It is important to point out the usefulness of high pressure for cold denaturation studies since at a pressure of 200 MPa, water can be maintained fluid to a temperature as low as - 20 ° C, resulting in the particular behavior of the water phase diagram (see Figure 3).

2.3. FLUORESCENCE SPECTROSCOPY

From a technical point of view, high pressure equipment for fluorescence spectroscopy is similar to that used for UV/vis spectroscopy. The difference is the geometry of the light beam where, for fluorescence spectroscopy, the optical excitation and optical detection beams are at 90 °.

Steady-state fluorescence and time-resolved methods together with data analysis were published by different research groups [see for example, 18-21]. Generally, the high pressure bomb has four window ports and one pressure inlet. The samples are placed in an inner cylindrical, bottle-shaped quartz cuvette. Correction factors are required to obtain fluorescence polarization values free of error due to the birefringence of the windows induced by hydrostatic pressure.

Currently in our laboratory, fluorescence measurements are carried out using a Bowman 2 fluorospectrophotometer (SLM Co) modified to measure fluorescence in the pressure range from 0.1 MPa to 700 MPa. According to the work of G. Weber *et al.*, the fluorescence spectra is quantified by specifying the center of spectral mass $< \upsilon >$:

$$< \upsilon > = \upsilon_i * F_i / F_i$$

where υ_i is the wavenumber and F_i the fluorescence intensity emitted at wavenumber υ_i . [19,20].

2.4. CIRCULAR DICHROISM UNDER HIGH PRESSURE

The first description of circular dichroism under high pressure was described twenty years ago and used for circular dichroism spectroscopy study of a gramicidin A derivative, up to 17 MPa [22]. The second description was given three years ago [23]. For the latter, in order to eliminate depolarization under the increased pressure, a metal slit having a 1 - 3 mm diameter hole was placed at the center of the 14 mm thick quartz or sapphire windows of a stainless steel high pressure cell, permitting CD spectra in a wavelength range 200 to 220 nm, at a pressure up to 200 MPa. Poly-L-glutamic acid and ribonuclease A were examined. For ribonuclease the results suggested that a significant part of the α-helical and β-sheet structures were altered at 200 MPa. After these attempts, few have survived to publication and in these cases with rather limited application.

We have tried to improve this method, which could be very useful for protein conformation studies at high pressure. Pressure CD would provide a complementary technique to the well established FTIR under pressure and the emerging high pressure NMR techniques.

However, CD is already sensitive to several optical problems and these are compounded on applying pressure. We have explored the viability of this method, in spite of technical difficulties [24].

2.4.1. The possibilities of artifacts.

To investigate the possibilities of CD under pressure, we have used our standard absorption cell described above. Since no further results were reported it was of major importance to eliminate the possibility of any artifacts, which would lead to distortions of the observed spectra and thus to misinterpretations of the observations. To our knowledge, two major technical causes can generate severe spectral modifications : *i)* induced birefringence of the quartz windows and *ii)* induced light depolarization. Birefringence has its origin in the non-planarity of the quartz windows, both those of the sample cuvette and those of the pressure cell itself, and these deviations from planarity will lead to the existence of a strong dichroic signal instead of a horizontal or nearly so base line. Depolarization will lead to a pressure dependent decrease in CD signal.

Measurement of baselines under various conditions showed that the main effects due to pressure arose due to distortion of the cell windows, dependent both on the quality of the window and on the surrounding support. The baseline signal varies dramatically with pressure, first increasing due to induced birefringence and then decreasing with the onset of depolarization. These variations are large compared with the signal and reversibility is not always assured. These types of artifacts severely limit the utility of pressure applied to CD.

2.4.2. Possible solutions to spectral artifacts

To improve the prospects, there are several developments that can be made to the design of the cell. Our cell is designed with tolerances appropriate for absorption. Better pressure distributing windows of fused silica are a primary requirement together with narrow apertures to minimize the distortion effects. A means to measure the baseline at the same time as the sample would improve reproducibility. To this end, work is in progress in the laboratory

2.5. STOPPED-FLOW METHOD

The stopped-flow apparatus operating at room temperature and atmospheric pressure is now a routine technique which was described many years ago. The first rapid-reaction apparatus was a continuous-flow system and is the ancestor of the present stopped-flow. The general design of a stopped-flow apparatus classically consists of a driving mechanism, two vertical syringes, a mixing chamber, an observation chamber and a waste syringe [25].

For recording rather fast reactions under high pressure, special requirements and special apparatus are necessary because the relatively long dead-time of high-pressure techniques using spectroscopic detection is the first important limitation to exploiting pressure enzyme kinetics. If the system under study can be characterized via optical detection (absorbance, fluorescence), the stopped-flow method is easily used. The second limitation deals with the temperature control which must be efficient to compensate the heat of compression. To solve these problems, devices were designed to mix samples under high-pressure, at controlled temperatures. The first apparatus was described by Grieger and Eckert [26] and modified by Sasaki *et al.* [27]. Both systems were thermostated and used the breakage of a foil diaphragm to force the mixing of two components. After different improvements [28], Heremans' group described a stopped-flow apparatus designed for spectroscopic detections of fast reactions at pressures up to 120 MPa by means of immersing a stopped-flow unit in a high-pressure bomb [29].

A new generation of stopped-flow apparatus was been recently described by P. Bugnon *et al.* which consists of one autoclave, including the static and dynamic components of the stopped-flow circuit; the drive mechanism and the high-pressure control unit. Measurements in absorbance or fluorescence mode can be performed at pressures up to 200 MPa, in a temperature range between - 40 and + 100 ° C. The sample circuit is chemically inert and there are no metallic parts in contact with sample solutions. The dead time of the instrument (i.e. the time needed for the reaction mixture to flow from the mixing chamber to the observation cell), is less than 2 ms in aqueous solution at room temperature up to 200 MPa [30].

Other high pressure stopped-flow instruments have been reported and a commercial unit is avaible from Hi-Tech Scientific, Salisbury, England.

To further reduce the dead-time with respect to the reactions studied, we have developed in the laboratory an apparatus which permits rapid mixing both at sub-zero temperatures and at high pressure (see Figure 4). The principle of our instrument incorporates certain features of previous stopped-flow systems described for cryo-enzymological studies and for investigations under high-pressure. The principle of this apparatus, already published [31,32], consists of a powerful pneumatic system driving the syringe mechanism, two vertical drive syringes containing the samples to be mixed, a mixing chamber, an observation chamber with quartz windows, and a waste syringe. Both temperature and pressure homogeneities are maintained by housing the whole apparatus in a high-pressure thermostated bomb. The stopped-flow apparatus can operate in absorbance or fluorescence mode over temperature and pressure ranges of + 40 to - 35 ° C and 1 to 300 MPa, respectively. The system is mounted either on an Aminco DW2 spectrophotometer permitting detection in a dual wavelength mode or on a spectrofluorometer specially designed in the laboratory (wavelength limits : 230 - 650 nm). The dead-time, nearly independent of pressure, is less than 5 ms when measuring in aqueous solutions at room temperature. In the presence of an organic solvent, this value increases to reach, for example, a value of 50 ms in a solvent system containing 40% ethylene glycol (v/v) at -15°C.

To be operative, the desiderata of a stopped-flow apparatus for cryo-baro-enzymological studies include :

1) efficient rapid mixing ;
2) good thermal and pressure equilibrium of the samples, the mixing chamber and the connecting parts ;
3) a recording system that can be used to follow rapid and very slow reactions with the same sample.

Neglect in any of these areas can lead to experimental artifacts or erroneous interpretation of the data . In our experience, spurious kinetic results can arise from :

1) incomplete mixing,
2) incomplete thermal equilibrium, mainly after compression, and
3) turbidity of samples.

These problems may be difficult to diagnose because each depends on the enzyme concentration.

Thus, it appears that extreme care must be exercised to avoid artifacts in optical spectroscopic measurements under these extreme conditions.

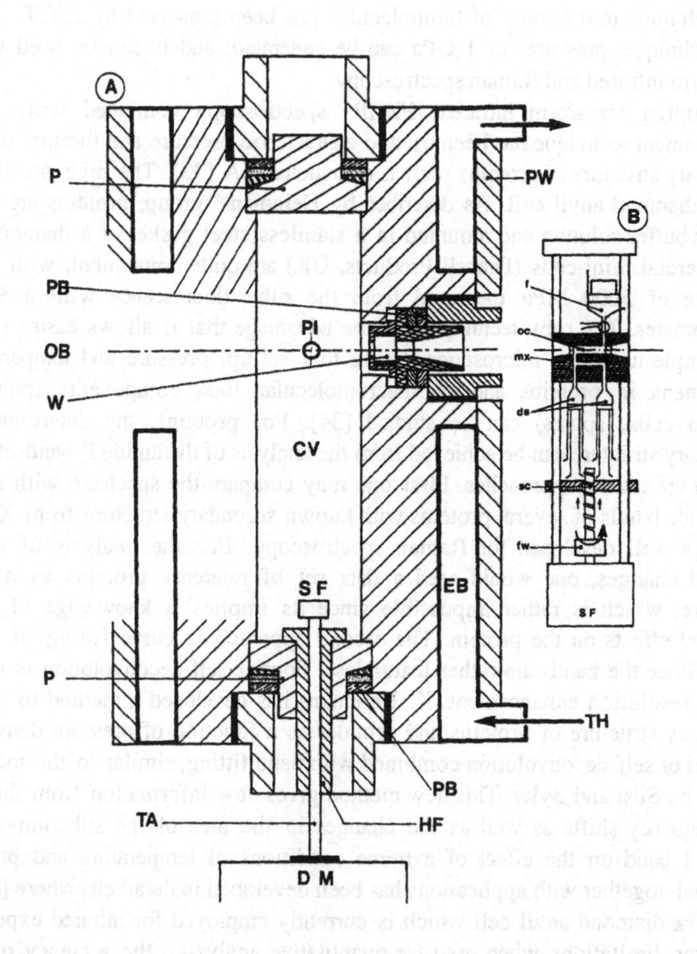

Figure 4. Schematic view of the high-pressure stopped-flow apparatus. (A) partial view of the bomb, P, plugs, CV, central volume containing the stopped-flow device; OB, optical beam; W, sapphire window; PB, packing, TH, thermostatization; PW, isolation, HF, high-pressure fitting; DM, driving mechanism, SF, stopped-flow adaptator. (B) Sttopped-flow module which is placed, when operate, in the central volume (CV) of the bomb (Adapted from ref. 7).

2.6. OTHER METHODS ADAPTED TO HIGH PRESSURE SPECTROSCOPIC DETECTION

2.6.1. *Infrared spectroscopy*

The aim of the present is not to extensively cover the wide range of optical methods adapted to high pressure. However, we must mention some specific techniques which are very well used in high-pressure experiments. The first is vibrational spectroscopy, which gives direct information about internal bonding in molecules. The application of this technique to the study of biomolecules has been pioneered by P.T.T. Wong. With this technique, pressures of 1 GPa can be generated, and it can be used with Fourier transform infrared and Raman spectroscopy.

Fourier transform infrared (FR-IR) spectroscopy combined with a resolution enhancement technique has been used to characterize pressure and thermal effects on the secondary structure of proteins such as ribonuclease A [33]. The high pressure cell used is the diamond anvil cell. As described by Heremans' group, proteins are dissolved in D_2O or buffer solution and mounted in a stainless steel gasket of a diamond anvil cell. Commercial mini-cells (Diacell Products, UK) are quite convenient, with a maximum pressure of 5000 MPa measured from the ruby fluorescence with a Spex Raman spectrometer. The ruby technique has the advantage that it allows easier inspection of the sample under the microscope. Using this set-up, pressure and temperature-induced phenomena in proteins and in macromolecular food components (polysaccharides, bacteria cells, spores) can be studied [34]. For proteins, the determination of the secondary structure can be achieved from the analysis of the amide I' band of the infrared spectra via several approaches. First one may compare the spectrum with a database of the amide bands of several proteins with known secondary structure from X-ray data, an analysis well developed for Raman spectroscopy. For the analysis of the pressure-induced changes, one would need a data set of reference proteins as a function of pressure, which is rather impossible since its implies a knowledge of all possible physical effects on the protein. The second approach is curve fitting of the amide I' band. Since the bands are rather featureless, Fourier self-deconvolution is often used to obtain resolution enhancement. K. Heremans has developed a method to determine the secondary structure of proteins and peptides as a function of pressure derived from the method of self-deconvolution combined with band fitting, similar to the method applied before by Susi and Byler. This new method gives new information from the analysis of the frequency shifts as well as the changes in the area of the subcomponents of the amide I band on the effect of extreme conditions of temperature and pressure. This approach together with applications has been developed in detail elsewhere [34].

The diamond anvil cell which is currently employed for infrared experiments has, however, limitations when used for quantitative analysis : the accuracy of pressure is low relative to the pressure range, accurately generating a pressure that is required is very difficult due to the pressure generation system, and the optical pathlength is not reproducible as a function of the applied pressure. Employing a traditional piston-cylinder type cell, the Taniguchi' group has developed a new cell with synthetic windows for quantitative infrared measurements of fluids [35]. This cell working up to 600 MPa has good accuracy in pressure control and has a constant optical pathlength.

The main advantage of the infrared technique is that it is an absorption technique where fluorescence does not interfere such as in the case of the Raman spectroscopy. However, because the strong absorption of water in the region of interest, infrared spectroscopy must be achieved in heavy water.

High-pressure infrared experiments also provide information on the dynamics of lipid hydrocarbon chains and their interactions with one another as shown for phospholipids [36].

2.6.2. Pressure perturbations

The use of pressure to perturb a system at equilibrium depends upon there being a difference in the volume of the components of the equilibrium mixture, according to :

$$A \xrightarrow{K_{eq}} B$$

where K_{eq} is the equilibrium constant.

If the molar volume of A differs from that of B by ΔV then, for a small pressure change ΔP:

$$\Delta \ln K = - \Delta V \cdot \Delta P / RT$$

where K is the pressure-induced change in equilibrium constant, R is the gas constant and T is the absolute temperature. The magnitudes of the perturbation induced in the concentrations of A and B depend not only on V but also on the magnitude of K [37].

Pressure-jump relaxation techniques with optical detection have been developed in several laboratories. After the rapid change in pressure the relaxation spectrum of the approach to the new equilibrium can be observed and analyzed. The pressure-perturbation has several advantages over temperature-perturbation as a technique for initiating chemical relaxation phenomena : wider choice of solvent composition, shorter time intervals between repeats and an extended time range. Moreover, pressure is a tool to initiate rapid reactions.

Different high-pressure systems have been described, but rather few for fast biochemical reactions. More than 20 years ago, Davis and Gutfreund have described an apparatus using the bursting of a disc to generate fast pressure-release [38]. A schematic illustration is shown in Figure 5. For this system, thought was given to obtaining higher operating pressures and shorter pressure-release times, whilst at the same time keeping cavitation and cell resonances at an acceptable minimum. A facility for using a variety of optical measuring techniques was also incorporated. Two different pressure-release devices can be fitted to the cell. One is a bursting disc valve (K). This device can relax the pressure in 25 μs. The alternative device is a mechanical valve (L) allowing a slower pressure-release on the order of 80 μs.

A number of optical detection techniques were incorporated in the design to enable as large a variety of signals as possible to be used for kinetics ; changes in turbidity, light absorption and fluorescence emission.

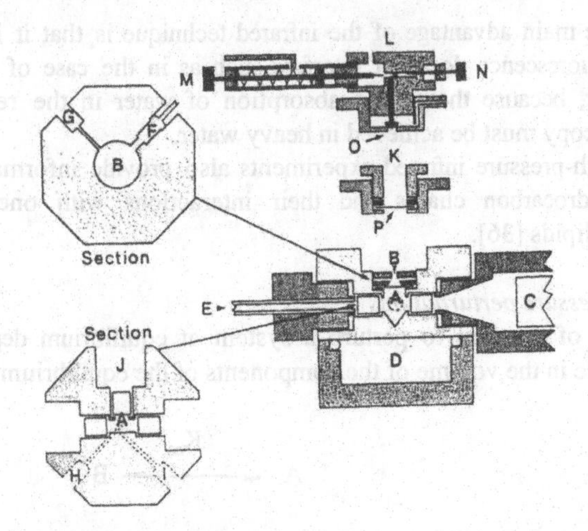

Figure 5. Schematic representation of the pressure-jump apparatus. A, observation cell; B, hydraulic chamber, C, absorbency photomultiplier; D, Thermostatted base; E, quartz fibre optic from light source; F, quartz pressure transducer; G, hydraulic pressure line; H & I, observation cell filling; J, fluorescence emission window; K, bursting disc pressure-release valve; N, reset mechanism; O, valve seat; P, bursting disc. (adapted from ref. 38).

In the laboratory, for slower reactions, we used manual equipment. The kinetics were induced by abrupt rises of pressure in 20 MPa steps. The dead time (about 2 s) corresponds to the time needed for raising the pressure and closing the valve. Nitrophenol absorbance at 398 nm was used to monitor temperature changes induced by pressure jumps. Immediately after the jump, the absorbance increased transiently for 1 min, corresponding to a temperature increase of approximately 0.75 ° C. This apparatus was used to study the pressure-induced spin-state transitions of cytochrome P-450scc [39].

3. Applications

The use of optical spectroscopic techniques in high pressure bioscience is so large that we will shown in the present article a selection of recent results obtained in our laboratory.

The first one concerns UV absorbance in the fourth derivative mode. We have applied this method to study the pressure induced unfolding of Sso7d from *S. solfataricus*. We use this very small protein (7 kDa), a DNA binding protein endowed with ribonuclease activity as a model to study the structural basis of protein stability. Indeed, this protein withstands pressure up to at least 2000 MPa and temperature to 90 °C. Furthermore, it is a well characterized protein which has been studied by FTIR, CD, NMR, scanning calorimetry and molecular dynamics simulation [15]. The analysis by the fourth derivative method of one of its mutants (F31A), (see Figure 6), shows that the polarity of tyrosine33 (the 283 nm band) changes in two distinct steps at increasing pressure. Between 1 and 120 MPa the amplitude of the tyrosine band decreases, and, between 200 and 330 MPa, it is blue-shifted, that its tyrosine 33 becomes exposed to the solvent. This stepwise change of the environment of tyrosine33, a residue which participates in the maintain of the hydrophobic core stability, indicates that pressure induced unfolding of Sso7d is not a simple two-state process. This very interesting reaction will be discussed in further details in the oral communication of E. Mombelli.

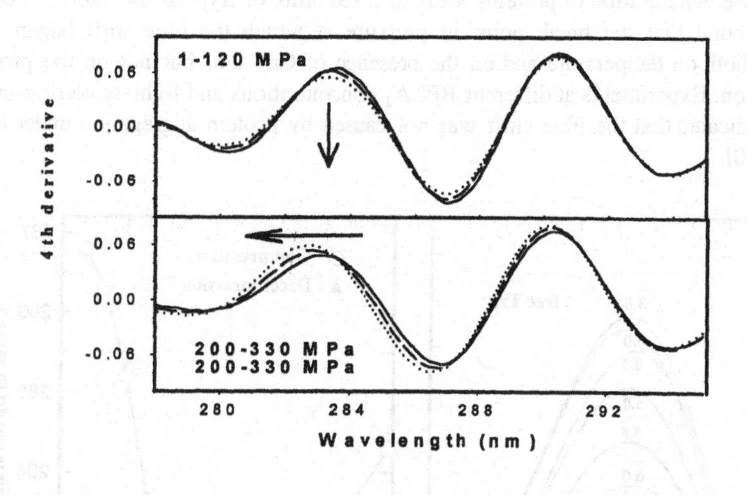

Figure 6. Effect of pressure on the 4th derivative UV absorbance spectrum of the F31A mutant of Sso7d from the thermophile archaebacterium *S. solfataricus*. The arrows indicate the shift of the tyrosine band (283 nm) at increasing pressure.

Another kind of result has recently being obtained in our laboratory by K. Ruan using fluorescence spectroscopy for protein denaturation study. According to our knowledge of protein denaturation caused by chemical denaturant, temperature and hydrostatic pressure, it is clear that the unfolding of protein is a step by step process depending on the increase of the strength of the denaturation condition. When the unfolding of protein is followed by monitoring the change in intrinsic tryptophan fluorescence, an increase in pressure induces generally a decrease in fluorescence intensity associated with a red shift. This results from a higher exposure of tryptophan residues to the solvent. Such a red shift has been observed in many pressure-induced denaturation studies of proteins. Recently, we studied the pressure denaturation of basic phospholipase A_2 ($BPLA_2$) which is one of three isoenzymes of phospholipase A_2 from the venom of *Agkistrodon Halys Pallas* (isoelectric point of 8.9). The primary structure of $BPLA_2$ has been reported, showing the existence of a single tryptophan residue (Trp 61) which is a particularly suitable situation for a fluorescence study. It was found that the pressure denaturation behavior of $BPLA_2$ is different from the above described general case. It was found that the tryptophan fluorescence emission spectra of the enzyme were significantly different in two pressure ranges : from 0.1 to 400 MPa and from 400 to 650 MPa. For increasing pressure, the spectra shifted to the red in the lower pressure range and to the blue in the higher pressure range. Whereas the red shift could be ascribed to the intrinsic pressure dependence of the fluorophore (trp), the blue shift indicated a pressure induced protein conformational change toward a structure where the single tryptophan is in a less polar environment, suggesting its burying deeper inside the protein. This is the first time that such a phenomenon has been observed. Generally, high pressure denaturation of proteins leads to a red shift of tryptophan fluorescence. It was also found that the break point in pressure at which the blue shift began was dependent both on temperature and on the presence of Ca^{++} ion, but not on the protein concentration. Experiments at different $BPLA_2$ concentrations and light scattering under pressure indicated that the blue shift was not caused by protein aggregation under high pressure [40].

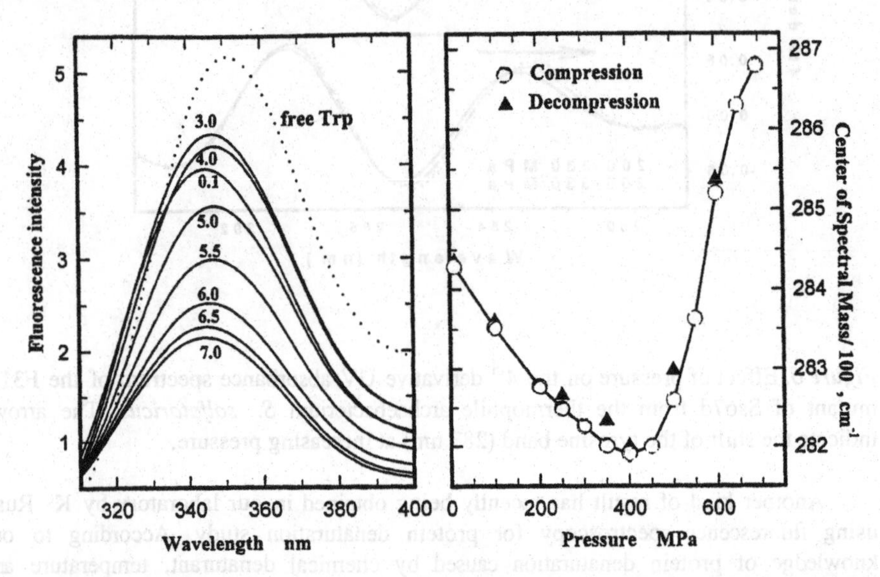

Figure 7. Pressure effect on fluorescence (left) and on center of spectral mass (right) of native basic phospholipase. The numbers on the spectrum were the applied pressure in kbar. Excitation wavelength : 295 nm (Adapted from ref. 40).

At the end, we would like give two examples of the potentiality of the stopped-flow method.

a) Using a combination of ultraviolet spectroscopy under pressure and stopped-flow kinetics under pressure, M.J. Kornblatt *et al.* have shown that the monomers of yeast enolase (dimeric in native form) produced by hydrostatic pressure are inactive [17]. Equilibrium constants and activation volumes for dissociation/inactivation produced by hydrostatic pressure have been determined under various conditions. The main result has been obtained by comparison of the pressure effects on the quaternary structure of enolase calculated from the fourth derivatives of spectra recorded under high pressure and the activity of the enzyme measured in stopped flow experiments (see Figure 7).

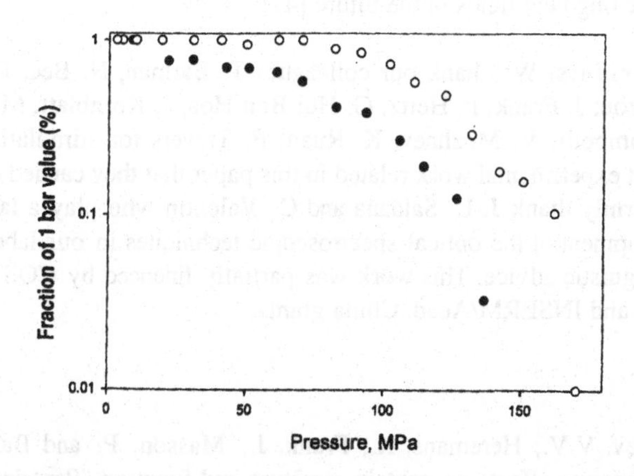

Figure 7. Effects of pressure on the quaternary structure and activity of yeast enolase. (•) The activity was measured in stopped-flow pressure experiments (the results are expressed as percentage of the 0.1 MPa value ; (o) quaternary structure (percentage dimeric) calculated from the 4th derivatves of spectra recorded under pressure. (Imidazole buffer, at 25 ° C). Enolase was assumed to be dimeric at atmospheric pressure. Enolase activity was assayed by following the production of phospho*enol*pyruvate at 240 nm. (adapted from ref. 17).

Removing the Mg^{2+}, either by adding EDTA or by preparing apoenzyme, displaces the equilibrium towards monomers and decreases both the activation volumes which occurs in the transition state for dissociation. It has been proposed that dissociation induced changes in the conformations of the 'mobile loopps' of enolase where a precise conformation of these loops is necessary for maintening subunit interaction.

b) Another example has been studied in our laboratory by Kunugi's group on the effects of high pressure on thermolysin activity and spectroscopic properties. This enzyme showed distinct pressure-induced activation with a maximum observed in the range 100 - 250 MPa depending of the substrate used [16]. The changes observed are explained by a simple two-state model accompanied by a large negative change in the volume of reaction.

4. Conclusion

It is not very hazarded to predict that the optical techniques adapted for high pressure experiments have a nice future. Structure-function and structure stability relationships in proteins associated with the study of mutant proteins and experimented using spectroscopic approaches will certainly contribute to increasing to our understanding of protein structure behavior, mainly with association with other structural methods.

Moreover, such complex problems can be best studied using molecular dynamics simulation associated with these optical methods [15]. Additional interest will come also from the use of more "extreme experimental conditions" as shown with recent publications opening new fields in the future [41].

Acknowledgements. We thank our colleagues T. Barman, N. Bec, J. Connelly, N. Dahan, P. Douzou, J. Frank, F. Heitz, G. Hui Bon Hoa, J. Kornblatt, M. Kornblatt, S. Kunugi, E. Mombelli, V. Mozhaev, K. Ruan, F. Travers for stimulating discussions and for the most experimental work related in this paper that they carried out. We should also like to warmly thank J.-L. Saldana and C. Valentin who play a large part in the technical development of the optical spectroscopic techniques in our laboratory and Dr. G. Yang for linguistic advice. This work was partially financed by COST D10 project, INSERM/SJPS and INSERM/Acad. China grants.

References

1. Mozhaev, V.V., Heremans, K., Frank, J., Masson, P. and Balny, C. (1996) High pressure effects on protein structure and function, *Proteins: Struc. Func. Gen.* **24**, 81-91.
2. Gross, M. and Jaenike, R. (1994) Proteins under pressure. The influence of high hydrostatic pressure on structure, function and assembly of proteins and protein complexes, *Eur. J. Biochem.* **221**, 617-630.
3. Silva, J.L. and Weber, G. (1993) Pressure stability of proteins, *Annu. Rev. Phys. Chem.* **44**, 89-113.
4. Jonas, J. and Jonas, A. (1994) High-pressure NMR spectroscopy of proteins and membranes, *Annu. Rev. Biophys. Biomol. Struct.* **23**, 287-318.
5. Paladini, A.A. and Weber, G. (1981) Absolute measurements of fluorescence polarization at high pressures, *Rev. Sci. Instr.* **53**, 419-427.
6. Shimizu, T. (1992) High pressure experimental apparatus with windows for optical measurements up to 700 MPa, in C. Balny et al. (eds.), *High Pressure and Biotechnology*, Colloque INSERM/J. Libbey Eurotext, London, vol. 224, pp. 525-527.
7. Balny, C., Masson, P. and Travers, F. (1989) Some recent aspects of the use of high pressure for protein investigations in solution, *High Press. Res.* **2**, 1-28.
8. Douzou, P. (1977) *Cryobiochemistry: an introduction*, Academic Press, New York.
9. Hui Bon Hoa, G., Douzou, P., Dahan, N. and Balny, C. (1982) High pressure spectrophotometry at sub-zero temperatures, *Anal. Biochem.* **120**, 125-135.
10. Talsky, G. (1994) *Derivative spectroscopy; low and high order*, VCH, Weinheim.
11. Kornblatt, J.A., Kornblatt, M.J. and Hui Bon Hoa, G. (1995) Second derivative spectroscopy of enolase at high hydrostatic pressure: an approach to the study of macromolecular interactions, *Biochemistry*, **34**, 1218-1223.

12. Lange R., Frank, J., Saldana, J.-L. and Balny, C. (1996) Fourth derivative UV-spectroscopy of proteins under high pressure. I. Factors affecting the fourth derivative spectrum of the aromatic amino acids, *Eur. Biophys. J.* **24**, 277-283.
13. Lange, R., Bec, N., Mozhaev, V.V. and Frank, J. (1996) Fourth derivative UV-spectroscopy of proteins under high pressure. II. Application to reversible structural changes, *Eur. Biophys. J.* **24**, 284-292.
14. Balny, C., Saldana, J.-L., Lange, R., Kornblatt, M.J. and Kornblatt, J.A. (1996) UV/vis biochemical spectroscopy under high pressure, in Ph. R. von Rohr and Ch. Trepp (eds.), *High Pressure Chemical Engineering*, Elsevier Science, pp. 553-558.
15. Mombelli, E., Afshar, M., Fusi, P., Mariani, M., Tortora, P., Connelly, J.P. and Lange, R. (1997) The role of phenylalanine 31 in the maintaining the conformational stability of ribonuclease P2 from *Sulfolobus solfataricus* under extreme conditions of temperature and pressure, *Biochemistry,* **36**, 8733-8742.
16. Kunugi, S., Kitayaki, M., Yanagi, Y., Tanaka, N., Lange, R. and Balny, C. (1997) The effect of high pressure on thermolysin, *Eur. J. Biochem.* **248**, 567-574.
17. Kornblatt, M.J., Lange, R. and Balny, C. (1998) Can monomers of yeast enolase have enzymatic activity ? *Eur J. Biochem.* **251**, 775-780.
18. Paladini, A.A. and Weber, G. (1981) Pressure-induced reversible dissociation of enolase, *Biochemistry*, **20**, 2587-2593.
19. Silva, J.L., Miles, E.W. and Weber, G. (1986) Pressure dissociation and conformational drift of the ß dimer of tryptophan synthase, *Biochemistry*, **25**, 5781-5786.
20. Ruan, K. and Weber, G. (1989) Hysteresis and conformational drift of pressure-dissociated glyceraldehydephosphate dehydrogenase, *Biochemistry*, **28**, 2144-2153.
21. Royer, C.A., Hinck, A.P., Loh, S.N., Prehoda, K.E., Peng, X., Jonas, J. and Markley, J.L. (1993) Effects of amino acid substitutions on the pressure denaturation of staphylococcal nuclease as monitired by fluorescence and nuclear magnetic resonance spectroscopy, *Biochemistry*, **32**, 5222-5232.
22. Harris, R.D., Jacobs, M., Long, M.M. and Urry, D.W. (1976) A high pressure sample cell for circular dichroism studies, *Anal. Biochem.*, **73**, 363-368.
23. Ozawa, S., Hayashi, R., Takahashi, S., Kawai, S. (1995) CD measurement of protein under high pressure in *Abstracts, International Conference on High Pressure Bioscience and Biotechnology*, Kyoto, A-3, p. 11.
24. Balny, C., Lange, R., Heitz, F., Connelly, J.P., Saldana, J.-L. and Ruan, K. (1998) High pressure as a tool to study protein conformation, *Rev. High Pressure Sci. Technol.* **7**, 1292-1296.
25. Markley, J.L., Travers, F. and Balny, C. (1981) Lack of evidence for a tetrahedral intermediate in the hydrolysis of nitroaniline substrates by serine proteinases, *Eur. J. Biochem.* **120**, 477-485.
26. Greiger, R.A. and Eckert, C.A. (1970) A new technique for chemical kinetics at high pressures, *AIChE J.* 766-770.
27. Sasaki, M., Amita, F. and Osugi, J. (1979) High pressure stopped-flow

apparatus up to 3 kbar, *Rev. Sci. Instr.* **50** 1104-1107.

28. Smith, P. Beile, E. and Berger, R. (1982) An observation chamber for high-pressure stopped-flow apparatus, *J. Biochem. Biophys. Methods*, **6**, 173-177.
29. Heremans, K., Snauwaert, J. and Rijkenberg, J. (1980) Stopped-flow apparatus for the study of fast reactions in solution under high pressure, *Rev. Sci. Instr.* **51**, 806-808.
30. Bugnon, P. Laurenczy, G. Ducommun, Y. Sauvageat, P.-Y., Merbach, E.A., Ith, R., Tschanz, R., Doludda, M. Bergbouer, R. and Grell, E. (1996) High-pressure stopped-flow spectrometer for kinetic studies of fast reactions by absorbance and fluorescence detection, *Anal. Chem.* **68**, 3045-3049.
31. Balny, C., Saldana, J.-L. and Dahan, N. (1984) High-pressure stopped-flow spectrometry at low temperatures, *Anal. Biochem.* **139**, 178-189.
32. Balny, C., Saldana, J.-L. and Dahan, N. (1987) High-pressure stopped-flow fluorometry at subzero temperatures : Application to kinetics of the binding of NADH to liver alcohol dehydrogenase, *Anal. Biochem.* **163**, 309-315.
33. Takeda, N., Kato, M. and Taniguchi, Y. (1995) Pressure and thermally-induced reversible changes in the secondary structure of ribonuclease A studied by FT-IR spectroscopy, *Biochemistry*, **34**, 5980-5987.
34. Heremans, K., Rubens, P., Smeller, L., Vermeulen, G. and Goossens, K. (1996) Pressure versus temperature behaviour of proteins: FT-IR studies with the diamond anvil cell, in R. Hayashi and C. Balny (eds.), *High Pressure Bioscience and Biotechnology*, Elsevier Science, pp. 127-134.
35. Kato, M. and Taniguchi, Y. (1995) A hydrostatic optical cell with synthetic diamond windows for quantitative infrared measurements of fluids, *Rev. Sci. Instrum.* **66**, 4333-4335.
36. Reis, O., Winter, R. and Zerda, T.W. (1996) The effect of high external pressure on DPPC-cholesterol multilamellar vesicles. A pressure-tuning Fourier transform infrared spectroscopy, *Biochim. Biophys. Acta*, **1279**, 5-16.
37. Geeves, M.A. (1991) The influence of pressure on actin and myosin interactions in solution and in single muscle fibres, *J. Cell Sci.*, **14**, 31-35.
38. Davis, J.S. and Gutfreund, H. (1976) The scope of moderate pressure changes for kinetic and equilibrium studies of biochemical systems, *FEBS Letters*, **72**, 199-207.
39. Bancel, F., Bec, N., Ebel, C. and Lange, R. (1997) A central role for water in the control of the spin state of cytochrome P-450scc, *Eur. J. Biochem.* **250**, 276-285.
40. Ruan, K., Lange, R., Zhou, Y. and Balny, C. (1998) Unusual affect of high hydrostatic pressure on basic phospholipase A_2 from venom of *Agkistrodon Halys Pallas*, *Biochem. Biophys. Res. Comm.* **249**, 844-848.
41. Luche, J.-L., Balny, C., Benefice, S., Denis, J.-M. and Petrier, C. (eds.) (1998) *Chemical Processes and Reactions under Extreme or non-Classic Conditions*, Off. Publ. Eur. Comm., Luxembourg.

HIGH HYDROSTATIC PRESSURE AND ENZYMOLOGY

Claude BALNY and Natalia L. Klyachko*
INSERM U 128, IFR 24, 1919, route de Mende
34293 MONTPELLIER Cedex 5, France
Email: balny@crbm.cnrs-mop.fr
*Department of Chemical Enzymology, Faculty of Chemistry,
Moscow State University, 119899 MOSCOW, Russia
Email: klyachko@enzyme.chem.msu.ru

Abstract

The recent interest in high pressure biochemistry requires more fundamental studies on the behavior of enzymes under such extreme conditions. In this paper we shortly review some basic knowledge on the high pressure effect on enzyme reactions, together with some new data concerning high pressure micellar enzymology taking as a model for enzyme studies in a membrane environment.

1. Introduction

In its simplest expression, an enzyme reaction is a series of elementary steps leading to the release of products. The effect of pressure on the overall reaction may be the consequence of its effects on the reaction itself, on enzyme conformational changes, on enzyme dissociation into subunits, or on substrate and product(s) level. The classical approach for the study of enzymatic reactions under high pressure may be the same as at atmospheric pressure: the steady state theory, giving k_{cat} and K_m values. This is very well documented by Morild [1], after the pioneering work of Laidler [2].

Another approach is to gather the information on the effects of pressure and/or temperature, two fundamental thermodynamic variables, on the individual rate constants of each elementary step of the enzyme reaction. This permits a clear and detailed analysis, while the measurement of a complex rate constant such as k_{cat}, can lead to ambiguous results [3].

In the present paper we would like mainly stress with the second approach which has been developed in our laboratory for many years [4,5]. Moreover, we will also present a rather new field: micellar enzymology under high pressure. Why the latter has been chosen?

Biological membranes are closely involved in enzymatic processes. In living cells, enzymes mostly act either on or near the "water/organic medium" interface, and the subcellular structure and the compartmentalization of enzymes play key roles in the regulation of metabolism. Moreover, the physical properties of water, such as, for

R. Winter and J. Jonas (eds.), High Pressure Molecular Science, 423–436.

example, polarity, viscosity, dielectric permeability, in proximity of the interface differ significantly from those of the bulk water. In this way, the high pressure parameter may be used as a new tool for better understanding biological processes.

2. Basic concepts

Enzymes fluctuate between different conformational states which may occur on a large scale or be restricted locally. According to Frauenfelder [6], these conformational states appear to be organized hierarchically in substates. The enzyme reaction which may involve one or a combination of several of these fluctuations often depends on the specific interaction with the surrounding solvent. Principles of protein dynamics and enzyme function under pressure is, however, still in its infancy and the only theory available is the classical transition state theory [7] which is an oversimplification of reactions carried out by complex macromolecules such as enzymes. In the absence of a better-adapted theory, we continue to use its formalism, in spite of the fact that the transition state must be characterized structurally [8]. Kinetic methods allows one to determine the activation volume of reaction, (i.e. the difference between the volume in the transition and the ground states) from the dependence of a rate constant, k, on pressure, according to :

$$\delta \ln k / \delta p = - \Delta V^{\neq}/RT \qquad (1)$$

By varying both pressure and temperature, one is able to obtain simultaneously kinetic (k), energetic ($\Delta H^{\neq}$, $\Delta S^{\neq}$) and structural ($\Delta V^{\neq}$) information of a transition state of the reaction.

Using the transition state theory, we have developed the elementary thermodynamic principles for enzyme reactions [3,4,9]. With temperature (T) and pressure (P) as variables, it is possible to integrate kinetic relations by including different parameters such as variations of heat capacity, compressibility or expandability, respectively. The final equations are not linear as a function of P or T. However, because of the nature of systems studied and the narrow ranges of P and T experimented, simplifications are possible, and for most systems the above eq. 1 can be used at constant temperature.

At the end, we must mention that for some systems where the viscosity is important, the treatment which may approach the real situation more closely is provided by Kramers' theory which take into account a pre-exponential term that includes the viscosity of the medium [10] (for details, see ref. 9).

In spite of the complexity of biochemical reactions, the quantities $\Delta V^{\neq}$, $\Delta G^{\neq}$, $\Delta S^{\neq}$ and $\Delta H^{\neq}$ for a given reaction as a function of T and P, respectively, are well related according to the Maxwell relationships:

$$- (\delta \Delta V^{\neq} / \delta T)_P = (\delta \Delta S^{\neq} / \delta P)_T \qquad (2)$$

$$\text{and} \quad (\delta \Delta H^{\neq} / \delta P)_T = \Delta V^{\neq} - T\,(\delta \Delta V^{\neq} / \delta T)_P \qquad (3)$$

$(\delta \Delta V^{\neq} / \delta T)_P$ is the temperature coefficient for the activation volume: expandability term.

$(\delta\Delta S^* / \delta P)_T$ and $(\delta\Delta H^* / \delta P)_T$ are the pressure coefficients for the activation entropy and enthalpy, respectively.

3. High pressure enzymology in bulk media

Literature provides a great number of studies concerning the effects of high pressure on enzyme reactions under steady-state conditions where pressure dependencies of k_{cat} and K_m were determined. In the following section we would like to present only some few specific examples studied in our laboratory.

3.1. TRANSIENT REACTIONS

Using high-pressure stopped-flow method presented in a preceding chapter of these Proceedings, it is possible to have an access to the elementary steps of the enzyme reactions. Different systems have been studied (creatine kinase [4], horseradish peroxidase complex I formation with peroxides [3], carbon monoxide binding to heme proteins [8,11], reduction of hydroxylamine oxidoreductase [12], etc. To increase the temporal resolution, cryoenzymology [5] has been used in addition to the fast kinetic detection. However, the addition of organic solvents (antifreezes) to the aqueous solution is necessary to maintain the media liquid at temperatures below 0° C. For many reactions, it appeared that the binding step (substrate or ligand to the enzyme) is a two step process: formation of a Michaelian compound followed by an "induced-fit" step resulting in a modification of the conformation of the enzyme according to:

$$E + S \underset{}{\overset{K_1}{\rightleftharpoons}} ES \xrightarrow{k_2} E^*S \qquad (4)$$

In order to determine the constants (and the thermodynamic parameters associated), it is possible to study the observed rate constant as a function of substrate (or ligand) concentration. One can then obtain K_1 and k_2, or only the second-order rate constant k_+ if a saturation plateau cannot be attained (the reaction is too fast or K_1 is too small).

For many examples, we found that the thermodynamic parameters measured were greatly modulated by the physical chemical parameters of the media (including temperature and solvent composition). These results suggest that the macroscopic thermodynamic response is mainly controlled by the solvent. By adjusting two variables (among T, P, solvent), it is possible either to amplify or to cancel out the effect of the third.

3.2. PSEUDO-STEADY STATE REACTIONS, A WAY TO APPROACH ENZYME CONFORMATIONAL STATE.

A characteristic example can be given with the reaction of carbamylation catalyzed by buty-rylcholinesterase. The effect of high-pressure on the reaction of butyrylcholinesterase with N-methyl-(7-dimethylcarbamoxy) quinolinium was determined under single-turnover conditions where the rate of carbamylation was monitored as the accumulation of a fluorescent ion. For this, P. Masson had used our high-pressure stopped-flow apparatus in fluorescence mode [13]. Elevated pressure favored the formation of the enzyme-substrate complex but inhibited carbamylation of the enzyme. Because a single reaction step was recorded, it was possible to interpret the data obtained under high pressure in terms of Michaelis-Menten equations. From these studies, it appeared that the substrate-induced conformational change and the change of water structure were dominant contributions to the overall volume change associated with substrate binding. The large positive activation volume measured (119 ml/mol) might also reflect extended structural and hydration changes. At pressures greater than 40 MPa, an additional change dependent on the substrate concentration occurred in a narrow pressure interval. This effect could be a result of appearing the substrate-induced pressure sensitive enzyme conformational state.

Using a combination of ultraviolet spectroscopy and stopped-flow kinetics under high pressure, conformational changes accompanying enzyme reactions have been also revealed for thermolysin [14] and yeast enolase [15].

3.3. THE USE OF HIGH-PRESSURE ENZYMOLOGY TO MODIFY CATALYTIC BEHAVIOR

Typical experiments were carried out by V.V. Mozhaev on the use of elevated pressures to increase the catalytic activity and thermal stability of α-chymotrypsin (CT) [16]. Under steady-state conditions, for an anilide substrate, an increase in pressure at 20° C, results in an exponential growth of the rate of hydrolysis, acceleration more pronounced at higher temperature (50° C). Elevated pressure is also efficient for the increasing stability of CT against thermal denaturation. This phenomenon, important for biotechnological applications, may be discussed in terms of the hypothesis which explain the action of external and medium effects on the protein structure, such as preferential hydration and osmotic pressure.

3.4. PRESSURE AND TEMPERATURE: FUNDAMENTAL PARAMETERS

Through the above examples, and with other data [17], it is clear that, for the enzyme reaction in bulk media (and, as we will show below, for micellar enzymology), it is important to take into consideration both parameters, temperature and pressure, which are related to the fundamental thermodynamic expressions. The omission of the temperature for the study of the effect of pressure on the enzyme reaction can give only a partial information since, for example, depending on the temperature, the sign of the activation volume can be different for the same reaction, as shown for hydroxylamine reductase [12] or myosin ATPase reactions [18], respectively.

4. Micellar enzymology under high pressure

4.1. MICELLAR ENZYMOLOGY

A new direction in enzymology, micellar enzymology, has been opened by using the systems of surfactant reversed micelles in organic solvents as a medium for chemical and enzymatic reactions (see [19-20] and reviews [21-24]).

Reversed micelles can be easily prepared by simple dissolving surfactants or lipids of different nature in organic solvents giving an optically transparent, thermodynamically stable medium. Surfactant molecules in the system arranged in spherical particles having a polar inner core formed by "head" groups and non-polar outer part formed by hydrocarbon "tails" of the surfactant (Figure 1). Reversed micelles represent dynamic formations that can exchange surfactant molecules with each other and with the solution at a high rate; the time being of the surfactant molecule in the micelle is about 10^{-7} s [25]. Moreover, surfactant molecules in the reversed micelle show a continuous vibration. Despite high mobility of surfactant molecules, the interface of micelles is rather well defined and non-permeable for an organic solvent surrounding the micelle. Micelles from anionic surfactants are characterized by a narrow distribution in terms of dimensions and form that do not depend on surfactant concentrations [26,27]. Surfactant reversed micelles in organic solvents can solubilize large amounts of water, other polar substances, and biomolecules (proteins, enzymes or antibodies), moreover, water content and the size of particles can be varied in wide limits. For example, in reversed micelles of anionic surfactant, sodium bis(2-ethylhexyl) sulfosuccinate, so-called Aerosol OT or AOT, in iso-octane, an increase in the water/surfactant molar ratio, w_o, from 0 to 50 results in a lengthening of the micelle hydrodynamic radius from 1.7 to 12 nm [26]. The water in the inner cavity of hydrated reversed micelles differs from bulk water by many physical chemical properties, such as acidity, microviscosity, polarity and so on [25, 26]. The difference depends on the degree of hydration becoming less and less with an increase in w_o.

Micellar enzymology is a promising approach to study and regulate the properties of biomolecules, and enzymes in particular. Many enzymes have been studied in the systems of reversed micelles in organic solvents [22-24]. Their catalytic properties can be widely regulated using several simple ways. One of them is variation of water content in the system. Most of the enzymes studied in surfactant reversed micelles exhibit a characteristic bell-shaped dependence of their catalytic activity on the surfactant hydration degree. Variations in this parameter are accompanied by changes in the size of the inner water cavity of reversed micelles. Maximal enzyme catalytic activity in reversed micelles is observed at a particular surfactant hydration degree when the size of the inner cavity corresponds to that of the entrapped protein [23]. Under conditions when such a complementarity is reached, the enzyme molecule experiences a close contact with the surrounding surfactant matrix. In this case the matrix can tightly "squeeze" the enzyme molecule, thereby "freezing" its rotational and vibrational motions that could lead to the fixation of the most catalytically active conformation. Indeed, the suggestion has been confirmed by parallel measurements of the catalytic activity of chymotrypsin and acid phosphatase, and the mobility of a spin label covalently attached to active sites of these enzymes. The results obtained show that under conditions when enzymes possess the highest catalytic activity, a minimum appears on the dependence of the rotational frequency of the spin label on the water content in the system, indicating an increase in the rigidity of the enzyme active site [23].

Figure 1 : Schematic representation of an enzyme-containing reversed micelle formed by surfactant (Aerosol OT) in non-polar solvent.

4.2. WHY MICELLAR ENZYMOLOGY UNDER HIGH PRESSURE ?

Can pressure give an additional advantages? What kind of effects one can expect using combination of these two factors of enzyme regulation?

We know from basic equation that activation volume, $\Delta V^{\neq}$, that represents the difference between the reagent volumes in the activation and ground states, is one of the main parameters that depends on pressure. It means, first, that pressure can play a key role in the enzyme regulation, if the volume changes occur upon pressure application. And, second, variation of water content that can be easily realized in the system of reversed micelles can bring an additional effect on the enzyme regulation at constant pressure.

4.2.1. Enzyme regulation.

Using α-chymotrypsin (CT), well-known model enzyme, hydrolysis of N-succinyl-L-phenylalanine p- nitroanilide (SPNA) and N-carbobenzoxy-L-tyrosine p-nitrophenyl ester (CBZTNP) was studied in reversed micelles of Aerosol OT in octane at different pressures [28-30]. These two substrates have different rate-limiting step in the enzyme catalysis; it is formation of the acyl-enzyme intermediate in the case of anilide substrates and deacylation in the case of ester substrates. Moreover, these two reactions have a different sign in activation volume values. α-Chymotrypsin hydrolysis of anilide and ester substrates is characterized by the negative and positive $\Delta V^{\neq}$ values, consequently. Hence, increasing pressure are accompanied by acceleration of the reaction in the case of anilide substrates and leads to its deceleration in the case of ester substrates. Figure 2 illustrates this effect in the system of reversed micelles of Aerosol OT in octane. As seen, an increase of pressure up to 200 MPa results in a sevenfold increase in the rate of anilide hydrolysis whereas it brings more than tenfold decrease in the rate of ester hydrolysis. The example demonstrates the strong possibility in changing the enzyme selectivity when difference between substrates increases hundred times upon pressure application.

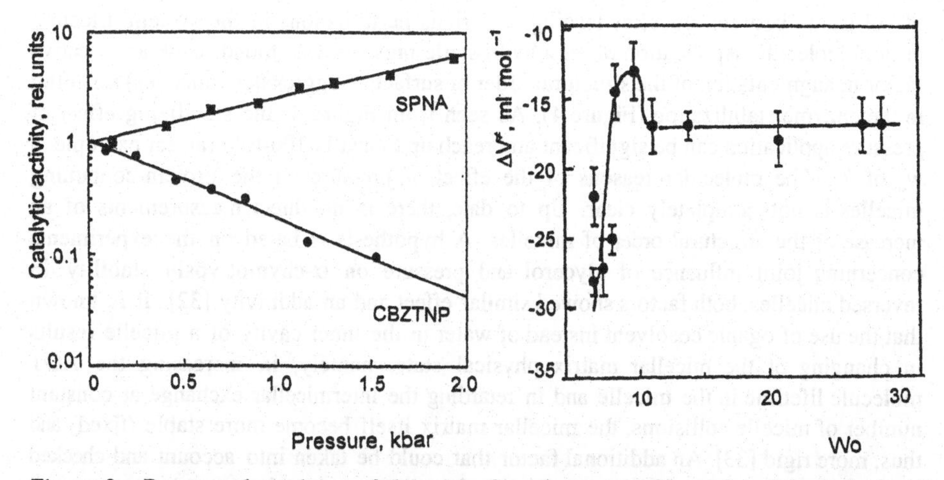

Figure 2 : Pressure dependence of the rate of α-chymotrypsin catalyzed hydrolysis of SPNA and CBZTNP in reversed micelles of Aerosol OT in octane [28-30].
Figure 3 : Dependence of the activation volume for α-chymotrypsin catalyzed hydrolysis of SPNA in reversed micelles of Aerosol OT in octane on surfactant hydration degree, w_o [28].

Variation of water content under pressure can also be an additional factor in enzyme regulation. Indeed, as found [28], activation volume in α-chymotrypsin catalyzed hydrolysis of SPNA also depended on surfactant hydration degree giving a bell-shaped curve, the same as for the enzyme catalytic activity changes (Figure 3). Maximum on the curve is observed at that hydration degree where the size of the micelle inner cavity is close to that of the enzyme ($w_o \approx 10$). As seen from Figure 3, the most pronounced changes in activation volume are observed at low water content ($w_o < 10$) where α-chymotrypsin creates its own micelle of an equal size and, therefore, protein-micelle interactions become another significant factor in the enzyme regulation which pressure can strongly affect.

4.2.2. Enzyme stabilization.
An important consequence and application of the effect of pressure on proteins via micellar matrix appears to be pressure stabilization of the enzyme in reversed micelles (i.e. an increasing the enzyme stability under pressure). In this respect, the thermal inactivation of α-chymotrypsin and the effect of pressure application on the enzyme stability have been studied in reversed micelles of AOT in octane [31]. At atmospheric pressure, an increase in temperature leads to a pronounced acceleration of inactivation of the enzyme; at 40° C, the catalytic activity drops more than 5 times in less than an hour whereas at 20° C, the enzyme remains 100 % active much longer. One of the possible reasons of such behavior could be weakening of the protein-micelle contacts (stabilizing action of the micellar matrix) with increasing temperature due to increasing the number

of collisions between micelles leading to serious perturbations in the system affecting protein molecule. Application of high hydrostatic pressure was found to be an effective factor in augmentation of the structural order of surfactant aggregates (micelles) resulting in the enzyme stabilization (Figure 4). As seen from Figure 4, the stabilizing effect of pressure application can be significant and reach up to nearly 100-fold (at, for example, a w_0 of 7). The molecular reasons of the effect of pressure on the protein-containing micelles is not completely clear. Up to date, there is no direct measurements of an increase of the structural order of micelles. A hypothesis is based on the experiments concerning joint influence of glycerol and pressure on α-chymotrypsin stability in reversed micelles; both factors showed similar effect and an additivity [32]. It is known that the use of oganic cosolvent instead of water in the inner cavity of a micelle results in changing of the micellar matrix physical state, namely, in increasing the AOT molecule lifetime in the micelle and in retarding the intermicellar exchange at constant number of micelle collisions, the micellar matrix itself become more stable (fixed) and thus, more rigid [33]. An additional factor that could be taken into account and checked in future is water molecules that forced out of the inner cavity when protein is entrapped into micelle of an equal or less size. According to NMR data, water molecules occupy void volume existing in spatial structure of AOT molecule [34]. Pressure can somehow affect the redistribution of these water molecules.

4.2.3. Significance of protein-micelle contacts.

The significance of protein-micelle contacts (interactions) and an important role of a micellar matrix in pressure effects on protein-containing reversed micelles have been demonstrated in the study of oligomeric enzymes: dimeric porcine heart malic dehydrogenase (MDH) [35], tetrameric rabbit muscle lactic dehydrogenase (LDH) and D-glyceraldehyde-3-phosphate dehydrogenase (GPDH) [36].

The catalytic activity of MDH in aqueous solution does not change much in wide range of pressure applied (from 0.1 to 100 MPa), slightly decreasing when pressure is increasing ($\Delta V^{\neq}_{app}$ is equal to 5). The picture changes dramatically when reversed micelles are used as a medium for the enzyme. The pronounced maximum appeared both on the dependence of the catalytic activity on water content and pressure. Figure 5 shows pressure dependence of MDH catalytic activity in reversed micelles of AOT in iso-octane at different surfactant hydration degrees. This maximum is observed at moderate pressure, 30-50 MPa, nevertheless, the difference between lowest and highest levels of the enzyme catalytic activity is about 10 times. Such phenomenon can be attributed to the pressure effect on the enzyme molecule via micellar matrix structure. Our experiment in water can be an indication that pressure does not affect the enzyme structure directly. MDH is an oligomeric enzyme consisting of two equal subunits, and one of hypotheses could be that pressure stabilizes a dimeric form of MDH in reversed micelles enduring (strengthening) protein-protein contacts and leading by this to a more compact form of the enzyme molecule existing in reversed micelles which appears then to be more sensitive to external effects. Another possibility could be that pressure application leads to the shift of the equilibrium between monomeric and dimeric forms of MDH to a monomeric one by its stabilization. The discrimination between these two possibilities cannot be done now.

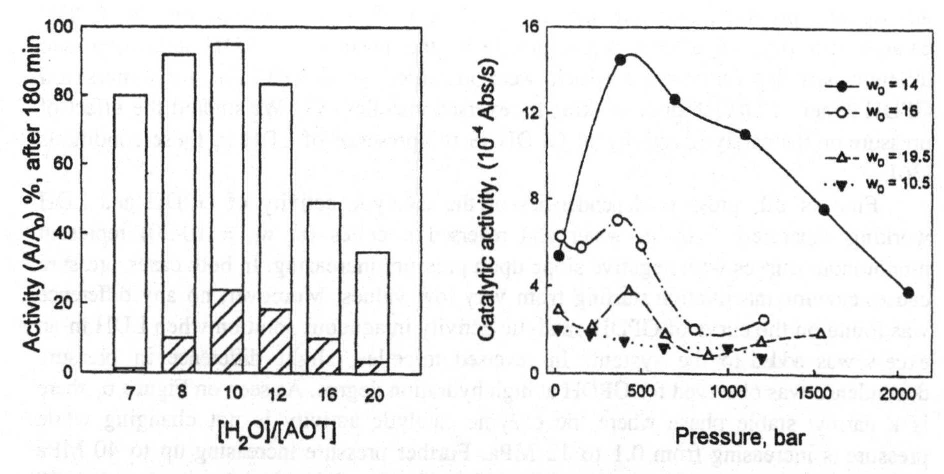

Figure 4 : Pressure stabilization of α-chymotrypsin (expressed as a ratio of the residual catalytic activity after 180 min of incubation at 35 ° C to the initial activity) in reversed micelles of Aerosol OT in octane at different hydration degrees; total bars - at 1000 bar, shaded parts - at 1 bar [31].

Figure 5 : Pressure dependence of the catalytic activity of malic dehydrogenase in reversed micelles of AOT in iso-octane at different hydration degrees [35].

4.3. HIGH PRESSURE AND SUPRAMOLECULAR PROTEIN COMPLEXES.

At atmospheric pressure, the principle of geometric fit was applied to the study of oligomeric enzymes consisting of several subunits [37,38]. The study of structure-function relationship realized in supramolecular protein complexes is one of the important tasks of modern molecular enzymology. However, an experimental study of these phenomena is limited by the absence of reliable approaches permitting to vary and control the supramolecular (oligomeric) composition of protein complexes. Using reversed micelles of Aerosol OT in octane the possibility to regulate the supramolecular structure of enzymes has been shown on a number of examples [37,38]. The dependence of the catalytic activity of oligomeric enzymes on hydration degree reveals several maxima. Each optimum represents formation and functioning of the appropriate oligomeric form. The structure of protein- containing micelles can be controlled by an independent non-kinetic method - ultracentrifugation (sedimentation analysis) [39]. The method allows one to determine the molecular weight of proteins or different oligomeric associates in reversed micelles at different hydration degrees.

In order to understand how pressure can affect protein-protein interactions in reversed micelles, LDH and GPDH were studied [36]. What was interesting for us to study in the case of GPDH and LDH is their work in a complex. Both enzymes are involved in the system of glycolysis possessing a natural potential to a complex formation. Indeed, in reversed micelles, bienzymic complexes were found existing. At atmospheric pressure, two maxima at w_o 21 and 36 were observed on the GPDH catalytic activity - hydration degree profile in the presence of LDH indicating two different particles functioning which were identified as GPDH-LDH monomers and GPDH dimer - LDH tetramer working in reversed micelles (38). We studied the effect of pressure on the catalytic activity of GPDH in the presence of LDH in these conditions [36].

First of all, pressure dependencies of the catalytic activity of GPDH and LDH working separately both in water and reversed micelles (at w_o = 10-30) represent monotonous curves with negative slope upon pressure increasing. In both cases pressure causes enzyme inactivation starting from very low values. Moreover, no any difference was found on the curve of GPDH catalytic activity in aqueous solution when LDH in an excess was added to the system. In reversed micelles, slight difference in pressure dependence was observed for GPDH at high hydration degree. As seen on Figure 6, there is a narrow stable phase where the enzyme catalytic activity is not changing while pressure is increasing from 0.1 to 12 MPa. Further pressure increasing up to 40 MPa leads to the profound decrease in the GPDH catalytic activity that drops more than 10 times. The phase transition occurs in the system at higher pressures followed by the turbidity increasing.

The significant effect of pressure on GPDH catalytic activity in the presence of LDH was revealed in reversed micelles in particular conditions (Figure 6). As seen, pressure stabilization and/or activation of GPDH is taking place at w_o = 21 (Figure 6a) and 36 (Figure 6c) where maxima are observed on the dependence of the enzyme catalytic activity on hydration degree [38] and complex formation occur between GPDH and LDH. At w_o = 28 (Figure 6b) (minimum on the dependence [38]), the addition of LDH does not bring any difference in pressure dependence of GPDH catalytic activity. As seen from Figure 6a, at w_o = 21, the activation of GPDH occurs in the presence of LDH starting from an atmospheric pressure, and the complex between two monomeric forms of enzymes appears to be more stable upon pressure increasing up to 50 MPa. Even more pronounced effect of GPDH activation upon pressure application can be found at w_o = 36 (Figure 6c). At this hydration degree where the complex between GPDH dimer and LDH tetramer can be formed in reversed micelles, the enzyme catalytic activity - pressure profile was found to be a bell-shaped curve. The catalytic activity of GPDH in the presence of LDH increases two times when pressure is increasing up to 20 MPa after that decreases being always higher than that in the absence of LDH. Whereas at 40 MPa the GPDH catalytic activity decreases dramatically, the GPDH catalytic activity in the complex with LDH retains at the initial level being 10 times higher than that in the absence of LDH.

5. The future of high pressure enzymology.

One actual challenge for industry is to use high pressure for biotechnological processes [40-42]. If some high pressure treated products are available on the market (mainly food products), further progress will require refinement of the theory of pressure effects on biosystems, including enzymology, which is one key of many processes such as, for example, polyphenoloxidases which are involved in browning of fruits and vegetables or

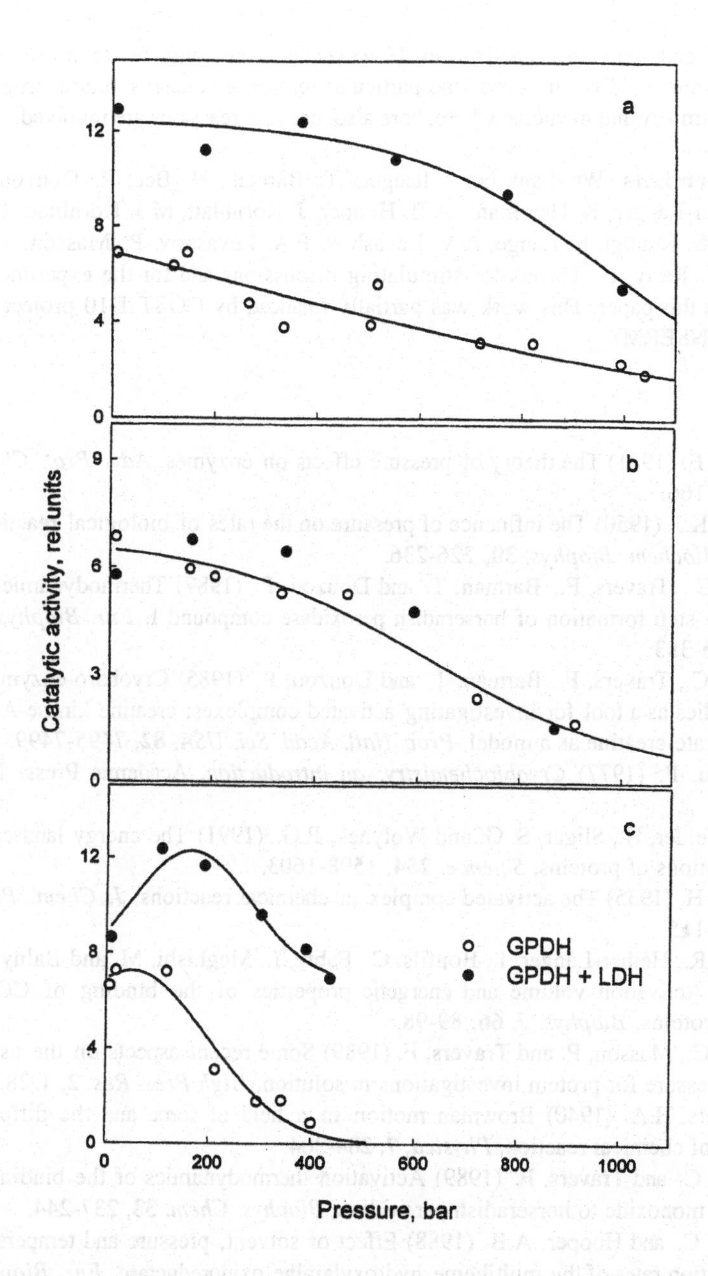

Figure 6 : **Pressure dependence of the catalytic activity of D-glyceraldehyde-3-phosphate dehydrogenase in the presence and absence of lactic dehydrogenase in reversed micelles at different w_0; a - 21, b - 28, c - 36 [36].**

the problem of endo-amylases inactivation. However, it is not only for food processes that high pressure could be involved, and particular relevance concerns future progress expected in pharmacy and medicine where, here also, enzyme reactions are involved.

Acknowledgements. We thank our colleagues T. Barman, N. Bec, P. Douzou, J. Frank, I. Heiber-Langer, K. Heremans, A.B. Hooper, J. Kornblatt, M.J. Kornblatt, E.V. Kudryashova, S. Kunugi, R. Lange, A.V. Levashov, P.A. Levashov, P. Masson, V.V. Mozhaev, R.V. Rariy, F. Travers for stimulating discussions and for the experimental work related in this paper. This work was partially financed by COST D10 project and PECO grant (INSERM).

References

1. Morild, E. (1981) The theory of pressure effects on enzymes, *Adv. Prot. Chem.* **34**, 93-166.
2. Laidler, K.J. (1950) The influence of pressure on the rates of biological reactions, *Arch. Biochem. Biophys.* **30**, 226-236.
3. Balny, C., Travers, F., Barman, T. and Douzou, P. (1987) Thermodynamics of the two step formation of horseradish peroxidase compound I, *Eur. Biophys. J.* **14**, 375-383.
4. Balny, C., Travers, F., Barman, T. and Douzou, P. (1985) Cryobaro-enzymologic studies as a tool for investigating activated complexes: creatine kinase-ADP-Mg-nitrate-creatine as a model, *Proc. Natl. Acad. Sci. USA*, **82**, 7495-7499.
5. Douzou, P. (1977) *Cryobiochemistry: an introduction*, Academic Press, New York.
6. Frauenfelder, H., Sligar, S. G. and Wolynes, P.G. (1991) The energy landscapes and motions of proteins, *Science*, **254**, 1598-1603.
7. Eyring, H. (1935) The activated complex in chemical reactions, *J. Chem. Phys.* **3**, 107-115.
8. Lange, R., Heiber-Langer, I., Bonfils, C., Fabre, I., Meghishi, M. and Balny, C. (1994) Activation volume and energetic properties of the binding of CO to hemoproteins, *Biophys. J.* **66**, 89-98.
9. Balny, C., Masson, P. and Travers, F. (1989) Some recent aspects on the use of high pressure for protein investigations in solution, *High Pres. Res.* **2**, 1-28.
10. Kramers, H.A. (1940) Brownian motion in a field of force and the diffusion model of chemical reaction, *Physica*, **7**, 284-304.
11. Balny, C. and Travers, F. (1989) Activation thermodynamics of the binding of carbon monoxide to horseradish peroxidase, *Biophys. Chem.* **33**, 237-244.
12. Balny, C. and Hooper, A.B. (1988) Effect of solvent, pressure and temperature on reaction rates of the multiheme hydroxylamine oxidoreductase, *Eur. Biophys. J.* **176**, 373-279.
13. Masson, P. and Balny, C. (1988) Effects of high pressure on the single-turnover kinetics of the carbamylation of cholinesterase, *Biochim. Biochem. Acta,* **954**, 208-215.

14. Kunugi, S., Kitayaki, M., Yanagi, Y., Tanaka, N., Lange, R. and Balny, C. (1997) The effect of high pressure on thermolysin, *Eur. J. Biochem.* **248**, 567-574.

15. Kornblatt, M.J., Lange, R. and Balny, C. (1998) Can monomers of yeast enolase have enzymatic activitiy ? *Eur. J. Biochem.* **251**, 775-780.

16. Mozhaev, V.V., Lange, R., Kudryashova, E.V. and Balny, C. (1996) Application of high hydrostatic pressure for increasing activity and stability of enzymes, *Biotech. Bioeng.* **52**, 320-331.

17. Mozhaev, V.V., Heremans, K., Frank, J., Masson, P. and Balny, C. (1996) High pressure effects on proteins structure and function, *Proteins: Struc. Func. Gene.* **24**, 81-91.

18. Saldana, J.L. and Balny, C. (1992) Device for optical studies of fast reactions in solution as a function of pressure and temperature, in C. Balny *et al.* (eds.), *High Pressure and Biotechnology*, J. Libbey, vol. **224**, pp. 529-531.

19. Martinek K., Levashov A.V., Klyachko N.L., Berezin I.V. (1977) Catalysis by water-soluble enzymes in organic solvents. Enzyme stabilization against denaturation through their inclusion in reversed micelles of surfactant. *Dokl. Acad. Nauk SSSR (Russ.)* **236**, 920-923.

20. Douzou P., Keh E., Balny C. (1979) Cryoenzymology in aqueous media: micellar solubilized water clusters. *Proc. Natl. Acad. USA*, **76**, 681-684.

21. Wolf R., Luisi P.L. (1979) Micellar solubilization of enzymes in hydrocarbon solvents. Enzymic activity and spectroscopic properties of ribonuclease in n-octane. *Biochem. Biophys. Res. Comm.* **89**, 209-217.

22. Luisi P.L., Giomini M., Pileni M.-P., Robinson B.H. (1988) Reverse micelles as hosts for proteins and small molecules. *Biochim. Biophys. Acta*, **947**, 209-246.

23. Martinek, K., Klyachko, N.L., Kabanov, A.V., Khmelnitsky, Yu. L., Levashov, A.V. (1989) Micellar enzymology: its relation to membranology. *Biochim. Biophys. Acta*, **981**, 161-172.

24. Tuena de Gomez-Puyou, M., Gomez-Puyou, A. (1998) Enzymes in low water systems. *Critical Rev. Biochem. Mol. Biol.* **33**, 53-89.

25. Fendler, J.H. (1982) *Membrane Mimetic Chemistry*, J. Wiley, New York

26. Eicke, H.F. (1980) Surfactants in nonpolar solvents: aggregation and micellization. *Top. Curr. Chem.* **87**, 85-145.

27. Kotlarchyk, M., Huang, J.S. and Chen, S.-H. (1985) Structure of AOT reversed micelles determined by small-angle neutron scattering. *J. Phys. Chem.* **89**, 4382-486.

28. Mozhaev, V.V., Bec, N. and Balny, C. (1994) Pressure effects on enzyme reactions in mainly organic media: α-chymotrypsin in reversed micelles of Aerosol OT in octane. *Biochem. Mol. Biol. Int.* **34**, 191-199.

29. Mozhaev, V.V., Heremans, K., Frank, J., Masson, P. and Balny, C. (1994) Exploiting the effects of high hydrostatic pressure in biotechnological applications. *TIBTECH*, **12**, 493-501.

30. Mozhaev, V.V., Bec, N. and Balny, C. (1995) Baroenzymology in reversed micelles, *Ann. N.Y. Acad. Sci.* **750**, 94-96.

31. Rariy, R.V., Bec, N., Saldana, J.-L., Nametkin, S.N., Mozhaev, V.V., Klyachko, N.L., Levashov, A.V. and Balny, C. (1995) High-pressure stabilization of α-chymotrypsin entrapped in reversed micelles of Aerosol OT in octane against thermal inactivation. *FEBS Lett.* **364**, 98-100.
32. Rariy, R.V., Bec, N., Klyachko, N.L., Levashov, A.V. and Balny, C. (1998) Thermostability of α-chymotrypsin in reversed micelles of Aerosol OT in octane solvated by water-glycerol mixtures. *Biotechnol. Bioeng.* **57**, 552-556.
33. Fletcher, P.D.I., Galal, M.F. and Robinson, B.H. (1984) Structural study of Aersol OT stabilised microemulsions of glycerol dispersed in n-heptane. *J. Chem. Soc. Faraday Trans.* **80**, 3307-3314.
34. Shapiro, Yu.E., Budanov, N.A., Levashov, A.V., Klyachko, N.L., Martinek, K. (1989) ^{13}C-NMR study of entrapping proteins (chymotrypsin) into reversed micelles of surfactants (Aerosol OT) in organic solvents (n-octane). *Collect. Czech. Chem. Comm.* **54**, 1126-1134.
35. Klyachko, N.L., Levashov, P.A., Levashov, A.V. and Balny, C. (1998) Pressure regulation of malic dehydrogenase in reversed micelles. *FEBS Lett.* (submitted).
36. Levashov, P.A., Klyachko, N.L., Levashov, A.V. and Balny, C. (1998) Pressure stabilises complexes between D-glyceraldehyde-3-phsphate dehydrogenase and lactic dehydrogenase in reversed micelles. *FEBS Lett.* (submitted).
37. Kabanov, A.V., Klyachko, N.L., Nametkin, S.N. and Levashov, A.V. (1991) Engineering of functional supramolecular complexes of proteins (enzymes) using reversed micelles as matrix microreactors. *Prot. Eng.* **4**, 1009-1017.
38. Levashov, A.V., Ugolnikova, A.V., Ivanov, M.V. and Klyachko, N.L. (1997) Formation of homo and heterooligomeric supramolecular structures by D-glyceraldehyde-3-phosphate dehydrogenase and lactade dehydrogenase in reversed micelles of aerosol OT in octane. *Biochem. Mol. Biol. Int.* **42**, 527-534.
39. Levashov, A.V., Khmelnitsky, Yu.L., Klyachko, N.L., Chernyak, V.Ya, Martinek, K. (1982) Sedimentation analysis of protein-Aerosol OT-water-octane system. *J. Colloid Interface Sci.* **88**, 444-457.
40. Balny, C., Hayashi, R., Heremans, K. and Masson, P. (eds.) (1992) *High Pressure and Biotechnology*, J. Libbey, London, **vol. 224**.
41. Hayashi, R. and Balny, C. (eds.) (1996) *High Pressure Bioscience and Biotechnology*, Elsevier Science, Amsterdam.
42. Heremans, K. (ed.) (1997) *High Pressure Research in the Bioscience and Biotechnology*, Leuven Univ. Press, Leuven.

THE PHASE DIAGRAM AND THE PRESSURE-TEMPERATURE BEHAVIOR OF PROTEINS

KAREL HEREMANS
Department of Chemistry, Katholieke Universiteit Leuven
B-3001 Leuven, Belgium

A fresh instrument serves the same purpose as foreign travel;
it shows things in unusual combinations.
A. N. Whitehead in: Science and the modern world.

Abstract.

The pressure and temperature behavior of proteins is discussed in the framework of the phase diagram. This gives unique information on the changes in heat capacity, thermal expansion and compressibility of protein unfolding. It also relates the cold, heat and pressure denaturation. The difference in pressure- and temperature-induced aggregation of unfolded proteins shows the unique features of pressure effects. A molecular interpretation of the thermodynamic quantities is not possible on the basis of model systems unless the packing defects are taken into account. High pressure molecular dynamic calculations contribute in a unique way to our understanding of pressure effects.

1. Introduction

Experimental data on the effect of temperature and pressure for physical, chemical and biological systems are usually interpreted at different levels of thinking. In biology one deals with very complex systems that are build up from a number of organelles embedded in an aqueous medium containing salts and other macromolecules. In biochemistry and biophysics one is studying dilute solutions of biomacromolecules under controlled conditions. Physicists tend to concentrate on pure simple compounds and on mechanistic aspects of phase transitions especially in condensed matter [1]. In order to get a deeper insight in these mechanisms, the study of mixtures is attempted such as is the case of N_2 and Ar, molecules that have a similar diameter. In such cases one looks for the behavior of the nitrogen to the presence of Ar [2]. The biophysical chemist may change his biomolecules by making mutants or he may look for different ligands that bind to the same active site on his protein [3,4]. The biologist has fewer options. However, in the case of bacteria, he may look for mutants that are more or less pressure and/or temperature sensitive [5].

The common strategy in the previous examples is to look for a response that results from a perturbation by changing the temperature, the pressure or, for solvent based systems, the solvent conditions. In the most favourable cases one looks for a relation

R. Winter and J. Jonas (eds.), High Pressure Molecular Science, 437–472.

between the structural changes of a biomolecule and his properties and or function. To this end a great variety of techniques are available that can follow the changes either on a local or on a global scale. The search is then for common principles that explain a wide variety of phenomena in terms of basic physical concepts. One might expect that biologists and microbiologists would tend to disagree with this approach by stressing the complexity of the biological systems and by putting emphasis on the reaction of the integrated system as a whole. In reality one notices that the tendency to interpret the behavior of very complex biological systems in molecular terms is getting more common. This can be illustrated by the rapidly developing field of the study of bacterial life under extreme conditions of temperature, pressure and solvent conditions such as salt, pH and water activity [6]. The significance of extremophiles and extreme conditions for biotechnology, medicine and research into the origin of life and early evolution has been discussed in a stimulating monograph [7].

One fascinating example of the crucial role of water in the behavior of organisms under extreme conditions is given by small organisms called Tardigrades. These animals, composed of about 40.000 cells, become immobile and shrink into a special state when the humidity of the surroundings decreases. In such a state they can survive temperature from –253°C up to 151°C and pressures up to 600 MPa [8]. In the normal state they are killed at 200 MPa. This behavior is quite similar to that of bacterial spores and dry proteins where pressures of more than 1 GPa are not able to provoke any changes [9,10]. The fact that the Tardigrades can undergo a transformation to an extreme dry state may be much more exceptional than the fact that they are resistant to extremes of temperature and pressure in the dry state.

An equally interesting observation is the correlation between the survival conditions of the deep-sea bacteria and the stability diagram of proteins in the temperature-pressure plane. Both diagrams can be schematically represented by an ellipse and Yayanos [11] has suggested that the killing of bacteria may be dictated by the denaturation of one crucial enzyme/protein in the metabolism of the bacteria.

The study of the behavior of proteins under extremes of temperature and pressure may therefore be extremely relevant for our understanding of biosystems under these conditions. For some proteins the effects of these parameters are reversible

$$\text{Native state (N)} \leftrightarrow \text{Denatured state (D)}$$

This representation indicates that the native conformation of a protein may undergo a transformation to another conformation that is not biologically active. The precise nature of the D state is a matter of considerable interest. A number of experimental approaches have shown the heat and the pressure denatured states are quite different [12,13,14]. A number of intermediate states have now been discovered on the reaction pathway between the native and the denatured state. For most proteins the changes are irreversible because of the strong tendency of the denatured state to undergo irreversible interactions with the formation of aggregates.

$$N \leftrightarrow D \rightarrow \text{Aggregates}$$

It is evident that the formation of these aggregates has for a long time been considered as a nuisance for the study of the thermodynamics of protein folding. However, recent research into the mechanism of molecular diseases has shown that the formation of aggregates of certain proteins may be the key to our understanding for at least some of these processes.

Most, if not all, experiments on the stability of proteins and enzymes are kinetic experiments. This rises the question as to the validity of a thermodynamic description and interpretation of the data. The advantage of the thermodynamic analysis is the generality of the physical picture that is used to describe the biotransformations in terms of volume and energy. The kinetic approach points towards possible mechanistic interpretations in terms of structural changes. In addition, it gives vital information on the feasibility of possible biotechnological applications [15].

2. Fundamental concepts

Before discussing the effects of temperature and pressure on dilute solutions of biological macromolecules, it is instructive to consider the effects on pure phases and mixtures of constant composition. The change in Gibbs free energy (G) of the system with pressure and temperature is given by the following expressions:

$$(\delta G/\delta P)_T = V \qquad (\delta G/\delta T)_P = -S$$

The first relationships allow us to compare the energy input in a given amount of water by applying a pressure of 1 GPa with heating of the same amount to 100°C. It turns out that the energy input on heating is about twice that on compressing. It is of interest to note that hexane the situation is almost exactly the opposite. More generally, a temperature increase of a system involves changes in energy and density whereas pressure affects primarily the density. This makes studies at constant density and variable energy important from a theoretical point of view.

The above equations can easily be applied to phase transitions of pure substances. When ΔG refers to the difference in free energy of the two phases, we get:

$$(\delta \Delta G/\delta P)_T = \Delta V \qquad (\delta \Delta G/\delta T)_P = -\Delta S$$

This procedure can be extended to the Maxwell equations. Changes in the free energy difference (ΔG) with temperature and pressure are given by:

$$d(\Delta G) = (\Delta V)\, dP - (\Delta S)\, dT$$

Integration of this equation may be performed under the condition that ΔV and ΔS are either temperature or pressure dependent. In the former case the phase boundaries are straight lines and a simple Clausius-Clapeyron equation is obtained. In the latter case reentrant phase behavior may be described as observed for the effect of pressure on the nematic-smectic A transition of certain liquid crystals [16].

Most biosystems are more complex than single component systems. They are dilute solutions of macromolecules in water and salts. In most cases we are interested in ΔG, the reaction free energy for the conversion of the native to the denatured state: G_D-G_N. Note that G_D and G_N are the partial molar Gibbs free energy, i.e. the chemical potentials of the substances. These are defined in a given medium with a defined composition and a given concentration of the macromolecule. The situation is similar for the partial molar volume of a substance. Here the effect of the medium composition is more familiar. The partial molar volume, V_i, of a solute molecule or ion is defined as the change in volume of the solution by the addition of a small amount of the solute over the number of moles of added solute keeping the amount of the other components constant.

$$V_i = \left(\frac{\partial V}{\partial n_i}\right)_{n_j,T,P}$$

In this definition V is the volume of the solution and n_i is the number of moles of the solute added. It is not equal to the volume of the molecule or the ions since it includes also the interaction with the solvent. This may be seen from the fact that the partial molar volume for salts such as $MgSO_4$ and Na_2CO_3 is negative at zero concentration because of the strong electrostriction of the solvent around the ions. For the same reason the volume of the uncharged glycolamide is larger (56.2 mL/mol) than that of the amino acid glycine (43.5 mL/mol).
The thermal expansion is defined as the relative change of the partial molar volume with temperature:

$$\alpha_i = \frac{1}{V_i}\left(\frac{\partial V_i}{\partial T}\right)_{P,j}$$

and the isothermal compressibility as the relative change of the volume with pressure:

$$\kappa_{T,i} = -\frac{1}{V_i}\left(\frac{\partial V_i}{\partial P}\right)_{T,J}$$

The isentropic compressibility is obtained from ultrasonic velocimetry and is defined as:

$$\kappa_{S,i} = -\frac{1}{V_i}\left(\frac{\partial V_i}{\partial P}\right)_{S,J}$$

Quantities derived from the volume (such as compressibility and thermal expansion) should be interpreted with great care. The results may depend on the sensitivity range of the method. Global measurements such as ultrasonics detect the whole molar volume, while some local probes may feel only the change of the protein interior volume.

Up to now time dependent phenomena have been assumed not to take place. In practice, however, this is almost never the case. For systems consisting of a pure substance this becomes evident in the formation of the glassy state in ice [17]. It is equally true for the solid –solid transitions of ice at low temperatures.

The structure of a protein is the result of a delicate balance between the intramolecular interactions in the polypeptide chain which compete with the solvent interactions. Proteins in the dry state are found to be extremely resistant to pressure unfolding an effect that may easily be studied in the diamond anvil cell [10]. These experiments emphasize the role of water as a plasticizer in biomacromolecules via the reduction of the glass-rubber transition temperature (T_g). These experiments also suggest that proteins in the dry state are below their glass transition temperature. The fundamental similarities between synthetic polymers and biomacromolecules has stimulated research in this field because of the importance of the T_g on all aspects of the physical chemical behavior. In view of the non-equilibrium nature of the glass transition temperature, this transition is governed by activation rather than equilibrium parameters. Limited studies on the effect of pressure on the glass transitions in polymers suggest a change of ca 22 K/100 MPa for nonhydrogen bonded systems. The effect of pressure on the hydrogen bonded system sorbitol shows a much weaker dependence of 4 K/100 MPa [18]. This suggests that similar orders of magnitude may be expected for glass transitions in proteins.

3. Phase (stability) diagram of proteins

The first systematic observation on the denaturation (coagulation) of proteins by high pressure was made by Bridgman [19]. He noticed that, at room temperature, the white of an egg coagulated when subjected to high hydrostatic pressure of 500 MPa. He carefully noted that at lower pressures the thickening is barely perceptible even after a long time, whereas at higher pressures a much shorter time is needed to obtain the same effect. At high pressures, the effect of temperature is small but interesting: The ease of coagulation increases contrary to what one might expect. Also when the egg white was taken to 1.2 GPa into the ice VI phase, the coagulated egg white did not seem to be affected by the freezing. Looking backwards from what we know today, it is clear that Bridgman observed the essential features of the phase diagram for the stability of proteins.

The next step forward was made by Suzuki [20] who determined the rate of denaturation of ovalbumin and carbonylhemoglobin as a function of temperature and pressure. He plotted his results in a p, T, k-diagram. He connected the points with the same rate constants at a given p and T. He noted that there was a changeover in sign from positive at high temperature and low pressure to negative at high pressure and low temperature for the activation energy ($\Delta H^{\#}$), entropy ($\Delta S^{\#}$) and volume ($\Delta V^{\#}$) as shown in Table I. The domains I, II and III are shown schematically in Figure 1. Suzuki explained the *negative activation energy* in domain I with the following mechanism:

$$P + n\,H_2O \leftrightarrow P\,(H_2O)n \rightarrow Pd \qquad mPd \rightarrow (Pd)m$$

In the first step water is pressed into the protein which then unfolds. In the second step the unfolded protein forms aggregates.

It should be pointed out that the negative activation energies observed by many investigators since then is not unique for the high pressure conditions. Negative activation energies are also observed for the denaturation of proteins by high urea concentrations [21]. To account for this the following scheme was proposed:

$$P + n\text{ urea} \leftrightarrow P(\text{urea})n \rightarrow Pd$$

TABLE 1. Pressure and temperature dependence of the sign of the activation parameters observed by Suzuki [20] for the denaturation of ovalbumin and hemoglobin.

	T (°C)	P (MPa)	$\Delta H^{\neq}$	$\Delta S^{\neq}$	$\Delta V^{\neq}$
I	< 30	> 400	-	-	-
II	> 40	> 300	+	±	-
III	> 60	< 300	+	+	+

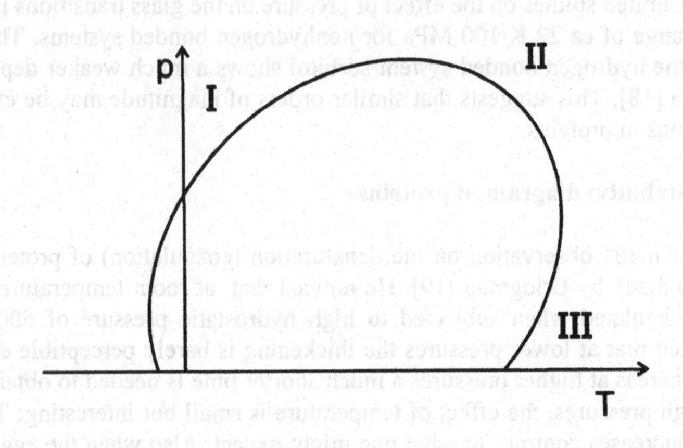

Figure 1. Kinetic diagram for the denaturation of ovalbumin. After Suzuki [20]. The domains I, II and III correspond to the conditions given in Table 1.

Thus it seems that the binding of urea at ambient pressure and the binding of water at high pressure may account for the negative activation energies that are observed in these processes.

A fairly complete list of all the experimental data kinetic data available on the protein denaturation/unfolding, enzyme inactivation is given in Table 2. The same table includes more complex systems that show similar behavior: the destruction of phages, viruses, bacteria and cells, the cloud formation in synthetic polymers and the gelation of starch. In all cases these are kinetic data although in a few cases a thermodynamic analysis is presented. However, it is also clear from these papers that under certain conditions of temperature and pressure the behavior of the system is controlled by the kinetics.

An important point that should be considered in the analysis of the data in the Table is the fact that water is a prerequisite for the re-entrant behavior. This is clear from the behavior of dry proteins and bacterial spores under pressure. Dry proteins, dry starch and dry bacterial spores may be compressed to very high pressures in the diamond anvil cell without showing any irreversible changes as judged from the infrared spectra. As soon as water is available, the systems behave in the normal way, i.e. a pressure-induced changes take place that are in most cases irreversible. This suggests that a possible mechanism includes at least one reversible step followed by an irreversible step as suggested above.

TABLE 2. Survey of experimental data that show re-entrant behavior in the P,T plane.

Process	System	Reference
Protein denaturation	Egg white	Bridgman [19]
	Ovalbumin	Suzuki [20]
	Hemoglobin	Suzuki [20]
	Ribonuclease	Brandts et al [22]
	Chymotrypsinogen	Hawley [23]
	Myoglobin	Zipp and Kauzmann [24]
	Egg White	Kanaya et al [25]
Enzyme inactivation	Chymotrypsin	Taniguchi and Suzuki [26]
	Lipoxygenase	Heinisch et al [27]
	Butylcholinesterase	Weingand-Ziadé et al [28]
	Lipoxygenase	Ludikhuyze et al [29]
	Polyphenoloxidase	Weemaes et al [30]
	Amylase	Ludikhuyze et al [31]
Protein unfolding	Amylase	Rubens et al [32]
	Lipoxygenase	Rubens et al [33]
Bacteria,Viruses and phages	*E. coli*	Ludwig et al [34]
	Deep sea bacteria	Yayanos [11]
	S. cerevisiae	Hashizume et al [35]
	E. coli	Sonoike et al [36]
	L. casei	Sonoike et al [36]
	L. innocua	Smelt et al [37]
	Bacteriophage T4	Gross and Ludwig [38]
Cell inactivation	Fruit fly eggs	Butz and Tauscher [39]
Reaction kinetics	Myosin + ATP	Saldana and Balny [40]
Cloud point	Synthetic polymers in H_2O	Kunugi et al [41]
Phase separation		Sun and King [42]
Starch gelation	potato	Thevelein et al [43]
	wheat	Douzals et al [44]
	rice	Rubens et al [45]

4. A physical model for the interpretation of the elliptic phase diagram

A general thermodynamic model for the analysis of the effect temperature and pressure on phase transition in pure compounds was first presented by Bridgman [19]. The first experimental data on the kinetics of the denaturation of proteins were presented by Suzuki [20]. The first thermodynamic analysis that is most often quoted in the biochemical literature is that given by Hawley [23]. It follows closely that given by Bridgman and was based on the ribonuclease data of Brandts et al. [22] and his own data on chymotrypsinogen.

Figure 2 shows the three common possible ways to denature a protein. These are the heat, pressure, and cold denaturation. In the thermodynamic description the denaturation is a reversible process. However, in most cases the denatured protein tends to produce a gel stabilized by an intermolecular network, which prevents the refolding to the native state. Since the equilibrium theory describes the denaturation by treating only two states, the irreversibility is not included in that model. The kinetic theory cannot distinguish between two denatured states, a metastable state where the protein is reversibly denatured (*D*) and an irreversibly inactivated (*I*) one:

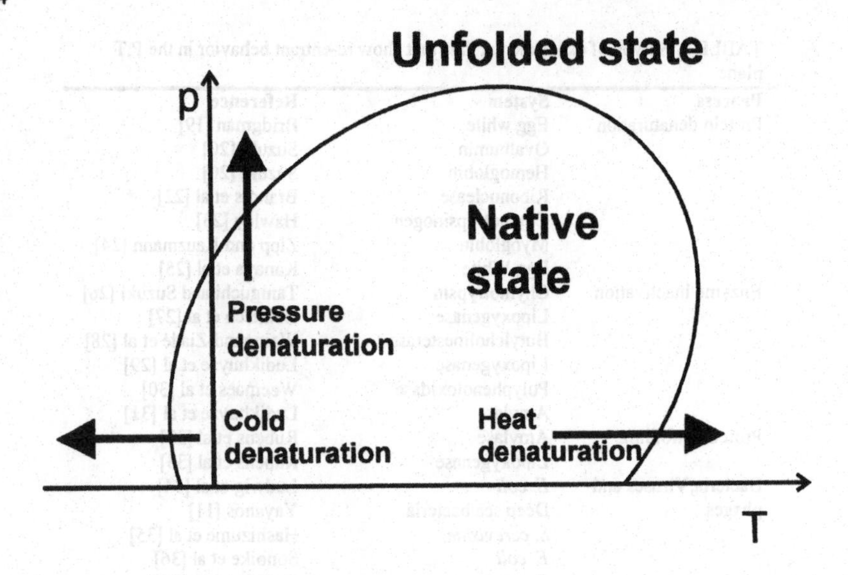

Figure 2. **The relation between heat, cold and pressure denaturation of proteins.**

$$N \underset{k_2}{\overset{k_1}{\rightleftharpoons}} D \xrightarrow{k_3} I$$

There are a few special cases of interest regarding the rates of the three conformational transitions. When the rate constant $k_3 << k_1, k_2$ then we have an equilibrium between the N and D states (with the equilibrium constant, $K = k_1/k_2 = D/N$), while the process $D \rightarrow I$ is slow. In this case the experiment reflects the kinetics of the process $D \rightarrow I$ with an apparent rate constant $k_{obs} = (k_1/k_2)\, k_3 = K\, k_3$. This case reflects the low temperature and low pressure conditions with a predominance of the native conformation. In the opposite limit, ($k_1, k_2 << k_3$) all the reversible denatured proteins will be irreversibly captured in an intermolecular network of the reversibly unfolded state. In this case the kinetics reflect the kinetics of the first step of the two-step denaturation process if the equilibrium is in favor of the native state. In the opposite case the observed kinetics give information on k_3. The latter case is rather exceptional since it would imply a measurement of the aggregation of the denatured state. However, these measurements are only accessible with specialized rapid reaction instrumentation [46].

It follows that under most circumstances kinetic data on the effect of temperature and pressure on the denaturation of proteins contains information that is a combination of thermodynamics and kinetics. It should be obvious that this model can easily be extended to more complex situations as the gelation of starch and the killing of bacteria.

Given the equilibrium assumption in the transition state theory, the thermodynamic theory can be extended to kinetic data. The terminology should then be adapted. Instead of a phase diagram in the P,T plane an isokinetic (isopleths) stability diagram is obtained P, T, k. Given the observation that the solvent also affects the temperature and pressure dependence of the kinetics, a P, T, k, n diagram is obtained. This terminology was suggested by Yayanos [11]. In the following we discuss the thermodynamics of the transformations. The advantage is the use of a simple physical model that gives the basis for a mathematical description.

The free energy change as function of pressure and temperature is given by:

$$d(\Delta G) = -(\Delta S)\,dT + (\Delta V)\,dP$$

where ΔG gives the difference in partial molar free energy between the denatured and the native state. An equivalent expression for the kinetic data would be:

$$d(\Delta G^{\#}) = -(\Delta S^{\#})\,dT + (\Delta V^{\#})\,dP$$

where $\Delta G^{\#}$ (= - RT ln k) expresses the difference between the transition state and the native state. In the following we only use the equilibrium expressions. Integration of the equation with the assumption that both ΔS and ΔV are temperature as well as pressure dependent leads to an expression of the following form:

$$\Delta G = a + b\,T + c\,T^2 + d\,P + e\,P^2 + f\,T\,P$$

Note that higher order terms in T^3, P^3, PT^2, P^2T, etc are neglected in this equation. Neglecting the higher order terms is equivalent to the assumption that the parameters c, e and f are independent of temperature and pressure. An equation of this type was obtained by Bridgman and applied to the unfolding of proteins by Hawley [23]. This equation has an advantage over purely empirical equations because the constants can be given a simple *physical* interpretation. It should be emphasised that the thermodynamic quantities are all partial molar quantities.

a: ΔG at given reference conditions: $\Delta G°$
b: $\Delta S°$ at reference conditions
c: T ($\delta\Delta S/\delta T$): change in the heat capacity (ΔCp)
d: $\Delta V°$ at reference conditions
e: $\delta\Delta V/\delta P$: change in the compressibility factor ($\Delta\kappa$)
f: $\delta\Delta V/\delta T$ (= - $\delta\Delta S/\delta P$): change in the thermal expansion factor ($\Delta\alpha$)

In view of the Maxwell equation ($\delta\Delta S/\delta P$) = - ($\delta\Delta V/\delta T$) the constant f has two equivalent interpretations. The equation may be written in the following final form

$$\Delta G = \Delta G_0 - \Delta S_0(T - T_0) - \frac{\Delta C_p}{2T_0}(T - T_0)^2 + \Delta V_0(P - P_0) + \frac{\Delta\kappa}{2}(P - P_0)^2 + \Delta\alpha\,(P - P_0)(T - T_0)$$

with the definitions of ΔC_p, $\Delta\kappa$ and $\Delta\alpha$ as given above. It should be noted that, except for the heat capacity, C_p, the compression and the thermal expansion do not have their convential definition as given in the section 2.

The conditions for the observation in such systems of an elliptical contour was first pointed out by Hawley [23]

$$(\Delta\alpha)^2 > \Delta\kappa\, \Delta C_p/T_o.$$

This condition is usually fulfilled since $\Delta\kappa$ and ΔCp have opposite sign for protein systems. An equation similar to the Clausius-Clapeyron equation can be obtained for the slope of the phase boundary

$$\frac{\partial T}{\partial P} = \frac{\Delta V_0 + \Delta\beta\,(P-P_0) + \Delta\alpha\,(T-T_0)}{\Delta S_0 - \Delta\alpha\,(P-P_0) + \Delta C_p (T-T_0)/T_0}$$

This expression reduces to the classical Clausius-Clapeyron equation when the difference in compressibility, thermal expansion and heat capacity vanish as is observed for most phase transitions in lipids [47]. The shape of the phase diagram for proteins is of considerable interest since, as pointed out previously, it contains information on the volume and entropy fluctuations and on the coupling between volume and entropy fluctuations. dT/dP and dP/dT become zero at two points in the diagram as can be seen from Figures 1 and 2. A more detailed discussion of the consequences of the thermodynamic model is given by Smeller and Heremans [48].

The free energy equation reduces to simpler expressions at constant temperature or pressure. At constant pressure we get the familiar expression that describes the heat as well as the cold denaturation:

$$\Delta G = \Delta G_0 - \Delta S_0 (T-T_0) - \frac{\Delta C_p}{2T_0}(T-T_0)^2$$

It can be shown that this is equivalent with Gibbs-Helmholtz equation

$$\Delta G = \Delta G_0 - \Delta S_0 \left(T-T_0\right) + \Delta C_p \left[\left(T-T_0\right) - T \ln\frac{T}{T_0}\right].$$

At constant temperature we obtain an equation that describes the pressure denaturation

$$\Delta G = \Delta G_0 + \Delta V_0 (P-P_0) + \frac{\Delta\kappa}{2}(P-P_0)^2 .$$

With the classical definitions of compressibility and thermal expansion a more complex equation was obtained by Haynes [49].

$$\Delta G = \Delta G_0 - \Delta S_0(T - T_0) - \frac{\Delta C_p}{2T_0}(T - T_0)^2 + \Delta V_0(P - P_0) + \frac{\Delta\kappa}{2}(P - P_0)^2$$
$$+ \Delta V(P_O, T_O)\left[1 + \Delta\alpha\,(T - T_0)\right]\left[1 - \exp(-\Delta\kappa\, P)\right] / \Delta\kappa$$

It is important to emphasise the power but also the limits of these equations. The power of the equation resides in the fact that it gives a consistent thermodynamic description of the heat, pressure and cold denaturation. The cross term in the equation finds its origin in the pressure and temperature dependence of the volume changes:

$$\Delta V(P,T) = \Delta V + \Delta\alpha\left(T - T_0\right) - \Delta\kappa\left(P - P_0\right)$$

The limits of the equation are that only differences in the thermal expansion, compressibility and heat capacity between the denatured and the native state are obtained. Additional data are needed which can be obtained from densitometry, calorimetry and ultrasonic velocimetry. Since the equation gives a relation between partial molar quantities, the usual difficulties arise for a molecular interpretation of these quantities. It is the usual practice to rely on low molecular mass model systems. However, in many model systems the cavity term is not considered or even neglected. Several papers in the literature have discussed the limitations of model systems. The other extreme in the interpretation can also be found in the literature. On the basis of NMR and circular dichroism work on a part of the lambda repressor, it has been suggested that the cold and the heat denatured states of this protein are thermodynamically and conformationally equivalent in the presence of 3 M urea [50]. It is argued that, given the importance of the role of the solvent in the folding process, there should be no reason to expect a difference between the heat and cold denatured state. Thus the strong temperature dependence of the properties of the water and the way it interacts with the denatured protein is thought to be the primary mechanism of the denaturation. If the interpretation is correct, then it follows that similar conformational changes are expected for pressure induced unfolding. This however seems not to be the case as is evident from studies on other proteins with different experimental approaches (see section 9). However, solely on the basis of thermodynamic arguments, as this interpretation neglects the thermodynamic changes in terms of partial molar quantities, this conclusion is not correct.

A final point is the applicability of the equation for the analysis of kinetic data on more complex systems than proteins. From Table 2 it can be seen that re-entrant P,T diagrams are observed for micro-organisms and starch. It has not been observed for nucleic acids and in the case of lipids it can be found under limited conditions were interdigitation of the hydrocarbon chains takes place [47]. It may then be concluded that the observation of elliptic stability diagrams in the case of micro-organisms can be attributed to the denaturation of a critical enzyme in the metabolism. In the case of starch, a preequilibrium as proposed for proteins can be assumed. It is of particular interest that no other polysaccharide shows the re-entrant behavior. This suggests that the packing of amylose and amylopectin is an important part of the mechanism.

5. Heat capacity, thermal expansion and compressibility

Before considering the heat capacity, the thermal expansion and the compressibility of proteins it is useful to consider pure substances and small molecules in solution. The contribution from the strong intermolecular hydrogen bonding may be seen from the differences between water and glycerol and ethanol, benzene and hexane as given in Table 3. The distinction is less pronounced for the heat capacity.

TABLE 3. Thermal expansion, isothermal compressibility and heat capacity of some liquids, and polymers. The quantities for amino acids and proteins are for dilute solutions in water.

	thermal expansion 10^{-6}/K	compressibility 1/Mbar	heat capacity (Cp) J/mol K
Water	210	45.8	75
Glycerol	500	21.4	224
Ethanol	1120	109	112
Benzene	1220	96	136
Hexane	1380	166	195
Polythene	150	17	2.3 (Cv)J/gK
nylon 6	80	15	1.6 (Cv)J/gK
amino acids	1000 - 1550	- 0.28 - -0.62	0.78 - 2J/gK
proteins	40 - 110	2-15	0.32 - 0.36J/gK

Except for water, the temperature and pressure dependence of these parameters is less documented in the literature. Such data would be extremely valuable for the precise modelling of the effect of temperature and pressure on biomaterials.

5.1. *Statistical mechanical interpretation of C_p, α and κ: Fluctuations*

The compressibility is a thermodynamic quantity of interest not only from a static but also a dynamic point of view. Its relevance to the biological function of a protein can be understood through the statistical mechanical relation between the isothermal compressibility κ_T and volume fluctuations [51,52]:

$$\langle V-\langle V\rangle\rangle^2 = \kappa_T k_B T V$$

where the expression between brackets indicates an average for the isothermal-isopiestic ensemble. Because of the small size of the protein, the volume fluctuations are relatively large, they are of the order of the volume changes for the denaturation.

In a thermodynamic system with constant T and P, the isopiestic heat capacity can be regarded as a measure of the entropy fluctuations of the system:

$$\langle S-\langle S\rangle\rangle^2 = k_B C_p$$

The partial molar heat capacity has been considered to be composed of intrinsic and hydration contributions. The intrinsic component contains contributions from covalent and noncovalent interactions. It has been shown that about 85% of the partial molar heat capacity of the native state of a protein in solution is given by the covalent structure [53]. Changes in the heat capacity upon unfolding are primarily due to changes in the hydration. A physical picture of entropy fluctuations means changing the conformation between ordered and less ordered structures. This can occur by hindered internal rotations, low frequency conformational fluctuations and high frequency bond stretching and bending modes.

Like the compressibility and heat capacity, the thermal expansion can also be related to the fluctuations of the system, although this relation is not as widely known. The thermal expansion is proportional to the cross-correlation of the volume and entropy fluctuations:

$$\langle SV - \langle S \rangle \langle V \rangle \rangle = k_B T V \alpha$$

This is in accordance with the intuitive picture, that the thermal expansivity characterizes some kind of coupling between the thermal (T, S) and the mechanical (P, V) parameters.

The biological relevance of the volume, as well as the entropy and volume-entropy fluctuations can be illustrated by referring to a number of processes that are related to the dynamical properties of proteins and biopolymers in general. These include the opening and closing of binding pockets in enzymes, the allosteric effects, the conversion of chemical in conformational energy in muscle contraction, the biological synthesis of proteins and nucleic acids and the transport of molecules trough membranes. It should also be stressed that this includes also the fluctuations between different conformational substates as discussed by Frauenfelder et al. [54]. It is clear that the expansion and contraction of the cavities plays an important role in fluctuations [55].

Kharakoz and Bychkova [56] pointed out that the above considerations are valid strictly only for a system consisting of a constant number of particles. If one considers an unfolded molecule, where the conformational fluctuations change the surface and therefore the size of the hydration shell, the fluctuation of the number of water molecules in the hydration shell must be taken into consideration. For the native, folded protein the fluctuation in the number of particles can be neglected. This does not apply to intermediate states such as molten globule states and certainly not to the unfolded state.

5.2. *Experimental determination of α and κ for proteins*

In the following sections we concentrate on the thermal expansion and the compressibility. These are far less explored experimentally than the heat capacities. However, like heat capacities they are important in our understanding of the changes that occur upon protein unfolding. But unlike changes in the heat capacity that are primarily interpreted as due to changes in the hydration, changes in thermal expansion and compressibility contain also contributions from cavities.

5.2.1. *Thermal expansion.*

The partial molar thermal expansion of proteins can be obtained from the temperature dependence of the partial molar volumes with densitometry as shown by the work of Chalikian et al. [57]. The interpretation in that paper focuses on the inapplicability of the small molecular weight model systems to the understanding of the behavior of

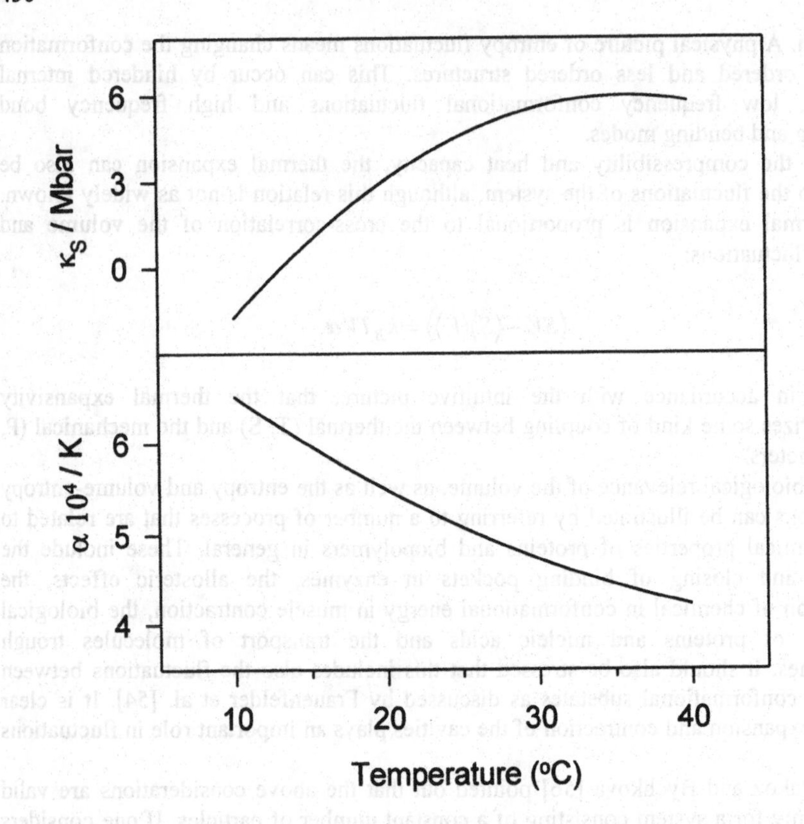

Figure 3. The effect of temperature on the thermal expansion [58] and the compressibility [59] of lysozyme.

proteins. It is suggested that mutual thermal motions of macromolecules and solvating waters involve modes that are absent in small molecules. More specifically, the hydration of non-polar groups on the surface of proteins is estimated to be different from the low molecular weight model systems. The same authors discussed this topic in more detail in terms of the "thermal volume" a volume that results from the thermally induced molecular vibrations. This volume is directly proportional to the isothermal compressibility [60].

In a seminal paper, Gekko and Hasegawa [59] have estimated the contribution of cavities and hydration to the temperature dependence of the isentropic compressibility of proteins. Bovine serum albumin and lysozyme show an increase in isentropic compressibility with increasing temperature. The authors attribute this mainly to a diminished hydration at higher temperature. At lower temperature the compressibility of both proteins becomes negative which is interpreted as a decreased contribution from the cavities. As can be seen from Figure 3, the temperature coefficient of the thermal expansion of lysozyme is negative. This is neither the behavior of a solid nor that of a liquid [58].

The isentropic compressibility of glycyldipeptides increases (becomes less negative) also with increasing temperatures [61]. This was interpreted as a decrease in electrostriction of the water surrounding the charged endgroups as the temperature is increased. Interestingly, the compressibility of dipeptides containing hydrophobic groups is less temperature dependent.

In both cases it is possible to present an alternative interpretation: The solvent structure changes with temperature thus accounting for the changes in volume change of the hydration.

Frauenfelder and coworkers [62] determined the thermal expansion of metmyoglobin from X-ray data. This refers to the expansion of the space inside the molecular surface and does not include the hydration water. The *linear* thermal expansion coefficient is estimated to be 115 10^{-6}/K, a value more than twice as that of liquid water (70 10^{-6}/K) but less than that of benzene (410 10^{-6}/K). It is concluded that the expansion is mainly due to the increase in volume of the numerous tiny packing defects in the protein, i.e. the subatomic free volumes between the atoms. The bulk of the overall effect is the separation between secondary structure units as the temperature increases. In the same temperature range the thermal expansion of ribonuclease A is about 2-3 fold smaller [63].

5.2.2. *Compressibility.*

Most experimental data on the compressibility of proteins have been obtained from ultrasonic velocimetry. The velocity of ultrasound (u) is related to the isentropic compressibility, κ_S, and the density, ρ, via the equation:

$$\kappa_S = 1/\rho u^2$$

The isothermal compressibility, κ_T, can be obtained from the following relation:

$$\kappa_T = \kappa_S + \alpha^2 T/\rho\, C_p$$

In this expression α is the thermal expansion, ρ, the density and C_p, the heat capacity. The definition of these parameters is as given in section 2. Sarvazyan [64] has drawn the attention to the potentialities of ultrasonic velocimetry as a method to obtain information on all molecular aspects related to compressibility. Recently, the methodology has been extended to the measurement of the compressibility of proteins as a function of pressure [65]. These effects should be clearly distinguished from the use of high power ultrasound to inactivate enzymes [66].

As pointed out before the compressibility of amino acids in aqueous solution is negative whereas the compressibility of proteins in the native state is positive. The negative compressibility of the unfolded state is usually interpreted as the consequence of the reduced compressibility of the hydration shell [67]. However, the disappearance of the cavities would also result in a decrease of the compressibility.

Chalikian et al. [68] studied the effect of pH on the temperature-induced compressibility change for chymotrypsinogen. Below pH 5.5, the compressibility of the unfolded protein is found to be higher than that of the folded state. At pH 6 and higher, the unfolded state aggregates and the compressibility decreases. It would be of particular interest to correlate these changes with the type of intermolecular aggregates formed.

Time resolved Brillouin difference spectroscopy has been used to measure the compressibility of lysozyme as a function of temperature [69]. Brillouin scattering results from the interaction of light with sound waves and the Doppler shift of the scattered light is proportional to the speed of sound. Upon denaturation by heat the compressibility increases but a further raise of the temperature gives a decrease in compressibility. This effect may be attributed to the aggregation of the protein. As indicated for the ultrasound experiments, the determination of the compressibility may be a sensitive probe to follow protein aggregation.

In principle it should be possible to obtain the compressibility of a protein matrix surrounding a chromophore, from the pressure-induced frequency shifts of the absorption spectrum of that chromophore. The approach can also be applied to fluorescence, phosphorescence and uv/visible spectroscopy as discussed in a recent review [70].

A more sophisticated luminescence method, *spectral hole burning* can be used to measure protein compressibility [71,72] in the low pressure regime. In this site selective spectroscopic technique a laser with very narrow bandwidth (10^8 Hz) is used to burn spectral hole into the inhomogeneously broadened absorption band. The hole arises because the laser light alters photochemically or photophysically a significant population of those molecules with transition frequencies corresponding to the laser (burning) frequency.

One can obtain the compressibility of a protein matrix surrounding the chromophore, from the pressure induced frequency shifts. This is based on a microscopic model that has been developed for the spectral hole burning technique which takes into account the long range *induced dipole-induced dipole* interactions between the protein matrix and the chromophore.

The compressibility values determined at 4 K for horseradish peroxidase are in the range of the compressibilities obtained for proteins at room temperature. Binding of an aromatic hydrogen-donor substrate to the free base mesoporphyrin substituted horseradish peroxidase was found to change the protein dynamics. This is reflected in the drastic increase of the compressibility [73]. Although this technique is restricted to chromoproteins and very low temperature conditions (*ca* 4 K), it is, in theory, possible to extend the method to the aromatic chromophores of the side chains of the proteins. A detailed review of the hole burning measurements can be found elsewhere [74].

It has to be mentioned that all of the above luminescence methods feel the compressibility of the interior of the protein molecule, which is different from the one determined by the ultrasonic measurements. By luminescence one measures the changes in the distances of the atoms of the protein. These distances change mainly because of the compression of the voids, while the ultrasonic measurements detect the effect of the hydration as well.

To correlate the effect of pressure on the frequencies in *infrared spectroscopy* ($d\nu/dP$) values with the compressibility is not as straightforward as in the fluorescence and visible/UV spectroscopy. The main difference is that the short range forces play a more important role than in the electronic transitions. The vibrational frequency is proportional to the second derivative of the potential with respect to the normal coordinate. The simplest calculation for the frequency shifts deals with a diatomic oscillator for which the vibrational frequency and equilibrium bond distance undergo slight changes. To the first order there is a simple relation between these changes [75]:

$$\frac{\Delta \nu}{\nu_0} = -a \frac{\Delta r}{r_e}$$

where the factor *a* depends on the leading terms in the series expansion of the potentials. Both the intrinsic potential of the oscillator and the interaction potential are to be taken into account.

The pressure effect depends on how the vibration affects the volume of the molecule. Gardiner et al. [76] presented an example of this effect in the case of toluene, where the dν/dP values for the ring breathing vibrations show the higher values. This means that a model that is based on long range interactions is not applicable here. The connection between the compressibility and dν/dP would have to include the nature of the vibrational mode. For proteins this compressibility would mainly reflect the compressibility along the C=O bond length, because the majority of the vibrational energy comes from the C=O stretching.

Fourier self-deconvolution can be used for resolution enhancement to separate the components of the amide I band characteristic for the different secondary structure [77]. The dν/dP values are typically in the range of -0.2 - -0.5 cm^{-1}/100 MPa for the α and β structure. The band corresponding to the unordered structure shows a positive (0.4 cm^{-1}/100 MPa) slope.

The pressure-induced changes of the folded structure of lysozyme was studied with *nuclear magnetic resonance* at 750 MHz [78]. The chemical shifts of 26 protons was followed up to 200 MPa. The main result is a compaction of the hydrophobic core consisting of bulky side chains. By contrast, it was found that the compaction is restricted to the α helical region in the crystal structure (see infra). The pressure effect on the structural dynamics of bovine pancreatic trypsin inhibitor was recently estimated by the same research group from the chemical shifts of the individual hydrogen bonds of the peptide bonds [79]. From the linear dependence of the chemical shifts on pressure, a pressure independent compressibility was assumed up to 200 MPa. This is a rather surprising result. But given the rather small pressure range, the change in compressibility might be under the limit of experimental detection.

The compressibility of the tetragonal crystal form of lysozyme up to 100 MPa has been obtained from *X-ray diffraction* by Kundrot and Richards [80]. As expected the compression is non-uniformly distributed. The domain of the α helices is found to be more compressible (5.7 $Mbar^{-1}$) than the other domain which is essentially β structure. It should be noted that this compressibility does not include the hydration water. Katrusiak and Dauter [81] have observed that the orthorhombic form of the same protein is stable up to 1 GPa. In solution lysozyme is known to unfold at about 0.5 GPa [82]. The reason for the high stability of the orthorhombic form compared to the solution and the tetragonal form deserves further study.

The compressibility of deoxymyoglobin has been estimated from *normal mode analysis calculations* up to 100 MPa by Yamato et al. [83]. In general, the helices are rigid but the interhelix regions are soft. The intrahelix rigidity is consistent with the findings of Kundrot and Richards for lysozyme [80]. Interestingly, the large cavities in the hydrophobic clusters do not make these clusters very compressible. Their relative compressibility depends on the size of the cavities, the larger cavities showing the largest compressibility. In agreement with NMR work of Morishima and Hara [84] the distal cavity is also found to be the most compressible.

6. The Grüneisen parameter

For an ionic solid such as rock-salt, it is possible to derive a simple relation between the thermal expansion, the compressibility and the specific heat. This is based on the existence of an asymmetrical potential energy curve for the whole crystal in terms of the separation between the atoms [85]. The relation is due to Grüneisen.

$$\gamma = \frac{V}{C_V} \frac{\alpha}{\kappa_T}$$

Pressure-induced vibrational wavenumber shifts in solids or polymers have been characterized by the volume-independent Grüneisen parameter [86].

$$\gamma_i = -\frac{d \ln \nu_i}{d \ln V} = -\frac{V}{\nu_i} \frac{d\nu_i}{dV}$$

If this parameter is assumed to be the same for all vibrations, one can obtain a bulk thermodynamic definition expressing the relation between the volume, the thermal expansion, the compressibility and the heat capacity at constant volume. The bulk Grüneisen parameter is found to be ca. 4 for polymers from the effect of pressure on the velocity of sound [87]. The data suggest that for the heat capacity only the interchain contribution should be taken into account. With this assumption, an order of magnitude calculation with the data available in Table I, shows that the bulk Grüneisen parameter for proteins is of the same order of magnitude as that of polymers. This suggests that the thermal expansion and the compressibility of proteins reflect primarily the movements of the domains between the secondary structures. These movements are reflected in the low frequency part of the vibrational spectrum. Unfortunately, no experimental data are available on the effect of pressure on these vibrations. The effect of pressure on the low frequency vibrations in liquid amides suggests differences between the hydrogen bonded and the non hydrogen bonded amides [88].

7. From model systems to proteins

Since Kauzmann [89] proposed micelles as a model for proteins, hydrophobic interactions have never been from the forefront in the interpretation of the stability of proteins. The fact however that the volume changes for pressure-induced protein denaturation are small (< 1%) compared to the total protein volume suggested that there is something missing in the picture of the protein model [22]. This apparent inconsistency between the observed volume change for protein denaturation and the expectations on the basis of low molecular weight compounds has been coined the "protein volume paradox" [60]. Several explanations have been offered the most recent being in terms of the role of packing forces in proteins [90], the nature of the unfolded state that retains a significant amount of structure [91] and the contribution of the "thermal volume" resulting from thermally induced molecular vibrations of the solute and the solvent leading to an additional expansion of the solvent [60]. But as pointed out by Murphy et al. [91], the partitioning of the solution space into solute and solvent volume may be performed in different ways and this leads to different conclusions on the volumetric properties.

The search for a correlation between hydration, compressibility, thermal expansion and heat capacity will always depend on a specific model that has to be used for the interpretation of thermodynamic data. However, a comparison between the simplest model for the partial molar volume of a small molecular weight compound:

$$V_i = V_{atoms} + \Delta V_{hydration}$$

and that for a protein:

$$V_i = V_{atoms} + V_{cavities} + \Delta V_{hydration}$$

suggests that apart from electrostatic interactions, hydrogen bonding and hydrophobic effects, cavities might play a crucial role in the volumetric properties of proteins. Their role has certainly neglected when compared to the more popular hydrophobic effect. However, the fact that elliptic phase diagram are observed for protein denaturation and starch gelation points to a more careful analysis of the role of other factors than hydrophobic interactions.

7.1 *Cavities in proteins.*

The role of cavities in the volumetric properties of proteins was probably first suggested by Weber and colleagues [92,93]. They suggested a common cause for the dissociation of oligomeric proteins and the unfolding of monomeric proteins: the imperfect packing of the atoms at the intersubunit surfaces and the restrictive effect of the covalent bond architecture of the protein.

Further examples as to the effect of cavities on the pressure behavior may be found in the studies of the F31A mutant of the 7 kDa protein Sso7d from S. solfataricus where the creation of a cavity drastically reduces the pressure stability [3]. Along the same lines one expects the volume changes for the unfolding of apomyoglobin to be larger than myoglobin because of the elimination of the void volume due to packing defects [94].

One possible technique that could be used to probe the role of cavities in proteins is *positron annihilation lifetime spectroscopy*. It is a powerful technique for determining the size of cavities and pores in materials. An estimate of the compressibility of the cavities in proteins should also be possible with technique. For an epoxy polymer, the free volume reduces exponentially with pressure with a reduction to half of the free volume at about 200 MPa [95]. The compressibility of the cavities is ca. 250 $Mbar^{-1}$ as estimated from the experimental data [96]. These numbers are much higher than the ones characteristic for proteins and suggest that there may be a considerable contribution of the reduction in cavity size to the compressibility of a protein.

7.2 The role of hydration.
Information on the hydration of molecules in solution may be obtained from several methods. Thermodynamic quantities such as partial molar volume, compressibility, thermal expansion and heat capacity probe the average of the population of water molecules whereas X-ray diffraction and NMR probe highly localised or highly immobilised water. Again a microscopic interpretation of the thermodynamic data needs a model for the interpretation [97].

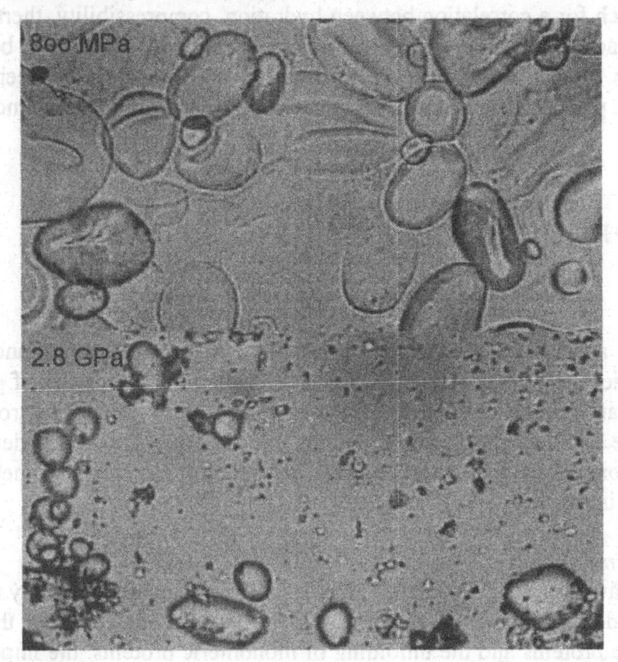

Figure 4. The uptake of water (top) and ethanol (bottom) by a suspension of potato starch granules at 800 MPa and 2.8 Gpa respectively. Note the absence of solvent uptake in the case of ethanol (Courtesy of Dr. J. Snauwaert).

The effect of hydration on the stability of proteins is well documented. Dry proteins can be compressed far beyond pressures that cause unfolding in aqueous solutions [10]. The absence of water explains also the extreme stability of bacterial spores [9] and small organisms to pressure treatment [8]. Similar effects are observed in the gelation of starch [98]. Whereas potato starch granules take up water at about 0.6 GPa pressures below the freezing point of ethanol (2 GPa) do not induce any uptake of solvent (see Figure 4).

The uptake of water during pressure-induced protein unfolding has been proposed earlier [92] and may also be observed from computer calculations at high pressure [99].

The phase diagram for protein unfolding is also affected by the composition of the solvent. Effects of pH on the phase diagram of metmyoglobin have been reported [24]. The effect of organic cosolvents upon the thermal stability of proteins has been reviewed recently by Timasheff [100]. Glycerol stabilises ribonuclease against temperature denaturation and affects the pressure-induced denaturation in a similar way.

Polyethyleneglycol stabilises both chymotrypsinogen and lactoglobulin against temperature denaturation. The same compound has a very small effect on the pressure-induced denaturation of chymotrypsinogen but destabilises lactoglobulin [77]. Evidently more systematic studies are needed in order to understand these effects. The effect of pressure on the catalytic activity of subtilisin Carlsberg has been studied in compressed gases and near the critical point [101]. It turns out that the activation volumes are much larger in the supercritical gases than in the liquid state. This was interpreted as resulting from the condensation of solvent to the solute.

8. Hydrostatic pressure and osmotic effects

The effects of organic cosolvents on protein stability and reactions have been interpreted in terms of the osmotic pressure of the solution with the implicated interpretation that these solvents avoid the protein surface. Thus water is considered to preferentially hydrate the protein surface [102]. In a few cases antagonistic effects have been observed of high hydrostatic pressure on osmotic pressure induced changes in protein conformation [103,104]. Volume changes are then calculated from the hydrostatic pressure effect and from the osmotic pressure effect:

$$\delta \ln K/\delta p_{hydrostatic} = - \Delta V_{hydrostatic}/RT$$

$$\delta \ln K/\delta p_{osmotic} = - \Delta V_{osmotic}/RT.$$

The volume changes obtained from osmotic pressure effects is then used to calculate the number of water molecules that are displaced when the activity of the water changes

$$\Delta V_{osmotic} = n\, V_{water}$$

This procedure raises interesting questions not only with regards to the interpretation of volume changes obtained from hydrostatic pressure experiments but also with regards to the molecular interpretation of osmotic pressure effects.

As pointed out in a previous section the partial molar quantities that play a role in volume changes are solvent as well as pressure dependent. Interpretation of the hydrostatic pressure effect in terms of density changes requires the knowledge of solvent as well as the pressure dependence of the partial molar volumes of all the reaction partners.

The osmotic pressure is not a pressure but an entropy (statistical) effect unless it is measured as a pressure with an osmometer. The changes in volume are due to mass movements between phases. The use of the term pressure in these situations creates confusion. From a thermodynamic point of view the term pressure should be restricted to the potential of volume changes. This definition reflects the analogy with temperature as the potential of heat flux and the chemical potential as the potential of matter flux [105].

The use of organic cosolvents for the removal of water from a membrane or a macromolecular system has been discussed by Parsegian et al. [106]. The term osmotic stress has been proposed but confusion has come in because of the use of the term osmotic pressure. In a recent paper Timasheff gives a strong warning against the possible misinterpretations of the role of cosolvents on protein reactions in disperse solutions [107]. In order to avoid further confusion Timasheff proposes the names cosolvent potential stress, preferential interaction stress or stress by Wyman linkage. The latter name refers to the competition between water and the cosolvent for binding places on the protein. In case of a chemical reaction such as the binding of a ligand or a change in conformation the following scheme has to be analysed:

Reactant (cosolvent)	↔	Product (cosolvent)
↑↓		↑↓
Reactant (water)	↔	Product (water)

For a detailed discussion of the effect of pressure and temperature on the binding partition function we refer to the book by Wyman and Gill [108]. In a separate communication in this volume we discuss the relation between osmotic pressure, hydration pressure, water activity and water potential and the various experimental configurations from which they are obtained [109].

The influence of various cosolvents on protein unfolding has been discussed by Timasheff [100]. Although pressure studies are limited, it seems that the stabilizing effect of organic cosolvents against temperature unfolding are also found against pressure unfolding [77]. Kinetic studies under pressure of the folding of staphylococcal nuclease in the presence of xylose, show that the sugar effect is primarily on the folding step suggesting that the transition state, a dry molten globule state, is close to the folded state [110]. The role of the cosolvent is also giving very useful insights into protein dynamics. Studies by Priev et al. [111] have shown that glycerol decreases the compressibility of the protein interior. These studies suggest a role for water as a lubricant for the conformational flexibility of proteins.

9. Cold, heat and pressure denaturation

A number of experimental approaches has been used to follow the heat, pressure and cold denaturation of proteins. These include NMR [12], UV spectroscopy [13] and DSC [112]. Whereas circular dichroism can be used to study cold and heat denaturation, for technical reasons, it cannot be used for the study of pressure denaturation. Then infrared spectroscopy, in combination with the diamond anvil cell, becomes the method of choice to follow the changes in secondary structure.

Information on the secondary structure can be obtained from the vibrations of the amide I band. 85 percent of the energy of this vibration comes from the C=O stretching. Since the oxygen atom is hydrogen bonded to the other part of the protein chain in a folded protein this is a conformation sensitive band. Its frequency is in the range of 1600-1700 cm^{-1} depending on the actual secondary structure. The usual effect of pressure on the vibrational bands is a blue shift (higher frequency), but hydrogen bonded systems are exceptions, showing a red shift. This is valid for the proteins as well. The increase in frequency with pressure demonstrates the strengthening of the hydrogen bond, which decreases the electron density of the C=O bond. Except for a few high pressure Raman studies [82,113] most of the vibrational spectroscopy was done

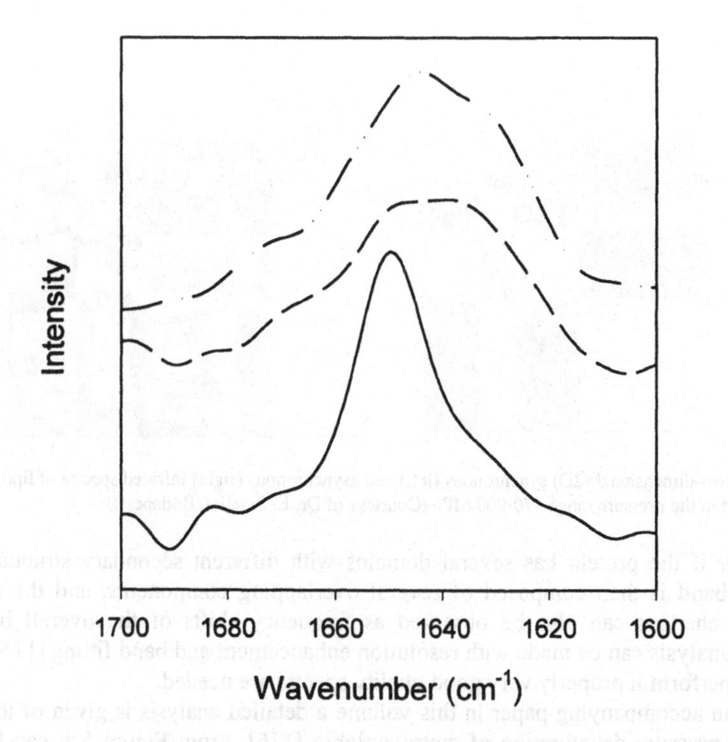

Figure 5. **The infrared spectra of metmyoglobin in the native state (full line) the pressure-denatured state at 11 kbar (broken line) and the pressure-assisted cold-denatured state at −16°C and 200 MPa (dash-dotted line). Meersman et al., unpublished.**

with FTIR (Fourier Transform Infrared) spectroscopy. The infrared spectroscopic experiments must be performed in D_2O because of the overlapping of the Amide I band with the strong water vibration at 1640 cm^{-1}. The use of D_2O involves the possibility of H/D exchange. While this effect becomes significant at the conformational changes, where new amino acids will be exposed to the solvent, it can also be used to probe the dynamics of proteins by estimating the internal accessibility of the polypeptide chain. D_2O has a slight stabilizing effect on the denaturation temperature, an effect that is not well understood [114].

For several proteins [10,77] a decreasing amide I frequency was found to be a general feature in the elastic pressure regime (typically below *ca.* 500 MPa). The maximum position of the amide I' band has a negative slope in the elastic region according to the above detailed picture. As such the band maximum can be a misleading

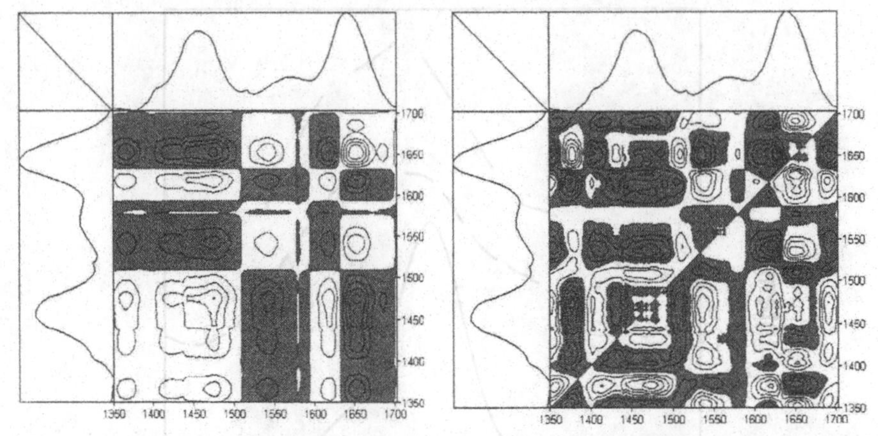

Figure 6. Two-dimensional (2D) synchronous (left) and asynchronous (right) infrared spectra of lipoxygenase investigated in the pressure range 470-900 MPa (Courtesy of Dr. L. Smeller, Budapest).

parameter if the protein has several domains with different secondary structure. The amide I band is then composed of several overlapping components, and the relative intensity changes can also be observed as frequency shifts of the overall band. A detailed analysis can be made with resolution enhancement and band fitting [115] but in order to perform it properly very good quality spectra are needed.

In an accompanying paper in this volume a detailed analysis is given of the cold, heat and pressure denaturation of metmyoglobin [116]. From Figure 5 it can be seen that the pressure and cold denatured protein have similar, but not identical, conformations. This is in sharp contrast with the heat denatured protein that shows the typical bands for intermolecular β-sheet formation [14].

The phase stability diagram of proteins may also be studied by infrared spectroscopy. For the enzyme lipoxygenase we have found an excellent correlation [33] between the changes in secondary structure and the inactivation of the enzyme as observed by Ludikhuyze et al. [29]. A similar correlation is found for the rate of hydrogen/deuterium exchange by following the changes in the amide II' region [33]. Hydrogen deuterium exchange experiments give important information on the protein dynamics as shown by the dependence of the accessibility of the protein interior and its pressure and temperature dependence. Even small proteins such as BPTI and ribonuclease P2 show considerable differences in their exchange rates and accessibility under normal conditions. Whereas P2 exchanges all the hydrogen under normal conditions, all hydrogen atoms in BPTI are exchanged after a pressure cycle of 500 MPa [3,10].

Two-dimensional IR spectroscopy.

The effects of H/D exchange and conformational changes can be separated by two-dimensional (2 D) infrared spectroscopy. The results for BPTI show that during the pressurization the first step is a small change in secondary structure followed by an increased rate of H/D exchange [117].

Pressure has multiple effects on proteins. Distortions of the spatial structure result in an increased hydrogen-deuterium exchange, which can further affect the secondary structure. Two-dimensional infrared spectroscopy (2D-IR) is a promising tool to distinguish these effects.

Figure 6 shows the 2D-IR synchronous and asynchronous spectra of lipoxygenase investigated in the pressure range 470-900 MPa. The amide I, II, and II' regions are shown. The negative (shadowed) cross-peaks on the synchronous plot show that changes in the amide II and II' are anticorrelated. The positive correlation between the 1653 cm^{-1} peak and the amide II vibrations shows that the helix structure disappears correlated with the H/D exchange. A more detailed analysis of both the synchronous and asynchronous spectra shows that the exchange is followed by further changes in the higher wavenumber part of the amide I region.

10. Unfolding and aggregation

The effect of pressure on protein-protein interactions is well documented [93]. At low pressures multimeric proteins dissociate reversibly into subunits. It is assumed that the free volumes at the interface of the subunits disappear at high pressure and are filled by the solvent. Under these conditions, pressure is used as a tool to study the assembly of viruses and macromolecular assemblages [118].

In many *in vitro* studies of protein folding the formation of aggregates is usually considered as an undesirable side effect which obscures the folding process as such. However, because of its important role in a number of diseases, the mechanism of the formation of aggregates, and the possible role of folding intermediates, deserves closer attention. Recent studies on lysozyme mutants have indicated that studies on model systems may help in obtaining a better insight into the molecular mechanisms underlying the formation of aggregates [119]. The general picture that is emerging from a number of studies is that under partial denaturation conditions proteins may acquire conformational intermediates that have a strong tendency to aggregate [120].

The phase diagram as we have discussed it in previous sections does not give any information about the mechanism of the unfolding and the aggregation of the unfolded state. Nor does it give information on the possible role of intermediates in the unfolding process which have been observed in a number of temperature and high pressure studies. Such a phase diagram, based on a two-state model, suggests that the heat-, pressure- and cold-unfolded state of the protein does not differ from each other qualitatively. This, however, is not supported by the recent NMR data for ribonuclease A, which indicate different structural changes for heat, cold and pressure unfolding [12].

The effect of pressure on protein-protein interactions has been studied with light scattering or turbidity methods [121,122]. These methods give information on the degree of association and aggregation. Information on conformational changes or on the nature of the intermolecular interactions may be expected from spectroscopic studies. As we have indicated the differences between the structural changes induced by temperature and pressure can be studied with FTIR spectroscopy. The infrared spectra of most heat unfolded proteins show two specific bands for the aggregated protein in the amide I' band range at 1615 and 1685 cm^{-1}. These aggregation specific bands are not observed in the case of the pressure unfolded protein [3,10,77]. These bands have been assigned to an intermolecular antiparallel β-sheet structure stabilized by hydrogen bonding [123]. Such a structure should not be confused with the classical antiparallel

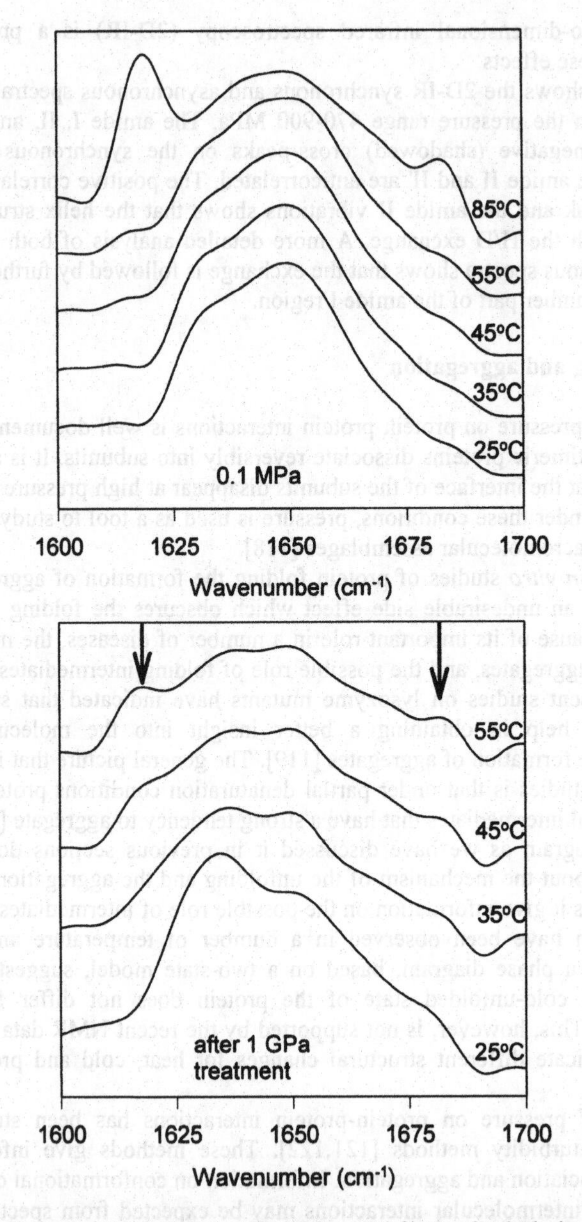

Figure 7. **The effect of a pressure pre-treatment on the development of intermolecular hydrogen bond network in lipoxygenase. The top figure shows that the hydrogen bond network develops at about 85°C. With a pressure pre-treatment at 1 Gpa the bands start to develop at 55°C (bottom). Rubens et al., unpublished.**

intramolecular β sheet secondary structure for which bands are expected in the region between 1630 and 1640 cm^{-1}. In the present case, the hydrogen bonds stabilize the intermolecular network of the heat-unfolded protein. The formation of the network after heat unfolding is an irreversible process, because the specific bands remain visible even after the cooling of the protein to room temperature. The FTIR study of many proteins leads to the conclusion that pressure unfolding can be reversible, while temperature unfolding is almost always irreversible.

We recently reported on the effect of temperature on the pressure-induced denaturation of myoglobin [124]. The results suggest that pressure induced partially unfolded states may play a very important role in the aggregation of proteins. The high pressure unfolding of horse heart metmyoglobin results in an intermediate form that shows a strong tendency to aggregate after pressure release. These aggregates are similar to those that are usually observed upon temperature denaturation.

We have observed similar effects for lipoxygenase as shown in Figure 7. In the upper part of the Figure the infrared spectra in the amide I' region indicate the formation of intermolecular antiparallel β-sheet stabilized by hydrogen bonding as the temperature is increased to 85°C. The formation of the aggregates is temperature dependent. The lower part shows the effect of temperature after a 1 GPa pressure treatment at room temperature. It is noted that intermolecular hydrogen bonding takes place well below 85°C. This suggests that pressure induces a conformation that is more prone to aggregate by temperature. It remains to be investigated whether the effects observed in myoglobin and lipoxygenase are a general property of proteins.

11. Unfolding in confined geometries

As discussed in previous sections, the temperature-induced denaturation of proteins gives rise to specific bands in the amide I part of the spectrum that can be assigned to intermolecular hydrogen bonding between anti-parallel β-sheets. For two proteins, myoglobin and lipoxygenase, we have shown in the previous section that pressure induces a state that gives rise to a temperature-induced aggregation at a lower temperature. These aggregates show also the characteristic bands that are found for heat denaturation. In order to study this phenomenon in more detail and to find out whether other conditions could provoke these specific structural features, we decided to study the unfolding in confined geometries. In a first set of experiments chymotrypsinogen was adsorbed on silica gel with an average pore diameter of 60 Å. Temperature induced denaturation of the adsorbed protein showed a marked decrease in the intensity of the bands attributed to intermolecular hydrogen bonding [125] This suggests that the presence of the intermolecular hydrogen bands indicate a fairly specific unfolding pattern that gives rise to very specific intermolecular hydrogen bonded structures.

An even more interesting result was obtained when chymotrypsin was incorporated in reversed micelles [126] Reversed micelles are formed when AOT is dissolved in octane. The size of the reversed micelles is determined by the ratio, w_0, of the total amount of water divided by the amount of surfactant. Figure 8 gives the result for w_0 = 38. It is noted that the spectrum of the protein incorporated in the reversed micelle is almost identical with that in solution suggesting that the conformation is the same. Increasing the pressure to 1.2 GPa denatures the protein irreversibly. However, the most surprising result is the spectrum that appears after the pressure is released. Surprisingly the spectrum is very similar to the spectrum that is obtained with a temperature induced denaturation. The origin of this effect is not exactly understood. Although the micelles, in the absence of protein, do not show any changes under pressure, it is not excluded that high molecular weight aggregates form in which the surfactant plays a specific role. Another observation that needs an explanation is the fact that the protein in the reversed

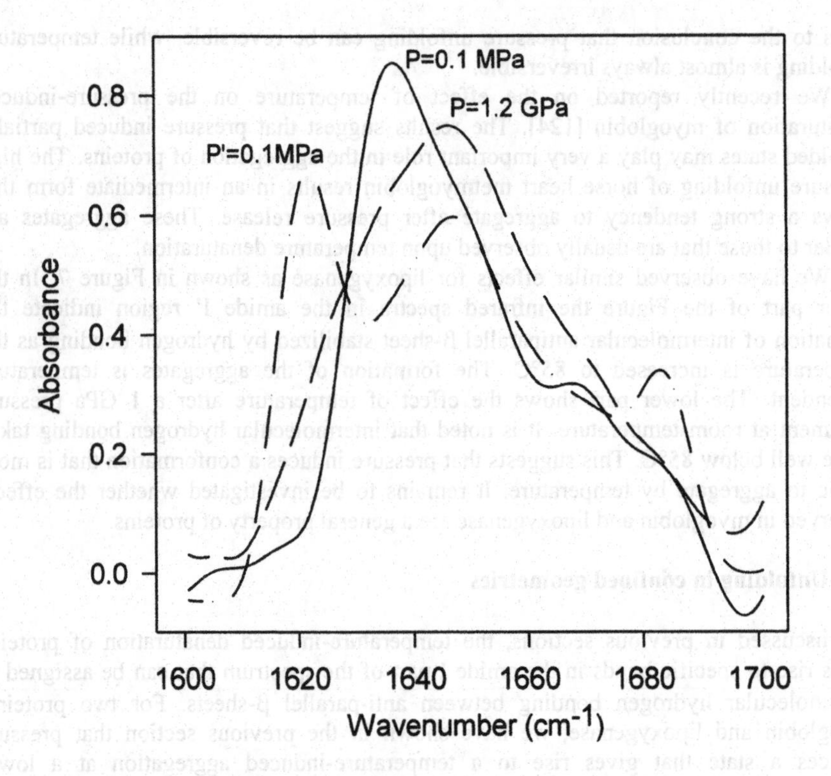

Figure 8. The effect of pressure on chymotrypsin incorporated in reversed micells of AOT in octane. Note the development of intermolecular hydrogen bonds after pressure release [125].

micelles denatures at a lower temperature and a higher pressure compared to the protein in solution.

The effect of pressure on the activity of enzymes in reversed micelles is discussed in the contribution by Balny and Klyachko in this volume [127].

12. Computer simulations of pressure effects

From a macroscopic point of view, the volume is a measure of that part of space that is inside a given surface. On the molecular and atomic level there is not precisely defined surface and it follows that a definition of the volume can use different approaches. The first one, the partial molar volume, is the phenomenological one and this is used in thermodynamics and in experimental work. The second one defines a surface such as the van der Waals or any other calculable surface, from which the volume is obtained. This approach is the one that is used in molecular dynamics simulations or other computer calculations. A central problem is the definition of the volume that gives the best agreements with experimental results. It was reported recently that the Voronoi volume and its related computed compressibility agree best with the experimental intrinsic compressibility [128].

With the exception of the normal mode analysis study of the behavior of deoxymyoglobin under pressure discussed in section 5, all high pressure computer simulations reported so far have been of the molecular dynamics (MD) type. The first high pressure MD simulations on BPTI were reported by Kitchen et al. [129]. No changes in the conformation were detected at 1 GPa. However, a subsequent MD simulation up to 2 GPa revealed changes in the secondary structure between 1 and 1.5 GPa [99]. These changes could be correlated with changes in the secondary structure observed with high pressure infrared studies [10].

Both papers reported the increased hydration of certain amino acids suggesting the penetration of solvent into the protein structure. This corroborates the water penetration models that have been proposed by Suzuki [20] and Weber et al. [92].

Van Gunsteren and coworkers [130] studied the unfolding of lysozyme. No net unfolding was observed at 1 GPa after 210 ps. However, fluorescence [92] as well as Raman [82] studies indicate that the protein unfolds at 500 MPa. Presumably, the origin of the absence of unfolding in the computer experiments might be kinetic. Paci and Marchi emphasize the difference in the physical nature of the structural effects by pressure and temperature [131]. For lysozyme the compression seems to be nonuniform in contrast to the dilation which is uniform as observed with X-ray diffraction [80].

A recent high pressure MD calculation by Floriano et al. [132] suggests that the unfolding of metmyoglobin is preceded by a molten globule intermediate. From an analysis of the hydrogen bonding pattern three stages are defined. Up to 600 MPa the structure remains native-like. Between 700 and 800 MPa, there is a substantial drop in the number of residues that are in the helical conformation. The tertiary structure is still largely as the native one but less rigid. This intermediate is suggested to be a molten globule state. Above 900 MPa less than 25% of the residues is part of a secondary structure. This is the denatured state. The authors note a qualitative agreement with the experimental studies of Zipp and Kauzmann [24]. It is of interest to compare the results with infrared studies that show a change in the amide I' band between 600 and 800 MPa [133]. No further changes are observed in the infrared spectrum between 0.9 and 1.2 GPa. Whether the structural changes observed with infrared spectroscopy are similar with those observed with the MD simulations is debatable. The observed spectrum at 800 MPa is not native-like when analysed in terms of secondary structure.

From the previous examples it should be clear that computer simulations will certainly assist in our understanding of pressure-induced protein unfolding. Of equal importance are simulations on the effect of pressure and temperature on the structure of water and on the non-covalent interactions that stabilise the structure of proteins. Silverstein et al. [134] used Monte Carlo simulations to study the anomalous properties of water and the hydrophobic effect. A 2D Mercedes Benz type molecule was used as a model for water. A number of features of water could be reproduced as well as features of the hydrophobic effect. It would be of interest to see whether this model reproduces the essential features of the effects of pressure. The pressure dependence of the hydrophobic interactions was calculated explicitly by Hummer et al. [135] by looking into the effect of pressure on the association of aliphatic residues in water. It is found that pressure stabilises the solvent-separated configuration of two aliphatic molecules compared to the contact configuration. This model supports the idea that pressure denaturation of proteins corresponds to the incorporation of water into the protein. It also emphasises the crucial role of water in the denaturation process.

Whereas the role of aliphatic residues is shown to play an important role in the pressure behavior of proteins, aromatic residues play also a role in some proteins [3]. It is known from experiments on model systems that the effect of pressure is such as to stabilise the stacked conformation [92,136]. It would therefore be of considerable interest to see whether the model calculations also predict such a behavior.

13. Why pressurize biomolecules?

Proteins can be unfolded by high hydrostatic pressure as well as by low or high temperature. This gives rise to an elliptic phase diagram in the temperature-pressure plane which describes the conditions under which the protein is in the native or the unfolded state at a given temperature and pressure with a specified solution composition. The elliptic shape of the phase boundaries follows from the temperature and pressure dependence of the entropy and volume change of the unfolding. The diagram gives also unique information on the changes in heat capacity, compressibility and thermal expansion of the unfolding process. These quantities are related to the changes in dynamics of the protein conformation. This information is difficult to obtain with other approaches.

The pressure-induced conformational transitions and unfolding can be studied with a number of spectroscopic techniques. FTIR spectroscopy in combination with the diamond anvil cell makes it possible to follow the changes in the secondary structure. In most cases the difference between the temperature and the pressure unfolding is very pronounced. This difference is not only reflected in the conformation of the polypeptide chain but also in the difference in intermolecular interactions of the proteins in the unfolded state. For the temperature unfolded proteins an extensive intermolecular hydrogen bond network develops. In many cases this network is absent in the pressure unfolded proteins. Some notable exceptions point to the use of pressure as a rather unique tool to unravel the interplay between protein unfolding and aggregation.

Clearly, a better understanding of the behavior of proteins under extreme conditions may be expected from a concerted application of a number of existing and new experimental approaches. But do proteins have simple strategies to cope with changes in the surroundings?

Behind (almost) every physical concept there is a *meta*-physical one. (To the reader who would disagree with this statement, the author replies that this proves the point!). Should we then look for another metaphysics (dogmas) to promote progress in high pressure molecular science? It has been suggested by A.N. Whitehead [137] that the whole of western philosophy can be considered footnotes to Plato. To take the analogy further, one could argue that high pressure bioscience as it developed during this century are footnotes to the Bridgman paper of 1914 in the sense that this paper described the basic ingredients.

Does this imply that there are no new developments? New developments imply different views (fresh looks) at existing experimental data. Two lines of developing concepts could play a role in the interpretation of pressure (and temperature) effects on proteins: There is first the new view of protein folding were the pathway concept is being gradually replaced by the assumed to be more realistic energy landscapes and folding funnels. This implies new interpretations of activation parameters [138]. The interesting point about this new view is that it equally rationalises the observations on glassforming liquids over a large range of behavior [139]. The point refers to the crucial contribution of water. Unless we understand in more detail the behavior of water we will probably never get close to a more detailed understanding of protein behavior. Progress along these lines is very exciting [17]. Secondly, there are the new developments in high pressure instrumentation which seem to be very strongly influenced by the new developments in physics. To mention only a few examples we refer to the combined use of FTIR and small-angle X-ray scattering [140] and the atomic force microscope that allows one to characterise soft materials under a wide variety of conditions [141]. But last but not least, we cannot but agree with a recent note by Kauzmann that the best experiments in science are those that lead to the least expected results [142].

Acknowledgements.
It is a great pleasure to acknowledge the contributions from coworkers and colleagues to the material and ideas discussed in this paper. The results from our research group presented in this paper were obtained with the support from the Research Fund of the K.U.Leuven, F.W.O Flanders, Belgium and the European Community.

14. References

1. Hemley, R.J. and Ashcroft, N.W. (1998) The revealing role of pressure in the condensed matter sciences, *Physics Today* **August**, 26-32.
2. Kooi, M.E. and Schouten, J.A. (1998) Raman spectra and phase behavior of the mixed solid N2-Ar at high pressure, *Phys. Rev.* **57B**, 10407-10413.
3. Fusi, P., Goossens, K., Consonni, R., Grisa, M., Puricelli, P., Vecchio, G., Vanoni, M., Zetta, L., Heremans, K., Tortora, P. (1997) Extreme heat- and pressure-resistant 7-KDa protein P2 from the Archeon *Sulfolobus solfataricus* is dramatically destabilized by a single-point amino acid substitution, *Proteins: Structure, Function and Genetics* **29**, 381-390.
4. Engelborghs, Y. (1998) General features of the recognition by tubulin of colchicine and related compounds, *Eur. J. Biophys.* **27**, 437-445.
5. Hauben, K., Bartlett, D., Soontjens, C., Cornelis, K., Wuytack, E., and Michiels, C. (1997) Escherichia coli mutants resistant to inactivation by high hydrostatic pressure. *Appl. and Environm. Microbiol.* **63**, 945-950.
6. Horikoshi, K. and Grant, W.D. (1998) *Extremophiles*, Wiley-Liss, New York.
7. Gross, M. (1998) *Life on the Edge*, Plenum Trade, New York.
8. Seki, K. and Toyoshima, M. (1998) Preserving tardigrades under pressure, *Nature* **395**, 853-854.
9. Sojka, B. and Ludwig, H. (1997) Effects of rapid pressure changes on the inactivation of *Bacillus subtilis* spores, *Pharm. Ind.* **59**, 436-438.
10. Goossens, K., Smeller, L., Frank, J. and Heremans, K. (1996) Conformation of bovine pancreatic trypsin inhibitor studied by Fourier transform infrared spectroscopy, *Eur. J. Biochem.* **236**, 254-262
11. Yayanos, A. (1998) Empirical and theoretical aspects of life at high pressure in the deep sea, in Horikoshi, K. and Grant, W.D. (1998) *Extremophiles*, Wiley-Liss, New York.
12. Zhang, J., Peng, X., Jonas, A. and Jonas, J. (1995) NMR study of the cold, heat, and pressure unfolding of ribonuclease A, *Biochemistry* **34**, 8361-8641.
13. Mombelli, E., Afshar, M., Fusi, P., Mariani, M., Tortora, P., Connelly, J.P. and Lange, R. (1997) The role of phenylalanine 31 in maintaining the conformational stability of ribonuclease P2 from *Sulfolobus solfataricus* under extreme conditions of temperature and pressure, *Biochemistry* **36**, 8733-8742.
14. Mozhaev, V., Heremans, K., Frank, J., Masson, P. and Balny, C. (1996) High pressure effects on protein structure and function, *Proteins: Structure, Function and Genetics* **24**, 81-91.
15. Hendrickx, M., Ludikhuyze, L., Van den Broeck, I. and Weemaes, C. (1998) Effects of high pressure on enzymes related to food quality, *Trends Food Sci. Technol.*, **8**, 197-203.
16. Klug, D.D. & Whalley, E. (1979) Elliptic phase boundaries between smectic and nematic phases. *J. Chem. Phys.* **71**, 1874-1877.

17. Mishima, O. and Stanley, H.E. (1998) The relationship between liquid, supercooled and glassy water, *Nature* **396**, 329-335.
18. Atake, T. and Angell, C.A. (1979) Pressure dependence of the glass transition in molecular liquids and plastic crystals, *J. Phys. Chem.*, **83**, 3218-3223.
19. Bridgman, P.W. (1914) The coagulation of albumen by pressure, *J. Biol. Chem.* **19**, 511-512.
20. Suzuki, K. (1960) Studies on the kinetics of protein denaturation under high pressure. *Rev. Phys. Chem. Japan*. **29**, 91-97.
21. Simpson, R.B. and Kauzmann, W. (1953) The kinetics of protein denaturation. I. The behavior of the optical rotation of ovalbumin in urea solutions. J. Am. Chem. Soc. 75, 5139-5152.
22. Brandts, J.F., Olivera, R.J. and Westort, C. (1970) Thermodynamics of protein denaturation. Effect of pressure on the denaturation of ribonuclease A. *Biochemistry* **9**, 1038-1047.
23. Hawley, S.A. (1971) Reversible pressure-temperature denaturation of chymotrypsinogen. *Biochemistry* **10**, 2436-2442.
24. Zipp, A. and Kauzmann, W. (1973) Pressure Denaturation of Metmyoglobin. *Biochemistry* **12**, 4217-4228.
25. Kanaya, H., Hara, K., Nakamura, A. and Hiramatsu, N. (1996) Time-resolved turbidimetric measurements during gelation process of egg white under high pressure, in *High Pressure Bioscience and Biotechnology*, Hayashi, R. and Balny, C. (Eds.) Elsevier, Amsterdam, pp. 343-346.
26. Taniguchi, Y. and Suzuki, K. (1983) Pressure inactivation of α-chymotrypsin, *J. Phys. Chem.* **87**, 5185-5193.
27. Heinisch, O., Kowalski, E., Goossens, K., Frank, J., Heremans, K., Ludwig, H. and Tauscher, B. (1995) Pressure effects on the stability of lipoxygenase: Fourier transform infrared spectroscopy and enzyme activity studies, *Z. Lebensm. Unters. Forsch.* **201**, 562-565.
28. Weingand-Ziadé, A., Renault, F. and Masson, P. (1997) Combined pressure/heat-induced inactivation of butyrylcholinesterase, *Biochim. Biophys. Acta* **1340**, 245-252.
29. Ludikhuyze, L., Indrawati, Van den Broeck, I., Weemaes, C. and Hendrickx, M. (1998) Effect of combined pressure and temperature on soybean lipoxygenase. 1. Influence of extrinsic and intrinsic factors on isobaric-isothermal inactivation kinetics, *J. Agr. Food Chem.* **46**, 4074-4080.
30. Weemaes PPO Weemaes, C.A., Ludikhuyze, L.R., Van den Broeck, I. and Hendrickx, M.E. (1998) Kinetics of combined pressure-temperature inactivation of avocado polyphenoloxidase, *Biotechnol. Bioeng.* **60**, 292-300.
31. Ludikhuyze, L.R., Van den Broeck, I., Weemaes, C.A., Herremans, C.H., Van Impe, J.F., Hendrickx, M.E. and Tobback, P.P. (1997) Kinetics for isobaric-isothermal inactivation of *Bacillus subtilis* α-amylase, *Biotechnol. Prog.* **13**, 532-538.
32. Rubens, P., Smeller, L. and Heremans, K. (1998) Pressure-temperature stability diagrams of proteins: α-amylases from *Bacillus* species, in *High Pressure food science, bioscience and chemistry*, Isaacs, N.S. (Ed), Royal Soc. Chem., Cambridge, pp. 411-416.
33. Rubens, P., Frank, J. and Heremans, K. (1999) Stability diagram of lipoxygenase as determined from H/D exchange kinetics and from conformational changes. This volume.
34. Ludwig, H., Scigalla, W. and Sojka, B. (1996) Pressure- and temperature-induced inactivation of microorganisms, in J. L. Markley, C. Royer & D. Northrup (Eds.) *High Pressure Effects in Molecular Biophysics and Enzymology*, Oxford University Press, pp. 346-363.
35. Hashizume, C., Kimura, K. and Hayashi, R. (1995) Kinetic analysis of yeast inactivation by high pressure treatment at low temperatures, *Biosci. Biotech. Comm.* **59**, 1455-1458.
36. Sonoike, K., Setoyama, T., Kuma, Y. and Kobayashi, S. (1992) Effect of pressure and temperature on the death rates of *L. casei* and *E. coli*, in C. Balny, R. Hayashi, K. Heremans, & P. Masson (Eds.), *High Pressure and Biotechnology*, John Libbey Eurotext Ltd, Montrouge, pp. 297-300
37. Hellemons, J.C. and Smelt, J.P.P.M. (1999) Building fail-safe models describing the effect of temperature and pressure on the kinetics of inactivation of infectious pathogens in foods, in *High pressure bioscience & biotechnology*, Ludwig, H. (Ed), Springer Verlag, Heidelberg, in press.
38. Gross, P. and Ludwig, H. (1992) Pressure-temperature-phase diagram for the stability of bacteriophage T4, in: C. Balny, R. Hayashi, K. Heremans, & P. Masson (Eds.), *High Pressure and Biotechnology*, John Libbey Eurotext Ltd, Montrouge, pp 57-59.
39. Butz, P. and Tauscher, B. (1995) Inactivation of fruit fly eggs by high pressure treatment, *J. Food Proc. Preserv.* **19**, 147-150.

40. Saldana, J.L. and Balny, C. (1992) Device for optical studies of fast reactions in solution as a function of pressure and temperature, in: C. Balny, R. Hayashi, K. Heremans, & P. Masson (Eds.), *High Pressure and Biotechnology*, John Libbey Eurotext Ltd, Montrouge, pp 529-531.
41. Kunugi, S., Takano, K., Tanaka, N., Suwa, K. and Akashi, M. (1997) Effects of pressure on the behavior of the thermoresponsive polymer poly(N-vinylisobutyramide) (PNVIBA), *Macromolecules* **30**, 4499-4501.
42. Sun, T. and King, H.E.Jr. (1996) Pressure-induced reentrant behavior in the poly(N-vinyl-2-pyrrolidone)-water system, *Phys. Rev.* **54** E, 2696-2703.
43. Thevelein, J., Van Assche, J.A., Heremans, K. and Gerlsma, S.Y. (1981) Gelatinisation temperature of starch as influenced by high pressure, *Carbohydrate Res.* **93**, 304-307.
44. Douzals, J.P., Perrier Cornet; J.M., Gervais, P. and Coquille, J.C. (1999) Hydration and pressure - temperature phase diagram of wheat starch, in *High pressure bioscience & biotechnology*, Ludwig, H. (Ed), Springer Verlag, Heidelberg, in press.
45. Rubens, P. and Heremans, K. (1999) Pressure and temperature phase diagram of starch gelation followed by FTIR, manuscript in preparation.
46. Balny, C. and Lange, R. (1999) Optical spectroscopic techniques in high pressure bioscience, This volume.
47. Winter, R., Landwehr, A., Brauns, T.H., Erbes, J., Czeslik, C. and Reis, O. (1996) High pressure effect on the structure and phase behavior of model membrane systems, in *High Pressure Effects in Molecular Biophysics and Enzymology*, J. L. Markley, C. Royer & D. Northrup (Eds.), Oxford University Press, p. 274-297.
48. Smeller, L. and Heremans, K. (1997) Some thermodynamic and kinetic consequences of the phase diagram of protein denaturation, in: K. Heremans (Ed), *High Pressure Research in Bioscience and Biotechnology*, Leuven University Press, pp. 55-58.
49. Haynes, J.M. (1968) Thermodynamics of freezing in porous solids, in *Low temperature biology of foodstuffs*, Hawthorn, J. (Ed), Pergamon Press, Oxford, pp 79-104.
50. Huang, G.S. and Oas, T.G. (1996) Heat and cold denatured states of monomeric lambda repressor are thermodynamically and conformationally equivalent, *Biochemistry* **35**, 6175-6180.
51. Cooper, A. (1973) Thermodynamic fluctuations in protein molecules, *Proc. Nat. Acad. Sci. USA* **73**, 2740-2741.
52. Landau, L. and Lifshitz, E. (1969) *Statistical Physics*, Theoretical Physics Vol. 5, Pergamon Press Oxford.
53. Gomez, J., Hilser, V.J., Xie, D. and Freire, E. (1995) The heat capacity of proteins, *Proteins: Structure, Function and Genetics* **22**, 404-412.
54. Frauenfelder, H., Alberding, N.A., Ansari, A., Braunstein, D., Cowen, B.R., Hong, M.K., Iben, I.E.T., Johnson, J.B., Luck, S., Marden, M.C., Mourant, J.R., Ormos, P., Reinisch, L., Scholl, R., Schulte, A., Shyamsunder, E., Sorensen, L.B., Steinbach, P.J., Xie, A., Young, R.D. and Yue, K.T. (1990) Proteins and pressure, *J. Phys. Chem.* **94**, 1024-1037.
55. Kaminsky, S.M. and Richards, F.M. (1992) Reduction of thioredoxin significantly decreases its partial specific volume and adiabatic compressibility, *Protein Science*, **1**, 22-30.
56. Kharakoz, D.P. and Bychkova, V.E. (1997) Molten globule of human α-lactalbumin: hydration, density and compressibility of the interior, *Biochemistry* **36**, 1882-1890.
57. Chalikian, T.V., Totrov, M., Abagyan, R. and Breslauer, K.J. (1996) The hydration of globular proteins as derived from volume and compressibility measurements: cross correlating thermodynamic and structural data, *J. Mol. Biol.* **260**, 588-603.
58. Hiebl, M. and Maksymiw, R. (1991) Anomalous temperature dependence of the thermal expansion of proteins, *Biopolymers* **31**, 161-167.
59. Gekko, K. and Hasegawa, Y. (1989) Effect of temperature on the compressibility of native globular proteins, *J. Phys. Chem.* **93**, 426-429.
60. Chalikian, T.V. and Breslauer, K.J. (1996) On volume changes accompanying conformational transition of biopolymers, *Biopolymers* **36**, 619-626.
61. Hedwig, G.R., Høiland, H. and Høgseth, E. (1996) Thermodynamic properties of peptide solutions. Part 15. Partial molar isentropic compressibilities of some glycyl dipeptides in aqueous solution at 15 and 35°C, *J. Sol. Chem.* **25**, 1041-1053.
62. Frauenfelder, H., Hartmann, H., Karplus, M., Kuntz, Jr. I.D., Kuriyan, J., Parak, F., Petsko, G.A., Ringe, D., Tilton, Jr. R.F., Connelly, M.L. and Max, N. (1987) Thermal expansion of a protein, *Biochemistry* **26**, 254-261.

63. Tilton Jr., R.F., Dewan, J.C. and Petsko, G.A. (1992) Effects of temperature in protein structure and dynamics: X-ray crystallographic studies of the protein ribonuclease-A at nine different temperatures from 98 to 320 K, *Biochemistry* **31**, 2469-2481.
64. Sarvazyan, A.P. (1991) Ultrasonic velocimetry of biological compounds, *Annu. Rev. Biophys. Biophys. Chem.* **20**, 321-342.
65. Benolenko, V.N., Chalikian, T., Funck, Th., Kankia, B. and Sarvazyan, A.P. (1997) High resolution ultrasonic measurements as a tool for studies on biochemical systems under variation of pressure, in: K. Heremans (Ed.), *High Pressure Research in Bioscience and Biotechnology*, Leuven University Press, pp. 147-150.
66. Froment, M.Th., Lockridge, O. and Masson, P. (1998) Resistance of butyrylcholinesterase to inactivation by ultrasound: effects of ultrasound on catalytic activity and subunit association, *Biochim. Biophys. Acta*, **1387**, 53-64.
67. Kharakoz, D.P. (1997) Partial volumes and compressibilities of extended polypeptide chains in aqueous solution: additivity scheme and implication of protein unfolding at normal and high pressure, *Biochemistry* **36**, 10276-10285.
68. Chalikian, T., Völker, J., Anafi, D. and Breslauer, K.J. (1997) The native and the heat-induced denatured states of α-Chymotrypsinogen A: Thermodynamic and Spectroscopic studies, *J. Mol. Biol.* **274**, 237-252.
69. Doster, W., Simon, B., Schmidt, G. and Mayr, W. (1985) Compressibility of lysozyme in solution from time-resolved Brillouin difference spectroscopy, *Biopolymers*, **24**, 1543-1548.
70. Heremans, K. and Smeller, L. (1998) Protein structure and dynamics at high pressure, *Biochim. Biophys. Acta*, **1386**, 353-370.
71. Zollfrank, J., Friedrich, J., Fidy, J. and Vanderkooi, J.M. (1991) Photochemical holes under pressure: Compressibility and volume fluctuations of a protein, *J. Chem Phys.* **94**, 8600-8603.
72. Friedrich, J. (1995) Hole burning spectroscopy and the physics of proteins, *Methods in Enzymology* 246, 226-259.
73. Fidy, J., Vanderkooi, J.M., Zollfrank, J. and Friedrich, J. (1992) Softening of the packing density of horseradish peroxidase by a H-donor bound near the heme pocket, *Biophys. J.* **63**, 1605-1612.
74. Köhler, M., Friedrich, J. and Fidy, J. (1998) Proteins in electric fields and pressure fields: basic aspects, *Biochim. Biophys. Acta* **1386**, 255-288.
75. Sandroff, C.J., King, Jr. H.E., and Herschbach, D.R. (1984) High pressure study of the liquid/solid interface: Surface enhanced Raman scattering from adsorbed molecules, *J. Phys. Chem.* **88**, 5647-5653.
76. Gardner, D.J., Walker, N.A. and Dare-Edwards, M.P. (1987) Density and temperature effects on relative Raman intensities in liquid toluene, *Spectrochimica Acta* **43A**, 1241-1247.
77. Heremans, K., Goossens, K. and Smeller, L. (1996) Pressure-tuning spectroscopy of proteins: Fourier transform infrared studies in the diamond anvil cell, in: J. L. Markley, D.B. Northrop, C. A. Royer (Eds), *High pressure effects in molecular biophysics and enzymology*, Oxford University Press, N.Y., pp. 44-61.
78. Akasaka, K. Tezuka, T. and Yamada, H. (1997) Pressure-induced changes in the folded structure of lysozyme, *J. Mol. Biol.* **271**, 671-678.
79. Li, H., Yamada, H. and Akasaka, K. (1998) Effect of pressure on individual hydrogen bonds in proteins. Basic Pancreatic Trypsin Inhibitor, *Biochemistry* **37**, 1167-1172.
80. Kundrot, C.E. and Richards, F.M. (1987) Crystal structure of hen egg-white lysozyme at a hydrostatic pressure of 1000 atmospheres, *J. Mol. Biol.* **193**, 157-170.
81. Katrusiak, A. and Dauter, Z. (1996) Compressibility of lysozyme protein crystals by X-ray diffraction, *Acta Cryst.* **D52**, 607-608.
82. Heremans, K. and Wong, P.T.T. (1985) Pressure effect on the Raman spectrum of proteins: Pressure induced changes in the conformation of lysozyme in aqueous solutions, *Chem Phys. Letters* **118**, 101-104.
83. Yamato, T., Higo, J., Seno, Y. and Go, N. (1993) Conformational deformation in deoxymyoglobin by hydrostatic pressure, *Proteins: Structure, Function and Genetics* **16**, 327-340.
84. Morishima, I. and Hara, M. (1983) High-pressure nuclear magnetic resonance studies of hemoproteins. Pressure-induced structural changes in the heme environments of ferric low-spin metmyoglobin complexes, *Biochemistry* **22**, 4102-4107.
85. Tabor, D. (1991) *Gases, liquids and solids and other states of matter*, Cambridge University Press, Cambridge.
86. Flores, J.J. and Chronister, E.L. (1996) Pressure-dependent Raman shifts of molecular vibrations in poly-(methyl methacrylate) and polycarbonate polymers, *J. Raman Spectrosc.* **27**, 149-153.

87. Wada, Y., Itani, A., Nishi, T. and Nagai, S. (1969) Grüneisen constant and thermal properties of crystalline and glassy polymers, *J. Polym. Sc. A2* 7, 201-208.
88. Goossens, K., Smeller, L. and Heremans, K. (1993) Pressure tuning spectroscopy of the low-frequency Raman spectrum of liquid amides, *J. Chem. Phys.* **99**, 5736-5741.
89. Kauzmann, W. (1959) Some factors in the interpretation of protein denaturation, *Adv. Prot. Chem.* **14**, 1-63.
90. Harpaz, Y., Gerstein, M. and Chothia, C. (1994) Volume changes on protein folding, *Structure* **4**, 641-649.
91. Murphy, L.R., Matubayasi, N., Payne, V.A. and Levy, R.M. (1998) Protein hydration and unfolding – insights from experimental partial specific volumes and unfolded protein models, *Folding and Design* **3**, 105-118.
92. Weber, G. and Drickamer, H.G. (1983) The effect of high pressure upon proteins and other biomolecules, *Q. Rev. Biophys.* **16**, 89-112.
93. Silva, J.L. and Weber, G. (1993) Pressure stability of proteins, *Ann. Rev. Phys. Chem.* **44**, 89-113.
94. Vidugiris, G.J.A. and Royer, C.A. (1998) Determination of the volume changes for pressure-induced transitions of apomyoglobin between the native, molten globule, and unfolded states, *Biophys. J.* **75**, 463-470.
95. Deng, Q. and Jean, Y.C. (1993) Free volume distributions of an epoxy polymer probed by positron annihilation: pressure dependence, *Macromolecules* **26**, 30-34.
96. Deng, Q., Sundar, C.S. and Jean, Y.C. (1992) Pressure dependence of free-volume hole properties in an epoxy polymer, *J. Phys. Chem.* **96**, 492-495.
97. Chalikian, T. and Breslauer, K.J. (1998) Thermodynamic analysis of biomolecules: a volumetric approach, *Curr. Opin. Struct. Biol.* **8**, 657-664.
98. Snauwaert, J., Rubens, P., Vermeulen, G., Hennau, F. and Heremans, K. (1998) *In situ* microscopic observations of pressure-induced gelatinization of starch in the diamond anvil cell, in *High Pressure food science, bioscience and chemistry*, Isaacs, N.S. (Ed), Royal Soc. Chem., Cambridge, pp. 457-464.
99. Wroblowski, B., Diaz, J.F., Heremans, K. and Engelborghs, Y. (1996) Molecular mechanisms of pressure induced conformational changes in BPTI, *Proteins: Structure, Function and Genetics* **25**, 446-455.
100. Timasheff, S.N. (1998) Control of protein stability and reactions by weakly interacting cosolvents: The simplicity of the complicated, *Adv. Prot. Chem.* **51**, 355-432.
101. Fontes, N., Nogueiro, E., Margarida Elvas, A., Correa de Sampio, T. and Barreiros, S. (1998) Effects of pressure on the catalytic activity of subtilisin Carlsberg suspended in compressed gases, *Biochem. Biophys. Acta* **1383**, 165-174.
102. Reid, C. and Rand, R.P. (1997) Probing protein hydration and conformational states in solution, *Biophys. J.* **72**, 1022-1030.
103. Robinson, C.R. and Sligar, S.G. (1995) Heterogeneity in molecular recognition by restriction endonucleases: Osmotic and hydrostatic pressure effects on *Bam*HI, *Pvu* II and *Eco*RV specificity, *Proc. Nat. Acad. Sci. USA* **92**, 3444-3448.
104. Di Primo, C., Deprez, E. Hui Bon Hoa, G. and Douzou, P. (1995) Antagonistic effects of hydrostatic pressure and osmotic pressure on Cytochrome P-450_{cam} spin transition, *Biophys. J.* **68**, 2056-2061.
105. Callen, H.B. (1985) Thermodynamics and an introduction to thermostatistics, John Wiley, New York.
106. Parsegian, V.A., Rand, R.P., Fuller, N.L. and Rau, D.C. (1986) Osmotic stress for the direct measurement of intermolecular forces, Methods in Enzymology 127, 400-416.
107. Timasheff, S.N. (1998) In disperse solution, "osmotic stress" is a restricted case of preferential interactions, *Proc. Nat. Acad. Sci. USA* **95**, 7363-7367.
108. Wyman, J. and Gill, S.J. (1990) *Binding and linkage: Functional chemistry of biological macromolecules*, University Science Books, California.
109. Pfeiffer, H. and Heremans, K. (1999) On the use of term osmotic pressure. This volume.
110. Frye, K.J. and Royer, C.A. (1997) The kinetic basis for the stabilization of staphylococcal nuclease by xylose, *Protein Science* **6**, 789-793.
111. Priev, A., Almagor, A., Yedgar, S. and Gavish, B. (1996) Glycerol decreases the volume and compressibility of protein interior, *Biochemistry* **35**, 2061-2066.
112. Privalov, P.L. (1990) Cold denaturation of proteins, *Crit. Rev. Biochem. Molec. Biol.*, **25**, 281-305.
113. Remmele, Jr., R.L., McMillan, P. and Bieber, A. (1990) Raman spectroscopic studies of hen egg-white lysozyme at high temperatures and pressures, *J. Prot. Chem.* **9**, 475-486
114. Kuhlman, B. and Raleigh, D.P. (1998) Global analysis of the thermal and chemical denaturation of the N-terminal domain of the ribosomal protein L9 in H_2O and D_2O. Determination of the thermodynamic

parameters, ΔH°, ΔS°, and ΔCp° and the evaluation of solvent isotope effects, *Protein Science* **7**, 2405-2412.
115. Smeller, L., Goossens, K., and Heremans, K. (1995) Determination of the secondary structure of proteins at high pressure, *Vibrational Spectrosc.* **8**, 199-203.
116. Meersman, F., Smeller, L. and Heremans, K. (1999) FTIR as a tool to study cold, heat and pressure denaturation of myoglobin. This volume.
117. Smeller, L. and Heremans, K. (1999) 2D FT-IR spectroscopy analysis of the pressure-induced changes in proteins, *Vibrational Spectrosc.*, in press.
118. Silva, J.L., Foguel, D., Da Poian, A.T., and Prevelige, P.E. (1996) The use of hydrostatic pressure as a tool to study viruses and other macromolecular assemblages, *Curr. Opin. Struct. Biol.* **6**, 166-175.
119. Booth, D.R., Sunde, M., Bellotti, V., Robinson, C.V., Hutchinson, W.L., Fraser, P.E., Hawkins, P.N., Dobson, C.M., Radford, S.E., Blake, C.C.F., and Pepys, M.B., (1997) Instability, unfolding and aggregation of human lysozyme variants underlying amyloid fibrillogenesis, *Nature*, **385**, 787-793.
120. Fink, A.L. (1998) Protein aggregation: folding aggregates, inclusion bodies and amyloid, *Folding and design* **3**, R9-R23.
121. Payens, T.A.J., and Heremans, K. (1969) Effect of pressure on the temperature-dependent association of β-casein, *Biopolymers* **8**, 335-345.
122. Gorovits, B.M. and Horowitz, P.M. (1998) High hydrostatic pressure can reverse aggregation of protein folding intermediates and facilitate acquisition of native structure, *Biochemistry* **37**, 6132-6135.
123. Clark, A. H., Saunderson, D.H. and Sugget, A. (1981) Infrared and laser-Raman spectroscopic studies of thermally-induced globular protein gels, *Int. J. Pept. Res.* **17**, 353-364.
124. Smeller, L., Rubens, P. and Heremans, K. (1999) Pressure effect on the temperature induced unfolding and tendency to aggregate of myoglobin, *Biochemistry*, in press.
125. Vermeulen, G. (1999) PhD thesis, Leuven (in Dutch).
126. Vermeulen, G. and Heremans, K. (1997) FTIR study of pressure and temperature stability of proteins in emulsions and reversed micelles, in: K. Heremans (Ed), *High Pressure Research in Bioscience and Biotechnology*, Leuven University Press, pp. 67-70.
127. Balny, C. and Klyachko, N.L. (1999) High hydrostatic pressure and enzymology, This volume.
128. Paci, E. and B. Velikson, B. (1997) On the volume of macromolecules, *Biopolymers* **41**, 785-797.
129. Kitchen, D.B., Reed, L.H. and Levy, R.M. (1992) Molecular dynamics simulation of solvated protein at high pressure, *Biochemistry* **31**, 10083-10093.
130. Hünenberger, P.H., Mark, A.E. and van Gunsteren, W.F. (1995) Computational approaches to study protein unfolding: hen egg white lysozyme as a case study, *Proteins: Structure, Function and Genetics* **21**, 196-213.
131. Paci, E. and Marchi, M. (1996) Intrinsic compressibility and volume compression in solvated proteins by molecular dynamics simulation at high pressure, *Proc. Nat. Acad. Sci. USA* **93**, 11609-11614.
132. Floriano, W.B., Nascimento, M.A.C., Domont, G.B. and Goddard III, W.A. (1998) Effects of pressure on the structure of metmyoglobin: molecular dynamics predictions for pressure unfolding through a molten globule intermediate, *Protein Science* **7**, 2301-2313.
133. Smeller, L., Rubens, P. and Heremans, K. (1996) High pressure FTIR studies on hemoproteins, in *High Pressure Science and Technology*, Trzeciakowski, W.A. (Ed.), World Scientific, Signapore, pp. 863-865.
134. Silverstein, K.A.T., Haymet, A.D.J. and Dill, K.A. (1998) A simple model of water and the hydrophobic effect, *J. Am. Chem. Soc.* **120**, 3166-3175.
135. Hummer, G., Garde, S., Garcia, A.E., Paulaitis, M.E. and Pratt, L.R. (1998) The pressure dependence of hydrophobic interactions is consistent with the observed pressure denaturation of proteins, *Proc. Nat. Acad. Sci. USA* **95**, 1552-1555.
136. Heremans, K. (1982) High pressure effects on proteins and other biomolecules, *Ann. Rev. Biophys. Bioeng.* **11**, 1-21.
137. Whitehead, A.N. (1979) *Process and Reality*, The Free Press, N.Y.
138. Dill, K.A. and Chan, H.S. (1997) From Levinthal to pathways to funnels, *Nature Structural Biology* **4**, 10-19.
139. Angell, C.A. (1997) Landscapes with megabasins: Polyamorphism in liquids and biopolymers and the role of nucleation in folding and folding diseases, *Physica D* **107**, 122-142.
140. Panick, G., Malessa, R., Winter, R., Rapp, G., Frye, K.J. and Royer, C.A. (1998) Structural characterization of the pressure-denatured state and unfolding/refolding kinetics of Staphylococcal nuclease by synchrotron small-angle X-ray scattering and Fourier-transform infrared spectroscopy, *J. Mol. Biol.* **275**, 389-402.
141. Weisenhorn, A.L., Khorsandi, M., Kasas, S., Gotzos, V. and Butt, H.J. (1993) Deformation and height anomaly of soft surfaces studied with an AFM, *Nanotechnology* **4**, 106-113.
142. Kauzmann, W. (1993) Reminiscences from a life in protein physical chemistry, *Protein Science* **2**, 671-691.

PRESSURE DENATURATION OF PROTEINS

Catherine A. Royer
Centre de Biochimie Structurale
INSERM Unité 414
Faculté de Pharmacie
15, ave. Charles Flahault
34060 Montpellier Cedex 02, France
royer@tome.cbs.univ-montp1.fr

Abstract
The study of pressure effects on protein stability has occupied a relatively marginal position in the field of protein folding, with very few thorough thermodynamic, structural and kinetic studies of this phenomenon. Moreover, theoretical treatment of the issue with a few recent exceptions, has been limited to declarations of its complexity and lack of concordance with the results from other approaches. This paucity of data and theory notwithstanding, understanding the fundamental physical basis for pressure effects on proteins is essential to progress in the field of protein folding. Moreover, pressure presents certain advantages as a perturbation methodology that render it an important, useful and complementary approach. In the present review, the issue of the fundamental basis for the effects of pressure is discussed. Reference is made to studies in the literature, but I have concentrated the detailed presentation on the body of work in pressure-induced protein unfolding carried out by my research group and collaborators on staphylococcal nuclease (Snase) over the past 5 years. The origins of the value of the change in volume upon unfolding must be understood prior to any thorough theoretical analysis of pressure effects. The various arguments for the multiple contributing factors are discussed and then recent studies from my research group designed to probe this question are presented, the overall conclusion being that the existence of packing defects in the folded structure represents the most likely candidate for the negative change in volume upon unfolding. Moreover, the results of the temperature dependence of the volume change for unfolding of Snase implicate the difference in thermal expansivity in the temperature dependence of the value of the volume change of unfolding. Next I present results of a characterization of the physical properties of the pressure denatured state of Snase, and compare these to studies on a number of other pressure denatured proteins. Finally, the results of a series of pressure-jump kinetic studies on the folding/unfolding reactions of this protein are discussed. It is too early to conclude whether the results from these pressure studies on Snase stability and their interpretations are general. For this, many more studies on a number of small, reversibly folding proteins will be required.

R. Winter and J. Jonas (eds.), High Pressure Molecular Science, 473–496.

Introduction

Understanding the mechanisms by which polypeptide chains of particular sequences spontaneously adopt specific, stable three dimensional native protein structures remains one of the most interesting and important problems in modern molecular biophysics. While random polypeptide sequences do not fold to unique structures, protein sequences have evolved in order to adopt specific three-dimensional conformations, which in addition, present specific thermodynamic stability and dynamic properties that are essential to their function. Interestingly, the viewpoint of protein chemists concerning the mechanisms of folding has evolved from the rather general oil-drop model, highlighting the hydrophobic effect [1], passing through a period of detailed description of folding kinetics and observable intermediates [2], back to a more general theory embodied by the energy landscape[3], based on a global overview of the protein's energy surface. Although the energy landscape or folding funnel theory is a global, statistical mechanical approach, the specificity of the final folded structure is incorporated essentially through the competition between the ruggedness of the landscape and the energy gap between the native and the mean of misfolded states[4]. The landscape theory for folding has been shown through lattice simulations to account very well for the temperature dependence of protein folding [5], experimentally well-understood as being based on the large decrease in heat capacity upon folding [6]. Pressure effects on protein structure however, are only now beginning to be considered in the context of the energy landscape theory. While it is clear than an understanding of the fundamental basis for the effect of pressure on protein structure will be essential to the development of a more complete theory for the folding process, experimentation on the effects of pressure in protein folding has remained relatively marginal to the folding field. The present review first examines the underlying basis for pressure effects on protein structure, i.e., the origins of the decrease in system volume upon unfolding. Secondly, the physical nature of the pressure denatured state is discussed. Finally, the kinetic aspects of pressure denaturation are presented in terms of the activation parameters for the folding and unfolding reactions and their physical interpretation.

The volume change upon unfolding

In early seminal work on the pressure denaturation of proteins, Brandts and coworkers [7] found that pressure lead to the unfolding of ribonuclease A, and that the volume changes for unfolding (ΔV_u) deduced from the equilibrium unfolding profiles fell in the range of -45 to -5 mL/mol, depending upon the pH and temperature at which the experiment was carried out. A few years later, Zipp & Kauzmann [8]) studied the pressure denaturation of metmyoglobin over a wide range of temperature, pH and pressure. They found the ΔV_u to be negative for nearly all conditions studied and on the order of -100 mL/mol in good agreement with dilatometric measurements of Katz [9]. Only at very high temperatures did the phase diagram for protein stability require a negative volume change for folding. Moreover they found that the volume change was independent of pressure. Negative values of ΔV_u have also been determined for ovalbumin and HSA[10], lysozyme [11,12], chymotrypsinogen [11, 13, 14], myoglobin

[15], Rnase A[16,17,18] and staphylococcal nuclease and a series of mutants of this protein [19-25]. The general picture of the temperature/pressure phase diagram that first emerged from the metmyoglobin study of Zipp and Kauzmann [8] is shown in Figure 1 below. It can be seen that the application of pressure over most of the temperature range results in unfolding of the protein ($\Delta V_u < 0$), although at very high temperature, this trend is reversed, and pressure stabilizes the folded structure (see arrow below). We will first consider the structural basis for the decrease in volume upon unfolding. Also apparent in the figure is the maximum in protein stability among the temperature axis, due to the large decrease in heat capacity upon folding.

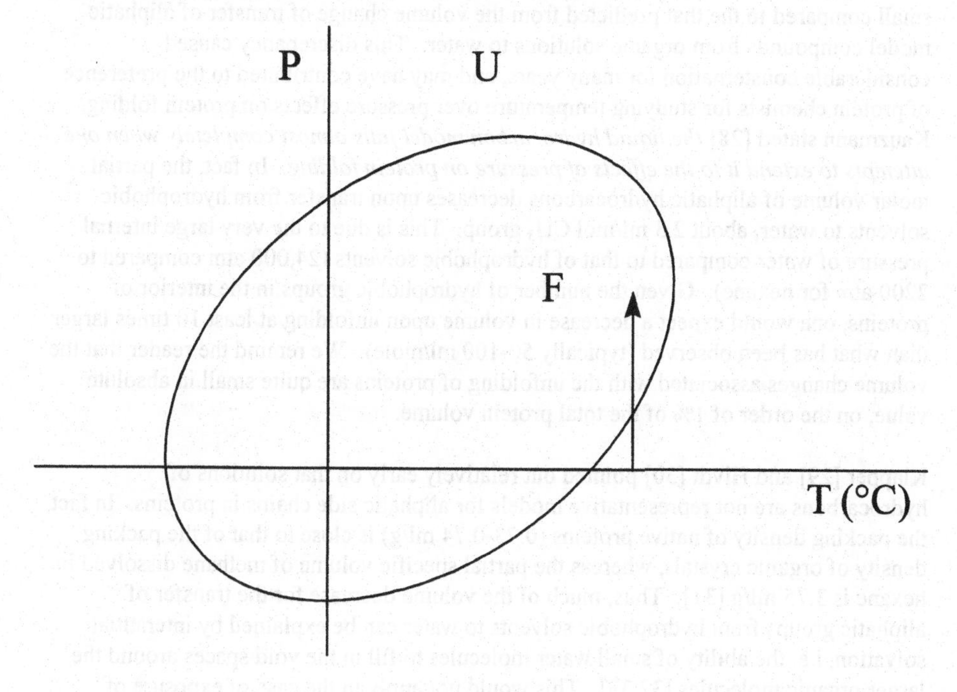

Figure 1. General temperature-pressure phase diagram for protein stability. U stands for the unfolded, and F for the folded state of the protein.

Of course, underlying pressure effects on protein structure is a difference in system (protein+solvent) volume between protein conformations. Generally, three factors are thought to contribute to the value of the volume change upon unfolding, ΔV_u:

- electrostriction of charged and polar groups that become exposed to solvent upon unfolding,
- the volume effects of transfer of hydrophobic groups from the protein interior to water,
- and the elimination of packing defects.

There is general agreement that the electrostriction and elimination of voids should provide a negative contribution to ΔV_u, but their magnitudes are not well-defined. And both the magnitude and the sign of the contribution of hydrophobic hydration have been subjects of debate [26]. Mozhaev and coworkers [27] point out that the pressure-induced unfolding of proteins over the past four decades has been interpreted in terms of the dominating view at the time of the forces contributing to protein stability. Thus, early interpretations imputed the volume change of unfolding to a decrease in the specific volume of the hydrophobic residues upon their transfer from the protein interior to water[8]. However, the magnitude of the volume change upon unfolding was quite small compared to the that predicted from the volume change of transfer of aliphatic model compounds from organic solutions to water. This discrepancy caused considerable consternation for many years, and may have contributed to the preference of protein chemists for studying temperature over pressure effects on protein folding. Kauzmann stated [28] *the liquid hydrocarbon model fails almost completely when one attempts to extend it to the effects of pressure on protein folding*. In fact, the partial molar volume of aliphatic hydrocarbons decreases upon transfer from hydrophobic solvents to water, about 2.3 ml/mol CH_3 group. This is due to the very large internal pressure of water compared to that of hydrophobic solvents (24,000 atm compared to 2200 atm for hexane). Given the number of hydrophobic groups in the interior of proteins, one would expect a decrease in volume upon unfolding at least 10 times larger than what has been observed (typically 50-100 ml/mole). We remind the reader that the volume changes associated with the unfolding of proteins are quite small in absolute value, on the order of 1% of the total protein volume.

Klapper [29] and Hivdt [30] pointed out relatively early on that solutions of hydrocarbons are not representative models for aliphatic side chains in proteins. In fact, the packing density of native proteins (0.73-0.74 ml/g) is close to that of the packing density of organic crystals, whereas the partial specific volume of methane dissolved in hexane is 3.75 ml/g [31]. Thus, much of the volume decrease for the transfer of aliphatic groups from hydrophobic solvents to water can be explained by interstitial solvation, i.e. the ability of small water molecules to fill in the void spaces around the larger organic molecules [32,33]. This would not apply in the case of exposure of tightly packed hydrophobic moieties of proteins. Hvidt [30] suggested that the volume change associated with exposure of hydrophobic residues of proteins actually proceeds with an increase in volume due to the increase in specific volume of the water molecules that form clathrate structures around the exposed hydrophobic groups [34]. In fact, the specific volume of water increases significantly when dissolved in organic solvents[35].

Although proteins are densely packed [36], the packing of proteins is constrained nonetheless by the covalent polypeptide backbone. This makes native protein structures relatively incompressible, about 5-10 times less compressible than water. It also results in the existence of small cavities and void volumes within the packed structure that cannot be compressed significantly without rotations about bonds. Rashin and co-workers [37] have estimated these cavities to represent up to 2% of the total protein volume, on the order of the observed volume changes for unfolding. Dilatometric studies on the helix-coil transition of polyglutamic acid [38] revealed a decrease in volume of about 1.0 ml/mol amino-acid moiety. These authors suggested that this was a *measure of the voids formed by the more ordered structure*. Weber and Drickamer [39] discussed the possible factors (electrostriction, hydrophobic hydration and elimination of voids) contributing to the decrease in system volume upon protein unfolding. Given the question of which effect is likely to dominate, they dismissed electrostriction since most charged residues are exposed already in native proteins, and agreed with Hvidt [30] on the inappropriateness of the liquid hydrocarbon model. They concluded thus, that increased hydration of the protein surface that accompanies the elimination of packing defects upon unfolding, rather than hydrophobic hydration or electrostriction, provided the basis for the volume decrease. They noted however that a quantitative answer to this question would require more data than was available at that time.

While pressure continued to be used as a tool to study protein structure and function by a number of groups, for several years after the appearance of the review articles of the early 1980's [39-41] the basis for the decrease in system observed for the unfolding of proteins was not revisited. Mozhaev and co-workers, in their recent review [27], again provide the list of possible contributing factors to the decrease in volume and note that, *pressure effects are especially pronounced in processes in which the hydration state of the system components changes significantly*. Chalikian and Breslauer [42] suggest that the small negative volume changes of unfolding results from simple compensation of hydration (<0) and thermal volume (>0).

Very recently, Hummer and coworkers [43] have presented an elegant theoretical explanation for the effects of pressure on proteins. Based on a simple model of two methane molecules in the *folded* (contact) state and in the *unfolded* (solvent separated) state, they used the information theory model [44] for hydrophobic interactions to calculate the effects of pressure on the Potential of Mean Force for methane association. In Figure 2 are shown the results of their calculations. It can be seen that pressure destabilizes the contact minimum with respect to the solvent separated minimum,

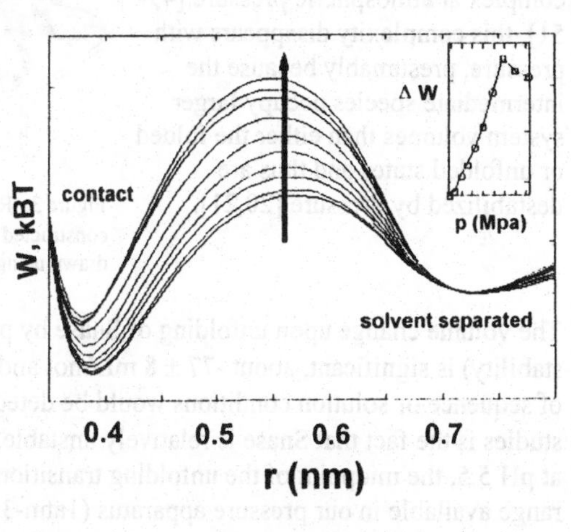

Figure 2. Pressure dependence of methane association [43]

thus pushing water between the two methane molecules. Essentially, the dominant factor acting to alter the PMF is the interstitial solvation (or excluded volume) discussed by Assarson and Eirich [32]. However, it was not apparent to us when we began our studies of pressure-induced unfolding of proteins in the early 1990's, that the dominant factor contributing to the decrease in system volume upon unfolding had been clearly identified. And while the recent theoretical considerations of Hummer *et al.*, [43] are quite convincing, in 1990 there was a clear lack of experimental results on the question. We reasoned at the time that a better quantification of the factors contributing to the volume change of unfolding would lead to a more in-depth understanding of the factors required for specific protein folding, much in the same manner as has the parameterization of the heat capacity change upon unfolding [45].

We therefore set out to test each of the possible contributions, electrostriction, hydrophobic hydration and void volume, to the overall volume change of unfolding. The protein system we chose to study was staphylococcal nuclease (Snase). Snase is a small (149 amino acids), single domain monomeric protein with no disulfide bridges. Its three dimensional structure has been determined and shows the protein to be composed of two subdomains, a β-barrel region and an α-helical domain (Figure 3) [46]. It has been shown by comparison of fluorescence, CD and calorimetric data that Snase unfolds reversibly in a two-state manner in equilibrium unfolding assays [47,48]. Moreover, although the kinetics of Snase folding are complex at atmospheric pressure [49-51], this complexity disappears with pressure, presumably because the intermediate species occupy larger system volumes than either the folded or unfolded states and thus are destabilized by pressure [20,21].

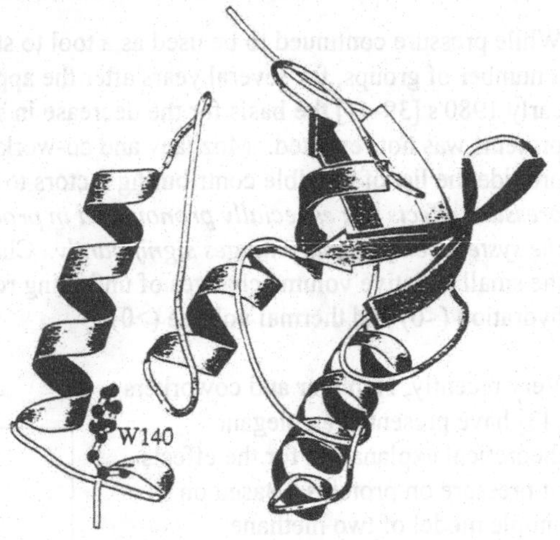

Figure 3. Ribbon diagram of the structure of Snase constructed from the coordinates determined in [46] and drawn using the program Molscript [75].

The volume change upon unfolding of Snase by pressure at 21 °C (near the maximal stability) is significant, about -77 ± 8 mL/mol and changes in its value due to alterations of sequence or solution conditions would be detectable. Particularly useful to our studies is the fact that Snase is relatively unstable, such that with a ΔV_u of -77 mL/mol, at pH 5.5, the midpoint of the unfolding transition is 1.5 kbar, well within the pressure range available in our pressure apparatus (1atm-3.0 kbar). Finally, Snase conveniently has a single tryptophan residue at position 140 in the α-helical domain, the fluorescence intensity of which is severely quenched (~50% at 360 nm) upon unfolding. Thus, monitoring this fluorescence signal change provides a simple, reproducible observable for detecting the unfolding transition. In Figure 4 is shown a typical high pressure equilibrium unfolding profile for Snase obtained by monitoring fluorescence intensity above 360 nm. The inset shows the spectral shift to the red that also accompanies unfolding, due to increased solvent relaxation around the exposed tryptophan residues in the pressure unfolded state.

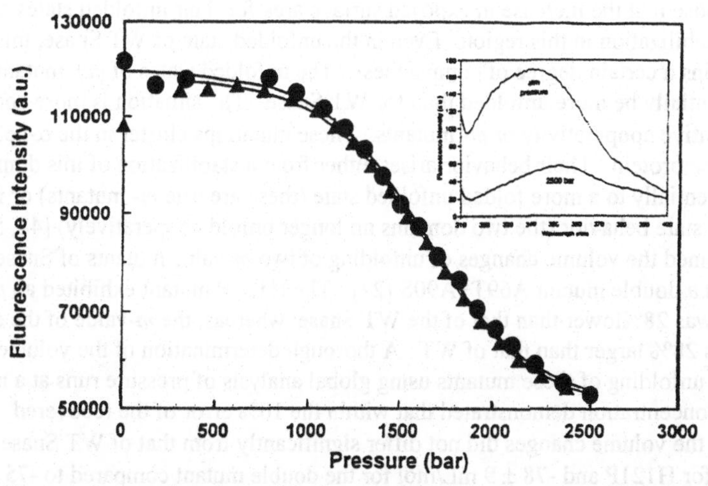

Figure 4. Pressure dependence of the fluorescence emission spectrum and intensity of Snase.

In order to test the possible contributions of electrostriction to the value of ΔV_u, profiles of the unfolding of Snase such as that shown in Figure 4 were obtained as a function of pH between pH 5.5 and pH 3.5 [19]. Further experiments were later carried out at pH 7.0 [21]. Fits of the data for the free energy and volume change of unfolding revealed that within the uncertainty of the recovered parameters, the volume change of unfolding of Snase is independent of pH over this range. Since the range from 7 to 3.5 includes the pKa values of histidine and glutamate and aspartate residues, the lack of pH effect on the volume change of unfolding provides evidence for the notion, proposed by Weber [39], that electrostriction does not play an important role in the mechanism pressure induced unfolding of this protein. We also noted that in this and subsequent studies, in which the midpoint of unfolding was adjusted between 500 and 2000 bar using chemical additives (denaturant, sugars) or pH, the volume change of unfolding was found to be constant, demonstrating that the any differences in compressibility between the folded and unfolded states are too small to influence the pressure dependence of unfolding.

We next wanted to probe the contributions of the exposure of hydrophobic surface area to the volume change of unfolding of Snase. For this we were fortunate to dispose of site-specific mutants of this protein for which the change in solvent accessible surface area between the folded and unfolded states differed from that of WT. This class of mutants, brought to light by Shortle and co-workers [52] and termed *m*-value mutants, exhibits either more or less pronounced cooperativity in denaturant unfolding profiles. In fact, the stability of many single domain proteins has been found to be a linear function of the concentration of denaturant [53-55]. The slope of this linear dependence is the cooperativity of the transition, and it has been found to be well-correlated with the amount of surface area (and hence the number of binding sites for denaturant) exposed to solvent upon unfolding [47, 54, 56, 57]. For Snase, the positive cooperativity or *m*+ mutations are all found in the β-barrel subdomain, leading Privalov and co-workers [48, 58] to propose that the increase in exposed surface area for their unfolded states arises from a destabilization in this region. Even in the unfolded state of WT Snase, this region retains a certain degree of compactness. The unfolded states of *m*+ mutants would essentially be more unfolded than for WT Snase. The situation is more complex for the negative cooperativity or *m*- mutants. These mutations cluster in the α-helical region of the protein. Their behavior arises either from a stabilization of this domain, leading essentially to a more folded unfolded state (these are true *m*- mutants) or from a loss of two state behavior (the two domains no longer unfold cooperatively) [48, 58]. We determined the volume changes of unfolding of two *m*-value mutants of Snase, H121P and a double mutant A69T+A90S [22]. The H121P mutant exhibited an *m*-value that was 28%lower than that of the WT Snase, whereas, the *m*-value of the double mutant was 28% larger than that of WT. A thorough determination of the volume change for unfolding of these mutants using global analysis of pressure runs at a number of xylose concentration demonstrated that within the 10% error of the recovered parameter, the volume changes did not differ significantly from that of WT Snase (-71 ± 3 mL/mol for H121P and -78 ± 9 mL/mol for the double mutant compared to -75 ± 5 for WT Snase). We later demonstrated that the lower *m*-value for the H121P mutant arises from a loss of 2-state behavior. However, the increased *m*-value for the double mutant resides in the difference in the amount of exposed surface area in the denatured state. Since we did not observe a large increase in the absolute value of the volume change for unfolding of this mutant, we concluded that it was not likely that hydrophobic hydration contributes significantly to the volume change of unfolding. Alternately, it could be that a positive contribution from hydrophobic hydration is essentially offset by a negative contribution from hydration of the polar groups. This could account for the discrepancy between the ΔV_u measured by hydrostatic as opposed to osmotic pressure. In either case, we conclude that the observed volume change of unfolding does not arise from exposure of buried amino acid residues to water.

Having eliminated compressibility, electrostriction and hydrophobic hydration as significant factors in determining the value of the volume change of unfolding of Snase, we next turned our attention to packing defects. In order to probe the role of internal voids in determining the value of the volume change of unfolding, pressure denaturation was carried out on three cavity mutants of Snase. All three mutations were made at position 66 which is a valine residue in the WT Snase. Two of the mutations were to smaller amino acid side chains (glycine, V66G, and alanine, V66A) while the third was mutated to a larger amino acid side chain (leucine, V66L). The crystal structure has been solved for the V66G mutant, bearing the largest cavity, and no crystallographically determined water molecules were found in the cavity (personal communication, Ed Lattman). Moreover, the main chain atoms in the WT Snase and V66G structures superimpose with a mean-square deviation of only 0.776 Å, and only minor readjustments of the sidechains around residue 66 are observed, indicating that the chain does not collapse into the cavity formed by substitution of V66 by G. If internal cavities make a significant contribution to the value of the volume change of unfolding then it would be expected that the V66G and V66A mutations, which introduce a cavity in the core of Snase, should result in a larger negative value for ΔV_u, with respect to WT Snase, while the V66L should decrease the absolute value of ΔV_u.

The value of ΔV_u for WT Snase is -75 ± 8 mL/mol [20, 22, 23]. Using the van der Waals volumes of the amino acid side chains determined by Richards [59] one can calculate crudely, the change in volume expected for the substitution of the valine at position 66 by glycine, alanine, and leucine. The predicted value of ΔV_u for the three mutants based on side chain van der Waals volumes would be -111, -100, and -66 mL/mol for V66G, V66A, and V66L, respectively. The actual experimental values for V66G and V66A are -112 +/- 8 and -104 +/- 6 mL/mol, respectively, while that for V66L was -94 +/- 12 mL/mol. The first two are in good agreement with the predicted values, while that of the V66L mutant is larger in absolute value than expected. A slightly larger negative ΔV_u for the V66L mutant as compared to the WT Snase might arise from local rearrangements in the structure which may actually create a new small cavity or small packing defects, as has been observed for lysozyme [60] and for unnatural amino acid substitutions at the cysteine in the V23C variant of Snase, exactly opposite V66 [61]. Accepting this caveat for the V66L mutant, it would appear that packing defects provide a significant contribution to the volume change of unfolding of this protein.

Our goal in these studies probing the fundamental basis for pressure denaturation of proteins was to define and if possible quantify the various contributing factors. We have found that difference in compressibility between the native and unfolded states is insignificant over the temperature range examined to date. Likewise, neither electrostriction nor hydrophobic hydration appears to play any significant role. The only modifications that altered the recovered value of the volume change of unfolding were those that modified the internal packing of the protein. Of course, Gregorio Weber deduced all of this from much less information nearly 15 years ago [39]. Our cumulative results on Snase simply serve to support this position. They also agree with the recent theoretical work of Hummer and co-workers [43] which implicated interstitial solvation (i.e., excluded volume) as the fundamental mechanism underlying pressure-induced unfolding of proteins.

The pressure denatured state

We and others have studied, for the most part, pressure effects on proteins for which the native state three dimensional structure has been determined by crystallographic methods or by NMR. A full interpretation of the volume change upon unfolding and of pressure effects on folding kinetics (discussed below) requires a reasonable description of the structural perturbations brought about by pressure. Quantitative studies of pressure effects on single chain proteins were first based on experiments in optical absorption spectroscopy [7, 8, 13]. These early studies indicated disruption of the tertiary contacts made by the aromatic residues of Rnase A and chymotrypsinogen, or the loss of the heme group in metmyoglobin, respectively, upon the application of pressure. Later, intrinsic tryptophan fluorescence measurements on lysozyme and chymotrypsinogen confirmed the disruption of tertiary structure by pressure in these two protein systems [11]. More recently, high pressure FTIR experiments on chymotrypsinogen [14] and Rnase A [16] have indicated that pressurization also disrupts the secondary structure of these proteins. On the other hand Goosens *et al.*[62] concluded that distinct high pressure FTIR spectra are indicative of retention of the secondary structure of BPTI under pressure. Unpublished data by Scarlata, Drickamer and Weber on the pressure dependence of the tyrosine emission of BPTI also indicate that the structure of this protein is relatively insensitive to pressure up to 12 kbar. One-dimensional NMR studies under pressure on lysozyme and staphylococcal nuclease [12, 19] have both indicated loss of native tertiary structure, whereas two-dimensional high pressure NMR on *arc* repressor revealed non-native intra-molecular tertiary interactions suggestive of a compact high pressure state for this protein [63]. This conclusion is consistent with the fact that at high pressure *arc* undergoes relatively fast rotational diffusion [64] and retains a high affinity for anilinonaphthalene sulfonate (ANS), a fluorescent dye that binds to compact denatured states of proteins. Thus, the particular properties of the pressure-denatured forms of proteins apparently depend upon the protein itself, as one might expect.

In order to ascertain to the greatest extent possible the nature of the conformation of pressure-denatured Snase, its pressure-induced unfolding was studied using small-angle X-ray scattering (SAXS) and Fourier-transform infrared (FT-IR) spectroscopy, which monitor changes in the compactness and secondary structural properties of the protein upon pressurization [65].

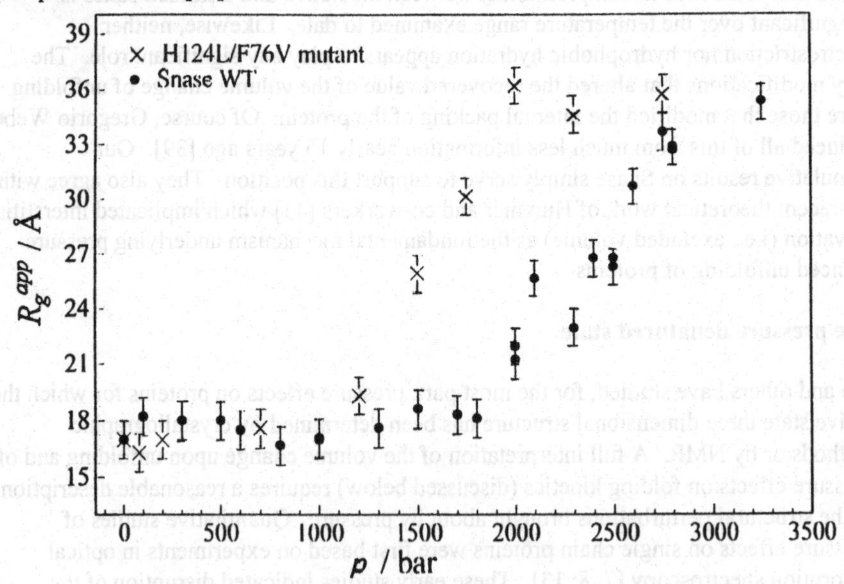

Figure 5. Results of the Guinier analysis of the SAXS profiles of Snase vs pressure [65].

Analysis of the high pressure SAXS data revealed that over a pressure range from atmospheric pressure to approximately 3 kbar the radius of gyration R_g of the protein increased from a value near 17 Å found for native Snase approximately two-fold to a value near 35 Å (Figure 5). Moreover, a large broadening of the pair-distance-distribution function was observed over this same range, indicating a transition from a globular to an ellipsoidal or dumbbell like structure (Figure 6). The radius of gyration of the pressure-denatured Snase is equivalent to that observed upon urea denaturation [66, 67].

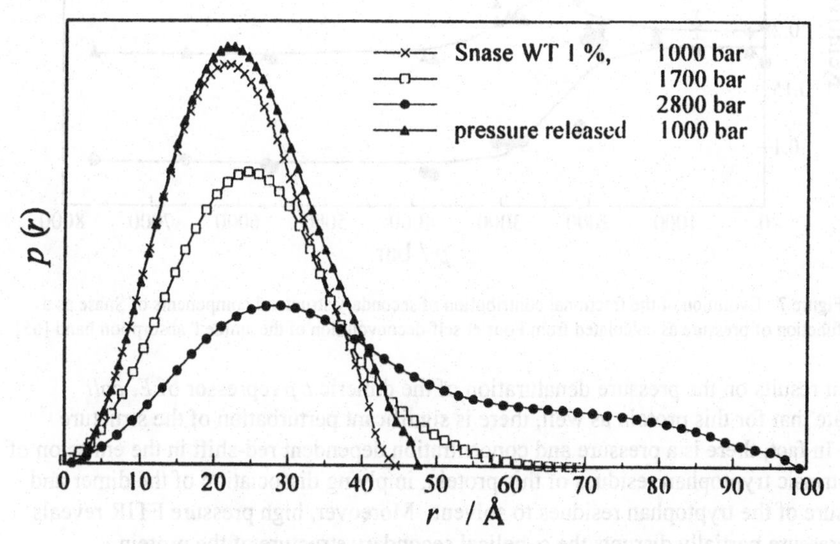

Figure 6. Pairwise atomic distance distribution for Snase as a function of pressure [65].

Deconvolution of the FTIR amide I' absorption band (Figure 7) reveals a pressure-induced denaturation process over the same pressure range as for fluorescence, NMR and SAXS that is evidenced by an increase in disordered and turn structures and a drastic decrease in the content of β-sheets and α-helices. The pressure-induced denatured state above 3 kbar retains nonetheless some degree of β-like secondary structure and the molecule cannot be described as a fully extended random coil. Temperature-induced denaturation involves a further unfolding of the protein molecule which is indicated by a larger R_g value of 45 Å (the theoretical value for a random coil polymer of this length) and significantly lower fractional intensities of IR-bands associated with secondary structure elements. Taken together with the red shift and quenching of the tryptophan fluorescence and the shifts in the histidine $^{\varepsilon}$H1 protons [19], we may conclude that the tertiary structure of Snase is significantly disrupted by pressure. The pressure-denatured state is swollen, beyond the level usually associated with molten globules, but does not assume a random coil configuration. Some contacts are retained, giving rise to the residual secondary structural signal in the FTIR experiments. These are likely due to fluctuating contacts in the β-barrel region.

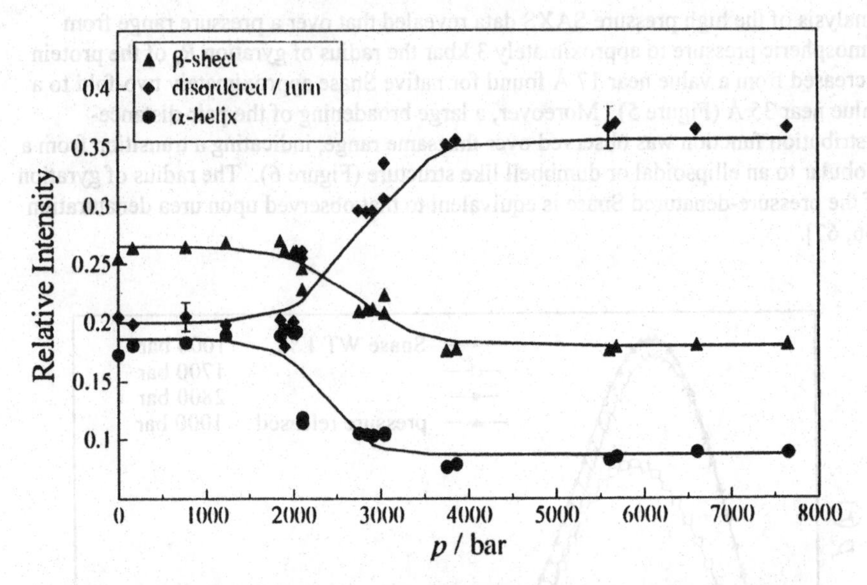

Figure 7. Evolution of the fractional contribution of secondary structural components of Snase as a function of pressure as calculated from Fourier self-deconvolution of the amide I' absorption band [65].

Recent results on the pressure denaturation of the dimeric *trp* repressor of *E. coli* indicate that for this protein as well, there is significant perturbation of the structure [68]. In fact, there is a pressure and concentration dependent red-shift in the emission of the intrinsic tryptophan residues of this protein, implying dissociation of the dimer and exposure of the tryptophan residues to solvent. Moreover, high pressure FTIR reveals that pressure partially disrupts the α-helical secondary structure of the protein. Conformational changes such as those observed for Snase and *trp* repressor are reminiscent of that described for pressure-denatured lysozyme [11] about which Weber & Drickamer [39] stated that the red shift in the tryptophan emission of this protein *points to considerable penetration of the structure by water*, a viewpoint recently supported by theoretical calculations [43].

Pressure effects on folding kinetics

While a reasonable number of studies pertaining to the equilibrium unfolding of proteins by pressure have appeared in the literature over the years and have for the most part been cited above, there has been very little done in the way of monitoring the effects of pressure on the kinetics of protein folding. An early study on the effects of pressure on the irreversible temperature-induced aggregation of [69] provided hints of these fascinating effects, and Zipp and Kauzmann [8] mention results of kinetics experiments on metmyoglobin without presenting the data. The only other data available in the literature is a study in the 1980's by Marden and coworkers [15] on the pressure denaturation of myoglobin. As we undertook to study the pressure denaturation of Snase, we noticed, as was the case for myoglobin and albumin, that as pressure was increased, the time necessary to achieve equilibrium following the application of pressure increased significantly. At 2 kbar, equilibration of the Snase fluorescence signal required 20 minutes, compared to less than 1 minute at atmospheric pressure.

The effect of pressure on the Gibbs free energy of unfolding necessarily arises from pressure effects on the rates of unfolding, refolding, or both. Following Eyring, a reaction rate k_p at a given pressure p can be expressed in terms of the rate at atmospheric pressure k' and the activation volume $\Delta V^{\ddagger}$ for the formation of the transition state [70]:

$$k_p = k' e^{(-p(\Delta V^{\ddagger})/RT)} \tag{1}$$

where R is the gas constant and T is the Kelvin temperature. Depending upon the relative volumes of the initial and transition states, increasing pressure can result either in an increase or a decrease in the reaction rate. For a two-state unfolding reaction such as that exhibited by Snase, the observed relaxation time τ at a given pressure is represented by the inverse of the sum of the two individual rate constants for the forward and backward reactions, unfolding and folding,

$$\tau = 1/(k_u + k_f) \tag{2}$$

and as such should be independent of the sign of the perturbation [71].

The pressure-jump fluorescence relaxation profiles for Snase (Figure 8) fit very well to single exponential decay kinetics and their analysis with this model yielded values for the relaxation time τ at each pressure [20]. Pressure-jump FTIR and SAXS experiments [65] were also single exponential functions. The pressure and time dependence of the loss of secondary structure and compactness were equivalent to that observed by fluorescence, confirming the two-state behavior for the transition (Figures 9 & 10). While Eyring theory may be too simplistic for application to such a complex reaction the results of the analysis are surprisingly straightforward.

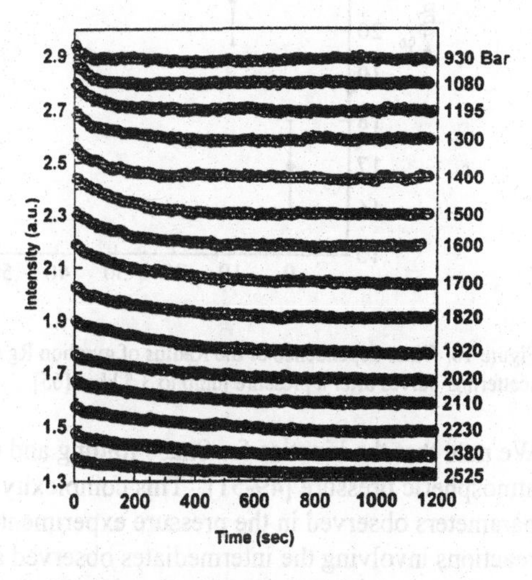

Figure 8. Pressure-jump fluorescence relaxation profiles for Snase as a function of pressure [20].

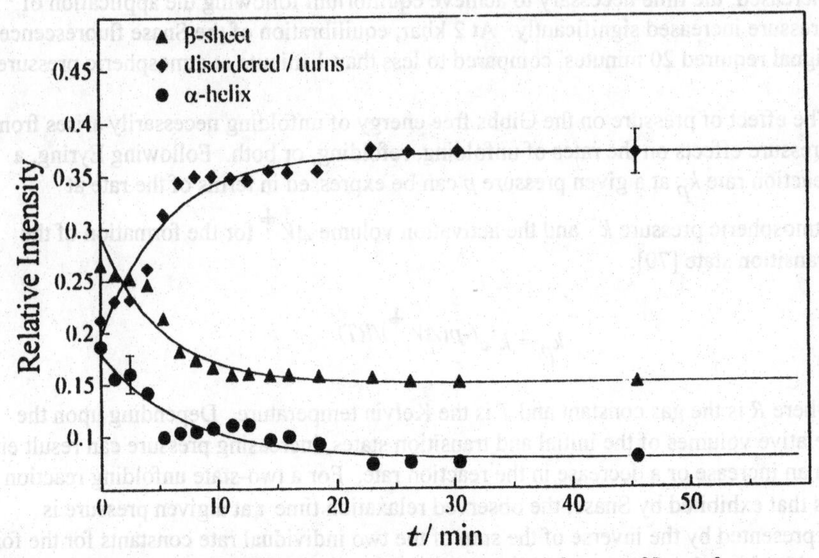

Figure 9. Time dependence of the evolution of secondary structural elements of Snase after a pressure jump to 4 kbar [65].

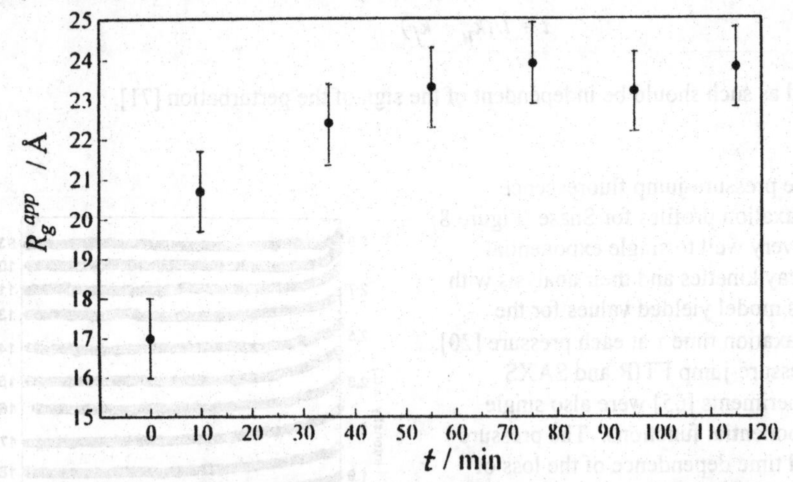

Figure 10. Time dependence of the Radius of gyration Rg of Snase obtained from Guinier analysis of the scattering curves after a pressure jump to 3.5 kbar [65].

We note that the kinetics for Snase folding and unfolding are non-two state at atmospheric pressure [49-51]. This complexity is not apparent in any of the order parameters observed in the pressure experiments. We can only conclude that the reactions involving the intermediates observed at atmospheric pressure speed up at high pressures, such that they are no longer partially rate-limiting, allowing observation of the main folding transition.

The plots of ln τ vs pressure (Figure 11) could be analyzed in terms of the activation volumes ($\Delta V^{\ddagger}_{f}$ and $\Delta V^{\ddagger}_{u}$) and the rates at atmospheric pressure (k^{o}_{f} and k^{o}_{u}) for the folding and unfolding reactions:

$$ln\tau = ln\,[1/((k^{o}_{f}exp(-p\Delta V^{\ddagger}_{f}/RT) + (k^{o}_{u}exp(-p\Delta V^{\ddagger}_{u}/RT))] \quad (3).$$

We found that the activation volume for folding was large and positive (+92 mL/mol) and that for unfolding was small and positive (+20 mL/mol). The difference between the two (72 ml/mol) was found to be within error if the value of the equilibrium volume change for folding (+75 mL/mol), as would be expected for a two-state reaction. Thus, the volume of the protein solvent system in the transition state is significantly larger than in the unfolded state and somewhat larger than in the folded state.

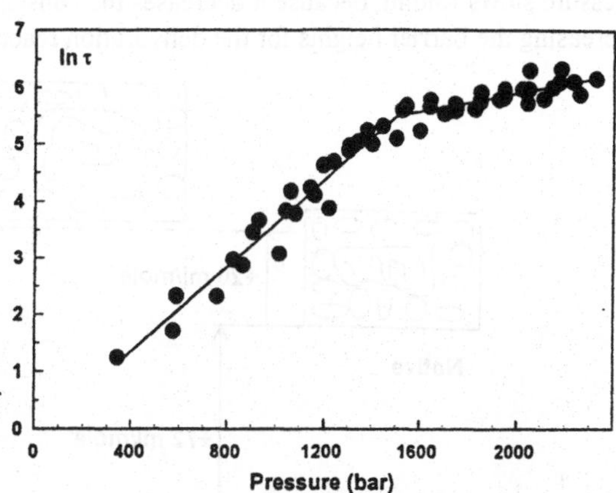

Figure 11. Pressure dependence of the natural logarithm of the relaxation time (fluorescence detected) [20].

The fact that the activation volume for folding $\Delta V^{\ddagger}_{f}$ is greater than that of unfolding $\Delta V^{\ddagger}_{u}$, means that high pressure unfolds proteins because the rate of folding is slowed significantly more by pressure than that of unfolding. Recent studies (Panick *et al.*, submitted) on the temperature dependence of the pressure-jump kinetics of the folding/unfolding of Snase demonstrate that the kinetic basis for the heat and cold denaturation of this protein, like others [72], lies in the non-Arrhenius behavior of the folding rate constant.

What image of the structure of the transition state do these volume differences provide? If we accept our previous argument pertaining to the basis for the volume changes of unfolding, then pressure-induced unfolding arises from the existence of solvent-excluded voids in the folded state. It is reasonable therefore to interpret the large increase in volume between the unfolded and transition states as resulting from collapse of the unfolded polypeptide chain to a loosely packed globule from which solvent is largely excluded and which contains more void than the unfolded chain (Figure 12). In the case of Snase, the small decrease in volume observed between the transition state and the folded state could arise from more efficient packing in the final folded structure, although recent analyses indicate that this change may not be significant [73]. Referring to the theoretical model of methane association in Figure 2 [43], it can be seen that pressure has similar effects on the activation barrier for association, which represents a solvent exclusion barrier. The authors pointed out that their activation volumes scaled well to those observed for Snase, given the number of hydrophobic contacts likely to be present in the transition state ensemble. Linear free energy relationships between the folding and unfolding rates for Snase and the stability of the protein [74] controlled by increasing osmolyte concentration [23] confirmed that over half of the hydrophobic contacts are made in the transition state. In the language of the energy landscape theory, pressure slows folding because it decreases the constant for diffusion down the funnel by increasing the barrier heights for the dehydration reactions associated with folding.

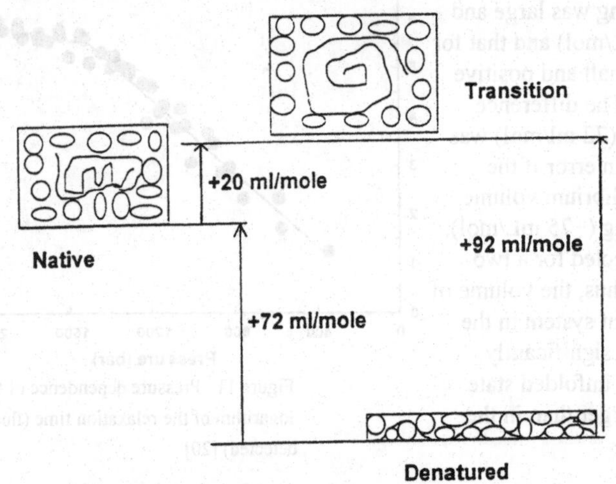

Figure 12. Volume profile for Snase folding/unfolding reactions [20].

In the case of *trp* repressor, a dimeric protein, a large increase in volume is observed as well for the transition from the unfolded state to the transition state, whereas a significant decrease in volume is observed between the folded and transition state. In this case the transition state lies between the native and unfolded states in terms of system volume. Our results on both proteins are consistent with a model in which collapse with concomitant desolvation constitutes the rate limiting step in folding. The pressure-induced changes in secondary structure for Snase exhibit the same relaxation behavior as do the changes in the collapse and the tryptophan fluorescence, implying that all of these order parameters are dependent upon the same rate limiting step.

Returning to the temperature-pressure phase diagram in Figure 1, we note that the studies presented here can provide reasonable explanations for the decrease in volume upon unfolding observed over most of the temperature range, but not for the stabilization of proteins by pressure at high temperature. Recently we have carried out studies on the temperature dependence of the equilibrium and kinetic folding/unfolding reactions for Snase and *trp* repressor [68]. In these studies we have found for both protein systems that the volume change for unfolding decreases significantly in absolute value with increasing temperature. This result is consistent with those of Zipp and Kauzman [8], although they had fewer data points and complications due to pressure-induced changes in pH. Kauzmann [8] remarked that the change in the volume change of unfolding with temperature simply represents the difference in thermal expansivity between the folded and unfolded states of the proteins. For both Snase and *trp* repressor this difference is on the order of 1 mL/mol deg, with the unfolded state exhibiting a larger thermal expansivity. This can be explained by the larger thermal volume of the residues in the unfolded state which exhibits a larger heat capacity and fewer constraints upon its motions. A diagram of temperature dependence of the partial molar volumes of the folded and unfolded states can be found in Figure 13. As the temperature increases the partial molar volume of the unfolded state will increase more than that of the folded state and as a consequence, the difference between the two partial molar volumes (i.e. the volume change upon unfolding) will diminish in absolute value, eventually reaching zero and then changing sign at high temperature. Thus the larger thermal volume for the unfolded state eventually compensates and then exceeds the packing defects in the folded state. At high temperature the re-entry into the unfolded state at high pressure (Figure 1), can be explained by the larger change in compressibility at high temperature.

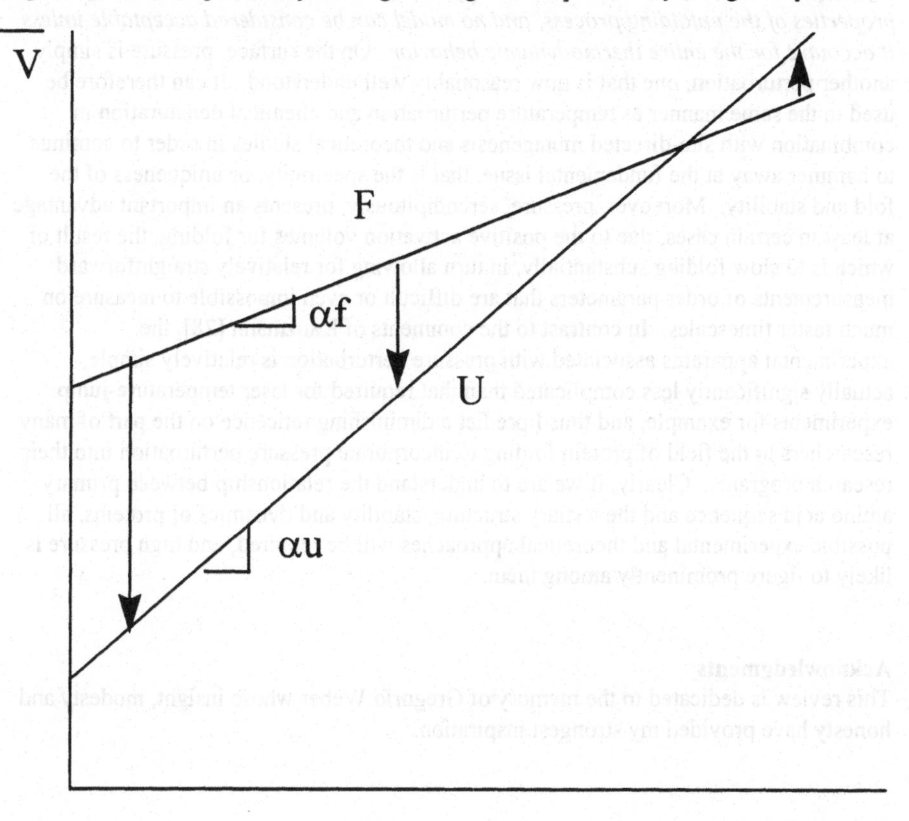

Figure 13. Schematic representation of the temperature dependence of the specific volumes of the folded and unfolded states of proteins, and the consequences of this dependence on the value and sign of the volume change upon unfolding.

Concluding Remarks

It would appear from the recent experimental and theoretical studies discussed above that, not surprisingly, Gregorio Weber's view of the basis for pressure's effects on protein structure was correct. The combination of packing defects and very low compressibility of the native structure lead to an unfolding of the protein by pressure over most of the available temperature range since this unfolded state is more efficiently solvated and thus occupies the smallest system volume. This understanding of the fundamental effect of pressure has allowed us to identify desolvation and collapse as being associated with the rate-limiting step in the folding of at least two proteins. Current studies on several more small reversibly folding proteins will help to verify the generality of this observation. What further consequences will the understanding and application of pressure effects have on the field of protein folding? Kauzmann [28] rightly pointed out that *volume and enthalpy changes are equally fundamental properties of the unfolding process, and no model can be considered acceptable unless it accounts for the entire thermodynamic behavior*. On the surface, pressure is simply another perturbation, one that is now reasonably well understood. It can therefore be used in the same manner as temperature perturbation and chemical denaturation in combination with site-directed mutagenesis and theoretical studies in order to continue to hammer away at the fundamental issue, that is the specificity, or uniqueness of the fold and stability. Moreover, pressure, serendipitously, presents an important advantage at least in certain cases, due to the positive activation volumes for folding, the result of which is to slow folding substantially, in turn allowing for relatively straightforward measurements of order parameters that are difficult or even impossible to measure on much faster timescales. In contrast to the comments of Kauzmann [28], the experimental apparatus associated with pressure perturbation is relatively simple, actually significantly less complicated than that required for laser temperature-jump experiments for example, and thus I predict a diminishing reticence on the part of many researchers in the field of protein folding to incorporate pressure perturbation into their research programs. Clearly, if we are to understand the relationship between primary amino acid sequence and the tertiary structure, stability and dynamics of proteins, all possible experimental and theoretical approaches will be required, and high pressure is likely to figure prominently among them.

Acknowledgments

This review is dedicated to the memory of Gregorio Weber whose insight, modesty and honesty have provided my strongest inspiration.

REFERENCES

1. Kauzmann, W (1959) Some factors in the interpretation of protein denaturation *in Advances in Protein Chemistry in* J C B Anfinsen, K Bailey, M L Anson, and J T Edsall, eds. Academic Press, New York and London, 1-66.

2. Kim, P. S. & Baldwin, R. L. (1990) Intermediates in the folding reactions of small proteins, *Annu. Rev. Biochem.* **59**, 631-660.

3. Onuchic, J. N., Luthey-Schulten, Z. & Wolynes, P. G. (1997) Theory of protein folding: the energy landscape perspective, *Annu. Rev. Phys. Chem.* **48**, 545-600.

4. Onuchic, J. N., Wolynes, P. G., Luthey-Schulten, Z. & Socci, N. D. (1995) Toward an outline of the topography of a realistic protein folding funnel *Proc. Natl. Acad. Sci. USA* **92**, 3626-3630.

5. Socci, N. D., Onuchic, J. N. & Wolynes, P. G. (1996) Diffusive dynamics for the reaction coordinate for protein folding funnels, *J. Chem. Phys.* **104**, 5860-5868.

6. Privalov, P L & Gill, S J (1988) Stability of protein structure and hydrophobic interaction, *Adv. Prot. Chem.* **39**, 191-234.

7. Brandts, J F, Oliveira, R J & Westort, C (1970) Thermodynamics of protein denaturation: Effects of pressure on the denaturation of ribonuclease A, *Biochemistry*, **9**, 1038-1047.

8. Zipp, A & Kauzmann, W (1973) Pressure denaturation of metmyoglobin, *Biochemistry*, **12**, 4217-4228.

9. Katz, S, Crissman, J K Jr. & Beall, J A (1973) Structure volume relationships of proteins: Dilatometric studies of the structural transitions engendered in serum albumin and myoglobin as a consequence of acid-base reaction in water and in denaturing media, *J. Biol. Chem.* **248**, 4840-4845.

10. Suzuki, K, Miyosawa, Y & Suzuki, C (1963) Protein denaturation by high pressure. Measurements of turbidity of isoelectric ovalbumin and horse serum albumin under high pressure, *Arch. Biochem. Biophys.*, **101**, 225-228.

11. Li, T M, Hook, T W, Drickamer, H G & Weber, G (1976) Plurality of pressure denatured forms in chymotrypsinogen and lysozyme, *Biochemistry*, **15**, 5572-5581.

12. Samarasinghe, S D, Campbell, D M, Jonas, A & Jonas, J (1992)High resolution NMR study of the pressure-induced unfolding of lysozyme, *Biochemistry*, **31**, 7773-7778.

13. Hawley, S A (1971) Reversible temperature-pressure denaturation of chymotrypsinogen, *Biochemistry*, **10**, 2436-2442.

14. Wong, P T T & Heremans, K (1988) Pressure effects on protein secondary structure and hydrogen deuterium exchange in chymotrypsinogen: A Fourier transform infrared spectroscopy study, *Biochim. Biophys. Acta.*, **965**, 1-9.

15. Marden, M C, Hui Bon Hoa, G & Marden-Stetzkowski, F (1986) Heme protein fluorescence vs. pressure *Biophys. J.* **49**, 619-627.

16. Takeda, N, Kato, M & Taniguchi, Y (1995) Pressure and thermally induced reversible changes in the secondary structure of ribonuclease A, *Biochemistry*, **34**, 5980-5987.

17. Zhang, J., Peng, X., Jonas, A. & Jonas, J. (1995) NMR study of the cold, heat and pressure unfolding of ribonuclease A, *Biochemistry* **34**, 8631-8641.

18. Nash D., Lee B.S., & Jonas J. (1996) Hydrogen exchange kinetics in the cold denatured state of ribonuclease A, *Biochim. Biophys. Acta*, **1297**, 40-48.

19. Royer, C A, Hinck, A P, Loh, S N, Prehoda, K E, Peng, X, Jonas, J & Markley, J L (1993) Effects of amino acid substitutions on the pressure denaturation of staphylococcal nuclease as monitored by fluorescence and nuclear magnetic resonance spectroscopy, *Biochemistry*, **32**, 5222-5232.

20. Vidugiris, G J A, Markley, J L & Royer, C A (1995) Evidence for a molten globule-like transition state in protein folding, *Biochemistry*, **34**, 4909-4912.

21. Vidugiris, G J A, Truckses, D M, Markley, J L & Royer, C A (1996) High pressure denaturation of staphylococcal nuclease proline to glycine substitution mutants, *Biochemistry*, **35**, 3857-3864.

22. Frye, K J, Perman, C S & Royer, C A (1996) Testing the correlation between ΔA and ΔV in protein unfolding using m-value mutants of staphylococcal nuclease, *Biochemistry*, **35**, 10234-10239.

23. Frye, K J & Royer, C A (1997) The kinetic basis for the stabilization of staphylococcal nuclease by xylose, *Prot. Sci.*, **6**, 789-793.

24. Eftink, M R, Ghiron, C A, Kautz, R A, & Fox, R O (1991) Fluorescence studies with staphylococcal nuclease and its site directed mutant, *Biochemistry*, **30**, 1193-1199.

25. Eftink, M R &Ramsay, G D (1996) Temperature and pressure induced unfolding of a mutant of staphylococcal nuclease A, *In* High-Pressure Effects in Molecular Biophysics and Enzymology. J. L. Markley, D. B. Northrop, and C. A. Royer, editors. Oxford University Press, New York. 62-73.

26. Dill, K A (1990)Dominant forces in protein folding, *Biochemistry* **29**, 7133-7155.

27. Mozhaev, V V, Heremans, K, Frank, J, Masson, P & Balny, C (1996) High pressure effects on protein structure and function, *Proteins: Struct. Funct. Genet.*, **24**, 81-91.

28. Kauzmann, W. (1986) Protein stabilization: Thermodynamics of unfolding, *Nature* **325**, 763-764.

29. Klapper, M. H. (1971) On the nature of the protein interior, *Biochim. Biophys. Acta* **229**, 557-566.

30. Hvidt, A. (1975) A discussion of pressure-volume effects in aqueous protein solutions, *J. theor. Biol.* **50**, 245-252.

31. Masterton, W L (1954) Partial molar volumes of hydrocarbons in solution, *J. Chem. Phys.* **22**, 1830-1833.

32. Assarsson, P & Eirich, F R (1968) Properties of amides in aqueous solution. I. A. Viscosity and density changes of amide-water systems. B. An analysis of volume deficiencies of mixtures based on molecular size differences (Mixing of hard spheres), *J. Phys. Chem.* **72**, 2710-2719.

33. Prehoda, K E, & Markley, J L (1996) Use of partial molar volumes of model compounds in the interpretation of high pressure effects on proteins, *In* High-Pressure Effects in Molecular Biophysics and Enzymology. J L Markley, D B Northrop, and C A. Royer, eds. Oxford University Press, New York. 33-43.

34. Frank, H. S. & Evans, M. W. (1945) Free volume and entropy in condensed systems: III Entropy in binary liquid mixtures, partial molal entropy in dilute solutions, structure and thermodynamics of aqueous electrolytes, *J. Phys. Chem.* **13**, 507-532.

35. Masterton, W L & Seiler, H (1968) Apparent and partial molar volumes of water in organic solvents, *J. Phys. Chem.* **72**, 4257-4262.

36. Richards, F M (1977) Areas, volumes, packing and protein structure, *Ann. Rev. Biophys. Bioeng.* **6**, 151-176.

37. Rashin, A A, Iofin, M & Honig, B (1986) Internal cavities and buried waters in globular proteins, *Biochemistry* 25: 3619-3625.

38. Nogochi, H & Yang, JT (1963) Dilatometric and refractometric studies of the helix-coil transition of poly-glutamic acid in aqueous solutions, *Biopolymers* **1**, 359-370.

39. Weber, G & Drickamer, H G (1983) The effect of high pressure on proteins and other biomolecules, *Quart. Rev. Biophys.*, **16**, 89-112.

40. Jaenicke, R (1981) Enzymes under extreme conditions, *Ann. Rev. Biophys. Bioeng.* **10**, 1-67.

41. Heremans, K (1982) High pressure effects on proteins and other biomolecules, *Annu. Rev Biophy. Bioeng.* **11**, 1-21.

42. Chalikian, T V and Breslauer K J (1996) On volume changes accompanying conformational transitions of biopolymers, *Biopolymers* **39**, 619-626.

43. Hummer, G., Garde, S., Garcia, A.E. Paulaitis, M.E. & Pratt, L. R. (1998) The pressure dependence of hydrophobic interactions is consistent with the observed pressure denaturation of proteins, *Proc. Natl. Acad. Sci. USA* **95**, 1552-1555.

44. Hummer, G., Garde, S., Garcia, A. E., Pohorille, A. & Pratt, L. R. (1996) An information theory model of hydropobic interactions, *Proc. Natl. Acad. Sci. USA* **93**, 8951-8955.

45. Murphy, K P & Freire, E. (1992) Thermodynamics of structural stability and cooperative folding behavior in proteins, *Adv. Prot. Chem.* **43**, 313-361.

46. Hynes, T R & Fox, R O (1991) The crystal structure of staphylococcal nuclease refines at 1.7 Å resolution, *Proteins: Struct. Funct. Genet.*, **10**, 92-105.

47. Shortle, D & Meeker, A K (1986) Mutant forms of staphylococcal nuclease with altered patterns of guanidine hydrochloride and urea denaturation, *Proteins: Struct. Funct. Genet.*, **1**, 81-89.

48. Carra, J H & Privalov, P L (1995)Energetics of denaturation and m-values of staphylococcal nuclease mutants, *Biochemistry*, **34**, 2034-2041.

49. Chen, H M, You, J L, Markin, V S & Tsong, T Y (1991)Kinetic analysis of the acid and alkaline unfolded states of staphylococcal nuclease, *J. Mol. Biol.* **220**, 771-778.

50. Chen, H M, Markin, V S & Tsong, T Y (1992)pH-induced folding/unfolding of staphylococcal nuclease: Determination of kinetic parameters by the sequential jump method, *Biochemistry*, **31**, 1483-1491.

51. Su, Z D, Arooz, M T, Chen, H M, Gross, C J & Tsong, T Y (1996) Least activation path for protein folding: investigation of staphylococcal nuclease folding by stopped-flow circular dichroism, *Proc. Natl. Acad. Sci. USA, 93*, 2539-2544.

52. Shortle, D, Meeker, A K & Gerring, S L (1989) Effects of denaturant at low concentrations on the reversible denaturation of staphylococcal nuclease, *Arch. Biochem. Biophy.* **2721**, 103-113.

53. Tanford, C. (1970) Protein denaturation, *Adv. Protein. Chem.* **24**, 1-95.

54. Schellman, J. A. (1978) Solvent denaturation, *Biopolymers* **17**, 1305-1322.

55. Pace, C. N. (1986)Determination and analysis of urea and guanidine hydrochloride denaturation curves, *Meth. Enzymol.* **131**, 266-280.

56. Shortle, D, Meeker, A K & Freire, E. (1988) Stability mutants of staphylococcal nuclease: large compensating enthalpy-entropy changes for the reversible denaturation reaction, *Biochemistry* **27**, 4761-4768.

57. Myers, J. K., Pace, C. N. & Scholtz, J. M. (1995) Denaturant m-values and heat capacity changes: Relation of changes in solvent accessible surface areas of protein folding, *Protein Sci.* **4**, 2138-2148.

58. Carra, J. H., Anderson, E. A. & Privalov, P. L. (1994) Three-state thermodynamic analysis of the denaturation of staphylococcal nuclease mutants, *Biochemistry* **33**, 10842-10850.

59. Richards, F M (1974) The interpretation of protein structures: Total volume, group volume and packing densities, *J. Mol. Biol.* **82**, 1-14.

60. Eriksson, A E, Baase, W A, Zhang, X .J, Heinz, D W, Blaber, M, Baldwin, E & Matthews, B W (1992) The response of proteins structure to cavity creating mutations and its relationship to the hydrophobic effect, *Science* **255**, 178-183.

61. Wynn, R, Harkins, P C, Richards, F M & Fox, R O (1997) Mobile unnatural amino acid side chains in the core of staphylococcal nuclease, *Prot. Sci.* **6**, 1621-1626.

62. Goosens, K., Smeller, L., Frank, J. & Heremans, K. (1996) Pressure-tuning the conformation of bovine pancreatic trypsin inhibitor studied by Fourier transform infrared spectroscopy, *Eur. J. Biochem.*, **236**, 254-262.

63. Peng, X., Jonas, J. & Silva, J. L. (1994) High pressure NMR study of the dissociation of *arc* repressor, *Biochemistry*, **33**, 8323-8329.

64. Silva, J. L., Silveira, C. F., Correia Junior, A. & Pontes, L. (1992) Dissociation of a native dimer of a molten globule monomer: Effects of pressure and dilution on the dissociation equilibrium of *arc* repressor, *J. Mol. Biol.*, **223**, 545-555.

65. Panick, G, Malessa, R, Winter, R, Rapp, G, Frye, K J & Royer, C A (1998) Structural characterization of the pressure denatured state and unfolding/refolding kinetics of staphylococcal nuclease by synchrotron small angle X-ray diffraction and Fourier Transform infrared spectroscopy, *J. Mol. Biol.*, 389-402.

66. Flanagan, J. M., Kataoka, M., Shortle, D. & Engelman, D. M. (1992) Truncated staphylococcal nuclease in compact but disordered, *Proc. Natl. Acad. Sci.*, **89**, 748-752.

67. Kataoka, M., Flanagan, J. M., Tokunaga, F. & Engelman, D. M. (1994) Use of X-ray solution scattering for a protein folding study, In: *Synchrotron Radiation in the Biosciences*. pp. 187-194, Chance, B., Deisenhofer, J., Ebashi, S., Goodhead, D. T., Helliwell, J. R., Huxley, H. E., Iizuka, T., Kirz, J., Mitsui, T., Rubenstein, E., Sakabe, N., Sasaki, T., Schmahl, G., Stuhrmann, H. B., Wüthrich, K. & Zaccai, G. (Eds.), Calendon Press, Oxford.

68. Desai, G., Panick, G., Winter, R. & Royer, C. A. (1998) Pressure denaturation of *trp* repressor, (submitted for publication).

69. Lumry, R. & Biltonen, R. (1969) in *Structure and Stability of Biological Macromolecules*, G. D. Fasman and S. N. Timasheff, Eds. (Marcel, Dekker, Inc., New York), vol. 2,chap. 2.

70. Gladstone, S., Laidler, K. J. & Eyring, H. (1941) in *The Theory of Rate Processes*, (McGraw-Hill Book Co., New York.

71. Eigen, M. & de Maeyer, L. in *Techniques in Organic Chemistry*, (1963) A. Weissberger, Ed., (Wiley, New York), pp. 895-1054.

72. Scalley, M. L. & Baker. D. (1997) Protein folding kinetics exhibit an Arrhenius temperature behavior when corrected for the temperature dependence of protein stability, *Proc. Natl. Acad. Sci USA*. **94**. 10636-10640.

73. Panick. G., Vidugiris. G. J. A., Winter. R. & Royer. C. A. (1998) Exploring the temperature-pressure phase diagram of staphylococcal nuclease (submitted for publication).

74. Nymeyer, H. Frye. K. J.. Royer. C. A., Garcia. A. E. Hummer. G. & Onuchic. J. N. Pressure probes the protein folding landscape (manuscript in preparation).

75. Kraulis. P J (1991) Molscript: A program to produce detailed and schematic plots of protein structure. *J. appl. Cryst.* **24**. 946-950.

HYDROSTATIC PRESSURE AS A TOOL TO STUDY VIRUS ASSEMBLY: PRESSURE-INACTIVATION OF VIRUSES BY FORMATION OF FUSION INTERMEDIATE STATES

A. C. OLIVEIRA, A. P. VALENTE, F. C. L. ALMEIDA, S. M. B. LIMA, D. ISHIMARU, R. B. GONÇALVES, D. PEABODY*, D. FOGUEL AND J. L. SILVA

*Departamento de Bioquímica Médica, Instituto de Ciências Biomédicas, Centro Nacional de Ressonância Magnética Nuclear de Macromoléculas, Universidade Federal do Rio de Janeiro, 21941-590 Rio de Janeiro, RJ, Brazil. *Dept Cell Biology, University of New Mexico, NM 87131-522*

ABSTRACT. The contribution of protein folding and protein-nucleic acid interactions to virus assembly has been measured in several bacterial, plant and animal viruses, using hydrostatic pressure as thermodynamic variable. By comparing the pressure stability among native wild-type viruses, single-amino acid mutants or empty particles, we have gained new insights about virus assembly and disassembly. We find that the isolated capsid proteins and the assembly intermediates are not fully folded, and that association of 60 or more subunits into an icosahedral particle is coupled to progressive folding of the coat protein and also to changes in interactions with the nucleic acid. Using pressure, we have detected the presence of a ribonucleoprotein intermediate, where the coat protein is partially unfolded but bound to RNA. These intermediates are potential targets for antiviral compounds. Pressure studies on viruses have direct biotechnological applications. The ability of pressure to inactivate viruses has been evaluated with a view toward the applications of vaccine development and virus sterilization. We demonstrate that pressure causes virus inactivation while preserving the immunogenic properties. There are substantial evidence that a high pressure cycle traps a virus in the "fusion intermediate state", not infectious but highly immunogenic. Pressure inactivation has been successful with viruses that cause disease in animals, especially foot-and-mouth disease virus (FMDV) and bovine rotavirus and humans, such as rhinoviruses, adenoviruses, alphaviruses, influenza and retroviruses.

1. Introduction

The virus particle is composed of either a membrane enveloped or noneneveloped protein shell and nucleic acid [1]. Virions have evolved to move their genome between cells of a susceptible host and between hosts. Evolution has routed the protein shell to adopt multiple functions: shielding of the nucleic acid, participation in chemical reactions for particle maturation and ability to penetrate into the cell and undergo

R. Winter and J. Jonas (eds.), High Pressure Molecular Science, 497–513.

disassembly. The coat proteins are usually arranged in a shell with icosahedral shape. To pack an infectious genome, integral multiples of 60 subunits are required to form the shell, resulting in nonidentical contacts between subunits. The icosahedral capsid is composed of 60T or 60P quasi-equivalent units, where T or P denotes the number of identical or different polypeptide sequences, respectively, that occupy an asymmetric unit of the icosahedron [1, 2] (Fig. 1).

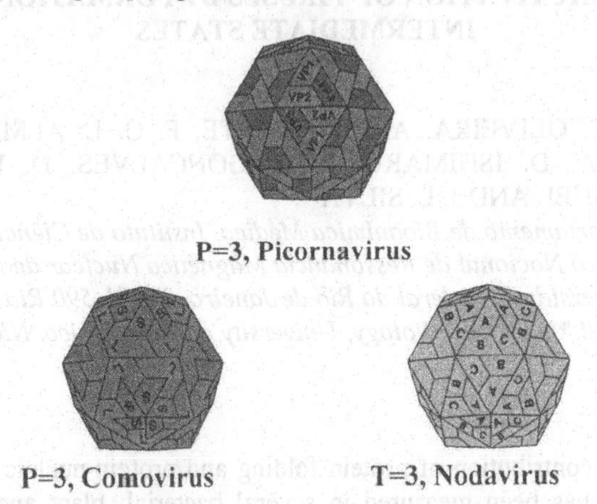

Figure 1. Schematic representation of T=3 / P=3 icosahedral virus particles. The particles are arranged on a T=3 icosahedral lattice (nodaviruses), or between three subunit type (picornaviruses and comoviruses) arranged on a pseudo T= 3 (P= 3) surface lattice. A portion of the icosahedral surface of a picornavirus is shown in detail (upper left). Each triangle represents an asymmetric unit that contains three protein subunits. The diagram at right shows how VP1 subunits are clustered around a fivefold axes, forming pentamers, whereas VP2 and VP3 alternate around a threefold axes, forming hexamers. The capsid proteins of picornaviruses, comoviruses and nodaviruses have tertiary folds similar to canonical viral β-sandwiches [1, 2].

Hydrostatic pressure has been used to study assembly of multimeric proteins and viruses [3-6]. Pressure has permitted to address the specific question of how the plasticity required for successful assembly of a virus particle is coded into the native conformation of a capsid protein subunit [6]. This combined thermodynamic and structural approach has been used to try to identify the general rules that govern virus assembly. In general, it has been found the capsid coat proteins (monomers or dimers) are much less stable to pressure than the assembled icosahedral particles [6]. The isolated capsid and the assembly intermediates assume different partially folded states in the assembly pathway [6-9]. Single-amino acid substitutions in the hydrophobic core of the coat protein of icosahedral viruses produce large decreases in stability against pressure and chemical denaturants. Recent studies also reveal the importance of the nucleic acid to scaffold the assembly of the particles [6]. High pressure has permitted us to trap a ribonucleoprotein intermediate (a potential target for antiviral drugs), where the coat protein is partialy unfolded but bound to RNA [6, 9, 10]. Intermediate states also appear in the dissociation of empty capsids, such as P22 procapsids [8].

In several animal viruses, pressure causes inactivation with no change or even increase in the immunogenicity, suggesting pressure as an alternative method to produce killed vaccines [11-14]. The use of pressure inactivation to produce human or veterinary vaccines may have important economic impact. This novel method could be feasible to produce inactivated or killed vaccines.

2. Free-energy stabilization in the formation of icosahedral shells

The dissociation of tobacco mosaic virus (TMV) by pressure was reported a half-century ago by Lauffer and Dow [15]. TMV has an helical structure which length is determined by the size of the RNA. More recently, Bonafe *et al.* [16] have extended the studies of TMV to low temperature under pressure. In 1988, we described the reversible pressure dissociation of an icosahedral virus, brome mosaic virus (BMV) [17]. Pressure studies have revealed differences between the stability of the whole virus particle and their capsid subunits. Monomeric capsid protein of P22 bacteriophage was very sensible to pressure and underwent denaturation at pressures below 2.0 kbar [8], whereas the T=7 procapsid shell did not dissociate with pressures up to 2.5 kbar. P22 procapsid shells dissociated only at high pressures and low temperatures, indicating that they are stabilized by entropy [8]. The dissociation was irreversible because the assembly of T=7 dsDNA viruses requires the scaffolding protein to act as a chaperone in the association reaction.

Several single-amino acid substitution mutants (G232D, W48Q and T294I) were utilized to dissect the factors that determine the free-energy stability of the icosahedral lattice of bacteriophage P22 [8, 18, 19]. In contrast to the wild type, the W48Q mutant shells could be easily dissociated by pressure at room temperature with little dependence on decreasing temperature, suggesting a smaller entropic contribution. On the other hand, the decreased stability found for the G232D and T294I mutants was associated with defective protein cavities (related to volume) [18, 19].

3. Isolation of a ribonucleoprotein complex

The contribution of protein-nucleic acid interactions to virus stabilization was measured by comparing pressure stability of empty capsids and ribonucleoprotein particles of comoviruses [7, 9, 10]. Empty particles of comoviruses were dramatically less stable than the RNA-encapsidated particles. Ribonucleoprotein components of several comoviruses reassembled into whole particles with the same hydrodynamic and spectroscopic properties as the control samples, whereas the pressure denaturation of empty particles was irreversible. A ribonucleoprotein intermediate was detected in the pressure-dissociation of comoviruses as based on light scattering, RNase resistance and bis-ANS binding data [6, 7, 9, 10]. As shown in Fig. 2, the ribonucleoprotein intermediate would serve as a core for ready regeneration of the particle when the pressure is reduced. An unusual feature of this ribonucleoprotein intermediate is the presence of partially unfolded coat proteins bound to the RNA. RNA apparently plays a chaperone-like role during assembly of the capsid. In the absence of RNA the subunits drift to a disorganized structure and cannot renature when the perturbation is withdrawn. Ribonucleoprotein intermediates have also been proposed to occur in R17 bacteriophage, nodaviruses and picornaviruses [6, 14], and, in animal viruses, they are potential target for antiviral drugs. The ribonucleoprotein intermediate seems to have a condensed structure [10], which demonstrates the high degree of plasticity of the coat-protein-RNA complex.

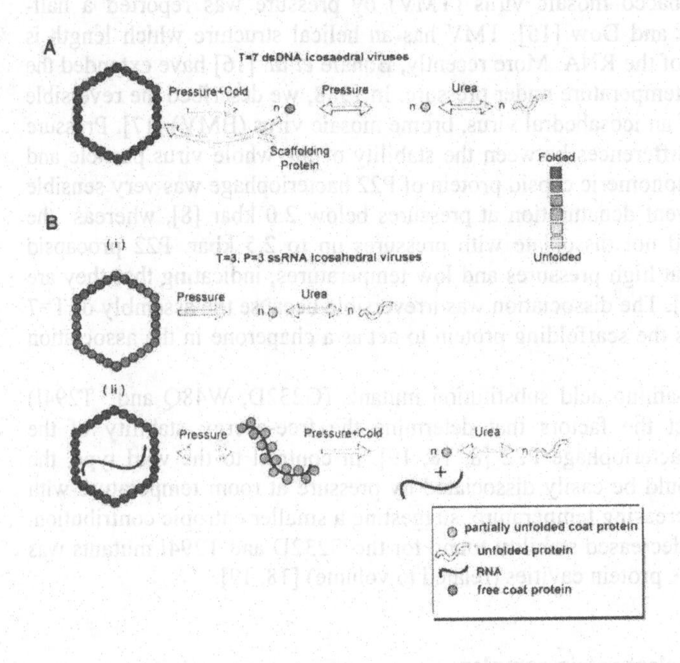

Figure 2. Folding gradient in the disassembly pathway of icosahedral viruses identified by high pressure and low temperature studies. Figure adapted from [6].

4. Protein-folding gradient in the assembly of icosahedral viruses

The general picture that has emerged for different viruses is a progressive decrease in folding structure in moving from assembled capsids to ribonucleoprotein intermediates (in the case of RNA viruses), free dissociated units (dimers or monomers), and finally unfolded monomers (Fig. 2). We propose a gradient of molten-globule states between the fully structured coat protein in the capsid and the unfolded monomers [6, 8, 10, 18]. High pressure would affect primarily the quaternary and tertiary structure of the capsid protein, leading to partially unfolded, molten globule conformations. In contrast, high urea concentrations would primarily disrupt the secondary structure.

All the viruses studied so far undergo cold dissociation and denaturation [6, 8-10, 14, 16, 18] similar to the cold denaturation observed with proteins [20, 21]. Whereas pressure dissociation are often completely reversible at room temperature, cold denaturation under pressure of viruses can lead to irreversibility. In the case of comoviruses, the coat protein dissociates from the RNA and assumes a conformation similar to that found in the pressure-denatured empty capsid. Therefore, entropy, especially due to the interaction between the coat protein and RNA, plays a major role to preserve the information necessary for virus assembly.

5. Assembly studies on MS2 phages

The linkage between protein-protein interactions and assembly is very remarkable in the T=3 RNA phages: R17 and MS2. Whereas coat protein dimers were fully dissociated into monomers at pressures below 2.2 kbar, whole phages were much more stable in this pressure range [22].

The MS2 bacteriophage is a member of single-stranded RNA phages that infect *Escherichia coli*. The icosahedral protein shell with triangulation number T=3 is composed of 180 copies of the coat protein and the structure is known at atomic resolution [23]. The thermodynamic stability of the wild type and different mutants of MS2 was studied (Fig. 3). Single-amino acid substitutions in the hydrophobic core of the coat protein produced large decreases in stability against pressure (Table 1, Fig. 3). The pressure effects were completely reversible.

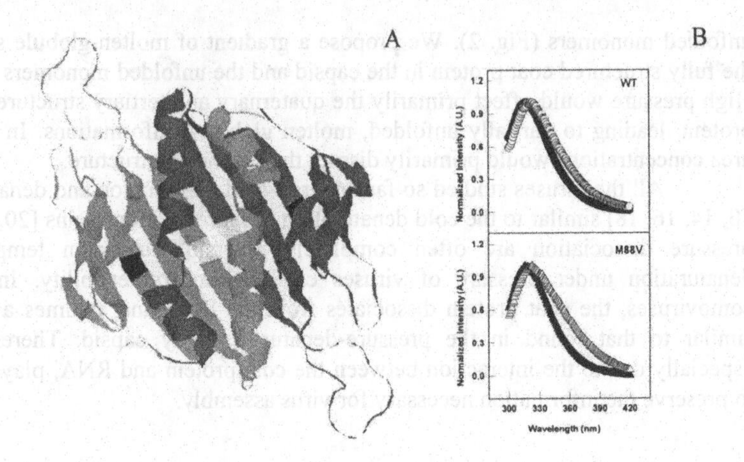

Figure 3. MS2 Bacteriophage structure. A. Ribbon representation of the dimeric structure of the MS2 coat protein. The residues in black show the position of the mutation T45S and in light grey the mutation M88V. PDB code: 1msc.pdb. B. Steady State fluorescence spectra of the wild type MS2 bacteriophage and the TS mutant M88V before (filled symbols) and under pressure treatment (3.4 kbar - hollow symbols).

Reversibility was tested by HPLC gel filtration chromatography, infectivity (Table 2) and by nuclear magnetic resonance (Figure 4). Although the fluorescence spectra, HPLC gel filtration and infectivity, indicated that the process was completely reversible, NMR spectroscopy showed that the particle gained more flexibility after a cycle of compression-decompression. Similar changes were elicited when the pH was decreased to 4.0. The sharp resonance lines that appear is a strong indication that segments of the coat protein (likely the AB and FG loops) that were immobile gained mobility after the compression at 3.4 kbar. This was based in the observation that almost no signal was observed at pH 7.0 before compression. ^{15}N relaxation measurements at the virus particle at lower pH (4.5) confirmed that the sharp lines are due to increased segmental mobility. The increase in flexibility are usually the characteristic of the particle after binding to the cellular receptor. As will be discussed below, the meaning of this result is particular important for the understanding of the pressure-induced inactivation of animal viruses where the pressure-treated particle is almost indistinguishable from the native one and presents properties similar to the cell-receptor bound particle, so called the "fusion intermediate state".

TABLE 1. Differences on MS2 bacteriophage stability produced by single-amino acid substitutions as followed by $P_{1/2}$ values.

MS2	$P_{1/2}$ (kbar) CM	$P_{1/2}$ (kbar) LS
WT	3.0	3.1
M88V	1.6	1.4
T45S	1.8	1.5

CM: center of spectral mass; LS: light scattering

TABLE 2. Pressure treatment of MS2 bacteriophage: reversibility measured by Infectivity assays and HPLC gel filtration chromatography.

MS2		INFECTIVITY TITER (PFU/mL)	GEL FILTRATION ELUTION PEAK AREA (%)
WT	Control	10^8	100
	Pressurized (3.4 kbar)	10^8	95
M88V	Control	10^8	100
	Pressurized (3.4 kbar)	10^7	89
T45S	Control	10^8	100
	Pressurized (3.4 kbar)	10^7	90

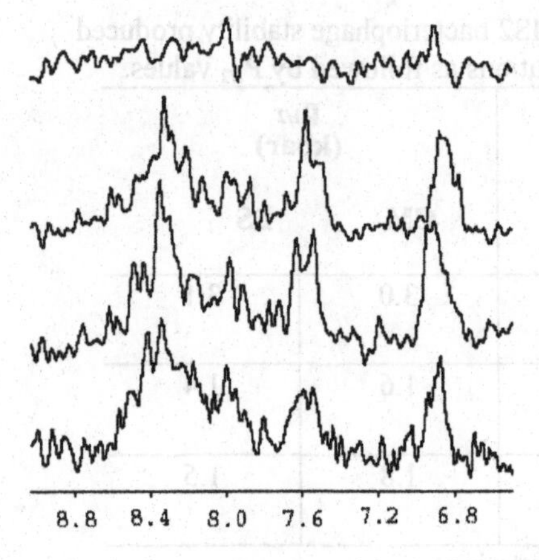

Figure 4. One-dimensional $^{15}N/^{1}H$ HMQC spectra of MS2 virus particle. Uniformly ^{15}N labeled MS2 was submitted to compression-decompression cycle and different pHs. The spectrum at pH 7.0 was done using a relaxation delay of 10 s. 3 s was used for all the others. All spectra was processed using 90° shifted sine bell.

Single-amino acid substitution in the hydrophobic core of the coat protein (M88V) or at the interface of the protein-RNA interaction (T45S) produced large decreases in stability against pressure (Fig. 3, Table 1). The diminished stability observed in both mutants can be explained by the large potential of these substitutions to create cavities ("cavity-creating mutants"). Pressure studies are now in progress with the isolated coat protein dimer and seems to reveal the importance of the RNA to scaffold the assembly of the particles, especially in those cases where the stability of the coat protein is put at a disadvantage with a mutation.

6. Structural transition in picornaviruses induced by pressure

The family *Picornaviridae* includes several viruses of great economic and medical importance [24]. Poliovirus replicates in the digestive tract, causing human disease that may range in severity from a mild infection to a fatal paralysis. The human rhinovirus is the most important etiologic agent of the common cold in adults and children. Foot-and-mouth disease virus (FMDV) causes one of the most economically important diseases in cattle. These viruses have in common a capsid structure composed of sixty copies of four different proteins, VP1 to VP4, and their 3D structures show similar general features [25-27]. Oliveira *et al.* [14] have described the differences in stability against high pressure and cold denaturation of these viruses. Both poliovirus and rhinovirus are stable to high pressure at room temperature - pressures up to 2.4 kbar are not enough to promote viral disassembly and inactivation. Within the same pressure range, FMDV particles are drastically affected by pressure, with a loss of infectivity of more than 4 log units. The dissociation of polio and rhino viruses can be observed only under high pressure at low temperatures in the presence of low concentrations of urea (1 to 2 M). The pressure and low temperature data reveal clear differences in stability among the three picornaviruses, FMDV being the most sensitive, polio the most resistant, and rhino having intermediate stability.

As mentioned above, dissociation of polio and rhino viruses could only be achieved by lowering the temperature under pressure (Fig. 5). At subzero temperatures both viruses exhibit cold-induced disassembly, as evidenced by the pronounced change in the Trp spectra – a decrease in center of spectral mass similar to the changes produced by 8 M urea. The thermodynamic parameters (Table 3) indicate again the importance of entropy to keep the stability of the particles. In comparison to rhinoviruses, the dissociation of poliovirus requires lower temperatures under pressure to be completed. However the change in entropy is much higher for rhinovirus (Table 3) than for poliovirus, which may arise from a more hydrophobic packing for rhino. On the other hand, polio has much smaller change in enthalpy.

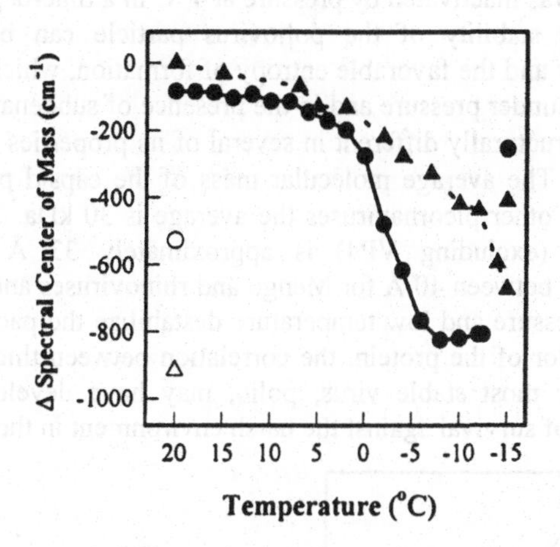

Figure 5. Cold denaturation under pressure (2.4 kbar) of polio (▲) and rhino (●) viruses in the presence of 1.5 M urea. The filled symbols in parentheses show the return to room temperature and atmospheric pressure. The hollow symbols at left show the data in the presence of 8 M urea (atmospheric pressure, room temperature).

TABLE 3. Enthalpy and entropy contributions to the free energy of association of polio and rhinovirus at 2.4 kbar[a].

Virus	ΔH, kcal/mol	$T\Delta S$ (0 °C), kcal/mol	ΔG (0 °C), kcal/mol
Poliovirus	479.00	767.40	-299.40
Rhinovirus 14	1822.00	2133.50	-311.50

a These parameters were obtained from van't Hoff plots.

FMDV is much less stable, since the pressure-induced changes that lead to inactivation can even occur at room temperature. While the rhino and poliovirus differ little in stability (less than 10 kcal/mol at 0 °C), the difference in free energy between these two viruses and FMDV was remarkable (more than 200 kcal per mol of particle). These differences are crucial to understanding the different factors that control assembly and disassembly of the virus particles during their life cycle.

Polioviruses was only fully inactivated after treatment under pressure (2.5 kbar) and low temperature in the presence of low concentrations of urea [14]. On the other hand, FMDV was inactivated by pressure at 4 °C in a time-dependent fashion (Fig. 6). The remarkable stability of the poliovirus particle can be explained by the unfavorable enthalpy and the favorable entropy of formation, which can only be broken by low temperatures under pressure and in the presence of subdenaturing concentrations of urea. FMDV is structurally different in several of its properties from other genera of picornaviruses [27]. The average molecular mass of the capsid protein is 24 kDa for FMDV, whereas for other picornaviruses the average is 30 kDa. The rms thickness of the FMDV capsid (excluding VP4) is approximately 33 Å whereas the other picornaviruses range between 40 Å for Mengo and rhinoviruses and 46 Å for poliovirus [24-27]. Because pressure and low temperature destabilize the packing of hydrophobic residues in the interior of the protein, the correlation between thickness of the shell is straightforward. The most stable virus, polio, may have developed thermodynamic stability as a means of survival against the harsh environment in the digestive tract.

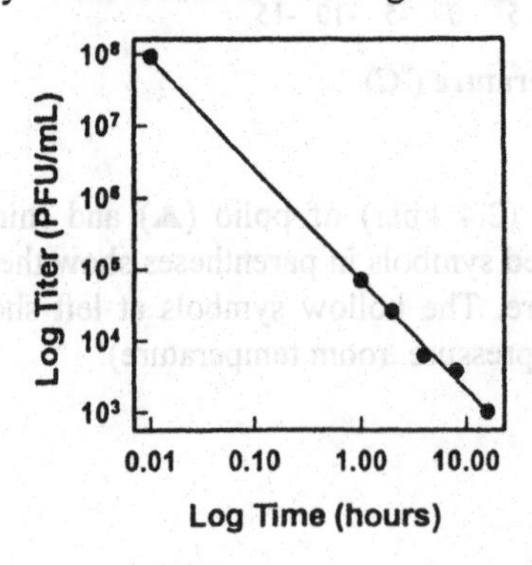

Figure 6. FMDV Infectivity under pressure for different times. FMDV samples were incubated at 2.5 kbar for 0, 1, 2, 4, 8 or 16 h and processed for plaque assay.

A scheme of the changes produced by pressure and low temperature on picornaviruses is presented in Fig. 7. The most important feature is that after a pressure cycle there is reassociation to a non-infectious particle (named P-particle). More drastic denaturation treatments, such as with high concentrations of urea, results in irreversible unfolding. The scheme also includes a possible explanation for loss of infection that has previously been proposed by other authors to account for heat treatment of some picornaviruses [28, 29], where the defective particle would lose VP4 and/or small molecules (pocket factors) bound to the canyon. The pressure-inactivated picornavirus may ressemble the A-particle detected in poliovirus and rhinovirus upon interaction with cell [24, 29] which also lack the internal capsid protein VP4. The A-particle is substantially less infectious than natural virions and is often identified as an intermediate in uncoating. These "fusion intermediate states" have been found in many nonenveloped and enveloped viruses [1]. During the fusion process, the conformations of the coat proteins and envelope glycoproteins change, which on one hand lead to non-infectious particles and on the other may lead to the exposure of previously occult epitopes, important for vaccine development. These irreversible conformational changes evoked by high pressure that resemble the changes that occur "*in vivo*" are discussed below for most of the viruses we have studied.

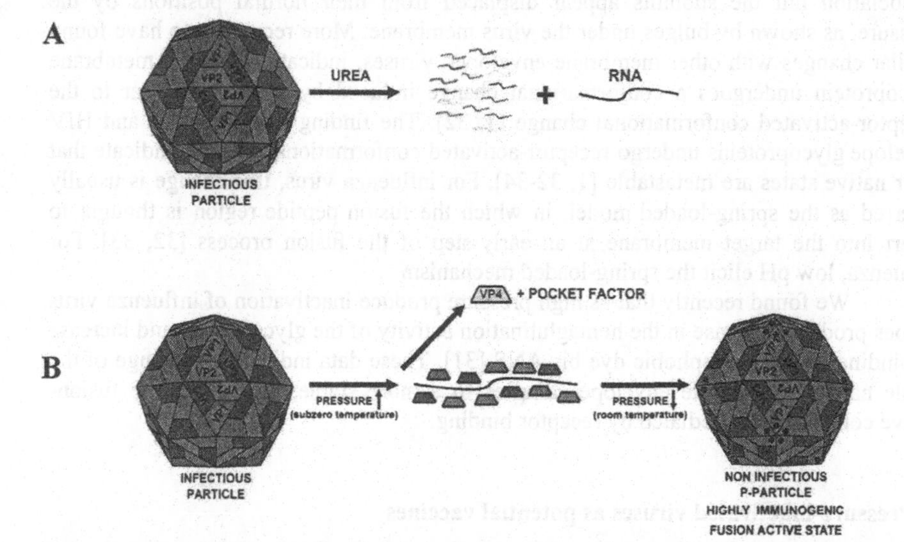

Figure 7. Proposed model for the disassembly of picornaviruses by denaturing high concentrations of urea (A) or by pressure and low temperature (B). In A, a high urea concentration elicits complete dissociation and denaturation, with separation of the RNA from the coat protein (urea-unfolded subunits are represented as elongated random coils) . In B, on contrary, pressure plus low temperature disrupt the icosahedral structure, but the capsid proteins (VP1, VP2 and VP3) still remain bound to the RNA. This particle loses infection on return to atmospheric pressure and room temperature. This infectivity loss may be due to release of VP4 and the "pocket factor".

7. Irreversible conformational changes in viruses

The survival of virus particles depends on the ability to protect for long periods of time the genetic material against the surroundings. Stability against many types of perturbations is required if the particles have to last long enough to reach their target hosts [2]. In addition, the virus needs to release the nucleic acid when bound or inside the host cells. The mechanisms that satisfy these contrary demands on the particle stability are only now starting to be elucidated. Like the multimeric proteins, the capsids of many viruses are dissociated by pressure in the range 1-3 kbar [6]. In Brome Mosaic Virus (BMV), the dissociation is completely reversible up to 1.4 kbar, that promotes 75% dissociation, but reassembly steeply decreases at higher pressures [17] owing to the formation of unspecific aggregates that remain on decompression.

The formation of non-infective particles after a cycle of compression and decompression has been demonstrated in many viruses like rotaviruses [13], adenoviruses [13], vesicular stomatitis virus (VSV) [11], simian immudeficiency virus [30], influenza [31] and picornaviruses [14]. In the case of VSV, which is a membrane enveloped virus, application of a pressure of 2.5 kbar for sufficient time abolishes the infectivity [11]. Electron microscopy of the compressed samples shows no detectable dissociation but the subunits appear displaced from their normal positions by the pressure, as shown by bulges under the virus membrane. More recently, we have found similar changes with other membrane-enveloped viruses, indicating that the membrane glycoprotein undergoes a conformational change induced by pressure similar to the receptor-activated conformational change [1, 32]. The finding that influenza and HIV envelope glycoproteins undergo receptor-activated conformational changes indicate that their native states are metastable [1, 32-34]. For influenza virus, this change is usually referred as the spring-loaded model, in which the fusion-peptide region is thought to insert into the target membrane at an early step of the fusion process [32, 33]. For influenza, low pH elicit the spring-loaded mechanism.

We found recently that as high pressure produce inactivation of influenza virus it does produce decrease in the hemaglutination activity of the glycoprotein and increase in binding of the hydrophobic dye bis-ANS [31]. These data indicate the change of the labile native state of the envelope complex to a more stable, mimicking the fusion-active conformation mediated by receptor binding.

8. Pressure-inactivated viruses as potential vaccines

Hydrostatic pressure can be used to inactivate animal viruses and result in particles that are good immunogens. The use of pressure to inactivate viruses has been evaluated with a view toward two potential applications: vaccine development and virus sterilization [11-14, 30]. Concerning viral vaccines, there are basically three types of immunization strategies: use of live (attenuated) particles; use of killed (inactivated) whole-virion particles; and use of subunit vaccines (hepatitis B). Immunization is the most efficient way of preventing infectious diseases in animals and humans [35, 36]. We have studied several viruses that cause disease in animals and humans. The antibodies against pressurized virus particles were as effective as those against the intact viruses when measured by their neutralization titer in plaque reduction assay.

Our studies on the pressure stability of viruses demonstrate the potential use of pressure to suppress virus infectivity, while preserving, or perhaps improving the immunogenic properties. The formation of non-infective particles after a cycle of compression and decompression has previously been demonstrated for rhabdoviruses [11], herpesviruses [37], immunodeficiency viruses [30], rotaviruses [13], adenoviruses [13] poliovirus and FMDV [14]. Pressure-inactivated FMDV retains its ability to produce high titers of neutralizing antibodies (Ishimaru *et al.*, manuscript in preparation). Table 4 summarizes the data on pressure inactivation of different animal and human viruses. Our results demonstrate the potential utilization of hydrostatic pressure for preparation of non-infectious whole virus particles. This is clearly worth considering because large losses in infectivity are obtained by a method that does not involve any covalent chemical reactions, and leads to production of an antigen with many of the chemical and physical properties of the intact viral particles. The high titers of the neutralizing antibodies elicited by pressure-inactivated viruses (Table 5) indicate that hydrostatic pressure can be used to prepare whole-virus immunogens. Effective immunization against viruses requires presentation of the whole virus particle to the immune system. This requirement coupled to the need for elimination of infectivity greatly limits the possibilities of preparation of appropriate vaccines. The use of hydrostatic pressure as a virus inactivation method may fulfill the two requirements described above. The employment of high pressure to prepare antiviral vaccines may have important advantages over other methods such as attenuation, chemical inactivation or isolated subunit vaccines. Attenuated live viruses can revert after a certain time and may cause the disease that they are intended to prevent, or worsen the real disease. Immunization with isolated subunits has several problems, especially because the immune system recognizes the isolated antigen less effectively than the whole virus. A physical method, such as high pressure, to prepare killed vaccines should not have the same problems.

TABLE 4. Pressure Inactivation of Viruses

Virus	Pressure (kbar)	Temperature (°C)	Decrease in Infectivity	Ref.
FMDV	2.4	4	10^5	14
VSV	2.6	20	10^5	11
Poliovirus	2.4	-15	$>10^4$	14
Bovine Rotavirus	2.5	20	10^5	13
Simian Rotavirus	2.5	20	10^4	13
Influenza	3.2	4	10^6	31
SIV and HIV	2.5	4	10^5	30
Adenovirus	2.2	20	10^5	13

The infectivity was assayed on cell monolayers.

TABLE 5. Immunogenicity of pressure-inactivated viruses as measured by neutralization titers

Virus		Neutralization Titer	Ref.
FMDV	Native	1:512	14
	Pressure-Inactivated	1:512	
VSV	Native	1:2000	11
	Pressure-Inactivated	1:2000	
Bovine Rotavirus	Native	1:1000	13
	Pressure-Inactivated	1:1000	
Simian Rotavirus	Native	1: 10^5	13
	Pressure-Inactivated	1: 10^5	

The neutralization titer is given by the reciprocal of highest dilution of antibodies (γ-globulin fraction) responsible for at least 90% reduction in the infectivity.

The reason why the pressurized viruses maintain the immunogenic potential probably resides in the fact that the structural changes are very subtle. As we discussed above pressure treatment seems to mimic the changes that are produced when viruses bind to cellular receptors. Recently, we found that pressure-inactivated VSV attaches to the cellular membrane, but it is not internalized by endocytosis [38]. The most reasonable explanation is that high pressure leads the envelope protein to the fusion conformation. This would be similar to the inhibition of an enzyme by a transition-state analog; the difference here is that a physical tool, pressure, freezes the viral protein in a conformation that cannot bind to the receptor preventing progression to endocytosis. Therefore, the other steps of the infection cycle are compromised. However, the attachment of inactivated virus, such as VSV, to the host cell [38] in a nonproductive way may be crucial to evoke immune response. First, because new epitope might be exposed which leads to efficient production of neutralizing antibodies; second, because the particle, although not infectious, can be processed by the cell resulting in a cellular response ($CD8^+$ cytotoxic T lymphocytes – CTLs).

The schematic representation for picornaviruses in Fig. 7B can be extended to enveloped viruses, the difference is that the main change after pressure release occurs in the membrane protein. The black circles represent the epitopes that seem to appear in the pressure-induced fusion-active state. As we reported in section 5, even the extremely simple particle of MS2 constituted by 180 copies of a single polypeptide chain undergo changes that are detected by heteronuclear NMR after a cycle of pressure dissociation and association. The changes in the NMR spectra were similar to those produced by lowering the pH to 4.5 which in many viruses cause the formation of the fusion-intermediate particle. It is tempting to suggest that we are detecting the dynamics of a loop important for binding and delivery of nucleic acid into the host bacterial cell. It may be a primordial switch for the fusion-active state which further evolved in animal viruses to more specialized structure due to the pressure of the immune system.

A recent study [39] describes the potential of using fusion complexes for HIV vaccine development. Using the idea that epitopes with superior immunogenicity might be exposed or created as HIV-1 begins to fuse with cell membranes, they tested fusion complexes for their ability to induce neutralizing antibodies in mice. Fusion-competent HIV vaccine immunogens were generated by fixing with formaldehyde the virus-cell complex. This preparation of fixed virus and whole cells elicited antibodies capable of neutralizing infectivity of 23 of 24 primary HIV isolates. The authors suggest that these fusion-dependent immunogens may lead to a broadly effective HIV vaccine. However, the major problem of their approach is the unclean preparation of a mixture of viruses and cells. Since high pressure is able to produce the same fusion-intermediate state, at least in some of the virus studied, it is a cleaner and more controllable way to produce antiviral vaccines.

9. References

[1] Harrison, S., Wiley, D.C., and Skehel, J.J. (1996) Virus Structure. In "Fields Virology" 3rd Edition (B. N. Fields, D. M. Knipe, P. M. Howley, et al., Eds.), Chapter 3, pp. 59-100, Lippincott-Raven Publishers, Philadelphia.

[2] Johnson, J. E (1996) Functional implications of protein-protein interactions in icosahedral viruses, *Proc. Natl. Acad. Sci. USA* **93**, 27-33.

[3] Weber, G. (1993) Pressure dissociation of the smaller oligomers: dimers and tetramers. In "*High Pressure Chemistry, Biochemistry and Material Sciences*", NATO ASI series C (R. Winter and J. Jonas, Eds), Kluwer Academic Publishers, Dordrecht, 401: 471-487.

[4] Silva, J.L. and Weber, G. (1993) Pressure stability of proteins, *Annu. Rev. Phys. Chem.* **44**, 89-113.

[5] Mozhaev, V. V., Heremans, K., Frank, J., Masson, P., and Balny, C. (1994). Exploiting the effects of hydrostatic pressure in biotechnological applications. *Trends Biotechnol.* **12**, 493-501.

[6] Silva, J. L., Foguel, D, Da Poian, A.T.and Prevelige, P.E. (1996) The use of hydrostatic pressure as a tool to study viruses and other macromolecular assemblages, *Curr. Opin. Struct. Biol.* **6**:166-175.

[7] Da Poian, A. T., Johnson, J. E. and Silva, J. L. (1994) Differences in pressure stability of the three components of cowpea mosaic virus: Implications for virus assembly and disassembly, *Biochemistry* **33**, 8339-8346.

[8] Prevelige, P.E., King, J. and Silva, J.L. (1994) Pressure denaturation of bacteriophage P22 coat protein and its entropic stabilization in the icosahedral shells, *Biophys. J.* **66**,1631-1641.

[9] Da Poian, A.T., Oliveira, A.C. and Silva, J.L. (1995) Cold denaturation of an icosahedral virus. The role of entropy in virus assembly, *Biochemistry* **34**, 2672-2677.

[10] Gaspar, L. P., Johnson , J.E., Silva, J.L. and Da Poian, A.T. (1997) Different Partially Folded States of the Capsid Protein of Cowpea Severe Mosaic Virus in the Disassembly Pathway, *J. Mol. Biol.* **273**, 456-466.

[11] Silva, J. L., Luan, P., Glaser, M., Voss, E.W. and Weber, G. (1992b) Effects of hydrostatic pressure on a membrane-enveloped virus: High immunogenicity of the pressure-inactivated virus, *J. Virol.* **66**, 2111-2117.

[12] Silva, J. L. (1993) Effects of pressure on multimeric proteins and viruses. In "*High Pressure Chemistry, Biochemistry and Material Sciences*", NATO ASI series C (R. Winter and J. Jonas, Eds), Kluwer Academic Publishers, Dordrecht, 401: 561-578.

[13] Pontes, L., Fornells, L.A., Giongo, V., Araujo, J.R.V., Sepulveda, A., Villas-Boas, M., Bonafe, C.F.S. and Silva, J.L. (1997). Pressure Inactivation of Animal Viruses: Potential Biotechnological Applications. High Pressure Research in the Bioscience and Biotechnology (Heremans, K., Ed.), Leuven University Press, Leuven, pp. 91-94.

[14] Oliveira, A. C., Ishimaru, D., Gonçalves, R. B., Mason, P., Carvalho, D., Smith, T. Silva, J. L. (1999). Low Temperature and Pressure Stability of Picornaviruses: Implication for Virus Uncoating, *Biophys. J.* **76**, 1270-1279.

[15] Lauffer M.A. and Dow, R.B. (1941) Denaturation of TMV at high pressure, *J. Biol. Chem.* **140**, 509-518.

[16] Bonafe, C.F., Vital, C.M., Telles, R.C., Goncalves, M.C., Matsuura, M.S., Pessine, F.B., Freitas, D.R., and Veja, J. (1998) Tobacco mosaic virus disassembly by high hydrostatic pressure in combination with urea and low temperature, *Biochemistry* **37**, 11097-11105.

[17] Silva, J. L., and G. Weber. (1988) Pressure-induced dissociation of brome mosaic virus, *J. Mol. Biol.* **199**, 149-161.

[18] Foguel , D., Teschke, C.M., Prevelige, P.E. and Silva, J.L. (1995) The role of entropic interactions in viral capsids: single-amino-acid substitutions in P22 bacteriophage coat protein resulting in loss of capsid stability, *Biochemistry* **34**, 1120-1126.

[19] Souza-Jr., P. C., Tuma, R., Prevelige, P. E., Silva, J. L., Foguel, D. (1999) Cavity defects in the procapsid of bacteriophage P22 and the mechanism of capsid maturation, *J. Mol. Biol.*, in press.

[20] Foguel, D. and Weber, G. (1995) Pressure-induced dissociation and denaturation of allophycocyanin at sub-zero temperatures, *J. Biol. Chem.* **270**, 28759-28766.

[21] Nash, D and Jonas, J. (1997) Structure of pressure-assisted cold denatured lysozyme and comparison with lysozyme intermediates, *Biochemistry* **36**, 14375-14383.

[22] Da Poian, A.T., Oliveira, A.C., Gaspar, L.P., Silva, J.L. and Weber, G. (1993) Reversible pressure dissociation of R17 bacteriophage: The physical individuality of virus particles, *J. Mol. Biol.* **231**, 999-1008.

[23] Valegard, K., Liljas, L., Fridborg, K. and Unge, T. (1990) The three-dimensional structure of the icosahedral bacterial virus MS2, *Nature (London)* **345**, 36-41.

[24] Rueckert, R. R. (1996) Picornaviridae: The Viruses and Their Replication. In "Fields Virology" 3rd Edition (B. N. Fields, D. M. Knipe, P. M. Howley, et al., Eds.), Chapter 21, pp. 609-645, Lippincott-Raven Publishers, Philadelphia.

[25] Rossmann, M. G., Arnold, E., Erickson, J.W., Frankenberger, E.A., Griffith, J.P., Hecht, H.J., Johnson, J.E., Kamer, G., Luo, M., Mosser, A.G., Rueckert, R.R.,

Sherry, B. and Vriend, G. (1985) Structure of a human common cold virus and functional relationship to other picornaviruses, *Nature* **317**, 145-153.
[26] Hogle, J. M., Chow, M. and Filman, D.J. (1985) Three-dimensional structure of poliovirus at 2.9 Å resolution, *Science* **229**, 1358-65.
[27] Acharya, R., Fry, E., Stuart, D.I., Fox, G., Rowlands, D. and Brow, F. (1989). The three-dimensional structure of foot and mouth disease virus at 2.9 Å, *Nature* **337**, 709-716.
[28] Giranda, V. L., Heinz, B.A., Oliveira, M.A., Minor, I., Kim, K.H., Kolatkar, P.R., Rossmann, M.G., Rueckert, R.R. (1992) Acid-induced structural changes in human rhinovirus 14: Possible role in uncoating, *Proc. Natl. Acad. Sci. USA* **89**, 10213-10217.
[29] Rossmann, M. G. (1994) Viral cell recognition and entry, *Protein Science* **3**, 1712-1725.
[30] Jurkiewicz E., Villas-Boas, M., Silva, J.L., Weber, G., Hunsmann, G. and Clegg, R.M. (1995) Inactivation of Simian Immunodeficiency Viruses by Hydrostatic Pressure, *Proc. Natl. Acad. Sci. USA* **92**, 6935-6937.
[31] Foguel, D., Dantas, V.S., Silva, A.C.B., Ano-Bom, A.P.D., Schwarcz, W.D., Silva, J.L. (1999) Mimicry of the fusion-active conformation of influenza virus by high pressure. Submitted
[32] Carr, C.M. and Kim, P.S. (1993) A spring-loaded mechanism for the conformational change of influenza hemagglutinin, *Cell* **73**, 823-832.
[33] Bullough, P.A., Hughson, F.M., Skehel, J.J., Wiley, D.C. (1994) Structure of influenza haemaglutinin at the pH of membrane fusion, *Nature* **371**, 37-43.
[34] Chan, D.C. and Kim, P.S. (1998) HIV Entry and Its Inhibition, *Cell* **93**, 681-684.
[35] Budowsky, E. I. (1991) Problems and prospects for preparation of killed antiviral vaccines, *Advances in Virus Research* **39**, 255-290.
[36] Bloom, B. R. (1996) A perspective on AIDS vaccines, *Science* **272**, 1888-1890.
[37] Nakagami, T., Shigehisa, T., Ohmori, T., Taji, S., Hase, A., Kimura, T. and Yamanishi, K. (1992) Inactivation of herpes viruses by hydrostatic pressure, *J. Virol. Methods* **38**, 255-261.
[38] Da Poian, A. T., Gomes, A.M.O., Oliveira, R.J.N. and Silva, J.L. (1996) Migration of Vesicular Stomatitis Virus Glycoprotein to the Nucleus of Infected Cells, *Proc. Natl. Acad. Sci. USA* **93**, 8268-8273.
[39] LaCasse, R.A., Follis, K.E., Trahey, M., Scarborough, J.D., Littman D.R., Nunberg, J.H. (1999) Fusion-competent vaccines: broad neutralization of primary isolates of HIV, *Science* **283**, 357-62

10. Acknowledgments:

This work was supported in part an International Grant from the Howard Hughes Medical Institute to JLS, by grants from Conselho Nacional de Desenvolvimento Científico e Tecnológico (CNPq), Financiadora de Estudos e Projetos (BID, PADCT and Pronex programs). J.L.S. is an International Scholar of the Howard Hughes Medical Institute

STRUCTURE AND STABILITY OF WILDTYPE AND F29W MUTANT FORMS OF THE N-DOMAIN OF AVIAN TROPONIN C SUBJECTED TO HIGH PRESSURES

A. YU, A. JONAS
Department of Biochemistry
University of Illinois at Urbana-Champaign
506 S. Mathews Avenue
Urbana, IL 61801, USA

J. JONAS, L. BALLARD
Department of Chemistry
University of Illinois at Urbana-Champaign
600 S. Mathews Avenue
Urbana, IL 61801, USA

L. SMILLIE, J. PEARLSTONE
MRC Group in Protein Structure and Function
Department of Biochemistry
University of Alberta
Edmonton, Alberta
CANADA

D. FOGUEL, J. SILVA
Departamento de Bioquímica Médica
Instituto de Ciências Biomédicas
Universidade Federal do Rio de Janeiro
Rio de Janeiro, BRAZIL

1. Abstract

The N-domain of troponin C (residues 1-90) regulates muscle contraction through conformational changes induced by Ca^{2+} binding. A mutant form of this domain of avian troponin C (F29W) has been used in previous studies to observe conformational changes that occur upon Ca^{2+} binding, and pressure and temperature changes. In this study we examined the effect of the point mutation on the protein structure and its stability to pressure. We performed 1-D and 2-D ^{1}H-NMR experiments at 300, 400, and 500 MHz on the wildtype and F29W mutant forms of the N-domain of chicken troponin C in the absence of Ca^{2+}. We found that the

R. Winter and J. Jonas (eds.), High Pressure Molecular Science, 515–521.

mutant protein at 5 kbar pressures had a destabilized β-sheet between the Ca^{2+}-binding loops, an altered environment near Phe 26, and reduced local motions of Phe 26 and Phe 75 in the core of the protein, probably due to a higher compressibility of the mutant. Under the same pressure conditions, the wildtype protein experienced little effect. These results suggest that the surface mutation (F29W) significantly destabilizes the N-domain of troponin C by altering the packing and dynamics of the hydrophobic core.

2. Introduction

Troponin C (TnC) is one of three members of the Troponin complex. Along with TnI and TnT, TnC functions by regulating muscle contraction in response to calcium. Calcium binds to TnC and causes conformational changes. Through protein-protein interactions, the conformational signal is propagated via TnI and TnT to the tropomyosin-actin filament affecting the latter's interaction with myosin heads of the thick filament and the contraction/relaxation cycle.

Avian TnC has been previously crystallized [1, 2] and shown to be a dumbbell shaped protein: The N-domain is postulated to have a regulatory role, while the C-domain is believed to have a more structural role. Both domains are highly helical, and each can bind two Ca^{2+} ions [3-5].

Because of its physiological role in muscle contraction, the conformational changes that accompany metal binding to TnC make this protein an attractive system for study. To simplify these studies, TnC has been separated into C- and N-domain fragments which retain the structural and Ca^{2+} binding properties of their respective counterparts in the intact protein [6]. In addition, since chicken skeletal TnC contains no natural tryptophans or tyrosines, spectral probes have been incorporated into the protein by mutagenesis. These mutants, namely F29W in the N-domain and F105W or F154W in the C-domain, have been previously characterized by fluorescence and other spectral methods [6-8] and permitted binding events in each domain to be monitored separately.

In the work described here, we have studied the N-domain of chicken TnC and its F29W mutant using our novel high pressure NMR approach in order to address two questions: How does a conservative (F→W) point mutation near the surface of the N-domain affect its stability, structure, and dynamics under pressure? And, does pressure affect different regions of the N-domain differently? The use of the high-pressure approach to perturb or unfold proteins is well justified by previous studies from our laboratory [9, 10] and by Foguel, Silva, and coworkers [11]. These studies have shown that high pressure is a controlled and gentle method for unfolding proteins and for stabilizing intermediates that are remarkably similar to protein folding intermediates [10].

3. Methods

The N-terminal domain of chicken TnC, comprising residues 1-90, and its F29W mutant were prepared by recombinant DNA techniques, expression in *E. coli*, and purification methods that were described previously [12]. Variable high-pressure

NMR experiments were performed at proton Larmor frequencies of 300 and/or 500 MHz using our unique high-pressure instrumentation, which has been described previously [13, 14]. The protein concentration in these experiments was typically 0.6 mM. The protein samples were dissolved in a pH*7.00±0.05 buffer composed of 20 mM TRIS-*d11*, 5 mM DTT, 2 mM EGTA, 100 mM KCl, and 1.5 mM TSP in D_2O. All studies were performed on the apo (Ca^{2+}-free form) of the TnC N-domain.

At 300 MHz, 1D PRESAT experiments were performed at 25°C with a 90° pulse width, up to a 4000 Hz sweep width, 16k complex points, a 6.25s total delay between scans (1.25s for a CW presaturation pulse), and 1024 scans. The measurements were taken in approximately 500 bar increments, to pressures as high as 5 kbar. The 300 MHz data was acquired with MacNMR software (Tecmag Inc., Houston, TX), and was later processed with NUTS software (AcornNMR, Fremont, CA) using 1.5 Hz exponential multiplication. At 500 MHz, 1D PRESAT experiments were performed in a similar manner, although with sweep widths of 5000-6000 Hz, 32k complex points, a 10s total delay (containing 1.5s of a CW presaturation pulse), and 128 scans. The 500 MHz data was collected with Varian VNMR software and, as with the 300 MHz data, was later processed using NUTS.

The 2D phase-sensitive NOESY experiments were performed at selected pressures on the 500 MHz system. Typical NOESY parameters consisted of a 5000 Hz sweep width, 32-64 scans, a 0.15s mixing time, a 2s recycle delay (2s of a CW presaturation pulse), and 1K x 256 complex points. The 2-D spectra were processed with NUTS software, using Gaussian multiplication in both dimensions and zero-filling to 2Kx2K complex points.

4. Results and Discussion

Prior to the analysis of high pressure NMR results, the resolved NMR peaks were assigned in the 5-8 ppm region by reference to the literature values [15, 16] (Figure 1, 1 bar, 25°C, pH* 7.0 spectrum). Figure 1 shows that the aromatic region (6.0 to 7.8 ppm) of the F29W mutant ^1H NMR spectrum undergoes various changes with increasing pressure. Three Phe peaks (F75δ, F26δ, and F78δ at 6.4, 6.6, and 7.0 ppm, respectively) (1 atm) are of special interest. These residues comprise part of the hydrophobic core of the N-domain of TnC. Although these residues are near each other in the protein structure, they demonstrate distinct trends with pressure. The F75δ proton peak first shifts upfield and then downfield with increasing pressure, while the F26δ proton peak moves upfield and broadens dramatically. In contrast to these peaks the F78δ and W29ε^3 peaks remain relatively unchanged up to 5 kbar pressures.

An explanation for these trends can be attempted from an examination of the core structure of the TnC N-domain. As shown by Herzberg, et al. [17] in the x-ray crystal structure of TnC, F75 and F26 are located near the fringes of the core while F78 is deeply buried within the protein. This is further supported by the solvent accessibility measurements performed on the x-ray structure [17]. Comparing the Ca^{2+}-free (closed) and Ca^{2+}-saturated (open) forms of TnC, Herzberg and coworkers found that F75 becomes more accessible to the solvent upon binding of Ca^{2+}, while F78 remains essentially unaffected. Similarly, our NMR data suggests a change in the environments of F75 and F26 at high pressures, but little effect on F78. The immutability of F78 at 5 kbar, furthermore, indicates that the core of the TnC F29W mutant is retained even at 5 kbar pressures and that, overall, the protein is highly resistant to denaturation by high pressures. The W29 residue is in a partially exposed region near loop I for the N-domain. The W29ε^3 chemical shift is close to that of a random coil Wε^3 resonance [16] and does not change significantly up to 5 kbar pressure. In previous studies, Foguel, et al. [11] observed no significant changes in Trp fluorescence when the TnC F29W N-domain was subjected to 2.2 kbar pressure at room temperature. Only at subzero temperatures (-11°C) under pressure did they find a large increase in fluorescence and lifetime of the Trp. They attributed the change to a conformational change which reduces quenching of the Trp fluorescence and exposes a hydrophobic region in the protein. Because low temperatures are known to decrease hydrophobic interactions and to destabilize proteins [9, 18], it is likely that the low temperature effects are much more important than the pressure effects in their study.

The D74α, D36α, and T72α proton resonance at 5.64, 5.57, and 5.42 ppm, respectively, arise from a short β-sheet in the native structure of TnC. As the pressure is increased to 5 kbar, these peaks shift upfield, broaden and disappear.

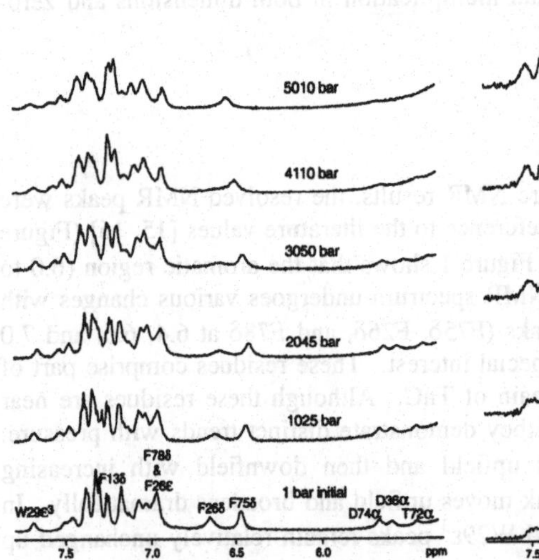

Figure 1. ^{1}H-NMR spectra of the F29W mutant of the N-domain of TnC as a function of increasing pressure at 25°C. The initial spectrum at 1 bar and the final spectrum after release of pressure (not shown) were identical. A 0.6 mM protein solution was used with a 500 MHz wide-bore Varian instrument.

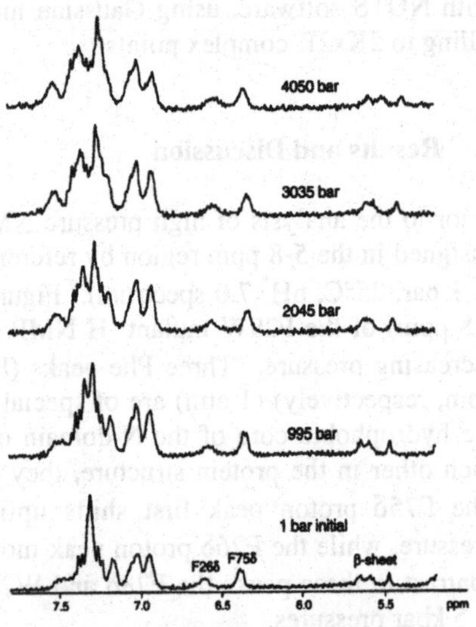

Figure 2. ^{1}H-NMR spectra of the wildtype form of the N-domain of TnC as a function of increasing pressure at 25°C. The spectral changes were completely reversible (not shown). A 0.6 mM protein solution was used with a 300 MHz wide-bore Tecmag system.

The upfield shift of the D74α, D36α, and T72α proton resonances indicates loosening of the β-sheet structure because they move towards their random coil conformation values of 4.76 and 4.35 ppm, respectively [16]. The broadening and decreasing intensity of the peaks further indicates a structural change and loss of β-sheet structure at high pressures. This interpretation is also supported by the 2D NOESY experiments. The close proximity of the D36 and D74 residues of the F29W N-domain at 1 atm was confirmed by a NOESY crosspeak between the D36α and D74α peaks (not shown). With increasing pressure, the intensity of the crosspeak weakened significantly and essentially disappeared by 4 kbar. This indicates that the β-sheet structure is disrupted at high pressures.

Figure 2 displays the wildtype N-domain ^{1}H-NMR spectra at increasing pressures. While there are some changes in the Phe peaks, especially in the range of 7.0 to 7.5 ppm, the resolved F78δ, F75δ, and F26δ peaks and the β-sheet proton peaks are not very susceptible to pressure effects. Likewise, 2D NOESY measurements of the β-sheet D36α-D74α crosspeak show that, in contrast to the mutant, the crosspeak appears to maintain its intensity to 4 kbar for the wildtype protein. Combined, these results show that the wildtype protein is more resistant to pressure than the F29W mutant as predicted by molecular dynamics simulations (results not shown).

It is of interest to note that the spectral region between 7 and 7.5 ppm shows changes with increasing pressure both for the wildtype and mutant proteins. This is not surprising because the hydrophobic core with stacked aromatic rings is expected to be sensitive to compression of the protein due to increasing pressure [19]. Compression of the stacked rings is likely to affect the dynamics and intermolecular interactions leading to spectral changes even at 1 kbar pressures.

The differences between the spectra for the two protein forms are illustrated in Figure 3, which shows the peak areas and peak widths of the F26δ and F75δ resonances as a function of pressure. While the peak areas and the line widths for the wildtype protein remain essentially constant over the entire pressure range, for the F29W mutant the peaks change significantly. The F26δ peak decreases in area and broadens about 2-fold, suggesting that the two F26δ protons become non-equivalent in the compressed protein and that local motions become more restricted upon increase of pressure. The F75δ peak, on the other hand, retains the same area but also broadens (2-fold) under pressure. Apparently, the two F75δ protons still experience the same chemical environment, but their motions are restricted. A reasonable explanation for the differences detected for these peaks, between the wildtype protein and the F29W mutant, is that the mutant has more free volume in the core, while the wildtype protein is probably optimally packed. Thus, the mutant may be compressed more under pressure, restricting further the motions of adjacent residues. In fact, Wagner [20] reported that the flipping motions of a Phe residue in the core of BPTI were restricted by subjecting that protein to 1.2 kbar pressures at 57°C.

Thus, in this study we have shown that under pressures of up to 5 kbar, the F29W mutant becomes destabilized in the β-sheet region and becomes compressed in the core near residues F26 and F75, altering the environment of F26. In contrast to the mutant protein, the wildtype form is quite insensitive to high pressures.

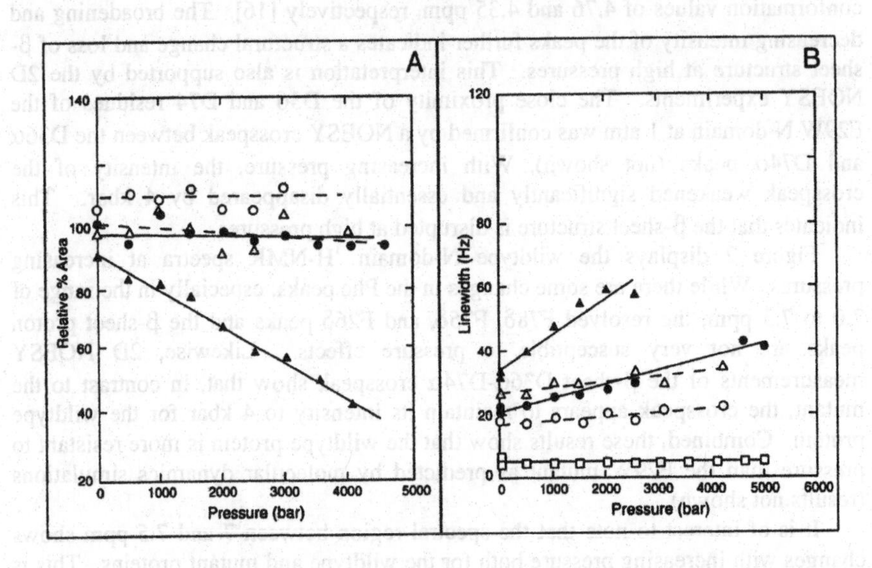

Figure 3. Resonance peak areas (A) and widths (B) for the wildtype and F29W mutant of the N-domain of TnC at 25°C. The data were obtained using NUTS software and manually from the spectra shown in Figures 1 and 2. The symbols correspond to: (•) F75δ, F29W mutant; (▲) F26δ, F29W mutant; (o) F75δ, wildtype protein; (Δ) F26δ, wildtype protein. Data for the TSP standard (□) are included to demonstrate the performance of the high resolution, high pressure probe for the 500 MHz system.

5. Acknowledgements

This work was supported in part by the National Institutes of Health under grant GM42452 (J. Jonas and A. Jonas) and by the Medical Research Council of Canada (L. Smillie).

6. References

1. Herzberg, O. and James, N.G. (1985) Structure of the calcium regulatory muscle protein troponin-C at 2.8 Å resolution, *Nature* **313**, 653-659.

2.. Satyshur, K.A., Pyzalska, D., Greaser, M., Rao, S.T., and Sundaralingam, M. (1994) Structure of chicken skeletal muscle troponin C at 1.78 Å resolution, *Acta Cryst.* **D50**, 40-49.

3. Li, M.X., Gagne, S.M., Tsuda, S., Kay, C.M., Smillie, L.B., and Sykes, B.D. (1995) Calcium binding to the regulatory N-domain of skeletal muscle troponin C occurs in a stepwise manner, *Biochemistry* **34**, 8330-8340.

4. Strynadka, N.C.J. and James, M.N.G. (1989) Crystal structures of the helix-loop-helix calcium-binding proteins, *Annu. Rev. Biochem.* **58**, 951-998.

5. Tobacman, L.S. (1996) Thin filament-mediated regulation of cardiac contraction, *Annu. Rev. Physiol.* **58,** 447-481.

6. Trigo-Gonzalez, G., Racher, K., Burtnick, L., and Borgford, T. (1992) A comparative spectroscopic study of tryptophan probes engineered into high- and low-affinity domains of recombinant chicken troponin C, *Biochemistry* **31,** 7009-7015.

7. Pearlstone, J.R., Borgford, T., Chandra, M., Oikawa, K., Kay, C.M., Herzberg, O., Moult, J., Herklotz, A., Reinach, F.C., and Smillie, L.B. (1992) Construction and characterization of a special probe mutant of troponin C: Application to analyses of mutants with increased Ca^{2+} affinity, *Biochemistry* **31,** 6545-6553.

8. Chandra, M., Fidalgo da Silva, E., Sorenson, M.M., Ferro, J.A., Pearlstone, J.R., Nash, B.E., Kay, C.M., and Smillie, L.B. (1994) The effects of N helix deletion and mutant F29W on the Ca^{2+} binding and functional properties of chicken skeletal muscle troponin C, *J. Biol. Chem.* **269,** 14988-14994.

9. Jonas, J. and Jonas, A. (1994) High-pressure NMR spectroscopy of proteins and membranes, *Annu. Revs. Biophys. Biomol. Structure* **23,** 287-318.

10. Nash, D. and Jonas, J. (1997) Structure of pressure-assisted cold denatured lysozyme and comparison with lysozyme folding intermediates, *Biochemistry* **36,** 14375-14383.

11. Foguel, D., Suarez, M.C., Barbosa, C., Rodrigues, Jr., J.J., Sorenson, M.M., Smillie, L.B., and Silva, J.L. (1996) Mimicry of the calcium-induced conformational state of troponin C by low temperature under pressure, *Proc. Natl. Acad. Sci. USA.* **93,** 10642-10646.

12. Li, M.X., Chandra, M., Pearlstone, J.R., Racher, K.I., Trigo-Gonzalez, G., Borgford, T., Kay, C.M., and Smillie, L.B. (1994) Properties of isolated recombinant N and C domains of chicken troponin C, *Biochemistry* **33,** 917-925.

13. Jonas, J., Koziol, P., Peng, X., Reiner, C., and Campbell, D.M. (1993) High-resolution NMR spectroscopy at high pressures, *J. Mag. Reson. B.* **102,** 299-309.

14. Ballard, L., Yu, A., Reiner, C., and Jonas, J. (1998) A high pressure, high resolution NMR probe for experiments at 500 MHZ, *J. Mag. Reson. A.* **133,** 190-193.

15. Findlay, W.A. and Sykes, B.D. (1993) ^{1}H-NMR resonance assignments, secondary structure, and global fold of the TR_1C fragment of turkey skeletal troponin C in the calcium-free state, *Biochemistry* **32,** 3461-3467.

16. Creighton, T.E. (1993) *Proteins, 2nd ed.*, W.H. Freeman and Company, New York.

17. Herzberg, O., Moult, J., and James, M.N.G. (1986) A model for the Ca^{2+}-induced conformational transition of troponin C, *J. Biol. Chem.* **261,** 2638-2644.

18. Zhang, J., Peng, X., Jonas, A., and Jonas J. (1995) NMR study of the cold, heat, and pressure unfolding of ribonuclease A, *Biochemistry* **34,** 8631-8641.

19. Kundrot, C.E. and Richards, F.M. (1987) Crystal structure of hen egg-white lysozyme at a hydrostatic pressure of 1000 atmospheres, *J. Mol. Biol.* **193,** 157-170.

20. Wagner, G. (1980) Activation volumes for the rotational motion of interior aromatic rings in globular proteins determined by high resolution ^{1}H NMR at variable pressure, *FEBS Letters* **112,** 280-284.

HIGH PRESSURE EFFECTS ON PROTEIN FLEXIBILITY AS MONITORED BY TRYPTOPHAN PHOSPHORESCENCE

P. CIONI
CNR, Istituto di Biofisica
Via S. Lorenzo 26, 56127 Pisa, ITALY

1. INTRODUCTION

Hydrostatic pressure is a modulator of biochemical processes: it inhibits bacterial growth, activates/inactivates enzymatic reactions. The application of high pressure above 3-4 kbar induces protein denaturation, whereas in the range between 1-2 kbar compression induces dissociation of oligomers[1]. It is commonly accepted that protein conformation is invariant to pressure and that inhibition of biological activity in the pre-denaturant range is due to the dissociation of oligomeric enzymes. But in recent years it has been proposed that inhibition of biological activity at moderate non-denaturating pressures might be due to the reduced flexibility of proteins rather than to the dissociation [2]. Up to date little is known about pressure effects on protein flexibility. In this paper we describe the use of the exquisite sensitivity of the phosphorescence lifetime of tryptophan to the fluidity of the environment to monitor changes in flexibility of several proteins under pressure. The results point out also important pressure-induced predissociational and post dissociational changes of protein structure

2. PHOSPHORESCENCE METHODOLOGY

Over the past two decades fluorescence spectroscopy has provided a wealth of both structural and dynamical information of biological macromolecules in solution. The potentialities of fluorescence methods are familiar to many. Less known and as yet little exploited is the phosphorescence emission from the internal tryptophan residue. The main difference between the two emission is the lifetime. Because phosphorescence is due to a spin forbidden transition, mostly between an excited triplet state and a singlet ground state, its lifetime is in the millisecond-second range and is 5 to 8 orders of magnitude longer than the lifetime of typical fluorescence states. The relative long triplet lifetime permits the investigation of dynamical events occurring in the ms-s range.

At room temperature dynamical information can be derived from Trp phosphorescence emission in two different ways: 1) from the bimolecular quenching

R. Winter and J. Jonas (eds.), High Pressure Molecular Science, 523–528.

rate constant of molecules that diffuse through the protein structure (recently it has been demonstrated that in the millisecond-second time scale of phosphorescence acrylamide can migrate through the macromolecule and that its rate is a measure of the frequency and amplitude of structural fluctuations underlying diffusional jumps [3]) and 2) from the dependence of the triplet lifetime (τ) on the local viscosity: investigations on the photophysics of the triplet state of the indole chromophore have emphasized the dramatic dependence of τ on microviscosity [4]. Indeed, the triplet lifetime of tryptophan residues varies from 6-6.5 s in glasses to 1 ms in aqueous solution at room temperature. Thus the triplet lifetime of Trp is exquisitely sensitive to the fluidity of the surrounding solvent/protein flexibility.

The unequalled sensitivity and wide dynamical range of the phosphorescence lifetime has been instrumental for detecting even minor changes in conformation of the macromolecules [5, 6].

3. PRESSURE EFFECTS ON PROTEIN FLEXIBILITY

The application of high pressure to protein in solution promotes any structural rearrangement of the macromolecule/solvent system that accommodates reduction in its volume. The compressibility (β) of proteins is determined mostly by two opposing contributions [7, 8] 1) reduction of internal cavities: this contribution is largest for inner regions of the macromolecule; 2) hydration of the polypeptide that will start from superficial regions. Reduction of cavities and hydration will exert opposing influences on protein dynamics. According to the correlation between compressibility, volume fluctuations and flexibility [9], compression will result in reductions in the mobility of internal regions. On the other hand, hydration which reduces the number of intramolecular H-bonds and concerns primarily peripheral regions of the polypeptide will tend to increase

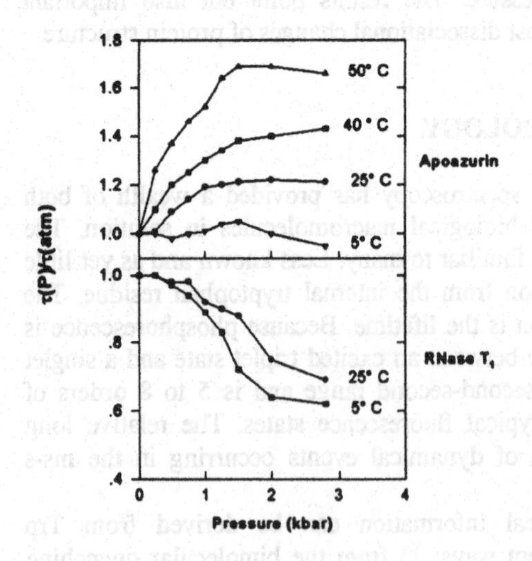

Figure 1. Relative change in the phosphorescence lifetime of Apoazurin and RNase T1 with applied pressure at different temperatures.

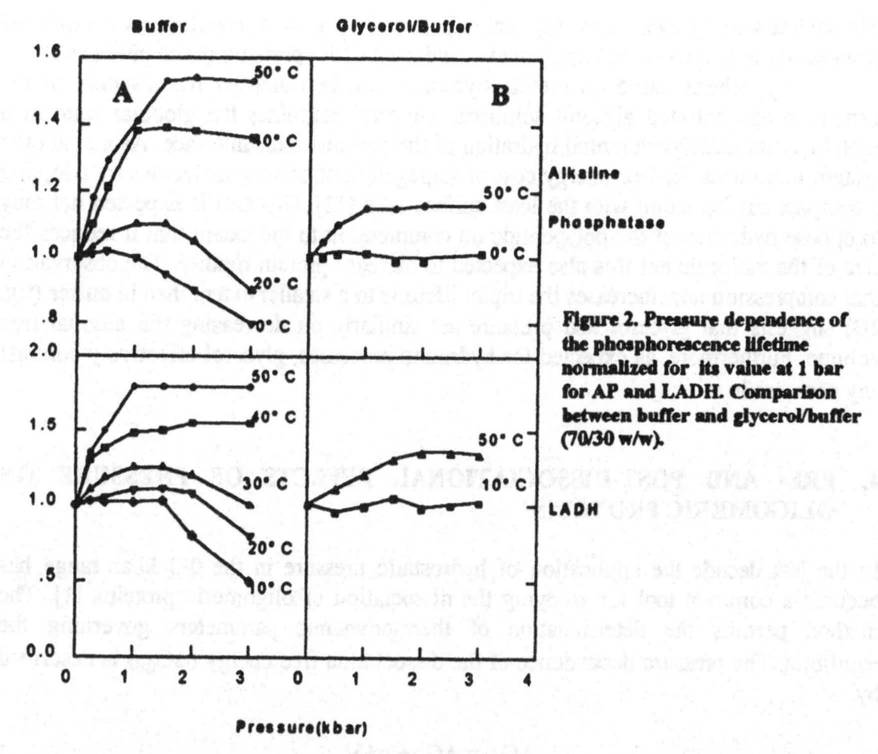

Figure 2. Pressure dependence of the phosphorescence lifetime normalized for its value at 1 bar for AP and LADH. Comparison between buffer and glycerol/buffer (70/30 w/w).

the flexibility of regions that in general are already more mobile. Since cavities are larger with thermal expansion and hydration increases at low temperature, the tightening/loosening effect of pressure will be more pronounced at high/low temperature, respectively. Our findings with apoazurin and Rnase T_1 (fig. 1) are consistent with the above expectations [10]. Trp48 of apoazurin is deeply buried and may be taken as a model for the effects of high pressure on the internal mobility of globular proteins. The pronounced lengthening of τ of Trp48 with pressure attests to a considerable decrease in flexibility of the chromophore' s environment, as expected from the reduction of internal cavities. The tightening effect increases with temperature, as cavity size increases with thermal expansion, and is less at low temperature where hydration of the polypeptide is favored.

In the case of RNase T_1 the triplet probe is nearest to the protein surface, only 2 Å from the solvent. High pressure reduces τ of this protein and an obvious explanation for the increased flexibility is given in terms of increased hydration of the peripheral region.

The same results have been obtained for other proteins with internal tryptophan residues as in the cases of alkaline phosphatase (AP) and liver alcohol dehydrogenase (LADH) (fig. 2A). For these two proteins we note that under conditions that strongly favor protein hydration, very high pressures and low temperatures, the lubricating action of the hydration water starts to dominate over any rigidity caused by the reduction in internal free volume even in the internal regions [11]. The same conclusions are inferred from acrylamide quenching experiments conducted under pressure (to be published).

Hypothesis based on protein hydration can be tested by the response of the system in concentrated glycerol solutions. Glycerol stabilizes the globular state of a protein by inducing preferential hydration of the protein-water interface. As a result the system minimizes the free energy cost of segregation of solvent molecules by adopting a compact conformation with the least surface area [12]. Glycerol is expected not only to oppose hydration of the polypeptide on compression to the extent that it reduces the size of the molecule but it is also expected to increase protein rigidity. The observation that compression now increases the triplet lifetime to a smaller extent than in buffer (fig. 2B) suggests that glycerol and pressure act similarly on decreasing the internal free volume. Furthermore, as expected for hydration processes, glycerol effectively contrast any decrease in τ.

4. PRE- AND POST-DISSOCIATIONAL EFFECTS OF PRESSURE ON OLIGOMERIC PROTEINS

In the last decade the application of hydrostatic pressure in the 0-3 kbar range has become a common tool for studying the dissociation of oligomeric proteins [1]. The method permits the determination of thermodynamic parameters governing the equilibria. The pressure dependence of the dissociation free energy ($\Delta G_{(p)}$) is described by

$$\Delta G_{(p)} = \Delta G_0 + p\Delta V$$

where ΔG_0 is the dissociation free energy at atmospheric pressure, p is the applied pressure and ΔV is the volume change of the reaction.

In the analysis of these equilibria it has always been assumed that pressure does not affect protein structure. But recent studies raise the concern that at the hydrostatic pressure required to cause subunit dissociation, pressure itself may directly influence the conformation of the oligomer and/or isolated subunit and that such perturbations might be responsible for some of the peculiarities of dissociation equilibria [11, 13, 14]. In this case the free energy of dissociation under pressure will contain a conformational term

$$\Delta G_{(p)} = \Delta G_0 + p\Delta V + \Delta G_{conf}$$

The sensitivity of τ to changes in the flexibility of the chromophore' s environment has revealed that high pressure can perturb the structure of both undissociated and dissociated species.

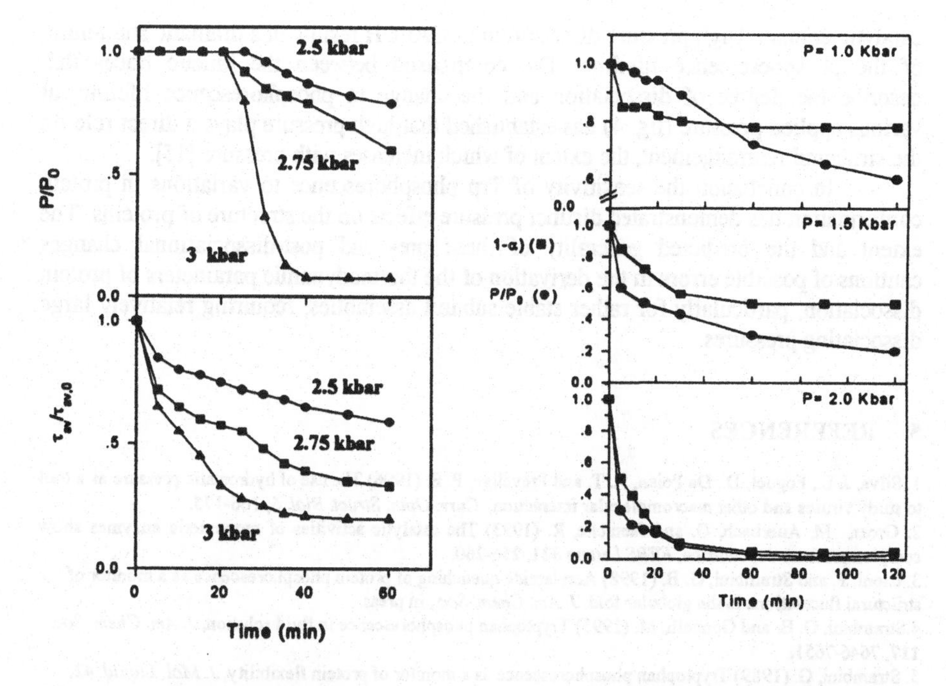

Figure 3. Time dependence of P/P0 and triplet lifetime of LADH at 0 °C under pressure.

Figure 4. Time dependence of the decrease in P/P_0 and in tetramer fraction of GAPDH at various dissociating pressures.

Dimeric LADH has the phosphorescent Trp (W314) buried at the subunit interface. Any change in the phosphorescence characteristics of LADH with pressure are to be attributed to alterations of the structure at the interface. In figure 3 is shown the comparison of the relative decrease of phosphorescence intensity (P/P_0) and τ/τ_0 profiles during the first hour of pressurization at 2.5, 2.75 and 3 kbar for LADH at 0 °C. In these conditions the LADH does not dissociate. The time dependent changes in phosphorescence characteristics distinguishes two consecutive steps. There is an initial stage characterized by a steady drop of τ/τ_0 that attests to a progressive loosening of the β sheet enveloping Trp314. After a lag period a second phase initiates in which deeper transformations of the protein structure cause quenching of the phosphorescence intensity that is complete at 3 kbar. The time evolution of the luminescence changes is strongly pressure-dependent. Therefore pressure perturbation results in a progressive hydration/loosening of the subunit interface region that can be responsible of a decrease of the subunit affinity under pressure [11].

An example of post-dissociational effects of pressure is given by gliceraldehyde-3-phosphate dehydrogenase (GAPDH). This tetrameric enzyme has the phosphorescent tryptophan in the catalytic domain at a site quite removed from the subunit interface. Consequently its environment is not altered by subunit dissociation per se, and changes in τ will necessarily reflect variation in the conformation of the catalytic domain. High pressure dissociation of GAPDH results in a dramatic shortening of the phosphorescence lifetime. The comparison between the kinetic traces that describe the degree of dissociation and the change in phosphorescence lifetime at various applied pressure (fig. 4) has established that high pressure plays a direct role on the structural rearrangement, the extent of which increases with pressure [15].

In conclusion the sensitivity of Trp phosphorescence to variations in protein conformation has demonstrated distinct pressure effects on the structure of proteins. The extent and the predicted generality of these pre- and post-dissociational changes cautions of possible errors in the derivation of the thermodynamic parameters of protein dissociation, particularly for rather stable subunit assemblies, requiring relatively large dissociating pressures.

5. REFERENCES

1. Silva, J. L., Foguel, D., Da Poian, A. T. and Prevelige, P. E. (1996) The use of hydrostatic pressure as a tool to study viruses and other macromolecular interaction, *Curr. Opin. Struct. Biol.* **6**, 166-175.
2. Gross, M, Auerbach, G. and Jaenicke, R. (1993) The catalytic activities of monomeric enzymes show complex pressure dependence, *FEBS Letters* **321**, 256-260.
3. Cioni, P. and Strambini, G. B. (1998) Acrylamide quenching of protein phosphorescence as a monitor of structural fluctuations in the globular fold, *J. Am. Chem. Soc.*, in press.
4. Strambini, G. B. and Gonnelli, M. (1995) Tryptophan phosphorescence in fluid solution, *J. Am. Chem. Soc.* **117**, 7646-7651.
5. Strambini, G. (1989) Tryptophan phosphorescence as a monitor of protein flexibility, *J. Mol. Liquid.* **42**, 155-165.
6. Strambini G. B. and Gabellieri E. (1996), Proteins in frozen solutions: evidence of ice-induced partial unfolding, *Biophys. J.* **70**, 971-976.
7. Gekko, K. And Hasegawa, Y. (1986) Compressibility-structure of globular proteins. *Biochemistry* **25**, 6563-6571.
8. Gekko, K. And Hasegawa, Y. (1989) Effect of temperature on the compressibility of native globular proteins, *J. Phys. Chem.* **93**, 426-429.
9. Cooper, A. (1976) Thermodynamic fluctuations in protein molecules, *Proc. Natl. Acad. Sci. USA* **73**, 2740-2741.
10. Cioni, P. and Strambini, G. B. (1994) Pressure effects on protein flexibility: monomeric proteins, *J. Mol. Biol.* **242**, 291-301.
11. Cioni, P. and Strambini, G. B. (1996) Pressure effects on the structure of oligomeric proteins prior to subunit dissociation, *J. Mol. Biol.* **263**, 789-799.
12. Timasheff, S. N. (1998) Control of protein stability and reactions by weakly interacting cosolvents: the simplicity of the complicated, *Adv. Prot. Chem.* **51**, 355-432.
13. Royer, C. A., Hinck, A. P. , Loh, S. N., Prehoda, K. E., Peng, X., Jonas, J. and Markley, J. L. (1993) Effects of aminoacid substitutions on the pressure denaturation of Staphylococcal nuclease as monitored by fluorescence and nuclear magnetic resonance spectroscopy, *Biochemistry* **32**, 5222-5232.
14. Peng, X., Jonas, J. and Silva, J. L. (1993) Molten-globule conformation of Arc repressor monomers determined by high-pressure ^{1}H NMR spectroscopy. *Proc. Natl. Acad. Sci. USA* **90**, 1776-1780.
15. Cioni, P. and Strambini, G. B. (1997) Pressure-induced dissociation of yeast glyceraldehyde-3-phosphate dehydrogenase: heterogeneous kinetics and perturbations of subunit structure, *Biochemistry* **36**, 8586-8593.

STABILITY DIAGRAM OF LIPOXYGENASE AS DETERMINED FROM H/D EXCHANGE KINETICS AND FROM CONFORMATIONAL CHANGES

P. RUBENS[1], J. FRANK[2] AND K HEREMANS[1]

[1]*Department of Chemistry, Katholieke Universiteit Leuven, B-3001 Leuven, Belgium.*

[2]*Kluyverlaboratorium for Biotechnology, Delft University of Technology, 2628 BC Delft, The Netherlands.*

Abstract

The stability of proteins under different conditions of pressure and temperature has already been the subject of many investigations. An elliptical outline of the stability curve can be suggested. In this study, we obtain the stability diagram of lipoxygenase by following the protein conformation with FTIR. Changes in the amide I (1700-1600 cm^{-1}) and amide II (1550 + 1450) region were monitored. Plotting midpoints of pressure and temperature induced cooperative changes gave an elliptical outline. In addition to the inactivation measurements of Ludikhuyze, we were interested to see if the typical behaviour could also be found in the hydrogen deuterium (H/D) exchange kinetics. Herefore, we followed the ratio of intensities at 1550 and 1450 cm^{-1}, which are proportional to the amount of H, resp. D in peptide bonds of the protein. A similar, but not identical, result was obtained. This shows that FTIR is a well suited technique to follow kinetics and thermodynamic equilibria in the protein under different physical conditions. The results of this work are compare with recent work on the inactivation kinetics at different pressures and temperatures.

1. Introduction

Protein stability is one of the main topics for many research groups. After decades of investigations on the influence of temperature and solvent conditions, it has become clear that pressure also can play an important role on the protein stability [1]. Screening protein stability under different physical conditions can be of interest in industrial application where food is treated under high pressure.

After Bridgman showed in 1914 that one was able to denature proteins of egg white [2], pressure has gained the interest of many researchers. It was shown that proteins denature under high pressure but on a different way as at high temperature. Moreover, combinations of temperature and pressure revealed a typical behaviour of

R. Winter and J. Jonas (eds.), High Pressure Molecular Science, 529–533.

proteins in a phasediagram. The outline of the stability diagram was mentioned by Suzuki [3]. It was found that in some cases, low pressure can stabilize protein against temperature denaturation while application of low temperatures can decrease the pressure stability. Hawley described in 1971 a theoretical model which permits to fit the pressure-temperature transition surface. It was shown that the surface can be fit to a relatively simple equation of state which shows approximately elliptical contours of constant free energy difference on the pressure-temperature plane [4,5].

Lipoxygenases catalyze the formation of fatty acid hydroperoxides, products used in further biochemical reactions leading to normal and pathological cell functions [6]. The stability against pressure and temperature denaturation have been investigated previously [7,8]. It was shown that protein denaturation was different in both cases. The major difference was the formation of intermolecular hydrogen bonds at high temperature. In recent work, Ludikhuyze et al. [9] showed results of inactivation measurements at different pressures and temperatures. Temperatures up to 35°C stabilized the protein against pressure denaturation. On the other hand, low pressures did not stabilize the protein against temperature denaturation. In this study, we will try to find a correlation between the inactivation data and the stability diagram that can be found when following the kinetics of hydrogen-deuterium exchange by using FTIR. Herefore, a thermostated diamond anvil cell (DAC) was used which allows to measure in a broad temperature (-20°C - 90°C) and pressure (1 bar - 30kbar) range.

2. Materials and methods

We used highly purified lipoxygenase form soybean that was dissolved in a 10mM TRIS-DCl buffer at pD=8.6 at a concentration of 100 mg/ml. The sample was stored overnight to ensure that exchange of the accessible protons has reached a static regime. The sample was mounted in a stainless steel gasket of a thermostated diamond anvil cell obtained from Diacell Products (Leicester, UK). The initial gasket thickness was 0.05 mm and the hole diameter 0.6 mm. The cell was placed in a Bruker IFS66 infrared spectrometer equipped with a liquid-nitrogen-cooled MCT detector. The sample compartment was continuously purged with dry air in order to avoid interference of atmospheric water vapour. $BaSO_4$ was added to determine the pressure. The time for each measurement started when the thermostated sample was brought at the appropriate pressure.

3. Results

The rate of hydrogen-deuterium exchange of the interior labile protons on the amide groups is dramatically increased by external pressure and is closely associated with the global conformational structure of the protein. These changes can be followed in the amide II region of the infrared spectrum. The band at ~1550 cm^{-1} is assigned to a mixed vibration involving N-H in plane bending and the C-N stretching. This band shifts to ~1455 cm^{-1} as a result of deuteration of labile protons on the amide group. By using a deuterated buffer, we were able to follow in-situ, the rate of exchange of the hydrogen atoms buried in the protein interior against deuterium atoms of the solvent. The ratio of intensities at 1550 and 1450 cm^{-1} was plotted in function of time. For each experiment, the exchange rate was followed during 3 hours. This resulted in an exponential decrease of the H/D ratio in the protein. First order kinetics were assumed and rate constants of the exchange processes were determined. This was done for a wide range of temperatures and pressures. Iso-rate-constant lines were plotted in a p,T-diagram (Fig.1). No data points were obtained for temperatures higher than 55°C as fast exchange kinetics occurred. At these temperatures, the remaining changes in the spectra happened beyond the spectral resolution of the spectrometer. For temperatures up to 35°C, a small stabilizing effect can be observed. Higher temperatures destabilize the protein and thereby accelerate the exchange process. These results are in good agreement with previously performed enzyme inactivation kinetics [9].

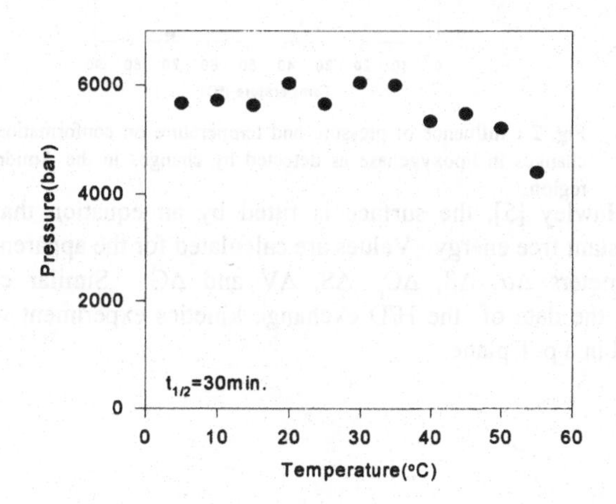

Fig. 1 : Influence of pressure and temperature on H/D exchange of lipoxygenase. Half life time of 30 min for H/D exchange was plotted at different conditions.

In addition to previous results, we tried to examine the stability of lipoxygenase on a different way. The pressure-temperature stability diagram has been determined from the conformational changes. The Amide I region in the infrared spectrum allows us to follow these changes in the protein. The bands in this region (1700-1600 cm^{-1}) are assigned to a conformational sensitive vibration that is mainly a C=O stretching. Each band in this region is associated with a secondary structure. Thus, the changes in the conformation will be reflected in the spectrum. The combination of the diamond anvil cell with an FTIR spectrometer allows us to follow, in-situ, the protein conformation under different condition of temperature and pressure. Plotting the spectral changes as a function of pressure and temperature results in a sigmoidal curve. At given conditions, a cooperative change takes place. The midpoints of these transitions were plotted in a p, T diagram (Fig.2)

Further analysis of this sigmoidal curve allows us to calculate ΔG values at different conditions. The K-values, obtained from the sigmoidal curve are transformed to ΔG values by using the equation : $\Delta G = -R\,T \ln K$ ($\Delta G^{\#} = -R\,T \ln K^{\#}$). This allows us to make a three dimensional plot of ΔG, p and T.

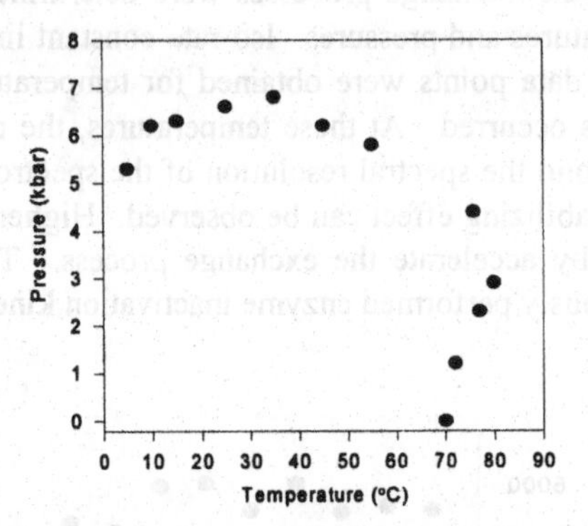

Fig. 2 : Influence of pressure and temperature on conformational changes in lipoxygenase as detected by changes in the Amide I region.

According to Hawley [5], the surface is fitted by an equation that gives elliptical contours of constant free energy. Values are calculated for the apparent thermodynamic transition parameters $\Delta\alpha$, $\Delta\beta$, ΔC_p, ΔS, ΔV and ΔG. Similar calculations were performed with the data of the H/D exchange kinetics experiment were the value of $\Delta G^{\#}$ was plotted in a p-T plane.

4. Discussion

FTIR is a well suited technique to follow kinetic and thermodynamic equilibria in the protein under different physical conditions. The combination of FTIR with a thermostated DAC has the major advantage that it allows in-situ observation of the sample at different physical conditions.

It is shown that the iso-rate constant line that is obtained by following the H/D exchange rate in a protein under different conditions presents the same outline as these from inactivation experiments. In contrast, the stability diagram, as determined from conformational changes shows a different outline. Especially the region of low pressure and high temperature reveals a significant difference. In the case of the H/D exchange kinetics, no stabilization with pressure against temperature denaturation is found. These results where in good agreement with the inactivation data. The difference between both results can be explained by the way these data points are obtained.

In the case of the H/D exchange experiment, one first increases the temperature and pressure is applied only after the solution is thermostated. By looking at the stability diagram as observed from conformational changes, on can see that this can lead to entering the region of the denatured state. By increasing the pressure at high temperature, it is possible to re-enter the region of the native state provided reversible changes can be observed. If one works with proteins that do not show a reversible transition between the native and the denatured state, it will not be possible to detect properties associated with the native protein. Thus, the stabilisation at low pressures against temperature denaturation is not observed.

In the case of the conformational change experiments, one first increases the pressure followed by a temperature scan. Hereby, the protein solution stays under conditions where the native state is stable. Only when reaching the denaturation temperature, the protein will lose its native conformation. By this way, it is not necessary that protein denaturation occurs in a reversible way to detect stabilization effects in the low pressure - high temperature region. As reversibility can play an important role, it should be clear that it is important to consider the different experimental procedures.

Acknowledgements

The results presented in this paper were obtained with the support from the Research Fund of the K.U.Leuven, F.W.O. Flanders, Belgium and the European Community.

References

[1] Heremans, K. and Smeller, L. (1998) Protein structure and dynamics at high pressure; *Biochim. Biophys. Acta.* **1386**, 353-370.

[2] Bridgman, P.W. (1914) The coagulation of albumen by pressure; *J. Biol. Chem.* **19**, 511-512.

[3] Suzuki, K. (1960) Studies on the kinetics of protein denaturation under high pressure; *Rev. Phys. Chem. Jpn.* **29**, 49-55.

[4] Hei, D.J. & Clark, D.S. (1994) Pressure stabilization of proteins from extreme thermophiles; *Appl. Env.Micr.* **60**, 932-939.

[5] Hawley, S.A., (1971) Reversible pressure-temperature denaturation of chymotrypsinogen; *Biochemistry* **13**, 2436-2442.

[6] Gaffney, B.J. (1996); Lipoxygenases: Structural principles and spectroscopy; *Annu. Rev. Biophys. Biomol. Struct.* **25**, 431-459

[7] Heinish, O., Kowalski, E., Goossens, K., Frank, J., Heremans, K., Ludwig, H. & Tausher, B. (1995) Pressure effects on the stability of lipoxygenase: Fourier transform-infrared spectroscopy (FTIR) and enzyme activity studies; *Z. Lebensm. Unters. Forsch.* **201**, 562-565.

[8] Mozhaev, V.V., Heremans, K., Frank, J., Masson, P. & Balny, C. (1996) High pressure effects on protein structure and function; *Proteins* **24**, 81-91.

[9] Ludikhuyze, L., Indrawati, I., Van den Broeck, I., Weemaes, C., Hendrickx, M. (1998) Effect of combined pressure and temperature on soybean lipoxygenase. 1. Influence of extrinsic factors on isobaric-isothermal inactivation kinetics. *J. Agric. Food Chem.* **46**, 1074-1080.

FTIR AS A TOOL TO STUDY COLD, HEAT AND PRESSURE DENATURATION OF MYOGLOBIN

F. MEERSMAN, L. SMELLER* and K. HEREMANS
Department of Chemistry, Katholieke Universiteit Leuven, B-3001 Leuven, Belgium
**Institute of Biophysics, Semmelweis University of Medicine, Budapest, H-1444 Hungary*

Abstract

We studied the denaturation of horse heart metmyoglobin in the diamond anvil cell (DAC) with Fourier transform infrared spectroscopy (FTIR). It is observed that the conformational changes due to the cold and the pressure denaturation are similar, but not identical and that both processes do not lead to a complete loss of the secondary structure. Heat denaturation distinguishes itself from the former two processes by the formation of two new bands at 1615 and 1683 cm^{-1} which are characteristic for intermolecular β-sheet aggregation. This also explains why a gel can be observed at the end of the experiment. A gel, however, can also be observed in the case of a pressure denaturation. Some aspects of pressure- and heat-induced aggregation are discussed as well.

1. Introduction

Cold, heat and pressure denaturation of proteins has been extensively studied using a variety of methods such as NMR [1], UV spectroscopy [2] and DSC [3]. We studied the denaturation of horse heart metmyoglobin in the diamond anvil cell (DAC) with Fourier transform infrared spectroscopy (FTIR). FTIR spectroscopy enables us to follow and assess, both qualitatively and quantitatively, the changes in the secondary structure upon denaturation. In combination with the DAC we can investigate the denaturation *in situ* in a broad pressure-temperature range (pressures up to 2GPa and temperatures between -200 and 100 °C). Here we report on the observed differences between these three ways of denaturation and the aggregation phenomenon.

2. Materials and Methods

Horse heart metmyoglobin was purchased from Sigma and used without further purification. The aqueous buffer was 25 mM deuterated acetate buffer, pD 4. Metmyoglobin was dissolved in the deuterated buffer and the sample was stored

R. Winter and J. Jonas (eds.), High Pressure Molecular Science, 535–539.

overnight to ensure a sufficient H/D-exchange. To study the cold denaturation we used the DAC (Diacell Products, Leicester, UK), which was placed in a cryocell. The pressure was increased up to 200 MPa by means of a helium driven membrane and the temperature was lowered using solid carbondioxide and acetone. In case of the pressure denaturation, the DAC was also used, but pressure was built up by means of a screw mechanism. Barium sulphate was used as an internal pressure standard in all cases. For the heat experiment a CaF_2 cell and a Graseby Specac Automatic Temperature Controller were used. The infrared spectra were obtained with a Bruker IFS66 FTIR spectrometer equipped with a liquid nitrogen cooled broad band MCT solid state detector. 250 Interferograms were coadded after registration at a resolution of 2 cm^{-1}. Fitting was done with a programme developped in our laboratory. The spectra were fitted after Fourier self-deconvolution by using a Lorentzian of 35 cm^{-1} half-bandwidth and a resolution enhancement factor (k value) of 2. The procedure and the above mentioned parameters are discussed elsewhere [4].

3. Results and discussion

The study of the cold denaturation of proteins in an aqueous solvent is limited by the formation of ice. It has been suggested that the protein might adsorb to the growing ice particles and consequently it might undergo a mechanical denaturation [5]. FTIR spectroscopy, however, shows no spectral changes in the general shape of the amide I' band upon ice formation. What can be observed is a change of background. In order to avoid this possible effect, several experimental setups have been described [6]. In this work we performed a pressure-assisted cold denaturation. The reason for this pressure assistance becomes clear when one takes the phase diagram of water into account [7]. The phase diagram of liquid water has a minimum at about 200 MPa. At that pressure water remains in the liquid phase down to –20 °C. On top of that we used an acetate buffer to assure a complete denaturation of the metmyoglobin before ice formation. The acetate buffer has the additional advantage that its pD does not change as a function of temperature.

Comparison of the spectral changes due to the cold denaturation with those due to the pressure denaturation shows that the changes in both cases are very similar, but not identical. This similarity together with the fact that the final denatured states are structurally comparable, could indicate that the mechanisms of these processes are quite alike. Furthermore, it seems that a significant amount of secondary structure, mainly intramolecular β-sheet, is still present in both the denatured states. This is in agreement with the NMR observations on ribonuclease [1] where a partially folded structure is still present in the denatured state. Some conformational data are shown in Table 1.

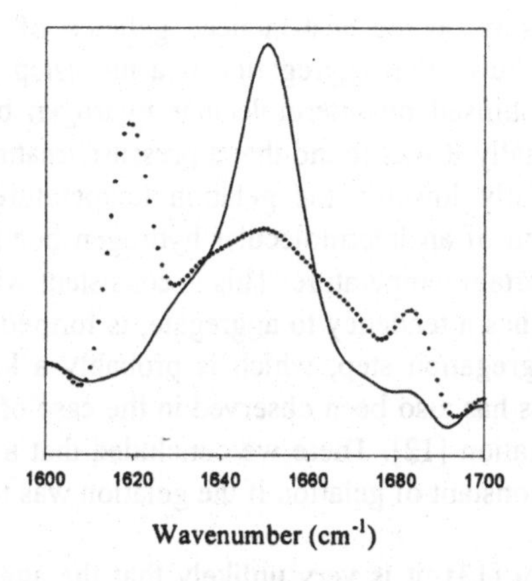

Fig. 1. Heat denaturation of metmyoglobin. Full line: native state, dotted line: denatured state

Heat denaturation, on the contrary, distinguishes itself from the former two processes by the formation and increase of two new bands at 1683 and 1615 cm^{-1} as the temperature increases. This is shown in Figure 1. These two bands are characteristic for intermolecular β-sheet aggregation [8]. This aggregation is probably the reason why heat denaturation is irreversible in most cases. It should be noted that these bands do not always appear at high temperatures. For instance, the heat denaturation of RNAse A, which is a common example of a reversible unfolding protein, does not lead to their formation [9]. The aggregation also explains why a gel can be observed at the end of the experiment. But a gel can also be observed in the case of a pressure denaturation where no such bands can be seen. This could point at the existence of different interactions leading to gel formation dependent on the mechanism of denaturation. These differences between pressure- and heat-induced gels have also been observed by rheological measurements for some proteins [10].

TABLE 1. The secondary structure content (%) of metmyoglobin at pD 4

	α-helix + random	Intramolecular β-sheet	Turn	Intermolecular β-sheet
Native	62	29	9	
Heat denatured	39	6.5	7	47
Cold denatured	50.5	44	5.5	
Pressure denatured	63	33	4	

Recently the effect of pressure on the heat-induced gelation of myoglobin has been studied [11]. There it was shown that aggregation is a multistep process and that the formation of a structure stabilised by intermolecular hydrogen bonds, i.c. a gel, is disfavoured by pressure. Finally it was found that a pressure treatment of the protein at room temperature significantly lowered the gelation temperature. This leads to the conclusion that the formation of an intermolecular hydrogen bonded network requires an unfolded protein and a certain temperature. This is consistent with the view that first a kind of intermediate, that has a tendency to aggregate, is formed, that, in a next step, will aggregate. For this aggregation step, which is probably a kinetic step, a higher temperature is required. This has also been observed in the case of trypsin at acidic pH for pressure-induced aggregation [12]. There we concluded that a higher pressure was needed to increase the rate constant of gelation if the gelation was to be observed within a reasonable timescale.

As stated before by Doi [13], it is very unlikely that the aggregating molecule is completely unfolded. This is also what we observe in case of the heat denaturation where the conformation of the protein just before the first appearance of the bands due to intermolecular β-sheet aggregation is still quite native like. When fitted, the structure of this state resembles the one of the cold denatured state. It was shown before by Jonas et al. [14] that the cold denatured state of lysozyme resembles an early folding intermediate. Is the intermediate state that will aggregate a molten globule?

4. Conclusions

From a conformational point of view the pressure and the cold denaturation are very alike but not identical, while the heat denaturation distinguishes itself by the formation of new bands. As for the aggregation process, it is plausible that a general scheme, which is valid for the heat-induced gelation as well as for the pressure-induced one, can be drawn up. The only difference between the two would be the nature of the interactions, that lead to the aggregation, on which the heat respectively the pressure act. In the case of heat denaturation the involvement of hydrogen bonds in intermolecular antiparallel β-sheet aggregation has been clearly shown, while those involved in the pressure-induced aggregation are still a topic of discussion.

Finally, if one assumes that the partially unfolded protein is incorporated in an aggregation network and thereby stabilized, one could wonder what the nature of the fully denatured state is.

Acknowledgements

The results presented in this paper were obtained with the support from the Research Fund of the K.U.Leuven, F.W.O Flanders, Belgium and the European Community.

References

[1] Zhang, J., Peng, X., Jonas, A. & Jonas, J. (1995) NMR study of the Cold, Heat, and Pressure Unfolding of Ribonuclease A, *Biochem.* **34**, 8631-8641.

[2] Mombelli, E., Afshar, M., Fusi, P., Mariani, M., Tortora, P., Connelly, J.P. & Lange,R. (1997) The role of Phe31 in maintaining the conformational stability of ribonuclease P2 from *Sulfolobus solfactarius* under extreme conditions of temperature and pressure, *Biochem.* **36**, 8733-8742.

[3] Privalov, P.L., Griko, Yu.V. & Venyaminov, S.Yu. (1986) Cold denaturation of myoglobin, *J. Mol. Biol.* **190**, 487-498.

[4] Smeller, L. Goossens, K & Heremans, K (1995) How to minimize certain artifacts in Fourier self-deconvolution, *Appl. Spectr.* **49**, 1538-1542.

[5] Strambini, G.E. & Gabellieri, E. (1996) Proteins in frozen solutions : Evidence of ice-induced partial unfolding, *Biophys. J.* **70**, 971-976.

[6] Franks, F. (1995) Protein destabilization at low temperatures, *Adv. Prot. Chem.* **46**, 105-139.

[7] Bridgman, P.W. (1935) The Pressure-Volume-Temperature Relations of the Liquid, and the Phase Diagram of Heavy Water, *J. Chem. Phys.* **3**, 597-605.

[8] Ismail, A.A., Mantsch, H.H. & Wong, P.T.T. (1992) Aggregation of chymotrypsinogen: portrait by infrared spectroscopy; *Biochem. Biophys. Acta* **1121**, 183-188.

[9] Fabian, H. & Mantsch, H.H. (1995) Ribonuclease A Revisited: Infrared Spectroscopic Evidence for the Lack of Native-like Secondary Structure in the Thermally Denatured State, *Biochem.* **34**, 13651-13655.

[10] Heremans, K., Van Camp, J. & Huyghebaert, A. (1997) High pressure effects on proteins, in *Food Proteins and Their Applications* (Ed. Damodaran, S. & Paraf, A.), pp. 473-502.

[11] Smeller, L., Rubens, P. & Heremans, K. (1999) Pressure effect on the temperature unfolding and tendency to aggregate of myoglobin, *Biochem.*, in press

[12] Meersman, F. (1998), unpublished results

[13] Doi, E. (1993) Gels and gelling of globular proteins, *Trends in Food Science & Technology* **4**, 1-5.

[14] Nash, D.P. & Jonas, J. (1997) Structure of the Pressure-Assisted Cold Denatured State of Ubiquitin, *Biochem. Biophys. Res. Com.* **238**, 289-291.

ON THE USE OF THE TERM OSMOTIC PRESSURE

H. PFEIFFER and K. HEREMANS
Department of Chemistry, Katholieke Universiteit Leuven,
Celestijnenlaan 200D, 3001 Leuven, Belgium

ABSTRACT: The use of thermodynamic quantities deviate occasional from their definitions given in standard textbooks. This could lead to misunderstandings and misinterpretations. This paper reviews some definitions and gives a classification of osmotic pressure and osmotic stress experiments as well as a critical view of the lipid and protein literature.

1. Introduction

The investigation of pressure and/or hydration effects of biological substances requires defined experimental conditions. In a pioneering publication, Parsegian et al. [1] gave a report about the "measured work of deformation and repulsion of lecithin bilayers". The authors describe three methods to achieve a controlled chemical potential of water that they call equivalent and complementary. In spite of the great merits of the introduction of those "osmotic stress methods" (which were extended to other bio-macromolecules) there are two aspects that should be mentioned. The first concerns the use of the term "osmotic pressure" as a synonym for the description of the variation of the chemical potential of water. Some interpretations of experimental results that were obtained by the application of the osmotic stress method could fail at the point where "osmotic pressure" is compared with the effects of hydrostatic pressure [2,3]. According to its original definition, osmotic pressure is a hydrostatic pressure. The second aspect is the equivalence of the "osmotic stress methods" that can only be valid for the calculation of the work of hydration if defined conditions are satisfied.

2. Definitions

2.1 OSMOSIS

The net flux of solvent molecules from one solution into another solution caused by different concentrations of solutes is called osmosis. This flux in the direction of the solution having the higher concentration is driven by diffusion. The concentration difference is maintained by a selective diffusion barrier that avoids the exchange of solute molecules.

R. Winter and J. Jonas (eds.), High Pressure Molecular Science, 541–545.

2.2 OSMOTIC PRESSURE

Osmotic pressure Π is defined as the smallest hydrostatic pressure that is required to stop or prevent osmosis (i.e. a kind of compensation pressure). In other words, the hydrostatic excess pressure ($\Pi = p - p_0$) which must act on the higher concentrating solution (phase 2) to enable the chemical equilibrium with the lower concentrating solution (phase 1=reference phase) is called osmotic pressure. If the molar volume V_w of solvent can considered being constant one can write:

$$\mu_1 = \mu_2(p_0) + V_w(p - p_0) \tag{1}$$

$$\Pi = \frac{\mu_1 - \mu_2(p_0)}{V_w} \tag{2}$$

The osmotic pressure (e.g. ideal solutions) arises from the dilution tendency of solvents, which is an entropic effect. However, in real systems there exist also non-entropic effects (e.g. specific solute-solvent interactions), which influence the difference of chemical potential and with it, osmotic pressure. If non-entropic effects begin to dominate the difference of chemical potential one should be careful to use the term "osmotic pressure". At least, the existence of osmotic pressure requires the existence of an appropriate two-phase system i.e. a single solution cannot have any osmotic pressure.

2.3 WATER ACTIVITY

The difference of the chemical potentials between the actual and reference state can be expressed in terms of the formalism made for ideal solutions ($\mu - \mu_0 = RT \ln a_W$). Then, the molar fraction x is replaced by an "effective" molar fraction called the activity a_W ($a_W = f x_W$ where f defines the activity coefficient). Note that the hydrostatic pressure enhances the activity coefficient and values of a>1 are possible [4].

2.4 WATER POTENTIAL

The difference of the chemical potentials between the actual and the reference state divided by the molar volume of the reference water V_w is called water potential, Ψ_W. [5]

$$\psi_W = \frac{\mu_W - \mu_{WO}}{V_w} \tag{3}$$

The water potential can be divided in different components as osmotic potential (entropic origin), turgid potential (hydrostatic pressure) and matrix potential (contains contributions from the interactions with colloids as e.g. in capillary and surface interactions).

	μ-μ_0	a	ψ
μ-μ_0		$\mu - \mu_0 = RT \ln a$	$\mu - \mu_0 = V_W \Psi$
a	$a = \exp\left(\frac{\mu - \mu_0}{RT}\right)$		$a = \exp\left(\frac{V_W \Psi}{RT}\right)$
ψ	$\Psi = \frac{\mu - \mu_0}{V_W}$	$\Psi = \frac{RT}{V_W} \ln a$	

Tab.1 A summary of the relations between chemical potential μ_W, activity a_W and water potential ψ_W, R is the gas constant, T the absolute temperature and V_W is the molar volume of water

3. Classification of the experimental configurations

3.1 CONFIGURATION I (piezotropic and non-isopotential)

As shown in Figure 1.I this experimental configuration represents one of the most used configurations to investigate high-pressure effects. The target dispersion is situated in a closed vessel that prevents a loss of solvent. Hydration means in that case an amplification or weakening of solvent/solute interactions due to hydrostatic pressure. Note, in this configuration the chemical potential of the components is enhanced if the mechanical pressure is increased.

3.2 CONFIGURATION II (piezotropic and isopotential)

As shown in Figure 1.II this corresponds e.g. to the original osmotic or hydration pressure experiments. The chemical potential of water is constant at every step of hydration because of the chemical equilibrium with the free water phase. The number of water molecules in the solution is correlated with the hydrostatic pressure.

3.3 CONFIGURATION III (isopiestic and non-isopotential)

As shown in Figure 1.III this configuration represents the isopiestic hydration using vapour pressure or appropriate osmotic solutions (see e.g. vapour pressure and osmotic stress method, Parsegian et al. [1]). One obtains functional pairs of the chemical potential and hydration. (Note that the use of "excluded" solutes to adjust the chemical potential was questioned by Timasheff [6].)

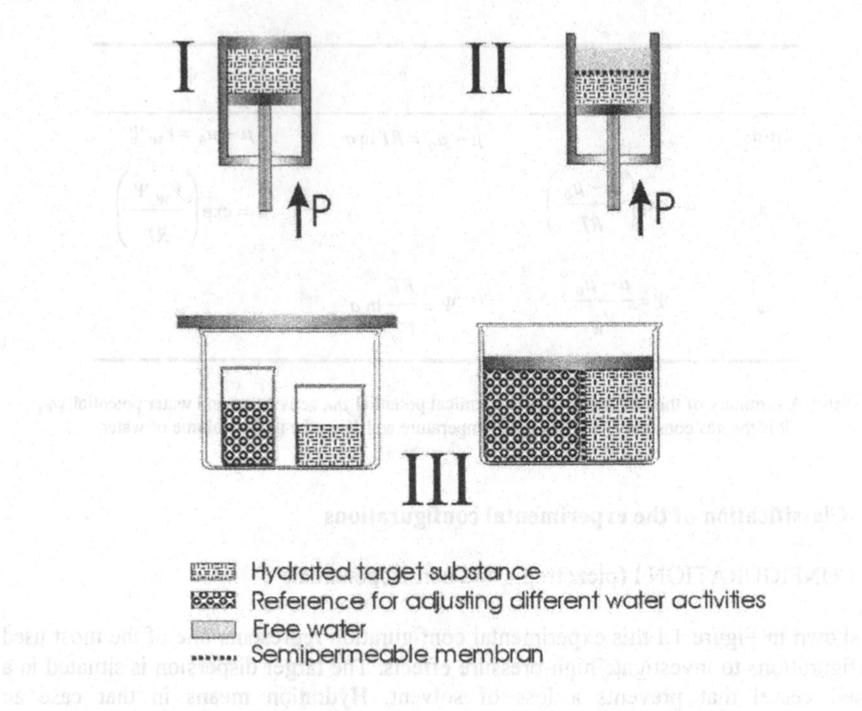

Fig. 1 Three different possibilities for the adjustment of pressure and/or hydration

	Constant	Variable
I	n_w	P, μ_w
II	μ_w	n_w, P
III	P	μ_w, n_w

Tab. 2 Relationship between the hydration hydrostatic pressure and chemical potential at different experimental configurations

4. Discussion

4.1 OSMOTIC PRESSURE

It appears that "osmotic pressure" is sometimes used as a synonym for the water potential ψ_W. This will be not a problem if one handles this "osmotic pressure" consequently as water potential (see e.g. [7]. In fact, Rand et al. clearly expressed that the volume changes due to "osmotic pressure" are different from those induced by high pressure (configuration I).

However, there are also other cases where the formalism seems to be applied in a different manner as shown in the following example [2]. The authors investigated the influence of "osmotic pressure" (water potential) and hydrostatic pressure on the low-spin - high-spin transition in case of cytochrom 450. They found that both "pressures" induce opposite or antagonistic effects i.e. hydrostatic pressure promotes the high spin to low spin transition and "osmotic pressure" promotes the low spin to high spin transition. If one correlate this effect with the activity of water, there remains indeed no opposite effect as the water activity is increased at hydrostatic pressure and decreased by "osmotic pressure". The opposite effects appear under this view less mysterious. Furthermore, it was correctly suggested that hydrostatic pressure and "osmotic pressure" probe different properties of the transition. However, this finding is presented as a conclusion as the different characters of the reaction volumes are a priori given by the different experimental configurations and they cannot be a matter of interpretation.

We think that in case of a more consequent use of an unequivocal nomenclature such confusing conclusions could be avoided.

4.2. EQUIVALENCE OF EXPERIMENTAL CONFIGURATIONS

Parsegian et al. [1] suggest that the experiments of type II and III are equivalent concerning the adjustment of chemical potential. This cannot hold exactly. If quantities owing pressure units as water potential and osmotic pressure have the same absolute magnitude at the same hydration degree, they are nevertheless related to different chemical potentials. Therefore, all microscopic and macroscopic properties that are correlated to the chemical potential should also be different.

For the determination of "osmotic pressure" resp. hydration pressure there are several methods proposed which correlate "osmotic pressure" to different thermodynamic or spectroscopic quantities. All this approaches assume that the hydrophilicity of the organic substances is independent of mechanical pressure i.e. not correlated with compressibility effects. However, this has to be proven in every case.

Acknowledgements

The results presented in this paper were obtained with the support from the Research Fund of the K.U. Leuven and F.W.O. Flanders, Belgium.

References

1. Parsegian, V.A., Fuller, N., and Rand, R.P. (1979) Measured work of deformation and repulsion of lecithin bilayers, *Proc. Natl. Acad. Sci. USA* **76**, 2750-2754
2. Di Primo, C., Deprez, E., Hui Bon Hoa, G., Douzou, P. (1995) Antagonistic effects of hydrostatic pressure and osmotic pressure on cytochrome P-450$_{cam}$ spin transition. *Biophys. J.* **68**, 2056-2061
3. Robinson, C.R., and Sligar, S.G. (1995) Heterogeneity in molecular recognition by restriction endonucleases: Osmotic and hydrostatic pressure effects on Bam HI, Pvu, and EcoRV specifity, *Proc. Natl. Acad. Sci. USA*. **92**, 3444-3448
4. Moore, W.J. (1972) *Physical Chemistry*, Longman, London, p. 122
5. Adam, J., Länger, P. and Stark, G. (1995) *Physikalische Chemie und Biophysik*, Springer Berlin, p.124
6. Timasheff, S.N. (1998) In disperse solution, "osmotic stress" is a restricted case of preferential interactions, *Proc. Natl. Acad. Sci. USA*, **95**, 7363-7367
7. Rand, R.P., Fuller, N.L., Butko, P. Francis, G. Nicholis, P. (1993) Measured change in Protein Solvation 'with Substrate Binding and turnover. *Biochemistry* **32**, 5923-5929

Subject Index

Author Index

List of Participants

R. Appel
Dept. of Physics & Astronomy
Texas Christian University
Fort Worth, Texas 76129
U.S.A.

Dr. C. Balny
U128 INSERM
Route de Mende
F-34033 Montpellier Cedex 1
France

Dr. L. Ballard
Department of Chemistry
University of Illinois
Urbana-Champaign, Il 61801
U.S.A.

Dr. W. Brasil
Physics Department
Cidade University
Rio de Janeiro
Brazil

Prof. Dr. M. Buback
Institut für Physikalische Chemie
Universität Göttingen
D-37077 Göttingen
Germany

O. Ces
Department of Chemistry
Imperial College
Exhibition Road
London SW 72 AY
U.K.

Dr. P. Cioni
CNR
Instituto di Biofisica
56127 Pisa
Italy

Dr. C. Czeslik
Department of Chemistry
University of Illinois
School of Chemical Sciences
Urbana, Il 61801
U.S.A.

Prof. Dr. H. Drickamer
Dept. of Chemical Engineering
University of Illinois
114 Roger Adams Lab.
Urbana, IL 61801
U.S.A.

Dr. D. J. Dunstan
Queen Mary and Westfield College
Department of Physics
University of London E1 4NS
U.K.

Dr. B. Efros
Donetsk Physics & Technology Inst.
of the Natl. Acad. of Sciences
340114 Donetsk
Ukraine

J. Eisenblätter
Physical Chemistry
University of Dortmund
D-44221 Dortmund
Germany

A. dias Ferrao-Gonzales
Dept. Bioquimica Medica-CCS
Rio de Janeiro CEP 21941-590
Brazil

U. Fekl
Department of Chemistry (Anorg. Chem.)
University of Erlangen-Nürnberg
D-91058 Erlangen
Germany

Dr. D. Foguel
Department of Biochemistry
Federal University of Rio de Janeiro
Rio de Janeiro CEP 21941-590
Brazil

Dr. A. Freiberg
Institute of Physics
University of Tartu
EE-2400 Tartu
Estonia

M. Frogley
Queen Mary and Westfield College
Physics Department
London E1 4N5
U.K.

A. Gabke
Department of Chemistry
University of Dortmund
D-44221 Dortmund
Germany

J. Greif
University of Firenze
Physics Department
Largo Enrico Fermi 2
I-50125 Firence
Italy

Prof. Dr. K. Heremans
Katholieke Universiteit Leuven
Department of Chemistry
B-3001 Leuven
Belgium

Prof. Dr. F. Hensel
Philipps-University of Marburg
Physical Chemistry
Hans-Meerwein-Straße
D-35032 Marburg
Germany

K. Hochgesand
Physical Chemistry
University of Dortmund
D-44221 Dortmund
Germany

Dr. G. Jenner
Laboratoire de Piezochimie Organique
Université Louis Pasteur
Institut de Chimie
1 rue Blaise Pascal
F-67008 Strasbourg Cedex
France

Prof. Dr. A. Jonas
Department of Biochemistry
University of Illinois
South Mathews Ave.
Urbana-Champaign, Il 61801
U.S.A.

Prof. Dr. J. Jonas
University of Illinois
School of Chemical Sciences
166 Roger Adams Lab.
Urbana, Il 61801
U.S.A.

Dr. Y.J. Kim
School of Chemical Sciences
University of Illinois
S. Mathews Ave.
Urbana-Champaign,Il 61801
U.S.A.

Dr. D. D. Klug
Steacie Institute for Molecular Sciences
National Research Council of Canada
Ottawa, Ontario K1A OR6
Canada

Dr. N.L. Klyachko
INSERM CNRS
1919 Route de Mende
F-34033 Montpellier Cedex 1
France

E. Kooi
Van der Waals-Zeeman Instituut
Universiteit van Amsterdam
1018 XE Amsterdam
The Netherlands

P. Krzeminski
Polish Academy of Science
Institute of Bioorganic Chemistry
61-704 Poznan
Poland

R. Kurzhöfer
Physical Chemistry
University of Dortmund
Otto-Hahn Str. 6
D-44221 Dortmund
Germany

Dr. J. Lee
Beckman Institute
405 N. Mathews Ave.
Urbana, Il 61801
U.S.A.

Dr. M.-H. Lemée-Cailleau
UMR CNRS 6626
University of Rennes I
F-35042 Rennes Cedex
France

H. Lotz
Van der Waals-Zeeman Instituut
Universiteit van Amsterdam
1018 XE Amsterdam
The Netherlands

Chr. Lorkowski
Department of Chemistry
University of Dortmund
D-44221 Dortmund
Germany

H. Lucas
LIMHP-CNRS
Institut Galilee
Université Paris
F-93430 Villetaneuse
France

Prof. Dr. H. Ludwig
Institut für Pharmazeutische Technologie
Universität Heidelberg
D-69120 Heidelberg
Germany

D. Mathieu
CEA Le Ripault, BP 16
37260 Monts
France

Dr. M.A. McLean
Beckman Institute
University of Illinois
405 N. Mathews
Urbana-Champaign, Il 61801
U.S.A.

F. Meersman
Department of Chemistry
Universiteit Leuven
B-3001 Leuven
Belgium

Prof. Dr. A. Misiuk
University of Electron Technology
Al. Lotnikow 46
02-668 Warsaw
Poland

Dr. E. Mombelli
INSERM CNRS
Route de Mende
F-34033 Montpellier
France

Dr. M.A.C. Nascimento
Physics Department
Cidade University
Rio de Janeiro
Brazil

A.C. de Oliveira
Department of Biochemistry
Federal Univers. of Rio de Janeiro
Rio de Janeiro CEP 21941-590
Brazil

A.M. de Oliveira Gomes
Department of Biochemistry
Federal Univers. of Rio de Janeiro
Rio de Janeiro CEP 21941-590
Brazil

Dr. P. Oszajca
Jagiellonian University
30060 Krakow
Poland

K.D. Patel
Department of Chemistry
University of North Carolina
Chappel Hill, NC 27599
U.S.A.

R.C. Pan
Department of Physics
University of Illinois
Urbana, IL 61801
U.S.A.

H. Pfeiffer
Department of Chemistry
Universiteit Leuven
B-3001 Leuven
Belgium

P. Rubens
Department of Chemistry
Universiteit Leuven
B-3001 Leuven
Belgium

Dr. J. A. Schouten
Van der Waals-Zeeman Laboratory
University of Amsterdam
1018 XE Amsterdam
The Netherlands

H. Seemann
Department of Chemistry
University of Dortmund
Otto-Hahn-Str. 6
D-44221 Dortmund
Germany

Dr. N.V. Shishkova
Donetsk Physics & Technology Inst.
of the Natl. Acad. of Sciences
R. Luxemburg Str.
340114 Donetsk
Ukraine

Prof. Dr. J. L. Silva
Departamento de Bioquimica Medica
Instituto de Ciencias Biomedicas
Universidade Federal do Rio de Janeiro
CEP 21941-590 Rio de Janeiro - RJ
Brazil

Dr. G. Stambini
Instituto di Biofisica
CNR
Via S. Lorenzo 26
I-56127 Pisa
Italy

M.M. Sun
Department of Chem. Engineering
201 Gilman Hall
Berkeley, CA 94720
U.S.A.

P. Teunissen
ATO-DLO
Agricultural Research Institute
P.O.-Box-17
6700 AA Wageningen
The Netherlands

Prof. Dr. R. van Eldik
Universität Erlangen-Nürnberg
Institut für Anorganische Chemie
D-91058 Erlangen
Germany

A. Wandel
Polish Academy of Science
Institute of Bioorganic Chemistry
UL Noskowskvego 12/14
61-704 Poznan
Poland

Prof. Dr. R. Winter
University of Dortmund
Physical Chemistry I
Otto-Hahn-Straße 6
D- 44221 Dortmund
Germany

Dr. J. Woenkhaus
Philipps-University of Marburg
Department of Chemistry
Hans-Meerwein-Straße
D-35032 Marburg
Germany

A. Yilmaz
Department of Chemistry
Middle East Technical University
06531 Ankara
Turkey

G.A. Yilmaz
Department of Chemistry
Middle East Technical University
06531 Ankara
Turkey

M. Zein
Physical Chemistry
University of Dortmund
Otto-Hahn-Str. 6
D-44221 Dortmund
Germany

Prof. Dr. W. Zerda
Dept. of Physics & Astronomy
Texas Christian University
TCU Box 298840
Fort Worth, Texas 76129
U.S.A.